Quarks, Nuclei and Stars

Memorial Volume Dedicated to Gerald E. Brown

Quarks, Nuclei and Stars

Memorial Volume Dedicated to Gerald E. Brown

Edited by

J. W. Holt
Texas A&M University, USA

Thomas T. S. Kuo
Stony Brook University, USA

K. K. Phua
Nanyang Technological University, Singapore

M. Rho
CEA Saclay, France

I. Zahed
Stony Brook University, USA

World Scientific

NEW JERSEY • LONDON • SINGAPORE • BEIJING • SHANGHAI • HONG KONG • TAIPEI • CHENNAI • TOKYO

Published by

World Scientific Publishing Co. Pte. Ltd.
5 Toh Tuck Link, Singapore 596224
USA office: 27 Warren Street, Suite 401-402, Hackensack, NJ 07601
UK office: 57 Shelton Street, Covent Garden, London WC2H 9HE

British Library Cataloguing-in-Publication Data
A catalogue record for this book is available from the British Library.

QUARKS, NUCLEI AND STARS
Memorial Volume Dedicated to Gerald E. Brown

ISBN 978-981-3220-64-5

Printed in Singapore

Contents

Preface

It is with great pleasure and years of memories that we write a short introduction to this memorial volume honoring Gerald E. Brown (1926–2013) or 'Gerry' as he was known to most of us. The planning of this volume was started in the summer of 2015, with the aim of publishing it in 2016 to commemorate Gerry's 90th birthday on July 26, 2016. The responses to the inquiries we sent out were overwhelmingly enthusiastic. As written in one contribution to this volume, Gerry was "an inspiring 'father figure' for generations of theoretical nuclear physicists and a great human being." We wholeheartedly agree.

Gerry devoted his life to physics. He spent most of his professional career at Stony Brook University (45 years from 1968–2013) and also held positions at Princeton University (1964–1968), NORDITA (Nordic Institute for Theoretical Physics, Copenhagen) (1960–1985), and the University of Birmingham, England (1950–1960). He studied at the University of Wisconsin (B.S. 1946) and Yale University (Ph.D. 1950). Gerry had worked closely with Gregory Breit (his Ph.D. thesis advisor), Rudolf Peierls (Birmingham) and Hans Bethe (Cornell) whom he had nicknamed his three 'eagles', guiding him to fly in the physics sky.

Gerry himself was also a great 'eagle', having taught, guided and taken personal care of about 100 Ph.D. students and also about 100 postdocs and colleagues. Most of the authors here were among them, making this volume like a reunion gathering. Gerry was interested in nearly all aspects of nuclear physics and astrophysics, both theoretical and experimental, and was quick to dive into new areas of research. This is reflected in the wide range of forefront subjects addressed in this memorial volume: nuclear structure, two- and three-body nuclear forces, strangeness nuclear physics, chiral symmetry, hadrons in dense medium, hidden local symmetry, heavy quark symmetry, cosmic neutrinos, nuclear double-beta decay, neutron stars, gravitational waves, renormalization group methods, exotic nuclei, electron ion collider, and much more.

We hope that readers will find this volume highly interesting and inspiring, not only in terms of physics content but also in the personal stories shared by the contributors that reveal the man behind the science.

We sincerely thank the authors of this volume for their enthusiastic support. We also sincerely thank Kim Tan, Chee-Hok Lim, Yubing Zhai (World Scientific) and Marlene Vera-Viteri (Stony Brook) for their excellent help in preparing this volume. Last but not least, the support from World Scientific is gratefully acknowledged.

To conclude, let us borrow a sentence from a paper of this volume, "Gerry is unforgettable".

J. W. Holt,[a] T. T. S. Kuo,[b] K. K. Phua,[c] M. Rho[d] and I. Zahed[b]
[a]*Texas A&M University*, [b]*Stony Brook University*,
[c]*Nanyang Technological University*, [d]*CEA Saclay*

Foreword I

Professor Gerry Brown passed away on May 31, 2013 in Stony Brook, New York. All of us who have made his acquaintance grieved over him.

He was not only a great physicist but also a great leader and a teacher of many of us including me. He was born in 1926 in Brookings, South Dakota which has a large population of Native Americans. Possibly because of this fact, he was a humanitarian and did never segregate anybody depending on his or her race. Therefore he attracted many Chinese and Japanese collaborators and students together with African Americans. Needless to say, he also had many American and European colleagues and students.

He was always very pleasant to talk with and teach us. He did not skimp his energy to advise anybody on his or her study and research. Furthermore he enjoyed cooperation with anybody. Because of his personality mentioned above, he accomplished many important researches which covered a broad range of nuclear physics from nuclear structure to astrophysics.

In this volume, readers will find how many collaborators cooperated with him and how broad the range of problems studied by him was.

It was in 1960 at Argonne National Laboratory that I met Gerry for the first time. I came to this laboratory in September 1959 as a research associate. Gerry Brown visited Argonne in 1960 and gave a talk on the dipole state in nuclei. He succeeded in explaining the E1 giant resonance as a collective vibration of nuclei from the effective two-body interaction. In 1982, he was awarded Tom W. Bonner Prize in Nuclear Physics because of this work. I was very pleased when listening to his talk and discussing with him. I worked those days mostly on the first order configuration mixing effect of nuclear interactions, namely the core-polarization effect on magnetic dipole and electric quadrupole moments.

I was also busy with the calculation of the second order correction from the tensor force in order to explain the spin–orbit interaction. I found that modifications to two-body interactions due to the nucleon–nucleon tensor force in second order were not negligible. I then told him those works of mine and asked his comments and criticism. He has been always open-minded and gave me advice several times.

In 1967–1968, I spent one year at Rutgers University as a visiting professor. During this time I visited Gerry in Princeton and worked under him for one or two days weekly. Those days, Gerry Brown and Tom Kuo were extremely active to develop their method to calculate effective two-body interaction systematically and quantitatively. They established thus the Kuo–Brown interaction. This was an epoch-making achievement. In the Kuo–Brown interaction the second order effect mentioned above is automatically taken into account.

I was invited as a professor by Gerry Brown to Stony Brook in 1971. I enjoyed very much cooperating with him. My first work there was to explain the second order corrections to magnetic dipole moments of nuclei with a LS closed shell plus or minus one nucleon, for example, ^{17}O, ^{17}F, ^{41}Ca and ^{41}Sc, and the Gamow–Teller (GT) β transitions among them. K. Shimizu, M. Ichimura and I in 1972 found that the second order effects can explain the quenching of $\tau\sigma$ moment in these nuclei. Namely large portions of the $\tau\sigma$ moment are spread over $2\hbar\omega$ excitations in those nuclei. Much later a majority of nuclear physicists tried to explain the quenching by Δ-hole admixture. Around Gerry Brown, my group sticked to the second order core polarization, but M. Rho and his collaborators insisted on the Δ-hole admixture. Gerry tried to be neutral. Frankly speaking, however, he seemed to favor slightly the Δ-particle picture until 1986.

In 1986, we celebrated the 60th birthday of Gerry in Stony Brook. On this occasion I was pleased to give a talk on the isoscalar magnetic moment in the relativistic mean field theory. In Tokyo, W. Bentz and collaborators had just finished a work on the use of the Ward–Takahashi identity in many-nucleon systems. They showed that the orbital part of the isoscalar magnetic moment is essentially the same as in the nonrelativistic theory, that is, not enhanced by a factor $M/M^\star$, where M is the free nucleon mass and $M^\star$ the effective nucleon mass. Gerry Brown developed very independently the similar explanation by using the idea of backflow current. We agreed with each other. In this way the magnetic moment problem in relativistic mean field theories was solved.

Gerry told me at that time that the second order core polarization was a more reasonable explanation of the quenching of the GT transitions.

By the way, in 1997, H. Sakai and his collaborators in Tokyo used the ^{90}Zr(p,n) reaction to measure the distribution of the GT strength in ^{90}Nb. After a careful analysis they arrived at the result that the GT strength summed up to 50 MeV excitation energy is $(93 \pm 5)\%$ of the Ikeda sum rule value. This clearly shows that the quenching of the GT strength due to the Δ-hole admixture must be very small.

Professor Gerry Brown has always encouraged us and given appropriate advices and criticisms. We will never forget him. When we encounter any difficulty, we will try to think what kind of advices would be given by him. Goodbye Gerry.

Akito Arima
Musashi University, Japan

Foreword II

This memorial volume is a tribute to the memory of an outstanding physicist. Gerry Brown's contributions in various fields, notably Many-Body Theory, Nuclear Physics and Astrophysics, are well known and will remain as a lasting proof of his scientific stature. This is well reflected in the large number of leading physicists contributing to this volume and in the wide range of subjects covered.

A foreword for a memorial volume contains in general a brief overview of the dedicatee's life and work. This need not be the case for this memorial. There is abundant literature online on the life and work of Gerry Brown. Gerry himself wrote an autobiographical essay "Fly with Eagles", published in *Annual Review of Nuclear and Particle Science* **51**, 1 (2001). This is pleasant reading from beginning to end and reflects the inimitable personality of his author. The "eagles" Gerry flew with are three of the greatest physicists of the last Century, Gregory Breit, Rudolph Peierls and Hans Bethe. Gerry had Breit as Ph.D. Advisor at Yale, then moved to Birmingham where he spent ten years in the theory group of Peierls. Gerry's flight with his last eagle, namely the intense collaboration with Bethe, came in only several years later, but lasted more than twenty years. A view of Gerry Brown "from the outside" is given by Sabine Lee in the chapter entitled "What is the Universe? G. E. Brown — His Life and Work", in *From Nuclei to Stars: Festschrift in Honor of Gerald E. Brown* (World Scientific, 2011).

When I was asked to write this Foreword, having realized that it was very hard to add something new about Gerry, I decided to tell a surely unknown story of a swimming relay race and of how this affected my own life.

I first met Gerry in August 1961 when I attended as a student the Enrico Fermi School on Nuclear Physics held in Varenna on Lake Como. The School was directed by Victor Weisskopf, and had as lecturers Gerry Brown, Amos de-Shalit, Torleif Ericson, Carl Levinson and Felix Villars. The subject of Gerry's lectures was "Collective motion and the Application of Many-body Techniques." The School lasted three weeks, so there was enough time for the students to become somewhat familiar with the lecturers. This was also favored by the enchanting beauty of the area. For those loving swimming, the best attraction was of course the lake, which is directly accessible from Villa Monastero, where the Schools take place. So, in the free time several participants, including Gerry and myself, were taking advantage of this opportunity. Varenna is a small village and our swimming activity did not pass unnoticed by local people. This brought up the idea of a swimming competition, physicists versus locals, on the occasion of an important local feast. It was then agreed on a 4×100 meter relay race, the prize for the winning team consisting of a case of high-quality wine bottles. Our team was formed on a voluntary basis and included Gerry, myself, and two more students from Germany and Sweden. The enthusiastic participation of Gerry is no wonder when knowing that at the age 17 he was a first-class seaman in the U.S. Navy. On the day of the competition, a crowd from the village gathered along the lake in Villa Monastero to see the triumph of the Varenna Natives over a poor amateur scientist team. To make a long story short, to the great disappointment of the local fans, we defeated the Natives by a large margin. Gerry kept a very good memory of this. I met Gerry for the last time in June 2006 in Darmstadt at the Symposium celebrating his 80th birthday . On that occasion, I gave him some large sized prints of photos taken 45 years before when he was swimming in Lake Como. The small story told above fits Gerry's bold personality, but there is a continuation of this story which has to do with another aspect of his personality.

In 1963 I had planned to spend the next academic year 1964/65 in Copenhagen. In December everything seemed to be arranged to fulfill my plans and I was of course very keen on the opportunity to join the nuclear theory group led by Aage Bohr and Ben Mottelson. On March 1964, however, I received a letter from Bohr telling me that he and Mottelson had decided to take a leave of absence during the coming academic year to concentrate on writing their book on nuclear structure. He therefore felt that it was most advisable for me to spend the coming year at some other institution and then come to Copenhagen for the year 1965/66. It was really a nice surprise for me to find together with the letter of Bohr a handwritten note from Gerry. He was telling me that he was going to Princeton next year and would have been glad to have me there. Since by that time the Department was full, he proposed me to try to go to the nearby Rutgers University with which he was planning to have a close collaboration. He ended by saying: "If this idea appeals to you, please write to Prof. Georges Temmer at Rutgers." I did so immediately and in a few weeks I was offered a research associateship at Rutgers. The end of the story is that I stayed three years at Rutgers.

These were exciting years for Nuclear Physics. In a short time Gerry gave a big boost to nuclear theory at Princeton. He also conducted a series of joint Princeton–Rutgers post-dinner "bull sessions". These were taking place on a weekly basis, starting with a seminar and going on with lively discussions, sometimes until past midnight. It was not unusual to have with us another eagle, Eugene P. Wigner. These bull sessions gave us the opportunity of learning a lot from Gerry, who was able to go to the heart of a problem often with a great sense of humor. It is worth recalling that it was in this nice context that the pioneering Kuo–Brown work on the derivation of the shell-model effective interaction from the nucleon–nucleon potential was conceived and performed.

We hope that this volume will help keeping the memory of Gerry Brown alive for younger generations.

Aldo Covello
Università di Napoli Federico II, Napoli, Italy

Foreword III

This volume presents a remarkable account of Gerry Brown's lasting scientific impact on several generations of physicists worldwide who had the pleasure and privilege to interact with him, as former students, postdocs, faculty colleagues, visting scientists and friends. Many of us vividly remember the stimulating environment that Gerry created with his Nuclear Theory Group in Stony Brook. Walking along the NTG corridor and into Gerry's office one could always sense that special atmosphere of "something new and exciting" in the air.

In the late sixties, as a graduate student at the University of Frankfurt, I discovered a book in the library that would keep me busy with excitement for some time: *Unified Theory of Nuclear Models and Forces* by G. E. Brown, Professor of Physics at Princeton University. The book was based on lectures that the author had given at a Nordic-Dutch winter meeting. Style and presentation were very different from what I had been familiar with in German academia. The Preface reported about the pleasures of skiing in the Norwegian mountains, and one could enjoy reading sentences like: "... let me say that I treat each problem the way that I enjoy treating it. Nature and I have achieved a reasonable modus vivendi in the past, and I have every reason to believe in the realization of further enjoyment in the future." Working through Chapters I to XIII was not an easy task. But it created a stimulating challenge for the young student and gave the motivation to try understanding the nuclear many-body problem from a microscopic point of view.

In 1973 I had the chance to meet the famous Professor Brown in person. It all started at a photonuclear physics conference in Asilomar, a peaceful place in coastal California. I had submitted an abstract, entitled "Discussion of possible modifications of the Gell-Mann–Goldberger–Thirring sum rule", as a short contribution to the conference. The celebrated GGT photonuclear sum rule was a basic frame of reference for analysing the enhancement of the nuclear dipole sum rule over the Thomas–Reiche–Kuhn sum and its relationship to mesonic exchange currents in nuclei. In its original version the sum rule rested on a simple assumption about the asymptotic behavior of the nuclear Compton scattering amplitude. It had occured to me that this assumption was inconsistent with high-energy photon–nucleus cross sections which had just been measured at DESY in Hamburg. It was important to incorporate the so-called nuclear shadowing effect into the GGT sum rule. My little abstract was upgraded to a full-scale talk and so a slightly nervous young Ph.D. unexpectedly found himself speaking in front of an audience that included Gerry Brown, Judah Eisenberg, Torleif Ericson, Claude Mahaux and Edward Teller. In the dispute following the talk, Gerry gave an illuminating interpretation of the physics behind what I had tried to explain. After the session Gerry and Torleif took me aside and continued the discussion. I can say that this experience changed my life. Gerry proposed that I should come to Stony Brook as a postdoc. This was the starting point of a highly productive collaboration and fruitful exchanges for many years to come.

Being able to work in Gerry's group was a real priviledge. Many postdocs and visiting scientists from all over the world came together and got involved in lively discussions at all times, especially in the legendary informal lunch seminars. In addition to working on projects together with Gerry himself, I had the chance to learn much about nuclear effective interactions from Tom Kuo, about two-pion exchange forces from John Durso and Andy Jackson, and about chiral symmetry and its role in nuclear physics from Mannque Rho. Sven-Olof Bäckman showed me how to handle Fermi liquid theory in practice. With Dan-Olof Riska we calculated in-flight pion absorption on the deuteron. Leonardo Castillejo visited regularly and always raised constructively critical questions. It was a great time.

One of the hot topics in the mid and late seventies was Migdal's proposal that pion condensates might be formed in nuclear and neutron matter. Gerry immediately put me on a project in order to find out under which conditions such a condensate could exist and what its consequences would be for nuclear physics. There was much exciting exchange about these issues with many masters in the field, including Migdal himself and Gordon Baym. As it turned out, a subtle competition between repulsive nuclear spin–isospin correlations and attractive pion exchange forces acted in such a way that the formation of a pion condensate was strongly inhibited. The quest for pion condensation became subject to observable constraints such as the properties of nuclear Gamow–Teller excitations. By that time pion-nuclear physics with all its implications had developed into an active field of research following pioneering work by Magda and Torleif Ericson. Gerry and I were particularly involved in studying the role of delta-isobars in nuclear many-body systems, work which culminated in a joint 1975 Physics Reports article that defined a framework for many activities thereafter.

This was also the time when the approximate chiral symmetry of QCD in its spontaneously broken (Nambu–Goldstone) phase began to be realized as a guiding principle not only for low-energy interactions of pions and nucleons but also for the nuclear many-body problem. The structure of the nucleon itself as a compact quark core surrounded by a Nambu–Goldstone boson field was a key topic in Stony Brook. I had the great chance to work with Gerry Brown and Mannque Rho on several aspects of the Chiral Bag model and its SU(3) extension to the baryon octet.

Regular visits to Stony Brook in later years further promoted the transatlantic long-range correlations with Gerry and his group. One of his last Ph.D. students, Jeremy Holt, spent several years as a postdoctoral research fellow with our group in Munich. This turned out once again to be a highly productive period in which the understanding of nuclear chiral dynamics and thermodynamics progressed significantly, especially through joint work with Norbert Kaiser at TUM. One can say that three generations of Gerry's "students" came together, and his legacy was with us. It is amazing how many such successful clusterings of Gerry's former collaborators were stimulated by his very special intuitive way of thinking. Building scientific and personal bridges, cheerfully ignoring national boundaries, was indeed one of Gerry Brown's basic messages to those working in the broad areas of quarks, nuclei and stars. What started with Gerry's creativity and transfer of pioneering ideas between Copenhagen and Stony Brook subsequently spread all over the world. And so these few lines of recollection end with a deep feeling of gratitude.

Wolfram Weise
Technical University of Munich, Germany

Gerry Brown, 1960s, Princeton, USA

Gerry Brown, 1980s, Stony Brook, USA

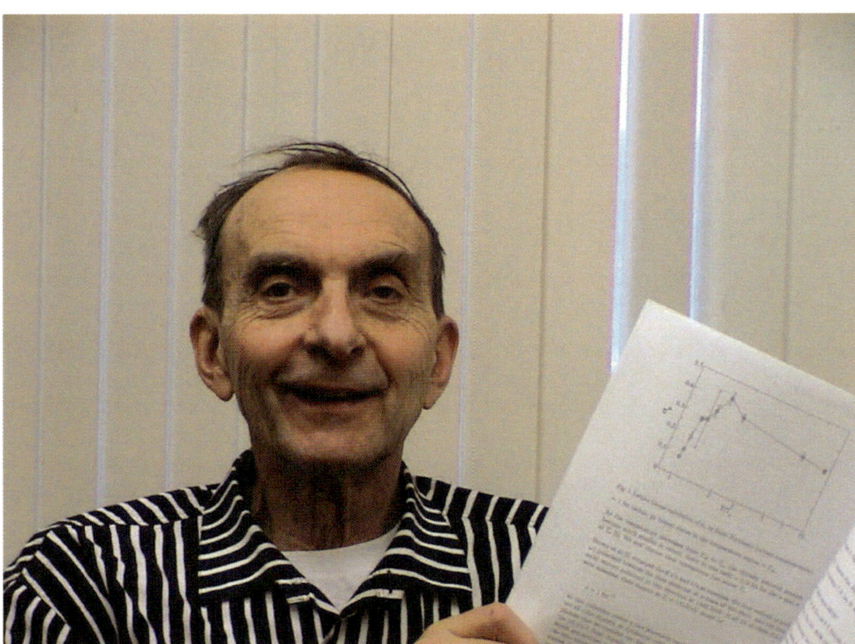

Gerry Brown, 2000s, Stony Brook, USA

Gerry and Betty Brown, 1980s, travelling

Gerry with Betty, Michael and Ann Anastasio, Thomas and Annette Kuo, 2007 at Stony Brook Distinguished Alumni Ceremony

Gerry with Achim Schwenk, 2006. Conference honoring Gerry's 80th birthday, Darmstadt, Germany

Gerry with friend, 2006. Conference honoring Gerry's 80th birthday, Darmstadt, Germany

Gerry with Paul Kienle, 2000s, Germany

The shapes of nuclei

G. F. Bertsch
Department of Physics and
Institute for Nuclear Theory,
University of Washington,
Seattle, WA 98195, USA
bertsch@uw.edu

Gerry Brown initiated some early studies on the coexistence of different nuclear shapes. The subject has continued to be of interest and is crucial for understanding nuclear fission. We now have a very good picture of the potential energy surface with respect to shape degrees of freedom in heavy nuclei, but the dynamics remain problematic. In contrast, the early studies on light nuclei were quite successful in describing the mixing between shapes. Perhaps a new approach in the spirit of the old calculations could better elucidate the character of the fission dynamics and explain phenomena that current theory does not model well.

Keywords: Nuclear structure; fission; nuclear deformation.

1. Introduction

In the early 1960s, Bohr and Mottelson pointed out some puzzling experimental data: some light nuclei thought to be spherical in shape had excited energy levels exhibiting characteristics of deformed nuclei. This was taken up first by Engeland[1] and then by Gerry Brown, who saw an opportunity to test the realistic nuclear interactions that were being developed at the time. The nuclei ^{16}O and ^{18}O were the first subject of study.[2] As his graduate student, I worked on a parallel study of Ca isotopes as part of my thesis project. Later, a more definitive study of the Ca nuclei was carried out by Gerace and Green.[3] As a general conclusion, one saw that the mixing between shapes could be understood with the realistic interactions derived from nucleon–nucleon scattering data.

Since those early days of nuclear structure physics, the subject of nuclear deformation has matured. First of all, we now know that the shape coexistence is ubiquitous in the low-energy spectra of nuclei across the periodic table, affecting even the fission properties of the heaviest nuclei. Also, we now have computational tools to describe and predict the static features of the landscape of nuclear shapes. However, the dynamics of shape change, i.e., how different shapes mix together, has been a challenging problem in the theory of heavy nuclei and is still not well understood. I shall describe some work I have been engaged in recently, to develop a new approach to fission dynamics in the spirit of the old studies on light nuclei.

2. Theory of Static Deformations

Nuclei are highly deformable, and the first task is to construct reliable models of the nuclear potential energy surface, i.e., the energy of configurations as a function of deformation coordinates such as the expectation values of quadrupole and higher moments. An immediate question is how to define both energy and shape of a configuration: the operators for these two quantities do not commute. The resolution of this conundrum is that we are only dealing with approximate wave functions, not the true eigenstates of the Hamiltonian. In practice, theory relies on the mean-field representation of wave functions as products of single-particle orbitals. The present-day calculational framework is very similar to the density functional theory (DFT) of condensed-matter physics. One defines an effective interaction, which may depend on the local density. The energies and shapes are determined by minimizing the energy expectation value just as in Hartree–Fock (HF) theory.

Several families of DFT are in use today, and all have considerable predictive power on static properties of nuclei. In the examples discussed below, I will show results obtained with the Gogny D1S functional.[4] It has 14 parameters, 3 of them fixed and 11 adjusted many years ago[5] to reproduce general nuclear properties. An example showing its considerable predictive power is recent systematic study of the low-energy spectroscopy of even–even nuclei.[4] I will come back to findings from this study later. Another success of the DFT approach is verification of additional shape minima at very high deformation in heavy nuclei. The potential energy surface that a fissioning nucleus traverses has at least two minima and perhaps more. Figure 1 shows the energy versus deformation in a typical fissile nucleus.

3. How Nuclei Change Shape

Up to now, the only practical approach for treating shape dynamics in heavy nuclei is the generator coordinate method (GCM) proposed by Hill and Wheeler.[6] Formally,

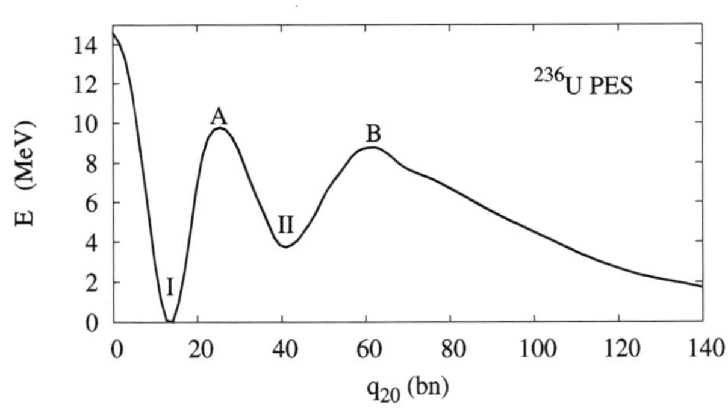

Fig. 1. Potential energy surface for ^{236}U covering the range of deformations up to the scission point. The curve was calculated by minimizing the D1S density functional with the mass quadrupole moment constrained to the value on the abscissa. The point marked "I" is the global minimum. The points marked "A" and "B" are saddle points on the fission path, with a second minimum "II" in between.

one generates a continuum of configurations by minimizing the Hamiltonian in the presence of an external field. Thus, the configuration $|\Psi_\lambda\rangle$ is defined by minimizing

$$\langle\Psi_\lambda|\hat{H} - \lambda\hat{Q}|\Psi_\lambda\rangle, \tag{1}$$

where $\hat{H}$ is the Hamiltonian or energy-density functional and $\hat{Q}$ is a one-body operator such as the axial quadrupole field:

$$\hat{Q}_{20} = \hat{z}^2 - \left(\hat{x}^2 + \hat{y}^2\right)/2. \tag{2}$$

Then one constructs an effective Hamiltonian from information about the diagonal and off-diagonal matrix elements $\langle\Psi_\lambda|\Psi'_\lambda\rangle$ and $\langle\Psi_\lambda|\hat{H}|\Psi'_\lambda\rangle$.

Computationally, the GCM can be implemented in two ways. In the first way, the moment $q = \langle\Psi_\lambda|\hat{Q}|\Psi_\lambda\rangle$ is treated as the position variable in a one-dimensional Schrödinger equation. It is straightforward to derive formulas for potential energy function $V(q)$ and a kinetic energy operator

$$-\frac{1}{2}\frac{\partial}{\partial q}\frac{1}{\mathcal{I}(q)}\frac{\partial}{\partial q} \tag{3}$$

in approximations such as the Gaussian overlap approximation.[7] This one-dimensional collective approach works well for treating the ground state tunneling under the barriers, but generalizing it to include excited states is cumbersome.[8] A more technical problem is necessity to define at least four or five deformation operators to completely specify the shape of the fissioning nucleus. It is possible, but not easy, to set up and solve the corresponding higher-dimension Schrödinger equation; a two-dimensional approximation was carried out in Ref. 9.

The second computational method is to discretize the GCM configurations with a mesh of λ values and minimize the Hamiltonian in that finite basis. There are obstacles to this approach as well. On a purely technical level, the fact that the configurations are not orthogonal leads to numerical instabilities. More serious on a fundamental level, it is not clear how to calculate interaction matrix elements between configurations when they are constructed via an energy-density functional. Several plausible prescriptions are possible, but the results can be unphysical if the wrong prescription is used.[10] Such problems never occurred in the early work on light nuclei — we used orthogonal bases and we had a pretty good idea of the interactions.

To build an alternative to the GCM approach, we should start by constructing an orthonormal basis within the mean-field framework. The dynamics can be developed later from the off-diagonal interactions in this basis. In the old work, the basis was constructed from the harmonic oscillator Hamiltonian by using the associated SU(3) group structure to organize the many-body configurations. This is obviously too crude for treating heavy nuclei, and it is better to use DFT to build the orbitals. In the nuclear context, one could use DFT orbitals and still preserve orthogonality by constraining them by their quantum numbers rather than by their deformations. For most nuclei, the mean-field potential is axially symmetric and has good parity at the DFT minima. This allows one to assign orbital quantum numbers K^π, where K is the angular momentum about the symmetry axis and π is the parity of the orbital. The DFT minimization can be carried out taking as a constraint the number of nucleons having given values of K, π, and isospin projection τ_z. Wave functions that differ in filling numbers are automatically orthogonal, so this should be very helpful for constructing an orthogonal basis. In the remaining sections of this paper, I will go through some examples illustrating the use of a K^π partitioning to define the configurations.

Before going on to the examples, it is instructive to see how the partitioning works in a semiclassical limit. Assume that the nucleus is spherical and the phase space density is uniform in spheres of radius R in coordinate space and k_F in momentum space. Then the number of nucleons having orbital quantum number M is given by

$$n_M = g\int^R d^3r \int^{k_F} \frac{d^3k}{(2\pi)^3}\delta(M-(\vec{r}\times\vec{k})_z), \tag{4}$$

where $g = 2$ is the spin degeneracy of the nucleon and M is the orbital angular momentum about the z-axis. The integrals can be carried out analytically; the result is

$$n_M = \frac{9N}{4k_FR}((1/2+x^2)\text{arccos}\, x - 3x(1-x^2)^{1/2}), \tag{5}$$

where $x = M/(k_FR)$. The same formula applies to ellipsoidally deformed nuclei with the radius R replaced by the transverse radius of the ellipsoid. Figure 2 shows the distribution for the neutrons in the spherical nucleus ^{208}Pb, comparing the mean-field fillings with the formula (5).

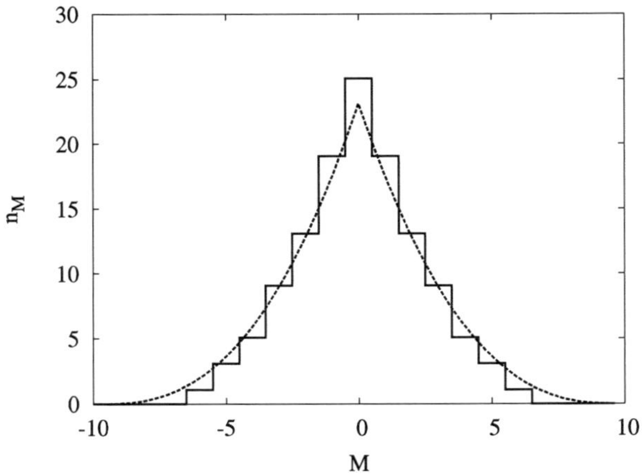

Fig. 2. M distribution of occupied neutron orbitals in the nucleus ^{208}Pb. Histogram: Spherical shell model. Dashed line represented Eq. (5) with $R = 7.1$ fm and $k_F = 1.35$ fm^{-1}.

4. Examples

The shells of the three-dimensional harmonic oscillator model are completely filled or empty at nucleon numbers $8, 20, 40, \ldots$. The first two examples here are nuclei at those possibly magic numbers: ^{40}Ca and ^{80}Zr. The third example deals with the early shape changes in a fissioning ^{236}U nucleus.

4.1. *The ^{40}Ca region*

The spectroscopy of ^{40}Ca has been known since the early 1960s. Figure 3 shows the states in the spectrum that are relevant to the discussion. The first excited state in the spectrum has angular moment zero, which is rare among the 600 or so-known even–even nuclei. The excited 0^+ together with the 2^+ and 4^+ above it form the lowest members of a rotational band. The evidence that these states are part of a band comes from the energy spacing and from the electromagnetic transition rates between states, indicated by arrows in the figure. The measured quadrupole transition strength between two lowest states in the band is $B(E2, 2_1^+ \to 0_2^+) = 250 \pm 35\,\mathrm{e}^2 \cdot \mathrm{fm}^4$.[11] On the scale of a single-particle quadrupole moment this is very large and would require an axis ratio of $a_z/a_\perp \approx 1.55$ to model the band as an ellipsoidal rotor. The corresponding intrinsic mass quadrupole moment in the band is

$$\langle \hat{Q}_{20} \rangle = 117\,\mathrm{fm}^2. \tag{6}$$

Now let us compare the GCM and the K^π-partition approaches to determine the structure of the band. We start with the GCM, applying a field $\lambda \hat{Q}_{20}$ to the spherical ground state configuration. The minimization is performed by small increments of λ up to the point where the configuration has the deformation (6) extracted from experiment. The results are shown in Fig. 4 as the open circles. The energy increases monotonically as the deformation increases. At the deformation corresponding to the one extracted from experiment, the excitation energy is 13 MeV. This is way off from the observed band head at 3.35 MeV. There will be a small gain of energy when one takes into account that the calculated energy is that of the band as a whole, but the energy gain is probably too small to approach the experimental value.

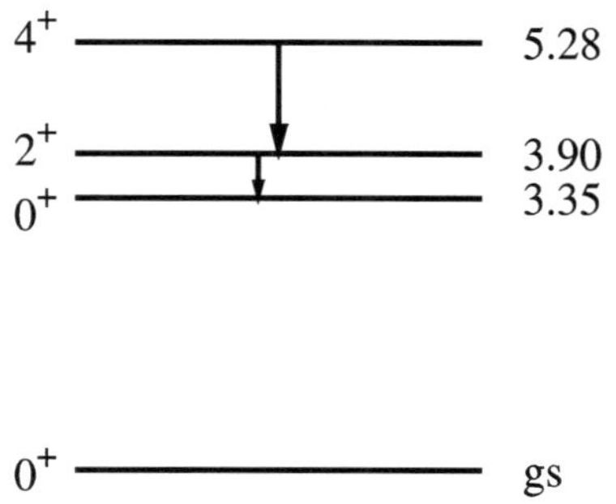

Fig. 3. The spectrum of ^{40}Ca, showing the first three levels of the deformed band and the gamma transitions that establish the deformation of the band.[11]

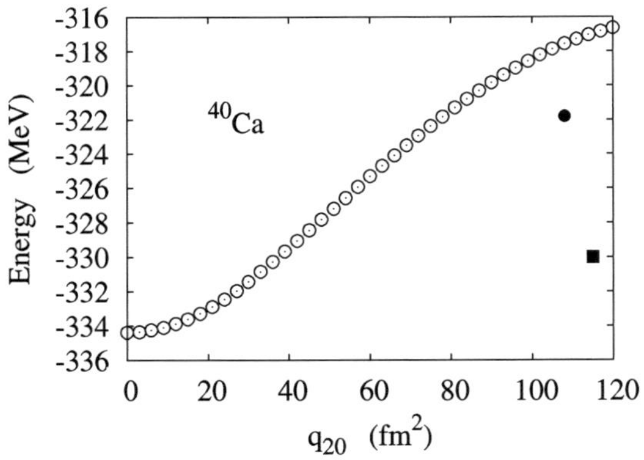

Fig. 4. Energies of ^{40}Ca configurations calculated by GCM (open circles) and by K^π-constrained minimization (black circle).

Anyway, a big surprise comes when we look at the shape of the nucleus, shown as in the right top panel of Fig. 5. One sees that the GCM-generated configuration no longer has good parity; there is a strong octupole component as well as the quadrupole deformation. This would imply that the band should have odd-parity members interleaved between the even-parity members. Since that is not the case, so we can conclude that GCM carried out in this way has failed.

Now lets try the K^π-constrained approach using the K^π partitions from Ref. 3. The ground state configuration has the fillings of the spherical shell model. The deformed configuration was built by taking the lowest four-particle four-hole excited state in a deformed harmonic oscillator potential. This is the Nilsson model; its diagram of orbital energies is shown in Fig. 6. We carry out the DFT minimization again, but now use the K^π quantum numbers of the occupied orbitals to constrain the minimization. The results for the spherical and deformed configurations are shown in Fig. 4 as the black square and black circle, respectively. The predicted quadrupole moment of the deformed configuration, 105 fm^2, agrees well with (6). The energy is still too high, but it is lower than the GCM energy and so is a better candidate for understanding the band structure. Of course, one can obtain the configuration by GCM minimization, but doing this would require a different starting point or additional shape-dependent constraints. One final point: as may be seen in the bottom panel of Fig. 5, the configuration found by starting from the four-particle four-hole configuration preserves the even parity of the band.

Ignoring the energy problem, one can try to calculate the mixing of deformed and spherical states, as was done by the early researchers. Unfortunately, as I mentioned earlier, there is no consistent way to extract configuration-interaction elements from a DFT. But DFT should still be useful to construct the configurations. The Hamiltonian matrix elements might be evaluated using these configurations as in the condensed-matter hybrid procedure.

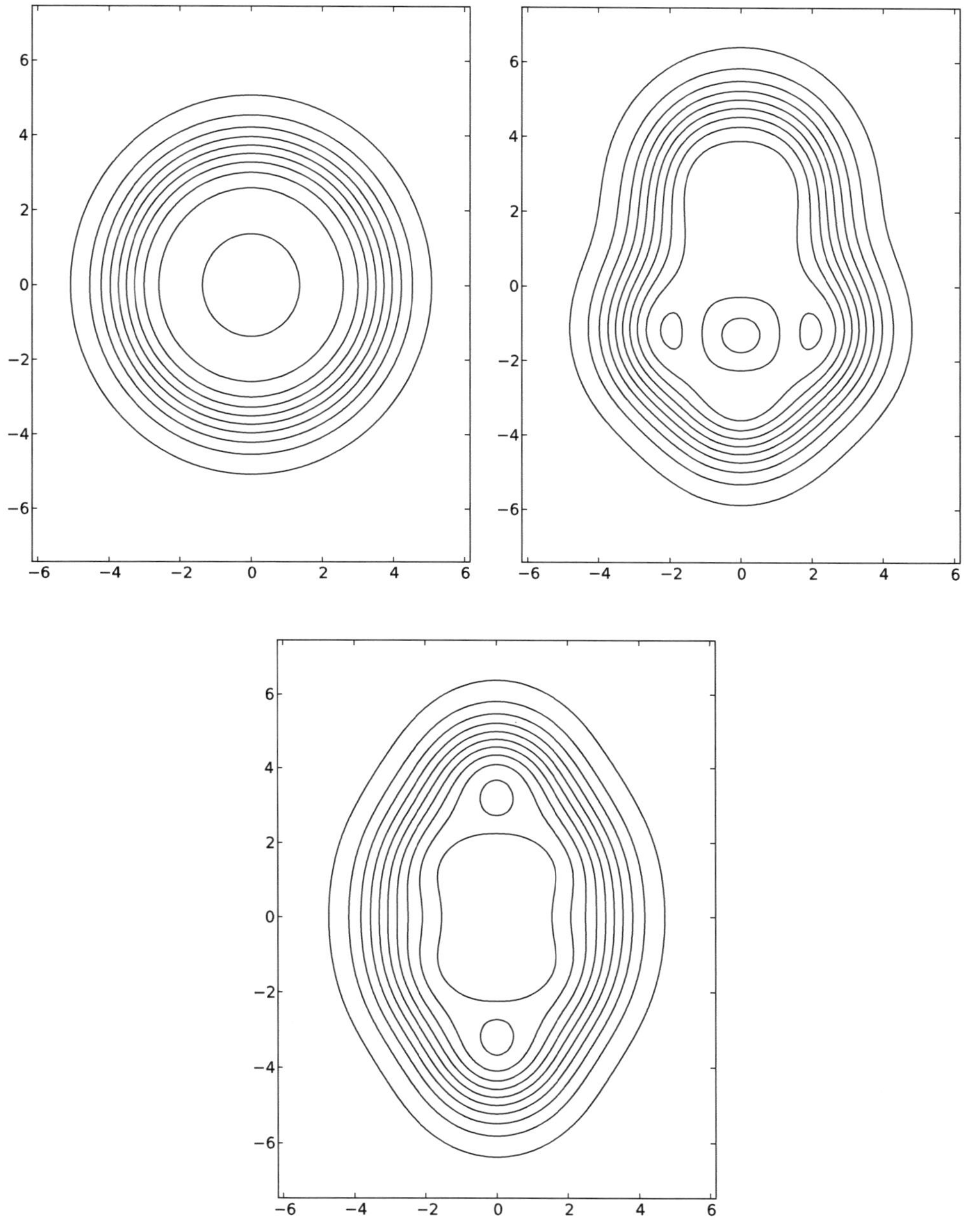

Fig. 5. Density distribution of the ^{40}Ca configurations. Left (top): ground state; right (top): GCM configuration at $q = 108\,\mathrm{fm}^2$; and bottom: four-particle four-hole K^{π}-constrained configuration.

4.2. *The ^{80}Zr region*

At the next harmonic oscillator shell closure ($N, Z = 20$ at ^{80}Zr), the competition between spherical and deformed configurations plays out differently. The coexistence question here was addressed in the DFT by Zheng and Zamick.[12] They tried

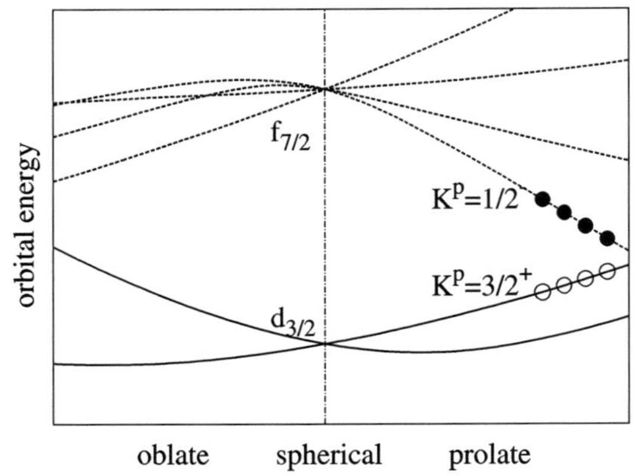

Fig. 6. Single-particle energies of orbitals around the Fermi level of ^{40}Ca as a function of deformation. The deformed band is attributed to four-particle four-hole excitation shown with solid and open circles (after Fig. 1 of Ref. 3).

several energy functionals in the Skyrme family and found that a deformed configuration came out lower than the spherical. While in ^{40}Ca there were four particle jumps needed to connect the configurations, 12 jumps were needed in ^{80}Zr. The Gogny D1S functional yields rather similar results. The corresponding potential energy surface is shown in Fig. 7. The filled square and circle show the minima for the spherical and deformed configurations, obtained by constraining the K^π partitions according to the spherical shell model and the 12-particle 12-hole Nilsson configuration. The latter has a very large deformation, $Q_{20} \approx 400\,\mathrm{fm}^2$, at an energy 3.5 MeV above the global minimum. The deformation is very robust with respect to

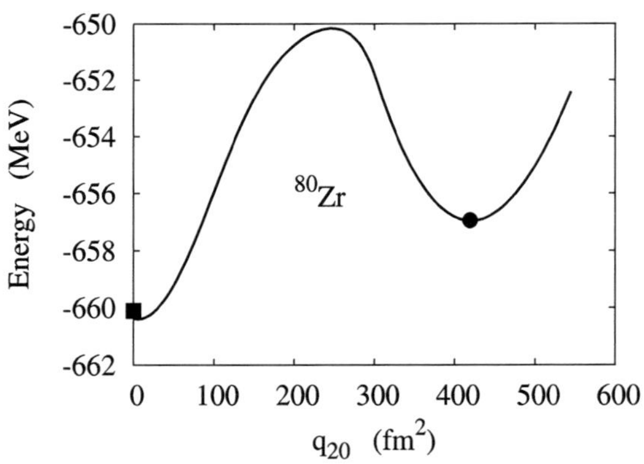

Fig. 7. The ^{80}Zr potential energy surface. The solid line is generated by the GCM procedure, starting from an octupole-deformed ground state minimum. The K^π-constrained minimization starting from the partition of the spherical shell model configuration is shown at the black square. Starting with the 12-particle 12-hole partitions gives the filled circle, coinciding with the second minimum of the GCM potential energy curve.

the choice of energy functional. In terms of the dimensionless deformation parameter β, the Skyrme functionals and the Gogny D1S all give $\beta = 0.44 \pm 0.01$. The authors of Ref. 12 also note that a simple formula derived earlier by myself is quite accurate.

The energetics are more complicated. First of all, the shell model configuration is not the lowest energy minimum at $q = 0$. The actual minimum has broken parity with a large octupole moment and a very small quadrupole moment. The entire potential energy surface up to high deformation can be generated by GCM procedure, provided one starts with a configuration that already has an octupole deformation. It is shown as the solid line in the figure.

While the 12-particle 12-hole is not the lowest configuration in the Gogny DFT, correlation effects can change the ordering. Recently[4] a systematic study of low-energy spectroscopy was carried out including all five quadrupole shape degrees of freedom in the GCM framework. In that way, rotational energies and some shape mixing effects are taken into account. As an example of the predictive power of the Gogny functional and the method, it was found that the calculated quadrupole transition moments of deformed nuclei agree with experiment upto 10% accuracy. For the ^{80}Zr nucleus, the authors found that the rotational energy was enough to bring the band head of the highly deformed configuration down to the ground state. Figure 8 compares their calculated 2^+ excitation energies across the chain of the heavy $N = Z$ even–even nuclei. One sees that 2^+ excitation energy in ^{80}Zr agrees very well with experiment. It is also the most highly deformed in the chain, judging by the excitation energy of the lowest 2^+ state. The heaviest measured nucleus in the figure, ^{96}Pd, is a real prediction, as the experimental measurement[13] was reported the following year. So, despite the mixed performance of DFT in ^{40}Ca, we find that the energy functionals become quite successful in a heavier region of nuclei.

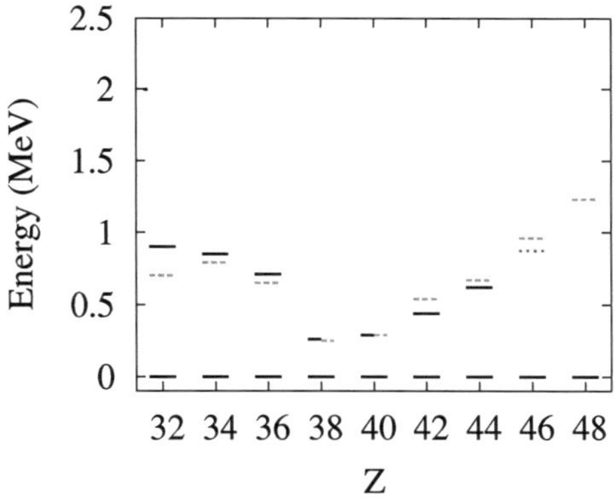

Fig. 8. (Color online) Systematics of 2^+ excitations in the heaviest known $N = Z$ nuclei. Solid black lines: older experiments; and red dashed line: Gogny/GCM theory.[4] Theory and experiment are visually indistinguishable for the nuclei ^{76}Sr and ^{80}Zr. The blue dotted line shows the new measurement at ^{96}Pd.[13]

5. Fission Dynamics

My ultimate goal is to gain a better understanding of the dynamics of fissioning nuclei. As mentioned at the beginning, the GCM approach works well for describing spontaneous fission as tunneling through a barrier in the potential energy surface. But when the excitation energy is above the barrier, it is far from clear what theoretical approach is justified. In any case, the spontaneous fission shows that the part of the interaction responsible for pairing is very important.[14] This suggests that the time-dependent Hartree–Fock–Bogoliubov (TD-HFB) approximation might be appropriate for above-barrier dynamics. Recently, computer power has become available to test the TD-HFB approximation for fission using current nuclear energy functionals. Such a study was carried out by Bulgac and collaborators.[15] Their starting point was a very deformed configuration past the second barrier and at a small excitation energy over the potential energy surface. They found that fission would take place, but the duration before the fragment formation could be very long.

A weak point of the TD-HFB approximation is that it takes into account only a very restricted set of interaction matrix elements to propagate the system from one configuration to another. Namely, configurations that can be generated as two-quasiparticle excitations of the starting configuration are included in some way, but all four-quasiparticle transitions are neglected. In the example of configuration mixing in ^{40}Ca, the HFB approximation would be useless. There is no pairing condensate, and even if there was one, the intermediate configurations are only partially represented by the allowed pair jumps.

While it would still be an enormous challenge to make a realistic treatment of fission in the discrete-configuration approach, we can at least see how the landscape could have traversed. We concentrate on the two minima and the barrier between them. Figure 9 shows the potential energy surface of ^{236}U again, expanding the

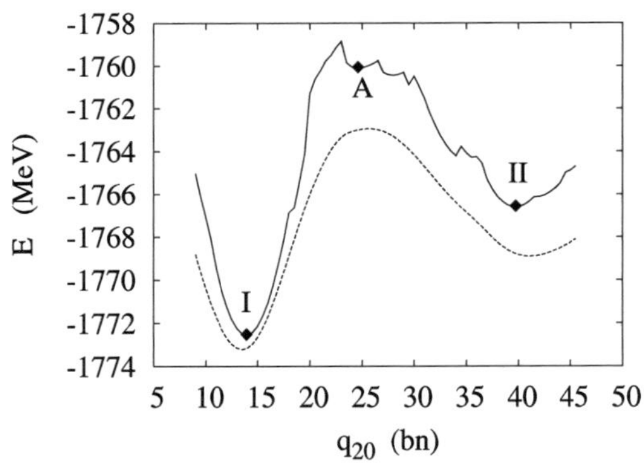

Fig. 9. ^{236}U HFB potential energy surface between the first and second minima and the energies of the partition-constrained HF wave functions. Solid and dashed lines are the HF and HFB potential energy surfaces, respectively. Diamonds show the energy and quadrupole moment of the partition-constrained minima.

horizontal scale in the region of the first barrier. The lower curve was obtained by GCM in the HFB approximation, which includes pairing effects. The more jagged curve above that is the GCM in HF, i.e., ignoring pairing effects. The HF configurations have definite partition of K^π quantum number. The black dots are three configurations generated by the K^π-constrained minimization, with the partitions taken from the GCM results. It is interesting to compare the K^π partitions found with the Gogny D1S energy function with other simpler ways for building a many-particle wave function. The simplest is the Nilsson model mentioned earlier. One finds that it gives exactly the same ground state K^π partition as the DFT. Another model simpler than the DFT is the finite-range liquid drop model.[16] Here the orbital energies are computed in a deformed potential well, but the shape of the well is determined by an energy functional based in part on the liquid drop model. It also gives the same K^π partition as the other models. This suggests that the K^π filling of the ground state is a quite robust property of the mean-field wave function. And it is certainly easier to specify, compared with many shape parameters needed to describe unambiguously the configurations obtained by GCM. However, at the points A and II along the fission path (Fig. 9), the partition obtained with the Gogny functional differs by one pair jump from the FRLD partition. It is likely that they are nearly degenerate, and would be mixed by the pairing interaction.

For dynamics, an important consideration is the number of steps it takes to get from one configuration to another via the two-particle interaction in the Hamiltonian. This suggests a distance measure between configurations as the number of pairing jumps needed to get from one to the other. Figure 10 shows a plot of the path from configuration I to II via the barrier top at A, showing both the quadrupole and hexadecapole coordinates of the intermediate configurations. The

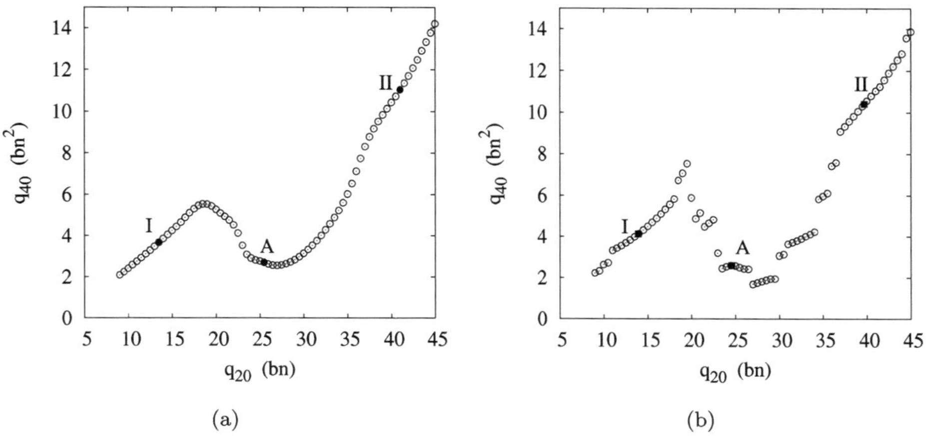

Fig. 10. Path from the first to the second minimum in the (q_{20}, q_{40}) plane. The left panel shows the HFB energies constrained by $q2$. The minima and intermediate maximum are marked by filled circles. The right panel shows the same path in the HF approximation. The filled diamonds are the partition-constrained minima. In principle, either method should produce the same global minimum.

left plot Fig. 10 is the GCM in the HFB approximation, which is seen to be smooth with very little structure. The right plot Fig. 10 shows the same path in the HF approximation. It is broken into a sequence of segments, each segment having a specific K^π partition. Analyzing these segments, one can determine that it takes six pair jumps to get from I to A and eight pair jumps to get from A to II. But it would be wrong to conclude that it takes 14 pair jumps to get from I to II. In fact, examination of end points reveals that the configuration II can be reached from I by changing the orbits of only six pairs. That of course requires surmounting a higher potential barrier. Which path is more important in the dynamics is not obvious; one needs to know interaction matrix elements between configurations as well as their energies to begin to address this question.

6. Conclusion

The configuration-interaction method has been very successful in light nuclei, and it should be possible, at least in a statistical way, to extend it to heavy nuclei. I have concentrated exclusively on the basis of wave functions, advocating the use of DFT constrained by K^π partitioning. This still leaves the interaction matrix elements to be determined. The same problem exists in condensed-matter theory, and there a hybrid approach has been quite successful.[17]

It would take a large effort to carry out this program in nuclear physics. However, there are good reasons, rooted in experimental findings, to undertake that effort. Empirically, there is strong evidence that the fissioning nucleus is close to a statistical equilibrium near the scission point.[18] The big open theoretical question is whether we can explain that in terms of Hamiltonian dynamics with realistic interactions.

Fluctuation phenomena also remain unexplained. Here are two examples. In 1978, Keyworth and collaborators measured the fission cross-sections for neutron-induced fission of ^{235}U separating out the individual angular momentum channels.

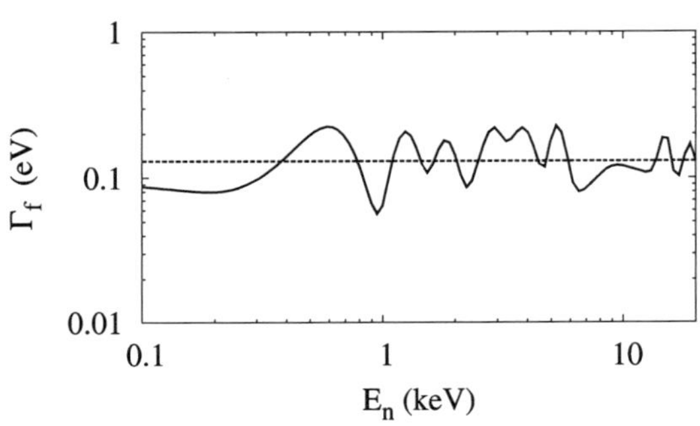

Fig. 11. Average fission width of the $J = 4$ compound nucleus states in ^{236}U as a function of excitation energy. The dashed line is the predicted width from the Bohr–Wheeler theory with one open channel. The solid curve is a fit to the experimental data in Ref. 19.

Besides the fluctuations associated with the compound nucleus states near the way-points I and II, they saw fluctuations on much larger energy scale.[19] The data for one of the angular channels is shown in Fig. 11. It may be that the discrete states near the barrier tops are responsible. A discrete basis would be very helpful here.

Another old experiment exhibiting unexplained fluctuations is the measurement of angular distributions of fission fragments by Huizenga and collaborators.[20] The distributions very close to threshold could be fitted with the usual theory, but on examination of a more extended range they found: "Further attempts to fit the data... were unsuccessful."

In summary, I think there is a good case for trying a discrete configuration approach to shape dynamics in heavy nuclei as an alternative to GCM.

References

1. T. Engeland, *Nucl. Phys.* **72** (1965) 68.
2. G. E. Brown and A. M. Green, *Nucl. Phys.* **75** (1966) 401.
3. W. J. Gerace and A. M. Green, *Nucl. Phys. A* **93** (1967) 110.
4. J. P. Delaroche *et al.*, *Phys. Rev. C* **81** (2010) 014303.
5. J. F. Berger, M. Girod and D. Gogny, *Comput. Phys. Commun.* **63** (1991) 365.
6. D. L. Hill and J. A. Wheeler, *Phys. Rev.* **89** (1953) 1102.
7. M. Girod and B. Grammaticos, *Phys. Rev. C* **27** (1983) 2317.
8. R. Bernard, H. Goutte, D. Gogny and W. Younes, *Phys. Rev. C* **84** (2011) 044308.
9. H. Goutte *et al.*, *Phys. Rev. C* **71** (2005) 024316.
10. L. M. Robledo, *J. Phys. G* **37** (2010) 064020.
11. J. R. MacDonald *et al.*, *Phys. Rev. C* **3** (1971) 219.
12. D. C. Zheng and L. Zamick, *Phys. Lett. B* **266** (1991) 5.
13. B. Cederwall *et al.*, *Nature* **469** (2011) 68.
14. R. Rodigruez-Guzman and L. M. Robledo, *Phys. Rev. C* **89** (2014) 054310.
15. A. Bulgac *et al.*, *Phys. Rev. Lett.* **116** (2016) 122504.
16. P. Möller *et al.*, *Phys. Rev. C* **79** (2009) 064304.
17. D. Sangalli *et al.*, arXiv:1603.00225 [physics.optics].
18. J. Randrup and P. Möller, *Phys. Rev. Lett.* **106** (2011) 132503.
19. M. S. Moore *et al.*, *Phys. Rev. C* **18** (1978) 1328.
20. A. Behkami *et al.*, *Phys. Rev.* **171** (1969) 1267.

The oxygen isotopes

B. Alex Brown
*Department of Physics and Astronomy
and National Superconducting Cyclotron Laboratory
Michigan State University, East Lansing,
Michigan 48824-1321, USA*

The properties of the oxygen isotopes provide diverse examples of progress made in experiments and theory. This chain of isotopes has been studied from beyond the proton drip line in ^{12}O to beyond the neutron drip line in ^{25,26}O. This short survey starts with the microscopic G matrix approach for ^{18}O of Kuo and Brown in the 1960's and shows how theory has evolved. The nuclear structure around the doubly-magic nucleus ^{24}O is particularly simple in terms of the nuclear shell model. The nuclear structure around the doubly-magic nucleus ^{16}O exhibits the coexistence of single-particle and collective structure.

Keywords: Shell model; oxygen isostopes.

1. Introduction

The properties of the oxygen isotopes provide diverse examples of progress made in experiment and theory. The known isotopes are shown in Fig. 1. There are three stable isotopes ^{16,17,18}O. Due to the temperature-dependent effect of the mass difference on evaporation and condensation, measurement of the changes in the isotopic ratio ^{18}O/^{16}O in ice cores provides information on historic climate changes.[1] With the advent of radioactive beams, there is now a nearly complete set of data known for all of the oxygen isotopes.[2] ^{12}O lies beyond the proton drip line and is observed to decay by emission of two protons ^{25}O lies beyond the neutron drip line and is observed to decay by the emission of one neutron. ^{26}O lies even further beyond the neutron drip line and is observed to decay by the emission of two neutrons. The root-mean-square (RMS) charge radii and charge form factors for the stable isotopes have been measured with electron scattering.[2] The RMS matter radii for ground states out to ^{24}O have been measured from interaction cross-sections[3,4] and elastic proton scattering.[5–7]

The properties of ^{16,17,18}O provided the first ingredients for the formulation of nuclear theory in terms of configuration–interaction (CI) models. $Z = 8$ is a magic number for all of the oxygen isotopes except ^{12}O. There are three strongly doubly-magic nuclei ^{14,16,24}O ($N = 6, 8, 16$) shown in dark red in Fig. 1, all with first excited states above 4 MeV, and a weaker doubly-magic nucleus, ^{22}O ($N = 14$), with a first excited state at 3.2 MeV shown in dark red. In the shell model, the

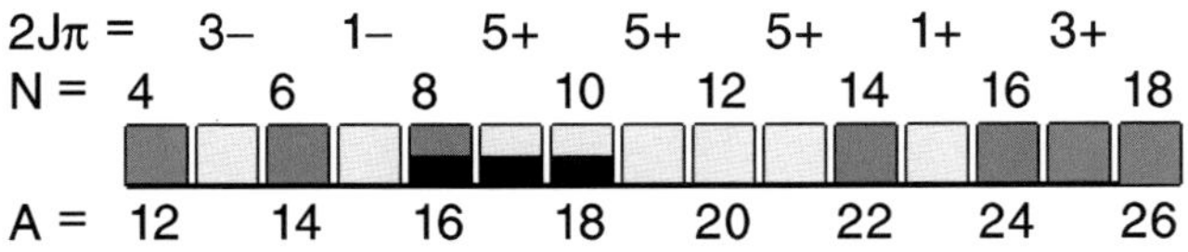

Fig. 1. (Color online) The oxygen isotopes. A is the mass number and N is the neutron number. $2J\pi$ is the ground state spin-parity for the ground states of the odd-neutron nuclei. The black rectangles show the three stable isotopes. The blue squares are those nuclei that are observed to be beyond the proton (left) and neutron (right) drip lines. The red squares are the doubly-magic even–even nuclei.

magic number arises when the neutron shell gaps at these neutron numbers are larger than two times the pairing gap.[8] This is the start of a rich doubly-magic number sequence observed up to ^{54}Ca.[9,10] The ground state spin of the odd–even nuclei shown in Fig. 1 are all explained by the sequential filling of the shell-model orbitals $0p_{3/2}(3/2^-)$, $0p_{1/2}(1/2^-)$, $0d_{5/2}(5/2^+)$, $1s_{1/2}(1/2^+)$ and $0d_{3/2}(3/2^+)$.

2. Structure for the *sd* Model Space Based on the Free Nucleon–Nucleon Interaction

I was a graduate student at Stony Brook from 1970–1973 working on experiments with my adviser Dave Fossen. During that time, I took several courses in nuclear theory, and had many interactions with Brown's nuclear theory group and their visitors. I will start the theory part of this review with the work of Kuo and Brown in the late 1960's.[11–15] Kuo and Brown planted many trees which have matured and left their seeds for the growth of nuclear theory. Their goal was to investigate the applicability of the free nucleon–nucleon potential determined by the scattering data for the shell-model description of finite nuclei. The first example they gave was for ^{18}O, where they started with a doubly closed-shell configuration for ^{16}O and calculated the spectrum of ^{18}O in the $(1s_{1/2}, 0d_{5/2}, 0d_{3/2})$ *sd* model space. The Brueckner reaction matrix G for the scattering of two nucleons evaluated with various methods was used to take into account the repulsive short-ranged part of the potential. The method of second-order core-polarization was used to renormalize the G-matrix elements into the *sd* model space. The end result was a set of two-body matrix elements (TBME) (63 of them) for the *sd* model space[15] that could be diagonalized to obtain the energies of ^{18}O relative to that of ^{16}O. The resulting energy spectrum is shown in Fig. 2 and compared to experiment. The agreement was excellent for some levels, but there are more levels observed in experiment compared to those calculated. These extra levels come from the excitation of protons from the p shell to the *sd* shell. There is mixing between these two sets,[16] but not enough to destroy the agreement for those dominated by *sd*. As one adds neutrons, these core-excited states move to higher energy. The Kuo–Brown TBME gave a starting point for the *sd* shell-model calculations beyond ^{18}O.

When the Kuo–Brown TBME are used for more neutron-rich oxygen isotopes, the agreement of the calculated spectra with experiment deteriorates. The reason for

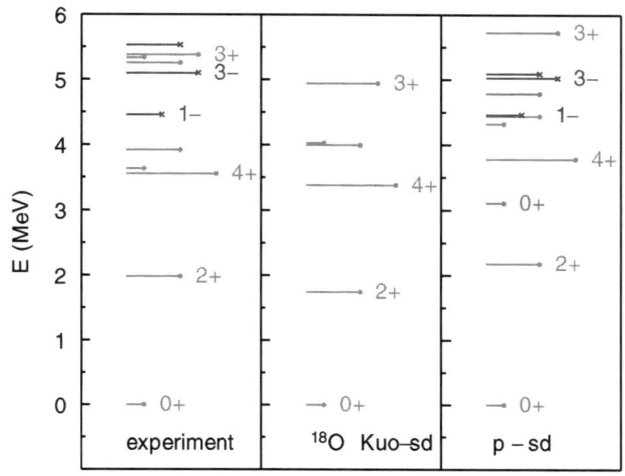

Fig. 2. (Color online) Energy levels for ^{18}O. Experiment is shown in the left-hand side with red lines for positive parity states and blue lines for negative parity states. The length of the line is proportional to the J value with a few lines labeled for reference. The middle panel shows the spectrum for positive parity states obtained with the Kuo–Brown Hamiltonian. The right-hand side shows the spectrum based on the WBP Hamiltonian from Ref. 17 that shows the sd-shell positive parity states (red), the $3p$-$1h$ negative parity states (blue) and the $4p$-$2h$ positive parity states (green).

this can be understood from a plot of the effective-single particle energies (ESPE) calculated relative to the closed-shell configurations $(0d_{5/2})^6$ for ^{22}O ($N = 14$), $(0d_{5/2})^6(1s_{1/2})^2$ for ^{24}O ($N = 16$) and $(0d_{5/2})^6(1s_{1/2})^2(0d_{3/2})^4$ for ^{28}O ($N = 28$). These are shown on the left-hand side of Fig. 3. The red dots indicate the position of the Fermi surface. The small gap at $N = 14$ means that the spectrum of the $(sd)^6$ calculation for ^{22}O does not have a closed-shell behavior in contradiction to experiment. The negative energy of the red dot at $N = 20$ means that ^{28}O is bound to neutron decay in contradiction to experiment.

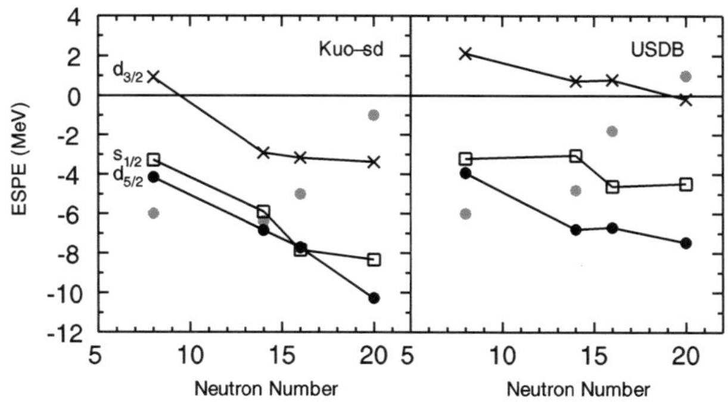

Fig. 3. (Color online) Effective single-particle energies obtained for the oxygen isotopes obtained with Kuo–Brown Hamiltonian[15] on the left and the USDB Hamiltonian[18] on the right. The orbital quantum numbers are shown on the left. The red dots indicate the position of the Fermi surface.

There have been two paths for correcting this problem. The first of these started in the 1980's by showing that one could find a universal set of TBME for the *sd*-shell Hamiltonian that unify the understanding of data related to states with an *sd* shell configuration. The second path which has started more recently involves the addition of three-body interactions together with improved techniques for renormalization. The first path has led to many similar local solutions involving the TBME such as those in the *pf* shell.[19] The second path should eventually lead to a universal solution for all nuclei.

3. The USD Hamiltonians

In the first method, known experimental data for binding energies and excitation energies are used to obtain empirical constraints on values of the TBME. The single-valued decomposition (SVD) method is used to find $n < 63$ linear combinations of TBME that are well determined by the data. The remaining linear combinations are fixed at the values given by the Kuo–Brown Hamiltonian. This first SVD fit was carried out by Chung[20] and Wildenthal.[21] Since one set of TBME could describe the properties of all states dominated by the *sd* shell configuration in nuclei with $A = 18$ to $A = 38$ this was called the universal *sd* (USD) interaction (W in the NuShellX Hamiltonian library[22]). The USD Hamiltonian was used to calculate observables for all *sd* shell nuclei.[23]

In 2006, an updated set of experimental energy data was used to obtain two new universal interactions, USDA and USDB.[18] This used about 600 data from $A = 17$ to $A = 40$. USDA is based on 30 SDV combinations, and USDB is based on 56 SDV combinations. The ESPE shown on the right-hand side of Fig. 3 are based on USDB. Compared to Kuo–Brown, one finds a shell gap at $N = 14$ and the Fermi surface for $N = 20$ is near zero. In the following, I will compare some experimental results to those obtained with USDB. The energies of states obtained with USDA are the same as those with USDB to within about 100 keV.

The binding energy of states in $^{22-28}$O (relative to ^{16}O) obtained with USDB are shown by the horizontal lines in Fig. 4. The data known in 2006 used to constrain the USDB TBME are shown by the three blue dots. All other horizontal lines are the energies predicted for states that were not known in 2006. All of these are unbound to neutron decay as shown by the arrows. With USDB, ^{24}O lies on the neutron drip line. Since 2006 many experiments have been carried out to measure these neutron decays. The experimental centroids of these decay resonances are shown by the red dots in Fig. 4. The interpretation of these results is particularly simple due to the fact that the *sd* wave functions for these nuclei near the doubly-magic nucleus ^{24}O are dominated by simple configurations, as we will discuss below.

In 2007, an unbound level in ^{23}O consistent with a $3/2^+$ assignment was observed at RIKEN[24] that is in good agreement with the prediction. Also in 2007 a low energy neutron from the decay of ^{23}O was observed at the NSCL which was associated with the predicted $5/2^+$ state just above the neutron-decay threshold.[25] The

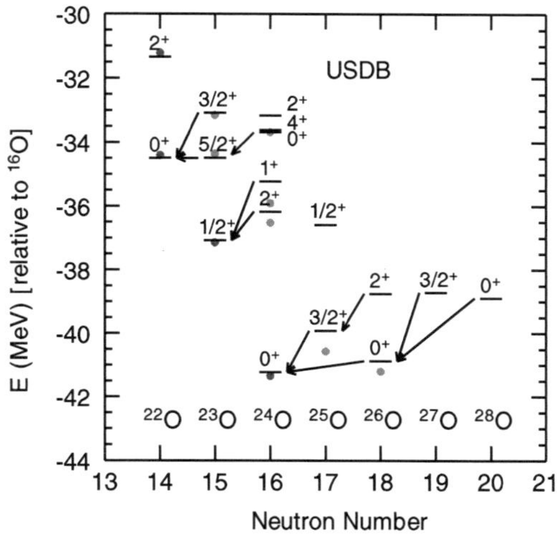

Fig. 4. (Color online) Energies of the states relative to ^{16}O as a function of mass number for the oxygen isotopes. The horizontal lines are the USDB predictions. The blue dots are experimental energies known before 2006. The arrows show the neutron decays discussed in the text. The red dots are the centroid energies observed for these neutron decays.

dominant configuration for this state is $(0d_{5/2})^5(1s_{1/2})^2$, and the calculated spectroscopic factor of 0.059 is small because it must go through the $(0d_{5/2})^4(1s_{1/2})^2$ admixture into the dominant $(0d_{5/2})^6$ configuration for the ^{22}O ground state. From the observed neutron-decay centroid energy of 40 keV one predicts a neutron-decay width of 10 eV (much smaller than the experimental resolution).

In 2009, the decay of two excited states of ^{24}O were observed[26] at an energy consistent with the 2^+ and 1^+ states in Fig. 4. This experiment showed that the first excited state in ^{24}O was high (4.0 MeV) confirming the doubly closed-shell nature of ^{24}O. The centroid of these two states observed are several hundred keV lower than those predicted (2^+ and 1^+). The configuration of these two excited states is $(0d_{5/2})^6(1s_{1/2})(0d_{3/2})$. In 2012, these two states were studied with proton scattering in inverse kinematics,[27] and the measured β_2 value for the 2^+ state was found to be in good agreement with the USDB calculations. In addition a state near 7.3 MeV was observed to neutron decay which indicates it may have the configuration $(0d_{5/2})^6(1s_{1/2})(pf)$ and have negative parity. Further experimental support for the USDB 2^+ and 1^+ predictions was obtained in 2015.[28] In 2015, the two-neutron decay of a state at 7.65(20) MeV was observed.[29] This is probably one of the states shown in Fig. 2 near this energy that have the dominant configuration $(0d_{5/2})^5(1s_{1/2})^2(0d_{3/2})$. From this configuration, one predicts a sequential decay through the $(0d_{5/2})^5(1s_{1/2})^2 5/2^+$ state in ^{23}O discussed above, in agreement with experiment.[29] The main configuration of the 0^+ state near 7.3 MeV (Fig. 3) is $(0d_{5/2})^6(0d_{3/2})^2$. It should be observed in a ^{22}O(t,p) experiment.

In 2008, the neutron decay of ^{25}O was observed[30] which was associated with the predicted $3/2^+$ state. The experimental centroid energy of this state of 770 keV is about 500 keV lower than predicted by USDB. The calculated spectroscopic factor $C^2S = 0.95$ results in a calculated neutron decay width of 79 keV (using the experimental decay Q value). The experimental width of 172(30) keV obtained in Ref. 30 is larger than that calculated, but a more recent experiment with higher statistics obtained is smaller of 88(6) keV,[31] in agreement with the prediction.

Predictions have been made for low-lying $3/2^-$ and $7/2^-$ states in ^{25}O coming from the pf shell orbitals.[32,33] The small gap between the $0d_{3/2}$ and the $1p_{3/2}-0f_{7/2}$ single-particle energies near ^{24}O is a consequence of the tensor interaction[34] together with a weak-binding effect that lowers the $1p_{3/2}$ single particle energy relative to that of $0f_{7/2}$.[35] The island of inversion[36] for nuclei near $N = 20$ and $Z \leq 12$ is due to this reduced energy gap. For these nuclei, the configurations of the type $(sd)^{n-2}(pf)^{m+2}$ come lower in energy than those for $(sd)^n(pf)^m$. A similar situation happens for $N = 8$ and $Z \leq 4$, where the shell gap between $0p_{1/2}$ and $1s_{1/2} - 0d_{5/2}$ is reduced leading to a configuration inversion in ^{12}Be (Ref. 37). This configuration inversion is also manifested in the properties[38–40] of the mirror nucleus ^{12}O that lies just beyond the proton drip line (see Fig. 1). At present, there is only a limit of 72 keV[39] on the two-proton decay width of ^{12}O.

The two neutron decay of ^{26}O–^{24}O has been observed.[31,41–43] In Ref. 42 a lifetime of 4.5(3.0) ps was obtained for this decay. In Ref. 31 the two neutrons were observed to be only 18(7) keV above threshold. The USDB prediction was 360 keV. The predictions obtained with many other theoretical models are shown in Ref. 44. Grigorenko has calculated the three-body decay of ^{26}O (Ref. 45). Based on his model, the lifetime for a decay energy of 18 keV should be on the order of 10^{-5}–10^{-3} ps. This three-body decay can provide unique insights into the two-neutron correlations. A better understanding of this three-body decay will be a challenge for future studies.

^{27,28}O have not been observed in radioactive beam experiments[46] indicating that they are neutron unbound, consistent with the *sd* shell calculations shown in Fig. 4 as well as with the calculations discussed below. But there are several low-lying resonances in ^{25}O (Ref. 33) that require a basis with the continuum. Since ^{26}O is unbound to two-neutron decay by only 18 keV one could speculate that there might be a four-neutron cluster bound state of ^{28}O. The tetra-neutron is unbound,[47] but the induced correlations in the nucleus could result in a state of ^{28}O with an extremely small binding that would make its production cross-section smaller than the limit observed experimentally.[46]

The overall agreement of experiment with the predictions of the USDB CI calculations for $^{23-26}$O is excellent. This success is due to the ability of all of the data related to *sd* shell configurations to be described in terms of a universal set of TBME. The TBME needed for $^{23-26}$O come from experimental energy data for $Z > 8$ as well as from excited states in $^{19-22}$O. These excited state configurations become important for the low-lying states of $^{23-26}$O. Experiments on $^{19-20}$O such as $^{18}\mathrm{O}(t,p)^{20}$ (Ref. 48), and from projectile fragmentation reactions (Ref. 49), were an important intermediate step in establishing the success of the CI calculations.

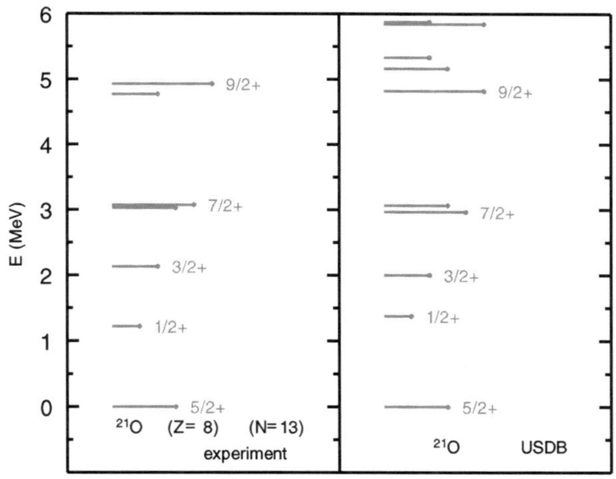

Fig. 5. Energies of the states for ^{21}O. The length of the line is proportional to the J value. The experimental results from Ref. 49 (left) are compared to the USDB predictions (right).

The experimental results for ^{21}O obtained from Ref. 49 are compared to the USDB calculations in Fig. 5. The agreement between experiment and theory is excellent except that the $3/2^+_2$ state is about 0.5 MeV too high in theory. The $9/2^+$ state is unbound to neutron decay by 1.1 MeV. Its observation in gamma decay puts a limit of 0.029 on the spectroscopic factor for this $\ell = 4$ neutron decay (in the sd model space the spectroscopic factor is zero). The spectroscopic factors obtained from an ^{20}O$(d,p)^{21}$O experiment in inverse kinematics[50] are in good agreement with the USDB predictions.

4. The Three-Nucleon Interaction and the Development of New Theoretical Methods

In the last decade, a new class of theoretical models has been developed. Because of the relatively simple shell structure of the neutron-rich oxygen isotopes, this has been the primary testing ground for these calculations. These are similar in spirit to the original Kuo–Brown methods but expand upon them in several ways. (1) The input for the basic interaction is based upon a consistent expansion from chiral EFT in terms of two- and three-nucleon interactions.[51,52] (2) Better methods to treat short-range correlations have been developed with the $V_{\text{low}k}$ and free-space similarity renormalization group (SRG) approaches.[53] (3) Due to advances in the computation techniques one can include up about 15 oscillator shells in the basis. (3) Better perturbative and non-perturbative methods have been formulated. (4) The continuum can be taken into account.[54–56] Calculations including the continuum[57] have been used to partially understand the quenching for the spectroscopic factors observed for deeply bound orbitals based upon the ratio of experimental knock-out cross-sections compared to those calculated from eikonal reaction theory with shell-model spectroscopic factors.[58]

Three-nucleon interactions are required to reproduce the magic number at ^{22}O ($N = 14$) and the position of the oxygen drip line at ^{24}O ($N = 16$).[59–62] When evaluated for an ^{16}O core, the main three-body terms are those involving the interaction of one valence neutron with two nucleons on the core, and the interaction of two valence neutrons and one nucleon in the core. This explains why effective one- plus two-body Hamiltonians of the USD type have been so successful.

Another place in the oxygen isotopes where the three-nucleon interaction is important is in the Gamow–Teller beta-decay of ^{14}O to the ground state of ^{14}N where the B(GT) is very small [B(GT) $= 3 \times 10^{-4}$].[63] The B(GT) is even smaller in the mirror transition from ^{14}C [B(GT) $= 4 \times 10^{-6}$],[63] and this results in the anomalously long half-life of 5700 years for ^{14}C that makes it so usefull for radiocarbon dating. Brown and others[64] showed that this small B(GT) can be explained by the Brown–Rho scaling model,[65,66] where the strength of the tensor interaction between nucleons is density dependent. Later it was shown that the density-dependent effects from the Brown–Rho model are similar to those from a three-nucleon force.[67] In large-basis CI calculations for ^{14}C, it was shown that the three-body interaction is important for obtaining the small B(GT) values.[68]

A commonly used input Hamiltonian due to Entem and Machleidt (EM)[51,52] is based upon an N^{3}LO two-nucleon plus a N^{2}LO three-nucleon interaction with low-energy coupling constants derived from nucleon–nucleon scattering and the properties of the $A = 3$ and $A = 4$ nuclei. The binding energies for the oxygen isotopes obtained with the EM input based on several types of non-perturbative methods[69–74] are consistent with each other and with experiment within 1–2 MeV (1% of the total binding energy). Excited states for neutron-rich nuclei from these calculations agree with each other and with experiment at the level of about 0.5 MeV. However, the intruder states in ^{18}O are not in the basis.

The RMS radii obtained with the EM interaction input is about 10% smaller than experiment for all of the oxygen isotopes.[75] This has lead to the development of a new Hamiltonian called NNLO$_{\text{sat}}$ that is simpler than EM, but where the determination of the low-energy coupling constants includes data on the binding energies and charge radii of ^{14}C and ^{16,22,24,25}O as well as the charge radii of ^{14}C and ^{16}O.[76,77] The matter radii for the neutron-rich oxygen isotopes with NNLO$_{\text{sat}}$ is much improved compared to experiment relative to those based on the EM input.[75] From the interaction cross-section experiments,[3,4] there is sharp increase in the matter radii between ^{22}O and ^{24}O where the $1s_{1/2}$ orbital is being filled that is only partly explained by the NNLO$_{\text{sat}}$ calculations as well as by energy-density functional calculations.[78] An increase is also observed in charge radii in calcium isotopes just after $N = 28(1p_{3/2})$,[79] the strontium isotopes just after $N = 50(1d_{5/2}$, the xenon isotopes just after $N = 82(2f_{7/2})$ and the lead isotopes just after $N = 126(2g_{9/2})$, where the given orbital is the one that starts to be filled after that magic number. All of these orbitals have an interior node whose self-consistent interactions with the core nucleons are determined by the density dependence of the equation-of-state.

Work is currently being carried out to use these non-perturbative methods to generate nucleus-dependent valence space Hamiltonians for the *sd* shell using the reference states from the targeted valence space (TVS) IM-SRG method.[80,81] It remains to be seen how or if a universal valence space Hamiltonian such as USD emerges from these calculations. Work is also started on the calculations for the nuclear moments and electromagnetic decay properties. In the CI calculations, the electric quadrupole moments and B(E2) for the oxygen isotopes depend upon the neutron effective charge which empirically is about 0.45 e.[82] [The B(E2) for ^{18}O are not well described by the *sd* shell wave functions due to the mixing with the intruder states[16,83]]. A question for the future will be to see if a universal set of low-energy coupling constants can found to reproduce experimental properties of all nuclei. If this is not the case, it may be due to real and induced four-body and higher interactions that are at present not included in the non-perturbative calculations.

5. Beyond the *sd* Shell

The methods discussed in the last section are mostly based on simple Hartree–Fock configurations for the starting point (the reference state). For example, a closed shell for ^{16}O or the $(0d_{5/2})^6$ configuration for ^{22}O. This cannot account for states whose major configuration involves core excitations. The first excited state of ^{16}O is a 0^+ state at 6.05 MeV. Its configuration is dominated by four-particles and four-holes (4*p*–4*h*) relative to the closed-shell configuration. The 0^+ state at 3.63 MeV in ^{18}O shown in Fig. 1 is dominated by a 4*p*–2*h* configuration.[16] Due to the large number or orbitals involved in these particles and holes, these intruder states are often deformed in the CI basis, and they were first discussed in terms of the Brown–Green model for coexisting spherical and deformed configurations.[84–88] Calculations have been made using the weak-coupling model where the correlations between nucleons in the same major shell are dominant and the cross-shell interaction is treated as a small perturbation.[83,89] The deformed configurations are often truncated in an SU_3 basis.[90]

Fortune used the Bansal–French–Zamick weak coupling model[91] for the 6*p*–4*h* states in ^{18}O (Ref. 92), which he suggests starts with the 0^+ state at 7.11 MeV. The success of the methods based on the spherical Hartree–Fock starting points must be that the intruder states largely decouple (coexist) with the spherical *sd* shell configurations. But this is an approximation, and ultimately one must consider both kinds of states at the same time.

Beyond the Brown–Green model, several empirical shell-model-based approaches have been made. Use of just the three orbitals $(0p_{1/2}, 0d_{5/2}, 1s_{1/2})$ closest to the Fermi surface of ^{16}O (the ZBM model space) gives a starting point and some insights.[93,94] The empirical WBP and WBT Hamiltonians[17] were developed for pure (unmixed) particle-hole configurations involving the orbitals $(0p_{3/2}, 0p_{1/2}, 0d_{5/2}, 1s_{1/2}, 0d_{3/2})$ for the mass region $A = 10$–20. The results for the

$2p$, $3p$–$1h$ and $4p$–$2h$ configuration of ^{18}O are shown on the right-hand side of Fig. 2. The $2p$ (red) and $4p$–$2h$ (green) states mix,[16] leading to the experimental spectrum on the left-hand side of Fig. 2. The problem for mixing of these states due to the truncation in the np–mh sequence is discussed in Ref. 95. For the future, one might be able to use the new many-body methods to derive a Hamiltonian for the full p–sd model space as used in Ref. 96. One should also consider alpha clustering such as with the algebraic cluster model for ^{16}O which describes the strong electron scattering transitions from the ground state to the members of a $3^-, 4^+, 6^+$ (A_1) band.[97]

In summary, I have shown that the observed properties of the oxygen isotopes have provided a rich, and in some cases simple, set of observables which have been used over the past 50 years to develop and test the theories for nuclear structure. There are still problems to solve.

Acknowledgments

I acknowledge useful discussions with Heiko Hergert and support from NSF grant PHY-1404442.

References

1. K. K. Andersen *et al.*, *Nature* **431** (2004) 147.
2. http://www.nndc.bnl.gov/ensdf/.
3. A. Ozawa, T. Susuki and I. Tanihata, *Nucl. Phys. A* **693** (2001) 32.
4. R. Kanungo *et al.*, *Phys. Rev. C* **84** (2011) 061304(R).
5. E. Fabrici, S. Micheletti, M. Pignanelli, F. G. Resmini, R. De Leo, G. D'Erasmo and A. Pantaleo, *Phys. Rev. C* **21** (1980) 844.
6. E. Khan *et al.*, *Phys. Lett. B* **490** (2000) 45.
7. MUST Collab. (E. Becheva *et al.*), *Phys. Rev. Lett.* **96** (2006) 012501.
8. B. A. Brown, *Phys. Rev. Lett.* **111** (2013) 162502.
9. B. A. Brown and W. A. Richter, *Phys. Rev. C* **72** (2005) 057301.
10. D. Steppenbeck *et al.*, *Nature* **502** (2013) 207.
11. T. T. S. Kuo and G. E. Brown, *Nucl. Phys.* **85** (1966) 40.
12. T. T. S. Kuo, *Nucl. Phys. A* **90** (1967) 199.
13. R. D. Lawson, M. H. Macfarlane and T. T. S. Kuo, *Phys. Lett.* **22** (1966) 168.
14. G. E. Brown and T. T. S. Kuo, *Nucl. Phys. A* **92** (1967) 481.
15. T. T. S. Kuo, *Nucl. Phys. A* **103** (1967) 71.
16. R. L. Lawson, F. J. D. Serduke and H. T. Fortune, *Phys. Rev. C* **14** (1976) 1245.
17. E. K. Warburton and B. A. Brown, *Phys. Rev. C* **46** (1992) 923.
18. B. A. Brown and W. A. Richter, *Phys. Rev. C* **74** (2006) 034315.
19. M. Honma, T. Otsuka, B. A. Brown and T. Mizusaki, *Phys. Rev. C* **69** (2004) 034335.
20. W. Chung, Empirical renormalizations of shell-model Hamiltonians and magnetic dipole moments of sd-shell nuclei, Ph.D. thesis, Mich. State University, East Lansing (1976).
21. B. H. Wildenthal, *Prog. Part. Nucl. Phys.* **11** (1984) 5.
22. B. A. Brown and W. D. M. Rae, *Nucl. Data Sheets* **120** (2014) 115.
23. B. A. Brown and B. H. Wildenthal, *Ann. Rev. Nucl. Part. Sci.* **38** (1988) 29.
24. Z. Elekes *et al.*, *Phys. Rev. Lett.* **98** (2007) 102502.

25. A. Schiller *et al.*, *Phys. Rev. Lett.* **99** (2007) 112501.
26. C. R. Hoffman *et al.*, *Phys. Lett. B* **672** (2009) 17.
27. K. Tshoo *et al.*, *Phys. Rev. Lett.* **109** (2012) 022501.
28. W. F. Rogers *et al.*, *Phys. Rev. C* **92** (2015) 034316.
29. M. D. Jones *et al.*, *Phys. Rev. C* **92** (2015) 051306(R).
30. C. R. Hoffman *et al.*, *Phys. Rev. Lett.* **100** (2008) 152502.
31. Y. Kondo *et al.*, *Phys. Rev. Lett.* **116** (2016) 102503.
32. A. Lapailleur *et al.*, *Phys. Rev. C* **92** (2015) 054309.
33. K. Hagino and H. Sagawa, *Phys. Rev. C* **93** (2016) 034330.
34. T. Otsuka, R. Fujimoto, Y. Utsuno, B. A. Brown, M. Honma and T. Mizusaki, *Phys. Rev. Lett.* **87** (2001) 082502.
35. I. Hamamoto, *Phys. Rev. C* **76** (2007) 054319.
36. E. K. Warburton, J. A. Becker and B. A. Brown, *Phys. Rev. C* **41** (1990) 1147.
37. H. Sagawa, B. A. Brown and H. Esbensen, *Phys. Lett. B* **309** (1993) 1.
38. D. Suzuki *et al.*, *Phys. Rev. C* **93** (2016) 024316.
39. M. F. Jager *et al.*, *Phys. Rev. C* **86** (2012) 011304(R).
40. H. T. Fortune and R. Sherr, *J. Phys. G* **40** (2013) 055102.
41. E. Lunderberg, *Phys. Rev. Lett.* **108** (2012) 142503.
42. Z. Kohley *et al.*, *Phys. Rev. Lett.* **110** (2013) 152501.
43. C. Caesar *et al.*, *Phys. Rev. C* **88** (2013) 034313.
44. M. Thoennessen *et al.*, *Acta Phys. Polon. B* **44** (2013) 543.
45. L. V. Grigorenko, I. G. Mukha and M. V. Zhukov, *Phys. Rev. Lett.* **111** (2013) 042501.
46. O. Tarasov *et al.*, *Phys. Lett. B* **409** (1997) 64.
47. K. Kisamori *et al.*, *Phys. Rev. Lett.* **116** (2016) 052501.
48. S. LaFrance *et al.*, *Phys. Rev. C* **20** (1979) 1673.
49. M. Stanoiu *et al.*, *Phys. Rev. C* **69** (2004) 034312.
50. B. Fernandez-Dominguez *et al.*, *Phys. Rev. C* **84** (2011) 011301(R).
51. R. Machleidt and D. R. Entem, *Phys. Rep.* **503** (2011) 1.
52. D. R. Entem and R. Machleidt, *Phys. Rev. C* **68** (2003) 041001.
53. S. K. Bogner, R. J. Furnstahl and A. Schwenk, *Prog. Part. Nucl. Phys.* **65** (2010) 94.
54. A. Volya and V. Zelevinsky, *Phys. Rev. C* **74** (2006) 064314.
55. N. Michel, W. Nazarewicz, M. Ploszajczak and T. Vertse, *J. Phys. G* **36** (2009) 013101.
56. K. Tsukiyama, M. Hjorth-Jensen and G. Hagen, *Phys. Rev. C* **80** (2009) 051301(R).
57. O. Jensen, G. Hagen, M. Hjorth-Jensen, B. A. Brown and A. Gade, *Phys. Rev. Lett.* **107** (2011) 032501.
58. A. Gade *et al.*, *Phys. Rev. C* **77** (2008) 044306.
59. T. Otsuka, T. Suzuki, J. D. Holt, A. Schwenk and Y. Akaishi, *Phys. Rev. Lett.* **105** (2010) 032501.
60. G. Hagen, M. Hjorth-Jensen, G. R. Jansen, R. Machleidt and T. Papenbrock, *Phys. Rev. Lett.* **108** (2012) 242501.
61. J. D. Holt, J. Menendez and A. Schwenk, *Phys. Rev. Lett.* **110** (2013) 022502.
62. A. Cipollone, C. Barbieri and P. Navratil, *Phys. Rev. Lett.* **111** (2013) 062501.
63. W.-T. Chou, E. K. Warburton and B. A. Brown, *Phys. Rev. C* **47** (1993) 163.
64. J. W. Holt, G. E. Brown, T. T. S. Kuo, J. D. Holt and R. Machleidt, *Phys. Rev. Lett.* **062501** (2008).
65. G. E. Brown and M. Rho, *Phys. Rev. Lett.* **66** (1991) 2720.
66. G. E. Brown and M. Rho, *Phys. Rep.* **396** (2004) 1.
67. J. W. Holt, N. Kaiser and W. Weise, *Phys. Rev. C* **79** (2009) 054331.
68. P. Maris, J. P. Vary, P. Navratil, W. E. Ormand, H. Nam and D. J. Dean, *Phys. Rev. Lett.* **106** (2011) 202502.

69. G. R. Jansen, J. Engel, G. Hagen, P. Navratil and A. Signoracci, *Phys. Rev. Lett.* **113** (2014) 142502.
70. S. Binder, J. Langhammer, A. Calci and R. Roth, *Phys. Lett. B* **736** (2014) 119.
71. A. Cipollone, C. Barbieri and P. Navratil, *Phys. Rev. C* **92** (2015) 014306.
72. H. Hergert, S. K. Bogner, S. Binder, A. Calci, J. Langhammer, R. Roth and A. Schwenk, *Phys. Rev. C* **87** (2013) 034307.
73. H. Hergert, S. Binder, A. Calci, J. Langhammer and R. Roth, *Phys. Rev. Lett.* **110** (2013) 242501.
74. S. K. Bogner, H. Hergert, J. D. Holt, A. Schwenk, S. Binder, A. Calci, J. Langhammer and R. Roth, *Phys. Rev. Lett.* **113** (2014) 142501.
75. V. Lapoux, V. Soma, C. Barbieri, H. Hergert, J. D. Holt and S. R. Stroberg, *Phys. Rev. Lett.* **117** (2016) 052501.
76. A. Ekstrom, G. R. Jansen, K. A. Wendt, G. Hagen, T. Papenbrock, B. D. Carlsson, C. Forssn, M. Hjorth-Jensen, P. Navrtil and W. Nazarewicz, *Phys. Rev. C* **91** (2015) 051301.
77. B. D. Carlsson, A. Ekstrom, C. Forssn, D. F. Strmberg, G. R. Jansen, O. Lilja, M. Lindby, B. A. Mattsson and K. A. Wendt, *Phys. Rev. X* **6** (2016) 011019.
78. B. A. Brown, S. Typel and W. A. Richter, *Phys. Rev. C* **65** (2002) 014612.
79. R. F. Garcia Ruiz *et al.*, *Nat. Phys.* **12** (2016) 594.
80. S. R. Stroberg, H. Hergert, J. D. Holt, S. K. Bogner and A. Schwenk, *Phys. Rev. C* **93** (2016) 051301(R).
81. S. R. Stroberg, A. Calci, H. Hergert, J. D. Holt, S. K. Bogner, R. Roth and A. Schwenk, *Phys. Rev. Lett.* **118** (2017) 032502.
82. W. A. Richter and B. A. Brown, *Phys. Rev. C* **80** (2009) 034301.
83. P. J. Ellis and T. Engeland, *Nucl. Phys. A* **144** (1970) 161; **A181** (1972) 368.
84. T. Engeland, *Nucl. Phys.* **72** (1965) 68.
85. G. E. Brown and A. M. Green, *Nucl. Phys.* **75** (1966) 401; **85** (1966) 87.
86. G. E. Brown and A. M. Green, *Phys. Lett.* **15** (1965) 168.
87. T. Erikson, K. F. Quader, G. E. Brown and H. T. Fortune, *Nucl. Phys. A* **465** (1987) 123.
88. T. Erikson, *Nucl. Phys. A* **170** (1971) 513; *Nucl. Phys. A* **211** (1973) 105.
89. A. Arima, H. Horiuchi and T. Sebe, *Phys. Lett. B* **24** (1967) 129.
90. J. P. Elliott, *Proc. R. Soc. A* **245** (1958) 129; J. P. Elliott and M. Harvey, *Proc. R. Soc. A* **272** (1963) 557.
91. R. Bansal and J. B. French, *Phys. Lett.* **11** (1964) 145; L. Zamick, *Phys. Lett.* **19** (1965) 580.
92. H. T. Fortune, *Phys. Rev. C* **18** (1978) 1053.
93. A. P. Zuker, B. Buck and J. B. McGrory, *Phys. Rev. Lett.* **21** (1968) 39.
94. A. P. Zuker, *Phys. Rev. Lett.* **23** (1969) 983.
95. E. K. Warburton, B. A. Brown and D. J. Millener, *Phys. Lett. B* **293** (1992) 7.
96. Y. Utsuno and S. Chiba, *Phys. Rev. C* **83** (2011) 021301(R).
97. R. Bijker and F. Iachello, *Nucl. Phys. A* **957** (2017) 154.

Lessons on doing science from my father, Gerry Brown

C. Titus Brown
Population Health and Reproduction,
School of Veterinary Medicine,
University of California, Davis, CA 95616, USA
ctbrown@ucdavis.edu

My father Gerry Brown practiced science in a certain way during his life, but much of his approach to work and research was not clearly articulated. In this article I convey some of the lessons I learned growing up with him.

Keywords: Mentoring; career; physics.

Dr. Gerald E. Brown was a well-known nuclear physicist and astrophysicist who worked at Stony Brook University from 1968 until his death in 2013. He was internationally active in physics research from the late 1950's onwards, ran an active research group at Stony Brook until 2009, and supervised nearly a hundred Ph.D students during his life. He was also my father.

It's hard to write about someone who is owned, in part, by so many people. I came along late in my father's life (he was 48 when I was born), and so I didn't know him that well as an adult. However, growing up with a senior professor as a father had a huge impact on my scientific career, which I can recognize even more clearly now that I'm a professor myself.

Gerry (as I called him) didn't spend much time directly teaching his children about his work. When I was invited to write something for his memorial book, it was suggested that I write about what he had taught me about being a scientist. I found myself stymied, because to the best of my recollection we had never talked much about the practice of science. When I mentioned this to my oldest brother, Hans, we shared a good laugh — he had exactly the same experience with our father, 20 years before me!

Most of what Gerry taught me was taught by example. Below are some of the examples that I remember most clearly, and of which I'm the most proud. While I don't know if either of my children will become scientists, if they do, I hope they take these to heart — I can think of few better scientific legacies to pass on to them from my father.

Publishing work that is interesting (but perhaps not correct) can make for a fine career

My father was very proud of his publishing record, but not because he was always (or even frequently) right. In fact, several people told me that he was somewhat notorious for having a 1-in-10 "hit rate" — he would come up with many crazy ideas, of which only about 1 in 10 would be worth pursuing. However, that 1 in 10 was enough for him to have a long and successful career. That this drove some people nuts was merely an added bonus in his view.

Gerry was also fond of publishing controversial work. Several times he told me he was proudest of the papers that caused the most discussion and collected the most rebuttals. He wryly noted that these papers often gathered many citations, even when they turned out to be incorrect.

The best collaborations are both personal and professional friendships

The last twenty-five years of Gerry's life were dominated by a collaboration with Hans Bethe on astrophysics, and they traveled to Pasadena every January until the early 2000s to work at the California Institute of Technology. During this month, they lived in the same apartment, worked together closely, and met with a wide range of people on campus to explore scientific ideas; they also went on long hikes in the mountains above Pasadena (chronicled by Chris Adami in Ref. 1). These close interactions not only fueled his research for the remainder of the year, but also emanated from a deep personal friendship. It was clear that, to Gerry, there was little distinction between personal and professional in his research life.

Science is done by people, and people need to be supported

Gerry was incredibly proud of his mentoring record, and did his best to support his students, postdocs, and junior colleagues both professionally and personally. He devoted the weeks around Christmas each year to writing recommendation letters for junior colleagues. He spent years working to successfully nominate colleagues to the National Academy of Sciences. He supported junior faculty with significant amounts of his time and sometimes by forgoing his own salary to boost theirs. While he never stated it explicitly, he considered most ideas somewhat ephemeral, and thought that his real legacy — and the legacy most worth having — lay in the students and colleagues who would continue after him.

Always treat the administrative staff well

Gerry was fond of pointing out that the secretaries and administrative staff had more practical power than most faculty, and that it was worth staying on their good side. This was less a statement of calculated intent and more an observation that many students, postdocs, and faculty treated nonscientists with less respect than they deserved. He always took the time to interact with them on a personal level, and certainly seemed to be well liked for it. I've been told by several colleagues who worked with Gerry that this was a lesson that they took to heart in their own interactions with staff, and it has also served me well.

Hard work is more important than brilliance

One of Gerry's favorite quotes was "Success is 99% perspiration, 1% inspiration", a statement attributed to Thomas Edison. According to Gerry, he simply wasn't as smart as many of his colleagues, but he made up for it by working very hard. I have no idea how modest he was being — he was not always known for modesty — but he certainly worked very hard, spending 10–14 h a day writing in his home office, thinking in the garden, or meeting with colleagues at work. While I try for more balance in my work and life of myself, he demonstrated to me that sometimes grueling hard work is a necessity when tackling tricky problems: for example, my Earthshine publications (e.g., Ref. 2) came after a tremendously unpleasant summer working on some incredibly messy and very tedious analysis code, but without the resulting analysis we wouldn't have been able to advance the project (which continues today, almost two decades later).

Experiments should talk to theory, and vice versa

Steve Koonin once explained to me that Gerry was a phenomenologist — a theorist who worked well with experimental data — and that this specialty was fairly rare because it required communicating effectively across sub disciplines. Gerry wasn't attracted to deep theoretical work and complex calculations, and in any case liked to talk to experimentalists too much to be a good theorist — for example, some of our most frequent dinner guests when I was growing up were Peter Braun- Munzinger and Johanna Stachel, both experimentalists. So he chose to work at the interface of theory and experiment, where he could develop and refine his intuition based on competing world views emanating from the theorists (who sought clean mathematical solutions) and experimentalists (who had real data that needed to be reconciled with theory). I have tried to pursue a similar strategy in computational biology.

Computers and mathematical models are tools, but the real insight comes from intuition

Apart from some early experience with punch cards at Yale in the 1950's, Gerry avoided computers and computational models completely in his own research (although his students, postdocs and collaborators used them, of course). I am told that his theoretical models were often relatively simple approximations, and he himself often said that his work with Hans Bethe proceeded by choosing the right approximation for the problem at hand — something at which Bethe excelled. Their choice of approximation was guided by intuition about the physical nature of the problem as much as by mathematical insight, and they could often use a few lines of the right equations to reach results similar to complex computational and mathematical models. This search for simple models and the utility of physical intuition in his research characterized many of our conversations, even when I became more mathematically trained.

Teaching is largely about conveying intuition

Once a year, Gerry would load up a backpack with mason jars full of thousands of pennies, and bring them into his statistical mechanics class. This was needed for one of his favorite exercises — a hands-on demonstration of the law of large numbers and the central limit theorem, which lie at the heart of thermodynamics and statistical mechanics. He would have students flip 100 coins and record the average, and then do it again and again, and have the class plot the distributions of results. The feedback he got was that this was a very good way of viscerally communicating the basics of statistical mechanics to students, because it built their intuition about how averages really worked. This approach has carried through to my own teaching and training efforts, where I always try to integrate hands-on practice with more theoretical discussion.

Benign neglect is a good default for mentoring

Gerry was perhaps overly fond of the concept of "benign neglect" in parenting, in that much of my early upbringing was at the hands of my mother with only occasional input from him. However, in his oft-stated experience (and now mine as well), leaving smart graduate students and postdocs to their own devices most of the time was far better than trying to actively manage (or interfere in) their research for them. I think of it this way: if I tell my students what to do and I'm wrong (which is likely, research being research), then they either do it (and I suffer for having misdirected them) or they don't do it (and then I get upset at them for ignoring me). But if I don't tell my students what to do, then they usually figure out something better for themselves, or else get stuck and then come to me to discuss it. The latter two outcomes are much better from a mentoring perspective than the former two.

Students need to figure it out for themselves

One of the most embarrassing (in retrospect) interactions I had with my father was during a long car ride where he tried to convince me that when x was a negative number, $-x$ was positive. At the time, I didn't agree with this at all, which was probably because I was a stubborn 7 years old. While it took me a few more years to understand this concept, by the time I was a math major I did have the concept down! Regardless, in this, and many other interactions around science, he never browbeat me about it or got upset at my stupidity or stubbornness. I believe this carried through to his interactions with his students. In fact, the only time I ever heard him express exasperation was with colleagues who were acting badly.

A small nudge at the right moment is sometimes all that is needed

A pivotal moment in my life came when Gerry introduced me to Mark Galassi, a physics graduate student who also was the systems administrator for the UNIX systems in the Institute for Theoretical Physics at Stony Brook; Mark found out I was interested in computers and gave me access to the computer system. This was one of the defining moments in my research life, as my research is entirely computational! Similarly, when I took a year off from college, my father put me in touch with Steve Koonin, who needed a systems administrator for a new project; I ended up working with the Earthshine project, which was a core part of my research for several years, and when I was trying to decide what grad schools to apply to, Gerry suggested I ask Hans Bethe and Steve Koonin what they thought was the most promising area of research for the future — their unequivocal answer was "biology!" This drove me to apply to biology graduate schools, get a Ph.D in biology, and ultimately led to my current faculty position. In all these cases, I now recognize the application of a light touch at the right moment, rather than the heavy-handed guidance that he must have desperately wanted to give at times.

Conclusion

There are many more personal stories that could be told about Gerry Brown, including his (several, and hilarious) interactions with the East German secret police during the cold war, his (quite bad) jokes, his (quite good) cooking, and his (enthusiastic) ballroom dancing, but I will save those for another time. I hope that his friends and colleagues will see him in the examples above, and will remember him fondly.

Acknowledgments

I thank Chris Adami, Erich Schwarz, Tracy Teal, and my mother, Elizabeth Brown, for their comments on drafts of this paper.

References

1. C. Adami, "Three weeks with Hans Bethe," in *Hans Bethe and His Physics* (World Scientific, Singapore, 2006), pp. 45–111.
2. P. Goode, J. Qiu, V. Yurchyshyn, J. Hickey, M. Chu, E. Kolbe, C. Brown and S. Koonin, *Geophys. Res. Lett.* **28** (2001) 1671–1674.

Thermal effects in dense matter beyond mean field theory

Constantinos Constantinou
Institute for Advanced Simulation, Institut für Kernphysik,
and Jülich Center for Hadron Physics, Forschungszentrum Jülich,
D-52425 Jülich, Germany
c.constantinou@fz-juelich.de

Sudhanva Lalit* and Madappa Prakash†
Department of Physics and Astronomy,
Ohio University, Athens, Ohio 45701, USA
**sl897812@ohio.edu*
†prakash@ohio.edu

The formalism of next-to-leading order (NLO) Fermi Liquid Theory (FLT) is employed to calculate the thermal properties of symmetric nuclear and pure neutron matter in a relativistic many-body theory beyond the mean field level which includes two-loop (TL) effects. For all thermal variables, the semi-analytical NLO corrections reproduce results of the exact numerical calculations for entropies per baryon up to 2. This corresponds to excellent agreement down to subnuclear densities for temperatures up to 20 MeV. In addition to providing physical insights, a rapid evaluation of the equation of state (EOS) in the homogeneous phase of hot and dense matter is achieved through the use of the zero-temperature Landau effective mass function and its derivatives.

Keywords: Thermal effects in dense matter; next-to-leading order Fermi liquid theory; field-theoretical models.

1. Introduction

Core-collapse supernovae, neutron stars from their birth to old age, and binary mergers involving neutron stars all pass through stages in which there are considerable variations in the baryon density, temperature, and lepton content. Simulations of these astrophysical phenomena involve general relativistic hydrodynamics and neutrino transport with special relativistic effects. Convection, turbulence, magnetic fields, etc., also play crucial roles. The macroscopic evolution in each case is governed by microphysics involving strong, weak and electromagnetic interactions. Depending on the baryon density n, temperature T, and the lepton content of matter (characterized by $Y_{Le} = n_{Le}/n$ when neutrinos are trapped or by the net electron concentration $Y_e = n_e/n$ in neutrino-free matter), various phases of matter are encountered. For sub-nuclear densities ($n \lesssim 0.1$ fm^{-3}) and temperatures $T \lesssim 20$ MeV, different inhomogeneous phases are encountered. A homogeneous phase of nucleonic and leptonic matter prevails at near- and supra-nuclear densities ($n \gtrsim 0.1$ fm^{-3}) at all temperatures. With progressively increasing density, homogeneous matter may contain hyperons, quark matter and Bose condensates.

Central to an understanding of the above astrophysical phenomena is the equation of state (EOS) of matter as a function of n, T, and Y_{Le} (or Y_e) as it is an integral part of hydrodynamical evolution, and controls electron capture and neutrino interactions in ambient matter. The EOS of dense matter has been investigated in the literature extensively, but for the most part those for use in the diverse physical conditions of relevance to astrophysical applications have been based on mean field theory (MFT) in both nonrelativistic potential and relativistic field-theoretical approaches. A recent paper honoring Gerry Brown reviews the current status and advances made to date in the growing field of neutron star research.[1]

The objective of this work is to assess the extent to which the model independent formalism of Fermi Liquid Theory (FLT)[2] is able to accurately describe thermal effects in dense homogeneous nucleonic matter under degenerate conditions for models beyond MFT. Recently, a next-to-leading order (NLO) extension of the leading-order FLT and its relativistic generalization in Ref. 3 was developed in Ref. 4. The FLT+NLO formalism was applied to nonrelativistic potential models with contact and finite-range interactions as well as to relativistic models of dense matter at the mean field level in Ref. 4. Excellent agreement with the results of exact numerical calculations for all thermal variables was found with the semi-analytical FLT+NLO results. In this contribution, we present similar excellent agreement with the exact numerical results of a relativistic field-theoretical model beyond the MFT level that includes two-loop (exchange) effects recently reported in Ref. 5. The gratifying result is that the FLT+NLO formalism extends agreement with the exact numerical results for all n and T for which the entropy per baryon $S \leq 2$. This means that, for $T \lesssim 20\,\mathrm{MeV}$, the method can adequately describe state variables down to a density of ~ 0.1 fm^{-3}. For densities below ~ 0.1 fm^{-3}, inhomogeneous phases occur for which a separate treatment is required. This development not only provides a check of time-consuming many-body calculations of dense matter at finite temperature, but also serves to accurately (and, to rapidly) calculate thermal effects from a knowledge of the zero-temperature single-particle spectra for S up to 2 for which effects of interactions are relatively important.

The organization of this contribution is as follows. In Sec. 2, we summarize the NLO formalism of FLT recently developed in Ref. 4. Section 3 contains a brief description of the relativistic field-theoretical model that extends MFT to include two-loop (TL) effects as implemented in Ref. 5. Working formulas required for the evaluation of the degenerate-limit thermal effects (in particular, expressions for the single particle spectra) are given in this section which also includes our results and associated discussion. A summary of our work along with conclusions are in Sec. 4. Personal tributes to Gerry Brown from two of the authors (Constantinou and Prakash) form the content of Sec. 5.

2. Next-to-Leading Order Fermi Liquid Theory

The thermodynamics of fermion systems entails evaluation of integrals of the type

$$I = \int_0^\infty dp\, g(p) \frac{1}{1 + \exp\left[\dfrac{\epsilon(p,n) - \mu}{T}\right]}, \tag{1}$$

where T is the temperature, μ is the chemical potential, and ϵ is the single-particle spectrum of the underlying model. The functional form of $g(p)$ is particular to the state property in question. Equivalently, we can write

$$I = \int_0^\infty dy \frac{\phi(y)}{1 + \exp(y - \eta)}, \tag{2}$$

where

$$y = \frac{\epsilon(p,n) - \mathcal{U}(n)}{T}, \quad \eta = \frac{\mu - \epsilon(p=0,n)}{T}, \tag{3}$$

$$\phi(y) = \frac{\mathcal{M}(p)Tg(p)}{p}, \quad \mathcal{M}(p) = p\left(\frac{\partial \epsilon}{\partial p}\right)^{-1}. \tag{4}$$

Above, $\mathcal{U}(n)$ is inclusive of all those terms in the spectrum which depend only on the density n. The Landau effective mass function, $\mathcal{M}(p)$, and its derivatives with respect to momentum p play crucial roles in determining the thermal effects.

In the degenerate limit, characterized by large values of the parameter η, Sommerfeld's Lemma

$$I \overset{\eta\gg 1}{\longrightarrow} \int_0^\eta \phi(y)dy + \frac{\pi^2}{6}\frac{d\phi}{dy}\bigg|_{y=\eta} + \frac{7\pi^4}{360}\frac{d^3\phi}{dy^3}\bigg|_{y=\eta} + \cdots \tag{5}$$

can be used for the approximate evaluation of such integrals. Truncation of the series at the first term recovers results for cold matter; the second term produces the familiar FLT corrections and the third term represents the NLO extension to FLT. Owing to the asymptotic nature of the Sommerfeld formula, the expansion will, in general, diverge at higher orders unless all terms are retained.

The number density of single-species fermions with γ internal degrees of freedom in three dimensions is (throughout we use units in which $\hbar = 1$)

$$n = \frac{\gamma}{2\pi^2} \int dp \frac{p^2}{1 + \exp\left(\dfrac{\epsilon - \mu}{T}\right)}. \tag{6}$$

In this context, we take n to be an independent variable as is appropriate for a system that does not exchange particles with an external reservoir but whose total volume is allowed to change. Thus

$$n(T=0) = n(p_F) = \frac{\gamma p_F^3}{6\pi^2} = n(T) \simeq n(p_\mu), \tag{7}$$

where p_F is the Fermi momentum and $n(p_\mu)$ is the result of Eq. (6) evaluated according to Eq. (5). Perturbative inversion of Eq. (7) leads to

$$p_\mu = p_F \left[1 - \frac{\pi^2}{6}\frac{m^{*2}T^2}{p_F^4}\left(1 + \frac{p_F}{m^*}\frac{d\mathcal{M}}{dp}\bigg|_{p_F}\right) + \cdots\right], \tag{8}$$

where

$$m^* = \mathcal{M}(p = p_F) \tag{9}$$

is the Landau effective mass. The combination of Eq. (8) with the Sommerfeld expansion of the entropy density, formally given by

$$s = \frac{\gamma}{2\pi^2}\int dp\, p^2 \left\{f(p)\ln f(p) - [1 - f(p)]\ln[1 - f(p)]\right\}, \tag{10}$$

$$f(p) = \frac{1}{1 + \exp\left[\dfrac{\epsilon(p) - \mu}{T}\right]} \tag{11}$$

yields an expression for s in terms of quantities defined on the Fermi surface:

$$s = \frac{\gamma p_F m^* T}{6} - \frac{\gamma\pi^2}{15}\frac{m^{*3}T^3}{p_F^3}(1 - L_F) \tag{12}$$

$$= 2anT - \frac{16}{5\pi^2}a^3 n T^3(1 - L_F), \tag{13}$$

where $a = \pi^2 m^*/(2p_F^2) = \pi^2/(4T_F)$ is the level density parameter with T_F denoting the Fermi temperature, and

$$L_F \equiv \frac{7}{12}\left(\frac{p_F}{m^*}\frac{\partial\mathcal{M}}{\partial p}\bigg|_{p_F}\right)^2 + \frac{7}{12}\frac{p_F^2}{m^*}\frac{\partial^2\mathcal{M}}{\partial p^2}\bigg|_{p_F} + \frac{3}{4}\frac{p_F}{m^*}\frac{\partial\mathcal{M}}{\partial p}\bigg|_{p_F}. \tag{14}$$

Then the entropy per particle is the simple ratio $S = s/n$ whereas the thermal energy, pressure and chemical potential are obtained via Maxwell's relations (the integrals below are performed at constant density):

$$E_{\rm th} = \int T\, dS = aT^2 - \frac{12}{5\pi^2}a^3T^4(1 - L_F), \tag{15}$$

$$P_{\rm th} = -n^2\int \frac{dS}{dn}\,dT = \frac{2}{3}anQT^2 - \frac{8}{5\pi^2}a^3nQT^4\left(1 - L_F + \frac{n}{2Q}\frac{dL_F}{dn}\right), \tag{16}$$

$$\begin{aligned}\mu_{\rm th} &= -\int \frac{ds}{dn}\,dT \\ &= -a\left(1 - \frac{2Q}{3}\right)T^2 + \frac{4}{5\pi^2}a^3T^4\left[(1 - L_F)(1 - 2Q) - n\frac{dL_F}{dn}\right],\end{aligned} \tag{17}$$

where

$$Q = 1 - \frac{3n}{2m^*}\frac{dm^*}{dn}. \tag{18}$$

Other quantities of interest such as the specific heats at constant volume and pressure, and the thermal index are given by standard thermodynamics:

$$C_V = T\left.\frac{dS}{dT}\right|_n = 2aT - \frac{48}{5\pi^2}a^3T^3(1-L_F), \tag{19}$$

$$C_P = T\left.\frac{dS}{dT}\right|_P = 2aT + \frac{T}{n^2}\frac{\left(\left.\frac{\partial P_{\text{th}}}{\partial T}\right|_n\right)^2}{\left.\frac{\partial P_{\text{total}}}{\partial n}\right|_T}, \tag{20}$$

$$\Gamma_{\text{th}} = 1 + \frac{P_{\text{th}}}{nE_{\text{th}}} = 1 + \frac{2Q}{3} - \frac{4}{5\pi^2}a^2nT^2\frac{dL_F}{dn}. \tag{21}$$

Note that while the NLO terms in the thermal quantities above have the same temperature dependences as those of a free Fermi gas, the accompanying density-dependent factors differ reflecting the effects of interactions.

3. Application to Models Beyond Mean Field Theory

In this work, we investigate the degenerate-limit thermodynamics of a relativistic field-theoretical model in which the nucleon-nucleon (NN) interaction is mediated by the exchange of σ, ω, ρ and π mesons (scalar, vector, iso-vector and pseudo-vector, respectively). Nonlinear self-couplings of the scalar field are also included. The model is described by the Lagrangian density[5–7]

$$\mathcal{L} = \mathcal{L}_N + \mathcal{L}_{\text{meson}} \tag{22}$$

$$\begin{aligned}\mathcal{L}_N = \bar{N}\Big[i\gamma^\mu\Big(\partial_\mu + i\frac{g_\rho}{2}\boldsymbol{\rho}_\mu\cdot\boldsymbol{\tau} + ig_\omega\omega_\mu \\ - i\frac{g_A}{2f_\pi}\gamma_5\boldsymbol{\tau}\cdot\partial_\mu\boldsymbol{\pi}\Big) - (M - g_\sigma\sigma)\Big]N,\end{aligned} \tag{23}$$

$$\begin{aligned}\mathcal{L}_{\text{meson}} = {}& \frac{1}{2}\partial_\mu\sigma\partial^\mu\sigma - \left(\frac{1}{2} + \frac{\kappa_3}{3!}\frac{g_\sigma\sigma}{M} + \frac{\kappa_4}{4!}\frac{g_\sigma^2\sigma^2}{M^2}\right)m_\sigma^2\sigma^2 \\ & - \frac{1}{4}V^{\mu\nu}V_{\mu\nu} + \frac{1}{2}m_\omega^2\omega_\mu\omega^\mu \\ & - \frac{1}{4}B_{\mu\nu}B^{\mu\nu} + \frac{1}{2}m_\rho^2\boldsymbol{\rho}_\mu\cdot\boldsymbol{\rho}^\mu \\ & + \frac{1}{2}\partial^\mu\boldsymbol{\pi}\cdot\partial_\mu\boldsymbol{\pi} - \frac{1}{2}m_\pi^2\boldsymbol{\pi}\cdot\boldsymbol{\pi},\end{aligned} \tag{24}$$

where

$$V_{\mu\nu} = \partial_\mu\omega_\nu - \partial_\nu\omega_\mu, \tag{25}$$

$$B_{\mu\nu} = \boldsymbol{\tau}(\partial_\mu\boldsymbol{\rho}_\nu - \partial_\nu\boldsymbol{\rho}_\mu) + i\frac{g_\rho}{2}[\boldsymbol{\rho}_\mu\cdot\boldsymbol{\tau}, \boldsymbol{\rho}_\nu\cdot\boldsymbol{\tau}] \tag{26}$$

are the field-strength tensors and τ are the $SU(2)$ isospin matrices. We use the masses $M = 939$ MeV, $m_\sigma = 550$ MeV, $m_\omega = 783$ MeV, $m_\rho = 770$ MeV and $m_\pi = 138$ MeV, the couplings $g_\sigma = 8.604$, $g_\omega = 7.522$, $g_\rho = 7.614$, $\kappa_3 = 4.84$ and $\kappa_4 = -4.47$, the pion decay constant $f_\pi = 93$ MeV and the nucleon axial current constant $g_A = 1.26$ as in Ref. 5.

All thermodynamic quantities of interest can be derived from the grand potential density Ω which is related to the pressure by $\Omega = -P$. For an isotropic system in its rest-frame, the pressure is obtained from the diagonal elements of the spatial part of the energy-momentum tensor $T_{\mu\nu} = \partial\mathcal{L}/\partial(\partial_\mu\phi)\partial_\nu\phi - g_{\mu\nu}\mathcal{L}$. In MFT, the result is

$$\begin{aligned} P &= \frac{1}{3}\langle T_{ii}\rangle \\ &= \frac{\gamma_{\rm spin}}{3}\sum_i \int \frac{d^3p}{(2\pi)^3}\frac{p^2}{E^*(p)}f_i(p) + \frac{g_\omega^2}{2m_\omega^2}n^2 + \frac{g_\rho^2}{8m_\rho^2}(n_i - n_j)^2 \\ &\quad - \frac{m_\sigma^2}{g_\sigma^2}(M - M^*)\left[\frac{1}{2} + \frac{\kappa_3}{6M}(M - M^*) + \frac{\kappa_4}{24M^2}(M - M^*)^2\right], \end{aligned} \tag{27}$$

where

$$E^*(p) = (p^2 + M^{*2})^{1/2}. \tag{28}$$

In the mean-field approximation, the spectrum $\epsilon_i(p)$ that enters the Fermi distribution function $f_i(p) = \{1 + \exp[(\epsilon_i(p) - \mu_i)/T]\}^{-1}$ is given by

$$\epsilon_i(p) = \pm E^*(p) + \frac{g_\omega^2}{m_\omega^2}n + \frac{g_\rho^2}{4m_\rho^2}(n_i - n_j). \tag{29}$$

The $+(-)$ sign corresponds to particles (antiparticles) and the subscripts i, j to the two nucleon species. The Dirac effective mass M^* results from the minimization of Ω with respect to the expectation value of the scalar field.

The leading corrections to the mean-field Ω arise from TL exchanges of the mesons involved in the model. These corrections are given by (see, e.g., Refs. 5 and 8)

$$\Omega_{{\rm ex},\sigma} = -\frac{\gamma_{\rm spin}}{4}g_\sigma^2\int d\tau_p d\tau_q\ f_s(p,q)D(k;m_\sigma^*)\sum_i f_i(p)f_i(q), \tag{30}$$

$$\Omega_{{\rm ex},\omega} = -\frac{\gamma_{\rm spin}}{4}g_\omega^2\int d\tau_p d\tau_q\ f_v(p,q)D(k;m_\omega)\sum_i f_i(p)f_i(q), \tag{31}$$

$$\begin{aligned} \Omega_{{\rm ex},\rho} = -\frac{\gamma_{\rm spin}}{16}g_\rho^2\int d\tau_p d\tau_q\ f_v(p,q)D(k;m_\rho) \\ \times \sum_i f_i(p)[f_i(q) + 2f_j(q)], \end{aligned} \tag{32}$$

$$\Omega_{\mathrm{ex},\pi} = -\frac{\gamma_{\mathrm{spin}}}{16}\left(\frac{g_A M^*}{f_\pi}\right)^2 \int d\tau_p d\tau_q\; f_{pv}(p,q) D(k;m_\pi) \times \sum_i f_i(p)[f_i(q) + 2f_j(q)], \tag{33}$$

where

$$d\tau_p = \frac{d^3p}{(2\pi)^3 2E^*(p)}, \quad f_s(p,q) = 4(p^\mu q_\mu + M^{*2}), \tag{34}$$

$$f_v(p,q) = 8(p^\mu q_\mu - 2M^{*2}), \quad f_{pv}(p,q) = 16(p^\mu q_\mu - M^{*2}), \tag{35}$$

$$D(k;m) = \frac{1}{k^\mu k_\mu - m^2}, \quad k^\mu = p^\mu - q^\mu, \quad p^\mu p_\mu = q^\mu q_\mu = M^{*2}. \tag{36}$$

The corresponding TL contributions to the single-particle spectrum (via $\epsilon^i_{\mathrm{ex}} = \delta\Omega_{\mathrm{ex}}/\delta n_i(p)$ with $\delta/\delta n_i(p) \int d^3p/(2\pi)^3 f_i(p) = 1$; $i =$nucleon species) are[5,8]:

$$\epsilon^i_{\mathrm{ex},\sigma} = -\frac{\gamma_{\mathrm{spin}}}{8}\frac{g_\sigma^2}{E^*(p)} \int d\tau_q\; f_s(p,q) D(k;m_\sigma^*) f_i(q), \tag{37}$$

$$\epsilon^i_{\mathrm{ex},\omega} = -\frac{\gamma_{\mathrm{spin}}}{8}\frac{g_\omega^2}{E^*(p)} \int d\tau_q\; f_v(p,q) D(k;m_\omega) f_i(q), \tag{38}$$

$$\epsilon^i_{\mathrm{ex},\rho} = -\frac{\gamma_{\mathrm{spin}}}{32}\frac{g_\rho^2}{E^*(p)} \int d\tau_q\; f_v(p,q) D(k;m_\rho)[f_i(q) + 2f_j(q)], \tag{39}$$

$$\epsilon^i_{\mathrm{ex},\pi} = -\frac{\gamma_{\mathrm{spin}}}{32E^*(p)}\left(\frac{g_A M^*}{f_\pi}\right)^2 \int d\tau_q\; f_{pv}(p,q) D(k;m_\pi)[f_i(q) + 2f_j(q)]. \tag{40}$$

Note that at the TL level, the self-interactions of the scalar field bestow upon it an effective scalar-meson mass

$$m_\sigma^* = m_\sigma \left[1 + \kappa_3\left(\frac{M - M^*}{M}\right) + \frac{\kappa_4}{2}\left(\frac{M - M^*}{M}\right)^2\right]^{1/2}, \tag{41}$$

which is used in all exchange terms involving the σ- meson.

3.1. *TL calculations of dense nucleonic matter*

The degenerate limit formalism delineated in Sec. 2 requires for its implementation, in principle, only the $T = 0$ parts of the spectrum (for $\mathcal{M}(p)$) and the pressure (for C_P and M^*). Note, however, that for cold matter the statements $dP/d\sigma = 0$ and $d\mathcal{E}/d\sigma = 0$ are equivalent (being that at $T = 0$, $P = -n\, d\mathcal{E}/dn$) and that the energy density $\mathcal{E}$ is somewhat easier to minimize with respect to σ in order to obtain M^*. We therefore opt to work with the latter. Confining ourselves to symmetric nuclear matter (SNM) and pure neutron matter (PNM) in the interest of simplicity, we have for the energy density (in the notation of Ref. 8)

$$\mathcal{E}_{\rm TL} = \mathcal{E}_{\rm MFT} + \mathcal{E}_{{\rm ex},\sigma} + \mathcal{E}_{{\rm ex},\omega} + \mathcal{E}_{{\rm ex},\rho} + \mathcal{E}_{{\rm ex},\pi}, \tag{42}$$

$$\mathcal{E}_{\rm MFT} = 2\gamma_{\rm iso}\int_0^{p_F} \frac{d^3p}{(2\pi)^3}E^* + \frac{1}{2}\frac{g_\omega^2}{m_\omega^2}n^2 + \frac{(1-\gamma_{\rm charge})}{8}\frac{g_\rho^2}{m_\rho^2}n^2$$
$$+ \frac{m_\sigma^2}{g_\sigma^2}(M-M^*)^2\left[\frac{1}{2} + \frac{\kappa_3}{3!}\left(\frac{M-M^*}{M}\right) + \frac{\kappa_4}{4!}\left(\frac{M-M^*}{M}\right)^2\right], \tag{43}$$

$$\mathcal{E}_{{\rm ex},\sigma} = \gamma_{\rm iso}\frac{g_\sigma^2}{(2\pi)^4}M^{*4}\left[\frac{1}{4}(x\eta - \ln\xi)^2 + \left(1 - \frac{w_\sigma^*}{4}\right)I(w_\sigma^*)\right], \tag{44}$$

$$\mathcal{E}_{{\rm ex},\omega} = \gamma_{\rm iso}\frac{g_\omega^2}{(2\pi)^4}M^{*4}\left[\frac{1}{2}(x\eta - \ln\xi)^2 - \left(1 + \frac{w_\omega}{2}\right)I(w_\omega)\right], \tag{45}$$

$$\mathcal{E}_{{\rm ex},\rho} = \frac{(\gamma_{\rm iso} + 4\gamma_{\rm charge})}{4}\frac{g_\rho^2}{(2\pi)^4}M^{*4}\left[\frac{1}{2}(x\eta - \ln\xi)^2 - \left(1 + \frac{w_\rho}{2}\right)I(w_\rho)\right], \tag{46}$$

$$\mathcal{E}_{{\rm ex},\pi} = \frac{(\gamma_{\rm iso} + 4\gamma_{\rm charge})}{4}\left(\frac{g_A M^*}{f_\pi}\right)^2\frac{M^{*4}}{(2\pi)^4}\left[(x\eta - \ln\xi)^2 - w_\pi I(w_\pi)\right] \tag{47}$$

and for the spectrum (via $\epsilon_{{\rm ex},i} = \delta\mathcal{E}_{{\rm ex},i}/\delta n$; $i =$ meson)

$$\epsilon_{\rm TL} = \epsilon_{\rm MFT} + \epsilon_{{\rm ex},\sigma} + \epsilon_{{\rm ex},\omega} + \epsilon_{{\rm ex},\rho} + \epsilon_{{\rm ex},\pi}, \tag{48}$$

$$\epsilon_{\rm MFT} = M^*e + \frac{g_\omega^2}{m_\omega^2}n + \frac{(1-\gamma_{\rm charge})}{4}\frac{g_\rho^2}{m_\rho^2}n, \tag{49}$$

$$\epsilon_{{\rm ex},\sigma} = \frac{g_\sigma^2}{(2\pi)^2}\frac{M^*}{2e}\left[\frac{1}{2}(x\eta - \ln\xi) + \left(1 - \frac{w_\sigma^*}{4}\right)\frac{2}{r}J(w_\sigma^*)\right], \tag{50}$$

$$\epsilon_{{\rm ex},\omega} = \frac{g_\omega^2}{(2\pi)^2}\frac{M^*}{2e}\left[(x\eta - \ln\xi) - \left(1 + \frac{w_\omega}{2}\right)\frac{2}{r}J(w_\omega)\right], \tag{51}$$

$$\epsilon_{{\rm ex},\rho} = \frac{(\gamma_{\rm iso} + \gamma_{\rm charge})}{4}\frac{g_\rho^2}{(2\pi)^2}\frac{M^*}{2e}\left[(x\eta - \ln\xi) - \left(1 + \frac{w_\rho}{2}\right)\frac{2}{r}J(w_\rho)\right], \tag{52}$$

$$\epsilon_{{\rm ex},\pi} = \frac{(\gamma_{\rm iso} + \gamma_{\rm charge})}{4}\left(\frac{g_A M^*}{f_\pi}\right)^2\frac{M^*}{2e(2\pi)^2}\left[2(x\eta - \ln\xi) - w_\pi\frac{2}{r}J(w_\pi)\right], \tag{53}$$

where

$$\gamma_{\rm iso} = \begin{cases} 1 \text{ PNM}, \\ 2 \text{ SNM}, \end{cases} \qquad \gamma_{\rm charge} = \begin{cases} 0 \text{ PNM}, \\ 1 \text{ SNM}, \end{cases} \tag{54}$$

$$x = \frac{p_F}{M^*}, \quad \eta = (1+x^2)^{1/2}, \quad \xi = x + \eta, \tag{55}$$

$$r = \frac{p}{M^*}, \quad e = (1+r^2)^{1/2}, \quad t = r + e, \tag{56}$$

$$w_i = \frac{m_i^2}{M^{*2}}, \quad i = \sigma^*, \omega, \rho, \pi, \tag{57}$$

$$I(w_i) = \frac{1}{4}\int_1^\xi \int_1^\xi dz\, dy \left(1 - \frac{1}{z^2}\right)\left(1 - \frac{1}{y^2}\right) \ln\left[\frac{(zy-1)^2 + w_i zy}{(z-y)^2 + w_i zy}\right], \tag{58}$$

$$J(w_i) = \frac{1}{4}\int_1^\xi dz \left(1 - \frac{1}{z^2}\right) \ln\left[\frac{(zt-1)^2 + w_i zy}{(z-t)^2 + w_i zy}\right]. \tag{59}$$

With the inclusion of the TL contributions and noting that $M^* = M - g_\sigma \sigma$, the Dirac effective mass M^* is determined by solving

$$\partial\mathcal{E}_{\rm TL}/\partial M^* = 0 \quad \text{with} \tag{60}$$

$$\frac{\partial \mathcal{E}_{\rm MFT}}{\partial M^*} = 2\gamma_{\rm iso}\int_0^{p_F} \frac{d^3p}{(2\pi)^3}\frac{M^*}{E^*} - \frac{m_\sigma^2}{g_\sigma^2}(M - M^*) \times \left[1 + \frac{\kappa_3}{2}\left(\frac{M-M^*}{M}\right) + \frac{\kappa_4}{6}\left(\frac{M-M^*}{M}\right)^2\right], \tag{61}$$

$$\frac{\partial \mathcal{E}_{\rm ex,\sigma}}{\partial M^*} = \frac{4\mathcal{E}_{\rm ex,\sigma}}{M^*} + \gamma_{\rm iso}\frac{g_\sigma^2}{(2\pi)^4}M^{*4} \times \left[-\frac{x^3}{M^*\eta}(x\eta - \ln\xi) + \left(1 - \frac{w_\sigma^*}{4}\right)\frac{\partial I_{\sigma^*}}{\partial M^*} + \frac{w_\sigma^*}{2M^*}I_{\sigma^*}\right], \tag{62}$$

$$\frac{\partial \mathcal{E}_{\rm ex,\omega}}{\partial M^*} = \frac{4\mathcal{E}_{\rm ex,\omega}}{M^*} + \gamma_{\rm iso}\frac{g_\omega^2}{(2\pi)^4}M^{*4} \times \left[-\frac{2x^3}{M^*\eta}(x\eta - \ln\xi) - \left(1 + \frac{w_\omega}{2}\right)\frac{\partial I_\omega}{\partial M^*} + \frac{w_\omega}{M^*}I_\omega\right] \tag{63}$$

$$\frac{\partial \mathcal{E}_{\rm ex,\rho}}{\partial M^*} = \frac{4\mathcal{E}_{\rm ex,\rho}}{M^*} + \frac{(\gamma_{\rm iso} + 4\gamma_{\rm charge})}{4}\frac{g_\rho^2}{(2\pi)^4}M^{*4} \times \left[-\frac{2x^3}{M^*\eta}(x\eta - \ln\xi) - \left(1 + \frac{w_\rho}{2}\right)\frac{\partial I_\rho}{\partial M^*} + \frac{w_\rho}{M^*}I_\rho\right], \tag{64}$$

$$\frac{\partial \mathcal{E}_{\rm ex,\pi}}{\partial M^*} = \frac{6\mathcal{E}_{\rm ex,\pi}}{M^*} + \frac{(\gamma_{\rm iso} + 4\gamma_{\rm charge})}{4}\left(\frac{g_A M^*}{f_\pi}\right)^2 \frac{M^{*4}}{(2\pi)^4} \times \left[-\frac{4x^3}{M^*\eta}(x\eta - \ln\xi) - w_\pi\frac{\partial I_\pi}{\partial M^*} + \frac{2w_\pi}{M^*}I_\pi\right], \tag{65}$$

where

$$\frac{\partial I(w_{\sigma^*})}{\partial M^*} = \frac{1}{2M^*}\left\{w_{\sigma^*}\left(1 - \frac{M^*}{m_\sigma^*}\frac{dm_\sigma^*}{dM^*}\right)\right. \times \int_1^\xi \int_1^\xi dy\, dz \frac{1}{yz}\frac{(1-z^2)^2(1-y^2)^2}{[(zy-1)^2 + w_{\sigma^*}zy][(z-y)^2 + w_{\sigma^*}zy]}$$

$$- \frac{x\xi}{\eta}\left(1-\frac{1}{\xi^2}\right) J(w_{\sigma^*}; t \to \xi)\Bigg\}, \tag{66}$$

$$\frac{dm_\sigma^*}{dM^*} = -\frac{m_\sigma^2}{2Mm_\sigma^*}\left[\kappa_3 + \kappa_4\left(\frac{M-M^*}{M}\right)\right], \tag{67}$$

$$\frac{\partial I(w_i)}{\partial M^*} = \frac{1}{2M^*}\Bigg\{w_i \int_1^\xi \int_1^\xi dy\, dz \frac{1}{yz} \frac{(1-z^2)^2(1-y^2)^2}{[(zy-1)^2 + w_i zy][(z-y)^2 + w_i zy]}$$

$$- \frac{x\xi}{\eta}\left(1-\frac{1}{\xi^2}\right) J(w_i; t \to \xi)\Bigg\}; \quad i = \omega, \rho, \pi. \tag{68}$$

To obtain Eqs. (66) and (68), the 2-dimensional Leibniz rule

$$\frac{d}{dt}\int_{x_0(t)}^{x_1(t)} \int_{y_0(t)}^{y_1(t)} F(x,y,t)dx\, dy$$

$$= \int_{y_0}^{y_1}\left[F(x_1)\frac{\partial x_1}{\partial t} - F(x_0)\frac{\partial x_0}{\partial t}\right] dy + \int_{x_0}^{x_1}\left[F(y_1)\frac{\partial y_1}{\partial t} - F(y_0)\frac{\partial y_0}{\partial t}\right] dx$$

$$+ \int_{x_0}^{x_1}\int_{y_0}^{y_1} \frac{\partial F}{\partial t} dx\, dy \tag{69}$$

was employed.

3.1.1. *Numerical notes*

The various integrals above are readily calculated by using the Gauss–Legendre quadrature method.[9] The results reported below were calculated using 32 points and weights in each dimension although use of 16 points and weights was found to be adequate. The derivatives of $\mathcal{M}(p)$ and $L_F(n)$ were calculated using the 5-point rule.[9] Root finding was accomplished by the Newton–Raphson scheme. All numerical results of our Fortran code were also verified by using Mathematica.

3.2. *Results and discussion*

The variational procedure $\partial\mathcal{E}_{TL}/\partial M^* = 0$ in Eq. (60) minimizes the energy density of the system and results in the optimal baryon (Dirac) effective mass M^* at each n. (This minimization condition also yields the expectation value of the scalar field $\sigma = (M - M^*)/g_\sigma$.) The values of M^* in SNM and PNM are shown in the upper left panel of Fig. 1. Use of these M^*'s in the expressions for the energy density in Eq. (42) allows for a calculation of the energy per particle, and are shown in the right panel of Fig. 1 for SNM and PNM at $T = 0$. These results yield good agreement with nuclear and neutron star phenomenology.[5] The TL contributions to the energy density play a significant role in determining $M^*(n)$. The pattern $M^*(\text{PNM}) \geq M^*(\text{SNM})$ for a given baryon density stems from the isospin-invariant nucleon–nucleon interactions employed in the model. An MFT calculation — that

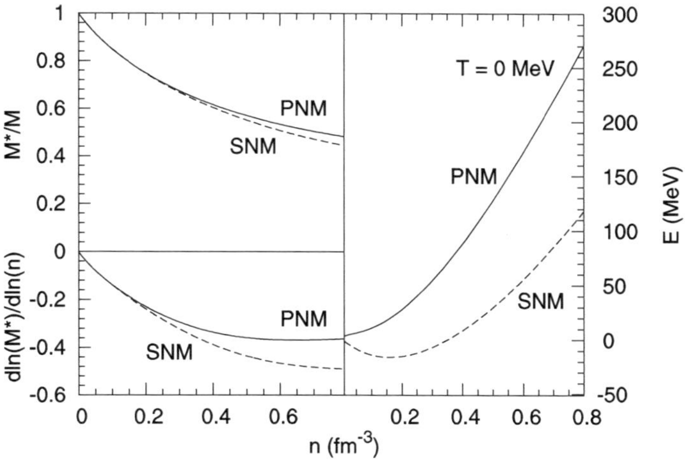

Fig. 1. Upper left panel: Dirac effective masses M^* (Eq. (60)) scaled with the vacuum nucleon mass versus density n in symmetric nuclear matter (SNM) and pure neutron matter (PNM) at temperature $T = 0$. Lower left panel: Logarithmic derivatives of M^* with respect to n. Right panel: Energy per particle $E = \mathcal{E}/n - M$ versus n in SNM and PNM at $T = 0$.

is, without the TL terms in Eq. (42) — that yields closely resembling E versus n curves as shown here through a readjustment of the various coupling strengths produces M^* curves that vary more steeply with density (not shown here, but see Ref. 5). As $M^*(n)$ and its logarithmic derivative with respect to n (lower left panel of Fig. 1) enter prominently in determining the thermal properties, contrasts between different levels of theoretical approximations (MFT versus MFT+TL in our case here) are afforded.

Under degenerate conditions for which $T/T_F \ll 1$, thermal effects depend sensitively on details of the single-particle spectrum near the Fermi surface. The various contributions to the $T = 0$ single-particle spectra in SNM and PNM are shown in Fig. 2 at $n = n_0 = 0.16$ fm^{-3}. Note that the dominant contribution from the MFT part in Eq. (48) has been divided by a factor of 20 to fit within the figure where contributions from the exchange of σ, ω, ρ, and π mesons from Eq. (48) are also shown. Although subdominant in their contributions to the spectra, the exchange contributions significantly alter the $M^*(n)$ curves from those of MFT and hence the MFT term of $\epsilon(p)$.

Depending on the density, the magnitude and slope of the single-particle spectrum are also altered from its MFT contribution as can be seen in Fig. 3 where results for $n = 3n_0$ are shown. Such differences will be reflected in the thermal properties, particularly in the NLO terms of FLT.

The Landau effective masses $m^*(n)$ from Eq. (9) scaled with the vacuum nucleon mass are shown in the left panel of Fig. 4 as functions of density in SNM and PNM. The associated logarithmic derivatives are in the right panel of this figure. The TL results are substantially larger than those of MFT for the same n (see Ref. 5).

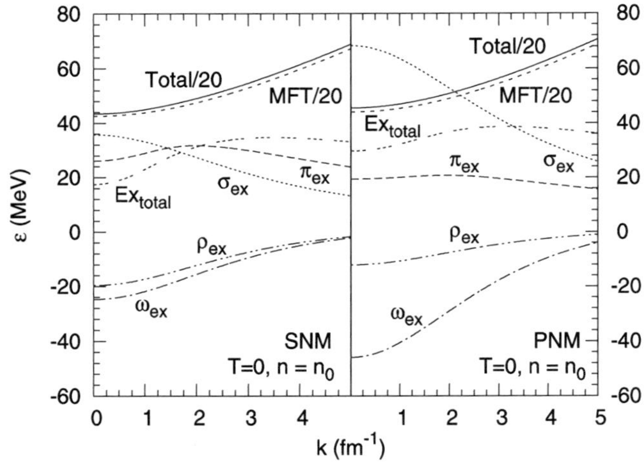

Fig. 2. Contributions from MFT and TL terms (from Eq. (48)) involving the exchange of σ, ω, ρ, and π mesons to the total single-particle energy versus wave number in SNM and PNM at the baryon density $n = n_0 = 0.16$ fm^{-3}.

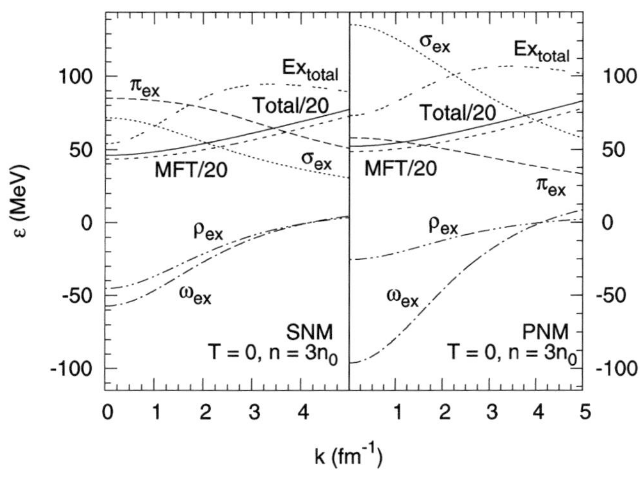

Fig. 3. Same as Fig. 2, but for $n = 3n_0$.

The nonmonotonic behaviors and the minima in the $m^*(n)$ curves are characteristic of relativistic field theoretical models in which $M^*(n)$ continually decreases with increase in n. Together with the derivatives of the Landau effective mass function required at NLO in FLT, $m^*(n)$ and its logarithmic derivative play important roles in improving the accuracy of the degenerate limit thermodynamics.

We turn now to compare the thermal properties from FLT and FLT+NLO with those from the exact numerical results of Ref. 5 for the TL calculations at temperatures of $T = 20$ and 50 MeV, respectively. In all cases, comparisons shown

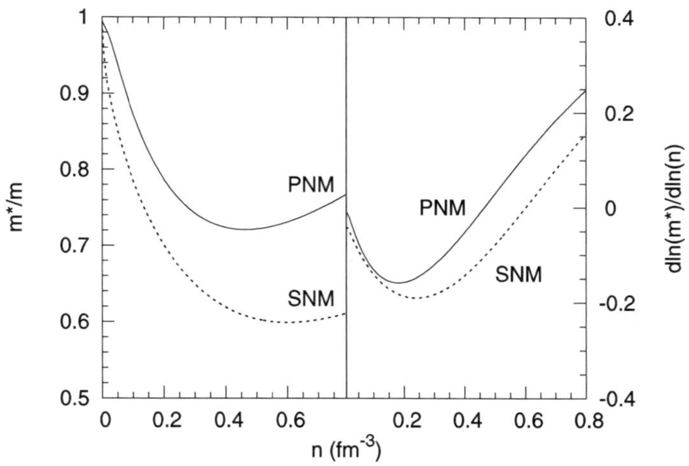

Fig. 4. Left panel: Landau effective masses from Eq. (9) scaled with the vacuum nucleon mass versus density in SNM and PNM. Right panel: Logarithmic derivatives of the Landau effective masses with respect to density.

for $T = 50$ MeV highlight the onset of semi- or nondegenerate regions in density for which results of degenerate limit FLT and FLT+NLO begin to become inadequate.

In astrophysical phenomena involving supernovae, neutron stars and binary mergers, the entropy per baryon S serves as a gauge to track hydrodynamical evolution and its consequences.[10] Figure 5 shows S versus n in SNM and PNM at $T = 20$ and 50 MeV, respectively. For both SNM and PNM, the NLO corrections substantially improve agreement with the exact results for S up to 2. For $T = 20$ MeV,

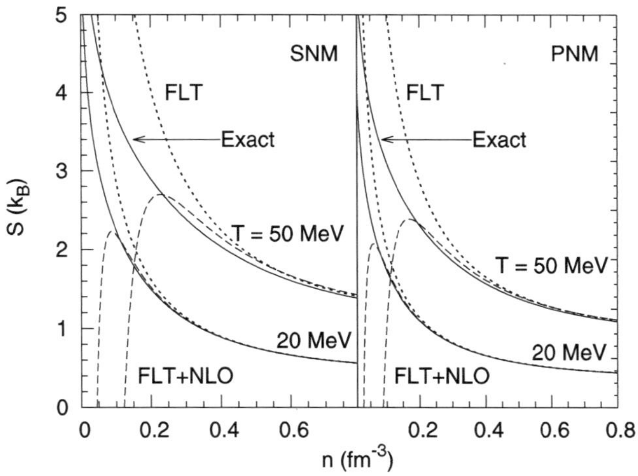

Fig. 5. Entropies per particle $S = s/n$ versus baryon number density in SNM and PNM. Results labeled "Exact" are from Ref. 5. The leading order FLT results are labeled "FLT" whereas "FLT+NLO" stands for results of NLO FLT with s from Eq. (13). Values of temperatures are as indicated in the figure.

the agreement extends to the subnuclear density of $n = 0.1\ \mathrm{fm}^{-3}$ for both SNM and PNM. This agreement is encouraging as for $n \lesssim 0.1\ \mathrm{fm}^{-3}$ and $T \lesssim 20$ MeV, matter exists in an inhomogeneous phase consisting of heavy nuclei, light nuclear clusters such as α particles, tritons and deuterons, and dripped nucleons (as also leptons and photons) for which a separate treatment is required. The lesson learned is that up to $S = 2$, the thermal properties of bulk homogeneous nucleonic matter are adequately described by a knowledge of the $T = 0$ spectra of nucleons from which all thermal properties can be obtained through the use of FLT carried up to NLO terms.

In Fig. 6, we show results for the thermal energies. As for S, the NLO corrections extend agreement with the exact results down to $n \simeq 0.1\ \mathrm{fm}^{-3}$ in both SNM and PNM. The agreement for PNM extends to somewhat lower densities because PNM is more degenerate than SNM at the same n. The results at $T = 50$ MeV indicate the densities for which matter is in the semi- or nondegenerate regions. The substantial improvement offered by the NLO corrections are, however, noteworthy.

Figure 7 contains results for the thermal pressures. As for S and $E_{\rm th}$, agreement of the FLT+NLO results with those of exact numerical calculations extend up to $n \simeq 0.1\ \mathrm{fm}^{-3}$ at $T = 20$ MeV. The situation with the results at $T = 50$ MeV is less satisfactory. The disagreement with the exact results at this temperature is partly owing to the fact that M^* begins to acquire a non-negligible temperature dependence as T increases,[11] not considered in the FLT+NLO treatment. Also at work is the fact that the thermodynamic identity $\mathcal{E} + P = Ts + \mu n$ cannot be satisfied even in principle beyond the Hartree level unless the theory is exactly solved.

The thermal parts of the chemical potentials are shown in Fig. 8. As for S, $E_{\rm th}$, and $P_{\rm th}$, the NLO corrections render significant improvement over the FLT results down to $n \simeq 0.1\ \mathrm{fm}^{-3}$ for $T = 20$ MeV. The agreement of the FLT+NLO results

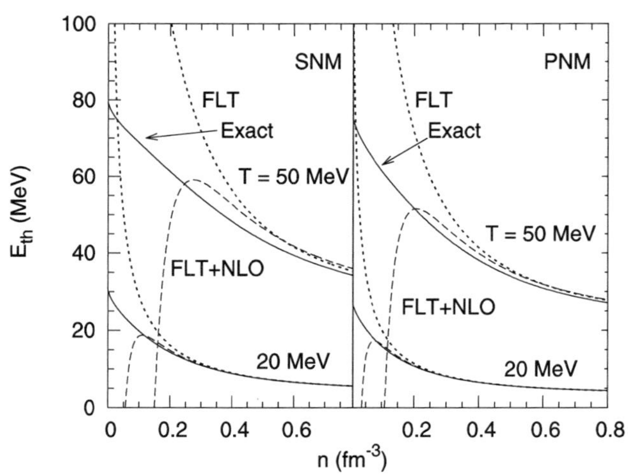

Fig. 6. Same as Fig. 5 but for thermal energies from Eq. (15).

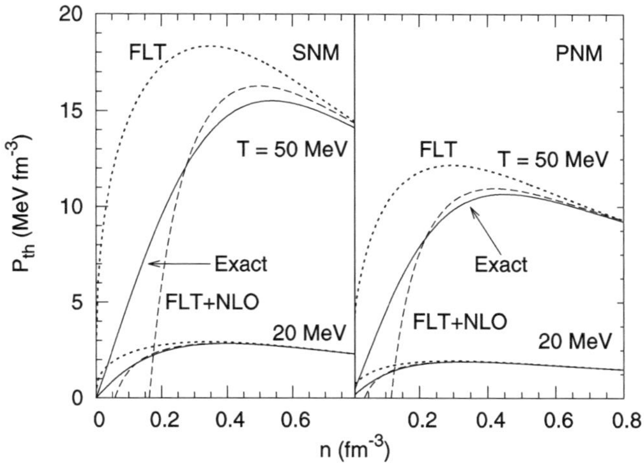

Fig. 7. Same as Fig. 5, but for thermal pressure from Eq. (16).

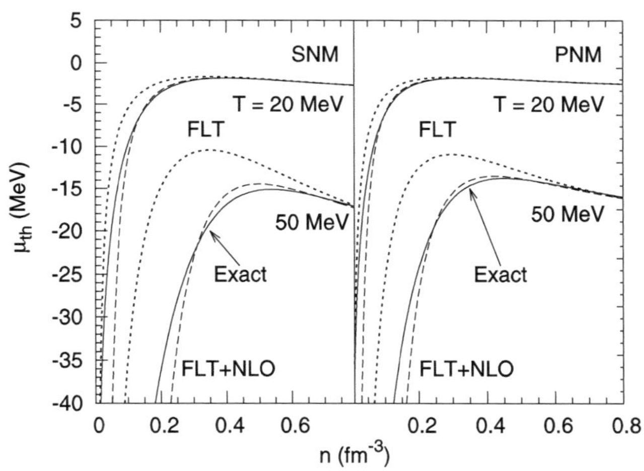

Fig. 8. Same as Fig. 5, but for thermal chemical potential from Eq. (17).

with the exact results is quantitatively better for PNM than for SNM because of its higher degeneracy at this temperature at the same density. The NLO improvements at $T = 50$ MeV are less striking than at 20 MeV, and suffer from the same maladies as the other thermal variables. Analytic expressions for Fermi integrals being asymptotic expansions, this disagreement is unavoidable particularly in the semi-degenerate region. A separate treatment as espoused in Ref. 11 is necessary in the nondegenerate region.

The specific heat at constant volume C_V is shown in Fig. 9. Exact results for C_V were not calculated in Ref. 5, but we can easily gauge the improvement from NLO corrections at near nuclear densities by recalling that at leading order in FLT,

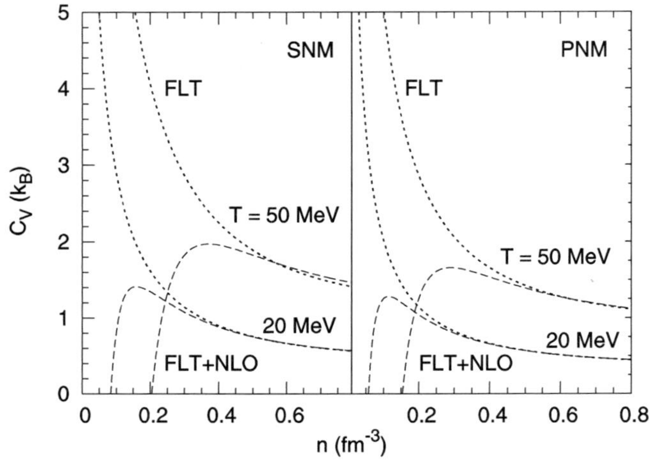

Fig. 9. Same as Fig. 5, but for specific heat at constant volume from Eq. (19).

$C_V = S$. The quantity C_V plays a major role in the long-term cooling of a neutron star. For example, the time for a star's center to cool by neutrino emission can be estimated by

$$\Delta t = -\int \frac{nC_V}{\epsilon_\nu}\, dT, \tag{70}$$

where ϵ_ν is the neutrino emissivity and T is the temperature. At low temperatures ($T \leq 1$ MeV), however, corrections to C_V arising from Cooper-pairing of nucleons must be considered.

Figure 10 shows results for the specific heat at constant pressure. As for C_V, exact numerical results for C_P are not yet available, hence only the FLT and

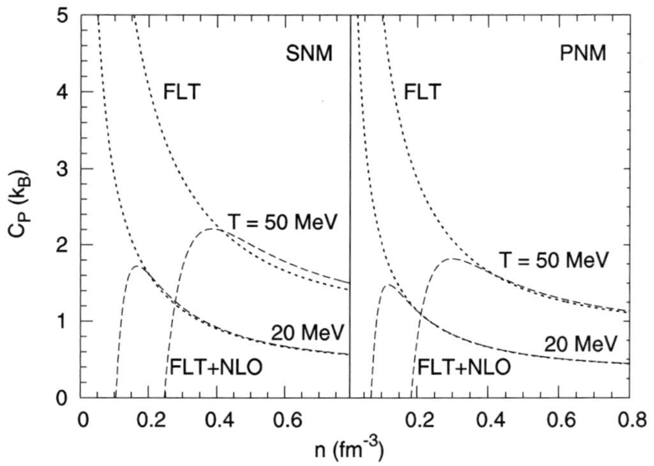

Fig. 10. Same as Fig. 5, but for specific heat at constant pressure from Eq. (20).

FLT+NLO results are shown. At leading order in FLT, $C_P = S$. It is intriguing that at $T = 20$ MeV, the NLO corrections do not alter the leading order FLT result down to near nuclear densities in both SNM and PNM. While the situation is similar for $T = 50$ MeV in PNM for $n \gtrsim 0.35$ fm^{-3}, NLO corrections are apparent in SNM. Exact calculations at $T = 50$ MeV would be necessary to confirm the extent to which NLO corrections improve the FLT results. A relation similar to Eq. (70) but with C_V replaced by C_P and ϵ_ν replaced by $\epsilon_{\gamma+\nu}$ is often used in the literature for time estimates in astrophysical phenomena.

Results for $\Gamma_{\rm th}$ are shown in Fig. 11. At leading order in FLT, $\Gamma_{\rm th} = \frac{5}{3} - \frac{n}{m^*}\frac{dm^*}{dn}$ and is independent of T. This feature is borne out by the results (short dashed curves) in both SNM and PNM, the differences between them stemming from differences in the logarithmic derivatives of the Landau effective masses (see Fig. 4). At NLO, $\Gamma_{\rm th}$ acquires a temperature dependence owing to terms proportional to T^4 in both $P_{\rm th}$ and $\epsilon_{\rm th}$. With increasing n, and hence degeneracy, the coefficient of the leading $\mathcal{O}(T^2)$ term in $\Gamma_{\rm th}$ decreases making the NLO corrections to diminish in magnitude. The agreement of the FLT+NLO results with the exact results extends to sub-nuclear densities at $T = 20$ MeV in both SNM and PNM. The $T = 50$ MeV results delineate the regions of density for which a semi-degenerate analysis is warranted. At very low densities, the exact $\Gamma_{\rm th}$ tends to $\frac{5}{3}$ the value for nonrelativistic ideal gases.

From the results shown above in Figs. 5–11, it is clear that the lowest density beyond which the FLT+NLO results reproduce the exact numerical results as a function of increasing temperature steadily increases owing to the semi-degenerate region being encountered. The case of $T = 30$ MeV is especially interesting as it happens to be the maximum temperature encountered in core-collapse supernova simulations. Lacking exact numerical results for MFT+TL calculations at $T = 30$ MeV,

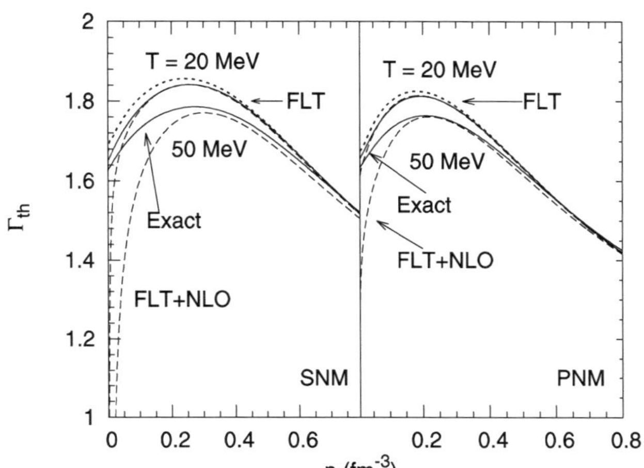

Fig. 11. Same as Fig. 5, but for the thermal index from Eq. (21).

we performed exact numerical calculations for MFT using the couplings in Ref. 5 and compared the ensuing results with those of FLT+NLO (not shown here). For all thermal variables, our findings are: (1) for SNM, very good agreement is found for $n \gtrsim 0.2$ fm^{-3}, and (2) for PNM, the agreement is very good starting from the nuclear density of $n_0 \gtrsim 0.16$ fm^{-3}. We expect a similar behavior for MFT+TL because the total single-particle spectra for both SNM and PNM are predominantly composed of their corresponding MFT parts. For values of Y_e intermediate to those of SNM and PNM, caution must be exercised in carrying the conclusions above as one or the other nucleonic species may be in the non- or semi-degenerate region.

4. Summary and Conclusions

In this work, the NLO extension of Landau's FLT developed in Ref. 4 was utilized to calculate the thermal properties of symmetric nuclear and pure neutron matter (SNM and PNM) for the relativistic model of Ref. 5 in which TL corrections to MFT were included. In FLT, the Landau effective mass m^* and its logarithmic derivative with respect to density n suffice to capture the leading order temperature (T) effects. The NLO corrections, which account for the next-higher-order effects in T, require up to second-order derivatives of the generalized Landau effective mass function $\mathcal{M}(p) = p(\frac{\partial\epsilon}{\partial p})^{-1}$, where $\epsilon \equiv \epsilon(n,p)$ is the density and momentum dependent part of the $T = 0$ single-particle spectrum. The explicit form of $\epsilon(n,p)$ depends on the specific nature of the $T = 0$ many-body calculation performed. Contrasting examples include models with contact or finite range interactions, MFT versus MFT+TL approximations, Bruekner–Hartree–Fock versus Dirac–Brueckner–Hartree–Fock, effective field-theoretical approaches at various levels of approximation, etc. For all these cases, the NLO extension enables the calculation of the entropy density and specific heats up to $\mathcal{O}(T/T_F)^3$ whereas the energy density, chemical potential and pressure to $\mathcal{O}(T/T_F)^4$ (where T_F is the Fermi temperature) extending the leading order results of FLT.

Our comparisons of FLT and FLT+NLO results with those of the exact numerical calculations reported in Ref. 5 for the relativistic model in which TL effects were included reveal that substantial improvements are achieved by the NLO corrections for all thermal variables (entropy, energy, pressure, chemical potential, and specific heats) for entropy per baryon S of up to 2. It is noteworthy that the NLO corrections extend agreement with the exact results to sub-nuclear densities of $n \sim 0.1$ fm^{-3} for $T = 20$ MeV, whereas the FLT results are valid for densities beyond ~ 0.2 fm^{-3}. Insofar as for $T \lesssim 20$ MeV and $n \lesssim 0.1$ fm^{-3}, an inhomogeneous phase consisting of heavy nuclei, light nuclear clusters, dripped nucleons, and pasta-like configurations exists which requires a separate treatment, the semi-analytical formulas of the FLT+NLO formalism enable a rapid evaluation of thermal effects in bulk homogeneous matter in addition to providing physical insights and checks of time-consuming exact numerical calculations.

Several areas for further investigation remain including an assessment of nonanalytic contributions arising from long-wavelength fluctuations, single particle-hole excitations and, collective and pairing correlations close to the Fermi surface.[2] Establishing their roles in astrophysical phenomena needs further work and will be reported elsewhere.

5. Personal Tributes to Gerry Brown

Constantinos Constantinou When the seminar room of the Nuclear Theory Group at Stony Brook was named in his honor, Gerry was playfully 'upset': "Are they telling me that I should retire?" He certainly had no such plans; it simply wasn't in the stars- collapsing, exploding or otherwise. He would come in (almost) every day full of energy and new ideas about problems to solve and just early enough to win his little battle with John Milnor for the #2 YITP parking spot; well... sometimes.

I was fortunate to have been under Gerry's tutelage for about two years. During this time he made it a point that I should learn FLT-among many other things which would be sprung upon me faster than I could dig out references for. In hindsight, I should have asked Gerry for those; he was much better than any search engine in this regard. In our contribution here, we apply an extended version of FLT to a relativistic model beyond the mean-field level. Perhaps Gerry would have liked it.

Madappa Prakash Gerry Brown was often fond of saying that although Landau's FLT originated in Russia, few Russians used it and it was left for others to exploit Landau's genius. Gerry, along with Kevin Bedell, taught me this subject which I have used whenever I can to gain physical insights. I recall Gerry struggling to satisfy Landau's forward-scattering sum rule using results from many-body calculations of nuclear matter. He exhorted all who could (particularly, Pandharipande) to provide him with Fermi-liquid parameters and was disappointed when they could not owing to the inherent difficulties in their many-body methods. The laments of his toil are recorded in a Physics Reports he wrote with Sven-Olaf Bäckman and Joni Niskanen.[12]

Our contribution to Gerry's 90th birthday memorial tribute here resulted from my collaboration with Gerry's last graduate student Constantinos Constantinou and my current graduate student Sudhanva Lalit. I like to think that in undertaking the work reported here, I am passing on the lessons Gerry taught me. He would have been pleased with the extensions to leading-order FLT that was developed by us in Ref. 4 and put to good use. Not one to praise anyone to his or her face, Gerry would have said "What about Landau's forward-scattering sum rule?". No doubt that would have annoyed us, but he always wanted to go forward.

A few "Gerry-sms" that I can never forget. "Don't be a scholar, do things". This one used to annoy me the most. What annoys me more is that I use it on my own students nowadays! "People keep saying they're consistent. But, are they right?" No arguments there. When warned to be careful, "I'm never careful! I want to get ahead". Bravado there. But, I have read his D.Sc. thesis in which he was super-careful. "I've no problem using results that I don't understand." He selected results of those he trusted. There are more, but for some other time and some other place.

Acknowledgments

We are grateful to Xilin Zhang for providing numerical results of the exact two-loop calculations in tabular form. This work was supported by the U.S. DOE under Grant No. DE-GG02-93ER-40756 (for S.L and M.P.).

References

1. J. M. Lattimer and M. Prakash, The equation of state of hot, dense matter and neutron stars, *Phys. Rep.* **621** (2016) 127.
2. G. Baym and C. J. Pethick, *Landau Fermi-Liquid Theory* (Wiley Interscience, New York, 1991).
3. G. Baym and S. A. Chin, Landau theory of relativistic Fermi liquids, *Nucl. Phys. A* **262** (1976) 527.
4. C. Constantinou, B. Muccioli, M. Prakash and J. M. Lattimer, Degenerate limit thermodynamics beyond leading order for models of dense matter, *Ann. Phys.* **363** (2015) 533.
5. X. Zhang and M. Prakash, Hot and dense matter beyond relativistic mean field theory, *Phys. Rev. C* **93** (2016) 055805.
6. J. D. Walecka, A theory of highly condensed matter, *Ann. Phys.* **83** (1974) 491.
7. J. Boguta and A. R. Bodmer, Relativistic calculation of nuclear matter and the nuclear surface, *Nucl. Phys. A* **292** (1977) 413.
8. S. A. Chin, A relativistic many-body theory of high density matter, *Ann. Phys.* **108** (1977) 301.
9. M. Abramowitz and I. A. Stegun, *Handbook of Mathematical Functions*, National Bureau of Standards, Applied Mathematics Series, Vol. 55, National Bureau of Standards, 10th printing (1972).
10. H. A. Bethe, G. E. Brown, J. Applegate and J. M.Lattimer, Equation of state in the gravitational collapse of stars, *Nucl. Phys. A* **324** (1979) 487.
11. M. Prakash, T. L. Ainsworth, J. P. Blaizot and H. Wolter, *Windsurfing the Fermi Sea*, Vol. II, eds. T. T. S. Kuo and J. Speth (Elsevier, Amsterdam, 1987).
12. S.-O. Bäckman, G. E. Brown and J. Niskanen, The nucleon-nucleon interaction and the nuclear many-body problem, *Phys. Rep.* **124** (1985) 1.

Core polarization and modern realistic shell-model Hamiltonians

L. Coraggio[*,§], A. Covello[†,¶], A. Gargano[*,||]
and N. Itaco[*,‡,**]

[*]*Istituto Nazionale di Fisica Nucleare, Complesso Universitario di Monte, S. Angelo, Via Cintia, I-80126 Napoli, Italy*

[†]*Dipartimento di Fisica "E. Pancini", Università di Napoli "Federico II", Complesso Universitario di Monte S. Angelo, Via Cintia, I-80126 Napoli, Italy*

[‡]*Dipartimento di Matematica e Fisica, Università della Campania "Luigi Vanvitelli", viale Abramo Lincoln 5, I-81100, Caserta, Italy*

[§]*coraggio@na.infn.it*
[¶]*covello@na.infn.it*
[||]*gargano@na.infn.it*
[**]*itaco@na.infn.it*

The understanding of the convergence properties of the shell-model effective Hamiltonian, within the framework of the many-body perturbation theory, is a long-standing problem. The infinite summation of a certain class of diagrams, the so-called "bubble diagrams," may be provided calculating the Kirson–Babu–Brown induced interaction, and provides a valid instrument to study whether or not the finite summation of the perturbative series is well-grounded. Here, we perform an application of the calculation of the Kirson–Babu–Brown induced interaction to derive the shell-model effective Hamiltonian for *p*-shell nuclei starting from a modern nucleon–nucleon potential, obtained by way of the chiral perturbation theory. The outcome of our calculation is compared with a standard calculation of the shell-model Hamiltonian, where the core-polarization effects are calculated only up to third-order in perturbation theory. The results of the two calculations are very close to each other, evidencing that the perturbative approach to the derivation of the shell-model Hamiltonian is still a valid tool for nuclear structure studies.

Keywords: Nuclear shell model; realistic nucleon–nucleon potentials; effective interactions.

1. Introduction

The work of Gerry Brown has greatly contributed to the development of many aspects of nuclear physics.[1] In particular, his 1966 paper with Tom Kuo "Structure of finite nuclei and the free nucleon–nucleon interaction. An application to ^{18}O and ^{18}F"[2] has laid the basis for a more fundamental approach to nuclear shell-model by linking the valence-space Hamiltonian to the underlying nuclear forces. In this paper, ^{18}O and ^{18}F were described as two valence nucleons outside an inert ^{16}O core, interacting through an effective interaction derived from the Hamada–Johnston nucleon–nucleon (NN) potential[3] and composed of the Brueckner G-matrix and second-order core polarization diagrams, which take into account particle-hole excitations of the core excluded from the chosen model space. The inclusion of these diagrams greatly improved the agreement of the calculated energy spectra with experiment, making the Kuo–Brown matrix elements for the *sd*-shell widely used for decades in nuclear-structure calculations.

Since the paper of Kuo and Brown, more solid theoretical foundations have been established for the realistic shell-model approach and significant progress has been made in the derivation of the effective valence-space Hamiltonian from the free NN potential. An account of the state of the art and the most relevant advances can be found in some review articles.[4,5] It is worth mentioning that along the trail of these improvements, a number of realistic shell-model calculations have been performed during the last 20-odd years, which have provided an accurate description of nuclear structure properties for nuclei in various mass regions both close to and far from the stability valley.

In a recent memorial volume dedicated to Brown,[6] we compared the results for the *sd*-shell nuclei reported in the Kuo–Brown paper with those obtained by using a modern shell-model effective Hamiltonian, showing that the original Kuo–Brown approach was able to catch the main aspects of the physics of many-nucleon systems within the framework of the shell-model.

Here, we focus on a more specific topic: the effects of core-polarization (CP) diagrams, whose second-order contribution to the interaction beyond the G matrix is crucial to obtain good agreement between theory and experiment for ^{18}O and ^{18}F.[2]

The importance of these effects in nuclear structure was first recognized by George Bertsch[7] for ^{18}O and ^{42}Sc using the Kallio–Kolltveit potential.[8] He showed that CP diagrams corresponding to one-particle-one-hole excitations at second-order in perturbation theory could lead to a renormalization of the G-matrix interaction as large as 30%. Soon afterwards, Kuo and Brown published the above mentioned paper, that marked a turning point in the development of the field.

This work was the start of a wide sequence of studies aimed at investigating the various problems related to the perturbative derivation of the shell-model effective interaction from the free NN potential, in the attempt to make this kind of calculation a reliable tool for a quantitative description of nuclear structure properties. In this context, one of the main issue has been the role of CP contributions beyond second-order, to which a great effort has been devoted.

For instance, third-order CP diagrams, including those with one fold, were studied in detail by Barrett and Kirson[9] for the *sd*-region nuclei, while Hjorth-Jensen *et al.*[4] carried out extensive investigations of the third-order CP diagrams for tin region. A main outcome of these studies is that third-order CP diagrams have to be explicitly taken into account, their contribution being not negligible. This seems to indicate that CP effects are so strong that also higher order diagrams cannot be disregarded. However, higher-order CP calculations are difficult to perform. As a matter of fact, order-by-order calculation of CP diagrams beyond third-order has never been carried out, mainly because the number of CP diagrams grows rapidly in the perturbative expansion. Alternatively, one may try to resort to a nonperturbative method to study the role of CP diagrams.

In a paper published in 2005 by Brown *et al.*[10] a method was proposed for efficiently summing CP diagrams to all orders in perturbation theory. The method is based on the effective and accurate induced interaction approach of Kirson[12] and Babu and Brown,[13] referred to as KBB method, where the vertex functions are obtained by solving self-consistently a set of equations. The application presented in Refs. 10 and 11 is similar to that performed few years before by Kirson to study ^{18}O and ^{18}F nuclei using a G-matrix interaction. It is worth to point out that thanks to the use of the $V_{\text{low}-\text{k}}$[14] vertices in place of the G-matrix ones, Brown and coworkers were able to include a wider class of diagrams when compared to that studied by Kirson, taking full advantage of the $V_{\text{low}-\text{k}}$ independence on both the starting energy and Pauli exclusion operator.

In the present paper, we have carried out all-order CP calculations for p-shell nuclei, starting from a chiral NN potential, along the same lines of Ref. 10 by including TDA diagrams (dubbed in the following "bubble diagrams"). We show in the next section that, at variance with that paper, where the other considered class of diagrams were limited to second-order, we extend the calculation to third-order diagrams, which means that third order RPA diagrams are explicitly taken into account.

A major merit of the present work is also the use of a NN potential derived within the chiral perturbation theory (ChPT). As a matter of fact, during the last decade these NN potentials have provided an approach to the problem of nuclear forces that is grounded in the quantum chromodynamics. The two main advantages of ChPT are that it generates nuclear two- and many-body forces on equal footing[15–17] and that the so obtained nuclear potentials are naturally tailored for the low-energy regime of the nuclear structure physics, thus avoiding the complications related to renormalizing the short-range repulsion.

In the next section, we give a few details about the chiral NN potential employed and a short description of the perturbative expansion of the effective shell-model Hamiltonian and of the KBB induced interaction approach. In Sec. 3, the results obtained for p-shell nuclei will be reported. Comments and conclusions are drawn in Sec. 4.

2. Outline of Calculations

2.1. *Realistic nucleon–nucleon potential*

We start this section with a description of a low-momentum chiral potential, which has been introduced for the first time to perform shell-model calculations in the sd-shell region.[18] The aim was to generate a chiral potential to be compared directly with a $V_{\text{low-k}}$ derived from the chiral potential of Entem and Machleidt,[19] renormalized within a cutoff momentum $\Lambda = 2.1$ fm^{-1}.[14]

As a matter of fact, within the $V_{\text{low-k}}$ approach, a realistic potential V_{NN} is smoothed by integrating out a cutoff momentum Λ the high-momentum modes of V_{NN}. This is achieved by the unitary transformation suggested by Lee and Suzuki.[14,20]

It is worth to recall that, the ChPT is characterized by the spontaneously broken chiral symmetry, and the degrees of freedom which are relevant for nuclear physics, nucleons and pions.

In order to reach a proper convergence rate, the ChPT expansion is valid only for momenta $Q < \Lambda_\chi \simeq 1$ GeV, where Λ_χ denotes the chiral symmetry breaking scale. Realistic NN potentials derived from ChPT are typically multiplied by a (nonlocal) regulator function $f(p', p) = \exp[-(p'/\Lambda)^{2n} - (p/\Lambda)^{2n}]$. Present chiral NN potentials[19,21] typically employ values for Λ around 2.5 fm^{-1}.

On these premises, in Ref. 18, we have constructed a low-momentum chiral N^{3}LO NN potential, using the regulator function given in the above expression with $n = 10$ and $\Lambda = 2.1$ fm^{-1}. We have dubbed this potential N^{3}LOW.

One advantage of this potential is that it is a smooth potential with a perturbative-like behavior as $V_{\text{low-k}}$, but it is given in analytic form, the same as for the "hard" N^{3}LO potential by Entem and Machleidt of 2003.[19]

The N^{3}LOW potential reproduces accurately the empirical deuteron binding energy, the experimental low-energy scattering parameters, and the empirical phase-shifts of NN scattering up to 200 MeV laboratory energy (see Fig. 1).

It is worth to point out that a great advantage of ChPT is that it generates nuclear two- and many-body forces on an equal footing. Most interaction vertices that appear in the three-nucleon force (3NF) and in the four-nucleon force (4NF) also occur in the two-nucleon force (2NF). The parameters attached to these vertices are fixed in the construction of the chiral 2NF. Consistency requires that for the same vertices the same parameter values are used in the 2NF, 3NF, 4NF.

We have employed the N^{3}LOW potential in calculations both for infinite nuclear matter[22,23] and — within the framework of the realistic shell model — finite nuclei.[6,18,24]

In this paper, we are going to deal with the latter, so a short introduction to the realistic shell model is in order.

2.2. *Shell-model effective Hamiltonian*

Within the realistic shell-model, the effective Hamiltonian, namely the single-particle energies (s.p.e.) and two-body matrix elements (TBME) of the residual interaction, is derived from a realistic NN force,[5] such as the N^{3}LOW potential, by means of many-body perturbation theory. To this end, we employ the method developed by Kuo and coworkers through the 1970s,[25] namely the well-known

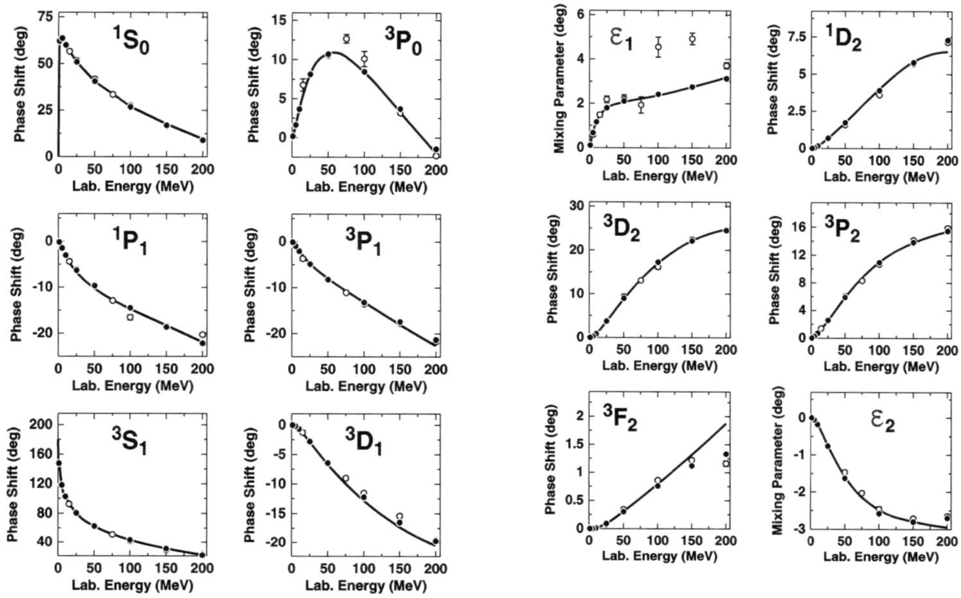

Fig. 1. Phase parameters of neutron–proton scattering up to 200 MeV laboratory energy. The solid lines show the predictions by the N^3LOW potential. The solid dots and open circles represent the Nijmegen multienergy *np* phase shift analysis and the GWU/VPI single-energy *np* analysis SM99, respectively.

$\hat{Q}$-box plus folded-diagram method,[5,25] where the $\hat{Q}$-box is a collection of one- and two-body irreducible valence-linked Goldstone diagrams. Within this framework the effective Hamiltonian H_{eff} can be written in an operator form as

$$H_{\text{eff}} = \hat{Q} - \hat{Q}' \int \hat{Q} + \hat{Q}' \int \hat{Q} \int \hat{Q} - \hat{Q}' \int \hat{Q} \int \hat{Q} \int \hat{Q} + \cdots , \tag{1}$$

where the integral sign represents a generalized folding operation, and $\hat{Q}'$ is obtained from $\hat{Q}$ by removing first-order terms. The folded-diagram series is summed up to all orders using the Lee–Suzuki iteration method.[20]

Our aim is to evaluate the impact of the KBB all-order summation of the bubble diagrams on the calculation of effective shell-model Hamiltonians, with respect to a standard order-by-order calculation of the $\hat{Q}$-box perturbative expansion. Some details on the nonperturbative method we have used, based on the induced interaction KBB method, are given in Sec. 2.3.

To do this, we compare in Sec. 3, the results of two different shell-model Hamiltonians, where the $\hat{Q}$-box is obtained

(a) by using a perturbative expansion with all diagrams up to third-order[5];
(b) by substituting the bubble diagrams belonging to the third-order with the KBB summation.

In both calculations of H_{eff}, we start from a purely intrinsic Hamiltonian, that can be written as follows:

$$H_{\text{int}} = \left(1 - \frac{1}{A}\right) \sum_{i=1}^{A} \frac{p_i^2}{2m} + \sum_{i<j=1}^{A} \left(V_{ij}^{\text{NN}} - \frac{\mathbf{p_i} \cdot \mathbf{p_j}}{mA}\right) = \left[\sum_{i=1}^{A} \left(\frac{p_i^2}{2m} + U_i\right)\right]$$
$$+ \left[\sum_{i<j=1}^{A} \left(V_{ij}^{\text{NN}} - U_i - \frac{p_i^2}{2mA} - \frac{\mathbf{p_i} \cdot \mathbf{p_j}}{mA}\right)\right] = H_0 + H_1, \quad (2)$$

where U is the harmonic oscillator one-body potential.

Since we deal with light nuclei belonging to the p-shell, it is needed to employ an intrinsic Hamiltonian especially if results of shell-model calculations are compared with those of *ab initio* nuclear-structure approaches.[5]

As regards the dimension of the intermediate-state space of our perturbative expansion, we have included states between successive vertices with excitation energy less than $E_{\max} = N_{\max}\hbar\omega$, where $\hbar\omega$ is the chosen harmonic oscillator parameter and $N_{\max}$ should be sufficiently large to ensure that both s.p.e. and TBME are independent, within a few keV, from $E_{\max}$. In our calculations, we have taken $\hbar\omega = 19\,\text{MeV}$ and $N_{\max} = 18$.

2.3. *Kirson–Babu–Brown method*

The KBB induced interaction method provides a summation at all orders of the bubble diagrams, and is based on the self-consistent solution of non linear coupled equations for the irreducible vertex functions.

Let us consider the particle-hole (ph) vertex function f, that is reported pictorially in Fig. 2 and is defined, in an operatorial form, by way of the self-consistent equation

$$f = V + \Sigma g_{ph} \Sigma + \Sigma g_{ph} f g_{ph} \Sigma + \Sigma g_{ph} f g_{ph} f g_{ph} \Sigma + \cdots , \quad (3)$$

where g_{ph} is the free ph Green function, and Σ denotes the vertex for particle-core and hole-core coupling.[10]

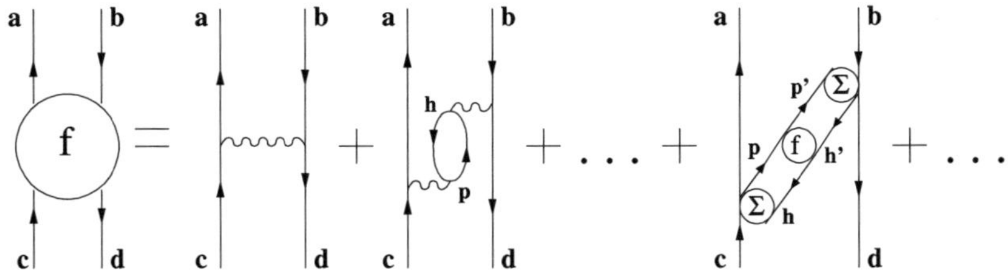

Fig. 2. Diagrammatic expansion of the ph vertex function f. Σ denotes the vertex for particle-core and hole-core coupling.[10]

By considering $\Sigma = V$ in Eq. (3), we can easily see that f represents an infinite summation of CP diagrams:

$$f = V + Vg_{ph}V + Vg_{ph}fg_{ph}V + Vg_{ph}fg_{ph}fg_{ph}V + \cdots . \tag{4}$$

From the presence of f on both left- and right-hand sides of Eq. (4), it is clear that an iterative solution of this integral equation corresponds to a summation at all orders of the bubble diagrams.

When dealing with nuclear many-body calculations, it is needed to derive also the particle–particle (pp) vertex function Γ, whose self-consistent expression is given in operatorial form as

$$\Gamma = V + \Sigma g_{ph}\Sigma + \Sigma g_{ph}fg_{ph}\Sigma + \Sigma g_{ph}fg_{ph}fg_{ph}\Sigma + \cdots . \tag{5}$$

The vertex functions f and Γ are then coupled together, defining a coupling vertex function Σ:

$$\begin{aligned} \Sigma &= V + \Sigma_{ph} + \Sigma_{pp} \\ \Sigma_{ph} &= Vg_{ph}V + Vg_{ph}fg_{ph}V + Vg_{ph}fg_{ph}fg_{ph}V + \cdots , \\ \Sigma_{pp} &= Vg_{pp}V + Vg_{pp}\Gamma g_{pp}V + Vg_{pp}\Gamma g_{pp}\Gamma g_{pp}V + \cdots , \end{aligned} \tag{6}$$

where g_{pp} is the free pp Green function.

The self-consistent vertex functions f and Γ are determined from Eqs. (3), (5), and (7).

It should be pointed out that the above coupled integral equations are similar to those reported by Kirson in his papers,[12] where only Σ_{ph} was taken into account, at variance with the present work including both Σ_{ph} and Σ_{pp}.

In order to solve the equations for f and Γ, we have followed the approach reported in Ref. 10, where the true ph and pp Green functions were used. The introduction of these functions

$$G_{ph} = g_{ph} + g_{ph}fG_{ph}, \tag{7}$$

$$G_{pp} = g_{pp} + g_{pp}\Gamma G_{pp}, \tag{8}$$

aims to partially sum and regroup our series, in order to give the self-consistent Eqs. (3), (5) and (7) a much simpler form:

$$\begin{aligned} f &= V + \Sigma G_{ph}\Sigma, \\ \Gamma &= V + \Sigma G_{ph}\Sigma, \\ \Sigma &= V + VG_{ph}V + VG_{pp}V. \end{aligned} \tag{9}$$

This procedure allows to follow an iterative method able to solve very efficiently the above coupled equations.[10]

3. Results

In this section, we present some selected results of shell-model calculations for p-shell nuclei, using realistic shell-model Hamiltonians derived from the chiral two-body potential N^3LOW following the prescriptions illustrated in Sec. 2. More precisely, we employ an $H_{\rm eff}$ derived at third-order in perturbation theory, as we have done in many other works,[5] and another one where all bubble diagrams up to third-order belonging to the $\hat{Q}$-box have been substituted with the KBB induced interaction. We refer to the calculations with the first $H_{\rm eff}$ as (I), the calculations with the second one are labelled as (II). The model space we have employed is the standard one, that is set up by the proton and neutron single-particle orbitals $0p_{3/2}, 0p_{1/2}$. It has to be mentioned that the s.p.e. and the TBME of $H_{\rm eff}$ (I) have been already published in a preceding paper.[5]

In Fig. 3, the calculated spectra of ^{6}He up to 7 MeV excitation energy are reported, and compared with the experimental bound states that have a firm spin-parity assignment.[26]

As can be seen, the eigenvalues obtained with Hamiltonians (I) and (II) are very close to each other, the largest discrepancy being 350 keV for the second excited 2^+ state and less than 200 keV for all other ones. The same degree of agreement can be found if one considers for calculations (I) and (II) only diagrams up to second-order in perturbation theory.

This evidences that, in the identical-particle channel, the $H_{\rm eff}$ provided by the partial summation of CP diagrams up to third order, included in the $\hat{Q}$-box, is in a good agreement with that obtained summing bubble diagrams at all orders.

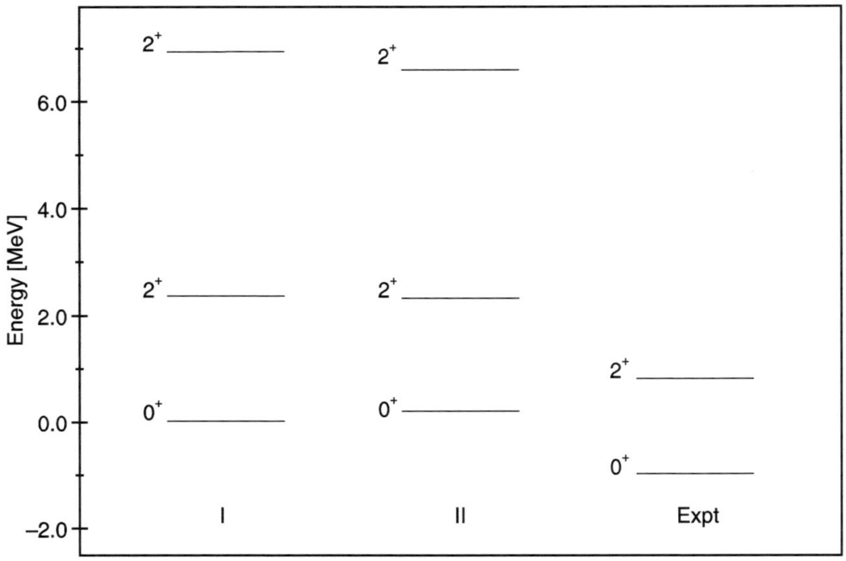

Fig. 3. Experimental[26] and theoretical spectra of ^{6}He calculated with $H_{\rm eff}$ (I) and (II) (see text for details).

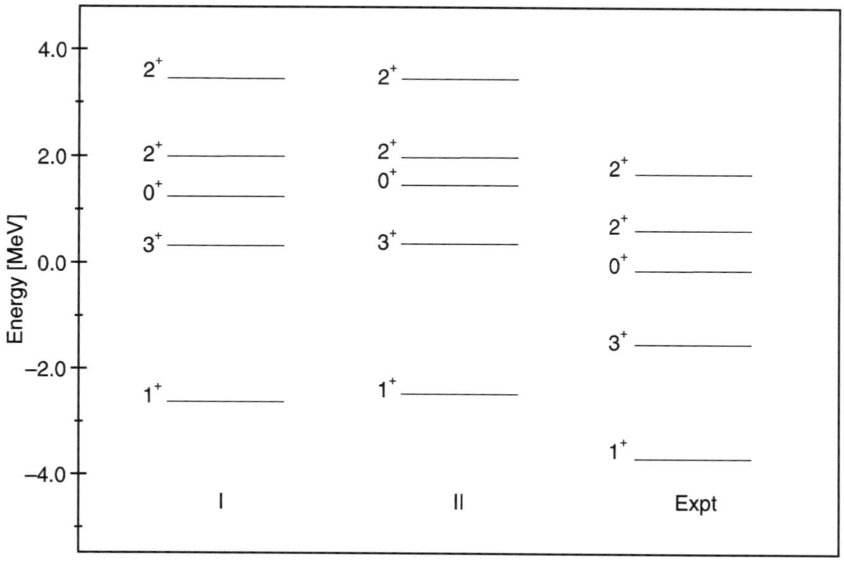

Fig. 4. Experimental[26] and theoretical spectra of ^{6}Li calculated with $H_{\rm eff}$ (I) and (II) (see text for details).

As regards the proton–neutron interaction channel, we have calculated the spectra of ^{6}Li up to 7 MeV excitation energy too, and they can be found in Fig. 4, where the comparison with the corresponding observed states is also shown.

It can be observed that for all calculated eigenvalues, except one, the discrepancy between calculation (I) and (II) do not exceed 220 keV. Actually, the agreement between the results of the two calculations is even better than that we have found in the identical-particle interaction channel.

We consider the above results a major achievement, since they may be of help in answering the long-standing question on the convergence of the perturbative expansion of the shell-model effective Hamiltonian. In this connection, it is worth mentioning, however, that calculations for the *sd*-region have shown a stronger effect of the KBB induced interaction[10] with respect to those reported here for the *p*-region. In fact, in the paper by Holt and coworkers,[10] where the perturbative expansion was limited to second-order, it was shown that for ^{18}O an all-order summation of the CP diagrams leads to an increase of about 500 keV in the 0^+ ground-state energy, which goes up to about 1 MeV for the 1^+ ground state of ^{18}F. As a matter of fact, the difference we have found for *p*-shell nuclei between partial and infinite summation of CP diagrams is about one order of magnitude smaller than that in Refs. 10 and 11. One possible reason may be the smaller hole space corresponding to the doubly closed core of ^{4}He with respect to ^{16}O, implying that particle-hole excitations play a minor role in the derivation of the shell-model effective Hamiltonian for *p*-shell nuclei.

Further studies along this direction are certainly needed.

4. Summary and Outlook

In this paper, we have presented the results of a study aimed at understanding the contribution of the infinite summation of the well-known bubble diagrams when calculating perturbatively the shell-model effective Hamiltonian.

To this end, we have derived two different H_{eff}s for the p-shell region, the first one being calculated including all diagrams up to third-order in perturbation theory, and the second one subtracting the finite order bubble diagrams and substituting them with the KBB induced interaction.

The results of shell-model calculations, starting from those H_{eff}s, have been compared for two-valence nucleon systems, namely the neutron–neutron and proton–neutron ones. The calculated spectra obtained within these two approaches show an excellent agreement, driving to the conclusion that the derivation of the shell-model effective Hamiltonian up to third-order in perturbation theory takes into account CP excitations excellently.

In previous studies[10,11] this comparison was less satisfactory, the testing ground being heavier mass nuclei in the ^{16}O region. This evidences the need of further investigations, that involve both p- and sd-shell nuclei and the implementation of different realistic forces as input NN potential.

The evolution of this study may shed light on the understanding of the perturbative properties of modern shell-model effective Hamiltonians, by a direct comparison with the infinite summation of a specific diagrammatic topology.

References

1. Special Issue Dedicated to the Memory of Gerald E. Brown, *Nucl. Phys. A* **928** (2014) 1; Memorial Volume in Honor of Gerald E. Brown, *Phys. Rep.* **621** (2016) 1.
2. T. T. S. Kuo and G. E. Brown, *Nucl. Phys.* **85** (1966) 40.
3. T. Hamada and I. D. Johnston, *Nucl. Phys.* **34** (1962) 382.
4. M. Hjorth-Jensen, T. T. S. Kuo and E. Osnes, *Phys. Rep.* **261** (2016) 125.
5. L. Coraggio, A. Covello, A. Gargano, N. Itaco and T. T. S. Kuo, *Ann. Phys. (NY)* **327** (2012) 2125.
6. L. Coraggio, A. Covello, A. Gargano, N. Itaco and T. T. S. Kuo, *Nucl. Phys. A* **928** (2014) 024311.
7. G. F. Bertsch, *Nucl. Phys.* **74** (1965) 234.
8. A. Kallio and K. Kolltveit, *Nucl. Phys.* **53** (1964) 87.
9. B. R. Barrett and M. W. Kirson, *Nucl. Phys.* **148** (1970) 145.
10. J. D. Holt, J. W. Holt, T. T. S. Kuo, G. E. Brown and S. K. Bogner, *Phys. Rev. C* **72** (2005) 041304(R).
11. T. T. S. Kuo, J. D. Holt, J. W. Holt, G. E. Brown and S. K. Bogner, *J. Phys. Conf. Ser.* **20** (2005) 1.
12. M. W. Kirson, *Ann. Phys. (NY)* **66** (1971) 624; M. W. Kirson, *Ann. Phys. (NY)* **68** (1971) 556; M. W. Kirson, *Ann. Phys. (NY)* **82** (1974) 345.
13. S. Babu and G. E. Brown, *Ann. Phys. (NY)* **78** (1973) 1.
14. S. Bogner, T. T. S. Kuo, L. Coraggio, A. Covello and N. Itaco, *Phys. Rev. C* **65** (2002) 051301(R).
15. S. Weinberg, *Phys. Lett. B* **295** (1992) 114.

16. U. van Kolck, *Phys. Rev. C* **49** (1994) 2932.
17. R. Machleidt and D. R. Entem, *Phys. Rep.* **503** (2011) 1.
18. L. Coraggio, A. Covello, A. Gargano, N. Itaco, T. T. S. Kuo, D. R. Entem and R. Machleidt, *Phys. Rev. C* **75** (2007) 024311.
19. D. R. Entem and R. Machleidt, *Phys. Rev. C* **68** (2003) 041001(R).
20. S. Y. Lee and K. Suzuki, *Phys. Lett. B* **91** (1980) 173.
21. E. Epelpaum, W. Glöckle and Ulf-G. Meissner, *Nucl. Phys. A* **747** (2005) 362.
22. L. Coraggio, J. W. Holt, N. Itaco, R. Machleidt and F. Sammarruca, *Phys. Rev. C* **87** (2013) 014322.
23. L. Coraggio, J. W. Holt, N. Itaco, R. Machleidt, L. E. Marcucci and F. Sammarruca, *Phys. Rev. C* **89** (2014) 044321.
24. L. Coraggio, A. Covello, A. Gargano and N. Itaco, *Phys. Rev. C* **81** (2010) 064303.
25. T. T. S. Kuo and E. Osnes, *Lecture Notes in Physics*, Vol. 364 (Springer-Verlag, Berlin, 1990).
26. Data extracted using the NNDC Online Data Service from the ENSDF database, file revised as of 27 July, 2016.

Science and status of the Electron Ion Collider

A. Deshpande
Department of Physics & Astronomy,
Stony Brook University,
Stony Brook, NY 11794-3800, USA
abhay.deshpande@stonybrook.edu

The US Nuclear Science Advisory Committee (NSAC) recently recommended the construction of a high-luminosity, high-energy Electron Ion Collider (EIC), with polarized beams capable of colliding polarized electrons with polarized proton and light ion beams, and with any nucleus. The $\sqrt{s}$ range between 40 GeV and 140 GeV, and luminosity range from 10^{33-34} cm^{-2} s^{-1} were recommended. It is anticipated that under the current guidance from the DOE, the collider could become operational in the second half of the 2020's. This paper summarizes its science and the scope of this over all project.

Keywords: DIS; Quarks; spin; nuclei; gluons.

1. My Communication with Gerry

Although I had a very short overlap with Gerry at Stony Brook, Gerry made a profound impression on me. On the very first formal day of my appointment at Stony Brook, I walked over to my office, and this pleasant old gentleman was waiting outside my office with a paper in his hand. After the introductions, with a warm smile Gerry said, he is so delighted that I joined the faculty at Stony Brook, and that he looking forward to *learning about nucleon spin from me.* He surely had ways to make young scientists feel important. At the end of that 10 min conversation, Gerry gave me the paper about nucleon spin that had appeared on the archive that morning. He wanted to know what I thought of it. About 2 h later Gerry appeared at my office again to ask if I had read the paper. I, somewhat embarrassed explained to him that I was just settling in and further that I had to go back to my experiment at BNL for the rest of the day. He said, "OK, then would you be ready to discuss the paper in the morning?". His childlike eagerness was overwhelming his perception of the surroundings and his interactions. I saw that in him in the next few years of my interactions and observations while he was present in seminars or other discussions. That first impression stayed with me for ever. After a few of our discussions, he invited me without any fixed time-line to write a review of nucleon spin and the RHIC spin program. I unfortunately, as luck would have it, neither he nor I followed up on completing that review. Perhaps, this paper can be considered as that paper and response to his invitation.

2. The Science of Electron Ion Collider

Under the umbrella of Nuclear Science, we study the origin and structure of the core of the atom, the nucleus and the nucleons (protons and neutrons) within it. Nuclei account for essentially all of the mass of the visible universe. Gerry Brown and his collaborators worked on understanding and characterizing the structures and interactions inside the nuclei for more than five decades. Their aim was to understand the observed universe including the birth, life, and death of stars.

Half a century of investigations have revealed that nucleons are themselves composed of more basic constituents called quarks, bound together by the exchange of gluons, and have led to the development of the fundamental theory of strong interactions known as Quantum Chromo-Dynamics (QCD). The force in QCD is mediated by color charge. Understanding these constituent interactions and the emergence of nucleons and nuclei from the properties and dynamics of quarks and gluons in QCD is a fundamental and compelling goal of nuclear science. Gluons interact with each other, generating a significant fraction of the nucleon mass and leading to a little-explored regime of matter, where gluons dominate its behavior. Hints of this regime manifest themselves when nucleons or nuclei collide at relativistic speeds in colliders such as HERA, RHIC and the LHC. The quantitative and systematic study of matter in this new regime requires a new experimental facility: an Electron Ion Collider (EIC).[1]

In the last decade, nuclear physicists have developed new phenomenological tools to enable remarkable tomographic images of the quarks and gluons inside unpolarized as well as polarized protons and neutrons. These tools will be further developed and utilized to study predominantly the valence quarks in the nucleon at the upgraded 12 GeV CEBAF at JLab and COMPASS Experiment at CERN. Applying these new tools to study the matter dominated by gluons will require the higher energy and beam polarization of an EIC.

As one increases the energy of the electron–nucleon collision, the process probes regions of progressively higher gluon density. However, the density of gluons inside a nucleon must eventually saturate to avoid untamed growth in the strength of the nucleon–nucleon interaction, which would violate the fundamental principle of unitarity.[8] To date this saturated gluon density regime has not been clearly observed, but an EIC could enable detailed study of this remarkable aspect of matter. This pursuit will be facilitated by electron collisions with heavy nuclei, where coherent contributions from many nucleons in a nucleus effectively amplify the gluon density being probed.

In 2015, Nuclear Physics Long Range Plan EIC was recommended to be "built in the US, as the highest priority for new facility construction"[2] after the completion of the existing construction project (FRIB) within the office of nuclear science. The EIC would extend the QCD science programs in the U.S. established at both the CEBAF accelerator at JLab and RHIC at BNL, and the one operational in Europe at CERN (COMPASS), in dramatic and fundamentally important ways.

When built the EIC will be the first and the only polarized electron–proton and electron–nucleus collider in the world. The most intellectually pressing questions that an EIC will address that relate to our detailed and fundamental understanding of QCD in this *frontier* environment are:

- How are the sea quarks and gluons, and their spins, distributed in space and momentum inside the nucleon? How are these quark and gluon distributions correlated with overall nucleon properties, such as spin direction? What is the role of the orbital motion of sea quarks and gluons in building the nucleon spin?
- Where does the saturation of gluon densities set in? Is there a simple boundary that separates this region from that of more dilute quark–gluon matter? If so, how do the distributions of quarks and gluons change as one crosses the boundary? Does this saturation produce matter of universal properties in the nucleon and all nuclei viewed at nearly the speed of light?
- How does the nuclear environment affect the distribution of quarks and gluons and their interactions in nuclei? How does the transverse spatial distribution of gluons compare to that in the nucleon? How does nuclear matter respond to a fast moving color charge passing through it? Is this response different for light and heavy quarks?

Answers to these questions are essential for understanding QCD and consequently, the nature of visible matter. An EIC is the ultimate machine to provide answers to these questions for the following reasons:

- A collider is needed to provide kinematic reach well into the gluon-dominated regime;
- Electron beams are needed to bring to bear the unmatched precision of the electromagnetic interaction as a probe;
- Polarized nucleon beams are needed to determine the correlations of sea quark and gluon distributions with the nucleon spin;
- Heavy ion beams are needed to provide undeniable access to the regime of saturated gluon densities and offer a precise dial in the study of propagation-length for color charges in nuclear matter.

The EIC would be distinguished from all past, current, and contemplated facilities around the world by being at the intensity frontier with a versatile range of kinematics and beam polarizations, as well as beam species. In particular, the EIC design exceeds the capabilities of HERA, the only electron–proton collider to date, by adding (a) polarized proton and light-ion beams; (b) a wide variety of heavy-ion beams; (c) two to three orders of magnitude increase in luminosity to facilitate tomographic imaging of nucleons and nuclei; and (d) wide energy variability to enhance the sensitivity to gluon distributions. Achieving these challenging technical improvements in a single facility will extend U.S. leadership in accelerator science and in nuclear science.

The scientific goals and the machine parameters of the EIC were delineated in deliberations at a community-wide program held at the Institute for Nuclear Theory (INT).[3] They were then summarized in a community wide EIC White Paper[1] that made the science case for the EIC in the NSAC Long Range Plan.[2] What follows is a high level excerpts from the EIC White Paper and its Science Goals followed by the Status of the EIC Since the Long Range Plan.

3. Physics with Polarized Beams at the EIC

Several decades of experiments on deep inelastic scattering (DIS) of electron or muon beams off nucleons have taught us about how quarks and gluons (collectively called partons) share the momentum of a fast-moving nucleon. They have not, however, resolved the question of how partons share the nucleon's spin and build up other nucleon intrinsic properties, such as its mass and magnetic moment. The earlier studies were limited to providing the longitudinal momentum distribution of quarks and gluons, *a one-dimensional view of nucleon structure.* The EIC is designed to yield much greater insight into the nucleon structure (Fig. 1, from left to right), by facilitating multi-dimensional maps of the distributions of partons in space, momentum (including momentum components transverse to the nucleon momentum), spin, and flavor.

The 12 GeV upgrade of CEBAF at JLab and the COMPASS at CERN will initiate such studies in predominantly valence quark region. However, these programs will be dramatically extended at the EIC to explore the role of the gluons and sea quarks in determining the hadron structure and properties. This will resolve crucial questions, such as whether a substantial "missing" portion of nucleon spin resides in the gluons. By providing high-energy probes of partons' transverse momenta, the EIC should also illuminate the role of their orbital motion contributing to nucleon spin.

An intensive and worldwide experimental program over the past two decades has shown that the spin of quarks and antiquarks is only responsible for ~30% of the

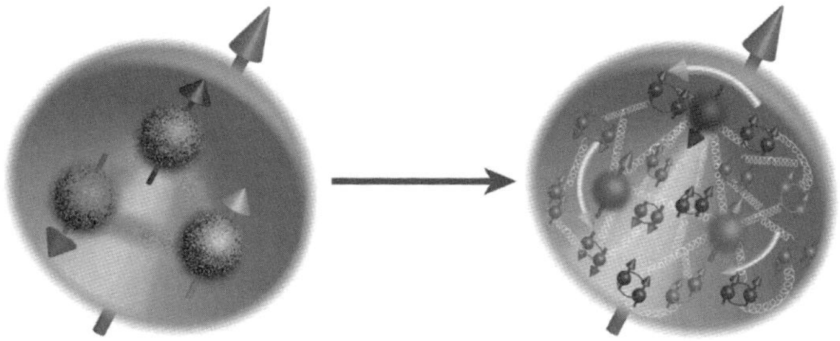

Fig. 1. Left shows a view of the proton as a one-dimensional object, ignoring the motion of partons inside the nucleon and also their relative position with respect to the presumed center of the nucleon. It also assumes that most observed properties of the nucleon originate from the valence quarks in it. The picture on the right acknowledges the possible existence of angular momentum, and possible correlation between the parton's momentum and position within the nucleon.

proton spin. Recent RHIC results indicate that the gluons' spin contribution in the currently explored kinematic region is nonzero, but not yet sufficient to account for the missing 70%. The partons' total helicity contribution to the proton spin is very sensitive to their minimum momentum fraction x (Bjorken "x") accessible by the experiments. With the unique capability to reach two orders of magnitude lower in x and to span a wider range of momentum transfer Q than previously achieved, the EIC would offer *the most powerful tool to precisely quantify* how the spin of gluons and that of quarks of various flavors contribute to the protons spin. The EIC would realize this by colliding longitudinally polarized electrons and nucleons, with both inclusive and semi-inclusive DIS (SIDIS) measurements.

Figure 2 (Right) shows the reduction in uncertainties of the contributions to the nucleon spin from the spin of the gluons, quarks and antiquarks, evaluated in the x range 0.001–1.0 and 0.0001–1.0 with EIC operation at two different center-of-mass energies indicated in the figure. The significant reduction in the uncertainties is apparent and dramatic compared to the current uncertainties. While the central values of the helicity contributions in Fig. 2 are derived from existing data, they could change as new data become available in the low-x region. The uncertainties calculated here are based on the state-of-the art theoretical treatment[4] of all available data related to the nucleon spin puzzle. Clearly, the EIC will make a huge impact on our knowledge of these quantities, unmatched by any other existing or anticipated facility.

3.1. *The confined motion of partons inside the nucleon*

SIDIS measurements are the most effective method to access the Transverse Momentum Distributions (TMDs) of partons within a nucleon.[5] Recent theoretical

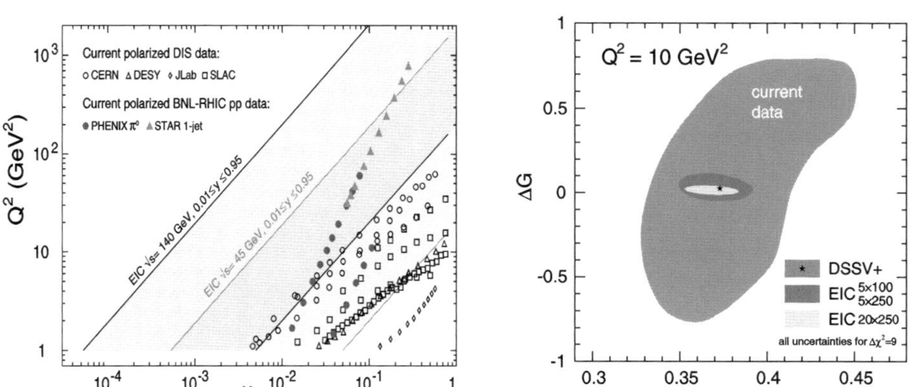

Fig. 2. Left: The range in parton momentum fraction x versus the square of the momentum transferred by the electron to the proton Q^2 accessible with the EIC in e+p collisions at two different center-of-mass energies, compared to existing data. Right: The projected reduction in the uncertainties of the gluon's helicity contribution ΔG versus the quark helicity contribution $\Delta\Sigma/2$ to the proton spin from the region of parton momentum fractions $x > 0.001$ that would be achieved by the EIC for different center-of-mass energies.

advances have led to a rigorous framework where information on the confined motion of the partons inside a fast-moving nucleon enables us to access the TMDs through the correlations between the motion of partons and their spin, as well as the spin of the parent nucleon. These correlations can arise from spin-orbit coupling among the partons, about which very little is known. TMDs thus allow us to investigate the full three-dimensional dynamics of the proton. With both electron and nucleon beams polarized at collider energies, the EIC will dramatically advance our knowledge of the motion of confined gluons and sea quarks in ways not achievable at any existing or proposed facility.

Figure 3 (Left) shows the TMD of up quarks inside a proton moving in the z-direction out of the page, with its spin polarized in the y-direction. The color code indicates the probability of finding the up quarks. The anisotropy in transverse momentum is described by the Sivers distribution function,[5] which is induced by the correlation between the proton's spin direction and the motion of its quarks and gluons. While the figure is based on a preliminary extraction of this distribution from current experimental data, nothing is known about the spin and momentum correlations of the gluons and sea quarks. The achievable statistical precision of the quark Sivers function from EIC kinematics is also shown in Fig. 3 (Right). Currently no data exist for extracting such a picture in the gluon-dominated region in the proton. The EIC will be crucial to initiate and realize such a program.

3.2. *The tomography of the nucleon — spatial imaging of gluons and sea quarks*

By choosing particular final states in electron+proton scattering, the EIC will probe the transverse spatial distribution of sea quarks and gluons in the fast-moving

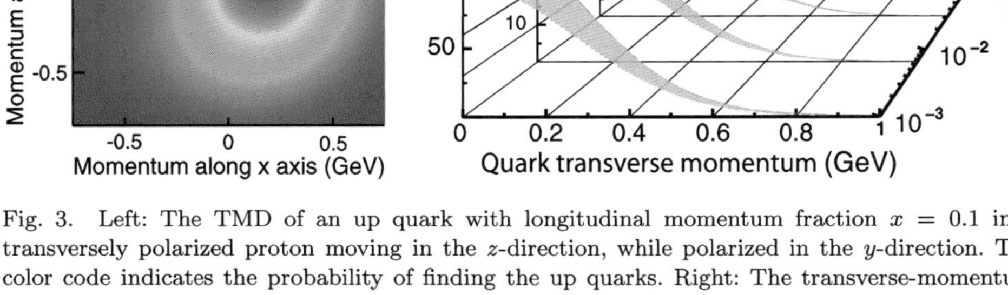

Fig. 3. Left: The TMD of an up quark with longitudinal momentum fraction $x = 0.1$ in a transversely polarized proton moving in the z-direction, while polarized in the y-direction. The color code indicates the probability of finding the up quarks. Right: The transverse-momentum profile of the up quark Sivers function at five x values accessible to the EIC, and corresponding statistical uncertainties.

proton as a function of the parton's longitudinal momentum fraction, x. This spatial distribution yields a picture of the proton that is complementary to the one obtained from the TMD of quarks and gluons, revealing aspects of proton structure that are intimately connected with the dynamics of QCD at large distances.

With its broad range of collision energies, its high luminosity and nearly hermetic detectors, the EIC could image the proton with unprecedented detail and precision from small to large transverse distances. The accessible parton momentum fractions x extend from a region dominated by sea quarks and gluons to one where valence quarks become important. This is illustrated in Fig. 4, and its expected precision for the spatial distribution of gluons from measurement of the exclusive process: electron + proton $\rightarrow$ electron + proton + J/ψ.[6]

The tomographic images obtained from cross-sections and polarization asymmetries for exclusive processes are encoded in generalized parton distributions (GPDs) that unify the concepts of parton densities and of elastic form factors.[7] They by definition include detailed information about spin-orbit correlations and the angular momentum carried by partons, including their spin and their orbital motion. The combined kinematic coverage of the EIC and of the upgraded CEBAF as well as COMPASS is essential for extracting quark and gluon angular momentum contributions to the proton's spin.

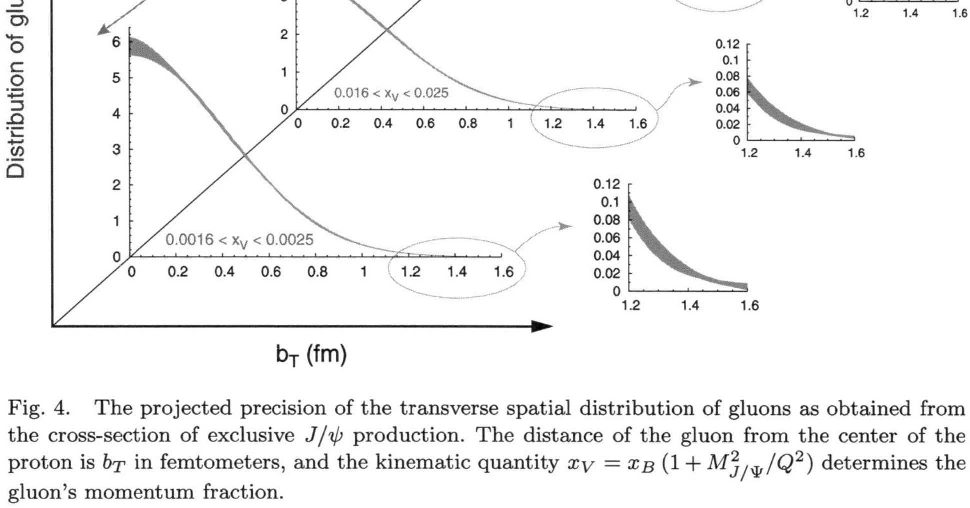

Fig. 4. The projected precision of the transverse spatial distribution of gluons as obtained from the cross-section of exclusive J/ψ production. The distance of the gluon from the center of the proton is b_T in femtometers, and the kinematic quantity $x_V = x_B\,(1 + M^2_{J/\Psi}/Q^2)$ determines the gluon's momentum fraction.

4. Physics with Nuclei

The nucleus is a QCD "molecule". Understanding the formation of nuclei in QCD is an ultimate long-term goal of nuclear physics. With its wide kinematic reach, as shown in Fig. 5 (Left), the capability to probe a variety of nuclei in both inclusive and SIDIS measurements, the EIC will be the first experimental facility capable of exploring the internal three-dimensional sea quark and gluon structure of a fast-moving nucleus. Furthermore, the nucleus itself is an unprecedented and probably the only QCD laboratory (for the foresee-able future) for discovering the collective behavior of gluonic matter at an unprecedented occupation number of gluons, and for studying the propagation of fast-moving color charges in a nuclear medium. Electron scattering off a nucleus is now established as the only unambiguous processes that may allow us to study these universal and very fundamental QCD phenomena systematically and cleanly.

4.1. *QCD at extreme parton densities*

In QCD, the large soft-gluon density enables (nonlinear) gluon–gluon recombination process to limit the otherwise seemingly infinite density growth. Such a QCD self-regulation mechanism generates a dynamic scale from the interaction of high density massless gluons, known as the saturation scale, Q_s, at which gluon splitting and recombination reach a balance. At this scale, the density of gluons is expected to saturate, producing new and universal properties of hadronic matter. The saturation scale Q_s separates the condensed and saturated soft-gluonic matter from the dilute, but confined, quarks and gluons in a hadron, as shown in Fig. 5 (Right).

The existence of such a state of saturated, soft gluon matter, often referred to as the Color Glass Condensate (CGC),[8] is a direct consequence of gluon self-interactions in QCD. It has been conjectured that the CGC of QCD has universal

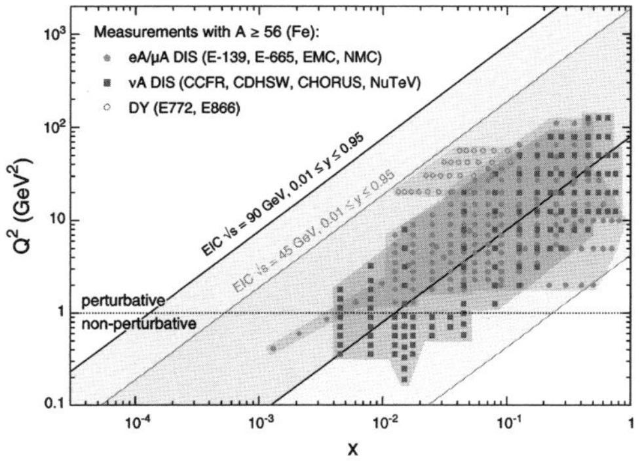

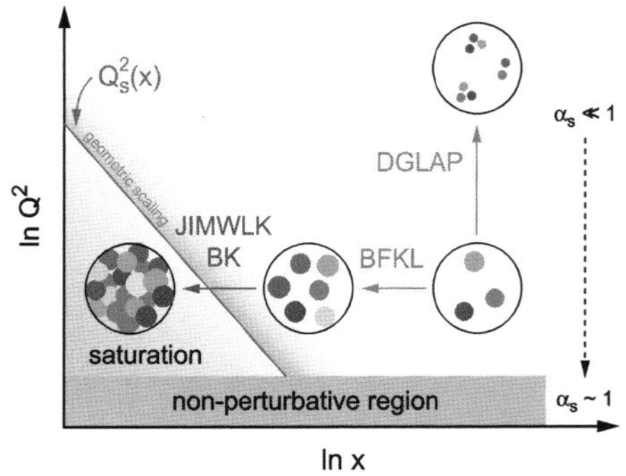

Fig. 5. Left: The range in the square of the transferred momentum by the electron to the nucleus, Q^2, versus the parton momentum fraction x accessible to the EIC in e-A collisions at two different center-of-mass energies, compared with the existing data. Right: The schematic probe resolution versus energy landscape, indicating regions of nonperturbative and perturbative QCD, including in the latter, low to high saturated parton density, and the transition region between them.

properties common to nucleons and all nuclei, which could be systematically computed if the dynamic saturation scale Q_s is *large*. However, such a semi-hard Q_s can only be unambiguously in electron–proton scattering at a multi-TeV proton beam. Heavy ion beams at the EIC could provide an alternative because the virtual photon in forward lepton scattering probes matter coherently over a characteristic length proportional to $1/x$, even exceeding the diameter of the Lorentz-contacted nucleus. Then, all gluons at the same impact parameter of the nucleus, enhanced by the nuclear diameter proportional to $A^{1/3}$ with the atomic weight A, contribute to the probed density, reaching saturation at far lower energies than would be needed in electron+proton collisions. While HERA, RHIC and the LHC have only found hints of saturated gluonic matter, the EIC would be in a position to seal the case, with its wide range of nuclei including the heaviest of them, completing the process started at those facilities.

Figure 6 illustrates some of the dramatic predicted effects of gluon density saturation in electron+nucleus versus electron+proton collisions at an EIC. The left frame considers coherent diffractive processes, defined to include all events in which the beam nucleus remains intact and there is a rapidity gap containing no produced particles. As shown in the figure, the fraction of such diffractive events are greatly

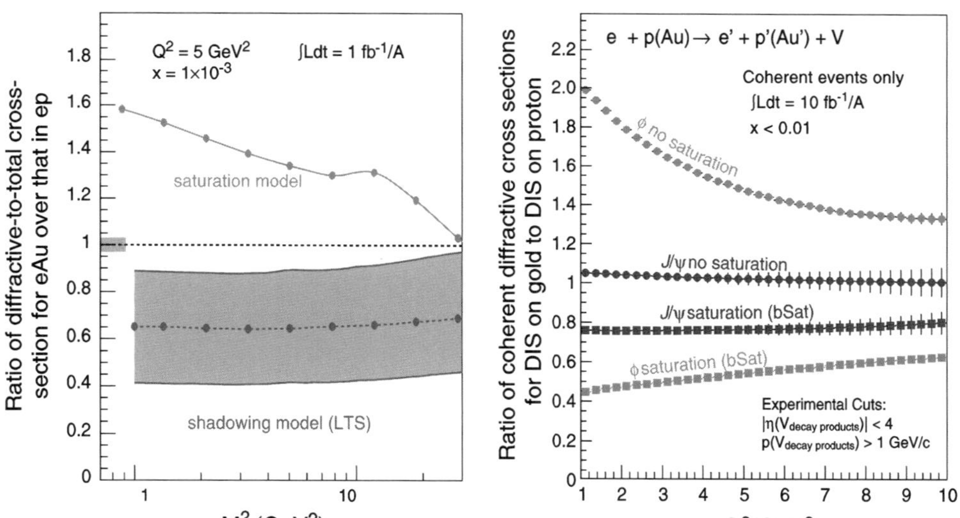

Fig. 6. (Color online) Left: The ratio of diffractive over total cross-section for DIS on gold normalized to DIS on proton plotted for different values of M_X^2, the mass squared of hadrons produced in the collisions for models assuming saturation and nonsaturation. The statistical error bars are too small to depict and the projected systematic uncertainty for the measurements is shown by the orange bar. The theoretical uncertainty for the predictions of the LTS model is shown by the grey band. Right: The ratio of the coherent diffractive cross-section in e+Au to $e+p$ collisions normalized by $A^{4/3}$ and plotted as a function of Q^2 for both saturation and nonsaturation models. The $1/Q$ is effectively the initial size of the quark–antiquark systems (ϕ and J/ψ) produced in the medium.

enhanced by gluon saturation (the red points) in comparison with the predictions of shadowing model (the blue points) in Fig. 6 (Left). An early measurement of coherent diffraction in $e + A$ collisions at the EIC would provide the first unambiguous evidence for gluon saturation.[1]

Figure 6 (Right) shows that gluon saturation is predicted to suppress vector meson production in $e+A$ relative to $e+p$ collisions at the EIC. The vector mesons result from quark–antiquark pair fluctuations of the virtual photon, which hadronize upon the exchange of gluons with the beam proton or nucleus. The magnitude of the suppression depends on the size (or color dipole moment) of the quark–antiquark pair, being significantly larger for produced ϕ (red points) than for J/Ψ (blue) mesons. An EIC measurement of the processes in Fig. 6 (Right) will provide a powerful probe to explore the properties of saturated gluon matter.[1]

4.2. *The tomography of the nucleus*

With its capability to measure the diffractive and exclusive processes with a variety of ion beams, the EIC will also provide the first three-dimensional images of sea quarks and gluons in a fast-moving nucleus with sub-femtometer resolution. For example, the EIC could obtain the spatial distribution of gluons in a nucleus by measuring the coherent diffractive production of J/Ψ in electron–nucleus scattering, similar to the case of electron–proton scattering shown in Fig. 4.

4.3. *Propagation of a color charge in QCD matter*

One of the key pieces of evidence for the discovery of the quark gluon plasma (QGP) at RHIC is jet quenching, manifested as a strong suppression of fast-moving hadrons produced in the very hot matter created in relativistic heavy-ion collisions. The suppression is believed to be due to the energy loss of colored partons traversing the QGP. It has been puzzling that the production is nearly identical for light as well as heavy quarks. We will get clues to understanding such mysteries at the EIC where one could study related phenomena in cold nuclear matter. For example, the variety of ion beams available for electron-nucleus collisions at the EIC would provide a *femtometer filter* to test and to help determine the correct mechanism by which quarks and gluons lose energy and hadronize in nuclear matter (see schematic in Fig. 7 (Left)). The process of DIS itself gives excellent control over the energy exchanged by the electron with the parton hit inside the nucleus. Together the two controls at the EIC will be unprecedented in their technical capability and the range of those control variables.

Figure 7 (Right) shows the ratio of the number of produced mesons in electron+nucleus and electron+deuteron collisions for pions (light mesons) and D^0-mesons (heavy mesons) at both low and high virtual photon energy ν, as a function of z.[1] The calculation of the lines and blue circle symbols assumes that the mesons are formed outside of the nucleus, as shown in the top sketch of Fig. 7 (Left), while the square symbols are simulated according to a model where a color neutral

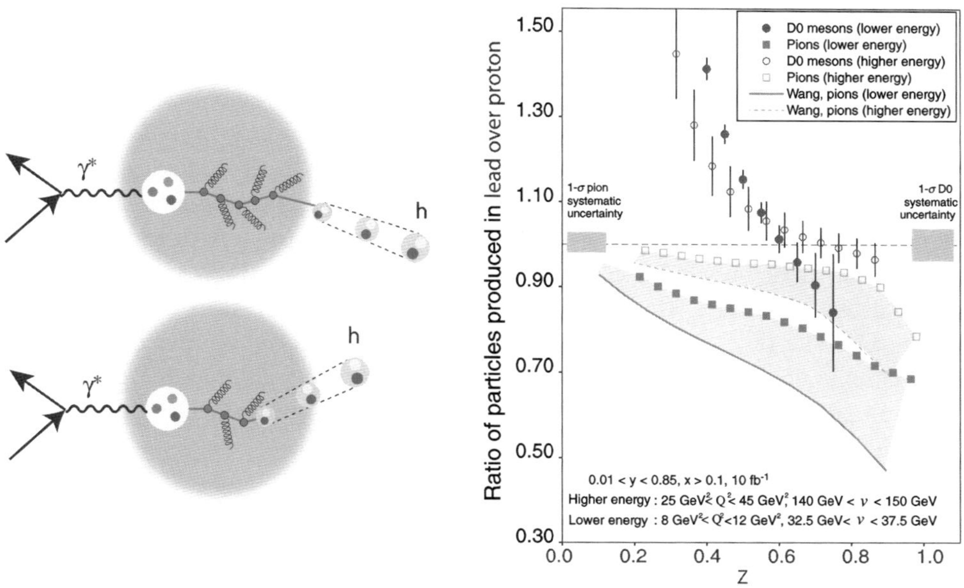

Fig. 7. (Color online) Left: A schematic of the interaction of a parton moving through cold nuclear matter: the hadron is formed outside (top) or inside (bottom) the nucleus. Right: The ratio of the semi-inclusive cross-section for producing a pion (red) composed of light quarks, and a D^0 meson (blue) composed of heavy quarks in e+lead collisions to e+deuteron collisions, plotted as a function of z, the ratio of the momentum carried by the produced hadron to that of the virtual photon (γ^*), as shown in the plots on the left.

pre-hadron was formed inside the nucleus, like in the bottom sketch of Fig. 7 (Left). The location of measurements within the shaded area would provide the first direct information on *when the mesons are formed.* Unlike the suppression expected for pion production at all z, the ratio of heavy meson production could be larger than unity due to very different hadronization properties of heavy mesons. The discovery of such a dramatic difference in multiplicity ratios between light and heavy mesons at the EIC will shed light on the hadronization process and on what governs the transition from quarks to hadrons — which is at the heart of the understanding of *confinement.*

4.4. *The distribution of quarks and gluons in the nucleus*

The EMC experiment at CERN and experiments in the following two decades revealed that the distribution of quarks in a fast-moving nucleus is a nontrivial function of Bjorken x, deviating significantly from unity, often referred to as nuclear shadowing. Because of unavailability of facilities to explore this in sufficiently low-x regions, there is to date no knowledge whether the same holds true for gluons. The EIC could measure the suppression in the region of gluon saturation. As a bonus while performing such measurements, the EIC could for the first time, reliably quantify the nuclear gluon distribution over a wide range of momentum fraction x.

5. Physics Possibilities at the Intensity Frontier

The subfield of Fundamental Symmetries in nuclear physics has an established history of key discoveries, enabled by either the introduction of new technologies or the increase in energy and luminosity of accelerator facilities. While the EIC is primarily being proposed for exploring new frontiers in QCD, it offers a unique new combination of experimental probes potentially interesting to the investigations in Fundamental Symmetries. For example, the availability of polarized beams at high energy and high luminosity, combined with a state-of-the-art hermetic detector, could extend Standard Model tests of the running of the weak-coupling constant far beyond the reach of the JLab12 parity violation program, namely toward the Z-pole scale previously probed at LEP and SLC. Figure 8 shows a recent study of what an EIC could achieve if the highest possible luminosity stated in the current machine designs can be achieved.[9]

6. Realization Challenges and Opportunities at the Frontier of Accelerator Science

Two independent designs for a future EIC have evolved in the United States. Both use the existing infrastructure and facilities available to the US nuclear science community. At Brookhaven National Laboratory (BNL), the eRHIC design (Fig. 9, top) utilizes a new electron beam facility based on an Energy Recovery LINAC (ERL) to be built inside the RHIC tunnel to collide with RHICs existing high-energy polarized proton and nuclear beams. An alternate design with is based on a Ring–Ring concept is also being pursued. At Jefferson Laboratory (JLab), the

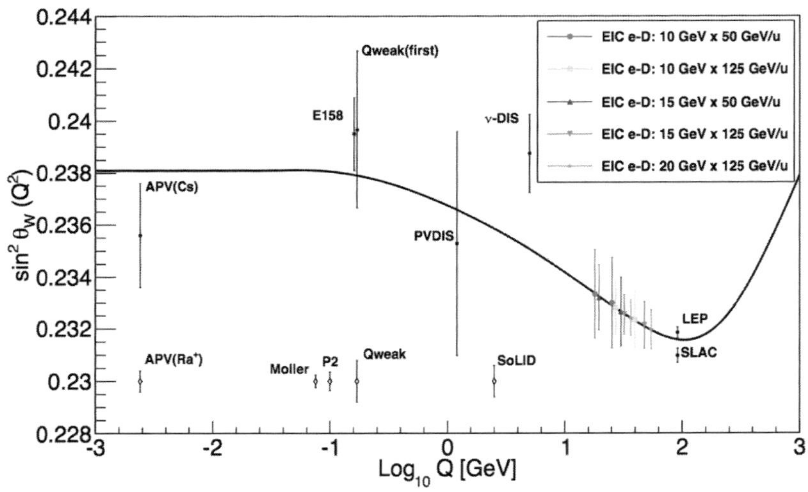

Fig. 8. The projected uncertainties on the weak mixing angle are shown along with other current and anticipated measurements of the parameter in the world. The EIC measurements assume ∼270 fb^{-1} integrated luminosity with e–D scattering. The colors indicate the energies at which EIC will have to be operated to make those measurements.

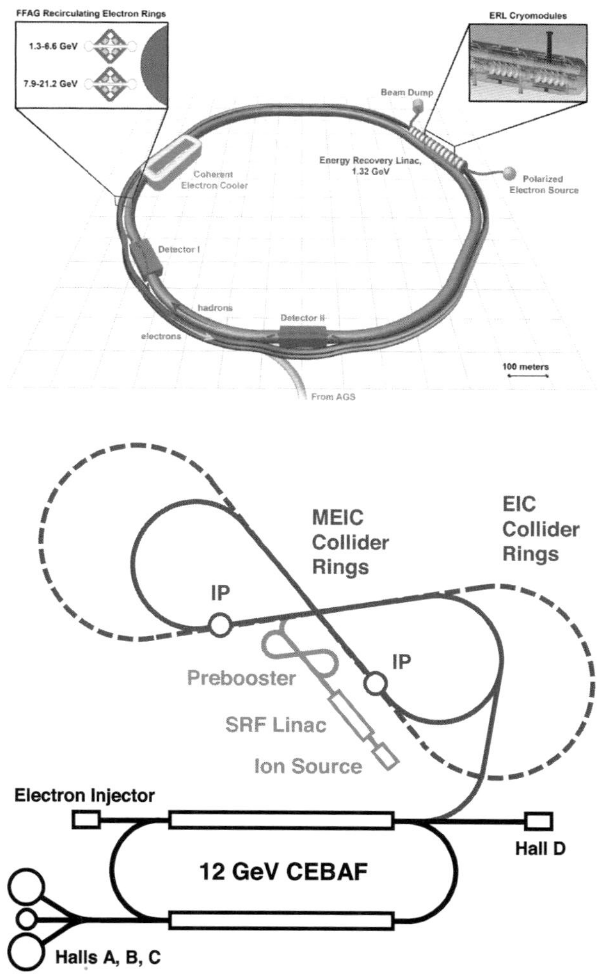

Fig. 9. (Color online) Top: The schematic of eRHIC at BNL, which would require construction of an electron beam facility (red) to collide with the RHIC blue beam at up to three interaction points. Botom: The schematic of JLEIC at JLab, which would require construction of an ion linac (red), and an electron-ion collider ring (blue) with at least two interaction points, around the 12 GeV CEBAF.

Medium energy Electron Ion Collider (JLEIC) design (Fig. 9, botom) employs a new electron and ion collider ring complex together with the 12 GeV upgraded CEBAF, now under construction, to achieve similar collision parameters. The EIC machine designs are aimed at achieving

- Highly polarized (∼70%) electron and nucleon beams
- Ion beams from deuteron to the heaviest nuclei (uranium or lead)
- Variable center of mass energies from ∼40–∼100 GeV, upgradable to ∼140 GeV
- High collision luminosity $\sim 10^{33-34}\,\mathrm{cm}^{-2}\mathrm{s}^{-1}$
- Possibilities of having more than one interaction region.

The EIC requirements will push accelerator designs to the limits of current technology, and will therefore need significant R&D which is now being supported by the DOE office of nuclear physics. Cooling of the hadron beam is essential to attain the luminosities demanded by the science. The development of coherent electron cooling is now underway at BNL, while the JLab design is based on conventional electron cooling techniques, but proposes to extend them to significantly higher energy and to use bunched electron beams for the first time.

An energy recovery linac at the highest possible energy and intensity are key to the realization of eRHIC at BNL, and this technology is also important for electron cooling in JLEIC at JLab. The eRHIC design at BNL also requires a high intensity polarized electron source that would be an order of magnitude higher in intensity than the current state-of-the-art, while the JLEIC design at JLab will utilize a novel Fig. 8 storage ring design for both electrons and ions.

The physics-driven requirements on the EIC accelerator parameters and extreme demands on the kinematic coverage for measurements makes integration of the detector into the accelerator a particularly challenging feature of the design. Lessons learned from past experience at HERA are being considered while designing the EIC interaction region. Driven by the demand for high precision on particle detection and identification of final state particles in both $e+p$ and $e+A$ programs, modern particle detector systems will be at the heart of the EIC. In order to keep the detector costs manageable, R&D efforts are under way on various novel ideas for: compact (fiber sampling and crystal) calorimetry, tracking (NaI coated GEMs, GEM size and geometries), particle identification (compact DIRC, dual radiator RICH

Fig. 10. Collaborating institutes in the EIC Users Group: 670 members from 150 institutions and 28 countries have joined the EIC Users Group as of December 2016.

and novel TPC) and high radiation tolerance for electronics. Meeting these R&D challenges will keep the U.S. nuclear science community at the cutting edge in both accelerator and detector technology.

In 2016, a world wide EIC Users Group (Fig. 10) was formed aimed at steering the EIC to reality by establishing a users group consisting of not only experimental physicists, but also theorists and accelerator scientists who would work closely with each other towards the most optimal designs of the machine, detector and the interaction region. The aim of this group is to channel the enthusiasm and efforts around the world in to a constructive movement to realize the EIC. We envision two detector collaborations evolving (at the least) to realize the physics promise at the EIC. The group has an organizational structure similar to that of large experimental collaboration, and currently consists of 670 users, 150 institutions, and a 10-member Steering Committee and its Chair. For more information about the EIC Users Group (EICUG) we refer you to its webpage.[10]

7. Outlook & Take Away Message

The Relativistic Heavy Ion Collider (RHIC) at BNL has revolutionized our understanding of hot and dense QCD matter through its discovery of the strongly-coupled quark gluon plasma that existed a few microseconds after the birth of the universe. Unprecedented studies of the nucleon and nuclear structure — including the nucleon spin, and the nucleon's tomographic images in the valence quark region — have been and will be possible with the high luminosity fixed target experiments at Jefferson Laboratory using the 6 GeV and 12 GeV CEBAF, respectively. The EIC promises to propel both programs to the next QCD frontier, by unraveling the three-dimensional sea quark and gluon structure of visible matter. Furthermore, the EIC will probe the existence of a universal state of saturated gluon matter and has the capability to explore it in detail. The EIC will thus enable the US to continue its leadership role in nuclear science research through the quest for understanding the unique gluon-dominated nature of visible matter in the universe.

Acknowledgments

The science case for the EIC was developed by a large number of scientists who participated in the INT workshop in 2010.[3] The science was then condensed in to the EIC White Paper.[1] I want to thank all those who contributed to those documents. The science presentation in this paper is their's. I would also like to thank Elke Aschenauer, Rolf Ent, Zein-Eddine Meziani, Richard Milner, Robert McKeown, Berndt Mueller, Jianwei Qiu, Thomas Ullrich, Raju Venugopalan, Steve Vigdor and Rik Yoshida for discussions, advice and collaboration on various aspects of the EIC project, including completion of various write-ups from which the material in this paper are taken. I would also like to thank the editors of this volume for being extra patient with me regarding delay in the submission of this manuscript.

References

1. A. Accardi *et al.*, *Eur. Phys. J. A* **52** (2016) 268, arXiv: 1212.1701.v3.
2. The 2015 Long Range Plan for Nuclear Science, http://science.energy.gov/np/nsac/.
3. D. Boer *et al.*, Gluons and the quark sea at high energies: Distributions, polarization and tomography, arXiv:1108.1713.
4. D. de Florian *et al.*, *Phys. Rev. Lett.* **101** (2008) 072001; *Phys. Rev. D* **80** (2009) 034030; *Prog. Part. Nucl. Phys.* **67** (2012) 251.
5. M. G. Perdekamp and F. Yuan, *Ann. Rev. Nucl. Part. Sci.* **65** (2015) 429.
6. E.-C. Aschenauer *et al.*, *J. High Energy Phys.* **1309** (2013) 093.
7. M. Diehl, *Phys. Rep.* **388** (2003) 41; X. Ji, *Ann. Rev. Nucl. Part. Sci.* **54** (2004) 413.
8. F. Gelis *et al.*, *Ann. Rev. Nucl. Part. Sci.* **60** (2005) 436.
9. Y. Zhao *et al.*, arXiv:1612.06927.
10. The EIC Users Group Web Page: EICUG.ORG.

Can one measure the Cosmic Neutrino Background?

Amand Faessler
Institute für Theoretische Physik der Universität Tübingen, D-72076 Tübingen, Germany
faessler@uni-tuebingen.de

Rastislav Hodák
Institute of Experimental and Applied Physics, Czech Technical University in Prague, CZ-128 00 Prague, Czech Republic

Sergey Kovalenko
Universidad Técnica Federico Santa María, Centro-Científico-Tecnológico de Valparaíso, Casilla 110-V, Valparaíso, Chile

Fedor Šimkovic
JINR, 141980 Dubna, Moscow Region, Russia
Faculty of Mathematics, Physics and Informatics, Comenius University, SK-842 48 Bratislava, Slovakia

The Cosmic Microwave Background (CMB) yields information about our Universe at around 380,000 years after the Big Bang (BB). Due to the weak interaction of the neutrinos with matter, the Cosmic Neutrino Background (CNB) should give information about a much earlier time of our Universe, around one second after the BB. Probably, the most promising method to "see" the CNB is the capture of the electron neutrinos from the Background by Tritium, which then decays into ^{3}He and an electron with the energy of the the Q-value $= 18.562\,\mathrm{keV}$ plus the electron neutrino rest mass. The "KArlsruhe TRItium Neutrino" (KATRIN) experiment, which is in preparation, seems presently the most sensitive proposed method for measuring the electron antineutrino mass. At the same time, KATRIN can also look by the reaction $\nu_e(\sim 1.95\,\mathrm{K}) +^3\mathrm{H} \rightarrow^3 \mathrm{He} + e^-(Q = 18.6\,\mathrm{keV} + m_{\nu e}c^2)$. The capture of the Cosmic Background Neutrinos (CNB) should show in the electron spectrum as a peak by the electron neutrino rest mass above Q. Here, the possibility to see the CNB with KATRIN is studied. A detection of the CNB by KATRIN seems not to be possible at the moment. But KATRIN should be able to determine an upper limit for the local electron neutrino density of the CNB.

Keywords: Neutrinos; Cosmic Neutrino Background; electron capture; KATRIN.

1. The Three Cosmic Backgrounds

The three Cosmic backgrounds

(1) Cosmic Gravitational Background (CGB; during the Big Bang (BB)),
(2) Cosmic Neutrino Background (CNB; 1 min after the BB) and
(3) Cosmic Microwave Background (CMB; $\sim$ 380,000 years after the BB)

can give information about the Universe at different times after the BB indicated above.

The inflationary expansion of the Universe by about a factor 10^{26} between roughly 10^{-35}–10^{-33}[s] after the BB couples according to the General Relativity to gravitational waves, which decouple after this time and their fluctuations are the seed for Galaxy Clusters and even Galaxies. These decoupled gravitational waves run since then with only very minor distortions through the Universe and contain a memory to the BB. The Evolved Laser Interferometer Space Antenna (eLISA) project of the European Space Agency[1] with three satellites may perhaps be able to see the CGB. Recently, the BICEP2 collaboration[2] claimed to have seen in the fluctuations and the polarization of the CMB fingerprints of the Gravitational Wave Background originating from the Inflationary Expansion during the BB. But this was probably a wishful interpretation of the data.

The CNB (often also called "relic neutrinos") decouples from matter about 1 sec after the BB in the radiation dominated era at a temperature of $10^{10}\,\mathrm{K} \equiv 1\,\mathrm{MeV}$. Today due to the expansion and cooling of the Universe, the relic neutrino temperature is 1.95 K and the average neutrino density in the Universe is 340 per cm^3 or 56 electron neutrinos per cm^3. Recently, Follin *et al.*[3] interpreted data about damping of acoustic oscillations of the CMB as an interaction with the CNB.

The CMB originates about 380,000 years after the BB at a temperature of around 3000 K and is well studied. It was detected by Penzias and Wilson[4] in 1964 (Nobel Prize 1978) as byproduct of their search for possible perturbations of the communication with satellites. Later on the CMB was confirmed and detailed by satellite observations.[5–7] The frequency distribution follows exactly Planck's black body formula and yields a surprisingly identical temperature up to four digits independent of the direction ($T_{0\gamma} = 2.7255(6)\,\mathrm{K}$).

$$\epsilon(f)df = \frac{8\pi h}{c^3} \cdot \frac{f^3 df}{\exp(hf/k_B T_0) - 1}[\text{Energy/Volume}]. \quad (1)$$

This work discusses the possibility of the detection of the CNB with the KATRIN experiment. This work is based on our publications,[8] the publication of Drexlin *et al.*[9] the KATRIN Design report[10] and numerous discussions with Drexlin and Weinheimer.[11]

Different methods have been discussed in the literature to search for these relic neutrinos.[12–18]

The search for the CNB with the induced beta decay[8,19] as originally proposed by Steven Weinberg[20] seems to be the most promising using the reaction:

$$\nu_e + {}^3\text{H} \rightarrow {}^3\text{He} + e^-. \tag{2}$$

The signal would show up by a peak in the electron spectrum with an energy of the neutrino mass above the Q value.

2. The Cosmic Photon or Microwave Background (CMB)

The Cosmic Background Radiation (CMB) was detected in 1964 by Penzias and Wilson.[4] The frequency distribution follows very accurately Planck's black body radiation law (1). As already mentioned, the temperature parameter T_0 fitted by (1) is independent of the observation direction up to the fourth digit. The average temperature of the photons in the CMB is about $3 \cdot T_0$. Today the temperature of the CMB is:

$$T_{0,\text{rad}} \equiv T_{0,\gamma} = 2.7255 \pm 0.0006\,[K]. \tag{3}$$

The Stefan–Boltzmann law for the energy density originates by integration over all frequencies f.

$$\epsilon_{\text{rad}} = \varrho_{\text{rad}} c^2 = \alpha T_0^4. \tag{4}$$

As soon as the electrons get permanently bound to the protons and the Helium nuclei, the Universe is electrically neutral and since then the photons move freely as the CMB.

3. When Decouple the Neutrinos from Matter?

Since the neutrinos interact with matter only by the weak interaction, they decouple much earlier than the photons, which interact by the electromagnetic force. The decoupling of the neutrinos from matter is determined by the competition of the expansion rate of the Universe (Hubble parameter) and the interaction rate of the neutrinos with matter. The Hubble parameter using the Friedmann equation[21,22] and the Stefan–Boltzmann law (4) is given by:

$$H = \frac{\dot{a}}{a} = \sqrt{\frac{8\pi G}{3}\varrho_{\text{total}}} = \sqrt{\frac{8\pi\varrho}{3M_{\text{Planck}}^2}} \propto T_0^2 \propto a^{-2}. \tag{5}$$

The Universe expansion rate is given by the Hubble parameter. It decreases with the square of the temperature. The variable "a" is any length scale describing the expansion of the Universe. The neutrino reaction rate reduces with the fifth power T^5 of the decreasing temperature.

$$\Gamma = n_\nu \langle \sigma v \rangle \approx T_0^3 G_{\text{Fermi}} T_0^2 = G_F T_0^5 \propto a^{-5}; \quad v \approx c = 1, \tag{6}$$

when the Hubble expansion rate (5) is getting faster or about equal to the neutrino reaction rate (6), the neutrinos decouple from matter.

$$\frac{\Gamma}{H} \doteq 1 = \left(\frac{k_B T_0}{1\,\mathrm{MeV}}\right)^3. \tag{7}$$

Thus, the decoupling temperature corresponds to:

$$T_0 = 1\,\mathrm{MeV} \equiv 10^{10}[\mathrm{K}]. \tag{8}$$

Today, the temperature of the neutrinos is 1.95 K.

Since in the radiation dominated era one has:

$$T_0 = \left(\frac{45\hbar^3 c^5}{32\pi^3 G}\right)^{1/4} \cdot \frac{1}{t^{1/2}} \equiv 1.3\left(\frac{t}{\mathrm{s}}\right)^{-1/2} [\mathrm{MeV}]. \tag{9}$$

Thus, the decoupling happens about 1 s after the BB.

4. Can KATRIN Detect the Cosmic Neutrino Background

Fermi's Golden Rule gives for the Tritium beta-decay probability:

$$\begin{aligned}\Gamma^{\beta}_{\mathrm{decay}}(^3\mathrm{H}) = \frac{1}{2\pi^3} \cdot \sum \int |\langle ^3\mathrm{He}|T|^3\mathrm{H}\rangle|^2 \cdot 2\pi\delta \\ \times (E_\nu + E_e + E_f - E_i)\frac{d\mathbf{p}_e}{2\pi^3} \cdot \frac{d\mathbf{p}_\nu}{2\pi^3}.\end{aligned} \tag{10}$$

$$\begin{aligned}\text{Theory:}\quad T^{\beta}_{1/2} &= \frac{\ln 2}{\Gamma^{\beta}_{\mathrm{decay}}(^3\mathrm{H})} = 12.32 \text{ years;} \\ \text{Experiment:}\quad T^{\beta}_{1/2} &= 12.33 \text{ years;}\end{aligned} \tag{11}$$

The beta-decay spectrum of the electron from the Tritium beta-decay has the form:

$$\frac{dN_e}{dE} = K \cdot F(E,Z) \cdot p_e E_e \cdot (E_0 - E_e) \cdot \sum_{j=1}^{3} |U_{ej}|^2 \cdot \sqrt{(E_0 - E_e)^2 - m^2_{\nu j}} \tag{12}$$

with:

$$K = \text{const}; \quad Q = 18.562\,\mathrm{keV}; \quad E_0 = Q + m_e;$$

$$E_e = \sqrt{m_e^2 + p_e^2}; \quad E = T_e = E_e - m_e \tag{13}$$

$$\nu_e = \sum_{j=1,3} U_{ej} \cdot m_{\nu j}, \tag{14}$$

ν_e flavor eigenstate; ν_j mass eigenstate. In the Kurie plot, the Tritium decay spectrum is for a massless neutrino a straight line as a function of the electron energy. The line hits the abscissa with the electron energy at the Q-value. The Kurie plot is obtained by dividing (12) by $K \cdot F(E,Z) \cdot p_e \cdot E_e$ and taking the square root. The neutrino mass modifies the Kurie plot in the interval $\langle Q - m_{\nu e} \cdot c^2 | Q = 18.562\,\mathrm{keV}\rangle$. The capture of the relic neutrinos from the CNB (2) should show as a peak in the electron spectrum at $Q + m_{\nu e} \cdot c^2$. The capture probability requires the same Fermi and Gamow–Teller matrix elements as for the decay. Thus, the theoretical prediction for the capture probability should be as accurate as the one for the decay.

$$\begin{aligned}\Gamma^{\beta}_{\text{capture}}(^3\text{H}) &= \frac{1}{\pi}(G_F \cos(\vartheta_C))^2 F_0(Z+1, T_e)\langle B_F(^3\text{H}) \\ &\quad + B_{GT}(^3\text{H})]p_e T_e \cdot \langle n_{\nu,e}\rangle \frac{n_{\nu,e}}{\langle n_{\nu,e}\rangle} \\ &= 4.2 \cdot 10^{-25} \frac{n_{\nu,e}}{\langle n_{\nu,e}\rangle} (\text{for one Tritium atom/year}); \\ \text{with: } \langle n_{\nu,e}\rangle &= 56\,\text{cm}^{-3}. \end{aligned} \tag{15}$$

The values for this effective strength of the Tritium source[23] and[8] have been reduced step by step. The correct value given by Drexlin[11] is 20 μg. This means $2 \cdot 10^{18}$ Tritium_2 molecules. The capture rate of relic neutrinos is then:

$$\text{Capture rate at KATRIN: } N_\nu(\text{KATRIN}) = 1.7 \cdot 10^{-6} \cdot \frac{n_{\nu,e}}{\langle n_{\nu,e}\rangle}. \tag{16}$$

If one uses the average relic neutrino number density $\langle n_{\nu,e}\rangle = 56\,\text{cm}^{-3}$, one predict only every 590,000 years a count. But there is the hope, that the local relic neutrino density in a galaxy increases by gravitational clustering. Ringwald and Wong[12] calculated, that relic neutrinos can cluster on the scale of a single galaxy and their halo and if one uses the proportionality to the baryon overdensity (see Ref. 19), then one can expect very optimistically an overdensities up to a factor $n_{\nu,e}/\langle n_{\nu,e}\rangle \leq 10^6$ in our neighborhood. With this very optimistic overdensity for the relic neutrinos of 10^6, one obtains from Eq. (16):

$$N_\nu(\text{KATRIN}) = 1.7 \cdot 10^{-6} \cdot \frac{n_{\nu,e}}{\langle n_{\nu,e}\rangle}[\text{year}^{-1}] \approx 1.7[\text{counts per year}]. \tag{17}$$

This seems not possible to measure for the moment. One way out would be to increase the effective activity of the tritium source. An effective mass of 2 mg Tritium would mean with the above optimistic estimate of the relic neutrino number overdensity $n_{\nu,e}/\langle n_{\nu,e}\rangle \approx 10^6$ about 170 counts per year, which should be feasible. But it should be possible with KATRIN, in its present form, to determine an upper limit for the local relic neutrino density of the CNB.

Is it possible to increase the Tritium source strength by factor 100? Increasing the source strength one has to keep in mind the following requirements.

Requirements:

(1) *Maximize detectable* 18.6 keV *electrons*, which contain the information on the neutrino mass and the capture of the relic neutrinos.
(2) 1 eV *energy resolution of the spectrometer.*
(3) *Conserve the orbital angular momentum of the electrons* for the cyclotron motion along the magnetic field lines from the source to the detector.
(4) *Conserve the magnetic flux.*

(5) *Focus a large number of* 18.6 keV *electrons into the* 9 cm *wide transport channel* to the spectrometer.

Discussion of these Requirements:

(i) *The increase of the Tritium source strength* is limited by the scattering of the emitted electrons by the tritium gas.[10] After the mean free path

$$\lambda_{\text{mean free path}} = \frac{1}{\rho \cdot \sigma(\text{electron-tritium})} = d_{\text{free}} \tag{18}$$

only about 37% decay electrons have not yet scattered. All the others also including electrons with the maximum energy, which contain the information on the neutrino mass and on the relic neutrino capture by Tritium, are lost for the measurement.[10] Thus, an increase of the Tritium source strength beyond the 20 μg reduces the number of counts of 18.6 keV electrons and is counter-productive.

(ii) If one request for the spectrometer, an *energy resolution of* 1 eV, one can afford to have only 1 eV of energy in the spectrometer in the perpendicular cyclotron motion. The spectrometer can be an opposing electric field of around 18,600 V control the longitudinal energy of the electron motion along the magnetic field lines and allow only electron to pass with energies around $Q = 18.562$ keV. The information on the neutrino mass and on the capture of the neutrinos from the CNB lies close to the upper end point of the electron spectrum in the interval $\langle Q - a \cdot m_{\nu e} \cdot c^2 | Q + a \cdot m_{\nu e} \cdot c^2 \rangle$ with a number "a" roughly between 2 and 20.

(iii) *The conservation of the angular momentum* in the circular cyclotron motion of an electron and also the corresponding magnetic moment of this ring current must be conserved along the trajectory of the electron from the source to the detector. This requires a fixed ratio of the perpendicular energy $E_\perp$ over the local magnetic field along the longitudinal orbit of the electrons.

$$|\mathbf{L}| = |\mathbf{r} \times \mathbf{p}| \propto \mu = \text{const} \propto \frac{E_{i\perp}}{B_i} = \frac{E_{f\perp}}{B_f}. \tag{19}$$

An energy resolution of about $\Delta E = 0.93$ eV thus requires:

$$\Delta E = 0.93 \text{ eV} = E_{f\perp} = \frac{B_f}{B_i} \cdot E_{i\perp} = \frac{3 \text{ Gauss}}{360 \text{ Gauss}} E_{i\perp} \tag{20}$$

with $B_i = 3.6$ Tesla and of the 18.6 keV at the source 10 keV in the perpendicular motion and a required resolution of $E_{f\perp} = 0.93$ eV one obtains for the magnetic field in the spectrometer a small value of about $B_f = 3$ Gauss.

(iv) *The magnetic flux* at the present source is given by

$$\text{Magnetic Flux(source)} = 53 \text{ cm}^2 \cdot 3.6 \text{ Tesla} \approx 190 \text{ Tesla} \cdot \text{cm}^2. \tag{21}$$

The flux must be the same in the spectrometer requiring a perpendicular area of the spectrometer of 63.6 m^2.

$$\text{Magnetic Flux(spectrometer)} = 63.6 \text{ m}^2 \cdot 3 \text{ Gauss} \approx 190 \text{ Tesla} \cdot \text{cm}^2. \tag{22}$$

The perpendicular area of 63.6 m^2 of the spectrometer required to conserve the magnetic flux, yields for an energy resolution of 0.93 eV a large size of the spectrometer with about 10 m diameter. See Fig. (2).

The low magnetic field in the spectrometer of only 3 Gauss is needed to transform the electron momenta, which are at the source almost perpendicular to the beam direction due to the cyclotron motion, into almost a translational direction (see Fig. 9 of the KATRIN design report Ref. 10 and Fig. 1 of this contribution), to reject with an electric opposing field all electrons almost up

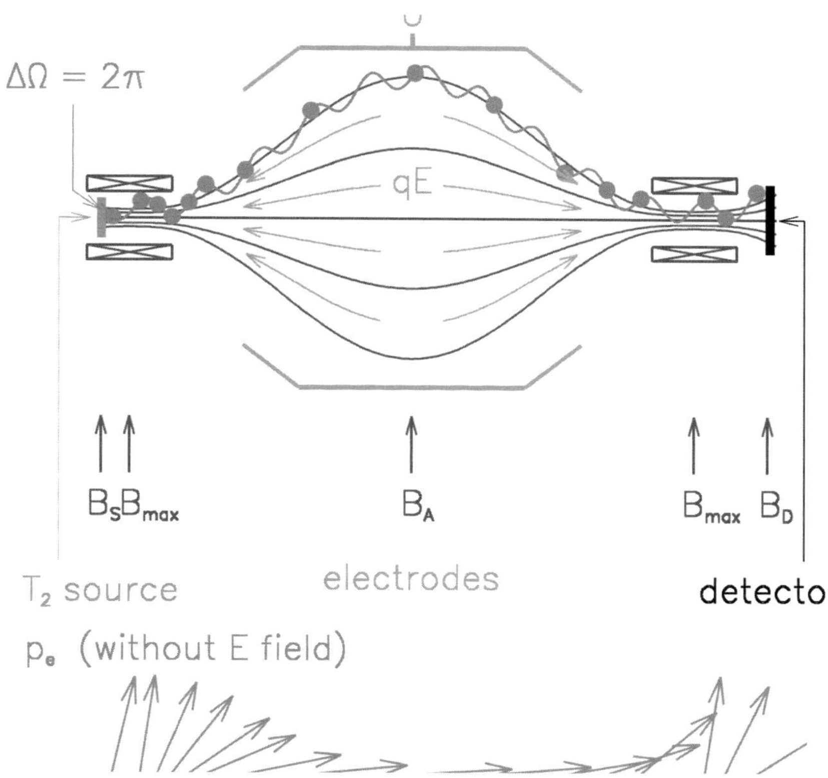

Fig. 1. Magnetic fields in the spectrometer. The magnetic field at the Tritium T_2 source B_S focuses ideally half of the the electrons in forward direction. In reality $B_{\max}$ contracts only part of the forward electrons into the transport channel with about 9 cm diameter. A to strong increase of the field $B_{\max}$ yields a magnetic mirror and reflects all electrons. The electrons follow in cyclotron rotation the magnetic field lines. Angular momentum conservation requests, that $E_\perp/B$ is constant along the electron trajectory. Thus, a small $E_\perp = 0.93\,\text{eV}$ in the middle of the spectrometer requires there a weak magnetic field of about 3 Gauss. Energy conservation then takes care, that the electron momentum in the middle of the spectrometer points forward into the longitudinal direction. The size and direction of the electron momentum is indicated at the lower end of the figure. This is an electric opposing field (not included in this figure) allows only electrons around the Q value containing the information about the neutrino mass and also about the CNB to reach the detector on the right. The electron momenta, direction and size, are indicated at the lower part of the figure. Figure by Christian Weinheimer, University Muenster,Germany, reproduced by his permission.

Fig. 2. KATRIN spectrometer. The conservation of the magnetic flux and energy resolution of 0.93 eV requires a spectrometer with a diameter of about 10 m. The figure is reproduced from the KATRIN collaboration with permission of the speakers Guido Drexlin, KIT Karlsruhe and Christian Weinheimer, University Muenster, Germany.

to the tritium Q-value ($Q = 18.562\,\text{keV}$). To increase the source strength by a factor 100 with a corresponding increase of the source area to $5000\,\text{cm}^2, 80\,\text{cm}$ diameter, one also needs to increase the spectrometer cross-section by a factor 100 or the diameter to about 90 m, which is not possible.

(v) *To focus a large number of* 18.6 keV *electrons at the source into the* 9 cm *wide transport channel* one needs the large magnetic field of 3.6 Tesla. This then requires by magnetic flux conservation and the requirement of 1 eV energy resolution the large size of the spectrometer. With the resolution of $\Delta E = 0.93\,\text{eV}$, KATRIN hopes to reduce the upper limit of the electron neutrino mass to about 0.2 eV (90%C.L.). Fitting at the upper end of the Kurie plot at $Q - m_{\nu,e}$ of the electron spectrum, the KATRIN collaboration hopes to determine the Q-value and the neutrino mass. The electron peak due to the capture of the relic neutrinos lies at $Q + m_{\nu,e}$. The neutrino mass and the energy resolution and the background remain the same as for the determination of the neutrino mass. One has only one additional fit parameter more (or two, if one counts the width of this peak), the counts in the peak at $Q+m_{\nu,e}$. At the moment, it does not seem possible to detect with a KATRIN type spectrometer the CNB. But one should be able to give an upper limit for the local relic neutrino overdensity $n_{\nu,e}/\langle n_{\nu,e}\rangle$ in our Galaxy.

5. Alternative Approaches

Two methods are discussed in the literature to go beyond KATRIN for the determination of the electron antineutrino mass[23] and the detection of the CNB.[24]

The "Project 8" at Santa Barbara and MIT wants to measure the electron antineutrino mass by the radiation emitted from the Tritium beta decay electrons moving in cyclotron resonances (rotations) along a longitudinal magnetic field.[23] This requires special antennas along the electron trajectory from the source to the detector and a detailed analysis of the radiation spectrum. The opinion of the experts of the feasibility of this approach to determine the neutrino mass varies.

The Princeton Tritium Observatory for Light, Early–Universe, Massive-Neutrino Yield (PTOLEMY) project[24] of the Princeton Plasma Physics Laboratory wants to measure neutrinos from the CNB. The planned Tritium source is exceptionally intense with a total mass of 100 g. A single graphene layer backed by a substrate can bind Tritium by sub-eV energies at each Carbon intersection on the surface directly facing vacuum. Maximal hydrogenation of the graphene surface achieves 3×10^{15} Tritium atoms per cm^2. Since the probability to scatter for one 18.6 keV electron on ^{12}C in one graphene layer is about 3.6%, one is limited to two or three stacked graphene layers (see Ref. 24, p. 2). This limits the weight of Tritium to about 1 μg per cm^2 with two layers. If one assumes two layers with 1 μg per cm^2 one needs for 100 grams of Tritium roughly 10^8 cm^2 = 10,000 m^2 of double graphene layers with a free way from the source to the detector. The PTOLEMY proposes a time-dependent system incorporating the Project 8 antenna technology to significantly reduce the phase space requirements, in additional to alternative distributed geometries for the MAC-E filter inspired by KATRIN. Although, the PTOLEMY collaboration discusses these problems in detail,[24] it is obvious that such a source is extremely difficult to realize.

6. Conclusion

- The average relic electron neutrino number density of $\langle n_{\nu,e}\rangle = 56\,\mathrm{cm}^{-3}$ KATRIN yields only every 590,000 years a count from the CNB with KATRIN. But the local overdensity due to gravitational clustering of the neutrinos in our galaxy will increase the counting rate. Estimates for this overdensity $n_{\nu,e}/\langle n_{\nu,e}\rangle$ vary widely from about 10^2 to 10^6 and depend on the mass of the relic neutrinos. The very optimistic value of the local overdensity of 10^6 yields with KATRIN 1.7 counts per year. Increasing the effective mass of the Tritium source from 20 μg to 2 μg would lead to 170 counts per year. Detection would be possible.

- The second problem could be the energy resolution of the KATRIN spectrometer of $\Delta E = 0.93\,\text{eV}$. With this resolution, KATRIN expects to extract from the electron spectrum at the upper end of the Kurie plot at $Q - m_{\nu,e}$ a very accurate Q-value with an error of milli-eV or less and an upper limit of the electron antineutrino mass of about $m_{\nu,e} \leq 0.2$ [eV] 90% C.L. If one can fit the Q value and the electron neutrino mass accurately enough, the position of the electron peak from the induced capture of the relic neutrinos is known to be at an electron energy of $Q + m_{\nu,e} \cdot c^2$. The energy resolution and the background is the same at $Q - m_{\nu,e} \cdot c^2$ and at $Q + m_{\nu,e} \cdot c^2$. One should with KATRIN at least be able to determine an upper limit for the local relic neutrino density.

Acknowledgments

We thank Christian Weinheimer and Guido Drexlin for discussions about the KATRIN spectrometer, Christian Tully for information about PTOLEMY and the Ministry of Education and Science of the Russian Federation (contract 12.741.12.0150). F. S. acknowledges the support by the VEGA Grant agency of the Slovak Republic under the contract No. 1/0876/12 and by the Ministry of Education, Youth and Sports of the Czech Republic under contract LM2011027. S. K. thanks for support by the FONDECYT Grant 1100582 and the Centro Cientifico-Tecnológico de Valparaíso PBCT ACT-028.

References

1. A.-S. Pau *et al.*, *Class. Quantum Grav.* **29** (2012) 124016, arXiv:1202.0839.
2. BICEP2 Collab., arXiv:1403.3985v2 [astro-ph].
3. B. Follin, L. Knox, M. Millea and Z. Pan, *Phys. Rev. Lett.* **115** (2015) 09301.
4. A. A. Penzias and R. W. Wilson, *Astrophys. J.* **142** (1965) 419.
5. N. W. Boggess *et al.*, *Astrophys. J.* **397** (1992) 420.
6. C. L. Bennett, arXiv:1212.5225 [astro-ph].
7. Planck Collab. *Astron. Astrophys.* **571** (2014).
8. A. Faessler, R. Hodak, S. Kovalenko and F. Simkovic, *J. Phys. G* **38** (2011) 075202, arXiv:1102.1799 [hep-ph]; *J. Phys. Conf. Ser.* **580** (2015) 012040; *EPJ Web Conf.* **71** (2014) 00044.
9. G. Drexlin, V. Hannen, S. Mertens and C. Weinheimer, *Adv. High Energy Phys.* **2013** (2013) 293986.
10. KATRIN Design Report 2004, FZKA Scientific Report 7090; MS-KP-0501.
11. G. Drexlin and Ch. Weinheimer, Private Communication 2013–2015.
12. A. Ringwald and Y. Wong, *J. Cosmol. Astropart. Phys.* **0412** (2004) 005.
13. R. Wigmans, *Astropart. Phys.* **19** (2003) 379.
14. A. Ringwald, arXiv:0301157 [hep-ph].
15. T. Weiler, *Phys. Rev. Lett.* **49** (1982) 234.
16. H. J. de Vega, O. Moreno, E. Moya de Guerra, M. Ramon Medrano and N. G. Sanchez, *Nucl. Phys. B* **866** (2012) 177, arXiv:1109.3452 [hep-ph].
17. F. Munyaneza and P. L. Biermann, *Astron. Astrophys.* **458** (2006) L9.
18. H. Päs and T. J. Weiler, *Phys. Rev. D* **63** (2001) 113015.
19. R. Lazauskas, P. Vogel and C. Volpe, *J. Phys. G.* **35** (2008) 1.

20. S. Weinberg, *Phys. Rev.* **128** (1962) 1457.
21. A. Friedmann, *Z. Phys.* **10** (1922) 377.
22. A. Liddle, *An Introduction to Modern Cosmology*, 2nd edn. (John Wiley, NY, 2005).
23. A. Kaboth, J. A. Formaggio and B. Monreal, *Phys. Rev. D* **82** (2010) 062001.
24. S. Betts *et al.*, arXiv:1307.4738v2.

Application of the hidden local symmetry in hadron physics

Masayasu Harada
Department of Physics, Nagoya University,
Nagoya 464-8602, Japan
harada@hken.phys.nagoya-u.ac.jp

About two decades ago, Gerry Brown pushed me to study hadron physics using hadronic models based on the hidden local symmetry (HLS). Since then, I have been involving many works related to the HLS. In this contribution to Gerry's memorial, I will summarize my recent analyses associated with the HLS. After a very brief summary of the HLS, I will show a way to determine parameters of the HLS model from holographic QCD models. This model provides a soliton solution which is identified with the nucleon. I will briefly introduce a model in which the solitons are put onto a crystal lattice for simulating nuclear matter. The results suggest existence of the chiral invariant mass of nucleon. I also briefly review my recent analyses on the nuclear matter by using a hadronic model including nucleon field together with the π, σ, ρ and ω mesons based on the parity doublet structure of nucleon combined with the HLS, in which the chiral invariant mass of the nucleon is naturally embedded. The result implies that studying effective masses of nucleons will give a clue for the chiral invariant mass. Finally, I will show our recent analysis on the medium modification of charmed meson masses using a chiral model based on the chiral doubling structure of charmed mesons. The result shows that the masses of the chiral partners of charmed mesons become closer in high density reflecting the particle chiral symmetry restoration in medium.

Keywords: Chiral symmetry; effective Lagrangian; vector meson; Skyrmions; nuclear matter; heavy quark symmetry.

1. Introduction

The first time I met Gerry was around the year of 1995 when I gave a seminar on the loop contribution in the hidden local symmetry (HLS) based on Refs. 1–3. Motivated by the discussion with Gerry and members in Strony Brook, I calculated corrections to the rho meson mass in the HLS at finite temperature,[4] which shows that the loop contribution increases the rho meson mass at finite temperature, on the contrary to the Brown–Rho scaling.[5] The agreement between the HLS calculation and the BR scaling is achieved when the notion of the intrinsic dependence of the HLS parameters reflecting the chiral restoration of QCD.[6–9] This vector manifestation (VM) is formulated based on the matching to QCD in a Wilsonian sense[7] together with the chiral perturbation theory (ChPT) with HLS.[1,10–13]

I spent three months at Stony Brook around the beginning of 2002, arranged by Gerry, slightly after Yamawaki and I started to write the manuscript of Physics Reports.[13] I gave a series of presentations concerning the ChPT with HLS and the

VM in Stony Brook. Gerry gave me many comments which were very useful to improve the manuscript. One year after I left Stony Brook, the manuscript was submitted and then accepted. After the submission, I made several works based on the VM, e.g., a study of the mass of a_1 meson,[14] a study of the effect of VM to the dilepton spectra,[15] an analysis based on the "hadronic freedom".[16] In particular, Ref. 16 is the only paper which I had a chance to be a co-author of Gerry. While we were writing the manuscript, I learned a lot from Gerry, such as a way of approximation to get a general idea.

About a decade ago, we proposed a way to determine the parameters of the HLS theory at $\mathcal{O}(p^4)$ from holographic QCD models by integrating out higher modes.[19,20] In Ref. 20, we have shown that integrating out heavy degrees is very powerful to obtain in particular form factors. At that time, a new experimental data for $\omega\pi^0$ transition form factor was provided by NA60,[24] which shows the violation of the ρ meson dominance. I would like to stress that the form factor obtained from the HLS nicely fit to the data, which may indicate that the contributions from heavier vector mesons are also important.

Using an HLS model in which parameters are determined from the holographic model of Sakai–Sugimoto,[21,22] we studied properties of the nucleon and the nuclear matter in the framework of Skyrme-crystal approach in Refs. 31–34. We confirmed the existence of the half-Skyrmion phase in which the space average of the chiral condensate vanishes. In particular in Refs. 31 and 32, we showed that the nucleon mass decreases with increasing density associate with decreasing chiral condensate in normal phase, while it is stable in the half-Skyrmion phase. This seems consistent with the existence of the chiral invariant mass of the nucleon which was pointed in Ref. 38 in a chiral model.

Very recently,[42] we study in-medium modification of effective masses of heavy–light mesons made of a charm quark and up or down quark. We construct an effective chiral model based on the chiral doubling structure proposed in Refs. 43 and 44, and ω meson can be naturally included by means of the HLS.

In this manuscript, I will briefly summarize some developments of the phenomenological analyses done in this decade using models based on the HLS. In Sec. 2, I give a brief summary of the basic notion of the HLS. Determination of parameters of the HLS from holographic QCD models is introduced in Sec. 3, where comparison of $\omega\pi^0$ transition form factor with experimental data is shown. Results of the Skyrme-type analysis using the HLS model are shown in Sec. 4. In Sec. 5, I give a brief summary of analysis of the nuclear matter based on a hadronic model based on the parity doublet structure of nucleons and the HLS. Then, the medium modification of charmed meson masses is presented in Sec. 6.

2. Hidden Local Symmetry

In this section, we briefly review the HLS model based on the $G_{\text{global}} \times H_{\text{local}}$ symmetry, where $G = \text{SU}(N_f)_\text{L} \times \text{SU}(N_f)_\text{R}$ is the global chiral symmetry and

$H = \mathrm{SU}(N_f)_\mathrm{V}$ is the HLS. The entire symmetry $G_\mathrm{global} \times H_\mathrm{local}$ is spontaneously broken down to a diagonal sum H which is nothing but the H of G/H of the nonlinear sigma model. Basic quantities for constructing an effective Lagrangian include two $\mathrm{SU}(N_f)$-matrix valued variables ξ_L and ξ_R which transform as

$$\xi_{\mathrm{L,R}}(x) \to \xi'_{\mathrm{L,R}}(x) = h(x) \cdot \xi_{\mathrm{L,R}}(x) \cdot g^\dagger_{\mathrm{L,R}}, \tag{1}$$

where $h(x) \in H_\mathrm{local}$ and $g_{\mathrm{L,R}} \in G_\mathrm{global}$. These variables are parameterized as

$$\xi_{\mathrm{L,R}} = e^{i\sigma/F_\sigma} e^{\mp i\pi/F_\pi} \quad [\pi = \pi^a T_a, \sigma = \sigma^a T_a], \tag{2}$$

where π denote the Nambu–Goldstone (NG) bosons associated with the spontaneous breaking of G chiral symmetry and σ denote the NG bosons absorbed into the gauge bosons. F_π and F_σ are relevant decay constants.

From the above ξ_L and ξ_R we can construct two Maurer-Cartan 1-forms:

$$\begin{aligned} \widehat{\alpha}_{\perp\mu} &= (D_\mu \xi_\mathrm{R} \cdot \xi_\mathrm{R}^\dagger - D_\mu \xi_\mathrm{L} \cdot \xi_\mathrm{L}^\dagger)/(2i), \\ \widehat{\alpha}_{\parallel\mu} &= (D_\mu \xi_\mathrm{R} \cdot \xi_\mathrm{R}^\dagger + D_\mu \xi_\mathrm{L} \cdot \xi_\mathrm{L}^\dagger)/(2i), \end{aligned} \tag{3}$$

where the covariant derivatives are given by

$$\begin{aligned} D_\mu \xi_\mathrm{L} &= \partial_\mu \xi_\mathrm{L} - i V_\mu \xi_\mathrm{L} + i \xi_\mathrm{L} \mathcal{L}_\mu, \\ D_\mu \xi_\mathrm{R} &= \partial_\mu \xi_\mathrm{R} - i V_\mu \xi_\mathrm{R} + i \xi_\mathrm{R} \mathcal{R}_\mu, \end{aligned} \tag{4}$$

with $\mathcal{L}_\mu$ and $\mathcal{R}_\mu$ being the external gauge fields introduced by gauging the chiral symmetry. An important basic building block of the Lagrangian beside the above 1-forms is the gauge field strength defined by

$$V_\mu = \partial_\mu V_\nu - \partial_\nu V_\mu - i[V_\mu,\ V_\nu]. \tag{5}$$

In addition, for including the effect of explicit chiral symmetry breaking due to the current quark masses, we define a field $\hat{\chi}$ as

$$\hat{\chi} = 2B\xi_L \mathcal{M} \xi_R, \tag{6}$$

where B is a constant of dimension-one and $\mathcal{M}$ is the quark mass matrix.

Now, by using the above building blocks, the leading order Lagrangian is given by[13,17,18]

$$\begin{aligned} \mathcal{L}_{(2)} &= F_\pi^2 \mathrm{tr}[\hat{\alpha}_{\perp\mu}\hat{\alpha}_\perp^\mu] + aF_\pi^2 \mathrm{tr}[\hat{\alpha}_{\parallel\mu}\hat{\alpha}_\parallel^\mu] \\ &\quad - \frac{1}{2g^2}\mathrm{tr}[V_{\mu\nu}V^{\mu\nu}] + \frac{1}{4}F_\chi^2 \mathrm{tr}[\hat{\chi} + \hat{\chi}^\dagger], \end{aligned} \tag{7}$$

where g is the gauge coupling constant, the parameter a is defined as $a = F_\sigma^2/F_\pi^2$ and F_χ in the fourth term is introduced to renormalize the quadratic divergence.[7]

In the HLS, it is possible to make a systematic derivative expansion including loop corrections, which we called the "ChPT with HLS" in Ref. 13. In the ordinary

ChPT,[45–47] derivatives and the external gauge fields are counted as $\mathcal{O}(p)$, and the quark mass matrix is counted as $\mathcal{O}(p^2)$:

$$\partial_\mu \sim \mathcal{R}_\mu \sim \mathcal{L}_\mu \sim \mathcal{O}(p), \quad \hat{\chi} \sim \mathcal{O}(p^2). \tag{8}$$

In addition to the above assignment, the gauge coupling g is counted as $\mathcal{O}(p)$ in the ChPT with HLS:

$$g \sim \mathcal{O}(p). \tag{9}$$

3. Holographic Models and HLS

In Ref. 19, we proposed a way to obtain a Lagrangian having the HLS as a low-energy effective model of a holographic QCD model such as the Sakai–Sugimoto model[21] by integrating out mesons heavier than the pion and rho meson. The method was extended by including external gauge fields in Ref. 20, where several form factors are predicted and compared with experimental data using the Sakai–Sugimoto model.[21,22] Here I just show one of main results in Refs. 19 and 20 by picking up the $\omega\pi^0$ transition form factor. At $\mathcal{O}(p^4)$ of the ChPT with the HLS, the form factor is expressed as

$$F_{\omega\pi^0}(q^2) = (1 - \tilde{r}) + \tilde{r}\frac{m_\rho^2}{m_\rho^2 - q^2}, \tag{10}$$

where $q^2 > 0$ in time-like region and $\tilde{r}$ is a parameter to be determined. In an effective HLS Lagrangian reduced from the Sakai–Sugimoto model, this $\tilde{r}$ is expressed as

$$\tilde{r}_{SS} = \frac{g_{\omega\rho_1\pi} g_{\rho_1}}{m_{\rho_1}^2} \Big/ \sum_{n=1}^{\infty} \frac{g_{\omega\rho_n\pi} g_{\rho_n}}{m_{\rho_n}^2}, \tag{11}$$

where $g_{\omega\rho_n\pi}$ is the coupling constant of the interaction among the pion and omega meson and the nth rho meson, g_{ρ_n} is the mixing strength between the photon and nth rho meson, and m_{ρ_n} is the mass of nth rho meson. The infinite sum in the denominator is done in the Sakai–Sugimoto model using the completeness condition, and the resultant value is $\tilde{r}_{SS} = 1.56$. In Fig. 1, the predicted form factor

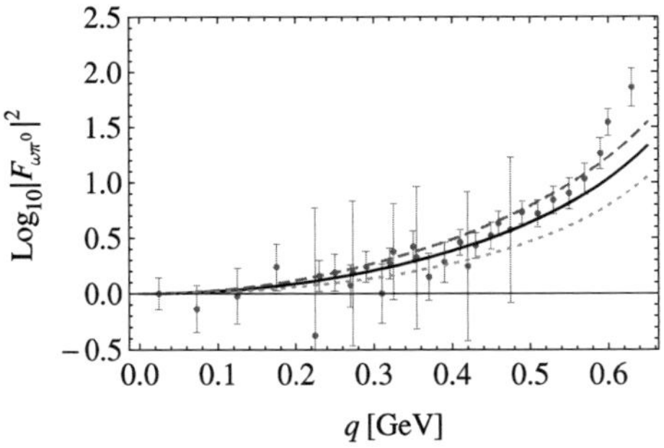

Fig. 1. Predicted $\omega\pi^0$ transition form factor compared with experiments. Figure is picked from Ref. 20. Experimental data are taken from Refs. 23 and 24.

is shown by the black solid curve together with the experimental data,[23,24] which yields $\chi^2/\text{d.o.f} = 45/31 = 1.5$. I also show the best fit curve with $\tilde{r}_{\text{best}} = 2.08$ ($\chi^2/\text{d.o.f} = 24/30 = 0.8$) and curve corresponding to the ρ meson dominance with $\tilde{r} = 1$ ($\chi^2/\text{d.o.f} = 124/31 = 4.0$) by blue dashed and red dotted curves, respectively. I note that the width of rho meson is not included in the above form factor, so that data for $q \gtrsim 0.6$ are not well fitted by the form factor obtained from the Sakai–Sugimoto model and the best fitted value of the HLS parameter.

In Ref. 25, nucleon electromagnetic form factors are studied in an HLS model with nucleon included, where the parameters are determined from an hQCD model with baryons proposed in Refs. 26–28. Similarly to the $\omega\pi^0$ transition form factor in Eq. (10), the EM form factors are expressed as a sum of two contributions: one is momentum-dependent part obtained by the exchange of rho and omega mesons, and another is a "core" contribution derived by integrating out infinite towers of vector mesons heavier than the rho and omega mesons.

I would like to stress that, from the goodness of the fit by the Sakai–Sugimoto model, the violation of rho meson dominance in the ω-π^0 transition form factor may indicate the existence of the contributions from the higher resonances.

4. Skyrme Models with HLS

In Refs. 29–33, we studied baryons described by a soliton existing in an HLS model including $\mathcal{O}(p^4)$, where all the low-energy constants are determined from the Sakai–Sugimoto model of holographic QCD as explained in the previous section. We confirmed the existence of the half-Skyrmion phase in the high density nuclear matter, where space average of the chiral condensate $\langle \bar{q}q \rangle$ vanishes although it is locally nonzero. Then, we showed that, in the low-density region where the half-Skyrmion phase is not achieved, the nucleon mass decreases against the increasing baryon number density, while it is stable in the half-Skyrmion phase. The density dependence of the nucleon mass is well fitted by[31]

$$\frac{m_N^*}{m_N} \simeq \begin{cases} \dfrac{1}{1+0.6(n/n_0)^2} & \text{for } n < n_{1/2}, \\ 0.63 & \text{for } n_{1/2} < n < n_c, \\ 0 & \text{for } n > n_c, \end{cases} \tag{12}$$

where n_0 is the normal nuclear matter density, $n_{1/2}$ is the critical density for the phase transition from the normal nuclear matter to the half-Skyrmion phase. By noting that the space average of the chiral condensate is zero in the half-Skyrmion phase, the above result may have a relation to the existence of the chiral invariant mass of nucleon.

5. Hadronic Model for Nuclear Matter with HLS

As I stated in the previous section, the existence of the chiral invariant mass of the nucleon is suggested by the Skyrme-crystal analysis. In addition, lattice analyses done in Refs. 35 and 36 might have relevance. Then, it would be interesting to study the nuclear matter using models in which the nucleon has a chiral invariant mass.

In Ref. 37, we constructed a hadronic model including nucleons with positive and negative parity together with the σ and π mesons based on the parity doublet structure.[38] We further included ρ and ω mesons in the framework of the HLS, and performed a mean field analysis of the nuclear matter. It was shown that for the wide range of the chiral invariant mass of nucleon m_0, $500 \le m_0 \le 900\,\text{MeV}$, the binding energy and the incompressibility at the normal nuclear mattter are reproduced in the model.

In Fig. 2, I show the density dependences of the mean fields, $\langle\sigma\rangle$ and $\langle\omega\rangle$. In Fig. 2(a), the blue curve shows the density dependence of $\langle\sigma\rangle$ for the chiral invariant mass $m_0 = 500\,\text{MeV}$, and the red curve shows the one for $m_0 = 700\,\text{MeV}$. This shows that the chiral condensate decreases faster for larger value of m_0. On the other hand, in our model, the mean field $\langle\omega\rangle$ is simply proportional to the baryon number density, $\langle\omega\rangle = \frac{g_{\omega NN}}{m_\omega^2}\rho_B$ for any value of m_0. Then, Fig. 2(b) shows just one straight line independently of the choice of m_0.

Using these density dependences of sigma and omega mean fields, we calculated effective masses of nucleons. Here, differently from our original paper,[37] following our recent work.[42] I define effective masses of nucleons by including effects of both σ and ω mean fields as

$$
\begin{aligned}
m_\pm^{(\text{eff})} &= \frac{1}{2}\left[\sqrt{(g_1+g_2)^2\langle\sigma\rangle^2 + 4m_0^2} \mp (g_2 - g_1)\langle\sigma\rangle\right] + g_{\omega NN}\langle\omega\rangle, \\
\overline{m}_\pm^{(\text{eff})} &= \frac{1}{2}\left[\sqrt{(g_1+g_2)^2\langle\sigma\rangle^2 + 4m_0^2} \mp (g_2 - g_1)\langle\sigma\rangle\right] - g_{\omega NN}\langle\omega\rangle,
\end{aligned}
\tag{13}
$$

where $m_+^{(\text{eff})}$ and $m_-^{(\text{eff})}$ are effective masses of nucleons with positive and negative parity, respectively, and $\overline{m}_\pm^{(\text{eff})}$ are those for anti-nucleons. Here, the form of the term with square bracket is actually same as the one at vacuum, and the density dependence appears through the mean field $\langle\sigma\rangle$. The second term proportional to $\langle\omega\rangle$ is added here. It should be noted that the omega contribution to the anti-nucleon is opposite to the one to the nucleon. I show the density dependence of

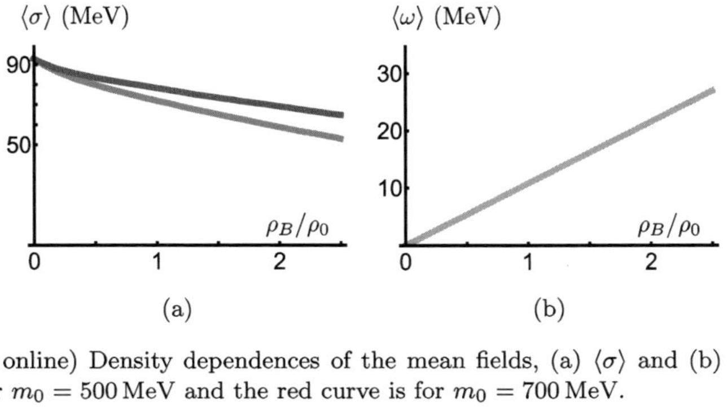

Fig. 2. (Color online) Density dependences of the mean fields, (a) $\langle\sigma\rangle$ and (b) $\langle\omega\rangle$. In (a), the blue curve is for $m_0 = 500\,\text{MeV}$ and the red curve is for $m_0 = 700\,\text{MeV}$.

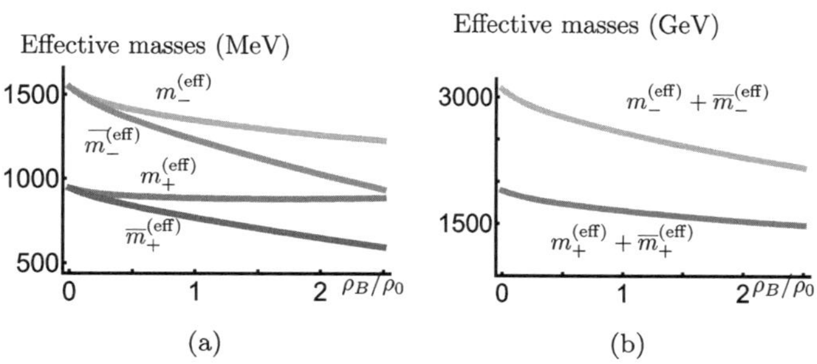

Fig. 3. Density dependences of effective masses for $m_0 = 700$. (a) shows the masses of $m_\pm^{(\text{eff})}$ and $\overline{m}_\pm^{(\text{eff})}$, and (b) shows the sums $m_+^{(\text{eff})} + \overline{m}_+^{(\text{eff})}$ and $m_-^{(\text{eff})} + \overline{m}_-^{(\text{eff})}$.

the effective masses of positive and negative parity nucleons and their anti-particles in Fig. 3. The left panel shows the masses of $m_\pm^{(\text{eff})}$ and $\overline{m}_\pm^{(\text{eff})}$ for $m_0 = 700\,\text{MeV}$. This shows that three masses ($m_-^{(\text{eff})}$ and $\overline{m}_\pm^{(\text{eff})}$) decrease with increasing density, while the ordinary nucleon mass ($m_+^{(\text{eff})}$) is rather stable by the balance between the decreasing effect from $\langle\sigma\rangle$ and the increasing effect of $\langle\omega\rangle$. This seems to indicate that it might be difficult to see the effect of partial chiral symmetry restoration from the density dependence of the ordinary nucleon mass only. The right panel of Fig. 3 shows the density dependence of the sum of a particle and its anti-particle. This shows that sum of the masses of nucleon and anti-nucleon decreases toward $2m_0$ reflecting the partial chiral symmetry restoration. Then, I would like to point out that studying effective masses will give a clue for m_0.

6. Medium Modification of Charmed Meson Masses

In Refs. 39–41, we studied medium modification of charmed meson masses using some chiral models, in which a model used in Refs. 40 and 41 is based on the chiral doubling structure proposed in Refs. 43 and 44. In the analysis done in Ref. 40, the nuclear matter is constructed by a Skyrme-crystal model mentioned in Sec. 4, and it was shown that mass difference of charmed mesons with negative parity and those with positive parity decreases with increasing baryon number density, reflecting the partial chiral symmetry restoration in nuclear matter. There the effect of nuclear matter is mediated by the σ and π mesons.

Recently, we study the medium modification of masses of charmed mesons including the effect mediated by the ω meson.[42] In the analysis, we include the ω meson into a chiral model based on the HLS. In this section, we briefly summarize some of main points.

In the analysis in Ref. 42, we work in the heavy quark limit and consider a simple chiral effective model for a heavy meson multiplet of charmed mesons with $J^P = 0^-$, 1^-, 0^+ and 1^+ based on the chiral doubling structure[43,44] as mentioned above. For studying the effective masses of charmed mesons in medium, we need to consider the interactions of these mesons with the σ and ω mesons. In Ref. 42, the σ is included together with π meson based on the chiral symmetry, and the ω meson is included based on the HLS.

Since we work in the heavy quark limit, D and D^* mesons form a heavy quark doublet which we call the H doublet, and D_0^* and D_1 does another doublet which we call the G doublet. Two masses in the H doublet as well as those in the G doublet are degenerated in the heavy quark limit, which we estimate by spin average of the physical masses as

$$m_H = \frac{m_D + 3m_{D^*}}{4}, \quad m_G = \frac{m_{D_0^*} + 3m_{D_1}}{4}. \tag{14}$$

By including the mean field contributions of the σ and ω mesons, the effective masses of H and G doublets are obtained as

$$\begin{aligned} m_H^{(\text{eff})} &= m - \frac{\Delta_M}{2f_\pi}\langle\sigma\rangle + g_{\omega DD}\langle\omega_0\rangle, \\ m_G^{(\text{eff})} &= m + \frac{\Delta_M}{2f_\pi}\langle\sigma\rangle + g_{\omega DD}\langle\omega_0\rangle, \end{aligned} \tag{15}$$

where m is the average mass of the H and G doublets as $m = (m_H + m_G)/2$. We obtain the effective masses of the anti-charmed mesons $\bar{D}$, $\bar{D}^*$, $\bar{D}_0^*$ and $\bar{D}_1$, by changing the sign in front of the coupling to the ω meson flipped, as

$$\begin{aligned} m_{\bar{H}}^{(\text{eff})} &= m - \frac{\Delta_M}{2f_\pi}\langle\sigma\rangle - g_{\omega DD}\langle\omega_0\rangle, \\ m_{\bar{G}}^{(\text{eff})} &= m + \frac{\Delta_M}{2f_\pi}\langle\sigma\rangle - g_{\omega DD}\langle\omega_0\rangle. \end{aligned} \tag{16}$$

In Ref. 42, we adopt a linear density approximation as a simple estimation, which leads to

$$\langle\omega_0\rangle = \frac{g_{\omega NN}}{m_\omega^2}\rho_B, \quad \frac{\langle\sigma\rangle}{\langle\sigma\rangle_0} = 1 - \frac{\sigma_{\pi N}}{m_\pi^2 f_\pi^2}\rho_B, \tag{17}$$

where $g_{\omega NN}$ is the omega meson coupling to the nucleon, m_ω is the mass of omega meson, ρ_B is the baryon number density, and $\sigma_{\pi N}$ is the coefficient of the π-N sigma term.

For making a numerical estimation, we use $m_G = 2.40\,\text{GeV}$, $m_H = 1.97\,\text{GeV}$, and $\Delta_M = m_G - m_H = 430\,\text{MeV}$ for masses, in addition to $m_\pi = 137\,\text{MeV}$ and $f_\pi = 92.3\,\text{MeV}$. As for other parameters, we use $\sigma_{\pi N} = 45\,\text{MeV}$, $|g_{\omega DD}| = 6$ and $|g_{\omega NN}| = 10$ as reference values. Note that different values are used in Ref. 42.

The density dependence of the masses is plotted in Fig. 4. Figure 4(a) shows that, for $g_{\omega DD}g_{\omega NN} < 0$, the masses of H and G decrease and those of $\bar{H}$ and $\bar{G}$ increase dominated by the ω contribution. For $g_{\omega DD}g_{\omega NN} > 0$, on the other hand, the density dependences are inverted.

In Fig. 5, I show density dependence of effective mass of charmed mesons where mean field values of $\langle\sigma\rangle$ and $\langle\omega_0\rangle$ are obtained from a model based on the parity doublet structure of the nucleon[37] explained in Sec. 5 for $m_0 = 700\,\text{MeV}$. Here I use $|g_{\omega DD}| = 6$ as a typical example. This shows density dependences similar to

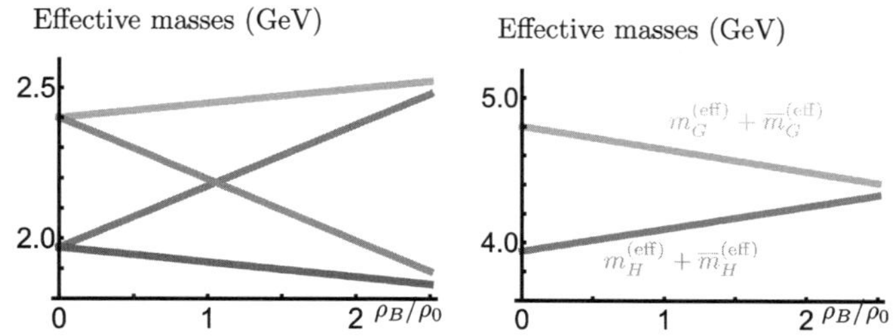

(a) (b)

Fig. 4. (Color online) Density dependence of the effective mass of charmed mesons. In (a) for $g_{\omega DD} g_{\omega NN} < 0$, the green, red, purple and blue curves show the masses of $\bar{G}$, G, $\bar{H}$ and H, respectively. For $g_{\omega DD} g_{\omega NN} > 0$, on the other hand, the green, red, purple and blue curves show the masses of G, $\bar{G}$, H and $\bar{H}$, respectively. In (b), the green and purple curves show the masses of $m_G^{(\text{eff})} + \overline{m}_G^{(\text{eff})}$ and $m_H^{(\text{eff})} + \overline{m}_H^{(\text{eff})}$, respectively, which do not depend on the sign of $g_{\omega DD} g_{\omega NN}$.

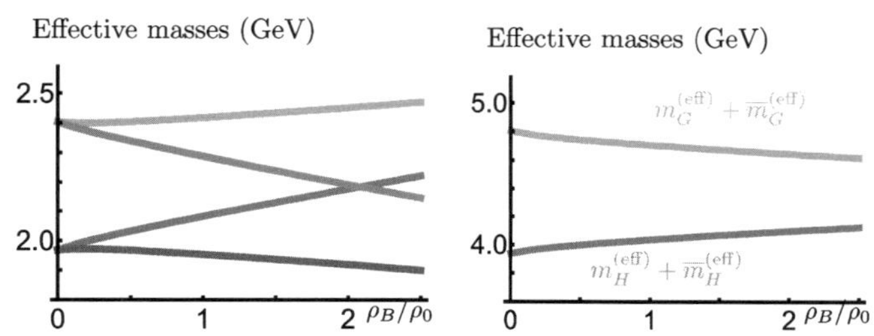

(a) (b)

Fig. 5. (Color online) Density dependence of the effective mass of charmed mesons where mean field values of $\langle \sigma \rangle$ and $\langle \omega_0 \rangle$ are obtained from the model in Ref. 37 for $m_0 = 700\,\text{MeV}$. In (a) for $g_{\omega DD} < 0$, the green, red, purple and blue curves show the masses of $\bar{G}$, G, $\bar{H}$ and H, respectively. For $g_{\omega DD} > 0$, on the other hand, the green, red, purple and blue curves show the masses of G, $\bar{G}$, H and $\bar{H}$, respectively. In (b), the green and purple curves show the masses of $m_G^{(\text{eff})} + \overline{m}_G^{(\text{eff})}$ and $m_H^{(\text{eff})} + \overline{m}_H^{(\text{eff})}$, respectively, which do not depend on the sign of $g_{\omega DD}$.

the ones obtained in the linear density approximation as shown in Fig. 4. But the dependences are a little milder. This is because, in Ref. 37, the value $g_{\omega NN} \simeq 6$ is obtained, which is smaller than the value used in the previous section.

The results in Figs. 4 and 5 indicate that increasing or decreasing of pseudoscalar D meson mass only is not enough for measuring the chiral symmetry restoration. One physical quantity which reflects the partial restoration is the mass difference between the pseudoscalar D meson and scalar D_0^* meson. The difference decreases with density showing the particel chiral symmetry restoration since the ω meson contribution is canceled. In addition to this mass difference, taking sum of masses of a particle and its anti-particle will give a clue as shown in Fig. 4(b): The sum of masses of H and $\bar{H}$ increases and that of G and $\bar{G}$ decreases with increasing density. This implies that the threshold energy for production of a pair of the D and anti-D mesons in medium is larger than vacuum reflecting the partial chiral symmetry restoration. I should note that the dependence on the choice of parameters of $\sigma_{\pi N}$ and $g_{\omega DD}$ is studied in Ref. 42.

7. Summary

As pointed by Gerrry, models based on the HLS are powerful to study the hadron physics closely related to the chiral symmetry property. I hope that I gave some feeling of their usefulness in this contribution.

Acknowledgments

I would like to thank Prof. Tom Kuo and Prof. Mannque Rho for giving me this opportunity to contribute to the book. This work is supported in part by the JSPS Grant-in-Aid for Scientific Research (c) No. 16K05345. This manuscript is dedicated to Professor Gerald E. Brown, who has been giving continuous encouragement on the author's works, in particular, on the hadron physics based on the HLS.

References

1. M. Harada and K. Yamawaki, *Phys. Lett. B* **297** (1992) 151.
2. M. Harada, T. Kugo and K. Yamawaki, *Phys. Rev. Lett.* **71** (1993) 1299.
3. M. Harada, T. Kugo and K. Yamawaki, *Prog. Theor. Phys.* **91** (1994) 801.
4. M. Harada and A. Shibata, *Phys. Rev. D* **55** (1997) 6716.
5. G. E. Brown and M. Rho, *Phys. Rev. Lett.* **66** (1991) 2720.
6. M. Harada and K. Yamawaki, *Phys. Rev. Lett.* **86** (2001) 757.
7. M. Harada and K. Yamawaki, *Phys. Rev. D* **64** (2001) 014023.
8. M. Harada and C. Sasaki, *Phys. Lett. B* **537** (2002) 280.
9. M. Harada, Y. Kim and M. Rho, *Phys. Rev. D* **66** (2002) 016003.
10. M. Tanabashi, *Phys. Lett. B* **384** (1996) 218.
11. H. Georgi, *Phys. Rev. Lett.* **63** (1989) 1917.
12. H. Georgi, *Nucl. Phys. B* **331** (1990) 311.
13. M. Harada and K. Yamawaki, *Phys. Rept.* **381** (2003) 1.
14. M. Harada and C. Sasaki, *Phys. Rev. D* **73** (2006) 036001.
15. M. Harada and C. Sasaki, *Phys. Rev. D* **74** (2006) 114006.
16. G. E. Brown, M. Harada, J. W. Holt, M. Rho and C. Sasaki, *Prog. Theor. Phys.* **121** (2009) 1209.
17. M. Bando, T. Kugo, S. Uehara, K. Yamawaki and T. Yanagida, *Phys. Rev. Lett.* **54** (1985) 1215.
18. M. Bando, T. Kugo and K. Yamawaki, *Phys. Rep.* **164** (1988) 217.
19. M. Harada, S. Matsuzaki and K. Yamawaki, *Phys. Rev. D* **74** (2006) 076004.
20. M. Harada, S. Matsuzaki and K. Yamawaki, *Phys. Rev. D* **82** (2010) 076010.
21. T. Sakai and S. Sugimoto, *Prog. Theor. Phys.* **113** (2005) 843.
22. T. Sakai and S. Sugimoto, *Prog. Theor. Phys.* **114** (2005) 1083.
23. CMD-2 Collab. (R. R. Akhmetshin *et al.*), *Phys. Lett. B* **613** (2005) 29.
24. NA60 Collab. (R. Arnaldi *et al.*), *Phys. Lett. B* **677** (2009) 260.
25. M. Harada and M. Rho, *Phys. Rev. D* **83** (2011) 114040.
26. D. K. Hong, M. Rho, H. U. Yee and P. Yi, *Phys. Rev. D* **76** (2007) 061901.
27. D. K. Hong, M. Rho, H. U. Yee and P. Yi, *J. High Energy Phys.* **0709** (2007) 063.

28. D. K. Hong, M. Rho, H. U. Yee and P. Yi, *Phys. Rev. D* **77** (2008) 014030.
29. Y. L. Ma, Y. Oh, G. S. Yang, M. Harada, H. K. Lee, B. Y. Park and M. Rho, *Phys. Rev. D* **86** (2012) 074025.
30. Y. L. Ma, G. S. Yang, Y. Oh and M. Harada, *Phys. Rev. D* **87**(3) (2013) 034023.
31. Y. L. Ma, M. Harada, H. K. Lee, Y. Oh, B. Y. Park and M. Rho, *Phys. Rev. D* **88** (2013) 014016, Erratum **88** (2013) 079904.
32. Y. L. Ma, M. Harada, H. K. Lee, Y. Oh, B. Y. Park and M. Rho, *Phys. Rev. D* **90** (2014) 034015.
33. M. Harada, H. K. Lee, Y. L. Ma and M. Rho, *Phys. Rev. D* **91** (2015) 096011.
34. M. Harada, Y. L. Ma, H. K. Lee and M. Rho, arXiv:1601.00058 [nucl-th].
35. G. Aarts, C. Allton, S. Hands, B. Jäger, C. Praki and J. I. Skullerud, *Phys. Rev. D* **92** (2015) 014503.
36. L. Y. Glozman, C. B. Lang and M. Schrock, *Phys. Rev. D* **86** (2012) 014507.
37. Y. Motohiro, Y. Kim and M. Harada, *Phys. Rev. C* **92** (2015) 025201.
38. C. E. Detar and T. Kunihiro, *Phys. Rev. D* **39** (1989) 2805.
39. D. Suenaga, B. R. He, Y. L. Ma and M. Harada, *Phys. Rev. C* **89**(6) (2014) 068201.
40. D. Suenaga, B. R. He, Y. L. Ma and M. Harada, *Phys. Rev. D* **91**(3) (2015) 036001.
41. D. Suenaga and M. Harada, *Phys. Rev. D* **93** (2016) 076005.
42. M. Harada, Y. L. Ma, D. Suenaga and Y. Takeda, arXiv: 1612.03496.
43. M. A. Nowak, M. Rho and I. Zahed, *Phys. Rev. D* **48** (1993) 4370.
44. W. A. Bardeen and C. T. Hill, *Phys. Rev. D* **49** (1994) 409.
45. S. Weinberg, *Physica A* **96** (1979) 327.
46. J. Gasser and H. Leutwyler, *Ann. Phys.* **158** (1984) 142.
47. J. Gasser and H. Leutwyler, *Nucl. Phys. B* **250** (1985) 517.

Partial restoration of chiral symmetry in hot and dense neutron matter

Jeremy W. Holt
Cyclotron Institute and
Department of Physics and Astronomy
Texas A&M University, College Station, TX, USA
holt@physics.tamu.edu

Keighley E. Rockcliffe
Department of Physics, Applied Physics, and Astronomy
Rensselaer Polytechnic Institute, Troy, NY, USA

We review efforts to describe the approach to chiral symmetry restoration in neutron matter from the low-energy realization of QCD, chiral effective field theory.

Keywords: Chiral symmetry restoration; chiral effective field theory; neutron matter.

1. Introduction

The restoration of chiral symmetry in hot and dense hadronic matter and the associated observable signatures in relativistic heavy ion collisions were favorite research interests of Gerry Brown for two decades.[1–5] Gerry attacked the problem with his characteristic directness, making an early prediction[1,2] with Mannque Rho that nucleons and heavy mesons shed mass in the approach to chiral symmetry restoration according to a universal scaling law. At the time that one of the authors (JWH) became Gerry's graduate student at Stony Brook in the mid-2000's, numerous experimental investigations of the Brown–Rho scaling conjecture were underway.[6–8] Despite the lack of clean experimental signatures[4,5] for Brown–Rho scaling at finite density and temperature, understanding the effects of medium-modified meson masses and couplings on the nucleon–nucleon (NN) potential was nevertheless an inspiring topic for a PhD thesis.

One of Gerry's favorite words of wisdom for young graduate students was, "Build simple models and don't let anyone tell you it has to be more complicated." The present contribution to this memorial volume will focus on a somewhat "more complicated" method to understand medium-modified NN interactions and the approach to chiral symmetry restoration in hot and dense neutron matter. This framework, utilizing microscopic chiral effective field theory (EFT) two- and three-body nuclear forces, in fact produces in-medium two-body interactions qualitatively similar to models incorporating Brown-Rho scaling.[9–15] When employed in many-body perturbation theory to calculate the neutron matter thermodynamic equation of state, the coarse-resolution chiral potentials used in the present work show excellent convergence properties. Here, again we are indebted to the work of Brown, who together with Kuo, Bogner and Schwenk, pioneered the use of low-momentum NN potentials[16] derived from effective interaction theory and the renormalization group.

Chiral symmetry generically appears in quantum field theories with massless fermions. In the chiral limit of QCD, the left- and right-handed quarks decouple and the Lagrangian is invariant under independent $SU(2)_{L,R}$ transformations in flavor space. In addition to the explicit breaking of chiral symmetry, due to the small but nonzero bare quark mass arising from coupling to the Higgs field, the strong attraction between quark–antiquark pairs leads to the formation of a scalar quark condensate $\langle 0|\bar{q}q|0\rangle$ in the QCD vacuum that spontaneously breaks chiral symmetry. At the high temperatures and/or densities encountered in core-collapse supernovae or neutron star mergers, however, chiral symmetry may be restored.[17,18]

At low densities, the quark condensate in nuclear matter decreases linearly[19,20] and proportional to the pion-nucleon sigma term $\sigma_{\pi N} = m_q \partial M_N/\partial m_q$, which encodes the small change in the nucleon mass M_N from the explicit breaking of chiral symmetry. Neglecting interaction contributions to the ground state energy, the linear term alone gives rise to chiral symmetry restoration in cold symmetric nuclear matter and pure neutron matter at a density $n \simeq 2.5n_0$, assuming a value of $\sigma_{\pi N} = 45\,\text{MeV}$[21] (note that a more recent chiral EFT analysis[22] has found the larger value of $\sigma_{\pi N} = 59 \pm 7\,\text{MeV}$), where $n_0 = 0.16\,\text{fm}^{-3}$. This is in the vicinity of the energy density at which chiral symmetry is restored at finite temperature and zero net baryon density from lattice QCD.[23] Corrections to the leading density dependence have been obtained from the quark-mass dependence of interaction contributions to the nuclear matter ground state energy in one-boson-exchange models,[24,25] in-medium chiral perturbation theory,[26,27] and by employing high-precision chiral nuclear forces at next-to-next-to-next-to-leading order (N3LO).[28] In all of these studies, nuclear mean fields and correlations arising from two- and three-body interactions consistently suppress chiral symmetry restoration beyond nuclear matter saturation density, relative to the noninteracting case.

At finite temperature, chiral perturbation theory was used to predict a phase transition to chiral symmetry restored matter at a critical temperature of $T_c \simeq 190\,\text{MeV}$,[29] which is somewhat larger than the presently accepted value of $T_c = 155 \pm 10\,\text{MeV}$ from lattice QCD.[30–32] In the present study, we compute both the temperature and density dependence of the scalar quark condensate in neutron matter from high-precision two- and three-nucleon interactions. Previous finite-temperature calculations of the quark condensate in symmetric nuclear matter in the framework of in-medium chiral perturbation theory[33] revealed that thermal effects wash out the interaction contributions that tend to delay chiral symmetry restoration at high densities. This leads to a nearly linear-density dependence of the condensate for temperatures greater than $T = 50\,\text{MeV}$. Hot proto-neutron stars born immediately after core-collapse supernovae or the hypermassive neutron stars that exist transiently after the merger of two neutron stars may therefore be more compelling candidate sites for quark-hadron phase transtions than cold neutron stars.

2. Neutron Matter at Finite Temperature from Many-Body Perturbation Theory

Chiral effective field theory is the appropriate tool to study hadronic matter at the scales relevant in nuclear astrophysics (well below the chiral symmetry breaking scale of $\Lambda_\chi \simeq 1\,\text{GeV}$). We start from a coarse-resolution chiral potential[34,35] with a momentum-space cutoff of $\Lambda = 414\,\text{MeV}$, which has been shown to exhibit good convergence properties[34–38] in many-body perturbation theory calculations of infinite nuclear matter, comparable to low-momentum potentials constructed via renormalization group methods.[16,39–41] In the present calculation, the free energy per particle of pure neutron matter at finite temperature is computed in the imaginary-time Matsubara formalism. The perturbation series for the grand canonical potential Ω reads

$$\Omega(\mu, T) = \Omega_0(\mu, T) + \lambda\Omega_1(\mu, T) + \lambda^2\Omega_2(\mu, T) + \cdots, \tag{1}$$

where λ is an arbitrary strength parameter, T is the temperature, and μ is the chemical potential. Equation (1) can be reformulated in the canonical ensemble through the Kohn–Luttinger–Ward prescription.[42,43] The result is a rearrangement of the perturbation series:

$$F(\mu_0, T) = F_0(\mu_0, T) + \lambda\Omega_1(\mu_0, T) + \lambda^2 \left(\Omega_2(\mu_0, T) - \frac{1}{2}\frac{(\partial\Omega_1/\partial\mu_0)^2}{\partial^2\Omega_0/\partial\mu_0^2}\right) + \cdots, \tag{2}$$

where μ_0 is the chemical potential of the noninteracting system. There is a one-to-one correspondence between the nucleon density n and the effective chemical potential μ_0 through

$$n(\mu_0, T) = -\frac{\partial\Omega_0}{\partial\mu_0}, \tag{3}$$

where the noninteracting grand canonical potential has the well-known form

$$\Omega_0(\mu_0, T) = -\frac{1}{\beta}\frac{1}{\pi^2}\int_0^\infty dp\, p^2 \log(1 + e^{-\beta(e(p)-\mu_0)}), \tag{4}$$

and $e(p) = p^2/2M_N$ is the single-particle energy. The first term in the expansion of the free energy is related to Ω_0 via $F_0(\mu_0, T) = \Omega_0(\mu_0, T) + n\mu_0$.

The first- and second-order perturbative contributions to the free energy of neutron matter are shown diagrammatically in Fig. 1. The wavy lines represent the sum of the free-space NN interaction and an in-medium NN potential constructed

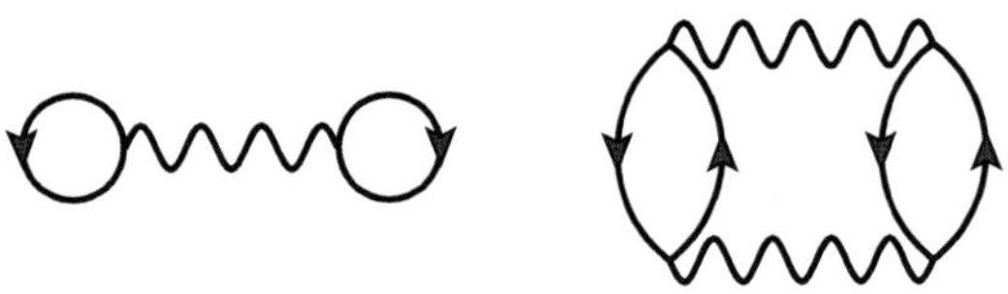

Fig. 1. First- and second-order diagrammatic contributions to the free energy of pure neutron matter from N3LO414 chiral two- and three-nucleon forces. The wavy line represents a density- and temperature-dependent NN interaction derived from the chiral three-body force.

from the N2LO chiral three-body force by summing one leg over the filled Fermi sea of noninteracting neutrons. This approximation can be improved upon by including three-body forces at N3LO[44] and by treating three-body forces explicitly at higher orders in perturbation theory.[45,46] In the present case, the effective interaction depends on the density and temperature of neutron matter. The explicit expressions for the diagrams shown in Fig. 1 are given by

$$\Omega_1(\mu_0, T) = \frac{1}{2}\sum_{\sigma_1,\sigma_2}\int \frac{d^3k_1 d^3k_2}{(2\pi)^6} n_{k_1} n_{k_2} \left\langle 12 \middle| \left(\bar{V}_{NN} + \frac{\bar{V}_{NN}^{\rm med}}{3}\right) \middle| 12 \right\rangle, \tag{5}$$

$$\Omega_2^n(\mu_0, T) = -\frac{1}{8}\prod_{i=1}^{4}\left(\sum_{\sigma_i}\int \frac{d^3k_i}{(2\pi)^3}\right)(2\pi)^3\delta(\mathbf{k}_1 + \mathbf{k}_2 - \mathbf{k}_3 - \mathbf{k}_4)$$
$$\times \frac{n_{k_1}n_{k_2}\bar{n}_{k_3}\bar{n}_{k_4} - \bar{n}_{k_1}\bar{n}_{k_2}n_{k_3}n_{k_4}}{e_3 + e_4 - e_1 - e_2}|\langle 12|(\bar{V}_{NN} + \bar{V}_{NN}^{\rm med})|34\rangle|^2, \tag{6}$$

where n_k is the Fermi distribution function, $\bar{n}_k = 1 - n_k$, and $\bar{V}$ is the antisymmetrized potential. We have written only the expression for the "normal" second-order contribution Ω_2^n and neglected the anomalous contribution, which was shown[36] to have a negligible effect when combined with the derivative term in Eq. (2). The free energy of neutron matter has been computed previously[37] employing the N3LO414 chiral potential. At low densities and high temperatures, the results were found to agree well with the model-independent virial expansion, and in the vicinity of nuclear matter saturation density the symmetry energy $E_{\rm sym} = 32.5\,{\rm MeV}$ and slope $L = 53.8\,{\rm MeV}$ are consistent with empirical constraints.[47]

From the Hellmann–Feynman theorem[20,33]

$$\left\langle \psi(\alpha) \middle| \frac{d}{d\alpha}H(\alpha) \middle| \psi(\alpha \right\rangle = \frac{d}{d\alpha}E(\alpha), \tag{7}$$

the chiral condensate in finite-density matter is related to the vacuum value by choosing $\alpha = m_q$:

$$\langle\psi|\bar{q}q|\psi\rangle - \langle 0|\bar{q}q|0\rangle = \frac{1}{2}\frac{d}{dm_q}[M_N n + \mathcal{E}], \tag{8}$$

where $|\psi\rangle$ is the neutron matter ground state and $\mathcal{E}$ is the energy density. Since, we do not have direct access to the quark-mass dependence of the neutron matter ground state energy, we employ the Gell-Mann–Oakes–Renner relation at leading order

$$m_\pi^2 f_\pi^2 \simeq -2\bar{m}_q \langle 0|\bar{q}q|0\rangle, \tag{9}$$

where $\bar{m}_q$ is the average light-quark mass and $f_\pi = 92.4\,\text{MeV}$ is the pion decay constant, to write the chiral condensate in terms of the pion-mass dependence of the energy per particle $\bar{E}$:

$$\frac{\langle \bar{q}q\rangle_n}{\langle \bar{q}q\rangle_0} = 1 - \frac{n}{f_\pi^2}\left[\frac{\sigma_{\pi N}}{m_\pi^2} + \frac{d\bar{E}}{dm_\pi^2}\right]. \tag{10}$$

This expression can be extended to finite temperature[33] by replacing the energy per particle in Eq. (10) with the free energy per particle $\bar{F}$,

$$\frac{\langle \bar{q}q\rangle_{n,T}}{\langle \bar{q}q\rangle_{0,0}} = 1 - \frac{n}{f_\pi^2}\frac{\partial \bar{F}(n,T)}{\partial m_\pi^2}, \tag{11}$$

and absorbing the nucleon mass and chemical potential μ_0 into $\bar{F}$.

3. Results

We begin by plotting in Fig. 2 the neutron matter thermodynamic equation of state (the free energy per particle as a function of temperature and density) obtained from the N3lO414 chiral two- and three-body potentials. The results are in good agreement with previous studies,[37,38] which employed the same two- and three-body potentials but included also self-consistent Hartree–Fock single-particle energies in the second-order diagrams. As we will show later, second-order perturbative contributions to the neutron matter equation of state and the chiral condensate are

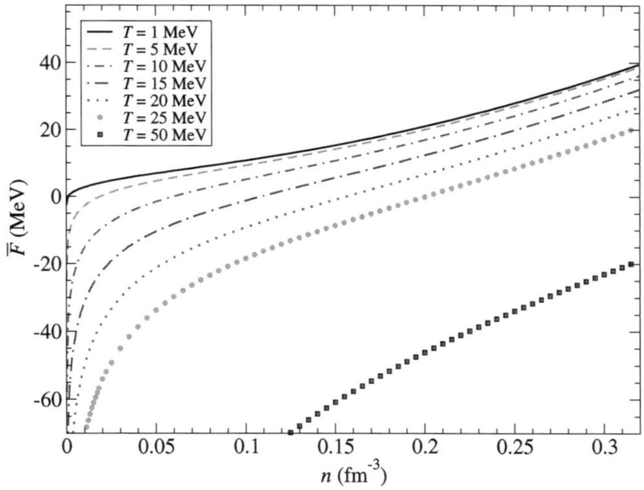

Fig. 2. Free energy per particle $\bar{F}$ of pure neutron matter as a function of density for different isotherms. Results are obtained at second order in many-body perturbation theory employing the coarse-resolution chiral potential N3LO414 described in the text.

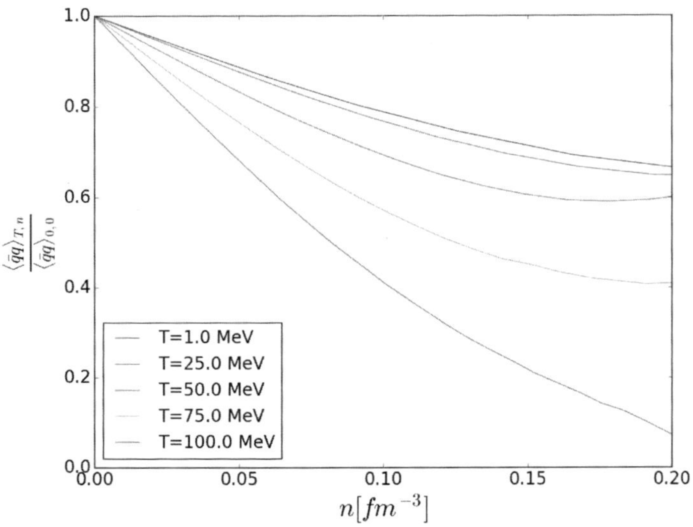

Fig. 3. Ratio of the chiral condensate in pure neutron matter to that in the vacuum as a function of density and temperature from the pion-mass dependence of the free energy computed in many-body perturbation theory with coarse-resolution chiral two- and three-body potentials.

relatively small. The inclusion of in-medium nucleon propagators will, therefore be postponed to future work and higher-order contributions to the nucleon single-particle potential will be addressed.

In Fig. 3, we show the ratio of the chiral condensate in neutron matter at finite temperature T to the chiral condensate in vacuum as a function of the nucleon number density n. In computing the numerical derivatives of the different contributions to the free-energy density with respect to the pion mass, we chose to vary the pion mass by 5%, 2% and 1%. We found only very small differences in the values of the numerical derivatives, and in the figure, the results from a 1% change in the pion mass are shown. We observe that at zero temperature, the density dependence of the chiral condensate falls within the uncertainty band obtained in Ref. 28, which included as well three-body forces at N3LO in the chiral power counting.

Temperatures up to 100 MeV are considered in the present work. In the highest temperature regime, the presence of thermal pions enhance the trend toward chiral symmetry restoration,[29,33,48] leading to a qualitative difference in the condensate ratio especially at low density. In the present work, we focus on nuclear interaction contributions to the medium dependence of the chiral condensate and will include effects due to thermal pions in future work. We also neglect the quark-mass dependence of the axial coupling strength g_A, pion decay constant f_π, and short-range low-energy constants in the 2N and 3N sectors. For zero-temperature neutron matter, these effects were estimated[28] to contribute on the order of 25% to the theoretical uncertainty in the chiral condensate. More detailed discussions can be found in the literature.[49,50]

As found in previous chiral effective field theory calculations[27,28] of the pion-mass dependence of the neutron matter ground state energy at zero temperature, we observe that interaction contributions result in a relatively small change in the leading linear decrease of the chiral condensate with increasing density. In particular, repulsive interactions delay the approach to chiral symmetry restoration and increase with nucleon density. At $n = 0.2\,\mathrm{fm}^{-3}$, nuclear interactions increase the chiral condensate by about 15%. This is a larger effect, though qualitatively similar, to what has been observed employing one-boson-exchange models of the nucleon–nucleon interaction.[24]

Increasing the temperature, on the other hand, is highly effective at promoting chiral symmetry restoration in neutron matter. This comes predominantly from the noninteracting contributions to the pion mass dependence of the neutron matter free energy, as shown in Fig. 4. Increasing the quark mass enhances both the value of the nucleon mass and the kinetic energy contribution to the neutron matter free energy. In Figs. 5 and 6, we plot the first-order, $\Delta^{(1)}\frac{\langle\bar{q}q\rangle_{T,n}}{\langle\bar{q}q\rangle_{0,0}}$, and second-order, $\Delta^{(2)}\frac{\langle\bar{q}q\rangle_{T,n}}{\langle\bar{q}q\rangle_{0,0}}$, perturbative contributions to the chiral condensate ratio from two- and three-nucleon chiral forces. For all densities and temperatures considered in the present work, both the first- and second-order contributions tend to delay the restoration of chiral symmetry for increasing density and temperature. This feature was observed already in Ref. 28 for neutron matter at zero temperature, and we note that it holds with increasing temperature. The leading Hartree–Fock contribution is consistently about three times larger in magnitude than the second-order correction. For temperatures up to $T = 50\,\mathrm{MeV}$ and densities below $n = 0.2\,\mathrm{fm}^{-3}$, the second-order interaction contribution to the scalar quark condensate gives less than a 10%

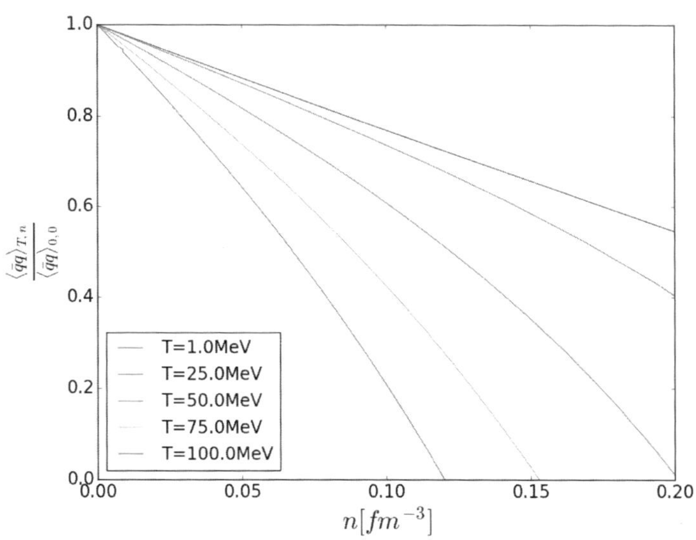

Fig. 4. Same as in Fig. 3, except that only the noninteracting contribution to the medium dependence of the chiral condensate is included.

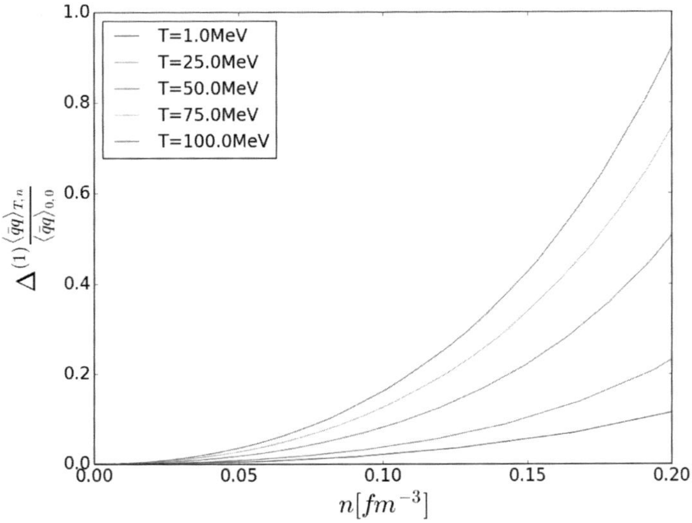

Fig. 5. Change in the chiral condensate ratio as a function of temperature and density including the first-order perturbative contribution to the free energy of neutron matter.

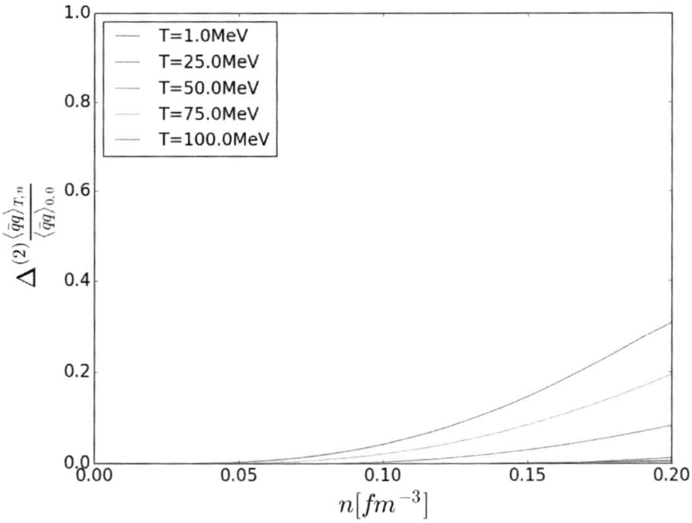

Fig. 6. Same as in Fig. 5, except that only the second-order perturbative contribution to the free energy has been included.

correction. This is in agreement with previous studies of the neutron matter equation of state where the use of coarse-resolution chiral potentials strongly reduced the effect of the second-order bubble diagram. Uncertainties associated with the pion-nucleon sigma term and low-energy constants in two- and three-nucleon forces are therefore expected to dominate the theoretical errors.[28]

4. Summary

Many-body perturbation theory with coarse-resolution chiral potentials now allows for the systematic study of infinite nuclear matter properties with reduced theoretical uncertainties. In the present work, we have calculated the scalar-quark condensate in neutron matter at finite temperature employing a recently developed chiral nuclear potential with a momentum-space cutoff of $\Lambda = 414\,\text{MeV}$. We considered separately the pion-mass dependence of noninteracting contributions to the free energy of neutron matter as well as interaction effects from two- and three-body nuclear forces. For all densities and temperatures, the noninteracting contributions are dominant and lead to a decrease of the chiral condensate with increasing density and temperature. In the absence of nuclear interactions, hot neutron matter around nuclear matter saturation density should already exhibit a transition to a chiral symmetry restored phase. However, interaction effects generically delay chiral symmetry restoration and increase in magnitude with both the density and temperature.

Acknowledgments

K. E. Rockcliffe acknowledges the kind hospitality of the Texas A&M Cyclotron Institute. Work supported in part by NSF grant PHY-1263281 and DOE grant DE-FG03-93ER40773.

References

1. G. E. Brown, *Nucl. Phys. A* **488** (1988) 689.
2. G. E. Brown and M. Rho, *Phys. Rev. Lett.* **66** (1991) 2720.
3. G. E. Brown and M. Rho, *Phys. Rept.* **363** (2002) 85.
4. G. E. Brown, J. W. Holt and M. Rho, *Phys. Rep.* **439** (2007) 161.
5. G. E. Brown, M. Harada, J. W. Holt, M. Rho and C. Sasaki, *Prog. Theor. Phys.* **121** (2009) 1209.
6. D. Trnka *et al.*, *Phys. Rev. Lett.* **94** (2005) 192303.
7. R. Nasseripour *et al.*, *Phys. Rev. Lett.* **99** (2007) 262302.
8. M. Kotulla *et al.*, *Phys. Rev. Lett.* **100** (2008) 192302.
9. B. Friman and M. Rho, *Nucl. Phys. A* **606** (1996) 303.
10. J. W. Holt, G. E. Brown, J. D. Holt and T. T. S. Kuo, *Nucl. Phys. A* **785** (2007) 322.
11. J. W. Holt, G. E. Brown, T. T. S. Kuo, J. D. Holt and R. Machleidt, *Phys. Rev. Lett.* **100** (2008) 062501.
12. J. W. Holt, N. Kaiser and W. Weise, *Phys. Rev. C* **79** (2009) 054331.
13. J. W. Holt, N. Kaiser and W. Weise, *Nucl. Phys. A* **870–871** (2011) 1.
14. J. W. Holt, N. Kaiser and W. Weise, *Nucl. Phys. A* **876** (2012) 61.
15. H. Dong, T. T. S. Kuo, H. K. Lee, R. Machleidt and M. Rho, *Phys. Rev. C* **87** (2013) 054332.
16. S. K. Bogner, T. T. S. Kuo and A. Schwenk, *Phys. Rep.* **386** (2003) 1.
17. M. Alford, M. Braby, M. Paris and S. Reddy, *Astrophys. J.* **629** (2005) 969.
18. T. Fischer, I. Sagert, G. Pagliara, M. Hempel, J. Schaffner-Bielich, T. Rauscher, F.-K. Thielemann, R. Käppeli, G. Martínez-Pinedo and M. Liebendörfer, *Astrophys. J. Suppl.* **194** (2011) 39.

19. E. G. Drukarev and E. Levin, *Nucl. Phys. A* **511** (1990) 679.
20. T. D. Cohen, R. J. Furnstahl and D. G. Griegel, *Phys. Rev. C* **45** (1992) 1881.
21. J. Gasser, H. Leutwyler and M. E. Sainio, *Phys. Lett. B* **253** (1991) 252.
22. J. M. Alarcón, J. M. Camalich and J. A. Oller, *Phys. Rev. D* **85** (2012) 051503.
23. P. Hegde and (BNL-Bielefeld-CCNU), *Proc. 32nd Int. Symp. Lattice Field Theory (Lattice 2014)* (2014), pp. 226.
24. G. Q. Li and C. M. Ko, *Phys. Lett. B.* **338** (1994) 118.
25. R. Brockmann and W. Weise, *Phys. Lett. B* **367** (1996) 40.
26. N. Kaiser, P. de Homont and W. Weise, *Phys. Rev. C* **77** (2008) 025204.
27. N. Kaiser and W. Weise, *Phys. Lett. B* **671** (2009) 25.
28. T. Krüger, I. Tews, B. Friman, K. Hebeler and A. Schwenk, *Phys. Lett. B* **726** (2013) 412.
29. P. Gerber and H. Leutwyler, *Nucl. Phys. B* **321** (1989) 387.
30. S. Borsányi, Z. Fodor, C. Hoelbling, S. D. Katz, S. Krieg, C. Ratti and K. K. Szabó, *J. High Energ. Phys.* **2010** (2010) 1.
31. A. Bazavov, T. Bhattacharya, M. Cheng, C. DeTar, H.-T. Ding, S. Gottlieb, R. Gupta, P. Hegde, U. M. Heller, F. Karsch, E. Laermann, L. Levkova, S. Mukherjee, P. Petreczky, C. Schmidt, R. A. Soltz, W. Soeldner, R. Sugar, D. Toussaint, W. Unger and P. Vranas, *Phys. Rev. D* **85** (2012) 054503.
32. T. Bhattacharya, M. I. Buchoff, N. H. Christ, H.-T. Ding, R. Gupta, C. Jung, F. Karsch, Z. Lin, R. D. Mawhinney, G. McGlynn, S. Mukherjee, D. Murphy, P. Petreczky, D. Renfrew, C. Schroeder, R. A. Soltz, P. M. Vranas and H. Yin, *Phys. Rev. Lett.* **113** (2014) 082001.
33. S. Fiorilla, N. Kaiser and W. Weise, *Phys. Lett. B* **714** (2012) 251.
34. L. Coraggio, J. W. Holt, N. Itaco, R. Machleidt and F. Sammarruca, *Phys. Rev. C* **87** (2013) 014322.
35. L. Coraggio, J. W. Holt, N. Itaco, R. Machleidt, L. E. Marcucci and F. Sammarruca, *Phys. Rev. C* **89** (2014) 044321.
36. C. Wellenhofer, J. W. Holt, N. Kaiser and W. Weise, *Phys. Rev. C* **89** (2014) 064009.
37. C. Wellenhofer, J. W. Holt and N. Kaiser, *Phys. Rev. C* **92** (2015) 015801.
38. C. Wellenhofer, J. W. Holt and N. Kaiser, *Phys. Rev. C* **93** (2016) 055802.
39. S. K. Bogner, A. Schwenk, R. J. Furnstahl and A. Nogga, *Nucl. Phys. A* **763** (2005) 59.
40. K. Hebeler and A. Schwenk, *Phys. Rev. C* **82** (2010) 014314.
41. K. Hebeler, S. K. Bogner, R. J. Furnstahl, A. Nogga and A. Schwenk, *Phys. Rev. C* **83** (2011) 031301.
42. W. Kohn and J. M. Luttinger, *Phys. Rev.* **118** (1960) 41.
43. J. M. Luttinger and J. C. Ward, *Phys. Rev.* **118** (1960) 1417.
44. I. Tews, T. Krüger, K. Hebeler and A. Schwenk, *Phys. Rev. Lett.* **110** (2013) 032504.
45. N. Kaiser, *Eur. J. Phys. A* **48** (2012) 58.
46. C. Drischler, A. Carbone, K. Hebeler and A. Schwenk, arXiv:1608.05615.
47. M. Dutra, O. Lourenço, J. S. Sá Martins, A. Delfino, J. R. Stone and P. D. Stevenson, *Phys. Rev. C* **85** (2012) 035201.
48. D. Toublan, *Phys. Rev. D* **56** (1997) 5629.
49. V. Bernard and U.-G. Meissner, *Phys. Lett. B* **639** (2006) 278.
50. J. C. Berengut, E. Epelbaum, V. V. Flambaum, C. Hanhart, U.-G. Meißner, J. Nebreda and J. R. Peláez, *Phys. Rev. D.* **87** (2013) 085018.

Compact star matter: EoS with new scaling law

Kyungmin Kim, Hyun Kyu Lee* and Jaehyun Lee
Department of Physics, Hanyang University,
Seoul 04763, Korea
**hyunkyu@hanyang.ac.kr*

In this paper, we present a simple discussion on the properties of compact stars using an EoS obtained in effective field theory anchored on scale and hidden-local symmetric Lagrangian endowed with topology change and a unequivocal prediction on the deformation of the compact star, that could be measured in gravitational waves. The objective is not to offer a superior or improved EoS for compact stars but to confront with a forthcoming astrophysical observable, the given model formulated in what is considered to be consistent with the premise of quantum chromodynamics (QCD). The model so obtained is found to satisfactorily describe the observation of a two-solar mass neutron star [P. B. Demorest *et al.*, *Nature* **467** (2010) 1081, J. Antoniadis *et al.*, *Science* **340** (2013) 1233232] with a minimum number of parameters. Specifically, the observable we are considering in this paper is the tidal deformability parameter λ (equivalently the Love number k_2), which affects gravitational wave forms at the late period of inspiral stage. The forthcoming aLIGO and aVirgo observations of gravitational waves from binary neutron star system will provide a valuable guidance for arriving at a better understanding of highly compressed baryonic matter.

Keywords: Compact star; equation of state; deformability; new scaling.

1. Introduction

The observation of two-solar mass neutron stars[1,2] seems to indicate that the equation of state (EoS) for compact stars needs to be sufficiently stiffer to accommodate the mass larger than 1.5-solar mass. Moreover, it requires the detailed information on the hadronic matter at higher density than the normal nuclear density, n_0, which seems to be, however, much higher beyond the reach of presently planned terrestrial laboratories. After the recent detections of gravitational waves from binary black holes,[3] the expectation of detecting gravitational waves from a binary neutron stars and/or a black hole-neutron star binary becomes very optimistic than ever. The gravitational waves emitted during the binary inspiral phase up to merging can provide us the information on the dense hadronic matter of higher density at the core of compact stars.

The nuclei with large atomic numbers are already obvious examples of highly dense matters composed of nucleons, $n \sim n_0 = 0.16/\text{fm}^{-3}$. The effective theories of nucleons have been developed and constrained by experimental data available up to and slightly above the normal nuclear density, n_0. Hence, they are fairly well controlled theoretically and experimentally. But the high density regime much above n_0 is more or less uncharted both experimentally and theoretically. Compact objects such as neutron stars are supposed to have higher core density than the normal nuclear density, n_0. Roughly for neutron star with mass $\sim 1.5M_\odot$ the relevant density at the center is likely around 2–$3n_0$ and for the mass $\sim 2M_\odot$ the density is supposed to be larger than $\sim 5n_0$. On top of the possibilities of getting high density nuclear matter at terrestrial laboratories, for example FRIB (USA), FAIR (Germany), J-Parc (Japan) and RAON (Korea) in near future, the possible detections of gravitational waves from binary neutron stars (or binary neutron star black hole) are believed to be promising probes of the high density interior of neutron stars. Theoretically, the success of low density effective theories univocally up to the normal nuclear density seems not to guarantee the similar success at higher density hadronic matter since the predictions of mass, radius, symmetry energy and deformability to name a few diverse from each other drastically beyond the normal nuclear density. In this sense, we are now at the very exciting period of foreseeing the opportunity of constraining theories at higher density by experiments and observations.

For the highly dense hadronic matter, recently we proposed a new approach[4–6] in which we try to formulate a field theory framework wherein both low and high density regimes are treated on the same footing. To cover both regimes in a consistent way in a unified field, theoretic approach seems like a tall order. In this work, we discuss the physical properties of stellar matter using the newly proposed scheme of a new scaling law Brown-Lee-Park-Rho (BLPR) in medium[7,8] all the way from normal nuclear density up to the higher density at the core of a neutron star. With a confrontation with the observed massive stars,[4] we discuss the physical properties of stellar matter with the EoS obtained therein.[a] The relevant quantities are mass, radius and deformability parameters, which could be constrained by the gravitational wave forms emitted during the binary inspiraling phase. The aim here is then to confirm or falsify the strategies taken and assumptions made in Dong *et al.*[4] and Lee *et al.*[5] and also to find the directions to be taken in constructing the correct effective theory at higher density.

In Sec. 2, the basic concept of unified approach in this work is discussed. Using the minimal effective Lagrangian in the frame work of relativistic mean field (RMF), the EoS of compact star with neutrons, protons, electrons and muon in weak equilibrium and charge neutrality condition is discussed in Sec. 3 and the mass and radius are estimated. Tidal deformation with new stiffer EoS is discussed with the observational possibility in gravitational wave detections at aLIGO[10] and aVirgo[11] in Sec. 4. The results are summarized in Sec. 5. We use units in which $c = G = 1$ and the notation in which Minkowski metric $\eta_{\mu\nu} = \text{diag}[-1, 1, 1, 1]$.

[a]This paper is an extended version of the previous work[9] prepared for Gerry Brown memorial volume.

2. Unified Approach with New Scaling

The unified field theoretic approach is formulated using an effective theory which has sufficient number of degrees of freedom and parameters to be able to implement desired symmetries in high density regime or at the critical density, $n_c > n_0$. The first step is to construct the functional forms of parameters of the effective theory in terms of the variables in quantum chromodynamics (QCD), for example, quark and gluon condensates. by matching procedures near at QCD scale. Then, if we can compute the density dependence of QCD condensates, they automatically determine the density dependence of parameters in the effective theory. In addition, the vacuum can be characterized by the expectation value of the relevant field, in this work scalar field $\langle \chi \rangle$, which is supposed to depend on the density. These are the intrinsic density dependencies (IDD) inherited from the quark–gluon condensates and scalar condensate. At low density (up to nuclear matter density), the success of chiral perturbation theory (ChPT) and low energy theorems implies that the QCD matching is implicitly taken care already. But at higher density, QCD matching or IDD will play a very crucial role since the naive extrapolation of ChPT, in which the QCD matching is implicit and therefore hidden, may go anywhere if not guided by the explicit QCD matching constraints.

The first step forward to that goal has been made using the effective Lagrangian with minimum number of degree of freedom, which are the pseudo-scalar mesons π, and scalar meson σ, vector mesons ρ and nucleon N. We adopt the RMF approach, which are found to work remarkably well for finite and nuclear matter as well as for highly dense matter inside compact stars. The success of RMF can be understood that RMF captures the physics of the Landau Fermi liquid fixed point at which nuclear matter is located. It is assumed that the symmetries hidden at lower density on top of underlying chiral symmetry are the hidden-local symmetry and the scale symmetry, which are supposed to be manifested explicitly eventually at the critical density higher than normal nuclear density. The details are reviewed in recent papers.[5,6,12]

When BR (Brown-Rho) scaling (old-BR)[13] is applied to the neutron-star calculation using realistic NN potentials,[14] the mass is estimated to be in the range, 1.2–$1.8M_\odot$, which is apparently less than $2M_\odot$. New BR/BLPR scaling has been proposed[4,15] to incorporate the change in topology of the crystal structure of skyrmions, skyrmion $\rightarrow$ half-skyrmion.[7] Suppose the threshold density, $n_{1/2}$, is higher but not so higher than the normal nuclear density, n_0, then we expect the physical effect of such topology change on nuclear matter at the density $n \geq n_{1/2}$. More elaborated effective Lagrangian including ω-meson has been discussed by Paeng *et al.*[6] but in this work, we use the simpler version discussed by Dong *et al.*[4]

The drastic change in the symmetry energy observed at the density $n_{1/2}$[7,8] can be translated into the parameter changes of the Lagrangian, a new scaling. With new scaling, new-BR/BLPR scaling, one can understand the origin of the drastic change in symmetry energy is due to the substantial change in tensor force, disappearance of ρ tensor component at higher density. The new scaling is incorporated into the $V_{\mathrm{low}\,k}$ — implemented EFT (Effective Field Theory) approach to calculate the EoS and the mass–radius relation of a compact object of a pure neutron matter. The stiffer EoS has been obtained as expected and the mass can be as large as $2.4M_{\odot}$,[4] which seems to be consistent with recently observed high mass neutron stars. In this work, we take a more realistic approach for the compact star with electrons, protons and neutrons, which are believed to be in weak equilibrium, rather than pure neutron matter. Near the surface of star, which is supposed to be in lower density region, $n < 0.5\,n_0$, we adopt the EoS used in Hebeler *et al.*[16]

It is assumed, in the range of density we are considering, the energy density of asymmetric nuclear matter ($n_p \neq n_n$ or $n_p/n \neq 1/2$) can be described by the conventional form in terms of symmetry energy, $S(n)$, as given by

$$\epsilon_{\mathrm{nuc}}(n, x) = \epsilon_{\mathrm{nuc}}(n, x = 1/2) + n(1 - 2x)^2 S(n) \tag{1}$$

where $x \equiv n_p/n$ is the fraction of proton density. Then the symmetry energy factor $S(n)$ can be obtained by

$$S(n) = \epsilon_{\mathrm{nuc}}(n, 0)/n - \epsilon_{\mathrm{nuc}}(n, x = 1/2)/n \tag{2}$$

which is equivalent to the difference of binding energy per nucleon between the symmetric nuclear matter ($x = 1/2$) and the neutron matter ($x = 0$). The pressure of nuclear matter is given by $p_{\mathrm{nuc}} = n^2 \frac{\partial \epsilon(n)/n}{\partial n}$. In this work, we use the corresponding binding energy and the symmetry energy factor obtained by Dong *et al.*[4] with the new scaling. In the weak equilibrium, the proton fraction, x, is determined essentially by the chemical potential difference between proton and neutron,

$$\mu_n - \mu_p = 4(1 - x)S(n) \tag{3}$$

together with charge neutrality condition, $n_p = n_e + n_\mu$.

Now given EoS for energy density, ϵ, and pressure, p, the radius, R, and mass, $m(R)$, can be determined by integrating the Tolmann–Oppenheimer–Volkoff (TOV) equations.[17,18] The equations are integrated up to the radius of the star, R, where $p(R) = 0$, and the mass of the star is determined by $m(R)$. The masses and radii, which depend on the EoS, are important in predicting the gravitational waves emitted from the coalescing binary neutron stars. During the inspiral period of binary neutron stars, tidal distortions of neutron stars are expected and the resulting gravitational wave is expected to carry the corresponding information of EoS.[19]

The tidal deformability of polytropic EoS, $p = K\epsilon^{1+1/n}$, where K is a pressure constant and n is the polytropic index, were evaluated by Flanagan and Hinderer[20,21] and by others in more detail.[22,23] Recent works by Reads *et al.*[24] and Hotokezaka *et al.*[25] demonstrate the measurability of the tidal deformation to constrain the EoS. In this work, we calculate the mass–radius relation and the tidal deformability using the stiffer EoS, which has been recently proposed with new scaling law (BLPR).[4,6]

3. Compact Star Composed of n, p, e and μ

The asymmetry of neutron and proton numbers at high density, dictated by the chemical potential difference, inevitably leads to the weak equilibrium configuration with electron and muon with neutrinos escaped. It can be summarized by the relation between chemical potentials given by

$$\mu_n - \mu_p = \mu_e = \mu_\mu. \tag{4}$$

The chemical potential difference between neutron and proton should be the same as the electron chemical potential. The last equality is due to the muon emergence at higher density when the chemical potential difference from neutron and proton becomes larger than the muon mass. With the charge carriers, proton, electron and muon, the local charge neutrality condition is given by

$$n_p = n_e + n_\mu. \tag{5}$$

The carrier densities at a given density can be solved using Eqs. (3)–(5). The density dependence of proton fraction, x, is shown in Fig. 1(a). One can see that the proton fraction increases significantly as density increases.

The total energy density and pressure are given by

$$\epsilon(n, x) = \epsilon_{\rm nuc} + \epsilon_{\rm lep}, \tag{6}$$

$$p(n, x) = p_{\rm nuc} + p_{\rm lep}. \tag{7}$$

The energy density, $\epsilon_{\rm lep}$, and the pressure, $p_{\rm lep}$ are given by the degenerate Fermi gas of the leptons (electron and muon) assuming a cold compact star ($T \sim 0$).

The resulting EoS is shown as a pressure-energy density diagram in Fig. 1. The causality limit, $c_s \leq c$, constrains the highest density, n_c, beyond which the stiffer EoS used is no more valid. For EoS used in this work, it is found to be $n_c \sim 5.7 n_0$.

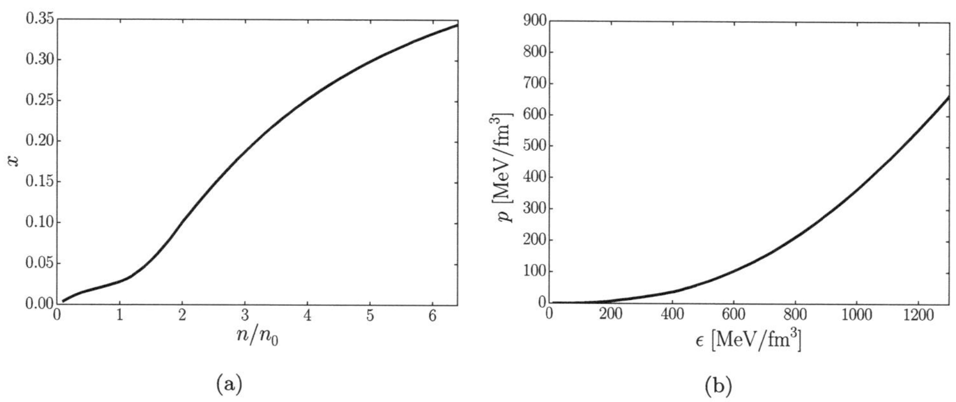

Fig. 1. (a) The proton ratio x as a function of nuclear density n. (b) The relation between energy density and pressure for an np asymmetric configuration.

For a static and spherically symmetric astrophysical compact star, the metric is given by

$$ds^2 = -e^{\Phi(r)}dt^2 + e^{\Lambda(r)}dr^2 + r^2 d\theta^2 + r^2 \sin^2\theta d\phi^2. \tag{8}$$

where Λ can be expressed in terms of a radial-dependent mass parameter, $m(r)$, introduced by

$$e^{\Lambda(r)} = \left(1 - \frac{2m}{r}\right)^{-1}. \tag{9}$$

Assuming a perfect-fluid stellar matter,

$$T_{\mu\nu} = (\epsilon + p)u_\mu u_\nu + p g_{\mu\nu}, \tag{10}$$

the relativistic hydrodynamic equilibrium is governed by the TOV equation

$$\frac{dm}{dr} = 4\pi r^2 \epsilon, \tag{11}$$

$$\frac{dp}{dr} = -(\epsilon + p)\frac{m + 4\pi r^3 p}{r(r - 2m)}, \tag{12}$$

$$\frac{d\Phi}{dr} = -\frac{1}{\epsilon + p}\frac{dp}{dr}, \tag{13}$$

where ϵ and p are energy density and pressure at r, respectively, and $u^\mu = dx^\mu/d\tau$ is the fluid's four-velocity. $m(r)$ is the mass enclosed inside the radius r. We can calculate the mass of compact star, M, and its radius, R, by integrating the TOV equation up to $p(R) = 0$ and we get the profile of $\Phi(r), m(r)$ and $p(r)$.

The EoS of np asymmetric configuration is used to solve the TOV equation resulting in the mass–radius curve shown in Fig. 2. For np asymmetric configuration, the possible maximum mass is estimated to be $M \sim 2.1 M_\odot$ with the radius $R \sim 11\,\text{km}$, where the central density is about $5.7\, n_0$. For pure neutron matter,[4] the

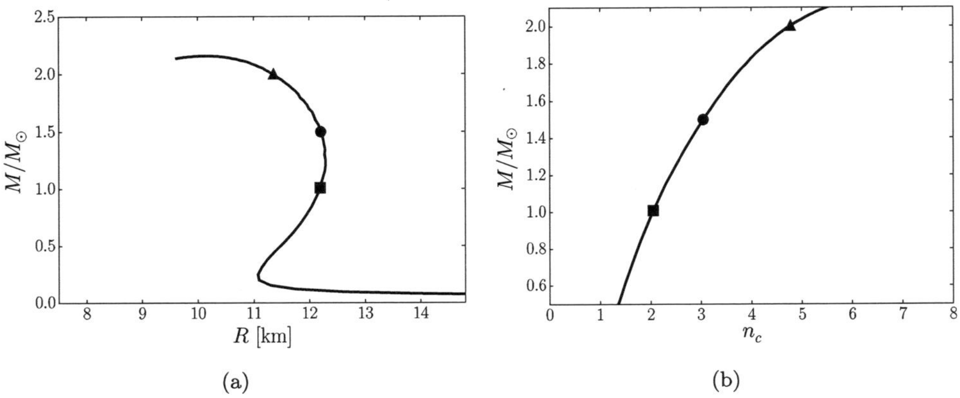

Fig. 2. (a) Mass(M)-radius (R) curve. The filled-square, filled-circle and filled-triangle correspond to $M = 1.0M_\odot$, $M = 1.5M_\odot$ and $M = 2.0M_\odot$ respectively. (b) n_c versus Mass.

possible maximum mass is approximately $M \sim 2.4M_\odot$ with the radius $R \sim 12\,\text{km}$ and $n \sim 4.7n_0$. In Fig. 2(a), the filled-square, filled-circle and filled-triangle correspond to $M = 1.0M_\odot$, $M = 1.5M_\odot$ and $M = 2.0M_\odot$, respectively. The compactness $C = \frac{M}{R}$ in the range of mass 1.0–2$M_\odot$ is found to be 0.12–0.26 and 0.14 for 1.4$M_\odot$, Fig. 2(b) shows the relation between central density n_c and the radius and between n_c and the mass of the star.

4. Tidal Deformability

When a compact star is placed in a static external field, we suppose a star in a spherically symmetric configuration is then deformed by the external field. The asymptotic expansion of the metric at large distances r from the star defines the quadrupole moment, Q_{ij}, and the external tidal field, $\mathcal{E}_{ij}$, as expansion coefficients[26] given by

$$-\frac{1+g_{00}}{2} = -\left[\frac{m}{r} + \frac{3}{2}\frac{Q_{ij}}{r^3}n^i n^j + \cdots\right] + \frac{1}{2}\mathcal{E}_{ij}r^2 n^i n^j + \cdots, \tag{14}$$

where $n^i = x^i/r$ and Q_{ij} and $\mathcal{E}_{ij}$ are both symmetric and traceless.[26]

The deformability parameter λ is defined by

$$Q_{ij} = -\lambda \mathcal{E}_{ij}, \tag{15}$$

which depends on the EoS of the nuclear matter and provides the information how easily the star is deformed. The deformability parameter can be reexpressed by the dimensionless Love number, k_2, defined by

$$\lambda = \frac{2k_2}{3}R^5. \tag{16}$$

In general, the linearized perturbations of the metric caused by an external field is given by,[27]

$$g_{\mu\nu} = g^{(0)}_{\mu\nu} + h_{\mu\nu}, \tag{17}$$

where $g^{(0)}_{\mu\nu}$ is a unperturbed metric,

$$g^{(0)}_{\mu\nu} = \text{diag}[-e^{2\Phi(r)}, e^{2\Lambda(r)}, r^2, r^2\sin^2\theta], \tag{18}$$

where $\Phi(r)$ and $\Lambda(r)$ will be determined by the stress–energy tensor configuration discussed below. $h_{\mu\nu}$ is a linearized perturbation, which carries the information of Q_{ij} and $\mathcal{E}_{ij}$ in Eq. (14). Since we will be considering the early stage of binary inspiral before the merging stage, the leading order tidal effects with even parity, $l = 2$, is dominated.[28] In the Regge–Wheeler gauge, the static- and even-parity perturbation with $l = 2$ denoted by $H(r)$ and $K(r)$ can be written in the following form[21]:

$$\begin{aligned} h_{\mu\nu} = \text{diag}[&-e^{2\Phi(r)}H(r)Y_{20}(\theta,\phi), -e^{2\Lambda(r)}H(r)Y_{20}(\theta,\phi), \\ &-r^2K(r)Y_{20}(\theta,\phi), -r^2\sin^2\theta K(r)Y_{20}(\theta,\phi)]. \end{aligned} \tag{19}$$

$K(r)$ is related to $H(r)$,

$$K'(r) = H'(r) + 2H(r)\Phi'(r), \tag{20}$$

where the prime $'$ denotes the differentiation d/dr.

On the other hand, the nonvanishing component of the perturbation of stress-energy tensor, $\delta T_{\mu\nu}$, due to the tidal deformation, corresponding to $l = 2$, $m = 0$ metric perturbation are given by

$$\delta T_0^0 = -\delta\epsilon(r)Y_{20}, \quad \delta T_i^i = \delta p(r)Y_{20}. \tag{21}$$

Using the linearized Einstein equations,

$$\delta G_{\mu\nu} = 8\pi\delta T_{\mu\nu}, \tag{22}$$

where $G_{\mu\nu}$ is Einstein tensor, we obtain the differential equation for $H(r)$:

$$H'' + \left(\frac{2}{r} + \Phi' - \Lambda'\right)H' \\ + \left\{2(\Phi'' - \Phi'^2) - \frac{6}{r^2}e^{2\Lambda} + \frac{3}{r}\Lambda' + \frac{7}{r}\Phi' - 2\Phi'\Lambda' + \frac{f}{r}(\Phi' + \Lambda')\right\}H = 0, \tag{23}$$

where we introduce $f(r)$ for $\delta T_{\mu\nu}$, given by

$$f(r) = \frac{d\epsilon}{dp}. \tag{24}$$

At the asymptotic distance from the stellar matter, $r \gg M$, where $T_{\mu\nu} = 0$, g_{00} can be approximated

$$-\frac{1 + g_{00}}{2} \to -\left[\frac{M}{r} - \frac{1}{2}H^{\text{asympt}}(r)Y_{20}\right], \tag{25}$$

where $H^{\text{asympt}}(r)$ is the solution of Eq. (23) at asymptotic distance given by

$$H^{\text{asympt}} = \frac{8}{5}\left(\frac{M}{r}\right)^3 c_1 + \left(\frac{r}{M}\right)^2 c_2 \cdots. \tag{26}$$

One can determine c_1 and c_2 using the continuity of $H(r)$ and $H'(r)$ at the boundary, $r = R$, both for interior and outside solutions of Eq. (23). Then we can read out the deformability parameter by comparing Eqs. (14) and (26)

$$\lambda = \frac{8}{15}M^5\frac{c_1}{c_2}. \tag{27}$$

Using the continuity condition, the $l = 2$ deformability λ can be written explicitly[21] in terms of the compactness $C = M/R$ and $y = RH'(R)/H(R)$,

$$\lambda_2 = \frac{16}{15}R^5C^5(1 - 2C)^2[2 + 2C(y - 1) - y]\{2C[6 - 3y + 3C(5y - 8)] \\ + 4C^3[13 - 11y + C(3y - 2) + 2C^2(1 + y)] \\ + 3(1 - 2C)^2[2 - y + 2C(y - 1)]\ln(1 - 2C)\}^{-1}. \tag{28}$$

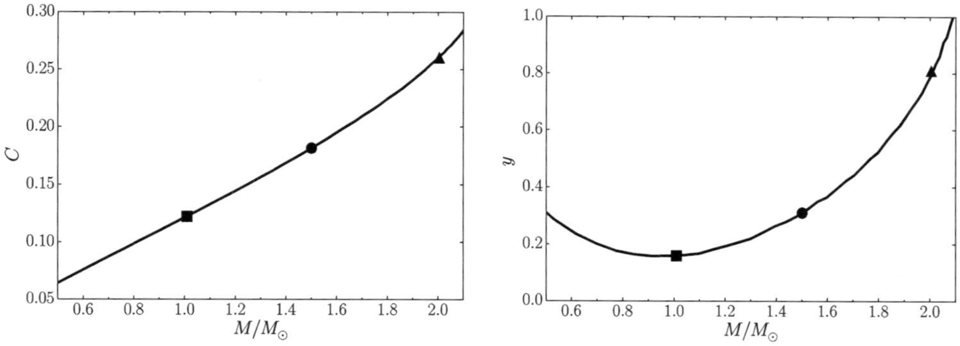

(a) (b)

Fig. 3. (a) Compactness versus mass. (b) y versus mass.

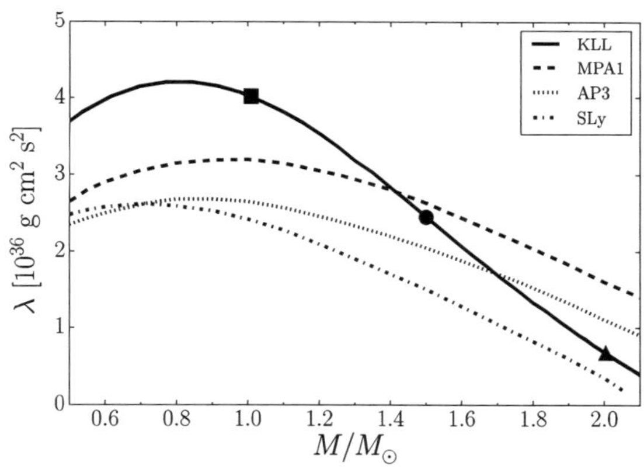

Fig. 4. The tidal deformability parameter λ in the mass range $0.5M_\odot$–$M_{\rm max}$.

By solving TOV equation and Eq. (23) together, we can then calculate y and the compactness C (see Fig. 3) for the interior solution and we obtain λ or Love number, Eq. (28), as shown in Fig. 4. The compactness $C = M/R$ in the range of mass 1.4–$2M_\odot$ is found to be 0.16–0.26 as shown in Fig. 3(a).

The deformability parameter for $1.4M_\odot$ is found to be 2.86. It can be compared with those of different EoS's with only $npe\mu$ matter. For example, the EoS's of SLy,[29] AP3[30] and MPA1[31] for the same mass give $\lambda = 1.70, 2.22$ and 2.79, respectively.[19] On the other hand, the slope of λ is found to be stiffer than those above in the mass range 1–$2M_\odot$. In the lower mass region around $\lesssim 1M_\odot$, the deformability parameter is found to be relatively higher than those of above EoS's ($\lambda < 3$), with the maximum value of 4.2 at $0.84M_\odot$. At higher mass region, the deformability parameter in this work is lower than those EoS above mentioned.

It is interesting to note that recent numerical analysis[24,25] demonstrated the measurability of tidal deformations[b] determined by the change of late inspiral wave forms for $\delta\Lambda > 400$.

5. Summary

We discussed the physical properties of stellar matter with a new stiffer EoS, which has been proposed recently using a new scaling law (new-BR/ BLPR) in medium caused by topology change at high density,[4] by extending Dong *et al.*'s work for pure neutron matter to a realistic nuclear matter of n, p, e and μ. The mass–radius and the tidal deformability were calculated.

The calculated maximum mass of compact star is found to be about $2M_\odot$ with its radius about 11 km. The radius for the mass range of 1–$2M_\odot$ is found to be 11.2–12.2 km. The calculated deformability parameter for the stiffer EoS employed in this work is in the range 4.0–0.68 $\times$ 10^{36} g cm^2 s^2.

What characterizes the approach presented in this work is the stiffening of the EoS due to topology change predicted in the description of baryonic matter with skyrmions put on crystal background to access high density. The change is implemented in the properties of the parameters of the effective Lagrangian anchored on chiral symmetry and manifests in nuclear EFT formulated in terms of RG-implemented $V_{\text{low }k}$. Given that the approach describes fairly well, the baryonic matter up to normal nuclear density, it is the changeover of skyrmions to half-skyrmions at a density $\sim(2\text{–}3)n_0$ that is distinctive of the model used. This topology change involves no change of symmetries — and hence no order parameters, therefore it does not belong to the conventional paradigm of phase transitions. But it impacts importantly on physical properties as described in various places in a way that is not present in standard nuclear physics approaches available in the literature. It is interesting to note that, as has been discussed recently,[32,33] there is another way to produce the stiffening in EoS to access the massive compact stars. It is to implement a smooth changeover from hadronic matter — more or less well-described — to strongly correlate quark matter, typically described in NJL model. By tuning the parameters of the quark model so as to produce a changeover at a density $\gtrsim 2n_0$, it has been possible to reproduce the features compatible with the properties of observed massive stars.

After the detections of gravitational waves binary black holes, the possible detection of gravitational wave signals from coalescing binary neutron stars are well expected during the next run of aLIGO and aVirgo and the detection will inform us of the detailed effect of the tidal deformation.[19,20,24,25] Recently, the tidally modified waveforms have been developed up to the high frequency of merger,[24,34] such that the deformability parameter λ, a function of the neutron-star EOS and

[b]In their analysis, the dimensionless form of deformability parameter, $\Lambda = G(c^2/GM)^5\lambda$ has been used. In this work, $\Lambda = 634$ for $M = 1.35M_\odot$.

mass, is measurable within the frequency range of the projected design sensitivity of aLIGO and aVirgo. It has been also demonstrated in Bayesian analysis that the tidal deformability can be measured to better than $\pm 1 \times 10^{36}\,\mathrm{g\,cm^2\,s^2}$ when multiple inspiral events from three detectors of aLIGO-aVirgo network[35,36] are analyzed.[37] They also show that the neutron star radius can be measured to better than $\pm 1\,\mathrm{km}$. Thus, the simultaneous measurement of mass, radius and deformability using gravitational wave detectors could present an exciting possibility to eventually pin down the highly uncertain EoS for the nuclear matter in the mass range of 1–$2M_\odot$. This would provide a probe for the state of baryonic matter at the high density that is theoretically the most uncertain. And the hope is whether one can confirm or falsify the strategies taken and assumptions made in Refs. 4 and 5 and whether the result would then help point the directions to be taken in the efforts described in Refs. 4 and 6.

Hyun Kyu Lee: "When Gerry invited me to Stony Brook in 1998 for my sabbatical year, he put me in a house just next to his. One late afternoon he came to our place with a big smile and a basket of potatoes he just dug out in his yard. He had keen interest in hearing the news of detecting gravitational waves, which was one of his favorite laboratories up in the sky. I am now missing his big smile and a basket of comments on recent detections of gravitational waves, GW150914 and GW151226."

Acknowledgments

The authors would like to thank Mannque Rho, Tom Kuo and Won-Gi Paeng for helpful discussions and acknowledge the hospitality at APCTP where a part of this work was done. The work was supported in part by WCU project of Korean Ministry of Education, Science and Technology (Grant No. R33-2008-000-10087-0) and HKL is also supported in Part by a Collaborative Research Program of the Construction and Operation for Large-Scale Science Data Center at KISTI (K-16-L01-C06-S01).

References

1. P. B. Demorest *et al.*, *Nature* **467** (2010) 1081.
2. J. Antoniadis *et al.*, *Science* **340** (2013) 1233232.
3. B. P. Abbott *et al.*, *Phys. Rev. Lett.* **116** (2016) 061102; B. P. Abbott *et al.*, *Phys. Rev. Lett.* **116** (2016) 241103.
4. H. Dong *et al.*, *Phys. Rev. C* **87** (2013) 054332.
5. H. K. Lee, W.-G. Paeng and M. Rho, *Phys. Rev. D* **92** (2015) 125033.
6. W.-G. Paeng, T. T. S. Kuo, H. K. Lee and M. Rho, *Phys. Rev. C* **93** (2016) 055203.
7. H. K. Lee, B. Y. Park and M. Rho, *Phys. Rev. C* **83** (2011) 025206.
8. H. K. Lee and M. Rho, *Int. J. Mod. Phys. E* **22** (2013) 1330005.
9. K. Kim, T. T. S. Kuo, H. K. Lee, J. Lee and M. Rho, arXiv:1412.5380.
10. See for example, the LIGO homepage, http://www.cascina.virgo.infn.it/advirgo.
11. See for example, the Virgo homepage, http://www.cascina.virgo.infn.it/advirgo.
12. M. Rho, arXiv:1604.02662.
13. G. E. Brown and M. Rho, *Phys. Rev. Lett.* **66** (1991) 2720.

14. H. Dong, T. T. S. Kuo and R. Machleidt, *Phys. Rev. C* **80** (2009) 065803.
15. H. K. Lee and M. Rho, *Eur. Phys. J. A* **50** (2014) 14.
16. K. Hebeler *et al.*, *Astrophys. J.* **773** (2013) 11.
17. R. C. Tolman, *Phys. Rev.* **55** (1939) 364.
18. J. R. Oppenheimer and G. M. Volkoff, *Phys. Rev.* **55** (1939) 374.
19. T. Hinderer *et al.*, *Phys. Rev. D* **81** (2010) 123016.
20. E. E. Flanagan and T. Hinderer, *Phys. Rev. D* **77** (2008) 021502.
21. T. Hinderer, *Astrophys. J.* **677** (2008) 1216.
22. T. Damour and A. Nagar, *Phys. Rev. D* **80** (2009) 084035.
23. T. Binnington and E. Poisson, *Phys. Rev. D* **80** (2009) 084018.
24. J. S. Read *et al.*, *Phys. Rev. D* **88** (2013) 044042.
25. K. Hotokezaka *et al.*, *Phys. Rev. D* **93** (2016) 064082.
26. K. S. Thorne, *Phys. Rev. D* **58** (1998) 124031.
27. K. S. Thorne and A. Campolattaro, *Astrophys. J.* **149** (1967) 591.
28. T. Regge and J. A. Wheeler, *Phys. Rev.* **108** (1957) 1063.
29. F. Douchin and P. Haensel, *Astron. Astrophys.* **380** (2001) 151.
30. A. Akmal, V. R. Pandharipande and D. G. Ravenhall, *Phys. Rev. C* **58** (1998) 1804.
31. H. Muther, M. Prakash and T. L. Ainsworth, *Phys. Lett. B* **199** (1987) 469.
32. K. Masuda, T. Hatsuda and T. Takatsuka, *Astrophys. J.* **764** (2013) 12, arXiv:1205.3621 [nucl-th].
33. T. Kojo, P. D. Powell, Y. Song and G. Baym, *Phys. Rev. D* **91** (2015) 045003.
34. S. Bernuzzi, A. Nagar, T. Dietrich and T. Damour, *Phys. Rev. Lett.* **114** (2015) 161103.
35. LIGO Scientific Collab. (G. M. Harry), *Class. Quantum Gravity* **27** (2010) 084006.
36. Virgo Collab. (F. Acernese *et al.*), Advanced virgo baseline design, Report VIR-027A-09 (2009).
37. B. D. Lackey and L. Wade, *Phys. Rev. D* **91** (2015) 043002.

Spinodal instabilities in baryon-rich quark matter

Che Ming Ko* and Feng Li†
Cyclotron Institute and Department of Physics and Astronomy,
Texas A&M University, College Station, TX 77843-3366, USA
**ko@comp.tamu.edu*
†fengli@comp.tamu.edu

For quark matter at finite baryon chemical potential, its density develops large fluctuations when it undergoes a first-order phase transition. Based on the Nambu–Jona–Lasinio (NJL) model, we have used the linear response theory to study the growth rate of density fluctuations and its dependence on the wavelength of unstable modes. Using the transport equation derived from the NJL model, we have also studied the time evolution of the unstable modes and the density fluctuations in a baryon-rich quark matter that is confined in a finite volume. Allowing the expansion of the quark matter, we have further studied the survivability of the density fluctuations as the density and temperature of the quark matter decrease. Possible experimental signatures of the density fluctuations have been suggested.

Keywords: NJL model; baryon-rich quark matter; spinodal instability; density fluctuations; heavy iron collisions.

1. Introduction

Studying the properties of baryon-rich quark–gluon plasma (QGP) is the main focus of the beam energy scan (BES) experiments[1–3] at the Relativistic Heavy Ion Collider (RHIC) as well as at the future Facility for Antiproton and Ion Research (FAIR). These experiments are expected to shed light on whether the phase transition from the baryon-rich QGP to the hadronic matter is a first-order one and the location of the critical end point in the QCD phase diagram if the phase transition is first-order. Gerry was once also interested in the hadronic matter to QGP transition, particularly about the role of the relevant degrees in the hadronic matter if masses of hadrons decrease with increasing temperature and density.[4,5] During his visit to Caltech in January, 1992 as a Sherman Fairchild Distinguished Scholar, he asked me to join him and take a look at this problem using the NJL model.[6,7] Although I did not find a relation between his approach and the NJL model, I did learn something about the NJL model. After about a quarter of century later, what I learnt at that time turns out to be very useful when my student Feng Li and I started to work on the spinodal instability of baryon-rich quark matter.[8] Using the NJL model and its extension, the Polyakov-Nambu-Jona-Lasinia (PNJL) model,[9]

*Corresponding author.

we have found via the linear response theory that the spinodal boundary in the temperature and density plane of the QCD phase diagram shrinks with increasing wave number of the unstable mode. In the small wave number or long wavelength limit, the spinodal boundary coincides with that determined from the isothermal spinodal instability in the thermodynamic approach. We have further found that the quark vector interaction suppresses unstable modes of all wave numbers. For the wave number dependence of the growth rate of unstable modes, it initially increases with the wave number but decreases when the wave number is large. We have further investigated how unstable modes would grow if one goes beyond the linear response or small amplitude limit by using the transport equation derived from the NJL model to study the time evolution of density fluctuations in a confined as well as in an expanding quark matter. We have found that the unstable modes have significant effects on the time evolution of higher-order density moments in the quark matter, the distribution of quark number in a subvolume of the quark matter, the quark momentum anisotropy and dilepton production rate from quark–antiquark annihilation. We have thus suggested to use these observables as signatures for a first-order phase transition of the baryon-rich quark matter produced in heavy ion collisions. In this paper, we give a brief summary of these results.

The paper is organized as follows: We first briefly review the NJL model in Sec. 2 and then discuss in Sec. 3 its application to the study of the spinodal instabilities in baryon-rich quark matter. In Sec. 4, we introduce the transport equation that is derived from the NJL model and then use it in Sec. 5 to study both the short and long time behaviors of the spinodal instability of a quark matter in a periodic box. The transport equation is further used in Sec. 6 to study how density fluctuations are affected by the expansion of the system as in heavy ion collisions. Finally, a summary is given in Sec. 7.

2. The NJL Model

The NJL Lagrangian for three quark flavors has the form[10]:

$$\begin{aligned}\mathcal{L} = {}& \bar\psi(i\not\partial - M)\psi + \frac{G}{2}\sum_{a=0}^{8}[(\bar\psi\lambda^a\psi)^2 + (\bar\psi i\gamma_5\lambda^a\psi)^2] \\ & + \sum_{a=0}^{8}\left[\frac{G_V}{2}(\bar\psi\gamma_\mu\lambda^a\psi)^2 + \frac{G_A}{2}(\bar\psi\gamma_\mu\gamma_5\lambda^a\psi)^2\right] \\ & - K[\det{}_f(\bar\psi(1+\gamma_5)\psi) + \det{}_f(\bar\psi(1-\gamma_5)\psi)], \end{aligned} \tag{1}$$

where $\psi = (\psi_u, \psi_d, \psi_s)^T$, $M = \mathrm{diag}(m_u, m_d, m_s)$ and λ^a is the Gell-Mann matrices with λ^0 being the identity matrix multiplied by $\sqrt{2/3}$. The last term is the Kobayashi–Maskawa–t'Hooft (KMT) interaction that breaks $U(1)_A$ symmetry[11] with $\det_f$ denoting the determinant in flavor space[12]:

$$\det{}_f(\bar\psi\Gamma\psi) = \sum_{i,j,k}\varepsilon_{ijk}(\bar u\Gamma q_i)(\bar d\Gamma q_j)(\bar s\Gamma q_k). \tag{2}$$

It gives rise to four-point interactions in two flavors and six-point interactions in three flavors. In the two-flavor case, the sum of scalar and pseudo-scalar interactions and the KMT interaction with $K = -G$ reduces to the original NJL model.[6,7] We note that for sufficient large G_V, such as $G_V = G_S$, the first-order phase transition induced by the attractive scalar interaction would disappear.[8,13]

3. Spinodal Instability in Baryon-rich Quark Matter

The spinodal instability in a quark matter occurs when its pressure decreases with increasing density either at constant temperature or constant entropy, i.e., $(\partial_n p)_T < 0$ (isothermal spinodal instability) or $(\partial_n p)_S < 0$ (isentropic spinodal instability). For a quark matter with temperature and density inside the spinodal region, small density fluctuations will develop into large fluctuations as a result of instability. The spinodal instability can also be studied in the linear response theory based on the mean-field approximation to the NJL model. In this case, one obtains a set of homogeneous linear algebraic equations for the density fluctuations. Fourier transforming these equations then allows us to determine the collective modes from the equation, $\det|I - \chi(k)| = 0$, where $\chi(k)$ is a matrix with elements given by the Fourier transform of quark correlators. Unstable modes, which determine the boundary of the spinodal region, correspond to collective modes with imaginary frequencies. The results for the spinodal boundaries of unstable modes of different wave numbers are shown in Fig. 1 for $G_V = 0$ (left window) and $G_V = 0.2\,G_S$ (right window). In both cases, the spinodal instability region shrinks as the wave number of unstable modes increases or its wavelength becomes shorter. This indicates

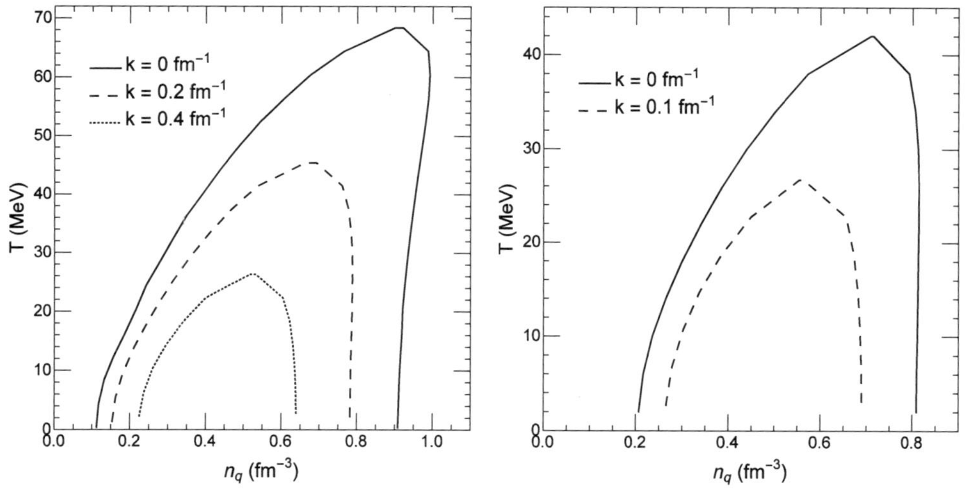

Fig. 1. Spinodal boundaries of unstable modes of different wave numbers in the temperature and net quark density plane from the NJL model with $G_V = 0$ (left window) and $G_V = 0.2\,G_S$ (right window).

that clumps of quark matter or the high density regions are more likely to merge into larger clumps, which correspond to modes of smaller wave number or longer wavelength, and this effect is larger at higher temperature. The vector interaction shrinks, however, the unstable region, particularly for unstable modes of larger wave number or shorter wavelength, as expected since a repulsive interaction drives particles away from density clumps and thus destroys the unstable modes. As shown in Ref. 8, including the Polyakov loop,[14] which takes into account the confinement–deconfinement transition of gluons, increases, on the other hand, the unstable region.

4. Transport Model Based on the NJL Model

For the long-time behavior of a quark matter or an expanding quark matter produced in a heavy ion collision, it can be studied using the Boltzmann or transport equation derived from the NJL Lagrangian for the phase-space distribution functions $f_a(X, \mathbf{p})$ of quarks and antiquarks of flavor a,[15] that is,[16]

$$\begin{aligned}\partial_{X^0} f_a(X,\mathbf{p}) + \frac{p^{i\pm}}{E_{\mathbf{p}^\pm}}\partial_{X^i} f_a(X,\mathbf{p}) - \partial_{X^i} V_a^S(X)\frac{M_a}{E_{\mathbf{p}^\pm}}\partial_{p_i} f_a(X,\mathbf{p})\\ \mp \partial_{X^i} V_0^V(X)\partial_{p_i} f_a(X,\mathbf{p}) \mp \partial_{X^i} V_j^V(X)\frac{p^{j\pm}}{E_{\mathbf{p}^\pm}}\partial_{p_i} f_a(X,\mathbf{p}) = \mathcal{C}[f_a]. \quad (3)\end{aligned}$$

In the above, $\mathbf{p}^\pm \equiv \mathbf{p} \pm \mathbf{V}^V$ is the kinetic momentum with the superscript "+" referring to quarks and "−" referring to antiquarks, V_μ^V is the vector potential, $M_a = m_{0a} - V_a^S$ is the effective quark mass with V_a^S being the scalar potential, and the collision term

$$\begin{aligned}C[f_a] \equiv \sum_{bcd} \frac{1}{1+\delta_{ab}} \int \frac{d^3\mathbf{p}_b}{(2\pi)^3 2E_b} \frac{d^3\mathbf{p}_c}{(2\pi)^3 2E_c} \frac{d^3\mathbf{p}_d}{(2\pi)^3 2E_d} \frac{(2\pi)^4}{2E_a}\\ \times \delta^4(p_a + p_b - p_c - p_d)|\mathcal{M}_{ab}|^2\\ \times [f_c f_d (1-f_a)(1-f_b) - f_a f_b (1-f_c)(1-f_d)] \quad (4)\end{aligned}$$

that describes scatterings among quarks and antiquarks. The transport equation can be solved using the test particle method[17] by expressing the distribution function in terms of the density of test particles that follow the Newton's equations of motions determined by the left-hand side of Eq. (3) and undergo scatterings according to Eq. (4).[16,18]

5. Quark Matter in a Box

We first consider a quark matter that is confined in a cubic box with periodic boundary conditions. The system is prepared by distributing many test particles inside the box according to the density of the system with their momenta given by the Fermi–Dirac distribution at certain temperature.

5.1. *Small amplitude density fluctuations*

To study the growth of density fluctuations from an initial distribution with density and temperature corresponding to that inside the spinodal region, we introduce an initial density fluctuation that has a sinusoidal oscillation in the z direction, $\rho_{\rm ini} = \rho_0(1 + 0.1\sin(2\pi z/L))$, where ρ_0 is the average initial density and L is the length of the box, corresponding to a wave number $k = 2\pi/L$. The left window of Fig. 2 shows how the amplitude of the sinusoidal wave grows with time in the case of $L = 20\,\text{fm}$, the average density $\rho_0 = 0.7\,\text{fm}^{-3}$ and an initial temperature $T = 45\,\text{MeV}$. Since the amplitude of density fluctuation at early times is expected to grow exponentially, it can be approximated by a hyperbolic cosine function of time, i.e., $\delta\rho(t) = \delta\rho_0\cosh(\Gamma_k t)$, where Γ_k is the growth rate and can be extracted directly from the numerical results. Solid circles in the right window of Fig. 2 show the growth rates of unstable modes of various wave numbers. They are seen to agree very well with those obtained from an analytical calculation based on the linearized Boltzmann equation[8] after including the finite grid size effect, shown by the dashed and solid lines for the cases with and without the collision term in the Boltzmann equation, respectively.

5.2. *Large amplitude density fluctuations*

To study how density fluctuations emerge and grow, we compare results from two calculations based on the same initial conditions but with and without the spinodal instability in the equation of state, which is achieved by introducing a vector

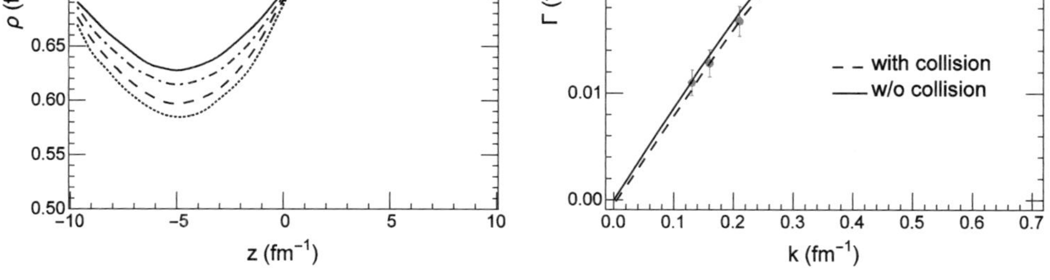

Fig. 2. Left window: Time evolution of an unstable density mode of wave number $k = 0.31\,\text{fm}^{-1}$. Right window: Growth rates extracted from numerically solving the Boltzmann equation for unstable modes of wave numbers $k = 0.63$, 0.31, 0.21, 0.16, $0.13\,\text{fm}^{-1}$ for quark matter of density $\rho = 0.7\,\text{fm}^{-3}$ and temperature $T = 45\,\text{MeV}$. Analytical results from the linearized Boltzmann equation of Ref. 8 are shown by dashed and solid lines for the cases with and without the collision term, respectively.

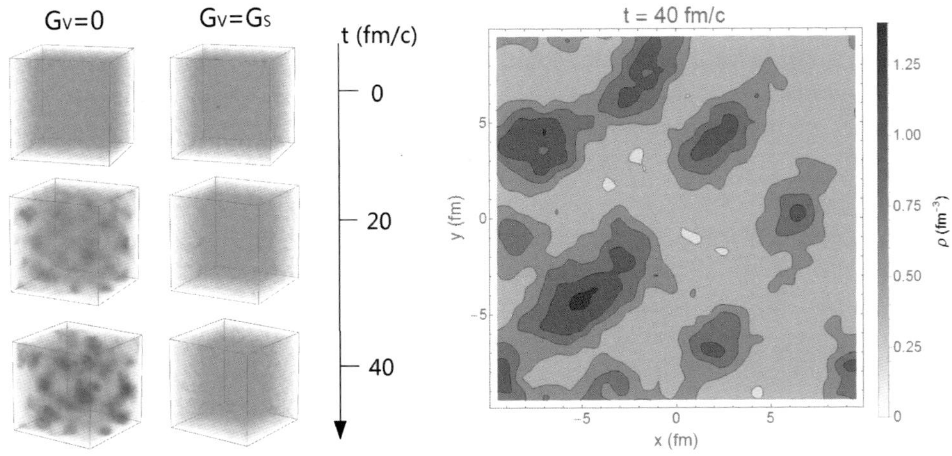

Fig. 3. Left window: Time evolution of density distribution in a quark matter of temperature $T = 20\,\text{MeV}$ and net quark density $n_q = 0.5\,\text{fm}^{-3}$ for the cases of $G_V = 0$ (left column) and $G_V = G_S$ (right column). Right window: Cross-sectional view of density distribution on the $z = 0$ plane at $t = 40\,\text{fm}/c$ for the case $G_V = 0$ with a first-order phase transition.

interaction in the NJL model. The left window of Fig. 3 shows the time evolution of the density distribution in a box of size $20 \times 20 \times 20\,\text{fm}^3$ for the two cases of $G_V = 0$ and $G_V = G_S$, with the darker color denoting the high density regions and the lighter color denoting the low density regions. Although the system is initially uniform in space, some dense spots are present due to statistical fluctuations as a result of finite number of test particles used in the calculation. In the case of $G_V = G_S$ without a first-order phase transition or spinodal instability, the density distribution in the box remains unchanged with time. This changes dramatically, however, for the case of $G_V = 0$. Due to the spinodal instability, the initial dense spots act like "seeds", which create several small low pressure regions that attract nearby partons and lead to the formation of many clusters at $t = 20\,\text{fm}/c$. These clusters further grow in size by connecting with each other and form stable large structures at $t = 40\,\text{fm}/c$, when the system clearly separates into two phases of matter with one of high density and the other of low density as shown in the right window of Fig. 3.

The density fluctuations can be quantified by the scaled density moments $\langle\rho^N\rangle/\langle\rho\rangle^N$,[19] where

$$\langle\rho^N\rangle \equiv \frac{\int d^3\mathbf{r}\rho(\mathbf{r})^{N+1}}{\int d^3\mathbf{r}\rho(\mathbf{r})}. \tag{5}$$

The scaled density moments are all equal to one for a uniform density distribution but become greater than one as the density fluctuations grow. In Fig. 4, we show by dotted, dashed and solid lines the scaled density moments for $N = 2$, 4 and

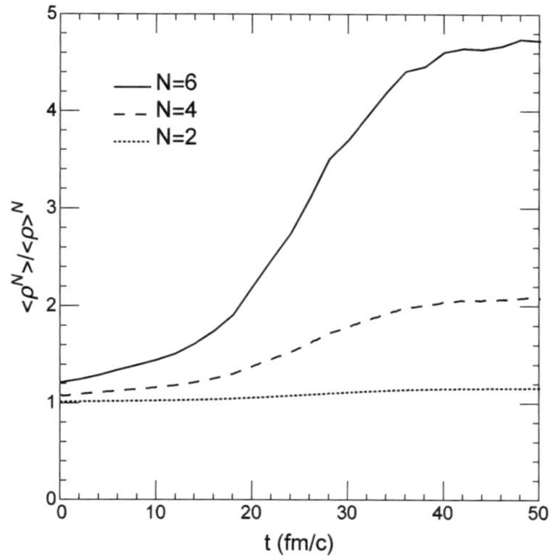

Fig. 4. Time evolution of the scaled density moments in a quark matter of temperature $T = 20\,\mathrm{MeV}$ and average net quark density $n_q = 0.5\,\mathrm{fm}^{-3}$ inside the spinodal region.

6, respectively. Our results show that the scaled density moments increase during the phase separation and reach their saturated values at about $t = 40\,\mathrm{fm}/c$, when the phase separation almost ends. Also, moments with larger N increase faster and saturate at larger values.

Other quantities of interest are the skewness and kurtosis of the particle multiplicity distribution, which have been proposed as possible signals for the critical phenomena[20] and have been studied in the BES experiments at RHIC.[1,2] They are defined as follows:

$$\text{skewness} \equiv \frac{\langle \delta N_q^3 \rangle}{\langle \delta N_q^2 \rangle^{3/2}}, \quad \text{kurtosis} \equiv \frac{\langle \delta N_q^4 \rangle}{\langle \delta N_q^2 \rangle^2} - 3. \tag{6}$$

Both quantities characterize how far an event-by-event multiplicity distribution deviates from a normal distribution. Theoretical calculations based on the grand canonical picture predict that both quantities diverge with the correlation length when a system approaches its critical point,[20] with the kurtosis diverging faster than the skewness. Therefore, they have been suggested as possible signals for the existence of a critical end point in the QCD phase diagram.

To be consistent with the grand canonical picture, we consider quarks in a subvolume of the box in our study and treat the remaining part as the reservoir. When the system is initially inside the spinodal instability region, quarks in the reservoir can sometimes move into the subvolume, but in most of the times quarks would leave from the subvolume to the reservoir. The number of quarks inside this subvolume thus varies drastically from event to event, leading to a large value for the skewness in its event-by-event distribution. In Fig. 5, we show the event-by-event distribution of the number of quarks in a subvolume from 1000 events at $t = 0$, 20 and $40\,\mathrm{fm}/c$ by the solid, dashed and dotted lines, respectively, for the

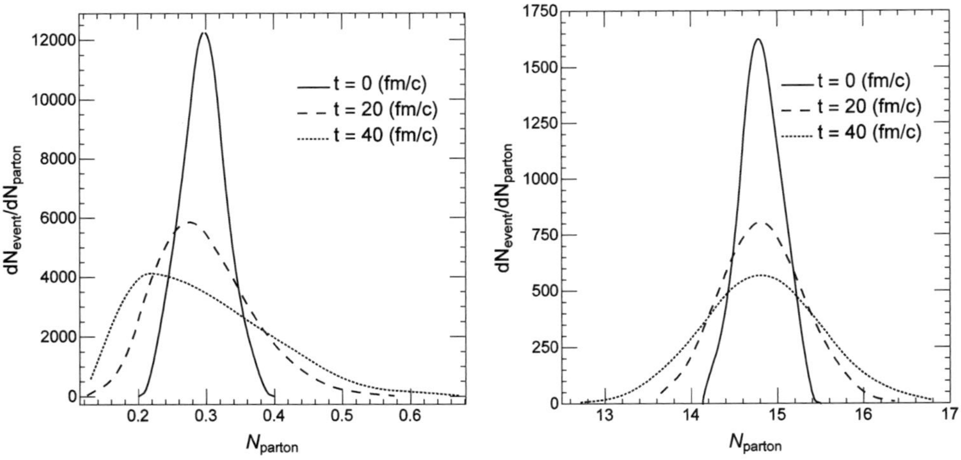

Fig. 5. Time evolution of the event-by-event distribution of the number of quarks in a subvolume of size $0.6\,\mathrm{fm}^3$ (left window) and $30\,\mathrm{fm}^3$ (right window) for a quark matter of temperature $T = 20\,\mathrm{MeV}$ and average net quark density $n_q = 0.5\,\mathrm{fm}^{-3}$ inside the spinodal region. The total number of events is 1000.

two cases of subvolume size of $0.6\,\mathrm{fm}^3$ (left window) and $30\,\mathrm{fm}^3$ (right window). The distribution in the case of small subvolume clearly becomes asymmetric as time increases, starting with an initial skewness of 0.11 and increasing to 0.60 at $20\,\mathrm{fm}/c$ and 0.75 at $40\,\mathrm{fm}/c$. This feature is absent in the case of large subvolume, where the distribution remains essentially symmetric with increasing time, with the skewness changing slowly from -0.001 ($t = 0$) to 0.086 ($t = 20\,\mathrm{fm}/c$) and 0.132 ($t = 40\,\mathrm{fm}/c$), and there is no apparent increase or decrease in the kurtosis.

6. Expanding Quark Matter

To study how large density fluctuations due to the spinodal instability as a result of a first-order phase transition obtained from the box calculation in the previous section are affected by the expansion of the system as in a heavy ion collision, we have carried out a dynamical calculation using the transport equation introduced in Sec. 4. For the initial quark and antiquark distributions, we use a blast-wave model by taking their positions to follow a spherical Wood–Saxon form:

$$\rho(r) = \frac{\rho_0}{1 + \exp((r - R)/a)} \tag{7}$$

with a radius $R = 5\,\mathrm{fm}$ and a surface thickness parameter $a = 0.5\,\mathrm{fm}$, similar to that expected from a central Au+Au collisions. The momenta of these partons are taken to be that of a Fermi–Dirac distribution at certain temperature.

We first study the time evolution of the temperature and net quark density in the central cell of volume of $\sim 43\,\mathrm{fm}^3$, which has an initial density $\rho_0 = 1.5\,\mathrm{fm}^3$ and temperature $T = 70\,\mathrm{MeV}$, and trace its trajectory in the QCD phase diagram as shown in the left window of Fig. 6 for the two cases with (solid line) and without

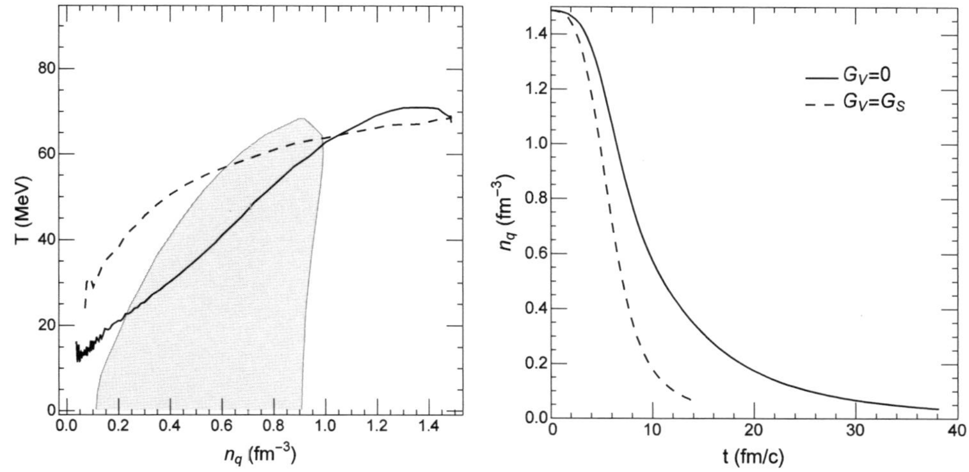

Fig. 6. Phase trajectory (left window) and time evolution (right window) of the central cell of an expanding quark matter for the two cases with (solid line) and without (dashed line) a first-order phase transition using the blast-wave initial conditions. The spinodal region is shown by the gray area.

(dashed line) a phase transition. Although the quark matter described by the transport equation may not always be in perfect thermal equilibrium, we approximate its temperature by that of an equilibrated one that has the same energy density and net quark density as in the NJL model. As expected, the quark matter with a first-order phase transition spends a longer time in the spinodal instability region, which is shown by the gray area, entering the region at about 6.5 fm/c and leaving the region at about 17.4 fm/c after spending about 10 fm/c inside this region. How the central density decreases with time is shown by the solid line in the right window of Fig. 6, which is seen to decrease slower than in the case without a first-order phase transition shown by the dashed line.

The density fluctuations can be seen from the density distribution on a plane such as the one at $z = 0$ shown in Fig. 7. The left window shows the density distribution at $t = 20\,\mathrm{fm}/c$ for the case with a first-order phase transition, while the right window shows that at $t = 10\,\mathrm{fm}/c$ for the case without a first-order phase transition, when the density of the central cell is about $0.2\,\mathrm{fm}^{-3}$ in both cases. Although density clumps appear in both cases, those in the one with a first-order phase transition are significantly larger. As in the case of quark matter in a box, we can quantify the density fluctuations by the scaled density moments.[19] They are shown in Fig. 8 by the black and red lines for the cases with and without a first-order phase transition, respectively. The dotted, dashed and solid lines are for $N = 2$, 4 and 6, respectively. In both cases, the scaled density moments first increase and then decrease with time. In the case without a first-order phase transition, this is caused by the fast increase of the surface of the quark matter and the quick deviation from its initial smooth Wood–Saxon density distribution. To the contrary,

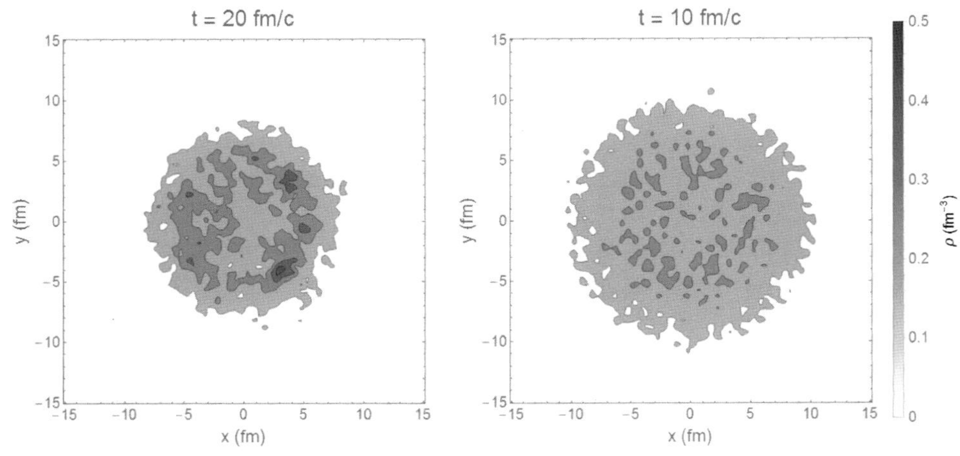

Fig. 7. Density distributions of an expanding quark matter on the $z = 0$ plane at $t = 20\,\text{fm}/c$ for the case with a first-order phase transition (left window) and at $t = 10\,\text{fm}/c$ for the case without a first-order phase transition (right window).

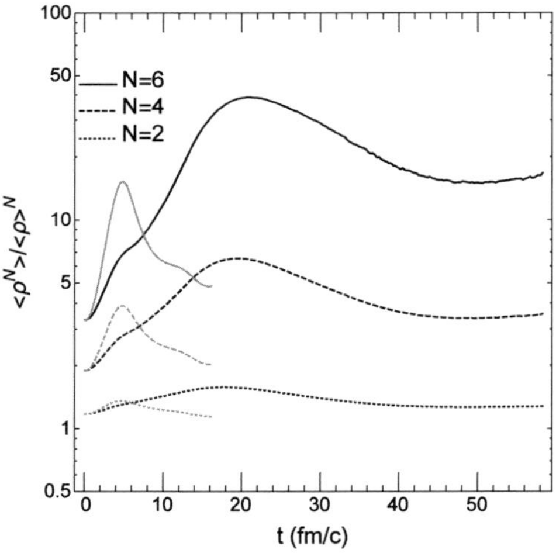

Fig. 8. (Color online) Scaled density moments as functions of time for the cases with (black lines) and without (red lines) a first-order phase transition.

the scaled density moments in the case with a first-order phase transition becomes much larger with time and only decreases slightly afterwards, reflecting the effect due to density clumps that distribute randomly inside the expanding quark matter. However, previous studies have indicated that possible signals for the enhanced scaled density moments are strongly diluted in experimental observables.[19]

Since density fluctuations can lead to spatial anisotropy even in central heavy ion collisions, it has been suggested that they may affect the anisotropic flows in the transverse plane.[21,22] The latter are defined by the coefficients v_n in the expansion of the transverse momentum distribution $f(p_T, \phi)$ as a Fourier series in the azimuthal angle ϕ,

$$f(p_T, \phi) = \frac{N(p_T)}{2\pi} \left\{ 1 + 2 \sum_{n=1}^{\infty} v_n(p_T) \cos[n(\phi - \psi_n)] \right\}, \tag{8}$$

where ψ_n is the event plane angle.[23] To calculate the anisotropic flow coefficients, we use the two-particle cumulant method,[24,25] namely, $v_n\{2\} = \sqrt{\langle \cos(n\Delta\phi) \rangle}$ by averaging over all particle pairs in an event. We have calculated $v_2\{2\}$ and $v_4\{2\}$ for 100 events of an expanding quark matter with the same blast-wave initial conditions, and their final event distributions are shown, respectively, in the left and right windows of Fig. 9 with the solid and dashed lines for the cases with and without a first-order phase transition, respectively. Both distributions peak at a larger value for the case with a first-order phase transition, particularly for v_4, thus providing a plausible signal for the first-oder phase transition. However, the values of the fluctuation induced v_2 and v_4 are much smaller than those in noncentral heavy ion collisions.

We have also studied the effect of density fluctuations on dilepton production from a quark matter. Since the dilepton production rate is proportional to the square of parton density, more dileptons are produced when the density fluctuation is large. Also, a longer partonic phase as a result of a first-order phase transition would increase the depletion yield as well. As usually done in studying dilepton production in heavy ion collisions,[26] we use the perturbative approach to calculate the dilepton yield from the quark–antiquark scattering by neglecting its effect on the dynamics of the expanding quark matter, and the results are shown in Fig. 10 by the solid and dashed lines for the cases with and without first-order phase transition, respectively. As expected, more dileptons are produced from the quark matter with a first-order

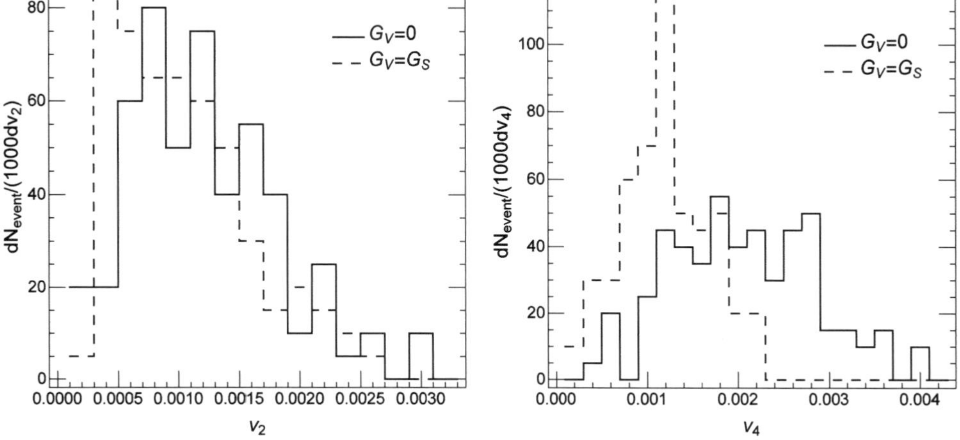

Fig. 9. Final anisotropic flow coefficients v_2 (left window) and v_4 (right window) distributions for 100 events of an expanding quark matter with the same blast-wave initial conditions.

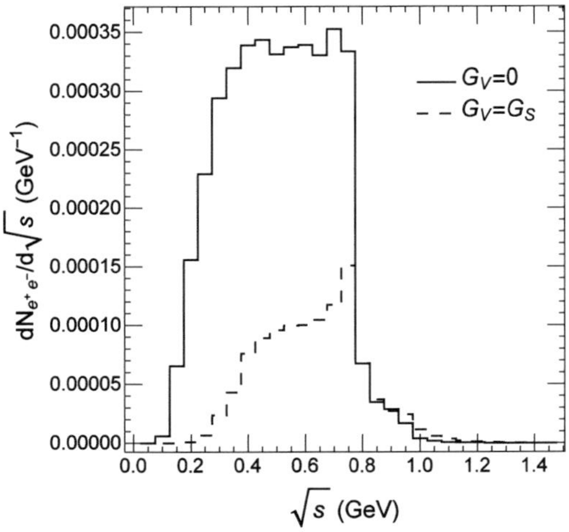

Fig. 10. Dilepton yield as a function of the invariant mass $\sqrt{s}$ for the cases with (solid line) and without (dashed line) a first-order phase transition in an expanding quark matter with the blast-wave initial conditions.

phase transition. We note the dilepton invariant mass spectrum peaks at $\sqrt{s} \approx 0.5\,\text{GeV}$ with the peak value being about $3.5 \times 10^{-4}\,\text{GeV}^{-1}$, which is comparable with the result obtained from a hadronic transport model.[27] This enhancement in dilepton production may thus be detectable in experiments.

7. Conclusions

The spinodal instability is a thermodynamic feature of a first-order phase transition in a many-body system. It occurs when its pressure in some parts decreases with increasing density. Using the linear response theory based on the NJL model in the mean-field approximation, we have determined the spinodal boundaries of unstable modes of different wavelengths in the temperature and density plane. In the long wavelength limit, the spinodal boundaries obtained in our study coincide with those determined from the isothermal spinodal instability in the thermodynamical approach. Also, the vector interaction is found to suppress the spinodal instabilities of unstable modes of all wavelengths as a result of its repulsive effect in baryon-rich quark matter. The spinodal instability can amplify the density fluctuations and lead to a phase separation in the system.

We have studied this phenomenon by solving the Boltzmann equations using the test particle method. For the case of a static box, we have found that the growth rates extracted from the early growth of a sinusoidal density fluctuation agree with the analytical results obtained from the linearized Boltzmann equation. We have also calculated the higher-order density moments of the quark matter and found them to increase and saturate at large values after phase separation. The skewness of the quark number event-by-event distribution in a small subvolume of the quark matter is also found to increase, but this feature disappears if the subvolume is large.

As for the expanding quark matter, we have used the blast-wave initial conditions and found that the expansion of the quark matter is slowed down by the presence of a first-order phase transition. Density clumps are found to appear and lead to an anisotropy in the momentum space, which can be characterized by the large scaled density moments and enhanced anisotropic flows v_2 and v_4, respectively. An enhancement in the dilepton yield is also observed. All these observables are potential probes of a first-order phase transition in the baryon-rich quark matter produced in heavy ion collisions.

Acknowledgments

This work was supported in part by the US Department of Energy under Contract No. DE-SC0015266 and the Welch Foundation under Grant No. A-1358.

References

1. T. K. Nayak, *Nucl. Phys. A.* **830** (2009) 555C–558C.
2. M. M. Aggarwal *et al.*, *Phys. Rev. Lett.* **105** (2010) 022302.
3. D. McDonald, *EPJ Web Conf.* **95** (2015) 01009.
4. G. E. Brown, H. A. Bethe and P. M. Pizzochero, *Phys. Lett. B* **263** (1991) 337–341.
5. G. E. Brown, A. D. Jackson, H. A. Bethe and P. M. Pizzochero, *Nucl. Phys. A* **560** (1993) 1035–1074.
6. Y. Nambu and G. Jona-Lasinio, *Phys. Rev.* **122** (1961) 345–358.
7. M. Asakawa and Y. Koichi, *Nucl. Phys. A* **504**(4) (1989) 668–684.
8. F. Li and C. M. Ko, *Phys. Rev. C* **93** (2016) 035205.
9. K. Fukushima, *Phys. Rev. D* **77** (2008) 114028, Erratum: *Phys. Rev. D* **78** (2008) 039902.
10. N. M. Bratovic, T. Hatsuda and W. Weise, *Phys. Lett. B* **719** (2013) 131–135.
11. G. 't Hooft, *Phys. Rev. D* **14** (1976) 3432–3450.
12. M. Buballa, *Phys. Rep.* **407** (2005) 205–376.
13. J. Xu, T. Song, C. M. Ko and F. Li, *Phys. Rev. Lett.* **112** (2014) 012301.
14. A. Polyakov, *Phys. Lett. B* **72**(4) (1978) 477–480.
15. S. P. Klevansky, A. Ogura and J. Hufner, *Ann. Phys.* **261** (1997) 37–73.
16. C. M. Ko, T. Song, F. Li, V. Greco and S. Plumari, *Nucl. Phys. A* **928** (2014) 234–246.
17. C.-Y. Wong, *Phys. Rev. C* **25** (1982) 1460–1475.
18. G. F. Bertsch and S. Das Gupta, *Phys. Rept.* **160** (1988) 189–233.
19. J. Steinheimer and J. Randrup, *Phys. Rev. C* **87** (2013) 054903.
20. M. A. Stephanov, *Phys. Rev. Lett.* **102** (2009) 032301.
21. C. Herold, M. Nahrgang, I. Mishustin and M. Bleicher, *J. Phys. Conf. Ser.* **509** (2014) 012065.
22. J. Steinheimer, J. Randrup and V. Koch, *Phys. Rev. C* **89** (2014) 034901.
23. B. H. Alver, C. Gombeaud, M. Luzum and J.-Y. Ollitrault, *Phys. Rev. C* **82** (2010) 034913.
24. S. Wang, Y. Z. Jiang, Y. M. Liu, D. Keane, D. Beavis, S. Y. Chu, S. Y. Fung, M. Vient, C. Hartnack and H. Stöcker, *Phys. Rev. C* **44** (1991) 1091–1095.
25. N. Borghini, P. M. Dinh and J.-Y. Ollitrault, *Phys. Rev. C* **64** (2001) 054901.
26. L. Xiong, Z. G. Wu, C. M. Ko and J. Q. Wu, *Nucl. Phys. A* **512** (1990) 772–786.
27. T. Galatyuk, P. M. Hohler, R. Rapp, F. Seck and J. Stroth, *Eur. Phys. J. A* **52**(5) (2016) 131.

Nucleon–anti-nucleon intruder state of Dirac equation for nucleon in deep scalar potential well

T. T. S. Kuo*,¶, T. K. Kuo†, E. Osnes‡ and S. Shu*,§

*Department of Physics and Astronomy, Stony Brook University, Stony Brook, NY 11794, USA

†Physics Department, Purdue University, West Lafayette, IN 47907, USA

‡Institute of Physics, University of Oslo, NO 0316 Oslo, Norway

§Department of Physics and Electronic Science, Hubei University, Wuhan 430062, P. R. China

¶thomas.kuo@stonybrook.edu

We solve the Dirac radial equation for a nucleon in a scalar Woods–Saxon potential well of depth V_0 and radius r_0. A sequence of values for the depth and radius are considered. For shallow potentials with $-1000\,\mathrm{MeV} \lesssim V_0 < 0$ the wave functions for the positive-energy states $\Psi_+(r)$ are dominated by their nucleon component $g(r)$. But for deeper potentials with $V_0 \lesssim -1500\,\mathrm{MeV}$ the $\Psi_+(r)$s begin to have dominant anti-nucleon component $f(r)$. In particular, a special intruder state enters with wave function $\Psi_{1/2}(r)$ and energy $E_{1/2}$. We have considered several r_0 values between 2 and 8 fm. For $V_0 \lesssim -2000\,\mathrm{MeV}$ and the above r_0 values, $\Psi_{1/2}$ is the only bound positive-energy state and has its $g(r)$ closely equal to $-f(r)$, both having a narrow wave packet shape centered around r_0. The $E_{1/2}$ of this state is practically independent of V_0 for the above V_0 range and obeys closely the relation $E_{1/2} = \frac{\hbar c}{r_0}$.

Keywords: Half-nucleon intruder state; Dirac equation for nucleon in scalar Woods–Saxon potential well.

1. Introduction

Intruder states have played an important role in nuclear structure physics. Two of us (TTSK and EO) first learnt this subject from Gerry Brown when we were his postdocs at Princeton and Nordita respectively long time ago. Gerry has taught, inspired and helped both of us a great deal; we are deeply grateful to him and remember well the pleasant time we had with him.

The nuclear shell model is a most successful model for nuclear structure. A desirable feature is its large energy gap between major shells. For example, the large energy gap between the $1s0d$ and $0p$ shells allows us to treat the nucleus ^{18}O using a small $2p0h$ model space, namely treating it simply as two $0s1d$ valence nucleons outside a closed ^{16}O core. Pioneered by Gerry, a large number of shell-model studies of nuclei using realistic nucleon–nucleon interactions have been carried out.[1–8]

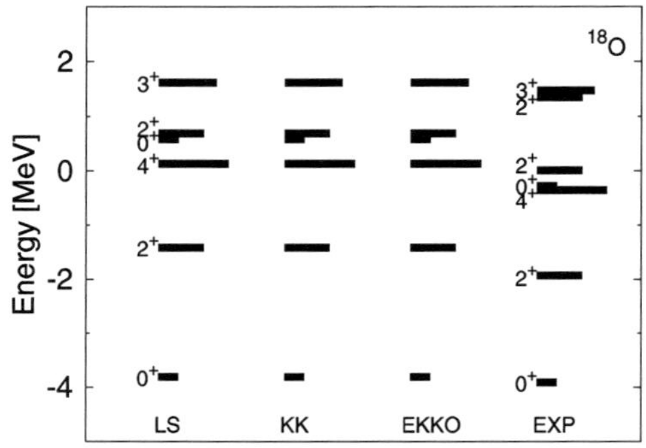

Fig. 1. Intruder states in low energy spectrum of ^{18}O.

The importance of intruder states in model-space effective interactions has been investigated by Schucan and Weidenmueller and others.[9,10]

In Fig. 1, we display the results of a typical calculation of ^{18}O.[7] In this calculation, a low-momentum interaction $V_{\mathrm{low}\,k}$ is employed, and the shell-model effective interaction has been calculated using three effective interaction methods denoted by LS, KK and EKKO[7] in the figure. As seen, calculations give only two 2^+ states while experiments have three. The missing one is an 'intruder' state whose wave function is mainly outside the $2p0h$ space which is the model space used for such calculation. The energies of the unperturbed $4p2h$ states are much higher than the $2p0h$ ones. But when the interaction is strong, some $4p2h$ states are pushed down, forming an intruder 2^+ state whose energy is as low as those states which are dominated by the $2p0h$ components. The presence of the intruder state is an indication that the Fermi sea (^{16}O-core) is no longer closed when the interaction is strong.

In the present short contribution dedicated to the memory of Gerry, we shall study the Dirac equation for a nucleon in a strong scalar potential well and discuss its possible intruder states. For a free nucleon, the negative-energy Dirac sea is closed. But when it is subject to a strong external field, it may be no longer closed and in analogy to the nuclear case described above, give rise to nucleon–anti-nucleon intruder states.

Before discussing our present work, we would like to mention briefly Gerry's style of physics; he was fond of and successful in using simple models to bring out the simple physics of complicated problems. He often said 'This is too complicated; I cannot check it with my 5-dollar calculator'. The Brown–Bolsterli schematic model for nuclear giant dipole resonance is a well known example.[11] The second-order core polarization diagram for nuclear effective interaction is another example, and has played an essential role in microscopic nuclear structure studies.[1–8] How would the properties of hadrons change with their surrounding medium is a quite complicated problem. But the Brown–Rho scaling[12,13] provides a remarkably simple answer, namely $m^*/m = 1 - Cn/n_0$ where m^* and m are respectively the in-medium and in-vacuum hadron mass. The medium density is n and the nuclear matter saturation density n_0. For low densities, $C \simeq 0.2$. Extensive applications of the Brown–Rho scaling on nuclear matter, neutron stars and the long life-time β-decay of ^{14}C have been carried out.[14–20]

In the following, we shall first describe some details for solving the Dirac radial equation with a scalar central potential V_s (Sec. 2). To mimic the extremely strong potential which may be present in highly compacted stellar objects, we shall consider the Dirac equation for a nucleon in super-deep scalar potentials in Sec. 3. The resulting nucleon–anti-nucleon intruder states will be discussed. In the final Sec. 4, we present a summary and conclusion.

In order to elucidate our calculation by a simple pedagogical example, we have repeated the calculation using a one-dimensional Woods-Saxon potential; this is included in an Appendix.

2. Dirac Radial Equation with Scalar Potential Well

We consider the Dirac equation for a nucleon

$$[c(\alpha \cdot \mathbf{p}) + \beta(mc^2 + V_s) + V_v]\psi = E\psi, \tag{1}$$

where m denotes its mass, and V_s and V_v respectively the scalar and vector potential. Its radial equation[22–24] is

$$\begin{aligned} \frac{dg(r)}{dr} &= -\frac{k}{r}g(r) + \frac{1}{\hbar c}[E - V_v(r) + V_s(r) + mc^2]f(r) \\ \frac{df(r)}{dr} &= +\frac{k}{r}f(r) - \frac{1}{\hbar c}[E - V_v(r) - V_s(r) - mc^2]g(r) \end{aligned} \tag{2}$$

with

$$k = -(l+1); \quad j = l + 1/2 = l; \quad j = l - 1/2 \tag{3}$$

and

$$\psi_{jlm} = \frac{1}{r}\begin{bmatrix} ig(r)y^j_{lm} \\ -f(r)y^j_{lm} \end{bmatrix}. \tag{4}$$

In the above, we use $mc^2 = 938\,\text{MeV}$ and $\hbar c = 197.3\,\text{MeV-fm}$. We shall consider in the present work the $l = 0$ and $j = 1/2$ state (namely $k = -1$).

To determine the bound-state eigenvalues of the above Dirac radial equations, a standard procedure is to integrate from $r = 0$ to r_{end} in two portions: (I) from $r = 0$ to r_{match} obtaining wave functions g_{I} and f_{I}, and (II) from r_{end} to r_{match} obtaining g_{II} and f_{II}, where $0 < r_{\text{match}} < r_{\text{end}}$. For an energy variable ω equal to a bound-state eigenvalue the logarithmic boundary conditions at r_{match} are satisfied, namely

$$\frac{d}{dr}\log(g_{\text{I}}(\omega, r_{\text{match}})) = \frac{d}{dr}\log(g_{\text{II}}(\omega, r_{\text{match}})), \quad \omega = E_1, E_2, \ldots, \tag{5}$$

and

$$\frac{d}{dr}\log(f_{\rm I}(\omega, r_{\rm match})) = \frac{d}{dr}\log(f_{\rm II}(\omega, r_{\rm match})), \quad \omega = E_1, E_2, \dots . \tag{6}$$

Note that the above two conditions are theoretically equivalent, but 'numerically' they are often not. High accuracy is generally needed for calculating the above quantities, and when this is not met they may give different and spurious eigenvalues.

In the present work, we employ the following alternative matching condition for determining the eigenvalues, namely

$$\frac{f_{\rm I}(\omega, r_{\rm match})}{g_{\rm I}(\omega, r_{\rm match})} = \frac{f_{\rm II}(\omega, r_{\rm match})}{g_{\rm II}(\omega, r_{\rm match})}. \tag{7}$$

This single condition is 'theoretically' equivalent to the previous two conditions, Eqs. (5) and (6). But 'numerically' we have found it being considerably more accurate and efficient. In our calculations, we shall mainly use this matching condition. We shall also use both Eqs. (5) and (6) and a visual inspection of the resulting wave functions to double check our results.

We consider a nucleon in a scalar Woods–Saxon potential well

$$V_s(r) = \frac{V_0}{1 + e^{(r-r_0)/\delta}} \tag{8}$$

with well depth V_0. We shall use a range of V_0 values from about -50 to minus few-thousand MeV. (Possible connection of our choices with dense stellar objects like black holes will be discussed in Sec. 4.) We shall also consider a range of r_0 values. For convenience, we shall use $\delta = 0.1\,\text{fm}$ for all calculations. The values for $r_{\rm match}$ and $r_{\rm end}$ are dependent on r_0. For example for $r_0 = 4\,\text{fm}$ we use $r_{\rm end} = 6\,\text{fm}$

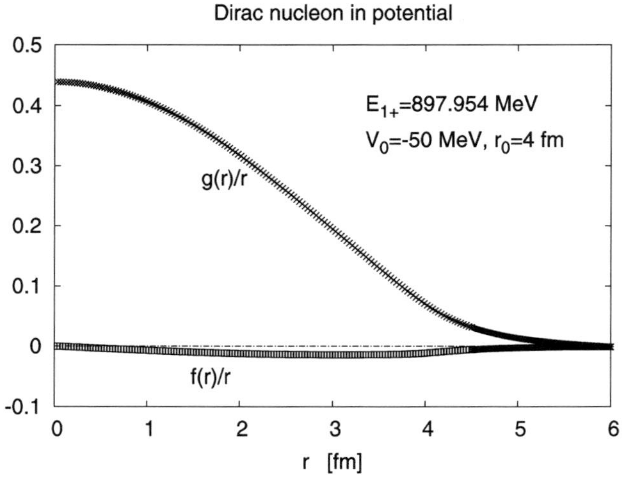

Fig. 2. Dirac wave functions for $E_{1+} = 897.952\,\text{MeV}$ and $V_0 = -50\,\text{MeV}$.

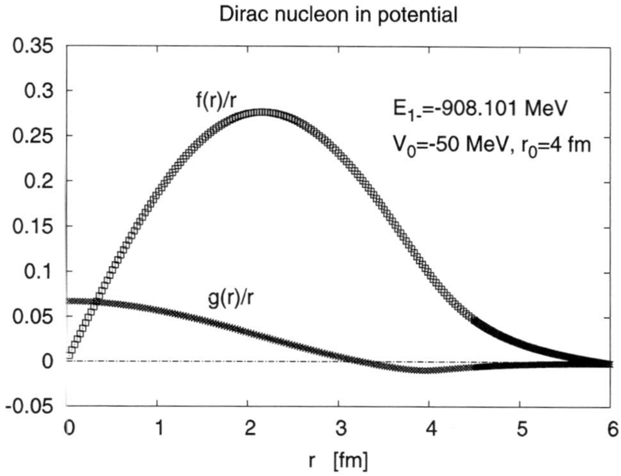

Fig. 3. Same as Fig. 2 but for $E_{1-} = -908.100\,\text{MeV}$.

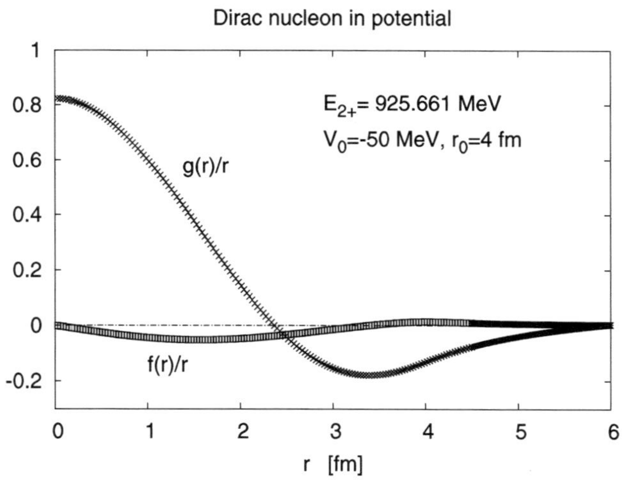

Fig. 4. Same as Fig. 2 but for $E_{2+} = 925.661\,\text{MeV}$.

and $r_{\text{match}} = 4.5\,\text{fm}$. For this case we have also used $r_{\text{match}} = 5.0\,\text{fm}$, which results in very good agreement (to fourth decimal place) with the former.

To illustrate our calculations, we first present some of our results with a $V_0 = -50\,\text{MeV}$ potential, $r_0 = 4\,\text{fm}$ and δ as given above. As seen in Fig. 2, the wave function of E_{1+} is largely dominated by its nucleon component $g(r)/r$. But the wave function for the negative energy E_{1-} is dominated by its anti-nucleon component $f(r)/r$ as seen in Fig. 3. In Fig. 4, we consider the wave functions corresponding of the E_{2+} state of the same potential. They are still dominated by $g(r)/r$ but have a slightly larger $f(r)/r$ component than E_{1+}. Note both of the wave functions in

Fig. 4 have a node. Our scheme for ordering the positive energy states is $E_{1+} < E_{2+} < E_{3+} < \cdots$ and $\cdots < E_{3-} < E_{2-} < E_{1-}$ for the negative energy ones.

These wave functions and those to be displayed later are normalized without the angular integration factor 4π, namely

$$\int (f(r)^2 + g(r)^2)dr = 1. \tag{9}$$

The above optical potential is similar to the empirical optical potentials used for a nucleon in a nucleus of mass number $A \simeq 40$.[21,22] Our results indicate that for such ordinary nuclear systems, the positive energy states of their nucleons are dominated by their nucleon component $g(r)$.

3. Nucleon–Anti-Nucleon Level Crossing and Intruder State

In Table 1, we present our results for a Dirac nucleon in the above potential well with different well depths V_0. Before discussing our results, let us mention that the matching conditions addressed earlier depends very sensitively on the energy variable ω. Thus, a very small change in ω can drastically change the wave functions f and g. As shown in Table 1, we have determined the energies E very accurately so as to satisfy the matching conditions. It is our hope that interested readers may check our numerical results.

Let us now discuss the results listed in Table 1. In Fig. 5, we plot the energies of the $1-$, $1+$ and $2+$ states versus V_0. As seen there is clearly a level-crossing behavior; the nucleon level E_{1+} descends with V_0 whereas the anti-nucleon E_{1-} level ascends. They 'cross' at a crossing potential of $V_0 \simeq -1000\,\text{MeV} \simeq -mc^2$. (In fact, they cannot actually cross for real V_0, but can do so for complex V_0.[9]) In Fig. 6, we plot the corresponding anti-nucleon fractions $\langle f \mid f \rangle$. As the well becomes deeper, the anti-nucleon fraction of the $1-$ state drops monotonically from ~ 1 at shallow well to ~ 0 at deep well of $V_0 \simeq -2000\,\text{MeV}$. In other words, this anti-nucleon is

Table 1. Energies and anti-nucleon fraction $\langle f \mid f \rangle$ of a nucleon in scalar Woods–Saxon potentials of depth V_0. The widths of these potentials are all $r_0 = 4\,\text{fm}$. The subscripts $1+$ and $1-$ refer to respectively the lowest positive- and highest negative-energy state. V_0 and E are both in units of MeV.

V_0	E_{1-}	$\langle f \mid f \rangle_{1-}$	E_{1+}	$\langle f \mid f \rangle_{1+}$
−50	−908.101	0.9908	897.954	0.004645
−100	−861.325	0.9883	849.499	0.005851
−300	−671.036	0.9780	654.283	0.01098
−500	−485.047	0.9570	461.309	0.02180
−700	−313.559	0.8594	275.994	0.05498
−900	−194.461	0.6487	120.856	0.2011
−1000	−191.321	0.3894	76.8122	0.3537
−1100	−236.603	0.1854	59.8212	0.4574
−1300	−397.210	0.05180	52.5618	0.4917
−1500	−583.955	0.02299	50.7829	0.4960
−1700	−777.233	0.01515	49.9197	0.4973

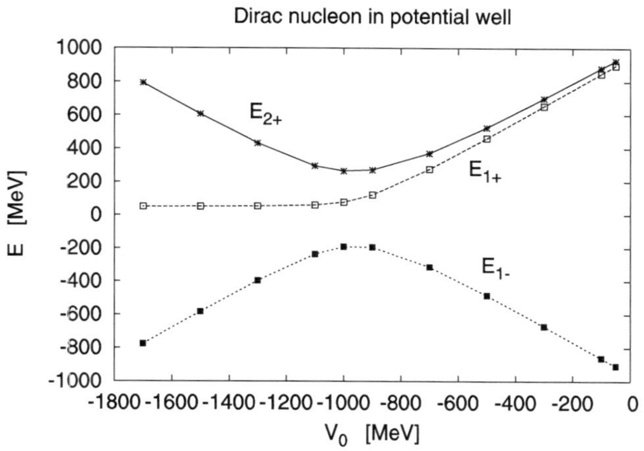

Fig. 5. Energy levels of nucleon in potential wells of different well depths V_0.

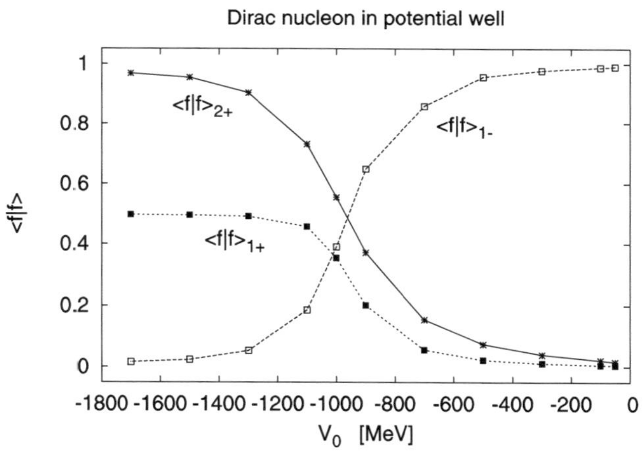

Fig. 6. Anti-nucleon fractions of nucleon in potential wells of different well depths V_0.

'transmuted' into a nucleon in the process. We note that the crossing potentials for E and $\langle f \mid f \rangle$ are approximately the same.

We now consider the evolution of the energy and anti-nucleon fraction of the 1+ state in Figs. 5 and 6. As the well becomes deeper, its energy drops till about $V_0 \simeq -1000\,\text{MeV}$. And until this strength is reached, this anti-nucleon fraction keeps on rising. But afterwards a plateau of energy $\simeq 50\,\text{MeV}$ and $\langle f \mid f \rangle \simeq 0.50$ is reached. It is of interest that this nucleon is approaching some sort of 'hybrid nucleon' composed of half nucleon and half anti-nucleon. It may be noted that, as indicated by Fig. 5, this hybrid state is the only bound state for $V_0 <\sim -2000\,\text{MeV}$, while all other states are diverging outward having energies $|E| > mc^2$. (By bound state we mean a state of energy being $|E| < mc^2$.) This hybrid state may be referred to as a half-nucleon intruder state.

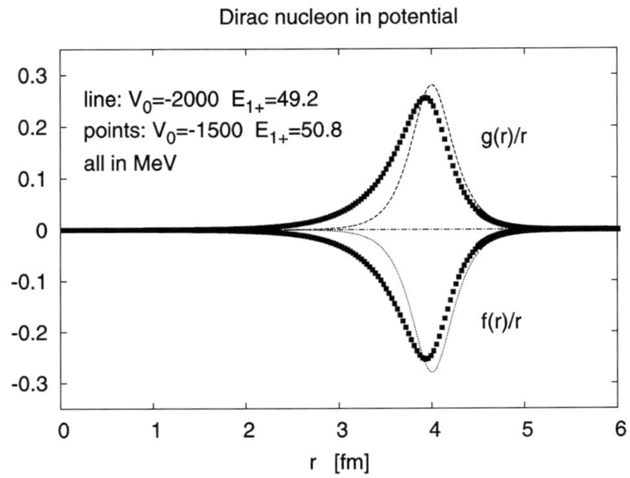

Fig. 7. Wave functions of a Dirac nucleon in a super-deep potential well of depth $V_0 = -1500\,\text{MeV}$ with energy $E_{1+} = 50.7829\,\text{MeV}$. Those of the corresponding state for $V_0 = -2000\,\text{MeV}$ are also shown.

In Fig. 7, we display the wave function of such a hybrid nucleon belonging to the $V_0 = -1500\,\text{MeV}$ potential. It is amusing that the radial distributions of its nucleon and anti-nucleon components are almost identical, except for a sign change in the wave function. Furthermore, this hybrid nucleon has the mass largely concentrated near its surface, making it a hollow nucleon. In this figure, the corresponding wave function for $V_0 = -2000\,\text{MeV}$ is also shown. It has a similar Gaussian wave packet shape, but of narrower width.

The wave function of the $1-$ anti-nucleon belonging to the same $-1500\,\text{MeV}$ potential is displayed in Fig. 8. As seen, it has almost no anti-nucleon content; it is a 'nucleon' evolved from its parent anti-nucleon. The wave function of the $2+$ state of this potential is shown in Fig. 9. Its $f(r)$ component is much larger than the $g(r)$ one, indicating this positive energy nucleon is actually predominantly composed of anti-nucleon.

Let us now discuss the results from potentials deeper than those shown in Table 1. In Fig. 10, we focus on the $1+$ state calculated with different well depths. Plotted are the anti-nucleon fraction $\langle f \mid f \rangle$ and the effective mass defined as $m^*/m \equiv E_{1+}/(mc^2)$ where m is the free nucleon mass. Near zero potential, we have $m^*/m \simeq 1$ and $\langle f \mid f \rangle \simeq 0$. As V_0 becomes more negative, the former descends and the latter ascends until $V_0 \simeq -1500\,\text{MeV}$. Afterwards, its $m * /m$ remains flat at about 0.05 (corresponding to $E_{1+} \simeq 50\,\text{MeV}$) and its $\langle f \mid f \rangle$ is flat at about 0.50. In the figure, we show results down to $V_0 \simeq -3500\,\text{MeV}$. We have tried deeper potentials, in fact as deep as $V_0 = -5400\,\text{MeV}$, and the flatness of these two curves remains unchanged. It seems that the above E_{1+} and $\langle f \mid f \rangle_{1+}$ are practically independent of V_0 for $V_0 <\sim -1500\,\text{MeV}$. The wave function and energy of this 'flat' state will be referred to as $\Phi_{1/2}$ and $E_{1/2}$, respectively.

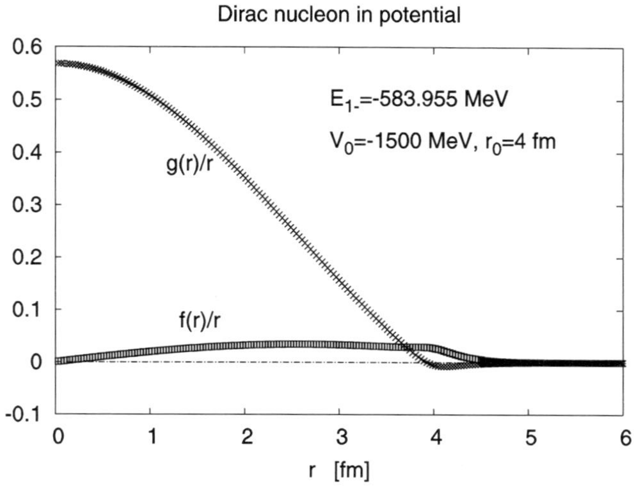

Fig. 8. Same as Fig. 7 but for energy $E_{1-} = -583.955\,\text{MeV}$.

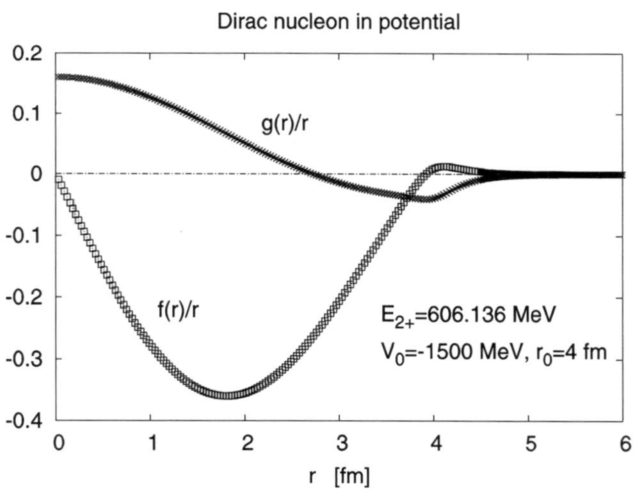

Fig. 9. Same as Fig. 7 but for energy $E_{2+} = 606.136\,\text{MeV}$.

Our results presented above are all obtained numerically. It would be nice if they can be obtained analytically. We have tried but so far have not succeeded. To partially understand the above 'flatness' behavior of Figs. 5, 6 and 10, let us rewrite Eq. (2) in matrix form as

$$\begin{bmatrix} -(V_s + mc^2)/\hbar c & \dfrac{d}{dr} + \dfrac{k}{r} \\ -\dfrac{d}{dr} + \dfrac{k}{r} & (V_s + mc^2)/\hbar c \end{bmatrix} \begin{bmatrix} f \\ g \end{bmatrix} = E \begin{bmatrix} f \\ g \end{bmatrix}. \tag{10}$$

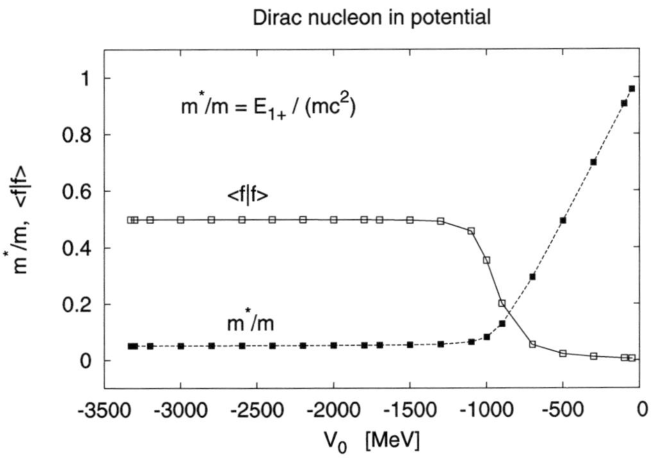

Fig. 10. Evolution of the E_{1+} energies and their anti-nucleon fractions with potential depth V_0. A potential radius of $r_0 = 4$ fm is used for all calculations shown.

Here, we do not have the V_v term of Eq. (2) as we include only the scalar potential V_s.

The diagonal terms $-(V_s + mc^2)$ and $(V_s + mc^2)$ play an important role in the above equation. When $V_s \simeq 0$, the wave functions of the positive- and negative-energy states are dominated by their nucleon and anti-nucleon component $g(r)$ and $f(r)$, respectively. But when V_s is largely negative, say $-2000\,\text{MeV}$, $\langle f|-(V_s+mc^2)|f\rangle$ becomes positive while $\langle g|(V_s+mc^2)|g\rangle$ becomes negative. Then the positive-energy state is dominated by the anti-nucleon component $f(r)$ while the negative-energy one by $g(r)$. This explains the level crossing seen in Fig. 5 for the 1− to 2+ right-to-left cross over and similarly the 2+ to 1− cross over.

The middle 1+ state of Fig. 5 is, however, a special case. Its energy E_{1+} for $V_0 <\sim -1500\,\text{MeV}$ is not only nearly independent of V_0 but is also rather small ($\sim 50\,\text{MeV}$). These properties suggest that there must be some cancellation among the contributions from V_s to this state, leading to a near-zero net contribution. A possible way to attain such cancellation is to have either $f = -g$ or $f = g$ where f and g are respectively the anti-nucleon and nucleon component of the eigen function of this 1+ state.

To study this possible cancellation, we rewrite Eq. (10) as

$$\left[-(V_s + mc^2)/(\hbar c) + \frac{d}{dr}\right](f+g) - \frac{k}{r}(f-g) = E(f-g) \tag{11}$$

and

$$\left[-(V_s + mc^2)/(\hbar c) - \frac{d}{dr}\right](f-g) + \frac{k}{r}(f+g) = E(f+g). \tag{12}$$

We have used the above equations to calculate the wave functions $(f+g)$ and $(f-g)$. These calculations can also be performed equivalently using Eq. (10). We have found for $V_0 \lesssim -1400\,\text{MeV}$ ($1.5\,mc^2 = 1407\,\text{MeV}$), the wave function $(f+g)$ is very much smaller than $(f-g)$ for the 1+ state. We have calculated for this state the norm $N_+ \equiv \langle f+g\,|\,f+g\rangle$ and $N_- \equiv \langle f-g\,|\,f-g\rangle$ with the normalization $N_+ + N_- = 1$. Our results are $(N_+, N_-) = (0.00004, 0.99996)$ for $V_0 = -1400\,\text{MeV}$ and $(0.00002, 0.99998)$ for $V_0 = -1500\,\text{MeV}$. In both calculations $r_0 = 4\,\text{fm}$ is used.

Thus for the 1+ state with $V_0 \lesssim -1500\,\text{MeV}$, we can drop the $(f+g)$ terms relative to the $(f-g)$ ones in the above equations. Then Eq. (11) gives the simple energy relation

$$E = -k\hbar c \frac{\langle g|1/r|g\rangle}{\langle g\,|\,g\rangle}, \tag{13}$$

for the energy of the 1+ state. Recall that $k = -1$ (see Sec. 2). To calculate the above energy, we need to have $\langle g|1/r|g\rangle$.

If $\langle g|1/r|g\rangle/\langle g\,|\,g\rangle$ is independent of V_0, then so is the above energy. In fact, we have found that for $V_0 \lesssim -1500\,\text{MeV}$, $(\langle g|1/r|g\rangle/\langle g\,|\,g\rangle)$ is closely equal to $1/r_0$ and is nearly independent of V_0. Thus the above energy is only weakly dependent on V_0 as indicated by Fig. 10 and Table 1. For the above V_0 range, g is generally of a Gaussian wave packet shape narrowly peaked around r_0. In Fig. 7, we have displayed such wave functions for $r_0 = 4\,\text{fm}$. We have found that the wave functions for other choices of r_0 are also of this shape. In Fig. 11, we display two such wave functions for $r_0 = 2$ and $7\,\text{fm}$. This special shape of the wave functions renders $(\langle g|1/r|g\rangle/\langle g|g\rangle) \simeq 1/r_0$ a good approximation. With this approximation, Eq. (13) becomes then

$$E \simeq \frac{-k\hbar c}{r_0}. \tag{14}$$

The accuracy of this approximation will be discussed later.

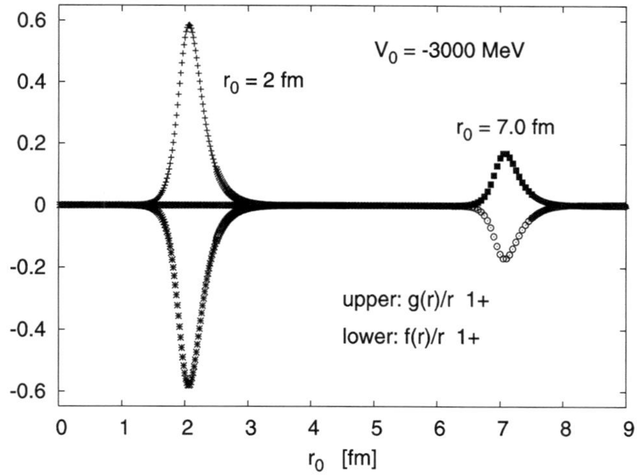

Fig. 11. Radial wave functions $g(r)/r$ and $f(r)/r$ for the 1+ state of potential wells $V_0 = -3000\,\text{MeV}$ and $r_0 = 2$ and $7\,\text{fm}$.

We have considered Eq. (11) for $f \simeq -g$. Let us now consider Eq. (12). In this situation, we can drop the $(f+g)$ terms there and have

$$\frac{dg}{dr} = -g(V_s + mc^2)/(\hbar c), \tag{15}$$

which is a simple differential equation for determining the wave function g. For the Woods–Saxon potential of Eq. (8), the above equation gives

$$g(r) = \exp\left[-\frac{V_0}{\hbar c}[(r-r_0) - \delta \log(1 + e^{(r-r_0)/\delta})] - \frac{mc^2}{\hbar c} r\right]. \tag{16}$$

Note that the above expression does not include the normalization constant. For $V_0 \lesssim -1500\,\text{MeV}$, the wave function calculated with the above equation agrees very well with the corresponding result from solving the full Dirac equation of Eq. (10).

The following approximation is helpful in understanding the structure of the $g(r)$ wave function. For r close to r_0, the Woods–Saxon potential of Eq. (8) can be approximated by a linear potential, namely

$$V_s(r) \to \frac{V_0}{2}\left[1 - \frac{r - r_0}{2\mu}\right]. \tag{17}$$

Then Eq. (15) gives

$$g(r) \propto \exp\left[\frac{V_0}{8\mu}(r - r_0)^2\right], \tag{18}$$

which is of Gaussian form, providing an explanation for the shape of the wave functions displayed in Figs. 7 and 11.

The parameter μ of Eq. (17) is determined by fitting the WS potential near r_0. As an example, we find $\mu = 0.144$ by fitting the $r_0 = 4\,\text{fm}$ and $V_0 = -2000\,\text{MeV}$ potential in the range of $r = 3.7$ to $4.3\,\text{fm}$. Then the wave function given by Eq. (18) is of the form $\exp[b(r-r_0)^2]$ with $b = V_0/(\hbar c 8\mu) = -8.80$. The same wave function obtained by solving the Dirac equations of Eq. (10) has been given in Fig. 7. Fitting it by $\alpha \cdot \exp[\beta(r - \gamma r_0)^2]$ gives $\alpha = 1.07$, $\beta = -8.37$ and $\gamma = 1.004$. Note that the β value is close to the above b value.

To estimate the accuracy of the energy approximation of Eq. (14), we have compared the energies so obtained with those from the full calculation of Eq. (10). Some sample comparisons are given in Fig. 12 where the energies for the 1+ state calculated from Eq. (10) using $V_0 = -3000\,\text{MeV}$ and $r_0 = 2$ to $7\,\text{fm}$ are compared with those given by Eq. (14). It is of interest that the two sets of energies agree remarkably well, indicating this approximation being highly accurate. Note that Eq. (14) is independent of V_0. This is consistent with Fig. 10 where the energies of the 1+ state are nearly independent of V_0 for $V_0 \lesssim -1500\,\text{MeV}$.

For particles of zero rest mass, the momentum-position uncertainty principle may be written as $\Delta E \Delta r \simeq \hbar c$; $E = pc$. It is of interest that the energy relation of this special state as given by Eq. (14) is similar to this uncertainty principle, suggesting that the nucleon in this special state has a near-vanishing effective mass as indicated by Fig. 10.

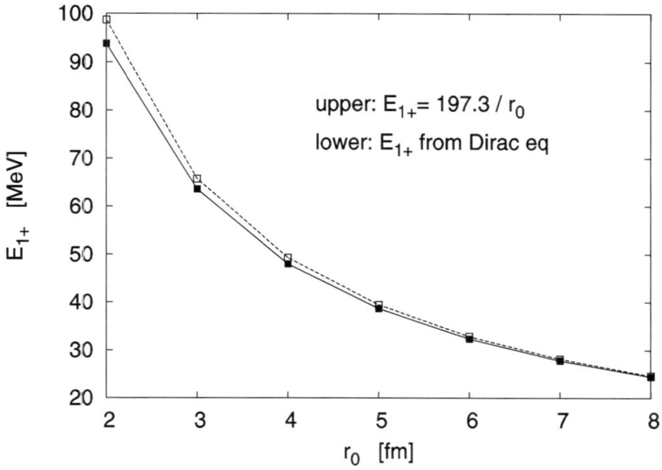

Fig. 12. Dependence of the 1+ energy on the range r_0 of the WS potential well. Common well depth of $V_0 = -3000\,\text{MeV}$ is used.

4. Summary and Discussion

In the present work, we have solved the Dirac radial equation for a nucleon in a scalar Woods–Saxon potential well with depth V_0 and radius r_0. For weak potentials such as $V_0 \simeq -50\,\text{MeV}$, the positive-energy states Ψ_+ of this equation have energies $\simeq +mc^2$ and wave functions dominated by the nucleon component $g(r)$, as indicated by Figs. 2 and 6. Similarly the negative-energy states Ψ_- have energies close to $-mc^2$ and wave functions dominated by the anti-nucleon component $f(r)$ as illustrated in Fig. 3.

As V_0 turns more negative (deeper), a qualitative reversal of the wave functions takes place. Namely, Ψ_- is generally becoming increasingly dominated by $g(r)$, and so is Ψ_+ by $f(r)$. Examples illustrating such tendencies are displayed in Figs. 8 and 9.

The lowest positive-energy state 1_+ is however an exemption, being a special intruder state. As indicated by Figs. 5 and 6, at weak potential strength this state has $E_{1+} \simeq mc^2$ and $\langle f \,|\, \Psi_{1+}\rangle^2 \simeq 0$. But as V_0 becomes $\lesssim -1500\,\text{MeV}$, this state is characterized by a special wave function $\Psi_{1/2}$ with constant composition of $f = -g$ where g and f are respectively its nucleon and anti-nucleon component and both are narrowly peaked around r_0. In addition, the energy of this state approaches a constant energy $E_{1/2} \simeq \hbar c/r_0$. These behaviors have been illustrated in Figs. 5, 6 and 10. For $V_0 \lesssim -2000\,\text{MeV}$, our results suggest that the above state, with wave function $\Psi_{1/2}$ and energy $E_{1/2}$, is the only bound state. It should be of interest if one could check this special intruder state experimentally. We note that a hypothetical scalar Woods–Saxon potential of very strong strength has been used in our work for calculating this and other intruder states. Is such a strong potential conceivable?

We have estimated that a nucleon near the surface of a medium-mass black hole (of mass $\sim 10^3$ solar mass and radius $\sim 10^3$ km[25,26]) would experience a gravitational potential of approximately -1400 MeV ($\cong 1.5\,mc^2$). There has been evidence for super-massive black holes such as the 4.3×10^6 solar-mass black hole in the Milky Way.[25,26] They may produce similar or stronger gravitational potential. Thus, the special nucleons indicated by this work may be present near such black holes.[a] This needs of course further study. With the presence of such strong gravitational field we may need to use a Dirac equation in curved space-time.[28,29] whereas we have been using the Dirac equation in flat space-time. It would indeed be of interest to obtain an estimate of the effects introduced by curved space-time.

Let us end by quoting a paragraph written by Gerry[6]:

"One of the authors, Gerry Brown, arrived at Princeton in early September, 1964. The next morning, as he came to the Palmer Physics Laboratory, Eugene Wigner, who just preceded him, opened the door for him. (It was a real contest to get ahead of Eugene and open the door for him which very few succeeded in doing.) Eugene asked Gerry, as we went into the building, what he planned to work on. 'I plan to work out the nucleon–nucleon interaction in nuclei.' Eugene said that it would take someone cleverer than him, to which Gerry replied that they probably disagreed what it meant to 'work out'. Gerry wanted to achieve a working knowledge, sufficiently good to be able to work out problems in nuclear physics."

In fact, it is our hope that the present calculations may serve as a warm-up exercise for 'working out' the Dirac equation for nucleons experiencing extremely strong scalar potential fields and its possible connection with dense stellar objects.

Acknowledgment

We thank Edward Shuryak, Ismail Zahed and Jeremy Holt for many helpful discussions. This work is supported by the US Department of Energy under contract DE-FG02-88ER4038. S. Shu is supported by the China Scholarship Council.

Appendix

In this Appendix the nucleon–anti-nucleon intruder states mentioned earlier are studied using a simpler and more transparent model. As indicated by Eq. (8), our earlier calculations are based on a model where a nucleon is placed in a spherical Woods–Saxon (WS) potential. When the depth of this potential is deeper than a certain value, the ground state of the nucleon becomes a special half-nucleon intruder state. One might ask if this holds true also for a one-dimensional Woods–Saxon potential.

[a]We have also made a similar estimate for a neutron star of mass $2.1 M_{\odot}$[27] and radius ~ 11 km.[20] The resulting potential is about -270 MeV. Thus, we do not expect the presence of such special nucleons near this neutron star, as indicated by Fig. 10.

To answer this question, we have repeated our earlier calculations using a one-dimensional WS potential

$$V_s(x) = V_0 \left[1 + \exp\left(\frac{xS(x) - x_0}{\delta}\right)\right]^{-1}; \quad S(x) = x/|x|. \tag{A.1}$$

The corresponding Dirac equation is

$$\begin{bmatrix} A & B \\ -B & -A \end{bmatrix} \begin{bmatrix} f \\ g \end{bmatrix} = E \begin{bmatrix} f \\ g \end{bmatrix} \tag{A.2}$$

with

$$A = -(V_s(x) + mc^2)/(\hbar c), \quad B = \frac{d}{dx}. \tag{A.3}$$

The solutions of this equation are paired, namely

$$E_{n+} = -E_{n-}; \quad (f_{n+}, g_{n+}) = \pm(g_{n-}, f_{n-}) \tag{A.4}$$

for all n. Equation (A.2) can be rewritten as

$$\begin{aligned} (A-B)(A+B)(f+g) &= E^2(f+g) \\ (A+B)(A-B)(f-g) &= E^2(f-g). \end{aligned} \tag{A.5}$$

The energies of these equations have been calculated as illustrated in Fig. A.1. (A common value of $\delta = 0.1$ fm is used in all calculations.) We note that the energies are paired, in agreement with Eq. (A.4).

As also shown in Fig. A.1, there is a clear level crossing behavior near $V_0 \simeq -1000\,\text{MeV}$. For potentials shallower than this value, the wave functions for the E_{n+} and E_{n-} states are dominated respectively by g and f. For potentials deeper, the energies of the E_{1+} and E_{1-} states are special, both being practically zero

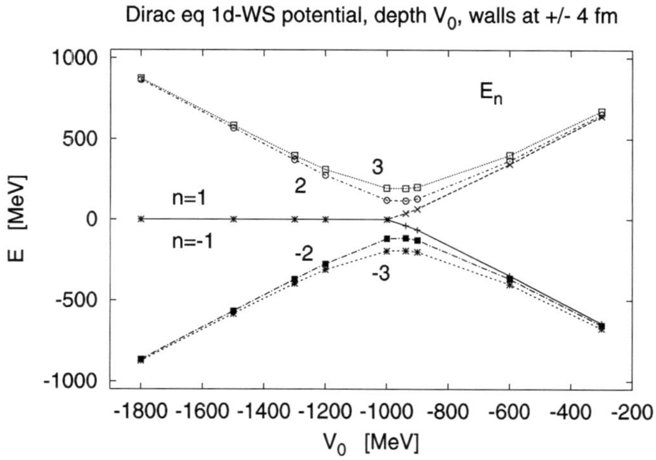

Fig. A.1. Energies of Dirac equation for a nucleon in a one-dimensional scalar WS potential. A similar plot for the spherical WS potential used earlier is shown in Fig. 3.

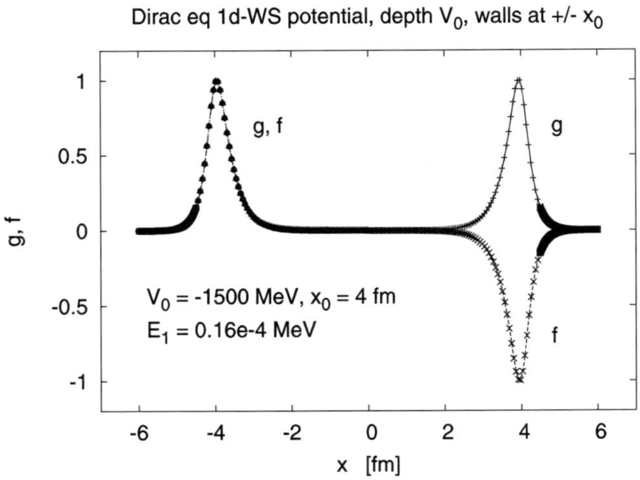

Fig. A.2. Wave function of a one-dimensional half-nucleon intruder state. Similar wave functions for the spherical WS potential used earlier are shown in Figs. 7 and 11.

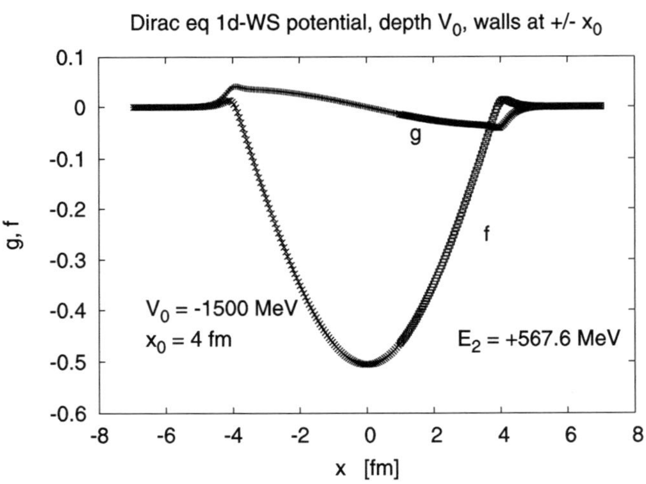

Fig. A.3. Wave function of a one-dimensional positive-energy state with f dominance. Similar wave functions for the spherical WS potential used earlier are shown in Fig. 9.

and independent of the potential depth, and their wave functions all have a special structure of either $f = -g$ or $f = g$ as illustrated in Fig. A.2. Also for this potential range, the wave functions for the $(E_{n+}, n > 1)$ states are dominated by f as illustrated in Fig. A.3. Similarly those for the $(E_{n-}, n > 1)$ states are dominated by g.

According to Eq. (A.2), the wave function (f, g) as a whole has a 'mixed parity' structure, in the sense that each of f and g is a function of definite parity but their parities are opposite. This is seen in Fig. A.2 where g has positive parity while f has negative parity. Similar parity structure is also seen in Fig. A.3. We note that

Eq. (A.5) provides a convenient way for obtaining approximate wave functions near the walls and near the center of the potential. This has been helpful in explaining the shapes of the wave functions of Eq. (A.2).

In short, the main features of the nucleon–anti-nucleon intruder states obtained from the spherical WS potential are well reproduced by the results from the one-dimensional WS potential.

References

1. G. E. Brown and T. T. S. Kuo, *Nucl. Phys.* **92** (1967) 481.
2. T. T. S. Kuo and G. E. Brown, *Nucl. Phys.* **85** (1966) 40.
3. G. E. Brown, *Unified Theory of Nuclear Models and Forces* (North-Holland, Amsterdam, 1971).
4. M. Hjorth-Jensen, T. T. S. Kuo and E. Osnes, *Phys. Rep.* **261** (1995) 126.
5. L. Coraggio, A. Covello, A. Gargano, N. Itaco and T. T. S. Kuo, *Prog. Part. Nucl. Phys.* **62** (2009) 135.
6. G. E. Brown, T. T. S. Kuo, J. W. Holt and S. Lee, *Nucleon–Nucleon Interactions and Nuclear Many-Body Problems* (World Scientific Publication, Singapore, 2010).
7. H. Dong, T. T. S. Kuo and J. W. Holt, *Nucl. Phys.* **930** (2014) 1.
8. T. T. S. Kuo, J. W. Holt and E. Osnes, *Phys. Scr.* **91** (2016) 033009.
9. T. H. Schucan and H. A. Weidenmueller, *Ann. Phys. (N.Y.)* **73** (1972) 103; **76** (1973) 483.
10. P. J. Ellis and E. Osnes, *Rev. Mod. Phys.* **49** (1977) 777; C. M. Vincent and S. Pittel, *Phys. Lett.* **47B** (1973) 327; P. A. Schaefer, *Ann. Phys.* **87** (1974) 375.
11. G. E. Brown and M. Bolsteri, *Phys. Rev. Lett.* **3** (1959) 377.
12. G. E. Brown and M. Rho, *Phys. Rev. Lett.* **66** (1991) 2720.
13. G. E. Brown and M. Rho, *Phys. Rep.* **396** (2004) 1; **253** (1991) 252.
14. J. W. Holt, G. E. Brown, J. D. Holt and T. T. S. Kuo, *Nucl. Phys. A* **785** (2007) 322.
15. J. W. Holt, G. E. Brown, T. T. S. Kuo, J. D. Holt and R. Machleidt, *Phys. Rev. Lett.* **100** (2008) 062501.
16. L. W. Siu, J. W. Holt, T. T. S. Kuo and G. E. Brown, *Phys. Rev. C* **79** (2009) 0540004.
17. H. Dong, T. T. S. Kuo and R. Machleidt, *Phys. Rev. C* **80** (2009) 065803.
18. H. Dong, T. T. S. Kuo and R. Machleidt, *Phys. Rev. C* **83** (2011) 054002.
19. H. Dong, T. T. S. Kuo, H. K. Lee, R. Macleidt and M. Rho, *Phys. Rev. C* **87** (2013) 054332.
20. W. G. Paeng, T. T. S. Kuo, H. K. Lee and M. Rho, arXiv:1508.05210; *Phys. Rev. C*, to be published (2016).
21. P. E. Hodgson, *The Nucleon Optical Potential* (World Scientific Publication, Singapore, 1994).
22. Y. Nogami and F. M. Toyama, *Phys. Rev. C* **42** (1990) 2449.
23. R. H. Landau, *Quantum Mechanics II* (John Wiley & Sons, New York, 1996).
24. R. R. Silbar and T. Goldman, *Eur. J. Phys.* **32** (2011) 212.
25. Black Hole-Wikipedia, https://en.wikipedia.org/wiki/Black_hole.
26. B. Schutz, *General Relativity* (Cambridge University Press, Cambridge, 2009).
27. J. Antoniadis *et al.*, *Science* **340** (2013) 6131.
28. B. Mukhopadhyay and S. Chakrabarti, *Nucl. Phys. B* **852** (2000) 629.
29. S. M. Carroll, *Spacetime and Geometry* (Addison and Wesley, US, 2004).

Neutron stars are gold mines

James M. Lattimer
Department of Physics and Astronomy,
Stony Brook University,
Stony Brook, NY 11794-3800, USA
James.Lattimer@StonyBrook.edu

Neutron stars are not only mines for clues to dense matter physics but may also be the auspicious sources of half of all nuclei heavier than $A = 60$ in the universe, including the auric isotopes. Although the cold dense matter above the nuclear saturation density cannot be directly explored in the laboratory, gilded constraints on the properties of matter from 1 to 10 times higher density can now be panned from neutron star observations. We show how upcoming observations, such as gravitational wave from mergers, precision timing of pulsars, neutrinos from neutron star birth and X-rays from bursts and thermal emissions, will provide the bullion from which further advances can be smelted.

Keywords: Neutron stars; dense matter; symmetry energy.

1. Introduction

A favorite discipline of Gerry's was the study of neutron stars: their birth in supernovae, their cooling by neutrino emissions, the possibility of meson condensates in their interiors and their collapse into black holes. Gerry was fond of spurring research by spinning provocative theories, such as the notion that the neutron star maximum mass is as small as $1.5M_\odot$, that "dropping masses" lead to kaon condensation at relatively low densities, and that black hole–neutron star binaries are far more abundant than neutron star–neutron star binaries. Gerry was frequently wrong in such assertions, but was also often proved correct, and there is no doubt that these provocations triggered important research. In the years since Gerry's last publications on these topics, a lot of progress has occurred. Gerry's thesis about the neutron star maximum mass evaporated with the discovery of higher, precisely measured, masses. On the other hand, Gerry's assertion that the equation of state (EOS) just above the saturation density ($n_s \simeq 0.16\,\mathrm{fm}^{-3}$, $\varepsilon_s \simeq 150\,\mathrm{MeV\,fm}^{-3}$) is relatively soft seems to be indicated by new theoretical studies of neutron matter as well as by laboratory measurements of neutron skin thicknesses, giant resonances and nuclear masses. (Technically, Gerry seems to be right that intermediate-density neutron star matter is relatively soft, but it now seems that symmetric matter is stiffer than Gerry preferred.) And with the recent discovery of binary black hole mergers by LIGO, we are on the verge of finding out if black hole–neutron star binaries are as abundant as Gerry believed.

Perhaps the most important discovery concerning neutron stars made in the last decade is that they can be relatively massive. The largest precision measurement is that of $2.01 \pm 0.04 M_\odot$[1] for the pulsar J0348 + 0432. This definitively rules out extensive softening beyond about $3n_s$. Indeed, single-handedly, it practically rules out the existence of pure quark stars and severely limits the possibilities for quark-hybrid stars.[2] We demonstrate below that, coupled with recent results from theoretical neutron matter calculations, it strongly suggests that neutron star radii are greater than about 11 km. The possibility of even larger masses is a definite possibility.[3]

We will also show that neutron matter studies, as well as nuclear experiments, imply that neutron stars in the physically relevant range 1–2 $M_\odot$ can be no larger than about 13 km. Therefore, even without astrophysical observations pertinent to radius determinations, the typical neutron star radius can now be confined to the range $11\,\text{km} < R < 13\,\text{km}$.

Upcoming observations of neutron stars from gravitational wave observations of neutron stars in mergers will yield information concerning the intermediate-density EOS and the neutron star radius through tidal deformations[4] immediately preceding their coalescence. Due to systematic uncertainties, the usefulness of this information, at first, will be rather limited. However, with an expanding catalog of detections, the limits to neutron star radii will become more precise with time. In the same time frame, it is also expected that continued observations of the pulsar binary[5] J0737 − 3039, a relativistic system containing two pulsars which already exhibits precession due to spin-orbit coupling, will yield a measurement of the moment of inertia of the more rapidly rotating pulsar (A) accurate to about 10%. This should translate into a radius measurement with an uncertainty of order 5%.

Observations of bursting and thermally cooling neutron stars will continue. Additional data from new probes hopefully will help to reduce existing systematic uncertainties in modeling. Neutron Star Internal Composition ExploreR[6] (NICER), beginning in February 2017, has a design goal of a measurement of a neutron star mass and radius to about 5%. We summarize how X-ray observations of these objects are able to simultaneously constrain masses and radii, and discuss some of the systematic uncertainties. We also show how the interpretation of these observations can be significantly strengthened by imposing reasonable priors involving causality, and general relativity, and even our limited knowledge of dense matter physics.

2. Neutron Star Structure

Assuming the validity of general relativity, the equations of stellar structure[7] are

$$\frac{dp}{dr} = -\frac{G}{c^2}\frac{(m + 4\pi p r^3/c^2)(\varepsilon + p)}{r(r - 2Gm/c^2)}, \qquad \frac{dm}{dr} = 4\pi\frac{\varepsilon}{c^2}, \tag{1}$$

where r is the radius and m is the gravitational mass interior to r. The only relevant aspects of the EOS are the pressure p and the total mass–energy density ε. Beginning from central ($r = 0$) values of the pressure (p_c) and energy density (ε_c), these equations are integrated to the surface R where the pressure vanishes. For normal matter, ε also vanishes at the surface.

These equations have multiple analytic solutions,[8] but we will make use only of the incompressible fluid case in which ε is uniform and equal to the constant ε_c. In this case, the solution becomes[7]

$$\frac{p}{\varepsilon_c} = \frac{\sqrt{1-2\beta(r/R)^2}-\sqrt{1-2\beta}}{3\sqrt{1-2\beta}-\sqrt{1-2\beta(r/R)^2}}, \qquad m = \frac{4\pi}{3}\frac{\varepsilon_c}{c^2}r^3, \tag{2}$$

where $\beta = 2GM/(Rc^2)$ is the compactness parameter, and M and R are the total mass and radius, respectively. Finite pressures require that $\beta \leq 4/9$ which sets its absolute upper limit. Imposition of the limit $p_c \leq \varepsilon_c$, required by causality, reduces this limit to $\beta \leq 3/8$. And, if as is widely believed to be likely, $p \leq \varepsilon/3$, the limit is further reduced to $\beta \leq 5/18$. In other words, a $1.4M_\odot$ neutron star would have a radius larger than 7.5 km.

2.1. *Maximally compact configurations*

However, the incompressible fluid automatically violates causality. The causal limit to the compactness can be determined, using the "minimal period", or maximal compactness conjecture of Koranda, Stergioulas and Friedman[9] in which $p = 0$ below ε_0 and $p = \varepsilon - \varepsilon_0$ above ε_0. In this case, the EOS has but a single parameter, ε_0, and solutions therefore scale. The maximum mass occurs when[10,11]

$$\begin{aligned} &\varepsilon_{c,\max} = 3.034\varepsilon_0, \quad p_{c,\max} = 2.0234\varepsilon_0, \quad n_{c,\max} = 2.251\varepsilon_0/\mu_0, \\ &M_{\max} = 4.09\sqrt{\varepsilon_s/\varepsilon_0}M_\odot, \quad R_{\max} = 17.07\sqrt{\varepsilon_s/\varepsilon_0}\ \text{km}, \quad \beta_{\max} = 0.354, \end{aligned} \tag{3}$$

where values at the star's center are indicated with the subscript c, and $\mu_0 \simeq 930.8\,\text{MeV}$ is the energy per baryon of iron nuclei (applicable to normal neutron stars with hadronic crusts). Using the largest precisely measured mass to represent a minimum value for $M_{\max}$, upper limits to thermodynamic quantities can be found:

$$\begin{aligned} &\varepsilon_{c,\max} \leq 50.8(M_\odot/M_{\max})^2\varepsilon_s, \quad p_{c,\max} \leq 34.0(M_\odot/M_{\max})^2\varepsilon_s, \\ &n_{c,\max} \leq 37.9(M_\odot/M_{\max})^2 n_s. \end{aligned} \tag{4}$$

PSR J0348 + 0432, with its mass of $2.01M_\odot$, therefore implies that

$$\varepsilon_0 \leq 4.14\varepsilon_s, \quad \varepsilon_{c,\max} \leq 12.6\varepsilon_s, \quad p_{c,\max} \leq 8.42\varepsilon_s, \quad n_{c,\max} \leq 9.39n_s. \tag{5}$$

Although these values apply to the largest densities and pressure in PSR J0348 + 0432, they must apply to all other known neutron stars.

The maximally compact EOS also implies minimum radii for any neutron star mass, not just the minimum radius corresponding to the maximum mass configuration. The maximally compact solution yields a mass–radius curve that closely resembles a self-bound configuration such as a pure quark matter star: in the bulk of the interior for low-mass configurations, the energy density is nearly constant so

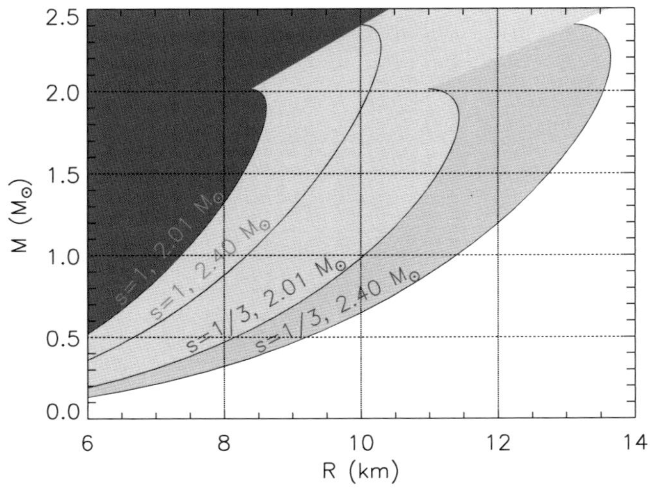

Fig. 1. Mass–radius diagram showing regions excluded by causality and the minimally compact conjecture. Boundaries are indicated for two values of $M_{\rm max}$ $(2.01M_\odot, 2.4M_\odot)$ and for two values of sound speed squared s (1, 1/3). Adapted from Ref. 12.

$M \propto R^3$. For masses approaching the maximum mass, the mass–radius trajectory develops greater curvature and ultimately its slope dM/dR becomes infinite and then negative. The maximum mass is reached when $dM/dR = 0$, as illustrated in Fig. 1. In this figure, the maximally compact M–R trajectory is scaled to achieve the observed minimum value of the maximum mass, $2.01M_\odot$; another trajectory utilizing $2.4M_\odot$ is shown for comparison. Regions with smaller radii than these trajectories are excluded by causality. It is apparent that the minimum compactness parameters for configurations with less than the maximum mass are greater than the value at the maximum, which is always $\beta_{\rm max} = 0.354$. For the case shown with $M_{\rm max} = 2.01M_\odot$, a $1.4M_\odot$ star must have $R \geq 8.15\,{\rm km}$, or $\beta \leq 0.254$.[a]

The figure also shows results when the speed of sound is limited instead by $c_s^2 = \partial p/\partial\varepsilon = s = 1/3$. In this case $\beta_{\rm max} = 0.271$ and $\varepsilon_{c,{\rm max}}/\varepsilon_0 = 4.826$. When $M_{\rm max} = 2.01M_\odot$, a $1.4M_\odot$ star must have $R \geq 11\,{\rm km}$, or $\beta \leq 0.188$. As we find below, nuclear physics information suggests that $s = 1/3$ provides a more realistic bound than $s = 1$ when applied to neutron star interiors.

2.2. *The structure of realistic stars*

In contrast to the M–R curves from the maximally compact conjecture are those predicted by "realistic" equations of state. Some examples are illustrated in Fig. 2. It is apparent that several of the illustrated equations of state are now ruled out by PSR J0438 and other neutron star mass measurements. Those that survive have the interesting, and apparently necessary, characteristic that the M–R trajectory

[a]In the limit that $M \to 0$, the minimum value of β also approaches 0.

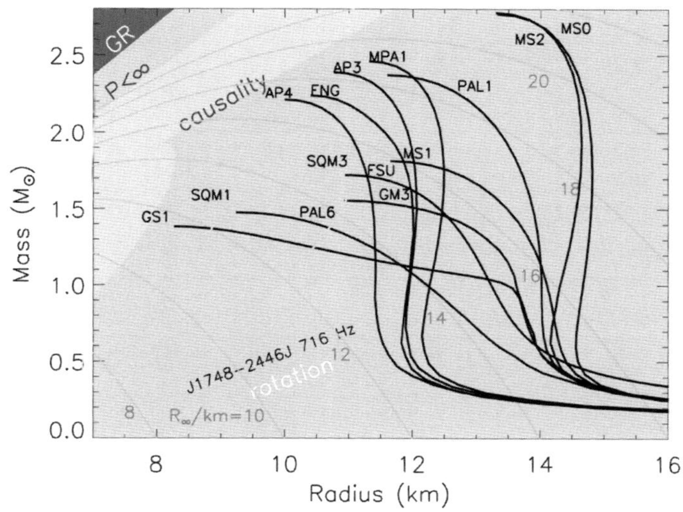

Fig. 2. (Color online) Mass–radius curves for a variety of popular EOSs (see Ref. 8 for descriptions). The green shaded region in the upper left is causally excluded; the green shaded region in the lower-right is excluded by the most rapidly spinning pulsar. Black curves are hadronic EOSs; green curves are for strange quark matter configurations. Lines of fixed $R_\infty = R/\sqrt{1-2\beta}$ are indicated as orange curves.

has a substantial segment with a relatively constant radius. Furthermore, the value of this radius is set by relatively low-density physics and not by the properties of matter near the central regions of neutron stars. A dimensionless analysis of the nonrelativistic structure equations (1) with a polytropic EOS (i.e., $p = K\rho^\gamma$, with K and γ constants) shows that

$$M \propto K^{1/(3\gamma-4)} R^{(\gamma-2)/(3\gamma-4)}. \tag{6}$$

Realistic equations of state have $\gamma \approx 2$ in the intermediate-density range $\rho_* = 1 - 3\rho_s$,[8] implying that $R \propto \sqrt{K}$ and is independent of mass as observed in Fig. 2. The value of $K \approx p_*/\rho_*^2$, where p_* is the pressure at ρ_*, suggesting that $R \propto \sqrt{p_*}$. Reference 8 demonstrated the existence of a phenomenological correlation between R and p_* involving a smaller power ($R \propto p_*^{1/4}$) for $\rho_s \lesssim \rho_* \lesssim 2\rho_s$, and that the smaller power is due to relativistic effects. This result is important, because it demonstrates a direct connection between the properties of neutron star matter in the vicinity of nuclear saturation density and the structure of neutron stars.

2.3. *Neutron star pressure and nuclear symmetry energy*

The pressure of degenerate neutron star matter near the nuclear saturation density n_s is, in large part, determined by the symmetry properties of the EOS. It is convenient to define the nuclear symmetry energy S as the energy difference between pure neutron matter and symmetric matter at a given density, $S(n) = E(n,0) - E(n,1/2)$, where $E(n,x)$ is the baryon energy per nucleon and x is the proton fraction. Because symmetric matter represents the optimal proton fraction at zero pressure, near $x = 1/2$, one can expand the energy of nuclear matter as

$$E(n,x) = E(n,1/2) + E_{\rm sym}(n)(1-2x)^2 + \cdots . \tag{7}$$

It has been found that the higher-order terms in Eq. (7) are not only small, but effectively cancel out even for values of x far from $1/2$ and near 0. In this case, $S(n) \simeq E_{\rm sym}(n)$. Introducing the incompressibility ($K_s \simeq 230\,\text{MeV}$) and the skewness ($K'_s \sim -2000\,\text{MeV}$) parameters, the nucleonic energy can be expanded about its value at n_s and $x = 1/2$, where x is the proton fraction:

$$E(n,x) = -B + \frac{K_s}{18}\left(\frac{n}{n_s}-1\right)^2 + \frac{K'_s}{162}\left(\frac{n}{n_s}-1\right)^3 + E_{\rm sym}(n)(1-2x)^2 \ldots . \tag{8}$$

The total energy includes leptonic contributions (mainly from electrons) as that from muons is small in the vicinity of n_s:

$$E_e = (3/4)\hbar c x (3\pi^2 n x^4)^{1/3}. \tag{9}$$

Because catalyzed matter in neutron stars is in beta equilibrium, i.e., $\mu_e = \mu_n - \mu_p = -\partial E/\partial x$, the equilibrium proton fraction at n_s is

$$x_s \simeq (3\pi^2 n_s)^{-1}(4S_v/\hbar c)^3 \simeq 0.04. \tag{10}$$

This small value of x_s enables the pressure at n_s to be expressed as

$$\begin{aligned} p(n_s,x_s) &= \left(n^2 \frac{\partial[E(n,x)+E_e(n,x)]}{\partial n}\right)_{n_s,x_s} \\ &= n_s(1-2x_s)\left[\frac{L}{3}(1-2x_s) + S_v x_s\right] \simeq \frac{L n_s}{3}, \end{aligned} \tag{11}$$

where we introduced the symmetry parameters $S_v \equiv E_{\rm sym}(n_s)$ and $L \equiv (3n\partial E_{\rm sym}/\partial n)_{n_s}$. The pressure at n_s thus depends primarily upon L and not S_v, because the terms containing x_s in Eq. (11) are relatively small. The values of K_s and K'_s are inconsequential. The equilibrium pressure at moderately larger densities is similarly insensitive to K_s and K'_s, but will depend on higher-order terms in an expansion of $E_{\rm sym}(n)$ about n_s. Evaluating the pressure for $n = 1.5n_s$, for example,

$$\frac{p(1.5n_s)}{n_s} = \frac{K_s}{8} + \frac{K'_s}{96} + \frac{3L}{4} + \frac{K_{\rm sym}}{8} + x\left(\frac{3S_v}{2} - \frac{9L}{2} - \frac{7K_{\rm sym}}{16}\right) + \cdots . \tag{12}$$

Here, the symmetry parameter $K_{\rm sym} \equiv (9n^2[\partial^2 E_{\rm sym}/\partial n^2])_{n_s}$. Note that the contributions from K_s and K'_s largely offset each other leading to the result that the pressure at intermediate densities ($1-3n_s$) is still dominated by the terms involving density derivatives of the symmetry energy.

3. Constraints on the Intermediate Equation of State from Nuclear Physics

3.1. *Nuclear experiments*

Information concerning the symmetry energy near the nuclear saturation density comes from a variety of nuclear observables, such as masses, giant dipole resonances and neutron skin thicknesses. It is very illustrative to consider the nuclear droplet model, which can surprisingly accurately describe the energetics. In this model the energy of a nucleus, neglecting shell and pairing energies, can be shown to be[13]

$$E(N,Z) = -BA + E_s A^{2/3} + \frac{3}{5}\frac{Z^2e^2}{R} + \frac{S_v(N-Z)^2}{A(1+S_sA^{-1/3}/S_v)} + \frac{Z(N-Z)e^2}{20R(S_vA^{1/3}/S_s+1)}, \tag{13}$$

where the terms represent, respectively, the bulk binding energy, the symmetric nucleus surface energy, the Coulomb energy, the symmetry energy and corrections to the Coulomb and symmetry energies due to polarization. $R \simeq (3A/(4\pi n_s))^{1/3} = r_o A^{1/3}$ is the nuclear radius, and S_s is the surface symmetry energy parameter. This model predicts[12] that

$$\frac{S_s}{S_v} \simeq \frac{a}{r_o}\left[\frac{L}{S_v} + \frac{1}{6}\left(\frac{L}{S_v}\right)^2 - \frac{K_{\rm sym}}{12S_v} + \cdots\right], \tag{14}$$

where $a \simeq 0.523\,\text{fm}$ is a surface thickness parameter. We kept terms up to quadratic order in the symmetry energy density expansion about n_s. The value of $K_{\rm sym}$ is largely unconstrained by nuclear experiments, but neutron matter studies lead to the relation (23) discussed in Sec. 3.2.

The symmetry energy can be extracted from the mass formula (13), in spite of the ignored shell and pairing corrections, by examining isobaric nuclear pairs. Consider the energy difference of an isobaric pair:

$$\begin{aligned} &E(N-1,Z+1) - E(N+1,Z-1) \\ &= \frac{12Ze^2}{5R}\left[1 - \frac{1}{12}\left(1+\frac{S_vA^{1/3}}{S_s}\right)^{-1}\right] \\ &\quad - 8(N-Z)\left(1+\frac{S_s}{S_vA^{1/3}}\right)^{-1}\left[\frac{S_v}{A} - \frac{e^2}{80R}\frac{S_s}{S_vA^{1/3}}\right], \end{aligned} \tag{15}$$

which, upon fitting to thousands of measured masses, results in a nearly linear correlation between S_v and S_s/S_v, and, because of Eq. (14), also forms a nearly linear correlation between L and S_v:

$$L \simeq 13S_v - 350\ \text{MeV}. \tag{16}$$

Unfortunately, the range of observed values of $(N-Z)/A$ is too small to permit separate values for L and S_v to be confidently determined. However, another nearly linear correlation results from observation of the dipole polarizability[14] (related to the giant dipole resonance):

$$\alpha_D \simeq \frac{AR^2}{20S_v}\left(1+\frac{5}{3}\frac{S_s}{S_vA^{1/3}}\right). \tag{17}$$

This correlation is similar to that derived from masses, except the predicted slope between S_s/S_v and S_v is 5/3 times larger; this translates also into a steeper slope for the $L - S_v$ correlation. The electric dipole polarizability of ^{208}Pb was recently determined by Ref. 15 to be $\alpha_D \simeq (20.1 \pm 0.6)\,\text{fm}^2$. For a value of $S_v \simeq 30.5\,\text{MeV}$, Eq. (17) then predicts that $S_s/S_v = 1.02 \pm 0.13$; use of Eqs. (14) and (23) then imply that $L \simeq (1.9 \pm 0.15)S_v \simeq (58 \pm 5)$ MeV. Alternatively, using Eq. (16) results in the consistent result $L \simeq (60 \pm 5.5)\,\text{MeV}$.

Yet another roughly linear correlation can be extracted from the neutron skin thickness, which in this model[13] becomes

$$r_{np} \equiv \frac{1}{N}\int n_n r^2 d^3r - \frac{1}{Z}\int n_p r^2 d^3r \simeq \sqrt{\frac{3}{5}}\frac{2r_o}{3}$$
$$\times\left(1+\frac{S_s}{S_vA^{1/3}}\right)^{-1}\left[\frac{N-Z}{A}\frac{S_s}{S_v} - \frac{3Ze^2}{140r_oS_v}\left(1+\frac{10}{3}\frac{S_s}{S_vA^{1/3}}\right)\right]. \tag{18}$$

In contrast to the mass and dipole polarizability correlations, this correlation has a small, negative, slope in the $S_s/S_v - S_v$ plane, making it nearly orthogonal to the mass and dipole correlations.[b] The neutron skin thickness of Pb^{208} was recently estimated by Ref. 16 to be $r_{np} \simeq (0.175 \pm 0.020)\,\text{fm}$. From Eq. (18) we estimate that $S_s/S_v \simeq 1.8 \pm 0.2$, which implies that $L/S_v \sim 3$ from Eqs. (14) and (23). The liquid droplet prediction should be corrected to account for the different diffusenesses of neutrons and protons,[17] which reduces L/S_v to about 2. In principle, a precise skin measurement could be extremely constraining on the preferred values of L and S_v; however, considering the relatively large spread in estimated neutron skin values over the last decade, the inherent systematic uncertainties are still large. Better experiments are required.

Figure 3 shows the experimental correlations between L and S_v. In addition, constraints from consideration of isobaric analog states[22] (IAS) and fragment distributions and flows in heavy ion collisions[23] (HIC) are displayed. The white region indicates concordance of all the experimental information (but is surely an understatement of the resulting constraint's uncertainty), with $S_v \simeq (31.5 \pm 0.5)\,\text{MeV}$ and $L \simeq (55 \pm 10)\,\text{MeV}$.

3.2. *Theoretical neutron matter studies*

Recent theoretical studies of pure neutron matter by multiple groups[20,24,25] are highly relevant for determination of symmetry energy parameters. Since the binding

[b]Note that without the polarization term in Eq. (18), the predicted slope would be zero; the polarization term introduces a negative correlation between S_s/S_v and S_v.

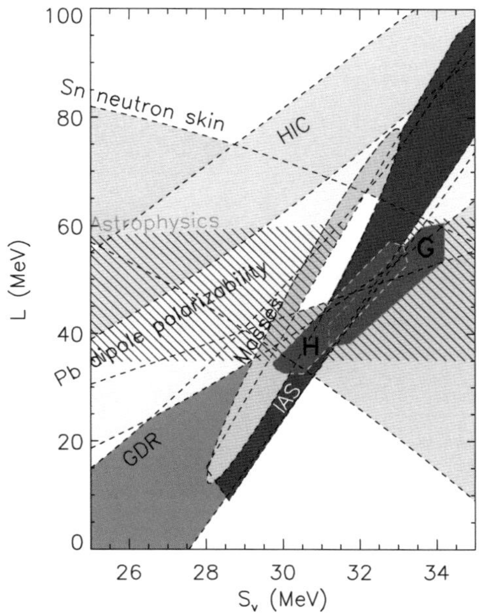

Fig. 3. Experimental constraints for symmetry energy parameters, adapted and revised from Ref. 18. See the text for further discussion and references to the experimental data and interpretation. The hatched region is an estimate derived from astrophysical considerations.[19] G and H refer to the neutron matter studies of Refs. 20 and 21, respectively. The white region is a simple concordance union of the experimental regions.

energy of uniform symmetric matter at n_s B is known to be about 16 MeV to within an MeV, the neutron matter energy at n_s yields S_v. And, as Eq. (11) shows, the pressure and its first derivative of pure neutron matter at n_s are equivalent to L and $K_{\rm sym}$, respectively. The major approaches are auxiliary field diffusion Monte Carlo studies[20] and chiral Lagrangian expansions.[21] Reference. 20 showed that a reasonable approximation to the neutron matter energy in the vicinity of n_s is

$$E(n,x) = au^\alpha + bu^\beta, \tag{19}$$

where $u = n/n_s$ and a, b, α and β are fitting parameters. We can immediately identify

$$S_v = B + a + b, \quad \frac{L}{3} = a\alpha + b\beta, \quad \frac{K_{\rm sym}}{9} = a\alpha(\alpha-1) + b\beta(\beta-1). \tag{20}$$

For recently published calculations,[20,24] the parameters have ranges

$$\begin{aligned} 12.6\,\text{MeV} < a < 13.4\,\text{MeV}, \quad 0.48 < \alpha < 0.51, \\ 3.2\,\text{MeV} < b < 5.6\,\text{MeV}, \quad 2.1 < \beta < 2.5. \end{aligned} \tag{21}$$

The predicted symmetry parameters thus lie in the ranges

$$\begin{aligned} 31\,\text{MeV} < S_v < 35\,\text{MeV}, \quad 39.8\,\text{MeV} < L < 61.7\,\text{MeV}, \\ 35.2\,\text{MeV} < K_{\rm sym} < 146.8\,\text{MeV}, \quad 0.65 < S_s/S_v < 0.89. \end{aligned} \tag{22}$$

It is interesting to note that the neutron matter calculations imply

$$K_{\rm sym}/S_v \approx 5L/S_v - 4.75, \tag{23}$$

and we proceeded to use this with Eq. (14) to estimate S_s/S_v. The average of values found by Ref. 20 are $S_v = 33\,\mathrm{MeV}$, $L = 48\,\mathrm{MeV}$, $K_{\rm sym} = 87\,\mathrm{MeV}$ and $S_s/S_v = 0.73$, while Ref. 24 find $S_v = 32\,\mathrm{MeV}$, $L = 45\,\mathrm{MeV}$, $K_{\rm sym} = 70\,\mathrm{MeV}$ and $S_s/S_v = 0.71$. The values for S_v and L are not only self-consistent but also compatible with those found from nuclear systematics involving masses, resonances and skin thicknesses (Fig. 3).

4. Extending the Equation of State to High Densities

The nuclear physics results for S_v and L allow for estimates of neutron star radii from their correlation with the pressure of neutron star matter (p_β)[8]:

$$R_{1.4} = (9.52 \pm 0.49)[p_\beta(n_s)/(\mathrm{MeV fm^{-3}})]^{1/4}\,\mathrm{km}, \tag{24}$$

where $R_{1.4}$ is the radius of a $1.4M_\odot$ star and the parameters were refined[18] to reflect the $2M_\odot$ minimum value for the neutron star maximum mass. We obtain the estimate $10.9\,\mathrm{km} < R_{1.4} < 13.5\,\mathrm{km}$ from the neutron matter results, including uncertainties reflected in Eq. (24).

The evidence is overwhelming that neutron stars have hadronic crusts. Thermal measurements (using X-ray observations) of neutron stars cooling on both long term (ages ranging from hundreds to a few millions of years) and short term (days to years after transient heating stemming from accretion of matter from binary companions) indicate cooling rates consistent with the existence of 0.5–1 km thick crusts containing masses of 0.02–$0.05M_\odot$.[26] Thinner crusts would imply large surface temperatures for young neutron stars; the fact that these are not observed also argues for moderately thick crusts. Bursting behavior on neutron star surfaces is also good evidence of hadronic crusts, as matter generally has to accumulate or else it burns continuously during accretion. In addition, observations of pulsar glitches are best explained by neutron $^1S_0$ superfluidity largely confined to a substantial hadronic crust.[27] Glitches are thought to involve the sudden transfer of excess angular momentum from matter spinning faster than the bulk of the star. The bulk of the star spins down due to the emission of dipole radiation producing observed pulses, but the superfluid component does not, at least until the differential in respective angular velocities becomes large enough to trigger a catastrophic transfer. The regularity of glitches argues in favor of a global phenomenon in which the entire excess angular momentum reservoir, estimated to involve a fractional moments of inertia $\Delta I/I \sim 0.05$, is transferred in each glitch.

Models of uniform nucleonic matter predict that below a density $n_0 \sim 0.4$–$0.5n_s$ matter becomes unstable to clustering, and this instability is the location of the core-crust boundary.[28] The crust contains, at its higher densities, nuclei arranged in a lattice immersed in a neutron fluid. Below a density of $0.002\,\text{fm}^{-3}$, called the neutron-drip density, matter consists of the nuclear lattice alone. Nearly all the crustal mass is in the denser region. Due to the fact that nuclei have internal densities near n_s, and consequently have almost zero internal pressure, the nucleonic contribution to pressure in the crust is very small. In fact, it is largely negative due to the negative lattice pressure. Stability of the crust, which requires positive pressures, is due to the presence of the relativistic degenerate electrons necessary to neutralize the positive nuclear charges. The pressure–density relation of these electrons is well-understood, so the crustal EOS is also well-understood. Small variations in the nuclear charges and masses, due to uncertainties in the nuclear incompressibility and symmetry energy, are not consequential.

The neutron matter studies suggest that the energy and pressure of neutron star matter higher than n_0 can be predicted to densities somewhat larger than n_s, perhaps to as high as $2n_s$, with moderate uncertainty. However, Eq. (4) shows that densities up to $9n_s$ could be achieved in neutron stars, given a minimum value for the maximum mass of $2M_\odot$. Nevertheless, there are two important constraints on the EOS at these densities. It should not violate causality, that is, $\partial p/\partial\varepsilon \leq 1$, yet the pressure should increase with increasing ε rapidly enough to support this minimum mass.

4.1. *Piecewise polytropes*

Read *et al.*[29] found that a piecewise polytrope scheme with three segments could accurately represent a large set of candidate hadronic equations of state for densities beyond the core-crust transition density $n_0 \simeq 0.37n_s$ and pressure $p_0 \simeq 0.24\,\text{MeV fm}^{-2}$. In such a scheme, there are polytropic regions, each with $p = K_i n^{\Gamma_i}$ with $i = 1$–3 within the density range $n_{i-1} < n < n_i$.[c] Because of the requirement of continuity of pressure at the segment boundaries n_i for $i = 1$–3, there are a total of six parameters: n_i and either Γ_i or p_i for $i = 1$–3. Moreover, they found that there exist optimum values for n_i that maintain accurate representations of all the candidate hadronic equations of state: $n_1 = 0.5n_2 = 0.25n_3 \simeq 1.85n_s$. Thus, the entire high-density EOS can be parametrized with only three parameters, which we choose to be p_1, p_2 and p_3. Reference 30 has also considered modeling the high-density EOS in this fashion, but with some differences (see below).

We note that the first polytropic segment with $n_0 < n < n_1$ is the region for which neutron matter calculations are valid. p_1, using Eq. (19), is

$$p_1 = n_s\left(a\alpha 1.85^{\alpha+1} + b\beta 1.85^{\beta+1}\right). \tag{25}$$

Neutron matter computations thus yield $8.1\,\text{MeV fm}^{-3} \lesssim p_1 \lesssim 20.9\,\text{MeV fm}^{-3}$. The pressures p_2 and p_3 will be much less constrained.

[c]For densities $n > n_3$, it is assumed that $p = K_3 n^{\Gamma_3}$.

Within the ith polytropic segment, assuming both pressure and energy density continuity, we have

$$n(p) = n_{i-1}\left(\frac{p}{p_{i-1}}\right)^{1/\Gamma_i}, \quad \Gamma_i = \frac{\ln(p_i/p_{i-1})}{\ln(n_i/n_{i-1})},$$
$$\varepsilon(p) = \left(\varepsilon_{i-1} - \frac{p_{i-1}}{\Gamma_i - 1}\right)\left(\frac{p}{p_{i-1}}\right)^{1/\Gamma_i} + \frac{p}{\Gamma_i - 1}, \tag{26}$$
$$c_s^2 = c^2\frac{\partial p}{\partial \varepsilon} = \frac{\Gamma_i p}{\varepsilon + p}.$$

Hydrodynamic stability requires $p_i \geq p_{i-1}$, which sets a lower limit to p_i for $i = 1$–3. Causality, i.e., $c_{s,i} \leq c$, sets upper limits to p_1 and p_2.[d] The causal limit to p_3 is somewhat arbitrary. Generally, if $\Gamma_3 > 2$, the causal limit will eventually occur, at the pressure

$$p_{cl} = p_2\left[\frac{(\Gamma_3 - 1)\varepsilon_2 - p_2}{\Gamma_3(\Gamma_3 - 2)p_2}\right]^{\Gamma_3/(\Gamma_3 - 1)}. \tag{27}$$

There seems to be two reasonable approaches. First one could require that $p_{cl} \geq p_{c,\max}$, where $p_{c,\max}$ is the central pressure of the maximum mass configuration (this configuration maximizes both the central pressure and sound speed for a given EOS). We call this model A. Alternatively, and more conservatively, within stars where $p_{cl} < p_c$, where p_c is the central pressure, one could impose a causal limit at higher pressures. Using $dp/(\varepsilon + p) + dn/n = 1$, we find in this region

$$\varepsilon = p + \varepsilon_{cl} - p_{cl}, \qquad n = n_{cl}\sqrt{\frac{2p + \varepsilon_{cl} - p_{cl}}{\varepsilon_{cl} + p_{cl}}}. \quad p \geq p_{cl}. \tag{28}$$

This alternative, which we call model B, seems rather arbitrary and less physically realistic, but we will consider it nonetheless. Analyses with models A and B have been summarized in Refs. 12 and 31. Yet another approach, considered by Ref. 30, where the upper limit to p_3 is set by the condition $c_{s,3} = c$, does not make sense because the pressure p_3 is often less than $p_{c,\max}$, in which case causality is violated in the interiors of many possible stars before the maximum mass is reached. This leads to underprediction of the most likely neutron star radius as a function of mass.

4.2. *Parameter constraints*

Figure 4 shows the ranges of allowed parameters for model A. An important feature is that the allowed parameter ranges are restricted by the assumed value of the minimum maximum mass, and are indicated by different colors. For

[d]These two constraints imply $p_0 \leq p_1 \leq 113.9\,\mathrm{MeV\,fm^{-3}}$ and $p_1 \leq p_2 \lesssim 500\,\mathrm{MeV\,fm^{-3}}$, so the upper limit $p_1 \lesssim 20.9\,\mathrm{MeV\,fm^{-3}}$ imposed by neutron matter calculations is far more constraining. This upper limit to p_1 imposes a new, lower causal limit $p_2 \lesssim 255\,\mathrm{MeV\,fm^{-3}}$.

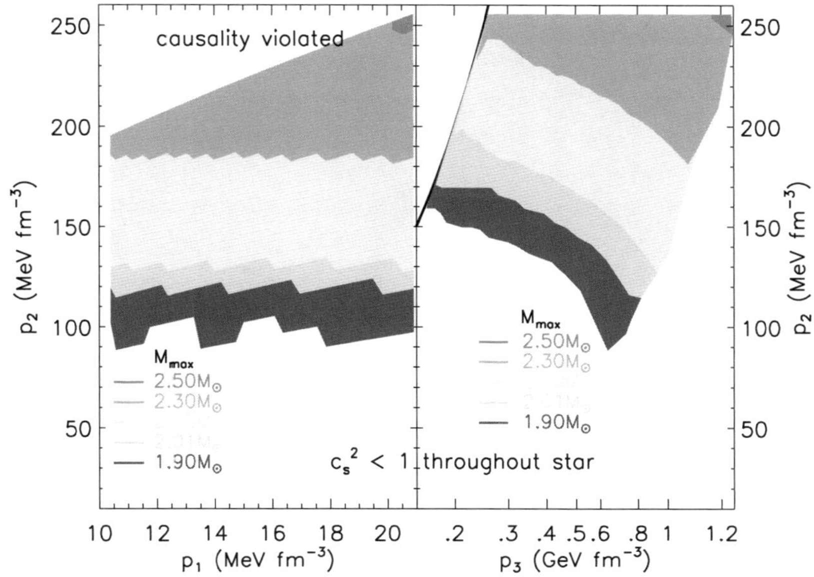

Fig. 4. The left panel shows allowed values of the EOS parameters p_1 and p_2 for various assumed values of $M_{\max}$ for model A. The allowed regions for lower values of $M_{\max}$ are partially obscured by those with larger values. The right panel shows allowed values for p_2 and p_3.

$M_{\max} = 1.90M_\odot$, for example, values of $p_2 > 90\,\mathrm{MeV\,fm^{-3}}$ are required, but this lower limit is continuously raised by increase in the assumed value of $M_{\max}$. When the maximum mass is raised from $1.90M_\odot$ to $2.01M_\odot$, there is relatively little additional restriction of permitted parameter ranges, but for $M_{\max} = 2.3M_\odot$, the permitted ranges of all the parameters are very tightly restricted. The case $M_{\max} = 2.5M_\odot$ seems to be forbidden.

On the other hand, when the causality restriction is relaxed, case B, the allowed parameter ranges increase for all values of $M_{\max}$, as displayed in Fig. 5. Note, in this case, that values of p_2 as small as the largest values of p_1 are now allowed. Similarly, in case B, the upper limits to p_3 are dramatically larger than in case A, and are nearly unbounded. In the case when $M_{\max} = 2.5M_\odot$, the parameter ranges are small but finite.

4.3. *Radius constraints*

Each legitimate set of parameters leads to a series of stellar models forming an individual M–R curve. It is interesting to explore the ranges of radii that result, and to compare this result with the expectations stemming from neutron matter theory and the correlation Eq. (24). The parameter with the most pronounced correlation with radius is expected to be p_1, given the phenomenological correlation Eq. (24). The robustness of the correlation, however, apparently depends on which piecewise polytropic approach we implement, as well as the choice of $M_{\max}$, as shown in Fig. 6.

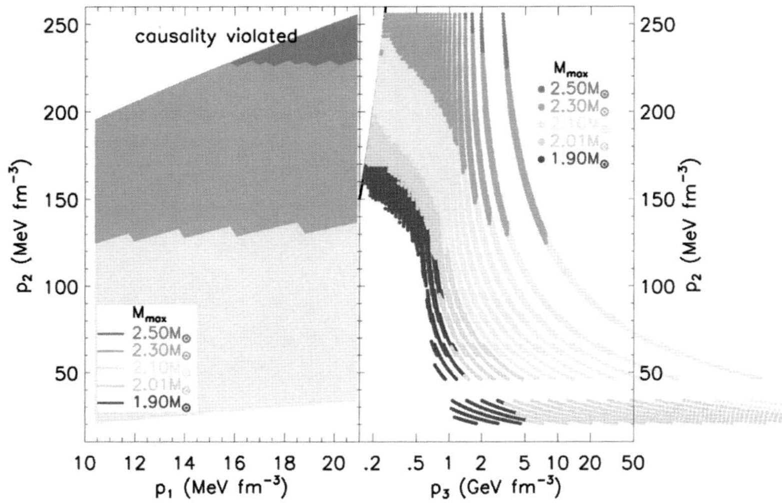

Fig. 5. The same as Fig. 5, but for model B.

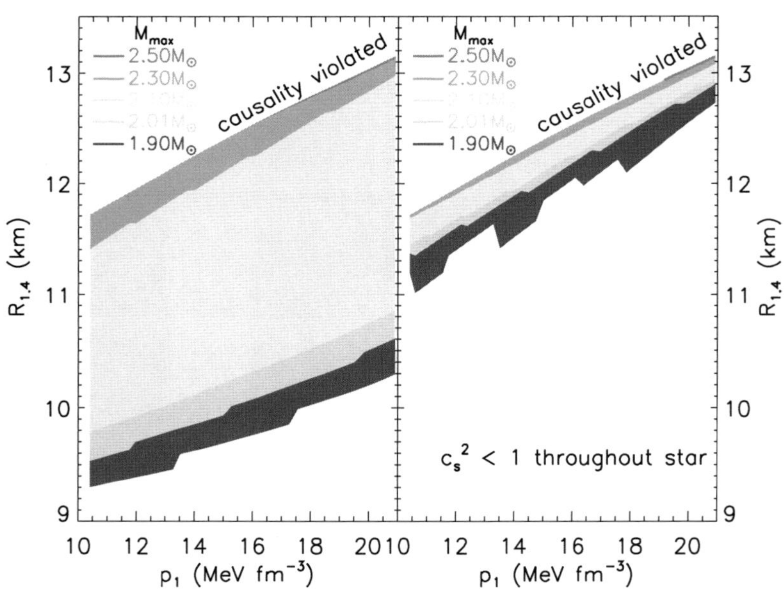

Fig. 6. (Color online) The radii of $1.4M_\odot$ stars as a function of the parameter p_1. The left panel represents approach B and the right panel approach A. For a given value of p_1, the spread in radius is due to varying values of p_2 and p_3. The colors indicate different assumptions about the minimum value of the maximum mass $M_{\max}$. Regions with larger values of $M_{\max}$ partially obscure regions with lesser values.

In the case of model B, the correlation between radius and p_1 is accurate to about ± 1 km, unless the maximum mass is greater than $2.1M_\odot$. On the other hand, in the case of model A, the correlation is accurate to about ± 0.2 km for $M_{\max} < 2.3M_\odot$, and to about ± 0.1 km for larger $\mathrm{M}_{\max}$.

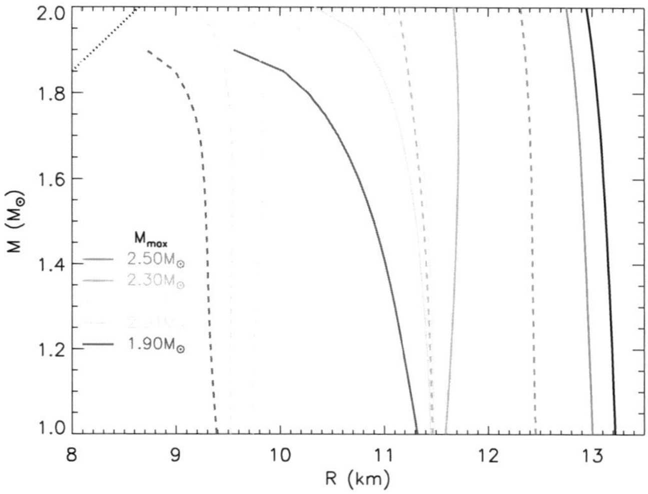

Fig. 7. (Color online) Allowed regions in the M–R plane as functions of the assumed minimum value of the maximum mass, $M_{\rm max}$. Colored solid lines are lower limits to R as a function of M for model A, while dashed lines are for model B. In both cases, upper limits to radii are shown by the solid black line.

Figure 7 compares the upper and lower limits to radii as a function of mass for the two approaches. As expected, the ranges of radii are small in model A, no more than about 2 km for $M \lesssim 1.8M_\odot$, but expand to about 4 km in the case of model B. The maximum mass becomes an important constraint if $M_{\rm max} \gtrsim 2.1M_\odot$ in both approaches. The lower limits in all cases are substantially larger than predicted by the maximally compact conjecture. The upper limits to radii are a result of neutron matter constraints on the upper limits to the value of p_1.

In a Bayesian analysis of observational data incorporating prior information constraining the EOS, including the existence of a hadronic crust, causality, the observed minimum value for $M_{\rm max}$, and symmetry energy information from theoretical neutron matter calculations and nuclear experiments, inferred radii will be strongly influenced by these priors. Randomly selecting parameters p_1, p_2 and p_3, and eliminating models not satisfying these prior assumptions, lead to a predicted distribution of neutron star radii. This can be displayed in a variety of ways, and we show two examples in Fig. 8. The first is the probability distribution of radii for $1.4M_\odot$ stars; the second is the average radii of stars for each modeled EOS, with masses chosen from a distribution similar to those observed in pulsar binaries. The latter is useful when considering astrophysical observations in which the stellar mass is not well-constrained, as is often the case for observations of quiescent low-mass X-ray binaries (qLMXBs) or Type I X-ray bursters. The influence of the lower and upper limits to radii from Fig. 7 are apparent.

Necessarily, model A does not allow for strong first-order phase transitions near the nuclear saturation density, i.e., in the range n_s–n_1. It can be shown that such phase transitions have a similar effect on radius constraints as model B even though

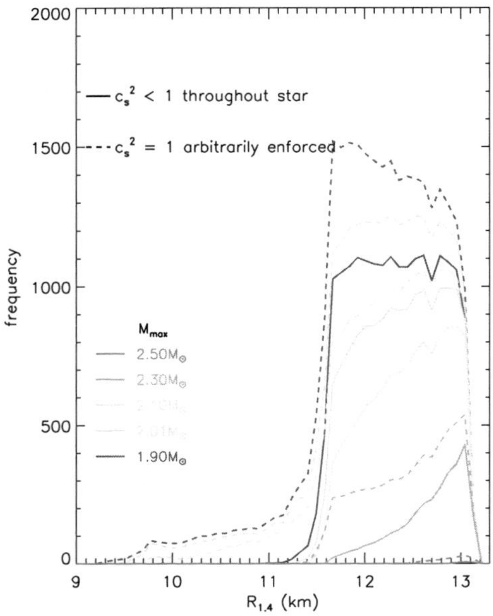

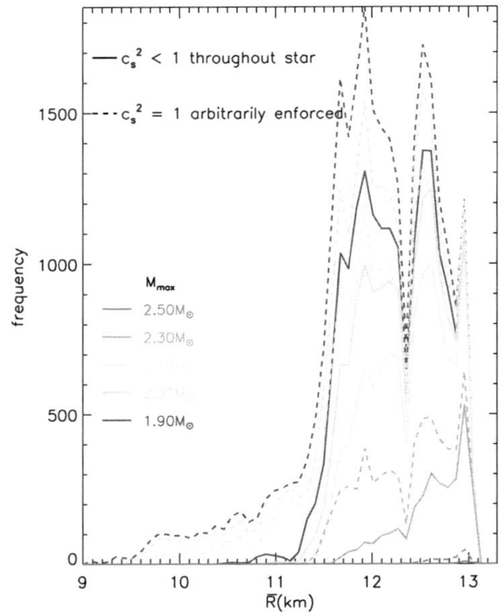

Fig. 8. Neutron star radius probability distributions subject to the constraints of a hadronic crust, causality, an observed minimum value of $M_{\max}$, correctness of general relativity, and neutron matter constraints on values of p_1. Left: Distribution of radii of $1.4M_\odot$ stars. Solid lines represent results for model A; dashed lines represent results for model B. The vertical scale is arbitrary. Right: Distribution of average radii from a mass distribution identical to that of measured pulsar masses in binary systems.

the causal limit to the sound speed is not approached in this density range (Sec. 4.8). As is clear, high accuracy observational data showing that neutron star radii are smaller than 11 km would be in serious tension with model A, and would therefore provide serious evidence in favor of one or more phase transitions near the nuclear saturation density, at least for extremely neutron-rich matter.

4.4. *Correlation involving the rotation rate*

One of the first nearly universal relations discovered connects the maximum spin frequency of a uniformly rotating neutron star and its average density[32,33]

$$f_K = 1.08 \left(\frac{M}{M_\odot}\right)^{1/2} \left(\frac{10\text{ km}}{R}\right)^{3/2} \text{ kHz}, \tag{29}$$

where f_K indicates the Keperlian frequency at which mass shedding occurs. The M and R in this formula refer to the mass and radius of the nonrotating configuration, not the spinning configuration. Although this formula was phenomenologically established from full axially-symmetric solutions of Einstein's equations, the dependence on mass and radius is identical to Newtonian gravity. The coefficient 1.08 kHz is very close to the 1.00 kHz of the analytic relativistic Roche model (which assumes the gravitational potential of the spinning star given by concentrating all its mass at the origin).[34,35] Other universal relations involving the shapes of rotating configurations can also be found (Sec. 5.4).

The highest frequency pulsar known is PSR J1748 − 2446ad[36] with $f = 716\,\text{Hz}$; its limiting radius as a function of mass is shown in Fig. 2. Unfortunately, the mass of this pulsar is unknown, so an upper limit to its radius is unable to be established. However, if most rapidly spinning pulsars, so-called millisecond pulsars, acquired their angular momentum by accretion from a companion, it is not out of the question that a rapidly rotating millisecond pulsar will eventually be found in a binary system with measured masses. Assuming that $M \lesssim 2M_\odot$, Eq. (29) implies $R \lesssim 16.6\,\text{km}$.

4.5. *Correlations involving the stellar binding energy*

One likely observable of a neutrino signal from a galactic supernova, i.e., a supernova with a distance of approximately 8.5 kpc, is the gravitational binding energy of the star, which could be measured to high precision. The details of flavor mixing among the emergent neutrinos are likely to be unimportant as, in the absence of mixing, neutrinos and antineutrinos are expected to be radiated in approximately equal numbers in all flavors, and the average energy difference among neutrinos and antineutrinos of different flavors is expected to be, at most, 1–2 MeV.[37] The various operating neutrino detectors, which are respectively sensitive to one or more flavors of neutrinos or antineutrinos, can then be used as calorimeters to estimate the total energy radiated in an event. Although neutrinos might continue to be detectable for tens of seconds for one or more detectors, almost all the energy is radiated in the first several seconds, when fluxes are relatively high. Therefore, the net radiated energy might be rather accurately determined. For SN 1987A, the binding energy was inferred from the neutrino energies and fluxes to be about $0.15M_\odot$, but with an uncertainty of about 50%.[38]

The stellar binding energy is defined as $\text{BE} = \mathcal{N}m_B - M$, where M is the total gravitational mass ($m(R)$), $\mathcal{N}$ is the total number of baryons, and m_B is the baryon mass. The definition of m_B is such that the binding energy of $\mathcal{N}$ baryons at a uniform baryon density of zero is itself zero. If it were to be assumed that the ground state energy per baryon at zero baryon density is $\mu_0 \simeq 930.8\,\text{MeV}$, then $m_B = \mu_0/c^2$. Alternatively, if m_B is taken to be the average baryon mass, about 939 MeV, the binding energy of a $1.4M_\odot$ configuration would differ from the previous assumption by about $0.088M_\odot c^2$, or about 5%. However, in a supernova collapse, the total neutrino energy released is essentially the binding energy difference between that of the progenitor's core and the supernova's remnant. Treated as a binding energy difference, the choice of energy zero, i.e., the value of m_B, is irrelevant. But since the binding energy of the progenitor's core is well-approximated by treating the core as composed of ground state iron, it is reasonable to compute the quantity BE by the choice $m_B = \mu_0/c^2$. Nevertheless, it remains necessary to take care in defining the stellar binding energy when confronting a precision measurement of the total released neutrino energy.

The net baryon number in the star is computed with the supplementary TOV equation

$$\frac{d\mathcal{N}}{dr} = 4\pi n r^2 \left(1 - \frac{2Gm}{rc^2}\right)^{-1/2}, \tag{30}$$

where n is the baryon density. For the piecewise polytrope equations of state, results are shown for model A in Fig. 9. The figure demonstrates the binding energy is highly correlated with the stellar compactness β, although that the correlation becomes less significant when $M \simeq M_{\rm max}$. It is also seen that the value of $M_{\rm max}$ plays an important role, in that the correlations become more significant as $M_{\rm max}$ is increased. These correlations had previously been noticed by Ref. 38 and in more recent works by other authors, but the role of the maximum mass has previously not been explored. We note, furthermore, that an even stronger correlation exists in model A between BE and M, rather than between BE and β. This correlation is shown in Fig. 10 for the case $M_{\rm max} \geq 2.01 M_\odot$. A least squares fit is given by

$$M = 0.01079 + 4.086\sqrt{{\rm BE}/M_\odot} - 1.073({\rm BE}/M_\odot)\ M_\odot, \tag{31}$$

which has a maximum error of 5.7% but a mean error of only 1.9% for all calculated models. This figure shows the extent to which the mass of the remnant can be determined from a binding energy measurement.

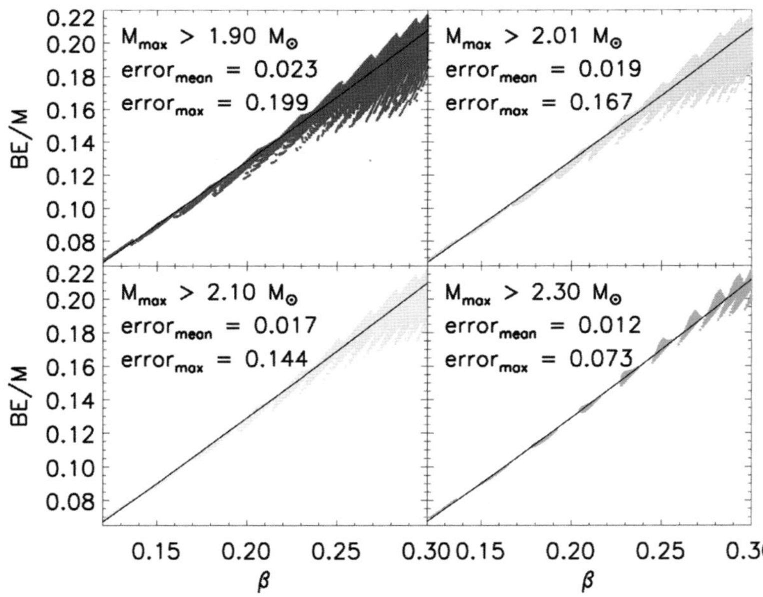

Fig. 9. Binding energy per unit mass as a function of stellar compactness $\beta = GM/Rc^2$ for model A. Four cases with different assumptions about the maximum mass are compared. Each point represents a stellar configuration. Also indicated are the mean and the maximum relative errors, compared to a least squares quadratic fit, for each case.

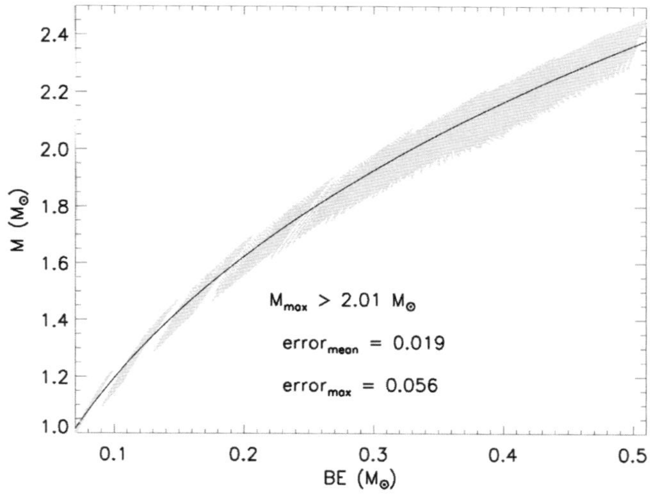

Fig. 10. Binding energy as a function of stellar mass for model A in the case that $M_{\rm max} \geq 2.01M_\odot$. Each point represents a stellar model of the indicated mass for parameters p_1, p_2 and p_3 spanning the ranges allowed by causality shown in Fig. 4. The solid curve is the least squares quadratic fit (Eq. (31)).

4.6. *Moment of inertia*

The pulsar binary PSR J0737 − 3039 has revealed the effects of spin-orbit coupling, having a precessional period of about 75 years. General relativity adds corrections to this precession leading to small changes in orbital inclination and also the advance of the orbit's periastron, both at the 2.5 order in a post-Newtonian expansion.[39,40] In the case of this system, whose orbit is observed edge-on, changes in the inclination will be nearly impossible to observe, but changes to periastron advance will be maximized. In the case of PSR J0737−3039 it is expected that the correction to periastron advance could be observed within a decade, and estimates to the moment of inertia I of the more rapidly spinning component, A, will be made. For a slowly spinning Newtonian star of uniform density, $I = (2/5)MR^2$. Since the stellar masses are known to quite high precision in this system, a measure of the moment of inertia translates into an even more precise measurement of the stellar radius.

In general relativity, for a slowly spinning configuration, the moment of inertia[41] is

$$I = \frac{8\pi}{3c^2\Omega}\int_0^R r^4\bar{\omega}(\varepsilon + p)e^{(\lambda-\nu)/2}dr, \tag{32}$$

where Ω is the rotation rate. Metric functions ν, λ and $\bar{\omega}$ are defined by

$$\frac{d\nu}{dr} = \frac{2G}{r^2c^4}(mc^2 + 4\pi pr^3)e^\lambda = -\frac{2}{\varepsilon + p}\frac{dp}{dr}, \quad e^{-\lambda} = 1 - \frac{2Gm}{rc^2},$$
$$\frac{d}{dr}\left(r^4 j\frac{d\bar{\omega}}{dr}\right) = -4r^3\bar{\omega}\frac{dj}{dr}, \tag{33}$$

where $j \equiv e^{-(\lambda+\nu)/2}$. The relevant boundary conditions are

$$\frac{d\nu}{dr} = \frac{d\lambda}{dr} = \frac{dj}{dr} = \frac{d\omega}{dr} = 0 \qquad r = 0,$$
$$\nu_R = -\lambda_R, \quad j_R = 1, \quad \frac{\bar{\omega}_R}{\Omega} = 1 - \frac{2GI}{R^3c^2}, \quad r = R. \tag{34}$$

We note, using the first two of Eq. (33), that

$$\frac{dj}{dr} = -\frac{j}{2}\left(\frac{d\nu}{dr} + \frac{d\lambda}{dr}\right) = -\frac{4\pi G}{c^4}(\varepsilon + p)rje^{\lambda}, \tag{35}$$

which follows from the first two of Eq. (33). Putting Eq. (35) in Eq. (32), and using the last of Eq. (33) and the outer boundary condition for j_R yield

$$I = -\frac{2c^2}{3G\Omega}\int_0^R r^3\bar{\omega}\frac{dj}{dr}dr$$
$$= \frac{c^2}{6G\Omega}\int_0^R d\left(r^4 j\frac{d\bar{\omega}}{dr}\right) = \frac{R^4c^2}{6G\Omega}\left(\frac{d\bar{\omega}}{dr}\right)_R. \tag{36}$$

Although $\bar{\omega}$ is defined by a second-order differential equation, computation of the moment of inertia only requires the solution of the first-order differential equation for $d\bar{\omega}/dr$. Substituting the variable $y = \bar{\omega}^{-1}d\bar{\omega}/dr$ into the last of Eq. (33), we find

$$\frac{dy}{dr} = -(y+4)\frac{dj}{jdr} - (3+y)\frac{y}{r}, \tag{37}$$

where $j^{-1}dj/dr$ is given by Eq. (35). Note that $y(r=0) = 0$. Then

$$I = \frac{R^3c^2}{6G}\frac{\bar{\omega}_R}{\Omega}y_R = \frac{R^3c^2}{2G}\frac{y_R}{3+y_R}, \tag{38}$$

using the last relation in Eq. (34) to eliminate $\bar{\omega}_R/\Omega$. Note that the moment of inertia, in this slowly rotating case, is independent of the rotation rate.

It has previously been noted[8] that the moment of inertia is highly correlated with stellar compactness, with the error in I for a given value of β being about $\pm 3\%$.[42] But when the constraints of a hadronic crust, neutron matter limits to p_1, causality and a minimum value for $M_{\max}$ are introduced, the correlation becomes much tighter. Figure 11 shows the resulting correlation for model A in the case that $M_{\max} \geq 2.01 M_\odot$. The least squares fit is given by

$$I/M^3 = 0.2239 + 0.7613\beta^{-1} + 0.2346\beta^{-2}, \tag{39}$$

which has a maximum error of 5.2% and a mean error of 0.9% for the calculated models. Note that the leading term is proportional to β^{-2}.

When applied to the pulsar binary PSR J0737 − 3039, these results indicate that a 10% accurate measurement of I of component A will translate into about a 6% uncertainty in A's radius. This is illustrated in Fig. 12 which utilizes the correlation between I/M^3 and β from Fig. 11 translated into the mass–radius plane.

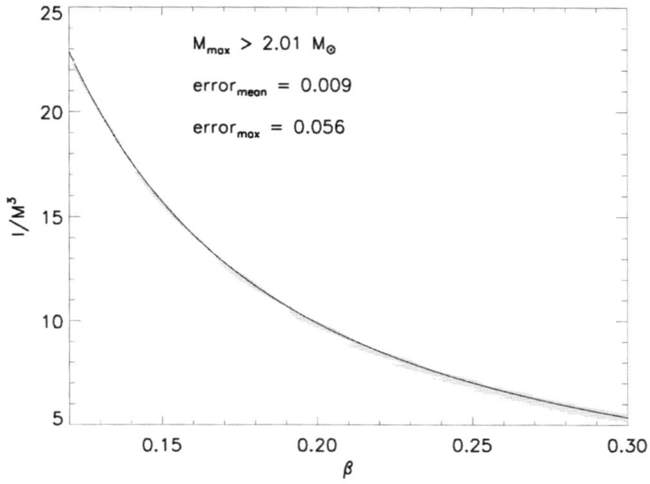

Fig. 11. The same as Fig. 10 except showing dimensionless moment of inertia I/M^3 as a function of compactness β. The solid curve is the least squares quadratic fit (Eq. (39)).

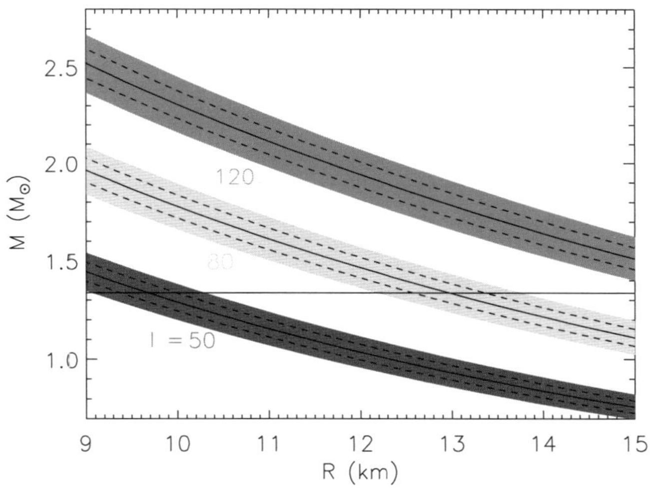

Fig. 12. (Color online) The potential of a 5% (dashed lines) or 10% (colored band) measurement of moment of inertia to constrain the mass and radius. Solid lines within the colored contours show Eq. (39) for three values of I in units of $M_\odot\,\mathrm{km}^2$. The case of PSR J0737 − 3039 with the well-measured mass $1.337M_\odot$ for component A is illustrated by the horizontal line.

4.7. *Tidal polarizability*

The merger of binary stars is expected to be the main source of gravitational wave signals observed with Advanced LIGO and VIRGO. Detection of a regular sinusoidal waveform prior to the merger allows the precise determination of the so-called chirp mass $\mathcal{M}_c = (M_1M_2)^{3/5}/(M_1+M_2)^{1/5}$. In the event at least one of the components is a neutron star, tidal effects during inspiral allow the possibility of breaking the remaining mass degeneracy through the measurement of the tidal deformability $\lambda = 2k_2R^5/(3G)$, where k_2 is the tidal Love number.[4] For a single star, the Love number is determined by[43]

$$\begin{aligned} k_2 &= \frac{8\beta^5}{5\mathcal{R}}[2 - w_R + 2\beta(y_R - 1)]e^{-2\lambda}, \\ \mathcal{R} &= 6\beta[2 - w_R + \beta(5w_R - 8)] - 3\lambda e^{-2\lambda}\left[2 - w_R + 2\beta(w_R - 1)\right] \\ &\quad + 4\beta^3\left[13 - 11w_R + \beta(3w_R - 2) + 2\beta^2(1 + w_R)\right], \\ \frac{dw}{dr} &= -\frac{w^2}{r} - \frac{w-6}{r}e^{\lambda} - \frac{4\pi Gr}{c^4}e^{\lambda}\left[5\varepsilon + 9p - w(\varepsilon - p) + \frac{d\varepsilon}{dp}(\varepsilon + p)\right] \\ &\quad + r\left(\frac{d\nu}{dr}\right)^2, \end{aligned} \tag{40}$$

where $\beta = GM/(Rc^2)$ and $w(0) = 2$ and $w_R = w(R)$. As in the case for the moment of inertia, only the solution of a single additional first-order differential equation for the variable w is required.

Although λ is highly correlated with β, the correlation between $\bar\lambda = \lambda/M^5$ and $\bar I = I/M^3$ is much tighter, as Ref. 44 found. For the case of model A, this correlation is shown in Fig. 13. The least squares fit shown is

$$\ln\bar\lambda = -35.52 + 48.85\ln\bar I - 24.17[\ln\bar I]^2 + 5.849[\ln\bar I]^3 - 0.5422[\ln\bar I]^4. \tag{41}$$

The maximum error is 2.5%, but the mean error is only 0.2% for the calculated models. It would be straightforward to combine Eqs. (39) and (41) to form a correlation between the tidal polarizability and compactness if one desired.

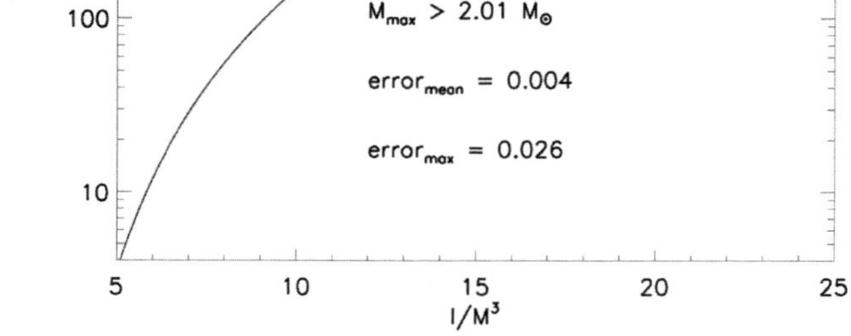

Fig. 13. The same as Fig. 10 except showing dimensionless tidal polarizability $\bar\lambda = \lambda/M^5$ as a function of dimensionless moment of inertia $\bar I = I/M^3$. The solid curve is the least squares fit (Eq. (41)).

For the case of a black hole–neutron star merger, the leading order correction to the gravitational waveform yields an estimate for $\bar{\lambda}$.[e] For an expected moderately strong typical detection, with a signal-to-noise ratio of 10, $\mathcal{M}_c$ can be determined to about 0.01%–0.02% accuracy, but both q and $\bar{\lambda}$ will be estimated to much lower accuracies.[45] It has also been found that for highly spinning black hole, estimated errors in the neutron star mass can approach 100%.[46] For neutron star masses ranging from $1.2M_\odot$ to $2.0M_\odot$ and black hole masses ranging from $4M_\odot$ to $12M_\odot$, $1.84 < \mathcal{M}_c/M_\odot < 3.97$ and $0.1 < q < 0.5$. Perhaps $\bar{\lambda}$ will be determined to about 20% – 25% accuracy, and the neutron star mass will be known to similar precision. Nevertheless it is interesting to examine the correlation between the tidal polarizability and neutron star mass. Figure. 14 displays $\bar{\lambda}$ as a function of the mass. Although the spread in $\bar{\lambda}$ for a given mass is broad, extending about a factor of 5 for $M \lesssim 2M_\odot$, the large radius sensitivity of $\bar{\lambda}$ (roughly proportional to R^5) implies that there should be a further correlation with radius or the EOS parameter p_1. This subcorrelation is also indicated in Fig. 14. Unfortunately, the expected large uncertainty in the neutron star mass would seemingly not permit a serious constraint on p_1 to be produced.

For the case of a binary neutron star merger, both components are tidally deformed, and the leading order correction to the gravitational waveform leads to the mass-averaged quantity

$$\bar{\Lambda} = \frac{16}{13}[\bar{\lambda}_1 q^4(12q+1) + \bar{\lambda}_2(1+12q)], \tag{42}$$

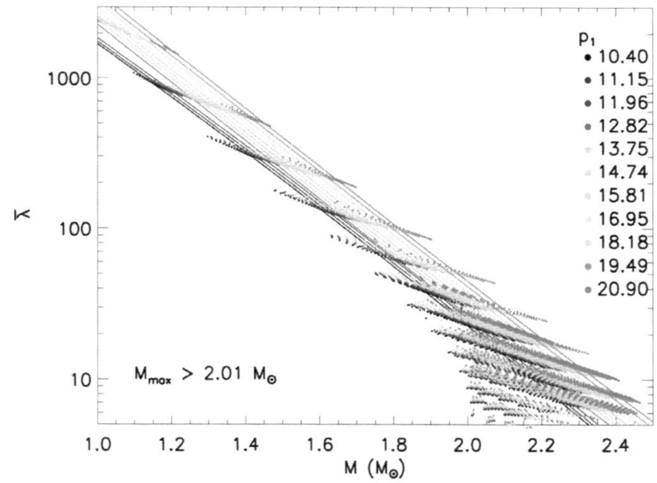

Fig. 14. (Color online) The correlation between the dimensionless tidal polarizability $\bar{\lambda}$ and the stellar mass, in the case of model A and $M_{\max} \geq 2.01M_\odot$. Each point corresponds to a calculated model with parameters p_1, p_2 and p_3 restricted to the values shown in Fig. 4. Individual models with differing values of p_1 within its assumed range are color-coded as indicated. Solid curves show least squares fits for each value of p_1.

[e]A black hole has zero tidal deformability.

where $q = M_1/M_2 \leq 1$ is the binary mass ratio. In the case of a binary neutron star merger with a signal-to-noise ratio of 20–30, once again the chirp mass will be determined to high precision but q will be found only to order 10%–15%.[47] However, in this case, since neutron star masses in a binary are quite likely to have $q \sim 1$, and their radii are likely to be nearly equal, their respective tidal polarizabilities will also be similar in value and, furthermore, correlated with their masses. Thus, $\bar{\Lambda}$ will be measurable to the same level of accuracy as in a black hole–neutron star binary. But in this case, the chirp mass will be known to comparatively high precision, and the possible range of q should be small. We therefore expect that a pronounced sensitivity to the average stellar radius, or equivalently to p_1, will survive the uncertainties in the component masses.

In Fig. 15, we show the binary tidal deformability as functions of $\mathcal{M}_c$ and q. There is an inherent spread in $\bar{\Lambda}$ values due to EOS variations. This spread is illustrated by computing the correlation between Λ and $\mathcal{M}_c$ for each value of p_1, then plotting the results in colors coded to p_1. It is evident from Fig. 15 that q is relatively unimportant, so that its intrinsically large uncertainty is irrelevant. Each series of contours correspond to $\bar{\Lambda}$ values differing by a factor $10^{1/4} \simeq 1.78$. The results are seen to be degenerate to a major extent in the values of $\mathcal{M}_c$ and p_1. Therefore, even a single moderately high signal-to-noise ratio observation of a binary neutron star merger could yield interesting information about the EOS and neutron star radii. This is quantitatively demonstrated in Fig. 16, which shows the degree to which p_1 can be constrained given the expected typical measurement errors in $\bar{\Lambda}$ and $\mathcal{M}_c$. It is interesting to note the striking global similarities between Figs. 16 and 14. Following years of LIGO runs, with an expected several dozen detections

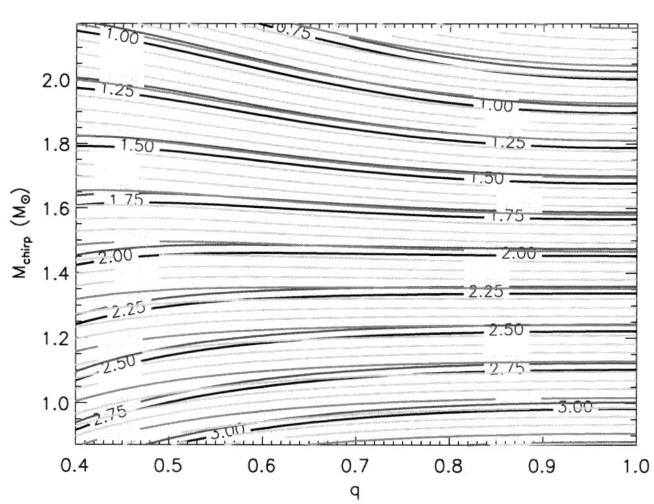

Fig. 15. (Color online) Contours of the binary tidal deformability $\log \bar{\Lambda}$ are displayed as functions of $\mathcal{M}_c = M_{\rm chirp}$ and q, for stellar configurations according to model A and with $M_{\rm max} \leq 2.01 M_\odot$. Colored contours indicate least squares fits of model results for selected values of p_1 corresponding to the color scheme shown in Fig. 14.

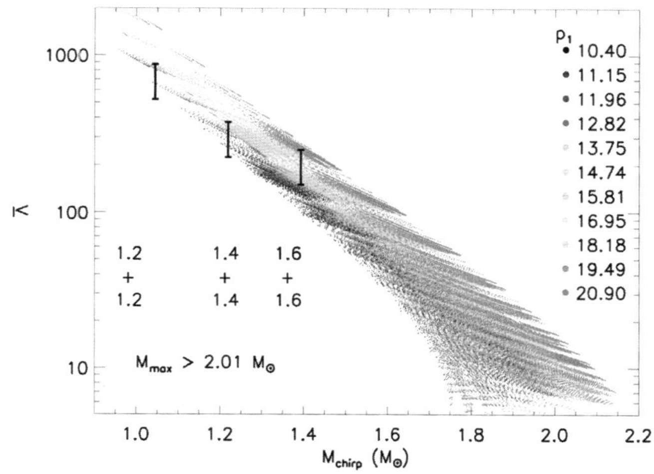

Fig. 16. The same as Fig. 14 except for the binary tidal deformability and the binary chirp mass rather than the individual tidal deformability and stellar mass. Typical observational errors for $\bar{\Lambda}$ and $\mathcal{M}_c = M_{\rm chirp}$ are indicated for the cases of three equal mass mergers involving components, respectively, of $1.2M_\odot$, $1.4M_\odot$ and $1.6M_\odot$. The vertical positions of the data points are chosen arbitrarily for purposes of illustration.

whose results can be statistically averaged, high levels of precision in estimates for p_1 can be anticipated.

4.8. *The effect of low-density strong phase transitions*

An important assumption we have made for models A and B is that the EOS remains without strong phase transitions at least up to the density $n_1 = 1.85n_s$. References 2 and 12 have explored in detail the possibility of phase transitions in this low-density region, and Refs. 21 and 13 noted that phase transitions near n_s have the potential to extend the range of $R_{1.4}$, with the widest range occurring when the phase transition begins at n_s. There is no evidence for a phase transition near n_s in symmetric matter, but in extremely neutron-rich matter the jury is still out. Reference 12 explored the role of phase transitions by allowing the lower and upper transition densities, n_L and n_U, respectively, to vary. For $n \geq n_U$, they assumed $p = s\varepsilon - \varepsilon_B$ with $s = 1$ or $1/3$ and ε_B a parameter, a scheme similar to that explored by Ref. 2. It is required that $p_U = p_L$ and $\mu_U = \mu_L = (\varepsilon_U + p_U)/n_U = (\varepsilon_L + p_L)/n_L$, which implies

$$n_U = \frac{(1+s)p_L + \varepsilon_B}{s\mu_L}, \quad \varepsilon_U - \varepsilon_L = \mu_L(n_U - n_L). \tag{43}$$

Therefore all possible transitions can be specified with the parameters n_L, n_U and s. Figure 17 shows that extremes occur when $n_L = n_s$: the smallest values then occur when $n_U \gg n_L$ and largest values occur when $n_U = n_L$. The ranges when $M_{\rm max} \geq 1.97M_\odot$ are, for $s = 1$, 8.4 km < 14.2 km, and for $s = 1/3$, 11.7 km < 12.6 km. It is also worth noting that, in the case $s = 1/3$, the lower limit to $R_{1.4}$ is 11.7 km, more

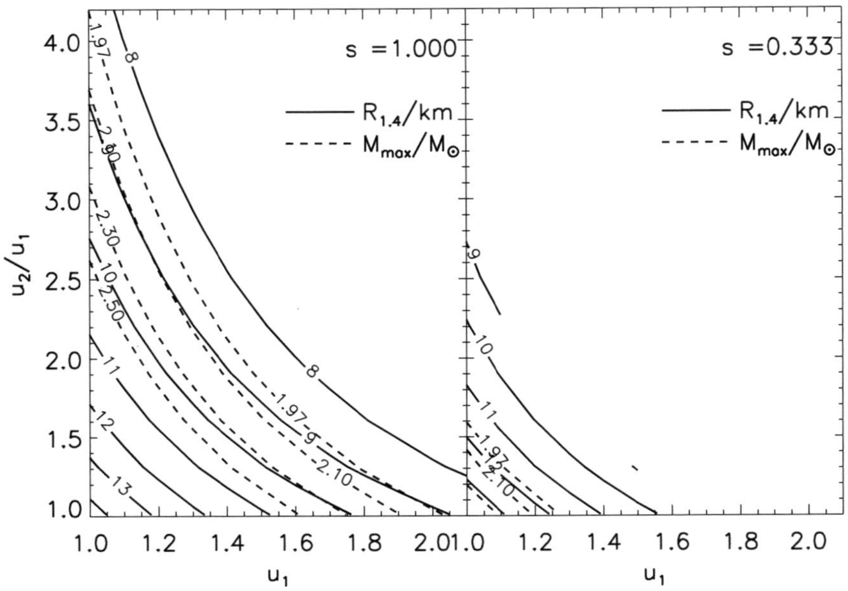

Fig. 17. Radius limits when low-density phase transitions occur. Solid (dashed) lines are contours of $R_{1.4}$ ($M_{\max}$). The left (right) panel is for $s = 1$ (1/3).

than 0.5 km larger than is the case without strong phase transitions. Furthermore, a phase transition remains compatible with $M_{\max} \geq 1.97M_{\odot}$ only if $n_L \lesssim 2n_s$. This dramatically reduces allowable models for strong phase transitions; nevertheless, they are not absolutely forbidden in either case even if $M_{\max} \gtrsim 2.3M_{\odot}$. There is thus some reasonable theoretical evidence to ignore the possibility of strong phase transitions.

5. Astrophysical Observations of Neutron Star Radii

5.1. *Photospheric radius expansion bursts*

Reference 48 proposed using photospheric radius expansion (PRE) X-ray bursts to obtain simultaneous mass and radius measurements in 1979, but the method did not become popular until about 2006. Type I X-ray bursts originate in accreting binaries when sufficient combustible mass accumulates onto a neutron star surface. The fuel ignites, with burning spreading quickly across the entire stellar surface. A subset of these bursts, PRE bursts, are energetic enough to lift the stellar photosphere to great heights. This requires luminosities exceeding the so-called Eddington luminosity[f]

$$L_{\mathrm{Edd},\infty} = \frac{4\pi cGM}{\kappa}\sqrt{1 - \frac{2GM}{R_{\mathrm{phot}}c^2}}, \tag{44}$$

[f]Quantities with the ∞ subscript are redshifted, as viewed at the Earth.

where κ is the atmosphere's mean opacity and $R_{\rm phot}$ is the distance of the photosphere from the star's center.[g] The bursts are short, lasting a matter of seconds, and the lifted material quickly loses opacity and falls back to the stellar surface. The burst decays as the stellar surface cools. Measurements of the flux and temperature during this decay permit the extraction of an angular area for the source

$$A = \frac{F_\infty}{\sigma_B T_{c,\infty}^4} = \frac{1}{f_c^4}\left(\frac{R_\infty}{D}\right)^2, \tag{45}$$

where the flux and color and effective temperatures at Earth are $F_\infty, T_{c,\infty}$ and $T_{\rm eff,\infty}$, respectively. $f_c = T_{\rm eff}/T_c$ is the color correction factor which accounts for the effects of a nonblackbody atmosphere, $R_\infty = R/\sqrt{1-2GM/(Rc^2)}$ is the apparent radiation radius and D is the distance.

Quantities extracted from observations include A, $F_{\rm Edd,\infty} = L_{\rm Edd,\infty}/(4\pi D^2)$, and D (see Ref. 49 for a summary and references). Quantities estimated from theory are κ and f_c. One can form two parameters from the observables:

$$\alpha \equiv \frac{F_{\rm Edd,\infty}}{\sqrt{A}}\frac{\kappa D}{f_c^2 c^3} = \beta(1-2\beta), \quad \gamma \equiv \frac{A}{F_{\rm Edd,\infty}}\frac{f_c^4 c^3}{\kappa} = \frac{R}{\beta(1-2\beta)^{3/2}}. \tag{46}$$

The quantity γ is independent of D, and $R_\infty = \alpha\gamma$ is independent of κ and $F_{\rm Edd,\infty}$. Note that $R \le \sqrt{27/3125}\ \gamma \simeq 0.093\gamma$ since $\beta(1-2\beta)^{3/2}$ has a maximum when $\beta = 1/5$. The solutions for mass and radius become

$$\beta = \frac{1}{4} \pm \frac{\sqrt{1-8\alpha}}{4}, \quad R_\infty = \alpha\gamma. \tag{47}$$

α should be less than 1/8 for real solutions to exist. In practice, however, observed values of α are greater than 1/8 to more than 1σ (Table 1). Monte Carlo sampling within the error ranges of the observables mostly leads to imaginary or acausal solutions (according to the *maximally compact* solution with $M_{\rm max} = 1.97M_\odot$). Restricting acceptances to valid trials, $\alpha \approx 1/8$ and $\beta \approx 1/4$ become strongly preferred. Preferred values for γ lie near their 1σ upper limits, γ_+. Then $R \approx \gamma_+/\sqrt{128}$,

Table 1. PRE bursters.

Source	α[a]	γ[a] (km)	$\gamma_+/\sqrt{128}$ (km)[b]	$\gamma\beta(1-2\beta)$ (km)[c]
4U 1820 − 30	0.182 ± 0.035	81.7 ± 18.0	8.8	10.2
SAX J1748.9 − 2021	0.170 ± 0.053	89.6 ± 28.6	10.4	11.1
EXO 1745 − 248	0.189 ± 0.060	70.9 ± 23.8	8.4	8.6
KS1731 − 260	0.187 ± 0.057	82.1 ± 24.8	9.4	10.0
4U1724 − 207	0.179 ± 0.055	86.6 ± 27.7	10.1	10.8
4U1608 − 52	0.213 ± 0.084	68.3 ± 22.0	8.0	10.0

Note: [a]Data from Ref. 49; 1σ uncertainties displayed.
[b]Inferred radii including Eddington flux redshift.
[c]Inferred radii without Eddington flux redshift.

[g]$R_{\rm phot} \gg R$ during the first part of the burst, and in principle $R_{\rm phot} = R$ when "touchdown" occurs and $L_{\rm Edd}$ is measured.

which ranges from 8 km to 10 km,[19] as shown in Table 1. That only extremely small fractions (i.e., about 4%) of trials are valid indicates serious problems with the model.

Possible solutions to the small acceptance are that the Eddington flux is measured while the photosphere is still well above the surface, so the redshift factor in Eq. (44) is absent,[19] or an accretion disk obscures part of the flux as seen from the Earth. In the former case, one finds

$$\alpha = \beta\sqrt{1-2\beta}, \quad \gamma = \frac{R}{\beta(1-2\beta)}, \tag{48}$$

whose real solution requires $\alpha \le 3^{-3/2} \simeq 0.192$ and $\beta \le 1/3$. All the sources but 4U1608 satisfy this. As a result, a much larger trial acceptance rate ensues, about 65%, after ensuring causality with $M_{\rm max} = 1.97 M_\odot$. Note now that $R = \gamma\beta(1-2\beta) \ge \gamma/9$, which is larger than for the previous model, as shown in Table 1. In the latter case, disk shadowing leads to an underprediction of A, an overprediction of $\alpha \propto 1/\sqrt{A}$, an underprediction of $\gamma \propto A$ and an underprediction of $R_\infty \propto \sqrt{A}$. In addition, the models assume uniform temperature over the source. It is straightforward to show that nonuniform temperatures generally lead to an underprediction of the effective emission area, or in this case, A.

Another approach is to limit the study to those bursts in which one can avoid the uncertainty of the location of the photospheric radius, i.e., choosing only "passive" bursts, and also using information from the whole cooling tail rather than just near touchdown and at late times. Reference 50 performed a study of three sources with this *cooling tail method* and showed that $F_{{\rm Edd},\infty}$ could be determined to at least 3% accuracy and $A' = D/R_\infty = (f_c^4 A)^{-1/2}$ to about 1%. Their sources all have nearly the same values for $\alpha \simeq 0.117$ and $\gamma \simeq 128$ km to about 2%. The major uncertainty is distance. Since α is smaller than 1/8 to about 1σ, most Monte Carlo trials produce real solutions. From Eq. (47), $\beta \simeq 0.185$ and $R_\infty \simeq 15$ km, leading to $M \simeq 1.5 M_\odot$ and $R \simeq 11.9$ km.

5.2. *Quiescent low-mass X-ray binaries*

qLMXBs produce thermal emission from neutron stars heated by recent episodes of accretion.[51] Reference 52 showed that models of thermal emission from pure hydrogen photospheres[53,54] give inferred radii of the correct order of magnitude. Continued observations and a consistent treatment of the surface gravity[55,56] led to estimates of radii between 8 km and 16 km for sources in globular clusters where D is reasonably well known.

Reference 57 combined new qLMXB observations and a "common radius" model (motivated by the generic vertical shape of M–R curves, cf. Fig. 2) and obtained a preferred value of 9.0 ± 1.4 km. However, individually analyzed, the sources each had radii in a wide range (7–20 km). Reference 58 pointed out that (i) some of the smaller radius neutron stars may have helium rather than hydrogen atmospheres, and (ii) the estimated amount of X-ray absorbing matter between the sources and Earth might be wrong. Reference 59 confirmed that the absorbing mass inferred for the source with the largest inferred radius was overestimated, leading to a radius reduction.

We note that the common radius model underpredicts the mean radius. Measurements of qLMXB spectra constrain R_∞ to a higher precision than the surface gravity $g = GM/(R^2\sqrt{1-2\beta})$, since the main global observables are F_∞, $T_{c,\infty}$ and D. For a blackbody

$$F_\infty = \left(\frac{R_\infty}{D}\right)^2 \sigma_B T_{c,\infty}^4. \tag{49}$$

Contours of g in the M–R plane are orthogonal to those of R_∞ and poorly constrained by available data for every source. Therefore, simultaneously refitting all sources under the constraint of a common radius is accommodated most easily by varying g rather than R_∞. As is clear from the R_∞ contours seen in Fig. 2, sources with widely varying estimated values of R_∞ can satisfy the constraint of a common radius only if that radius is smaller than the smallest R_∞ in the sample. Sources with large R_∞ can be reconciled by increasing their masses. For the analysis of Ref. 57, the mean R_∞ of the individually analyzed sources was 14.0 km; for the joint analysis with a common radius, it became 13.2 km. A red flag is that the range of individually predicted masses was $0.8M_\odot < M < 1.8M_\odot$ but in the common radius analyses it becomes $0.7M_\odot < M < 2.3M_\odot$.

5.3. *Bayesian analyses of observations with theoretical priors*

Reference 19 introduced Bayesian inference to combine M and R estimates from both qLMXBs and PRE X-ray bursts with information about neutron stars (hadronic crust, causality, general relativity, $M_{\rm max}$) to obtain constraints on the EOS. This work led to novel constraints on the density dependence of the nuclear symmetry energy and the nuclear incompressibility, as well as the EOS at high density. This work also demonstrated from the computation of Bayes factors that the PRE burst model without a redshift factor on $F_{\rm edd}$ is greatly preferred, and also demonstrated that helium, not hydrogen, is favored for at least one qLMXB source. The final result was a range of $R_{1.4}$ from 10.7 km to 12.5 km.

We can simulate this Bayesian analysis by combining the priors on radii, e.g., Fig. 8, with the astrophysical observations outlined above. For this exercise, we use the mass and radius information deduced for six PRE and six qLMXB sources compiled by Ref. 49. To simplify the analysis, Ref. 49 got rid of the "nuisance variable" M by averaging $R(M)$ for each source over the observed mass distribution of binary pulsars. The result is an overall average radius of 10.4 ± 0.7 km, with central values for the individual radii averaged over the neutron star mass distribution in the range 9.0 km$< \bar{R} <$ 12.5 km. Folding this distribution with the prior information from Fig. 8, one obtains the results shown in Fig. 18. Whereas the prior distributions

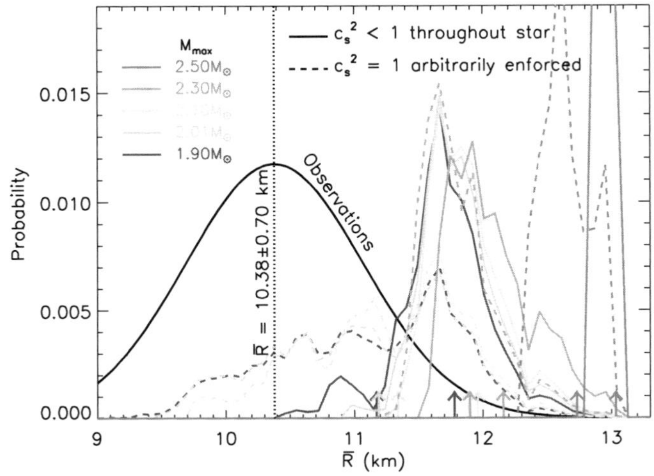

Fig. 18. (Color online) Mass-averaged radius probabilities. The black solid line shows the mass-averaged and source-averaged radius probability distribution for the six PRE and six qLMXB sources from Ref. 49. The solid colored curves show the results of folding the observed radius distribution with prior information from Fig. 8 in model A; dashed curves instead use prior information for model B. The colors code the assumed values of $M_{\rm max}$.

are nearly uniform in radii between 11 km and 13 km for model A, the fact that the observational average radius is less than the prior minimum (11 km) results in a preferred radius peaking near this minimum, at least unless $M_{\rm max}$ is very large. The preferred radius value increases with the assumed $M_{\rm max}$. The preferred radii for model A are naturally larger than those of model B by an amount that increases with increasing $M_{\rm max}$. Also note that the observed radius distribution increases the folded results for the probabilities of small radii for model B.

5.4. *Pulse-profile observations*

Surface emission from spinning neutron stars produces a periodic brightness change as hot and cold spots move in and out of the line of sight. Brightness variations could be due to magnetic fields, by nonuniform thermonuclear burning in an X-ray burster, or anisotropic accretion from a companion. The strong surface gravity of a neutron star affects the pulse shapes and amplitudes; generally, the more compact the star, the more of its surface is visible to a distant observer, and the smaller the amplitudes. Fortunately, using pulse profiles to measure compactness does not depend on the star's distance or intervening absorbing matter, unlike measurements from PRE bursts or qLMXBs. The method is one motivation for the funded NASA's *NICER*[6] and the proposed ESO Large Observatory For X-ray Timing (*LOFT*)[60] missions.

Early attempts to employ this technique generally studied bolometric emissions[61–64] which could not untangle degeneracies between M and R and produced estimated values for β that, while consistent with neutron stars, lacked enough accuracy to be useful. It has been suggested, however, that pulse profiles observed in two or more energy bands supply additional constraints that can break these degeneracies.[65–68] The method makes use of nearly universal relations, similar to the correlations shown in Fig. 13, among properties of rotating neutron stars. These include the surface ellipticity, the specific angular momentum $a = Ic/(GM^2P_s)$, where P_s is the spin period, the star's quadrupole moment, M, and $R_{\rm eq}$, the equatorial radius.[44,69,70] Models[63,67] show that the pulse profile is nearly independent of the angular size of a spot, and, using the aforementioned semi-universal relations, the profiles are determined essentially by M, $R_{\rm eq}$, the inclination i, and θ_s, the colatitude of the spot. M, $R_{\rm eq}$ and P_s are sufficient to determine R, the nonrotating radius for the same mass M, and, hence, will predict the EOS. The *NICER* mission was designed to acquire M and R to about 5% accuracy from at least one pulsing source, PSR J0437 − 4715.

6. Conclusions

We have probed the extent to which neutron star structure and the dense matter EOS are currently constrained by nuclear physics experiments, theoretical neutron matter calculations and astrophysical observations. We have been proactive in applying important restrictions stemming from assumptions that neutron stars have hadronic crusts, general relativity is the correct theory of gravity, a minimum observed value for $M_{\rm max}$, and causality is not violated. Analyses not including such assumptions about neutron stars are likely to predict neutron star radii, for example, that are much smaller or larger than the 11–13 km we suggest.

Acknowledgments

I gratefully acknowledge support from the Institute of Nuclear Theory as part of this work was inspired by and performed during the programs INT-16-2a and INT-16-2b. Partial support for this research was provided by the US DOE under grant DE-AC02-ER40317.

The author was supported in part by USDOE DEAC02-87ER40317.

References

1. J. Antoniadis *et al.*, *Science* **340** (2013) 448.
2. M. G. Alford and S. Han, *Eur. Phys. J. C* **52** (2015) 63.
3. J. M. Lattimer, *Ann. Rev. Nucl. Part. Sci.* **62** (2012) 488.
4. E. E. Flanagan and T. Hinderer, *Phys. Rev. D.* **77** (2008) 021502.
5. M. Kramer *et al.*, *Science* **314** (2006) 97.
6. K. Gendreau, Z. Arzoumanian and T. Okajima, *Proc. SPIE* **8443** (2012) 13.
7. J. R. Oppenheimer and G. M. Volkoff, *Phys. Rev.* **55** (1939) 374.
8. J. M. Lattimer and M. Prakash, *Astrophys. J.* **550** (2001) 426.
9. S. Koranda, N. Stergioulas and J. L. Friedman, *Astrophys. J.* **488** (1997) 799.
10. J. M. Lattimer and M. Prakash, *Phys. Rep.* **442** (2007) 109.

11. J. M. Lattimer and M. Prakash, "What a Two Solar Mass Neutron Star Really Means", in *From Nuclei to Stars*, ed. S. Lee (World Scientific, Singapore, 2011), p. 275.
12. J. M. Lattimer and M. Prakash, *Phys. Rep.* **621** (2016) 127.
13. J. M. Lattimer and A. W. Steiner, *Eur. Phys. J. A* **50** (2014) 40.
14. E. Lipparini and S. Stringari, *Phys. Lett. B* **112** (1982) 421.
15. A. Tamii *et al.*, *Phys. Rev. Lett.* **107** (2011) 062502.
16. L.-W. Chen, C. M. Ko, B.-A. Li and J. Xu, *Phys. Rev. C* **82** (2010) 024321.
17. M. Warda, M. Centelles, X. V. Nas and X. Roca-Maza, *Acta Phys. Pol. B* **43** (2012) 209.
18. J. M. Lattimer and Y. Lim, *Astrophys. J.* **771** (2013) 51.
19. A. W. Steiner, J. M. Lattimer and E. F. Brown, *Astrophys. J.* **722** (2010) 33.
20. S. Gandolfi, J. Carlson and S. Reddy, *Phys. Rev. C* **85** (2012) 032801.
21. K. Hebeler, J. M. Lattimer, C. J. Pethick and A. Schwenk, *Phys. Rev. Lett.* **105** (2010) 161102.
22. P. Danielewicz and J. Lee, *Nucl. Phys. A* **922** (2014) 1.
23. M. B. Tsang, Y. Zhang, P. Danielewicz, M. Famiano, Z. Li, W. G. Lynch and A. W. Steiner, *Phys. Rev. Lett.* **102** (2009) 122701.
24. C. Drischler, V. Somá and A. Schwenk, *Phys. Rev. D* **89** (2014) 02586.
25. F. Sammarruca, L. Coraggio, J. W. Holt, N. Itaco, R. Machleidt and L. E. Marcucci, *Phys. Rev. C* **91** (2015) 4311.
26. E. F. Brown, L. Bildsten and R. E. Rutledge, *Astrophys. J.* **504** (1998) 95.
27. B. Link, R. I. Epstein and J. M. Lattimer, *Phys. Rev. Lett.* **83** (1999) 3362.
28. G. Baym, H. A. Bethe and C. J. Pethick, *Nucl. Phys. A* **175** (1971) 22.
29. J. S. Read, B. D. Lackey, B. J. Owen and J. L. Friedman, *Phys. Rev. D* **79** (2009) 124032.
30. F. Özel and P. Freire, arXiv:1603.02698.
31. A. W. Steiner, J. M. Lattimer and E. F. Brown, *Eur. Phys. J. A* **52** (2016) 18.
32. J. M. Lattimer and M. Prakash, *Science* **304** (2004) 536.
33. P. Haensel, J. L. Zdunik, M. Bejger and J. M. Lattimer, *Astron. Astrophys.* **502** (2009) 605.
34. S. L. Shapiro, S. A. Teukolsky and I. Wasserman, *Astrophys. J.* **272** (1983) 702.
35. S. L. Shapiro, S. A. Teukolsky and I. Wasserman, *Nature* **340** (1989) 451.
36. J. W. T. Hessels, S. M. Ransom, I. H. Stairs, P. C. C. Freire, V. M. Kaspi and F. Camilo, *Nature* **311** (2006) 1901.
37. A. Burrows and J. M. Lattimer, *Astrophys. J.* **307** (1986) 178.
38. J. M. Lattimer and A. Yahil, *Astrophys. J.* **340** (1989) 425.
39. B. M. Barker and R. F. O'Connell, *Phys. Rev. D* **12** (1975) 329.
40. T. Damour and G. Schaeffer, *Nuovo Cimento B* **101** (1988) 127.
41. J. B. Hartle, *Astrophys. J.* **150** (1967) 1005.
42. J. M. Lattimer and B. F. Schutz, *Astrophys. J.* **629** (2005) 979.
43. S. Postnikov, M. Prakash and J. M. Lattimer, *Phys. Rev. D* **82** (2010) 024016.
44. K. Yagi and N. Yunes, *Phys. Rev. D* **88** (2013) 023009.
45. R. O'Shaughnessy, B. Farr, E. Ochsner, H.-S. Cho and C.-H. Lee, *Phys. Rev. D* **89** (2014) 064048.
46. H.-S. Cho, arXiv:1607.07118.
47. C. L. Rodriguez, B. Farr, V. Raymond, W. M. Farr, T. B. Littenberg, D. Fazi and V. Kalogera, *Astrophys. J.* **784** (2014) 119.
48. J. van Paradijs, *Astrophys. J.* **234** (1979) 609.
49. F. Ozel, D. Psaltis, T. Güver, G. Baym, C. Heinke and S. Guillot, *Astrophys. J.* **820** (2016) 28.

50. J. Nättilä, A. W. Steiner, J. E. Kajava, V. F. Suleimanov and J. Poutanen, *Astron. Astrophys.* **591** (2016) 23.
51. J. van Paradijs, F. Verbunt, R. A. Shafer and K. A. Arnaud, *Astron. Astrophys.* **182** (1987) 47.
52. R. Rutledge, L. Bildsten, E. Brown, G. Pavlov and E. Zavlin, *Astrophys. J.* **514** (1999) 945.
53. M. Rajagopal and R. W. Romani, *Astrophys. J.* **461** (1996) 327.
54. V. E. Zavlin, G. G. Pavlov and Y. A. Shibanov, *Astron. Astrophys.* **315** (1996) 141.
55. C. O. Heinke, G. B. Rybicki, R. Narayan and J. E. Grindlay, *Astrophys. J.* **644** (2006) 1090.
56. N. A. Webb and D. Barret, *Astrophys. J.* **671** (2007) 727.
57. S. Guillot, M. Servillat, N. A. Webb and R. E. Rutledge, *Astrophys. J.* **772** (2013) 7.
58. A. W. Steiner, J. M. Lattimer and E. F. Brown, *Astrophys. J. Lett.* **765** (2013) L5.
59. C. O. Heinke, H. N. Cohn, P. M. Lugger, N. A. Webb, W. C. G. Ho, J. Anderson, S. Campana, S. Bogdanov, D. Haggard, A. M. Cool and J. E. Grindlay, *Mon. Not. R. Astron. Soc.* **444** (2014) 443.
60. M. Feroci *et al.*, *Exp. Astron.* **34** (2012) 415.
61. G. G. Pavlov and V. E. Zavlin, *Astrophys. J.* **490** (1997) 91.
62. J. Poutanen and M. Gierlinski, *Mon. Not. R. Astron. Soc.* **343** (2003) 1301.
63. S. Bogdanov, G. B. Rybicki and J. E. Grindlay, *Astrophys. J.* **670** (2007) 668.
64. D. A. Leahy, S. M. Morsink and C. Cadeau, *Astrophys. J.* **672** (2008) 1119.
65. T. E. Strohmayer, W. Zhang and J. H. Swank, *Astrophys. J.* **487** (1997) 77.
66. M. C. Miller and F. K. Lamb, *Astrophys. J.* **499** (2009) 37.
67. D. Psaltis, F. Özel and D. Chakrabarty, *Astrophys. J.* **787** (2014) 136.
68. M. C. Miller and F. K. Lamb, *Astrophys. J.* **808** (2015) 31.
69. S. M. Morsink, D. A. Leahy, C. Cadeau and J. Braga, *Astrophys. J.* **663** (2007) 1244.
70. M. Bauböck, E. Berti, D. Psaltis and F. Özel, *Astrophys. J.* **777** (2013) 68.

Gravitational waves from neutron star binaries

Chang-Hwan Lee
Department of Physics,
Pusan National University,
Busan 46241, Korea
clee@pusan.ac.kr

With H. A. Bethe, G. E. Brown worked on the merger rate of neutron star binaries for the gravitational wave detection. Their prediction has to be modified significantly due to the observations of $2M_\odot$ neutron stars and the detection of gravitational waves. There still, however, remains a possibility that neutron star-low mass black hole binaries are significant sources of gravitational waves for the ground-based detectors. In this paper, I review the evolution of neutron star binaries with super-Eddington accretion and discuss the future prospect.

Keywords: Gravitational waves; neutron stars; black holes; accretion and accretion disks.

1. First Detection of Gravitational Waves: Binary Black Hole Coalescence

The first direct detection of gravitational waves, for which G. E. Brown had been waited for a long time, was announced on 12th February 2016,[1] even though the source is identified as a massive black hole–black hole binary instead of a neutron star–neutron star binary or a neutron star-black hole binary. According to the paper, on 14 September 2015, Laser Interferometer Gravitational-Wave Observatory (LIGO) had detected gravitational waves from a merging binary black holes composed of two black holes with masses $36^{+5}_{-4}M_\odot$ and $29^{+4}_{-4}M_\odot$. Recently, second gravitational-wave observation from a $22M_\odot$ binary black hole coalescence has been reported.[2] These unexpected observations of gravitational waves from black hole binaries open a new possibility in the field of gravitational-wave astronomy.

Before the first detection of gravitational waves, double pulsar binaries were expected to be the main source of gravitational waves. A binary pulsar PSR B1913+16 was observed in 1975,[3] and the change in their binary orbital period was consistent with the prediction based on the gravitational wave emission.[4] The first detection of gravitational waves generated by binary neutron star coalescence was expected to be around 2019. In this respect, the first detection of gravitation waves from binary black holes in 2015 gave a strong excitement to the community.

There is a possibility that the black hole binaries are most dominant gravitational-wave sources. However, neutron star binaries are still believed to be important sources for LIGO. In Sec. 2, I will summarize G. E. Brown's contribution to the evolution of neutron star binaries and the merger rates of the gravitational waves.[5,6] In Sec. 3, I will summarize the role of super-Eddington accretion in the evolution of neutron star binaries.[7] In Sec. 4, I will discuss the future prospect of neutron star binaries as sources of the gravitational waves.

2. Gravitational Waves and Neutron Star Binaries

In 1998, with H. A. Bethe, G. E. Brown estimated the merger rates of gravitational waves using the soft equation of state (EOS) in which kaon condensation is taken into account.[5] According to their original prediction, there are more neutron star-low mass black hole binaries than neutron star–neutron star or black hole–black hole binaries as sources of gravitational waves that can be detected by advanced LIGO. Their argument is based on the theoretical prediction that the maximum mass of neutron stars can be reduced to $1.5M_{\odot}$ if a soft EOS with kaon condensation is introduced.[8] These prediction was consistent with well-measured neutron star masses in neutron star–neutron star binaries.[9–11] However, due to the observation of $2M_{\odot}$ neutron stars in neutron star-white dwarf binaries,[12,13] their original prediction on the maximum mass of neutron stars has to be modified.

Lee *et al.*[6] introduced a possibility of supercritical accretion in the evolution of neutron star binaries. Their argument is based on the evolution of progenitors of neutron star binaries and is independent of neutron star EOS. They claimed that the mass distribution in neutron star–neutron star binaries is not the indication of maximum mass of neutron stars. They also discussed the possibility that neutron star masses can be significantly higher in neutron star-white dwarf binaries than in double neutron star binaries. These prediction turned out to be right by $2M_{\odot}$ neutron star observations in neutron star-white dwarf binaries,[12,13] even though they could not predict the value of maximum neutron star mass. Note that their prediction is before the discovery of $2M_{\odot}$ neutron stars. In the next section, I will summarize the consequences of the super-Eddington accretion in the evolution of neutron star binaries.

3. Evolution of Neutron Star Binaries with Super-Eddington Accretion

Following Lee *et al.*,[6] Lee and Cho discussed that there still remains a possibility that neutron star-low mass black hole binaries are significant sources of gravitational waves for the ground-based detectors.[7] In this section, we review their work.

Well-measured neutron star masses in neutron star–neutron star binaries are still below $1.5M_{\odot}$ within uncertainty ranges.[10,11] This observation was one of the main arguments for H. A. Bethe and G. E. Brown to draw a conclusion that maximum neutron star mass is about $1.5M_{\odot}$.[5] However, two $2M_{\odot}$ neutron stars have been found in neutron star-white dwarf binaries.[12,13] These observations ruled out lower maximum mass of neutron stars. The discrepancy in the mass distribution of neutron stars (between neutron star–neutron star and neutron star-white dwarf binaries) suggests that the masses of neutron stars depend on the evolutionary track of their progenitors.

In the evolution of a neutron star binary progenitor, the first-born neutron star can spiral into the expanding envelope of the companion star. During the in-spiral phase, the neutron star can accrete materials from the envelope of the companion. The amount of accretion is still quite uncertain and cannot be calculated from the first principles. Lee *et al.* suggested that the super-Eddington accretion occurs during the evolution of soft X-ray transient sources (black hole binaries)[14,15] and extended their argument to the neutron star binaries.[6,7] Eddington limit is the critical condition at which the photon pressure balances the gravitational force such that the ionized material cannot be accreted into the star. In the evolution of noncompact stars, the Eddington limit usually holds. However, for the accretion into neutron stars, the neutron star surface temperature can reach $\mathcal{O}(10^{10}\,\mathrm{K})$ above which the thermal neutrino emission wins against thermal photon emission. Since the neutrino interacts with ordinary matter very weakly, the produced thermal photons can carry energy and pressure out of the system allowing super-Eddington accretion.

In Fig. 1, neutron star mass distribution in the presence of supercritical accretion is summarized.[7] As in the figure, one can consider three different kinds of neutron star binaries.

(a) NS–NS binaries: If the difference in the progenitor (zero age main sequence; ZAMS) masses of a neutron star binary is less than 4%, two progenitors evolve

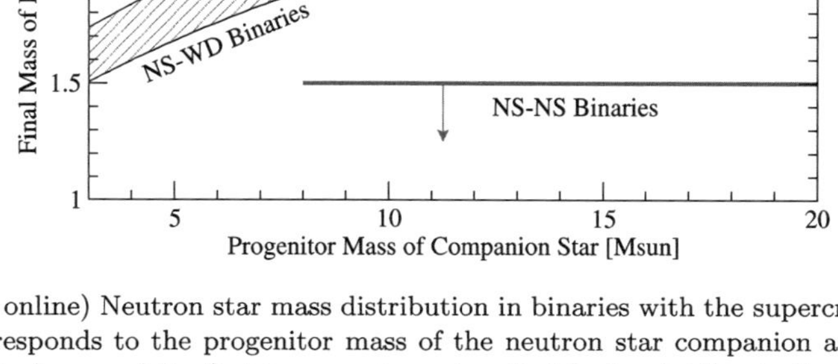

Fig. 1. (Color online) Neutron star mass distribution in binaries with the supercritical accretion. The x-axis corresponds to the progenitor mass of the neutron star companion and y-axis corresponds to the final mass of the first-born neutron star. Red horizontal line indicates the observed neutron star masses in neutron star-white dwarf (NS-WD) binaries. Blue horizontal line indicates the upper bound of neutron star masses in neutron star–neutron star binaries (NS–NS) binaries. When the progenitor mass of the companion star in x-axis is less (larger) than $\sim 8M_\odot$, the final companion star will result in a white dwarf (neutron star). The hatched area corresponds to the probability density of forming neutron star binaries indicating that the probability of forming high-mass neutron star–neutron star (HMNS–NS) or black hole-neutron star (BH-NS) binaries is similar to that of neutron star-white dwarf binaries. (This figure with caption is taken from Ref. 7.)

almost simultaneously and the first-born neutron star does not have enough time for the super-Eddington accretion. In this case, the observed final mass (y-axis) of the first-born (more massive) neutron star is the same as that at its birth. Since neutron star is formed as a result of Fe core collapse of a giant star (with $8M_\odot$–$20M_\odot$ as indicated in x-axis), the observed mass distribution is an indication of the Fe core mass distribution at the end of giant evolution. The upper mass limit ($\sim 1.5M_\odot$) shown in the observed neutron star–neutron star binaries indicates that the Fe core mass does not grow above $1.5M_\odot$ for the giants with mass range $8M_\odot$–$20M_\odot$.[16] Note that this argument is independent of neutron star EOS.

(b) NS-WD binaries: If the difference in ZAMS masses of a neutron star binary is bigger than 4%, the companion (less massive star) is still in main sequence when the more massive giant (with ZAMS mass in a range $8M_\odot$–$20M_\odot$) produces a primary neutron star as a result of core collapse. The primary neutron star spirals into the expanding envelope of companion when it expands, and can have super-Eddington accretion. If the companion is not massive enough to produce a neutron star, a white dwarf will be produced. The left part of the hatched area in Fig. 1 correspond to NS-WD binaries. Note that observed $2M_\odot$ neutron stars in neutron star-white dwarf binaries[12,13] are consistent with the mass distribution in Fig. 1.

(c) HMNS/BH-NS binaries: This class corresponds to the case in which the difference in ZAMS masses is bigger than 4% while both ZAMS masses are big enough ($>8M_\odot$) to produce neutron stars at the end of their evolution. In this case, as in Fig. 1, the primary (first-born) neutron star can have super-Eddington accretion and its final mass can be increased above $2.5M_\odot$. Note that the final fate of the primary neutron star, whether it will remain as a neutron star (high-mass neutron star; HMNS) or collapse into a black hole, depends on the neutron star EOS, i.e., the maximum mass of a neutron star.

Neutron stars have been detected mainly as pulsars and the recycling of the primary neutron star during the binary evolution increases the detectability of neutron star binaries because lower magnetic field with wider opening angle due to the recycling enhances the pulsar lifetime. On the other hand, if the primary neutron star collapses into a black hole, it cannot be observed as a pulsar and the secondary lower mass neutron star has shorter lifetime due to the stronger magnetic field. Hence, BH-NS binary may be very hard to be observed. In addition, there can be possible observational biases.

There is no observational evidence of the third class, HMNS/BH-NS binary, yet. However, if it exists, it can contribute to the gravitational wave detection.

4. Future Prospect

In the previous section, a possibility of new class of neutron star binaries, HMNS-NS or BH-NS binaries, is discussed. Its existence is still an open question. In order to have an answer, one should have better understanding on the detectability of the new class of neutron star binaries, in addition to the theoretical understanding on the mechanism of super-Eddington accretion and the maximum mass of neutron stars, i.e., neutron star EOS.

One possibility to test the existence of a new class is the detection of gravitational waves from neutron star binaries. Currently, only two gravitational wave sources have been detected and both of them are black hole binaries. In a few years, however, when the advanced LIGO reaches the design sensitivity, gravitational waves from neutron star binaries are expected to be detected. In principle, gravitational waves can carry information on the masses of a binary (Chirp mass) and the neutron star EOS.[17] Hence, one may be able to investigate the inner structure of neutron stars using gravitational waves and the existence of HMNS/BH-NS binaries.

Acknowledgments

I would like to dedicate this article to G. E. Brown who was a teacher, a collaborator, and a mentor not only for physics but also for my life. This work was supported by the National Research Foundation of Korea (NRF) grant funded by the Korea government (MSIP) (Nos. NRF-2015R1A2A2A01004238 and NRF-2016R1A5A1013277).

References

1. B. P. Abbott *et al.*, *Phys. Rev. Lett.* **116** (2016) 061102.
2. B. P. Abbott *et al.*, *Phys. Rev. Lett.* **116** (2016) 241103.
3. R. A. Hulse and J. H. Taylor, Discovery of a pulsar in a binary system, *Astrophys. J.* **195** (1975) L51.
4. J. M. Weisberg, D. J. Nice and J. H. Taylor, *Astrophys. J.* **722** (2016) 1030.
5. H. Bethe and G. E. Brown, *The Astrophys. J.* 506 (1998) 780.
6. C.-H. Lee, H.-J. Park and G. E. Brown, *Astrophys. J.* **670** (2007) 741.
7. C.-H. Lee and H.-S. Cho, *Nucl. Phys. A* **928** (2014) 296.
8. C.-H. Lee, *Phys. Rep.* **275** (1996) 255.
9. C.-H. Lee and G. E. Brown, *J. Korean Phys. Soc.* **49** (2006) 803.
10. J. Lattimer and M. Prakash, *Phys. Rep.* **442** (2007) 109.
11. M. Prakash, Proceedings of 8th International Workshop on Critical Point and Onset of Deconfinement, arXiv:1307.0397.
12. P. B. Demorest, T. Pennucci, S. M. Ransom, M. S. E. Roberts and J. W. T. Hessels, *Nature* **467** (2010) 1081.
13. J. Antoniadis *et al.*, *Science* **340** (2013) 448.
14. C.-H. Lee, G. E. Brown and R. A. M. J. Wijers, *Astrophys. J.* **575** (2002) 996.
15. C.-H. Lee, *J. Korean Phys. Soc.* **49** (2006) 1791.
16. G. E. Brown, A. Heger, N. Langer, C.-H. Lee, S. Wellstein and H. A. Bethe, *New Astron.* **6** (2001) 457.
17. K. Kiuchi, Y. Sekiguchi, K. Kyutoku and M. Shibata, *Class. Quantum Grav.* **29** (2012) 124003.

Baryons and chiral symmetry

Keh-Fei Liu
Department of Physics and Astronomy,
University of Kentucky, Lexington, KY 40506 USA
liu@pa.uky.edy

This manuscript is dedicated to the memory of Gerald E. Brown who was the author's Ph.D. thesis advisor, a mentor in his professional career and a lifelong friend.

The relevance of chiral symmetry in baryons is highlighted in three examples in the nucleon spectroscopy and structure. The first one is the importance of chiral dynamics in understanding the Roper resonance. The second one is the role of chiral symmetry in the lattice calculation of $\pi N \sigma$ term and strangeness. The third one is the role of chiral $U(1)$ anomaly in the anomalous Ward identity in evaluating the quark spin and the quark orbital angular momentum. Finally, the chiral effective theory for baryons is discussed.

Keywords: Baryons; chiral symmetry; Roper; quark spin; strangeness.

1. Introduction

It is well known that for three-flavor quantum chromodynamics (QCD), the chiral symmetry $\mathrm{SU}(3)_L \times \mathrm{SU}(3)_R$ is spontaneously broken to the diagonal $\mathrm{SU}(3)_V$ with the octet pseudoscalar mesons as the Goldstone boson. There is also a $U_A(1)$ anomaly which begets a heavy η' meson. As such, pions and chiral symmetry are important for low-energy hadron physics and the chiral dynamics has been successfully applied in subjects such as $\pi\pi$ scattering,[1,2] vector dominance,[3] KSRF relation,[4] low-energy πN scatterings,[1,5] πN scattering up to about 1 GeV with the skyrmion,[6] nucleon static properties,[7] electromagnetic form factors,[7] πNN form factor,[8] and the Goldberger–Treiman relation.[9]

The long range part of the nucleon–nucleon potential is due to one pion exchange and more modern NN potential has included the correlated two-pion exchange potential to account for the intermediate range attraction.[10,11] These are the major ingredients in the realistic phase-shift equivalent NN potentials.[12]

Realization and application of chiral symmetry in hadrons and nuclei has been a major research theme in Gerry Brown's scientific career. These include chiral symmetry in nucelon–nucleon interaction,[13] meson exchange currents,[14] and his joint research effort with Mannque Rho on little chiral bag,[15] Brown–Rho scaling,[16] and dense nuclear matter.[17]

With the advent of QCD and the lattice chiral fermion formulations in domain wall fermion[18] and overlap fermion,[19] lattice QCD has provided a new tool in addressing the role of chiral symmetry and the associated dynamics in first-principles calculation in terms of quarks and gluons.

In this memorial manuscript, I will discuss three examples in baryons where chiral symmetry plays a crucial role as a way to pay tribute to Gerry's teaching throughout the author's professional career and to echo his passion of chiral symmetry by extending the study of chiral symmetry from nuclear structure to nucleon structure.[20]

The three examples are the Roper resonance, the pion nucleon sigma term and strangeness, and the quark spin as representative cases for the importance of chiral symmetry in baryon spectroscopy and structure.

2. Roper Resonance

The Roper resonance has been studied extensively, but its status as the lowest excited state of the nucleon with the same quantum numbers is intriguing. First of all, it has been noted for a long time that it is rather unusual to have the first positive parity excited state lower than the negative parity excited state which is the $N^-_{1/2}(1535)$ in the $S_{11}\pi N$ scattering channel. This is contrary to the excitation pattern in the meson sectors with either light or heavy quarks. This parity reversal has been problematic for the quark models based on SU(6) symmetry with color–spin interaction between the quarks[21] which cannot accommodate such a pattern. Realistic potential calculations with linear and Coulomb potentials[22] and the relativistic quark model[23] all predict the Roper to be $\sim$100–200 MeV above the experimental value with the negative parity state lying lower. On the other hand, the pattern of parity reversal was readily obtained in the chiral soliton model like the Skyrme model via the small oscillation approximation to πN scattering.[6] Although the first calculation[24] of the original skyrmion gives rise to a breathing mode which is $\sim$200 MeV lower than the Roper resonance, it was shown later[25] that the introduction of the sixth-order term, which is the zero range approximation for the ω meson coupling, changes the compression modulus and produces a better agreement with experiment for both the mass and width in πN scattering.

Since the quark potential model is based on the SU(6) symmetry with residual color–spin interaction between the quarks; whereas, the chiral soliton model is based on spontaneous broken chiral symmetry, their distinctly different predictions on the ordering of the positive and negative parity excited states may well be a reflection of different dynamics as a direct consequence of the respective symmetry. This possibility has prompted Glozman and Riska to suggest[26] that the parity reversal in the excited nucleon and Δ, in contrast to that in the excited Λ spectrum, is an indication that the inter-quark interaction of the light quarks is mainly of the flavor–spin nature, which implies Goldstone boson exchange, rather than the color–spin nature due to the one-gluon exchange. This suggestion is supported in the lattice QCD study of "valence QCD",[27] where one finds that the hyperfine splitting between the nucleon and Δ and also between ρ and π are largely decimated when the Z-graphs in the quark propagators are removed. This is an indication that the color–magnetic interaction is not the primary source of the inter-quark spin–spin interaction for light quarks. (The color–magnetic part, being spatial in origin, should be unaffected by the truncation of Z-graphs in valence QCD, which only affects the time part.) Yet, it is consistent with the Goldstone-boson-exchange picture which requires Z-graphs and thus the flavor–spin interaction.

The failure of the SU(6) quark model to delineate the Roper and its photoproduction has prompted the speculation that the Roper resonance may be a hybrid state with excited glue[28] or a $qqqq\bar{q}$ five quark state.[29] Thus, unraveling the nature of Roper resonance has direct bearing on our understanding of the quark structure and chiral dynamics of baryons.

Lattice QCD is, in principle, the most desirable tool to adjudicate the theoretical controversy surrounding the issue. However, there is a complication. As shown in Fig. 1, the lattice calculations with the Clover fermions,[30–32] chirally improved fermions,[33] and twisted mass fermions[32] agree with the overlap fermion for the nucleon mass,[34] but their first excited nucleon state (i.e., the Roper) are mostly over 2 GeV and much higher than those of the overlap fermion in the pion mass range between 300 MeV and 600 MeV. Near the physical pion mass, they are $\sim$300 MeV above the experimental Roper mass at $\sim$1430 MeV, while the overlap fermion prediction agrees with experiment. This situation is basically a redux of the quenched calculations,[35,36] i.e., the overlap results are much lower than those of the Wilson-type fermions. Why is there such a discrepancy? Since the calculations with Wilson-type fermions, which breaks chiral symmetry, use the variational calculation, while the calculation with the chiral overlap fermion adopts the sequential

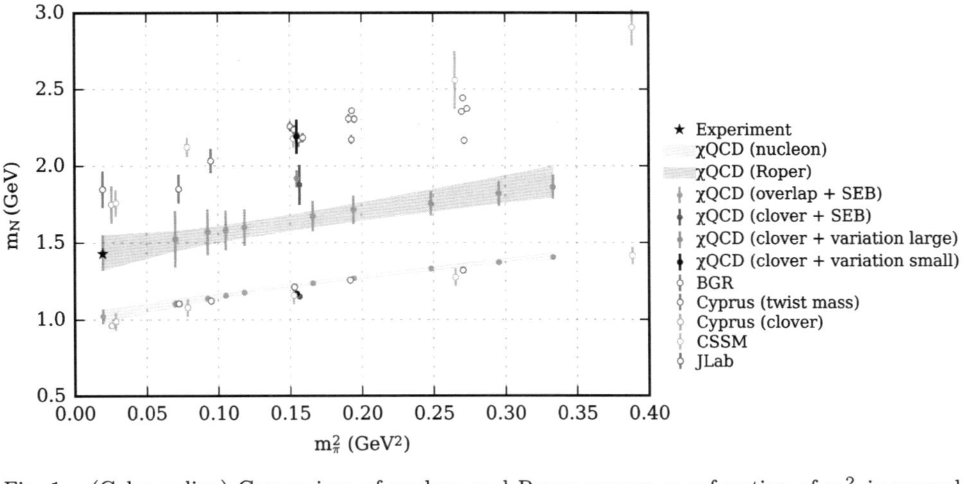

Fig. 1. (Color online) Comparison of nucleon and Roper masses as a function of m_π^2 in several dynamical fermion calculations with different fermion actions.

empirical bayes method (SEB)[37] to extract the Roper state, a question arises as to whether the discrepancy is due to different fitting algorithms or different actions at finite lattice spacing.

It was previously reported[34] that to check the different algorithmic approaches, the SEB method was used on the gauge configurations that are produced by HSC Collaboration[31] to calculate the nucleon and the Roper. These are 2+1 flavor Clover fermion gauge configurations on the anisotropic $24^3 \times 128$ lattice with $a_s = 0.123$ fm and the light u/d sea quark mass corresponds to a pion mass of ~ 390 MeV (Nota bene the HSC results in Fig. 1 were obtained on the $16^3 \times 128$ lattice with the same action and quark mass). The SEB fitting of the nucleon and Roper masses are shown in Fig. 1 together with the variational results from HSC.[31] We see that while the nucleon mass agrees with that from HSC; the Roper, on the other hand, is lower than that from HSC by ~ 300 MeV (blue point in Fig. 1) with a $\sim 3\sigma$ difference. To verify that this difference is not due to the fitting algorithm, we have carried out a variational calculation with different smearing sizes for the interpolation field. We see that when the r.m.s. radii of the Gaussian smeared source, include one as large as 0.86 fm, the Roper does appear lower at 1.92(6) GeV as shown by the second plateau in Fig. 2 (upper panel) and indicated by the red point in Fig. 1. This is in agreement with the SEB result which is shown as the blue point in Fig. 1. On the other hand, when the r.m.s. radii of all the Gaussian smeared sources are less than 0.4 fm, the nucleon excited state is higher — 2.19(11) GeV as shown in Fig. 2 (lower panel) and indicated by the black dot in Fig. 1, which is consistent with the HSC result indicated by the purple points in the same figure. This presumably confirms the speculation[34] that one needs a large enough source to have a better overlap with the $2S$ excited state which has a radial node at ~ 0.9 fm from the study of its Coulomb wave function[34]; whereas, smaller sources may couple to the $3S$ state strongly and results in a higher mass. More importantly, the fact that this new variational result agrees with that from the SEB method suggests that the SEB method is a legitimate approach and, consequently, its extraction of the Roper at ~ 1.6 GeV in this pion mass range from the overlap fermion calculation in Fig. 1 should be reliable.

The fitting algorithm issue being settled, this leads to the possibility that the difference is due to the different fermion actions at finite lattice spacing.

An extensive model has been constructed to study the N^* resonance in πN scattering partial waves.[38] The unperturbed states are the bare N and Δ and the meson–baryon reaction channels including $\pi N, \eta N$, and $\pi\pi N$ which has $\pi\Delta, \rho N$ and σN resonant components. This model fits the πN scattering data well in various channels below 2 GeV. It is found[39] in the P_{11} channel, the meson–baryon transition amplitude is strong, which shifts the bare $1/2^+ N^*$ at 1763 MeV down to $(\mathrm{Re}\,\mathrm{M_R}, -\mathrm{Im}\,\mathrm{M_R}) = (1357, 76)$ MeV which corresponds to the P_{11} pole of $N^*(1440)$. This is a shift of ~ 400 MeV in mass due to the meson–baryon coupling. Compared to the overlap fermion which has the lattice chiral symmetry which can have a larger

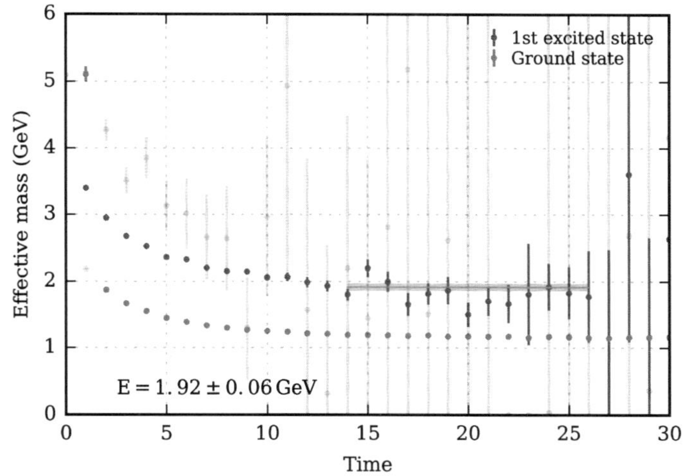

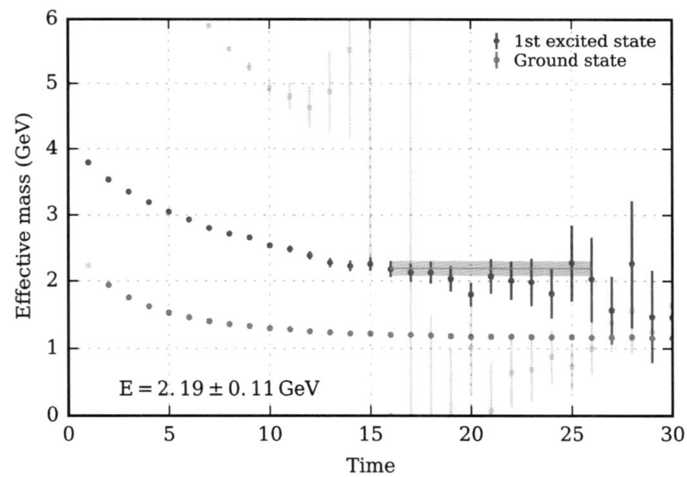

Fig. 2. Nucleon and Roper masses from the variational method including large smeared sources with radius as large as 0.86 fm (upper panel) and with the radius of the smeared sources limited to less than 0.4 fm (lower panel).

$\langle 0|\chi_{N,3q}|\pi N\rangle$ matrix element and a large Roper to πN coupling. These 3-quark to 5-quark coupling which invokes a pair creation or the Z-graph might be curtailed in the Wilson fermions like in "valence QCD" as we discussed above and consequently results in a higher Roper state. This is a plausible explanation which is verifiable with variational calculations of the Roper state with the Wilson–Clover fermion at smaller lattice spacings where chiral symmetry is better recovered. Given the pattern of level reversal, the πN scattering model and the lattice calculations, we believe the Roper resonance has a sizable πN component in its wave function[34] and is a showcase for the role of chiral dynamics.

3. πN and Strangeness Sigma Terms

As measures of explicit and spontaneous chiral symmetry breaking in the baryon sector, $\sigma_{\pi N}$, defined as

$$\sigma_{\pi N} \equiv \hat{m}\langle N|\bar{u}u + \bar{d}d|N\rangle, \tag{1}$$

where $\hat{m} = (m_u + m_d)/2$ is the averaged light quark mass, and f_s^N defined as the strangeness σ term as a fraction of the nucleon mass,

$$\sigma_{sN} \equiv m_s\langle N|\bar{s}s|N\rangle, \quad f_s^N = \frac{\sigma_{sN}}{m_N} \tag{2}$$

are fundamental quantities which pertain to a wide range of issues in hadron physics. They include the quark mass contribution in the baryon which is related to the Higgs contribution to the observable matter,[40,41] the pattern of SU(3) breaking,[40] πN and KN scatterings,[42,43] and kaon condensate in dense matter.[44] Using the sum rule of the nucleon mass, the heavy quark mass contribution can be deduced by that from the light favors, in the heavy quark limit and also in the leading order of the coupling.[41,45,46] At the same time, precise values of the quark mass term for various flavors, from light to heavy, are of high interest for dark matter searches,[47–49] where the popular candidate of dark matter (likes the weakly interacting mass particle) interacts with the observable world throughout the Higgs couplings, so that the precise determination of the $\sigma_{\pi N}$ and σ_{sN} can provide constraints on the dark matter candidates.

Phenomenologically, the $\sigma_{\pi N}$ term is typically extracted from the πN scattering amplitude. To lowest order in m_π^2, the unphysical on-shell isospin-even πN scattering amplitude at the Cheng–Dashen point corresponds to $\sigma(q^2 = 2m_\pi^2)$[42,43] which can be determined from πN scattering via fixed-q^2 dispersion relation.[43] $\sigma_{\pi N}$ at $q^2 = 0$ can be extracted through a soft correlated two-pion form factor.[50–52] Analysis of the πN scattering amplitude to obtain $\sigma_{\pi N}(0)$ from the Lorentz covariant baryon chiral perturbation and the Cheng–Dashen low-energy theorem are also developed.[53–55] They give $\sigma_{\pi N}$ values in the range $\sim$45–64 MeV, while the most recent analysis[55] gives 59.1(3.5) MeV.

Lattice calculations should be a good tool in giving reliable results to these quantities. Again, there is an issue about chiral symmetry. It was pointed out[56,57] that due to explicit chiral symmetry breaking, the quark mass in the Wilson-type fermions has an additive renormalization and the flavor-singlet and nonsinglet quark masses renormalize differently. In this case, the renormalized strange scalar matrix element $\langle N|\bar{s}s|N\rangle^R$ can be written as

$$\langle N|\bar{s}s|N\rangle^R = \frac{1}{3}[(Z_0 + 2Z_8)\langle N|\bar{s}s|N\rangle + (Z_0 - Z_8)\langle N|\bar{u}u + \bar{d}d|N\rangle], \tag{3}$$

where Z_0 and Z_8 are the flavor-singlet and flavor-octet renormalization constants, respectively. Z_0 differs from Z_8 by a disconnected diagram which involves a quark loop. In the massless renormalization scheme, one can calculate these renormalization constants perturbatively. For the massless case where $\bar{\psi}\psi = \bar{\psi}_L\psi_R + \bar{\psi}_R\psi_L$, a quark loop for the scalar density vanishes no matter how many gluon insertions there are on the loop, since the coupling involving γ_μ does not change helicity. Thus, the massless scalar quark loop is zero and $Z_0 = Z_8$. There is no mixing of the scalar matrix element with that of u and d. This is the same with the overlap fermion, since the overlap has chiral symmetry and the inverse of its massless quark propagator D_c anti-commutes with γ_5, i.e., $\{D_c, \gamma_5\} = 0$ as in the continuum.

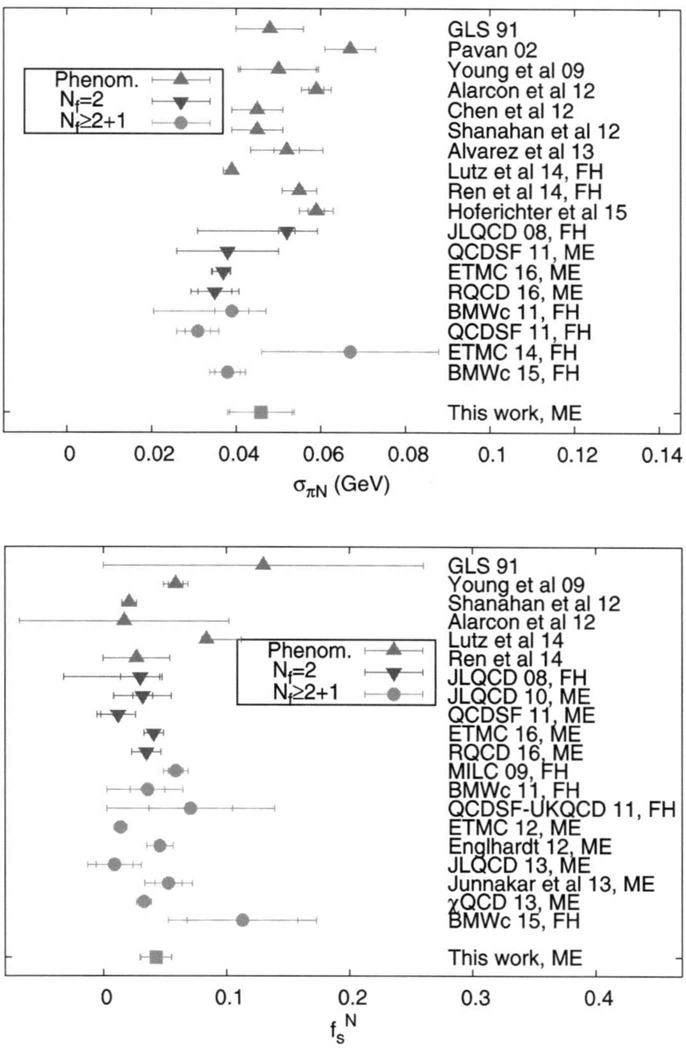

Fig. 3. (Color online) The results of $\sigma_{\pi N}$ (upper panel) and f_s^N (lower panel) from both phenomenology and lattice simulations. The narrow error bar for each data point is the statistical, and the broad one is that for the total uncertainty. The physical proton mass 938 MeV is used to obtain f_s^N in this work. They are color-coded in phenomenological and indirect approaches (green), $N_f = 2$ lattice calculations (blue), and $N_f = 2 + 1$ lattice calculations (red). Detailed references are given in Ref. 59.

This is not so for Wilson-type fermion, where its free quark propagator contains a term proportional to the Wilson r-term which violates chiral symmetry and will give a nonzero contribution to the scalar matrix element at the massless limit, leading to $Z_0 \neq Z_8$. Since the u and d matrix elements in the nucleon are not small, there can be a substantial flavor mixing at finite a. This lattice artifact due to nonchiral fermions can be removed by calculating Z_0 and Z_8.[41,58] Furthermore, the direct calculation of the matrix element with Wilson-type fermions faces the complication that the sigma term with bare quark mass is not renormalization group invariant. This can also be corrected with the introduction of various renormalization constants to satisfy the Ward identities.[41,58] All of these involve additional work and will introduce additional errors. On the contrary, there is no flavor mixing in the overlap fermion and the sigma terms are renormalization group invariant with bare mass and bare matrix element, since the renormalization constants of quark mass and scalar operator cancel, i.e., $Z_m Z_s = 1$ due to chiral symmetry. For the latest calculation with overlap fermion on $2 + 1$ flavor domain wall fermion gauge configurations for several ensembles with different lattice spacings, volume and sea masses including one at the physical pion mass, the global fit gives the prediction of $\sigma_{\pi N} = 45.9(7.4)(2.8)$ MeV and $\sigma_{sN} = 40.2(11.7)(3.5)$ MeV. They are shown in Fig. 3 together with those from phenomenological and indirect calculations, and also the $N_f = 2$ and $N_f = 2 + 1$ lattice calculations. This value of $\sigma_{\pi N}$ has a two-sigma tension with the recent results based on Roy–Steiner equations[55] which gives $\sigma_{\pi N} = 59.1(3.5)$ MeV.

To conclude, we believe that to calculate $\sigma_{\pi N}$ and σ_{sN} which are fundamental quantities reflecting both the explicit and spontaneous chiral symmetry, it is theoretically clean and straightforward procedure-wise to calculate them with chiral fermions on the lattice in order to obtain reliable results without the complication of renormalization and flavor-mixing as compared to nonchiral fermions.

4. Quark Spin and Orbital Angular Momentum

The quark spin content of the nucleon was found to be much smaller than that expected from the quark model by the polarized deep inelastic lepton–nucleon scattering experiments and the recent global analysis reveals that the total quark spin contributes only $\sim 25\%$ to the proton spin.[60] This is dubbed 'proton spin crisis' since no model seems to be able to explain it convincingly and, moreover, quantitatively.

Once again, first-principle lattice calculation should be able to address this issue. The ideal calculation would be to use the conserved axial-vector current of the chiral fermions which satisfies the anomalous Ward identity (AWI) on lattice at finite lattice spacing. However, it is somewhat involved to construct the current itself for the overlap fermion.[61] Before it is implemented, one can use the AWI as the normalization condition for the simpler local axial-vector current

$$\partial_\mu \kappa_A A^0_\mu = 2mP - 2iN_f q, \tag{4}$$

where $A^0_\mu = \sum_{i=u,d,s} \overline{\psi}_i i\gamma_\mu\gamma_5(1 - \frac{1}{2}D_{\rm ov})\psi_i$ is the local singlet axial-vector current and $mP = \sum_{i=u,d,s} m_i\overline{\psi}_i i\gamma_5(1 - \frac{1}{2}D_{\rm ov})\psi_i$ is the pseudoscalar density with $D_{\rm ov}$ being the massless overlap operator and q the local topological charge as derived in the Jacobian factor from the fermion determinant under the chiral transformation whose local version is equal to $\frac{1}{16\pi^2}{\rm tr}_c G_{\mu\nu}\tilde{G}_{\mu\nu}(x)$ in the continuum,[62] i.e.,

$$q(x) = {\rm Tr}\,\gamma_5 \left(\frac{1}{2}D_{\rm ov}(x,x) - 1\right) \xrightarrow[a\to 0]{} \frac{1}{16\pi^2}{\rm tr}_c\, G_{\mu\nu}\tilde{G}_{\mu\nu}(x), \tag{5}$$

where Tr refers to the trace over color and spin, whereas ${\rm tr}_c$ indicates the trace over color. κ_A in Eq. (4) is the finite lattice renormalization factor (often referred to as Z_A in the literature for the flavor nonsinglet case) needed for the local axial-vector current to satisfy the AWI on the lattice with finite lattice spacing, much like the finite renormalization for the vector and nonsinglet axial-vector currents. We shall call it lattice normalization. On the other hand, the mP and q defined with the overlap operators do not have multiplicative renormalization. There is a two-loop renormalization of the singlet A^0_μ and the topological charge q mixes with $\partial_\mu A^0_\mu$. It turns out that they are the same. Thus, the renormalized AWI is the same as the unrenormalized AWI (but normalized) to the α_s^2 order. To utilize the AWI, one needs to calculate the matrix elements of $2mP$ and $2q$ on the r.h.s. of the AWI and extrapolate to $q^2 = 0$. However, the smallest $|q^2|$ is larger than the pion mass squared on the lattices that we work on, the extrapolation to q^2 is not reliable. Instead, we shall match the form factors at finite $|q^2|$ from both sides, i.e.,

$$2m_N\kappa_A g^0_A(q^2) + q^2\kappa_A h^0_A(q^2) = 2mg_P(q^2) + N_f g_G(q^2), \tag{6}$$

where the singlet $g^0_A(q^2)$ and the induced pseodoscalar $h^0_A(q^2)$ are the bare form factors. $2mg_P(q^2)$ and $g_G(q)$ are the form factors for the pseudoscalar current and topology, respectively. From this normalization condition, one can determine κ_A and the normalized g^0_A is $\kappa_A g^0_A(0)$. This has been employed in the calculation of the strange quark spin to find $\Delta s + \Delta\bar{s} = -0.0403(44)(61)$.[63] This is more negative than the other lattice calculations with and axial-vector current, mainly because $\kappa_A = 1.36(4)$ is found to be larger than that of the flavor-octet axial-vector current. The lesson here is that, unless the conserved current is used to carry out the calculation, it is essential to adopt the AWI to obtain the normalization of the local axial-vector current. This is possible with the overlap fermion.

While the final numbers on the u and d spin fraction which include the connected insertion are still beng worked out, the initial results indicate that it is the larger negative $2mP$ matrix elements that cancel the positive topological charge term in the triangle anomaly in the disconnected insertions that lead to a small g^0_A.

There are various ways to decompose the proton spin in to quark and glue spins and orbital angular momenta.[64,65] From the symmetrized energy-mometum tensor of QCD (the Belinfante form), it is shown[66] that the proton spin can be decomposed as

$$\mathbf{J}_{\rm QCD} = \mathbf{J}_q + \mathbf{J}_g = \frac{1}{2}\mathbf{\Sigma}_q + \mathbf{L}_q + \mathbf{J}_g, \tag{7}$$

where the quark angular momentum $\mathbf{J}_q$ is the sum of quark spin and orbital angular momentum,

$$\mathbf{J}_q = \frac{1}{2}\mathbf{\Sigma}_q + \mathbf{L}_q = \int d^3x \left[\frac{1}{2}\overline{\psi}\gamma\gamma^5\psi + \psi^\dagger\{\mathbf{x} \times (i\mathbf{D})\}\psi\right], \tag{8}$$

and each of which is gauge invariant. The glue angular momentum operator

$$\mathbf{J}_g = \int d^3x[\mathbf{x} \times (\mathbf{E} \times \mathbf{B})], \tag{9}$$

is also gauge invariant. However, it cannot be further divided in to the glue spin and orbital angular momentum gauge invariantly with the Belinfante tensor.

Since it has a large finite volume effect to calculate the operator with a spatial $\mathbf{r}$ on the lattice with periodic boundary condition, one can instead calculate the quark and glue momentum and angular momentum from their form factors $T_1(q^2)$ and $T_2(q^2)$ and obtain the momentum and angular momentum fractions from their forward limits, i.e., $\langle x\rangle = T_1(0)$ and $J = \frac{1}{2}(T_1(0) + T_2(0))$, much like the electric charge and magnetic moment from the forward Dirac and Pauli form factors $F_1(0)$ and $F_2(0)$. After determining the quark angular momentum, the quark orbital angular momentum is obtained by subtracting the quark spin from it. This has been carried out in a quenched approximation.[67] The OAM fractions $2\langle L^q_{\rm kin}\rangle$ for the u and d quarks in the CI have different signs and add up to 0.01(10), i.e., essentially zero. This is the same pattern which has been seen with dynamical fermion configurations and light quarks, as pointed out earlier. The large OAM fractions $2\langle L^q_{\rm kin}\rangle$ for the u/d and s quarks in the DI is due to the fact that g^0_A in the DI is large and negative, about $-0.12(1)$ for each of the three flavors. All together, the quark OAM constitutes a fraction of 0.47(13) of the nucleon spin. The majority of it comes from the DI.

As far as the spin decomposition is concerned, it is found that the quark spin constitutes 25(12)% of the proton spin, the gluon total AM takes 28(8)% and the rest is due to the quark kinetic OAM which is 47(13)%. The pie charts of the quark spin, quark orbital angular momentum and glue angular momentum are given in Fig. 4.

Since this calculation is based on a quenched approximation which is known to contain uncontrolled systematic errors, it is essential to repeat this calculation with dynamical fermions of light quarks and large physical volume. However, we expect that the quark OAM fraction may still be large in the dynamical calculation.

In the naive constituent quark model, the proton spin comes entirely from the quark spin. On the other hand, in the skyrme model,[68] the proton spin originates solely from the OAM of the collective rotational motion of the pion field.[69] What is found in the present lattice calculation suggests that the QCD picture, aside from

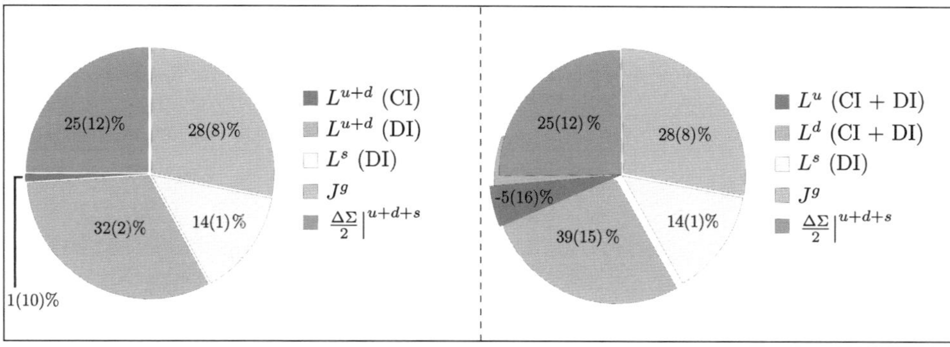

Fig. 4. Pie charts for the quark spin, quark orbital angular momentum and gluon angular momentum contributions to the proton spin. The left panel show the quark contributions separately for CI and DI, and the right panel shows the quark contributions for each flavor with CI and DI summed together for u and d quarks.

the gluon contribution, is somewhere in between these two models, indicating a large contribution of the quark OAM due to the meson cloud ($q\bar{q}$ pairs in the higher Fock space) in the nucleon.

5. Effective Theory of Baryons

Many estimates of quark spin and OAM contributions of the nucleon are based on quark models. However, quark models are not realistic effective theories of QCD, since they do not have chiral symmetry, a salient feature of QCD whose dynamics governs light-quark hadron structure, spectroscopy, and scattering at low energies. It is being learned quantitatively through lattice calculations of quark spin from the AWI[63,70,71] that the smallness of the quark spin contribution in the nucleon is related to the $U(1)$ anomaly, the same anomaly which is responsible for the large η' mass. This cannot be understood with quark models without the chiral $U(1)$ anomaly. Similarly, relativistic quark models do not explain the large OAM obtained from the lattice calculation in Sec. 4. Both the chiral quark model studies[26] and lattice calculation of valence QCD[27,34] reveal that the level reversal of the positive and negative parity excited states of the nucleon, i.e., $P_{11}(1440)$ (Roper resonance) and $S_{11}(1535)$, and the hyperfine splittings between the decuplet and octet baryons are dominated by the meson-mediated flavor–spin interaction, not the gluon-mediated color–spin interaction. All of these point to the importance of the meson degree of freedom ($q\bar{q}$ pairs in the higher Fock space) which is missing in the quark model.

To see how this comes about, one can follow Wilson's renormalization group approach to effective theories. It is suggested by Liu *et al.*[72] that the effective theory for baryons between the scale of $4\pi f_\pi(\sim 1\,\text{GeV})$, which is the scale of the meson size ($l_M \sim 0.2\,\text{fm}$), and $\sim 300\,\text{MeV}$, which is the scale of a baryon size ($l_B \sim 0.6\,\text{fm}$), should be a chiral quark model with renormalized couplings and renormalized meson, quark and gluon fields which preserve chiral symmetry. A

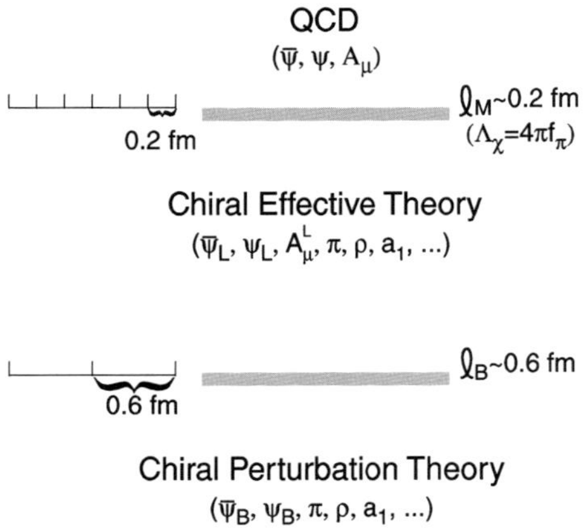

Fig. 5. A schematic illustration of the two-scale delineation of the effective theories. The shaded bars mark the positions of the cutoff scales l_M and l_B separating different effective theories.

schematic illustration for such division of scales[a] for QCD effective theories is illustrated in Fig. 5.

This suggestion is based on the observation that mesons and baryon form factor assume a monopole and dipole form, respectively. Since the πNN form factor is much softer than the $\rho\pi\pi$ form factor, it is suggested that the confinement scale of quarks in the baryon l_B is larger than l_M — the confinement scale between the quark and antiquark in the meson. This is consistent with the large-N_c approach to hadrons where the mesons are treated as point-like fields and the baryons emerge as solitons with a size of order unity in N_c.[74] Taking l_M from the $\rho\pi\pi$ form factor gives $l_M \sim 0.2\,\text{fm}$. This is very close to the chiral symmetry breaking scale set by $\Lambda_\chi = 4\pi f_\pi$. Considering them to be the same, operators of low-lying meson fields become relevant operators below Λ_χ. As for the baryon confinement scale, Liu *et al.* take it to be the size characterizing the meson–baryon–baryon form factors. Defining the latter from the respective meson poles in the nucleon pseudoscalar, vector, and axial-vector form factors in a lattice calculation[27] (see Fig. 17 in the reference), they obtained $l_B \sim 0.6 - 0.7\,\text{fm}$, satisfying $l_B > l_M$. Thus, in between these two scales l_M and l_B (corresponding to the scale of $\sim$300 MeV), one could have coexistence of mesons and quarks in an effective theory for baryons.

An outline is given[72] to show how to construct a chiral effective theory for baryons. In the intermediate length scale between l_M and l_B, one needs to separate the fermion and gauge fields in to long-range ones and short-range ones

$$\psi = \psi_L + \psi_S, \quad A^\mu = A^\mu_L + A^\mu_S, \tag{10}$$

[a]We should point out that although two scales are adopted here, they are distinct from those of Manohar and Georgi.[73] In the latter, the σ–quark model does not make a distinction between the quark fields in the baryons and mesons. As such, there is an ambiguity of double counting of mesons and $q\bar{q}$ states. By making the quark–quark confinement length scale l_B larger than the quark–antiquark confinement length scale l_M, one does not have this ambiguity.

where ψ_L/ψ_S and A_L^μ/A_S^μ represent the infrared/ultraviolet part of the quark and gauge fields, respectively, with momentum components below/above $1/l_M$ or Λ_χ. One adds irrelevant higher-dimensional operators to the ordinary QCD Lagrangian with coupling between bilinear quark fields and auxiliary fields as given in Ref. 75, interpreting these quark fields as the short-range ones, i.e., ψ_S and $\overline{\psi}_S$. Following the procedure by Li in Ref. 75, one can integrate out the short-range fields and perform the derivative expansion to bosonize ψ_S and $\overline{\psi}_S$. This leads to the Lagrangian with the following generic form:

$$\begin{aligned}\mathcal{L}_{\chi\,\mathrm{QCD}} &= \mathcal{L}_{\mathrm{QCD}'}(\overline{\psi}_L, \psi_L, A_L^\mu) + \mathcal{L}_M(\pi, \sigma, \rho, a_1, G, \ldots) \\ &\quad + \mathcal{L}_{\sigma q}(\overline{\psi}_L, \psi_L, \pi, \sigma, \rho, a_1, G, \ldots). \qquad (11)\end{aligned}$$

$\mathcal{L}_{\mathrm{QCD}'}$ includes the original form of QCD but in terms of the quark fields $\overline{\psi}_L, \psi_L$, and the long-range gauge field A_L^μ with renormalized couplings. It also includes higher-order covariant derivatives.[76] $\mathcal{L}_M$ is the meson effective Lagrangian, e.g., the one derived by Li[75] which should include the glueball field G. Finally, $\mathcal{L}_{\sigma q}$ gives the coupling between the $\overline{\psi}_L$, ψ_L, G and mesons. As we see, in this intermediate scale, the quarks, gluons and mesons coexist and meson fields couple to the long-range quark fields. Going further down below the baryon confinement scale $1/l_B$, one can integrate out $\overline{\psi}_L$, ψ_L and A_L^μ, resulting in an effective Lagrangian $\mathcal{L}(\overline{\Psi}_B, \Psi_B, \pi, \sigma, \rho, a_1, G, \ldots)$ in terms of the baryon and meson fields.[77] This would correspond to an effective theory in the chiral perturbation theory. In order for the chiral symmetry to be preserved, the effective theory of baryons at the intermediate scale necessarily involves mesons in addition to the effective quark and gluon fields. This naturally leads to a chiral quark effective theory.

Models like the little bag model with skyrmion outside the MIT bag,[78] the cloudy bag model[79] and quark chiral soliton model[80] have the right degrees of freedom and, thus, could possibly delineate the pattern of division of the proton spin with large quark OAM contribution. In particular, the fact that the u and d OAM in the MIT bag and to some extent the LFCQM in Table 1 have different signs from those of the lattice calculation may well be due to the lack of meson contributions as demanded by chiral symmetry in the effective theory of baryons.

6. Summary

We discuss three examples in lattice QCD to highlight the role chiral symmetry and chiral dynamics play in baryons. From the observation of parity reversal of the excited, the nucleon and Δ spectrum and the phenomenological model for the πN scattering and N^* states, it is suggested that it is the chiral dynamics that plays the leading role in the pattern of the low-lying baryon spectrum. This notion is supported by our study of the valence QCD and the fact that the lattice calculation of the Roper state by the nonchiral fermions at relatively coarse lattice spacing are substantially higher (by several hundred MeV's) than that of the chiral fermion which agrees with experiment.

All the differences of lattice calculations with different fermion formulation are supposed to go away as the lattice spacing approaches zero due to universality. However, at finite lattice spacing, the different results form nonchiral and chiral fermions serve to illustrate the role of chiral symmetry and confirm the observation that chiral dynamic seems to be ubiquitous in low-energy hadron physics. Thus, it is essential to incorporate Goldstone bosons in the effective theory of baryons in addition to quarks and gluons below the chiral scale of $4\pi f_\pi$.

Acknowledgment

This work is supported partially by US Department of Energy Grant No. DE-SC0013065.

References

1. S. Weinberg, *Phys. Rev. Lett.* **17** (1966) 616.
2. J. Gasser and H. Leutwyler, *Nucl. Phys. B* **250** (1985) 465, 517, 539.
3. J. J. Sakurai, *Phys. Rev. Lett.* **17** (1966) 1021; *Currents and Mesons* (The University of Chicago Press, Chicago, 1969).
4. K. Kawarabayashi and M. Suzuki, *Phys. Rev. Lett.* **16** (1966) 255; Fayyazuddin and Riazuddin, *Phys. Rev.* **147** (1966) 1071.
5. Y. Tomozawa, *Nuovo Cimento A* **46** (1966) 707.
6. M. Mattis, in *Chiral Solitons*, Phenomenology of the Meson–Skyrmion System, ed. K. F. Liu (World Scientific, Singapore, 1987), p. 171.
7. G. Adkins, C. R. Nappi and E. Witten, *Nucl. Phys.* **228** (1983) 552.
8. B. A. Li and K. F. Liu, *Chiral Solitons*, ed. K. F. Liu (World Publishing Co., Singapore, 1987), p. 421.
9. M. Goldberger and S. B. Treiman, *Phys. Rev.* **110** (1958) 1178.
10. G. E. Brown and J. D. Jackson, *The Nucleon–Nucleon Interaction* (North-Holland, Amsterdam, 1976).
11. C. W. Wong and K. F. Liu, Nucleon–nucleon interaction, in *Topics in Nuclear Physics*, eds. T. T. S. Kuo and S. S. M. Wong, Lecture Notes in Physics, Vol. 144 (Springer-Verlag, Berlin, 1981), pp. 1–174.
12. R. V. Mau, *Lect. Notes Phys.* **581** (2001) 1; R. B. Wiringa, V. G. J. Stoks and R. Schiavilla, *Phys. Rev. C* **51** (1995) 38, doi:10.1103/PhysRevC.51.38, arXiv:nucl-th/9408016; R. Machleidt and I. Slaus, *J. Phys. G, Nucl. Part. Phys.* **27** (2001) 69, arXiv:nucl-th/0101056; V. G. J. Stoks, R. A. M. Klomp, C. P. F. Terheggen and J. J. de Swart, *Phys. Rev. C* **49** (1994) 2950, arXiv:nucl-th/9406039.
13. G. E. Brown, Chiral symmetry and the nucleon–nucleon interaction, in *Mesons in Nuclei*, Vol. 1, eds. M. Rho and D. Wilkinson, (North-Holland, Amsterdam, 1979).
14. D. O. Riska and G. E. Brown, *Phys. Lett. B* **38** (1972) 193.
15. G. E. Brown and M. Rho, *Phys. Lett. B* **82** (1979) 177; G. E. Brown, A. D. Jackson, M. Rho and V. Vento, *Phys. Lett. B* **140** (1984) 285.
16. G. E. Brown and M. Rho, *Phys. Rev. Lett.* **66** (1991) 2720.
17. G. E. Brown and M. Rho, *Phys. Rep.* **363** (2002) 85, arXiv:hep-ph/0103102.
18. D. B. Kaplan, *Phys. Lett. B* **288** (1992) 342, arXiv:hep-lat/9206013; Y. Shamir, *Nucl. Phys.* **406** (1993) 90, arXiv:hep-lat/9303005.
19. H. Neuberger, *Phys. Lett. B* **417** (1998) 141.
20. K. F. Liu, *Nucl. Phys. A* **928** (2014) 99, arXiv:1404.3754 [hep-ph].

21. S. Capstick and W. Roberts, *Prog. Part. Nucl. Phys.* **45** (2000) S241, arXiv:nucl-th/0008028.
22. K. F. Liu and C. W. Wong, *Phys. Rev. D* **28** (1983) 170.
23. S. Capstick and N. Isgur, *Phys. Rev. D* **34** (1986) 2809.
24. K. F. Liu, J. S. Zhang and G. R. E. Black, *Phys. Rev. D* **30** (1984) 2015; J. Breit and C. R. Nappi, *Phys. Rev. Lett.* **53** (1984) 889.
25. U. B. Kaulfuss and U.-G. Meissner, *Phys. Lett. B* **154** (1985) 193.
26. L. Y. Glozman and D. O. Riska, *Phys. Rep.* **268** (1996) 263.
27. K. F. Liu, S. J. Dong, T. Draper, D. Leinweber, J. Sloan, W. Wilcox and R. Woloshyn, *Phys. Rev. D* **59** (1999) 112001.
28. T. Barnes and F. E. Close, *Phys. Lett. B* **123** (1983) 221; Z. Li, *Phys. Rev. D* **44** (1991) 2841; C. E. Carlson and N. C. Mukhopadhyay, *Phys. Rev. Lett.* **67** (1991) 3745.
29. O. Krehl, C. Hanhart, S. Krewald and J. Speth, *Nucl. Phys. C* **62** (2000) 025207; R. Jaffe and F. Wilczek, *Eur. Phys. J. C* **33** (2004) S38.
30. M. Mahbub, W. Kamleh, D. B. Leinweber, P. Moran and A. G. Williams, *Phys. Lett. B* **707** (2012) 389, arXiv:1011.5724.
31. R. G. Edwards, J. J. Dudek, D. G. Richards and S. J. Wallace, *Phys. Rev. D* **84** (2011) 074508, arXiv:1104.5152.
32. C. Alexandrou, T. Leontiou, C. N. Papanicolas and E. Stiliaris, *Phys. Rev. D* **91** (2015) 014506, arXiv:1411.6765 [hep-lat].
33. G. P. Engel, C. Lang, D. Mohler and A. Schfer, *Phys. Rev. D* **87** (2013) 074504, arXiv:1301.4318.
34. K. F. Liu, Y. Chen, M. Gong, R. Sufian, M. Sun and A. Li, *PoS LATTICE* **2013** (2014) 507, arXiv:1403.6847 [hep-ph].
35. N. Mathur, Y. Chen, S. J. Dong, T. Draper, I. Horváth, F. X. Lee, K. F. Liu and J. B. Zhang, *Phys. Lett. B* **605** (2005) 137, arXiv:hep-ph/0306199.
36. B. G. Lasscock, J. N. Hedditch, W. Kamleh, D. B. Leinweber, W. Melnitchouk, A. G. Williams and J. M. Zanotti, *Phys. Rev. D* **76** (2007) 054510.
37. Y. Chen *et al.*, arXiv:hep-lat/0405001.
38. B. Julia-Diaz, H. Kamano, T.-S. H. Lee, A. Matsuyama, T. Sato and N. Suzuki, *Phys. Rev. C* **80** (2009) 025207, arXiv:0904.1918 [nucl-th].
39. N. Suzuki, B. Julia-Diaz, H. Kamano, T.-S. H. Lee, A. Matsuyama and T. Sato, *Phys. Rev. Lett.* **104** (2010) 042302, arXiv:0909.1356 [nucl-th].
40. R. D. Young and A. W. Thomas, *Phys. Rev. D* **81** (2010) 014503, arXiv:0901.3310 [hep-lat].
41. RQCD Collab. (G. S. Bali *et al.*), *Phys. Rev. D* **93** (2016) 094504, arXiv:1603.00827 [hep-lat].
42. L. S. Brown, W. J. Pardee and R. D. Peccei, *Phys. Rev. D* **4** (1971) 2801.
43. T. P. Cheng and R. F. Dashen, *Phys. Rev. D* **4** (1971) 1561.
44. D. B. Kaplan and A. E. Nelson, *Phys. Lett. B* **175** (1986) 57.
45. M. A. Shifman, A. I. Vainshtein and V. I. Zakharov, *Phys. Lett. B* **78** (1978) 443.
46. XQCD Collab. (M. Gong *et al.*), *Phys. Rev. D* **88** (2013) 014503, doi:10.1103/PhysRevD.88.014503, arXiv:1304.1194 [hep-ph].
47. T. Falk, A. Ferstl and K. A. Olive, *Phys. Rev. D* **59** (1999) 055009; *Phys. Rev. D* **60** (1999) 119904, arXiv:hep-ph/9806413.
48. J. R. Ellis, K. A. Olive and C. Savage, *Phys. Rev. D* **77** (2008) 065026, arXiv:0801.3656 [hep-ph].
49. J. Giedt, A. W. Thomas and R. D. Young, *Phys. Rev. Lett.* **103** (2009) 201802, arXiv:0907.4177 [hep-ph].
50. J. Gasser, H. Leutwyler and M. E. Sainio, *Phys. Lett. B* **253** (1991) 252.

51. T. Becher and H. Leutwyler, *Eur. Phys. J. C* **9** (1999) 643, arXiv:hep-ph/9901384.
52. M. M. Pavan, I. I. Strakovsky, R. L. Workman and R. A. Arndt, *PiN Newslett.* **16** (2002) 110, arXiv:hep-ph/0111066.
53. J. M. Alarcon, J. Martin Camalich and J. A. Oller, *Phys. Rev. D* **85** (2012) 051503, arXiv:1110.3797 [hep-ph].
54. Y. H. Chen, D. L. Yao and H. Q. Zheng, *Phys. Rev. D* **87** (2013) 054019, arXiv:1212.1893 [hep-ph].
55. M. Hoferichter, J. Ruiz de Elvira, B. Kubis and U. G. Meiner, *Phys. Rev. Lett.* **115** (2015) 092301, doi:10.1103/PhysRevLett.115.092301, arXiv:1506.04142 [hep-ph].
56. UKQCD Collab. (C. Michael *et al.*), *Nucl. Phys. Proc. Suppl.* **106** (2002) 293, doi:10.1016/S0920-5632(01)01692-9, arXiv:hep-lat/0109028.
57. JLQCD Collab. (K. Takeda *et al.*), *Phys. Rev. D* **83** (2011) 114506, doi:10.1103/PhysRevD.83.114506, arXiv:1011.1964 [hep-lat].
58. QCDSF Collab. (G. S. Bali *et al.*), *Phys. Rev. D* **85** (2012) 054502, doi:10.1103/PhysRevD.85.054502, arXiv:1111.1600 [hep-lat].
59. Y. B. Yang, A. Alexandru, T. Draper, J. Liang and K. F. Liu, arXiv:1511.09089 [hep-lat].
60. D. de Florian, R. Sassot, M. Stratmann and W. Vogelsang, *Phys. Rev. D* **80** (2009) 034030, arXiv:0904.3821 [hep-ph].
61. P. Hasenfratz, V. Laliena and F. Niedermayer, *Phys. Lett. B* **427** (1998) 125, doi:10.1016/S0370-2693(98)00315-3, arXiv:hep-lat/9801021.
62. Y. Kikukawa and A. Yamada, *Phys. Lett. B* **448** (1999) 265, arXiv:hep-lat/9806013; D. H. Adams, *Ann. Phys.* **296** (2002) 131, arXiv:hep-lat/9812003; K. Fujikawa, *Nucl. Phys. B* **546** (1999) 480, arXiv:hep-th/9811235; H. Suzuki, *Prog. Theor. Phys.* **102** (1999) 141, arXiv:hep-th/9812019.
63. M. Gong, Y. B. Yang, A. Alexandru, T. Draper and K. F. Liu, arXiv:1511.03671 [hep-ph].
64. E. Leader and C. Lorc, *Phys. Rep.* **541** (2014) 163, doi:10.1016/j.physrep.2014.02.010, arXiv:1309.4235 [hep-ph].
65. K. F. Liu and C. Lorc, *Eur. Phys. J. A* **52** (2016) 160, doi:10.1140/epja/i2016-16160-8, arXiv:1508.00911 [hep-ph].
66. X. D. Ji, *Phys. Rev. Lett.* **78** (1997) 610, arXiv:hep-ph/9603249.
67. M. Deka *et al.*, *Phys. Rev. D* **91** (2015) 014505, doi:10.1103/PhysRevD.91.014505, arXiv:1312.4816 [hep-lat].
68. G. S. Adkins, C. R. Nappi and E. Witten, *Nucl. Phys. B* **228** (1983) 552.
69. B. A. Li, *AIP Conf. Proc.* **343** (1995) 802, doi:10.1063/1.48961.
70. Y. B. Yang, M. Gong, K. F. Liu and M. Sun, *PoS LATTICE* **2014** (2014) 138.
71. K. F. Liu, *Int. J. Mod. Phys. Conf. Ser.* **40** (2016) 1660005, doi:10.1142/S2010194516600053, arXiv:1504.06601 [hep-ph].
72. K. F. Liu, S. J. Dong, T. Draper, J. H. Sloan, W. Wilcox and R. M. Woloshyn, *Phys. Rev. D* **61** (2000) 118502.
73. A. Manohar and H. Georgi, *Nucl. Phys. B* **234** (1984) 189.
74. E. Witten, *Nucl. Phys. B* **160** (1979) 57.
75. B. A. Li, *Phys. Rev. D* **52** (1995) 5165, arXiv:hep-ph/9504304.
76. B. J. Warr, *Ann. Phys.* **183** (1988) 1.
77. Q. Wang, Y. P. Kuang, X. L. Wang and M. Xiao, arXiv:hep-ph/9910289.
78. G. E. Brown and M. Rho, *Phys. Lett. B* **82** (1979) 177.
79. A. W. Thomas, S. Theberge and G. A. Miller, *Phys. Rev. D* **24** (1981) 216.
80. M. Wakamatsu and Y. Nakakoji, *Phys. Rev. D* **74** (2006) 054006.

Heavy exotic molecules

Yizhuang Liu and Ismail Zahed
Department of Physics and Astronomy,
Stony Brook University, Stony Brook,
New York 11794-3800, USA

We briefly review the formation of pion-mediated heavy-light exotic molecules with both charm and bottom, under the general structures of chiral and heavy quark symmetries. The charm isosinglet exotic molecules with $J^{PC} = 1^{++}$ binds, which we identify as the reported neutral $X(3872)$. The bottom isotriplet exotic with $J^{PC} = 1^{+-}$ binds, and is identified as a mixed state of the reported charged exotics $Z_b^+(10610)$ and $Z_b^+(10650)$. The bound bottom isosinglet molecule with $J^{PC} = 1^{++}$ is a possible neutral $X_b(10532)$ to be observed.

Keywords: Heavy mesons; exotic molecules.

1. Introduction

A decade ago, the BaBar collaboration[1] and the CLEOII collaboration[2] have reported narrow peaks in the $D_s^+\pi^0$ (2317) and the $D_s^{*+}\pi^0$ (2460) in support of predictions from chiral and heavy-quark symmetry.[3–5] The heavy-light multiplet $(0^-, 1^-) = (D, D^*)$ has a chiral partner $(0^+, 1^+) = (\tilde{D}, \tilde{D}^*)$ that is about one constituent mass heavier.[3,4] More recently, the Belle collaboration[6] and the BESIII collaboration[7] have reported new multiquark exotics, outside the standard quark model classification. A key source for these exotics is $\Upsilon(10860)$ and its closeness to the $B\bar{B}^*\pi$ (10744) and $B^*\bar{B}^*\pi$ (10790) thresholds. The smallness of the phase space for the pion decay of $\Upsilon(10860)$ suggests that the decay process is slow, involving a molecular configuration on the way out. Heavy exotic molecules have been reported, such as the neutral $X(3872)$ and the charged $Z_c(3900)^\pm$ and $Z_b(10610)^\pm$. More of these exotics are expected to be unravelled by the DO collaboration at Fermilab,[8] and the LHCb collaboration at Cern.[9]

Theoretical arguments have predicted the possibility of molecular bound states involving heavy-light charm and bottom mesons through pion exchange.[10,11] Since, a number of molecular estimates were made by many.[11–16] Nonmolecular heavy exotics were also proposed using constituent quark models,[17] heavy solitonic baryons,[18,19] instantons[20] and QCD sum rules.[21]

In this contribution, we briefly review our recent analysis of the molecular configurations with heavy-light charm and bottom mesons and their chiral partners, under the general constraints of chiral and heavy-quark symmetry.[22] In Sec. 2, we outline the heavy-light effective action in leading order involving the $(0^\pm, 1^\pm)$ multiplets, and formulate the nonrelativistic bound state problem in the $J = 1$ channel. The results are summarized in Sec. 3. Our conclusions are in Sec. 4.

2. Molecules

The leading part of the heavy-light Lagrangian for the charmed multiplet $(0^-, 1^-)$ with pions reads[3,5]

$$\begin{aligned}\mathcal{L} &\approx +2i(\bar{D}\partial_0 D + \bar{\mathbf{D}}\cdot\partial_0\mathbf{D}) - \Delta m_D \bar{D}D - \Delta m_{\mathbf{D}}\bar{\mathbf{D}}\mathbf{D}\\ &\quad + i\frac{g_H}{f_\pi}\mathrm{Tr}\,\partial_i\pi(D_i D^\dagger - D D_i^\dagger + \epsilon_{ijk}D_k D_j^\dagger)\end{aligned} \tag{1}$$

with $\Delta m_i = m_i - m_C$ of the order of a quark constituent mass. The leading part of the heavy-light chiral doublers Lagrangian for the charmed $(0^+, 1^+)$ multiplet with pions reads[3]

$$\begin{aligned}\tilde{\mathcal{L}} &\approx +2i(\bar{\tilde{D}}\partial_0\tilde{D} + \bar{\tilde{\mathbf{D}}}\cdot\partial_0\tilde{\mathbf{D}}) - \Delta m_{\tilde{D}}\bar{\tilde{D}}\tilde{D} - \Delta m_{\tilde{\mathbf{D}}}\bar{\tilde{\mathbf{D}}}\tilde{\mathbf{D}}\\ &\quad + i\frac{g_H}{f_\pi}\mathrm{Tr}\,\partial_i\pi(i(\tilde{D}_i\tilde{D}^\dagger + \tilde{D}\tilde{D}_i^\dagger) + \epsilon_{ijk}\tilde{D}_k\tilde{D}_j^\dagger)\end{aligned} \tag{2}$$

with again $\Delta m_{\tilde{i}} = m_{\tilde{i}} - m_C$ of the order of a quark constituent mass. The $(0^+, 1^+)$ multiplet mixes with the $(0^-, 1^-)$ by chiral symmetry[3,4]

$$\delta\mathcal{L} = \frac{g_{HG}}{f_\pi}\mathrm{Tr}\,\partial_0\pi(\tilde{D}_i^\dagger D_i - i\tilde{D}^\dagger D + \text{c.c.}). \tag{3}$$

The molecular exotics of the type $D\bar{D}^*$, follows from (1) through one-pion exchange. The nonrelativistic character of the molecules yields naturally to a Hamiltonian description. Let $D_{0\bar{0}}(\mathbf{r})$ denote the wave function of the molecular scalar, and $\bar{Y}_{0\bar{i}}(\mathbf{r})$ and $Y_{i\bar{0}}(\mathbf{r})$ denote the wave functions of the molecular vectors, and $T_{i\bar{j}}(\mathbf{r})$ the wave function of the molecular tensors. Using (1)–(3) for the 2-body interactions, we have

$$\begin{aligned}(VT)_{k\bar{l}} &= C\epsilon_{kim}\epsilon_{\bar{l}\bar{j}n}\partial_{mn}V(r)T_{i\bar{j}},\\ (VT)_{0\bar{0}} &= C\partial_{i\bar{j}}\,VT_{i\bar{j}},\\ (V\bar{Y})_{k\bar{0}} &= -C\partial_k\partial_{\bar{j}}V(r)\bar{Y}_{0\bar{j}},\\ (VT)_{0\bar{k}} &= C\epsilon_{\bar{k}\bar{l}j}\partial_i\partial_j V(r)T_{i\bar{l}},\\ (VT)_{\bar{0}k} &= C\epsilon_{klj}\partial_{\bar{i}}\partial_j V(r)T_{l\bar{i}}\end{aligned} \tag{4}$$

with the isospin factor

$$C = \mathbf{I}_1\cdot\mathbf{I}_2 = \left(\frac{1}{4}\bigg|_{I=1}, -\frac{3}{4}\bigg|_{I=0}\right). \tag{5}$$

Here $V(r)$ is the regulated one-pion exchange using the standard monopole form factor by analogy with the pion–nucleon form factor.[23] It is defined with a core cutoff $\Lambda \gg m_\pi$[11,23]

$$V(r) = \left(\frac{g_H}{f_\pi}\right)^2 \frac{1}{4\pi}\left(\frac{e^{-m_\pi r}}{r} - \frac{e^{-\Lambda r}}{r} - (\Lambda^2 - m_\pi^2)\frac{e^{-\Lambda r}}{2\Lambda}\right) \tag{6}$$

Throughout, we will use $g_H = 0.6$[3,4] and $\Lambda = 1\,\mathrm{GeV}$. The choice of Λ is the major uncertainty in the molecular analysis. The one-pion exchange in (4) induces a D-wave admixture much like in the deuteron as a proton–neutron molecule.[23] It is very different from one-gluon exchange in heavy quarkonia.[17]

The pertinent projections onto the higher J^{PC} channels of the molecular wave functions in (4) require the use of both vector and higher tensor spherical harmonics.[24,25] For $J = 1$, we will use the explicit forms quoted in Ref. 25 with the ${}^SL_J$ assignment completely specified. For the $(1^\mp, 0^\mp)$ multiplets, there are four different 1^{PC} sectors

$$\begin{aligned}
&1^{++} : T^{2,2}_{i\bar{j}}({}^5D_1), Y^{0+}_i({}^3S_1), Y^{2+}_i({}^3D_1),\\
&1^{--} : T^{0,1}_{i\bar{j}}({}^1P_1), T^{2,1}_{i\bar{j}}({}^5P_1), T^{2,3}_{i\bar{j}}({}^5F_1), Y^{1-}_i({}^3P_1), D^1({}^1P_1'),\\
&1^{+-} : T^{1,0}_{i\bar{j}}({}^3S_1), T^{1,2}_{i\bar{j}}({}^3D_1), Y^{0-}_i({}^3S_1'), Y^{2-}_i({}^3D_1'),\\
&1^{-+} : T^{1,1}_{i\bar{j}}({}^3P_1), Y^{1+}_i({}^3P_1').
\end{aligned} \tag{7}$$

The normalized tensor harmonics are detailed in Refs. 22 and 25.

3. Results

In Fig. 1, we show the radial components (upper part) and percentage content (lower part) of the bound isosinglet charm wave function with energy $E = 3.867\,\mathrm{GeV}$, versus r in units of $\Lambda = 1\,\mathrm{GeV}$. The intra-coupling between the $(0^-, 1^-)$ and $(0^+, 1^+)$ multiplet causes the chiral partners $\tilde{D}\bar{\tilde{D}}^*$ to unbind. The ${}^SL_J$ assignments referring to the $(0^-, 1^-)$ multiplet, and the ${}^S\tilde{L}_J$ assignments referring to the $(0^+, 1^+)$ multiplet, are those listed in (7). The mixing results in a stronger binding in this channel are mostly an isosinglet ${}^1S_3$ contribution in the $(1^-, 0^-)$ multiplet with almost no D-wave admixture. This molecular state carries $J^{PC} = 1^{++}$. It is chiefly an isosinglet $D\bar{D}^*$ molecule, which we identify as the reported exotic $X(3872)$.

In Fig. 2, we show the radial components (upper part) and percentage content (lower part) of the bound isosinglet bottom wave function with energy $E = 10.532\,\mathrm{GeV}$, versus r in units of $\Lambda = 1\,\mathrm{GeV}$. Again, the ${}^SL_J$ and ${}^S\tilde{L}_J$ assignments refer to the $(0^\pm, 1^\pm)$ multiplets, respectively, as defined in (7). The 1^{++} mixed bound state is mostly a $B\bar{B}^*$ $({}^3S_1)$ molecule. A comparison of Figs. 1 and 2 shows that this neutral bottom molecular state is the mirror analogue of the neutral charm molecular state or $X_b(10532)$, yet to be reported.

In Fig. 3, we show the radial components (upper part) and percentage content (lower part) of the bound isosinglet bottom wave function with energy $E = 10.592\,\mathrm{GeV}$, versus r in units of $\Lambda = 1\,\mathrm{GeV}$. From the assignments given in (7), it

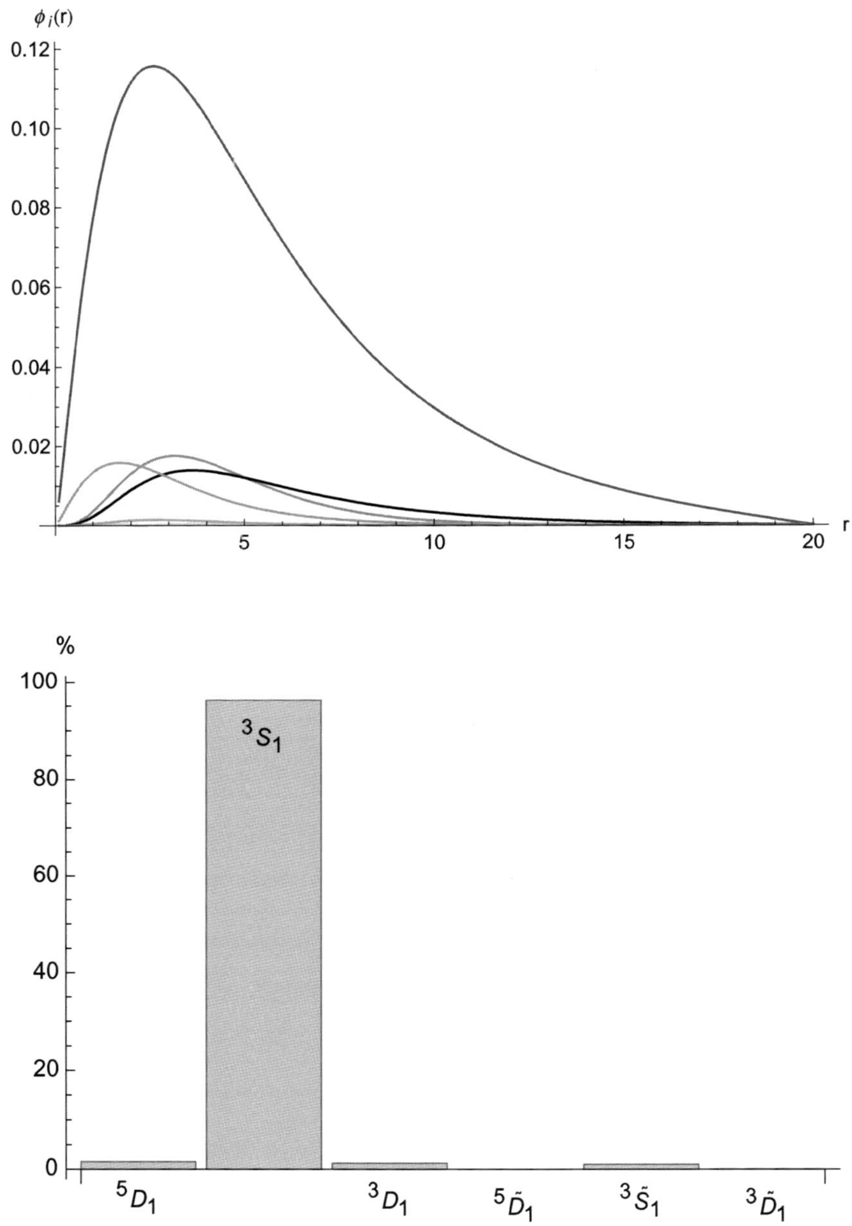

Fig. 1. $J^{PC} = 1^{++}$: radial wave functions (upper plot) and percentage content (lower plot) for the charm isosinglet exotic state ($C = -3/4$).

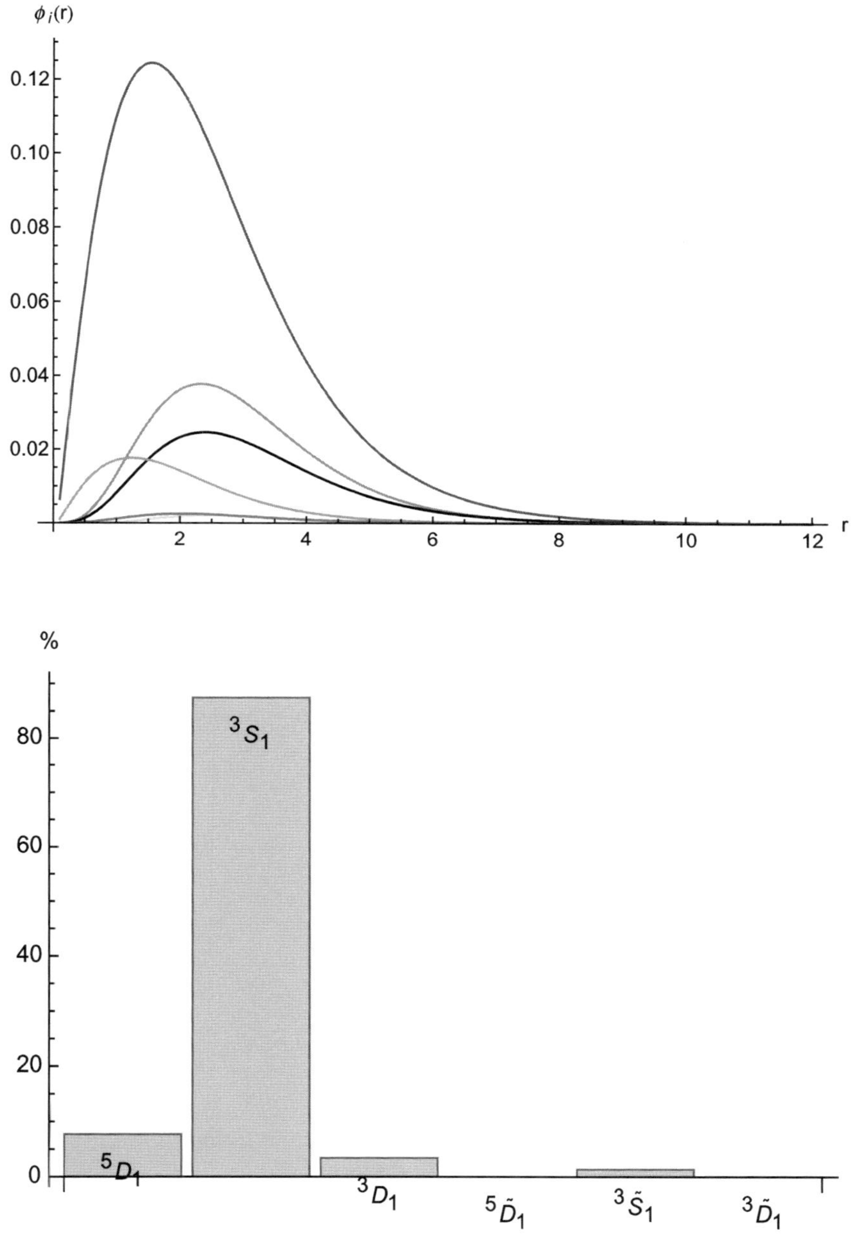

Fig. 2. $J^{PC} = 1^{++}$: radial wave functions (upper plot) and percentage content (lower plot) for the bottom isosinglet exotic state ($C = -3/4$).

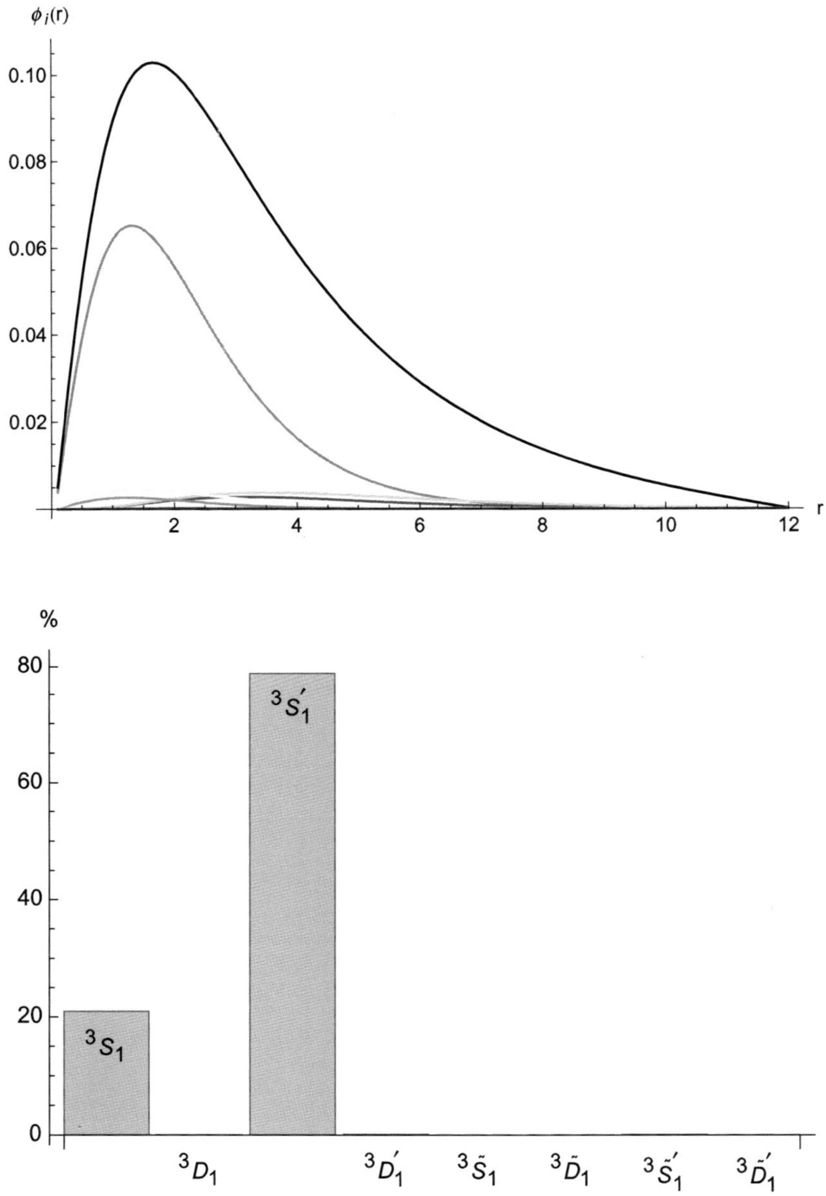

Fig. 3. $J^{PC} = 1^{+-}$: radial wave functions (upper plot) and percentage content (lower plot) for the bottom isotriplet exotic state ($C = +1/4$).

follows that 1^{+-} is a mixed isotriplet $B\bar{B}^*$ ($^3S_1'$) molecules, with a small admixture of $B^*\bar{B}^*$ ($^3S_1'$) molecule. This molecule is an admixture of the reported states $Z_b^+(10610)$ and $Z_b^+(10650)$.

4. Conclusions

We have briefly reported on the molecular states of doubly heavy mesons mediated by one-pion exchange for both the chiral partners $(0^\pm, 1^\pm)$ as a coupled channel problem, recently discussed in Ref. 22. The analysis complements and extends those presented in Refs. 11–16 by taking into account the structures of chiral and heavy quark symmetry, and by retaining most coupled channels between the $(0^-, 1^-)$ multiplet and its chiral partner $(0^+, 1^+)$. The key aspect of this coupling is to cause the molecules in the $(0^-, 1^-)$ multiplet to bind about twice more, and the molecules in the $(0^+, 1^+)$ multiplet to unbind. The charm isosinglet exotic molecules with $J^{PC} = 1^{++}$ is strictly bound for a pion-exchange cutoff $\Lambda = 1\,\text{GeV}$. This state is identified with the reported isosinglet exotic $X(3872)$ which in our case is mostly an isosinglet $D\bar{D}^*$ molecule in the $^1S_0$ channel with no D-wave admixture. The attraction in the isotriplet channel with $J^{PC} = 1^{+-}$ is too weak to bind the $D\bar{D}^*$ compound, suggesting that the reported isotriplet $Z_C(3900)^\pm$ is at best a threshold enhancement. The $Y(4260)$, $Y(4360)$ and $Y(4660)$ may point to the possibility of their constituents made of excited (D_1, D_2) heavy mesons and their chiral partners.[3,26] The isotriplet bottom exotic molecule with $J^{PC} = 1^{+-}$ which we have identified with the pair $Z_b^+(10610)$ and $Z_b^+(10650)$, which is a mixed state in our analysis. The isosinglet bottom exotic molecule with $J^{PC} = 1^{++}$ is a potential candidate for $X_b(10532)$, yet to be measured.

5. Tribute to Gerry Brown

Ismail Zahed Gerry has been a mentor and a colleague for the past three decades. He has taught many of us a great deal of how to approach physics from the bottom line point of view and encouraged us to pursue freely our physics interests, always in the quest of tangible results. Most of us here at Stony Brook and many elsewhere, owe so much to Gerry scientific leadership and personal friendship. We hope to pass his legacy to our students and collaborators.

Acknowledgments

This work was supported by the U.S. Department of Energy under Contract No. DE-FG-88ER40388.

References

1. BaBar Collab. (B. Aubert *et al.*), *Phys. Rev. Lett.* **90** (2003) 242001, arXiv:hep-ex/0304021.

2. CLEO Collab. (D. Besson *et al.*), *Phys. Rev. D* **68** (2003) 032002, Erratum: **75** (2007) 119908, arXiv:hep-ex/0305100.
3. M. A. Nowak, M. Rho and I. Zahed, *Phys. Rev. D* **48** (1993) 4370, arXiv:hep-ph/9209272; M. A. Nowak, M. Rho and I. Zahed, *Acta Phys. Polon. B* **35** (2004) 2377, arXiv:hep-ph/0307102.
4. W. A. Bardeen and C. T. Hill, *Phys. Rev. D* **49** (1994) 409, arXiv:hep-ph/9304265; W. A. Bardeen, E. J. Eichten and C. T. Hill, *Phys. Rev. D* **68** (2003) 054024, arXiv:hep-ph/0305049.
5. N. Isgur and M. B. Wise, *Phys. Rev. Lett.* **66** (1991) 1130; A. V. Manohar and M. B. Wise, *Heavy Quark Physics*, Cambridge Monographs on Particle Physics, Nuclear Physics and Cosmoloy, Vol. 10 (Cambridge University Press, 2000).
6. Belle Collab. (I. Adachi), arXiv:1105.4583 [hep-ex]; Belle Collab. (A. Bondar *et al.*), *Phys. Rev. Lett.* **108** (2012) 122001, arXiv:1110.2251 [hep-ex].
7. BESIII Collab. (M. Ablikim *et al.*), *Phys. Rev. Lett.* **110** (2013) 252001, arXiv:1303.5949 [hep-ex].
8. D0 Collab. (V. M. Abazov *et al.*), arXiv:1602.07588 [hep-ex].
9. LHCb Collab. (R. Aaij *et al.*), arXiv:1606.07895 [hep-ex]; LHCb Collab. (R. Aaij *et al.*), arXiv:1606.07898 [hep-ex].
10. M. B. Voloshin and L. B. Okun, *JETP Lett.* **23** (1976) 333; *Pisma Zh. Eksp. Teor. Fiz.* **23** (1976) 369.
11. N. A. Tornqvist, *Phys. Rev. Lett.* **67** (1991) 556; N. A. Tornqvist, *Z. Phys. C* **61** (1994) 525, arXiv:hep-ph/9310247; N. A. Tornqvist, *Phys. Lett. B* **590** (2004) 209, arXiv:hep-ph/0402237.
12. M. Karliner and H. J. Lipkin, arXiv:0802.0649 [hep-ph]; M. Karliner and J. L. Rosner, *Phys. Rev. Lett.* **115** (2015) 122001, arXiv:1506.06386 [hep-ph]; M. Karliner, *Acta Phys. Polon. B* **47** (2016) 117.
13. C. E. Thomas and F. E. Close, *Phys. Rev. D* **78** (2008) 034007, arXiv:0805.3653 [hep-ph]; F. Close, C. Downum and C. E. Thomas, *Phys. Rev. D* **81** (2010) 074033, arXiv:1001.2553 [hep-ph].
14. S. Ohkoda, Y. Yamaguchi, S. Yasui, K. Sudoh and A. Hosaka, *Phys. Rev. D* **86** (2012) 034019, arXiv:1202.0760 [hep-ph]; S. Ohkoda, Y. Yamaguchi, S. Yasui, K. Sudoh and A. Hosaka, arXiv:1209.0144 [hep-ph].
15. M. T. AlFiky, F. Gabbiani and A. A. Petrov, *Phys. Lett. B* **640** (2006) 238, arXiv:hep-ph/0506141; I. W. Lee, A. Faessler, T. Gutsche and V. E. Lyubovitskij, *Phys. Rev. D* **80** (2009) 094005, arXiv:0910.1009 [hep-ph]; M. Suzuki, *Phys. Rev. D* **72** (2005) 114013, arXiv:hep-ph/0508258; J. R. Zhang, M. Zhong and M. Q. Huang, *Phys. Lett. B* **704** (2011) 312, arXiv:1105.5472 [hep-ph]; D. V. Bugg, *Europhys. Lett.* **96** (2011) 11002, arXiv:1105.5492 [hep-ph]; J. Nieves and M. P. Valderrama, *Phys. Rev. D* **84** (2011) 056015, arXiv:1106.0600 [hep-ph]; M. Cleven, F. K. Guo, C. Hanhart and U. G. Meissner, *Eur. Phys. J. A* **47** (2011) 120, arXiv:1107.0254 [hep-ph]; T. Mehen and J. W. Powell, *Phys. Rev. D* **84** (2011) 114013, arXiv:1109.3479 [hep-ph]; X. W. Kang, Z. H. Guo and J. A. Oller, *Phys. Rev. D* **94** (2016) 014012, arXiv:1603.05546 [hep-ph].
16. E. S. Swanson, *Phys. Rep.* **429** (2006) 243, arXiv:hep-ph/0601110; Z. F. Sun, J. He, X. Liu, Z. G. Luo and S. L. Zhu, *Phys. Rev. D* **84** (2011) 054002, arXiv:1106.2968 [hep-ph].
17. A. V. Manohar and M. B. Wise, *Nucl. Phys. B* **399** (1993) 17, arXiv:hep-ph/9212236; N. Brambilla *et al.*, *Eur. Phys. J. C* **71** (2011) 1534, arXiv:1010.5827 [hep-ph]; M. B. Voloshin, *Prog. Part. Nucl. Phys.* **61** (2008) 455, arXiv:0711.4556 [hep-ph]; J. M. Richard, arXiv:1606.08593 [hep-ph].

18. D. O. Riska and N. N. Scoccola, *Phys. Lett. B* **299** (1993) 338.
19. M. A. Nowak, I. Zahed and M. Rho, *Phys. Lett. B* **303** (1993) 130.
20. S. Chernyshev, M. A. Nowak and I. Zahed, *Phys. Rev. D* **53** (1996) 5176, arXiv:hep-ph/9510326.
21. M. Nielsen, F. S. Navarra and S. H. Lee, *Phys. Rep.* **497** (2010) 41, arXiv:0911.1958 [hep-ph].
22. Y. Liu and I. Zahed, arXiv:1608.06535 [hep-ph].
23. G. E. Brown and A. D. Jackson, *The Nucleon–Nucleon Interaction* (North Holland, 1976).
24. A. R. Edmonds, *Angular Momentum in Quantum Mechanics* (Princeton University Press, 1974).
25. K. S. Thorne, *Rev. Mod. Phys.* **52** (1980) 299.
26. M. A. Nowak and I. Zahed, *Phys. Rev. D* **48** (1993) 356.

The nuclear force: Meson theory versus chiral effective field theory

R. Machleidt

Department of Physics, University of Idaho,
Moscow, Idaho 83844, USA
machleid@uidaho.edu

Gerry Brown has played a major role in initiating and stimulating systematic work on the theories of nuclear forces. It is the purpose of this paper to review those theories — in his memory.

Keywords: Nuclear forces; meson-theory; low-energy QCD; chiral effective field theory.

1. Introduction

Throughout his professional career, Gerry Brown has been concerned with the theory of nuclear forces.[1–7] He has discussed repeatedly and critically what a proper meson-theory for the two-nucleon force (2NF) should look like[2,5–7] and, early on, he has noticed the relevance of three-nucleon forces.[1,3,4]

Even though Gerry is a co-author of only a few papers on nuclear forces, he is the one who initiated and encouraged a lot of the research that became well known and established over time. For example, he invited Klaus Erkelenz[8] to Copenhagen and Stony Brook for extended research visits (in 1972 and 1973), and I myself have been a postdoc with him in Stony Brook in the years of 1976 and 1977. Thus, Gerry has been very supportive of the development of the Bonn potential.[8–11]

The dispersion theory-based Stony Brook potential[5,12] is essentially due to Gerry's initiative.[2] Besides this, he has always been very appreciative of the dispersion-theoretic work conducted by the Paris group,[13,14] which has led to the Paris Potential.[15]

Moreover, in all of his review papers and comments,[1–7] he has pointed out the importance of chiral symmetry as a crucial constraint for nuclear two- and three-body forces.

It is therefore appropriate to review, in Gerry Brown's memory, the traditional meson theory as well as the chiral effective field theory (EFT) of nuclear forces.

In Sec. 2, we will present the meson theory of the nucleon–nucleon (NN) interaction. In Sec. 3, we motivate the chiral EFT approach, for which we give an overview in Sec. 4, and NN potentials are presented in Sec. 5. Section 6 concludes the paper by conducting a systematic comparison between the two approached that are the subject of this review.

2. Meson Theory

In 1935, Yukawa introduced the concept of massive particle exchange to explain the (finite-ranged) nuclear force.[16] Yukawa's original derivation was done for scalar bosons. When finally a real meson was discovered in 1947/1948, it turned out to be pseudo-scalar with mass around 138 MeV and was dubbed the π-meson or pion. Consequently, in the 1950s, the attempts to derive the nuclear force focused on theories that included only pions. These "pion theories" had many problems and little success — for reasons we understand today: pion dynamics is constrained by chiral symmetry, a concept that was unknown in the 1950s. In the early 1960s, heavier (nonstrange) mesons were found in experiment, notably the vector (spin-1) mesons $\rho(770)$ and $\omega(782)$. Because of the problems with the pion theories, theoreticians were now happy to extend meson theory by including more and different species of mesons. This led to the one-boson-exchange (OBE) models, which were started in the 1960s and turned out to be very successful.

2.1. *The one-boson-exchange model*

OBE models include, in general, about half a dozen of bosons with masses up to about 1 GeV, Fig. 1. Not all mesons are equally important. The leading actors are the following four particles:

- The pseudoscalar pion with a mass of about 138 MeV and isospin $I = 1$ (isovector). It is the lightest meson and provides the long-range part of the potential and most of the tensor force.
- The isovector ρ meson, a 2π P-wave resonance of about 770 MeV. Its major effect is to cut down the tensor force provided by the pion at short range.
- The isoscalar ω meson, a 3π resonance of 783 MeV and spin 1. It creates a strong repulsive central force of short range ("repulsive core") and the nuclear spin-orbit force.
- The scalar–isoscalar $f_0(500)$ or σ boson with a mass around 500 MeV. It provides the crucial intermediate range attraction necessary for nuclear binding. The interpretation as a particle is controversial.[17] It may also be viewed as a simulation of effects of correlated S-wave 2π-exchange.

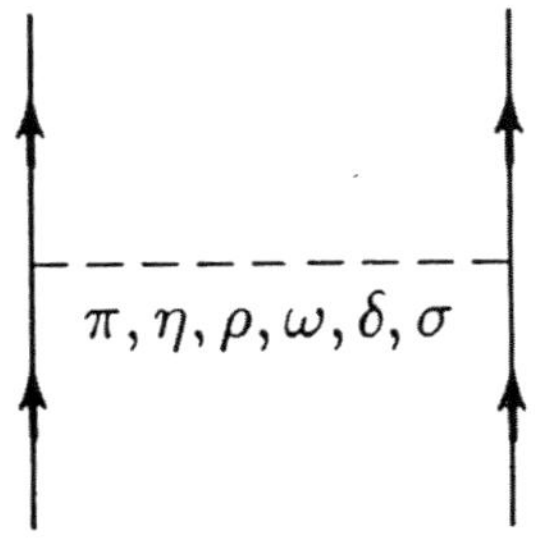

Fig. 1. The OBE model. Solid lines denote nucleons and the dashed line represents mesons.

Obviously, just these four mesons can produce the major properties of the nuclear force.[a]

A classic example for an OBE potential (OBEP) is the Bryan–Scott potential published in 1969.[19] Since it is suggestive to think of a potential as a function of r (where r denotes the distance between the centers of the two interacting nucleons), the OBEPs of the 1960s where represented as local r-space potentials. To reduce the original one-meson-exchange Feynman amplitudes to such a simple form, drastic approximations have to be applied. The usual method is to expand the amplitude in terms of p^2/M_N^2 (where M_N denotes the nucleon mass and p is an external momentum) and keep only terms up to first-order (and, in many cases, not even all of them). Commonly, this is called the *nonrelativistic OBEP*. Besides the suggestive character of a local function of r, such potentials are easy to apply in r-space calculations.

However, quantitatively, the *nonrelativistic* version of the OBE model has major drawbacks. It fails to describe certain partial waves correctly. The Bryan–Scott[19] and the Nijmegen[20] nonrelativistic OBEPs predict the phase shifts of the $^1P_1$ and $^3D_2$ states substantially above the data.[b]

An important advance during the 1970s has been the development of the *relativistic OBEP*.[8,9,22] In this model, the full, relativistic Feynman amplitudes for the various OBEs are used to define the potential. These nonlocal expressions do not pose any numerical problems when used in momentum space.[c] The quantitative deficiencies of the nonrelativistic OBEP disappear when the nonsimplified, relativistic, nonlocal OBE amplitudes are used. The very quantitative and successful CD-Bonn potential[23] is of this nature.

2.2. *Beyond the OBE approximation*

Historically, one must understand that, after the failure of the pion theories in the 1950s, the OBE model was considered a great success in the 1960s.

On the other hand, one has to concede that the OBE model is a great simplification of the complicated scenario of a full meson theory for the *NN* interaction. Therefore, in spite of the quantitative success of the OBEPs, one should be concerned about the approximations involved in the model. Major critical points include:

- The scalar isoscalar σ "meson" of about 500 MeV.
- The neglect of all noniterative diagrams.
- The role of meson–nucleon resonances.

[a]The interested reader can find a more detailed and pedagogical introduction into the OBE model in Secs. 3 and 4 of Ref. 18.

[b]A thorough discussion of this issue and figures illustrating the facts can be found in Ref. 21.

[c]In fact, in momentum space, the application of a nonlocal potential is numerically as easy as using the momentum-space representation of a local potential.

Two pions, when "in the air", can interact strongly. When in a relative P-wave ($L = 1$), they form a proper resonance, the ρ meson. They can also interact in a relative S-wave ($L = 0$), which gives rise to the σ boson. Whether the σ is a proper resonance is controversial, even though the Particle Data Group lists an $f_0(500)$ or $\sigma(500)$ meson, but with a width 400–700 MeV.[17] What is for sure is that two pions have correlations, and if one does not believe in the σ as a two-pion resonance, then one has to take these correlations into account. There are essentially two approaches to take care of these two-pion exchange (2PE) contributions to the NN interaction (which generates the intermediate range attraction): dispersion theory and field theory.

The "dispersion-theoretic" picture is described schematically in Fig. 2. In this approach one assumes that the total diagram (a) can be analyzed in terms of two "halves" (b). The hatched ovals stand for all possible processes which a pion and a nucleon can undergo. This is made more explicit in (d) and (e). The hatched boxes represent baryon intermediate states including the nucleon. (Note that there are also crossed pion exchanges which are not shown.) The shaded circle stands for $\pi\pi$ scattering. Quantitatively, these processes are taken into account by using empirical information from πN and $\pi\pi$ scattering (e.g., phase shifts) which represents the input for such a calculation. Dispersion relations then provide an on-shell NN

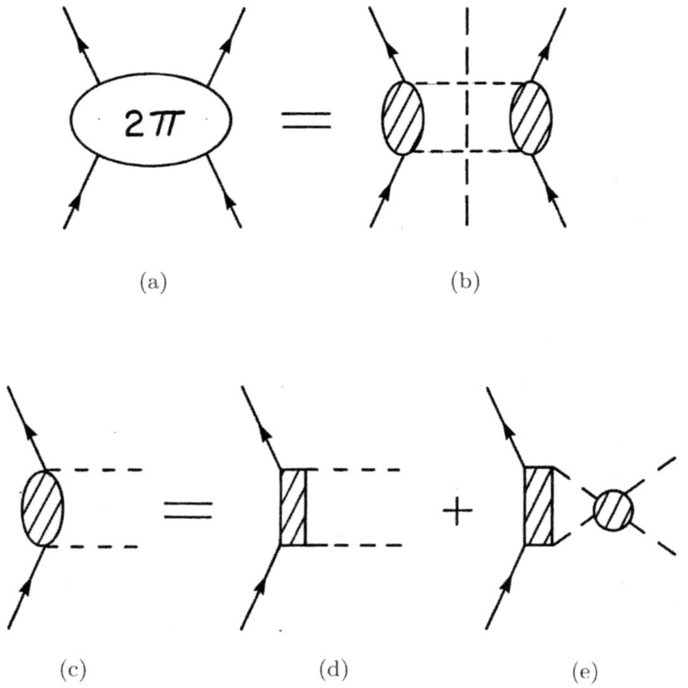

Fig. 2. The 2π-exchange contribution to the NN interaction as viewed by dispersion theory. Solid lines represent nucleons and dashed lines pions. Further explanations are given in the text.

amplitude, which — with some kind of plausible prescription — is represented as a potential. The Stony Brook[12] and Paris[13] groups have pursued this approach. They could show that the intermediate-range part of the nuclear force is, indeed, described correctly by the 2π-exchange as obtained from dispersion integrals. To construct a complete potential, the 2π-exchange contribution is complemented by one-pion and ω exchange. In addition to this, the Paris potential[15] contains a phenomenological short-range part for $r < 1.5\,\text{fm}$. For further details, we refer the interested reader to a pedagogical paper by Vinh Mau.[14]

A "field-theoretic model" for the 2π-exchange contribution is shown in Fig. 3. The model includes contributions from isobars as well as from $\pi\pi$ correlations. This can be understood in analogy to the dispersion relations picture. In general, only the lowest-lying πN resonance, the so-called Δ isobar (spin 3/2, isospin 3/2, mass 1232 MeV), is taken into account. The contributions from other resonances have proven to be small for the low-energy NN processes under consideration. A field-theoretic model treats the Δ isobar as an elementary (Rarita–Schwinger) particle. The six upper diagrams of Fig. 3 represent uncorrelated 2π exchange. The crossed (noniterative) two-particle exchanges (second diagram in each row) are important. They guarantee the proper (very weak) isospin dependence due to characteristic

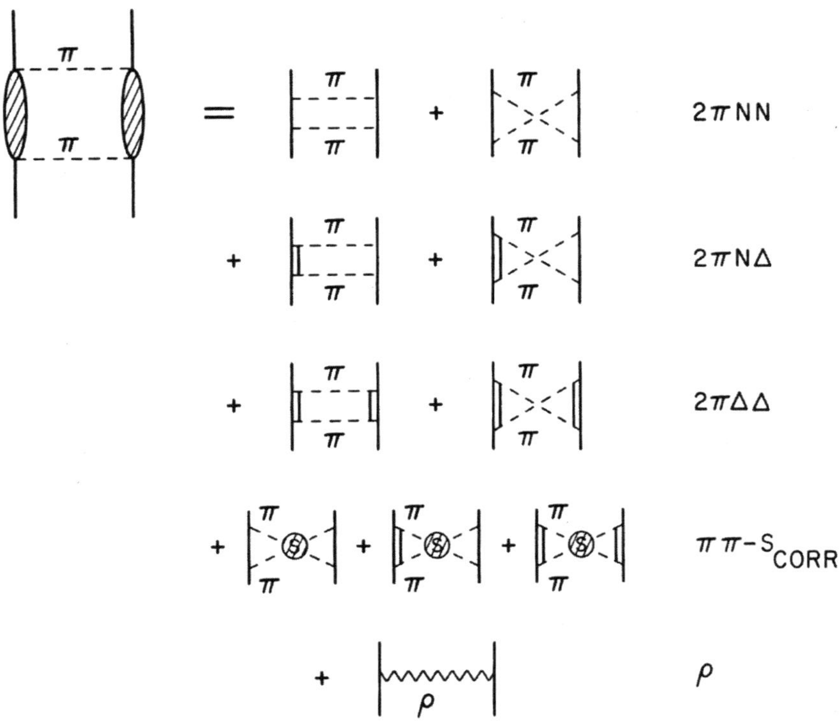

Fig. 3. Field-theoretic model for the 2π-exchange. Notation as in Fig. 2. Double lines represent isobars. The hatched circles are $\pi\pi$ correlations. Further explanations are given in the text.

cancelations in the isospin-dependent parts of box and crossed box diagrams. Furthermore, their contribution is about as large as the one from the corresponding box diagrams (iterative diagrams); therefore, they are not negligible. In addition to the processes discussed, also correlated 2π exchange has to be included (lower two rows of Fig. 3). Quantitatively, these contributions are about as sizable as those from the uncorrelated processes.

Besides the contributions from two pions, there are also contributions from the combination of other mesons. The combination of π and ρ is particularly significant, Fig. 4. This contribution is repulsive and important to suppress the 2π exchange contribution at high momenta (or small distances), which is too strong by itself.

The most developed meson-theoretic model for the *NN* interaction is the Bonn Full Model,[11] which includes all the diagrams displayed in Figs. 3 and 4 plus single π and ω exchange.

Having highly sophisticated models at hand, like the Paris and the Bonn potentials, allows to check the approximations made in the simple OBE model. As it turns out, the complicated 2π exchange contributions to the *NN* interaction tamed by the $\pi\rho$ diagrams can well be simulated by the single scalar isoscalar boson, the σ, with a mass around 550 MeV. Retroactively, this fact provides justification for the simple OBE model. To illustrate this point, we show, in Figs. 5 and 6, the phase shift predictions from the Bonn[11] and Paris[15] potentials as well as a relativistic OBEP.[18]

This could have been the happy end of the theory of nuclear forces. However, with the rise of QCD to the ranks of the authoritative theory of strong interactions, meson theory is demoted to the lower level of a model (even though a beautiful one), and we have to start all over again — in the next section.

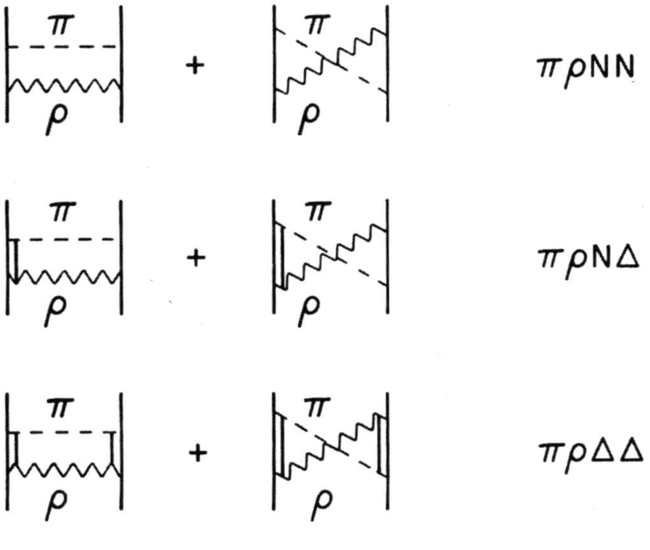

Fig. 4. $\pi\rho$ contributions to the *NN* interaction.

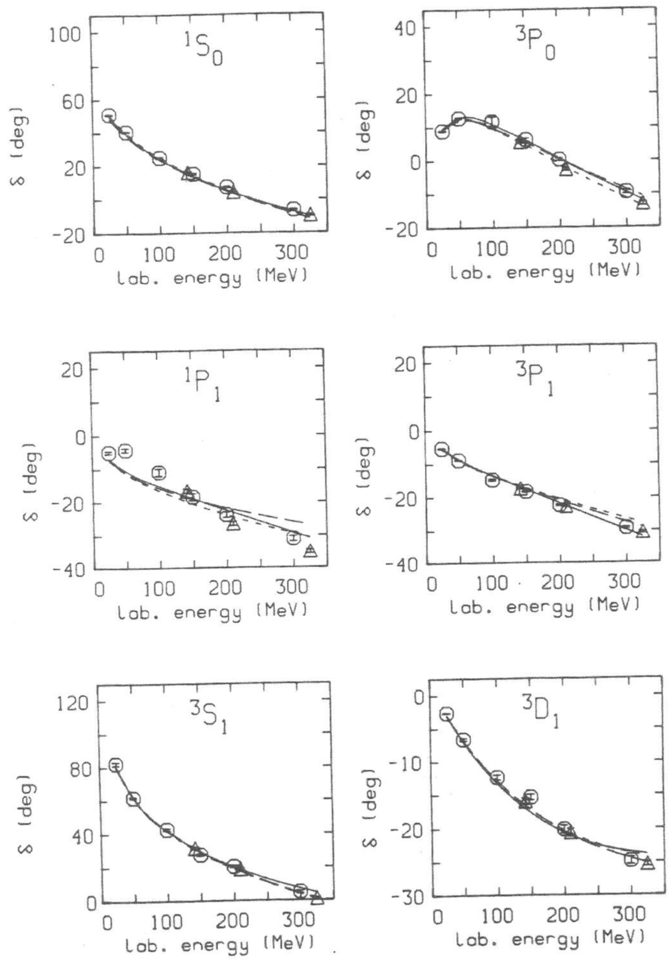

Fig. 5. Phase shifts of *np* scattering from some "classic" meson-exchange models for the *NN* interaction. Predictions are shown for the Bonn Full Model[11] (solid line), the Paris potential[15] (dashed) and a (relativistic) OBEP[18] (dotted). Phase parameters with total angular momentum $J \leq 1$ are displayed. Symbols represent results from *NN* phase shift analyses.

3. Effective Field Theory for Low-Energy QCD

QCD is the theory of strong interactions. It deals with quarks, gluons and their interactions and is part of the Standard Model of Particle Physics. QCD is a non-Abelian gauge field theory with color SU(3) the underlying gauge group. The non-Abelian nature of the theory has dramatic consequences. While the interaction between colored objects is weak at short distances or high momentum transfer ("asymptotic freedom"); it is strong at long distances ($\gtrsim 1$ fm) or low energies, leading to the confinement of quarks into colorless objects, the hadrons. Consequently, QCD allows for a perturbative analysis at large energies, whereas it is highly nonperturbative in the low-energy regime. Nuclear physics resides at low energies and the force between

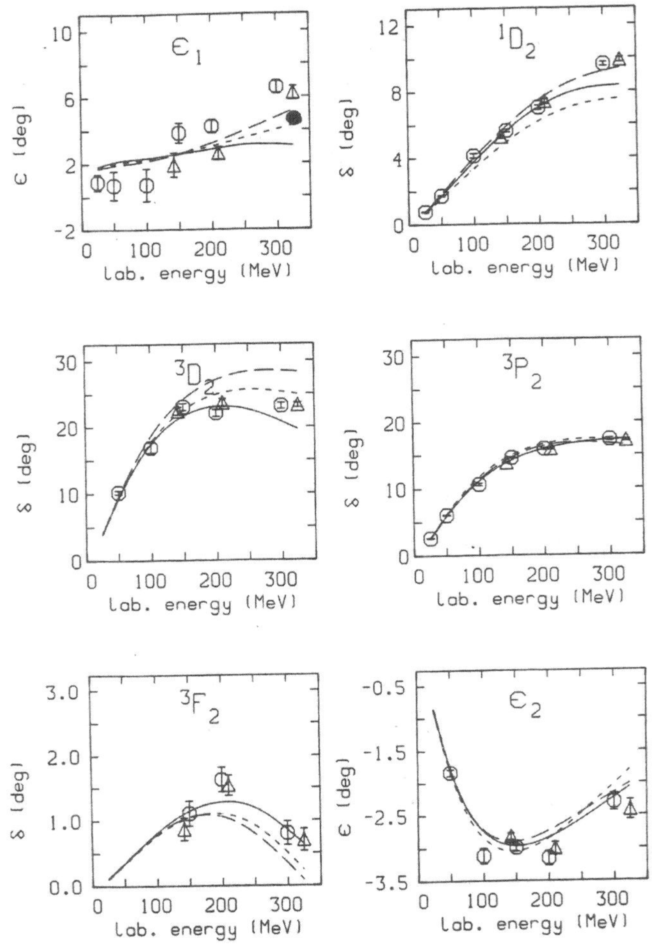

Fig. 6. Same as Fig. 5, but $J = 2$ phase shifts and $J \leq 2$ mixing parameters are shown.

nucleons is a residual color interaction similar to the van der Waals force between neutral molecules. Therefore, in terms of quarks and gluons, the nuclear force is a very complicated problem that, nevertheless, can be attacked with brute computing power on a discretized, Euclidean space-time lattice (known as lattice QCD). In a recent study,[24] the neutron–proton scattering lengths in the singlet and triplet S-waves have been determined in fully dynamical lattice QCD, with a smallest pion mass of 354 MeV. This result is then extrapolated to the physical pion mass with the help of chiral perturbation theory (ChPT). The pion mass of 354 MeV is still too large to allow for reliable extrapolations, but the feasibility has been demonstrated and more progress can be expected for the near future. In a lattice calculation of a very different kind, the NN potential was studied.[25] The central part of the potential shows a repulsive core plus attraction of intermediate range. This is a very promising result, but it must be noted that also in this investigation still rather large pion masses are being used. In any case, advanced lattice QCD calculations are under way and continuously improving. However, since these calculations are very time-consuming and expensive, they can only be used to check a few representative key-issues. For everyday nuclear structure physics, a more efficient approach is needed.

The efficient approach is an EFT. For the development of an EFT, it is crucial to identify a separation of scales. In the hadron spectrum, a large gap between the masses of the pions and the masses of the vector mesons, like $\rho(770)$ and $\omega(782)$, can clearly be identified. Thus, it is natural to assume that the pion mass sets the soft scale, $Q \sim m_\pi$, and the rho mass the hard scale, $\Lambda_\chi \sim m_\rho$, also known as the chiral-symmetry breaking scale. This is suggestive of considering an expansion in terms of the soft scale over the hard scale, Q/Λ_χ. Concerning the relevant degrees of freedom, for conventional nuclear physics, quarks and gluons are ineffective degrees of freedom, while nucleons and pions are the appropriate ones. To make sure that this EFT is not just another phenomenology, it must have a firm link with QCD. The link is established by having the EFT observe all relevant symmetries of the underlying theory. This requirement is based upon a "folk theorem" by Weinberg[26]:

> If one writes down the most general possible Lagrangian, including *all* terms consistent with assumed symmetry principles, and then calculates matrix elements with this Lagrangian to any given order of perturbation theory, the result will simply be the most general possible S-matrix consistent with analyticity, perturbative unitarity, cluster decomposition and the assumed symmetry principles.

In summary, the EFT program consists of the following steps:

(1) Identify the soft and hard scales, and the degrees of freedom appropriate for (low-energy) nuclear physics.
(2) Identify the relevant symmetries of low-energy QCD and investigate if and how they are broken.
(3) Construct the most general Lagrangian consistent with those symmetries and symmetry breakings.
(4) Design an organizational scheme that can distinguish between more and less important contributions: a low-momentum expansion.
(5) Guided by the expansion, calculate Feynman diagrams for the process under consideration to the desired accuracy.

This program works well for π–π[27] and π–N,[28] since in these cases the physical amplitudes can be calculated perturbatively. But, as it turns out, the *NN* scenario is more complicated. It involves a bound state (the deuteron) and large scattering lengths (in S-waves), which cannot be explained in terms of perturbation theory. Therefore, Weinberg suggested[29] to use the above (perturbative) scheme to derive the *NN potential* and to apply this potential then in a Schrődinger or Lippmann–Schwinger (LS) equation to obtain the *NN amplitude*. Notice that this procedure implies a nonperturbative resummation of the potential. Consequently, even when the potential is calculated at a well-defined order, the exact order of the NN amplitude is not clear due to the resummation. This will also obscure the error estimate of the predicted observables. An exception is NN scattering in peripheral partial waves, which can be calculated perturbatively. Therefore, later in this paper, we will first consider peripheral NN scattering before we discuss NN potential construction with all its conceptual problems (Sec. 5, particularly, Sec. 5.3).

To proceed, we will now elaborate on the above-listed steps, one by one. Since we discussed Step (1) already, we will address now Step (2).

3.1. *Symmetries of low-energy QCD*

In this section, we will give a brief introduction into (low-energy) QCD, its symmetries and symmetry breakings. More detailed presentations of this topic are provided in Refs. 30–32.

3.1.1. *Chiral symmetry*

The QCD Lagrangian reads

$$\mathcal{L}_{\rm QCD} = \bar{q}(i\gamma^\mu \mathcal{D}_\mu - \mathcal{M})q - \frac{1}{4}\mathcal{G}_{\mu\nu,a}\mathcal{G}^{\mu\nu}_a \tag{1}$$

with the gauge-covariant derivative

$$\mathcal{D}_\mu = \partial_\mu - ig\frac{\lambda_a}{2}\mathcal{A}_{\mu,a} \tag{2}$$

and the gluon field strength tensor[d]

$$\mathcal{G}_{\mu\nu,a} = \partial_\mu \mathcal{A}_{\nu,a} - \partial_\nu \mathcal{A}_{\mu,a} + g f_{abc}\mathcal{A}_{\mu,b}\mathcal{A}_{\nu,c}. \tag{3}$$

In the above, q denotes the quark fields and $\mathcal{M}$ the quark mass matrix. Further, g is the strong coupling constant and $\mathcal{A}_{\mu,a}$ are the gluon fields. The λ_a are the Gell–Mann matrices and the f_{abc} the structure constants of the SU(3)$_{\rm color}$ Lie algebra $(a, b, c = 1, \dots, 8)$; summation over repeated indices is always implied. The gluon–gluon term in the last equation arises from the non-Abelian nature of the gauge theory and is the reason for the peculiar features of the color force.

The masses of the up (u), down (d) and strange (s) quarks are[17]:

$$m_u = 2.3 \pm 0.7\,\text{MeV}, \tag{4}$$

$$m_d = 4.8 \pm 0.5\,\text{MeV}, \tag{5}$$

$$m_s = 95 \pm 5\,\text{MeV}. \tag{6}$$

These masses are small as compared to a typical hadronic scale, i.e., a scale of low-mass hadrons which are not Goldstone bosons, e.g., $m_\rho = 0.78\,\text{GeV} \approx 1\,\text{GeV}$.

[d]For SU(N) group indices, we use Latin letters, $\dots, a, b, c, \dots, i, j, k, \dots$, and, in general, do not distinguish between subscripts and superscripts.

It is therefore of interest to discuss the QCD Lagrangian in the limit of vanishing quark masses:

$$\mathcal{L}^0_{\text{QCD}} = \bar{q} i\gamma^\mu \mathcal{D}_\mu q - \frac{1}{4}\mathcal{G}_{\mu\nu,a}\mathcal{G}^{\mu\nu}_a. \tag{7}$$

Defining right- and left-handed quark fields,

$$q_R = P_R q, \quad q_L = P_L q, \tag{8}$$

with

$$P_R = \frac{1}{2}(1+\gamma_5), \quad P_L = \frac{1}{2}(1-\gamma_5), \tag{9}$$

we can rewrite the Lagrangian as follows:

$$\mathcal{L}^0_{\text{QCD}} = \bar{q}_R i\gamma^\mu \mathcal{D}_\mu q_R + \bar{q}_L i\gamma^\mu \mathcal{D}_\mu q_L - \frac{1}{4}\mathcal{G}_{\mu\nu,a}\mathcal{G}^{\mu\nu}_a. \tag{10}$$

This equation reveals that *the right- and left-handed components of massless quarks do not mix* in the QCD Lagrangian. For the two-flavor case, this is $\text{SU}(2)_R \times \text{SU}(2)_L$ symmetry, also known as *chiral symmetry.* However, this symmetry is broken in two ways: explicitly and spontaneously.

3.1.2. *Explicit symmetry breaking*

The mass term $-\bar{q}\mathcal{M}q$ in the QCD Lagrangian equation (1) breaks chiral symmetry explicitly. To better see this, let us rewrite $\mathcal{M}$ for the two-flavor case,

$$\begin{aligned}\mathcal{M} &= \begin{pmatrix} m_u & 0 \\ 0 & m_d \end{pmatrix} \\ &= \frac{1}{2}(m_u+m_d)\begin{pmatrix} 1 & 0 \\ 0 & 1 \end{pmatrix} + \frac{1}{2}(m_u-m_d)\begin{pmatrix} 1 & 0 \\ 0 & -1 \end{pmatrix} \\ &= \frac{1}{2}(m_u+m_d)I + \frac{1}{2}(m_u-m_d)\tau_3. \end{aligned} \tag{11}$$

The first term in the last equation in invariant under $\text{SU}(2)_V$ (isospin symmetry) and the second term vanishes for $m_u = m_d$. Thus, isospin is an exact symmetry if $m_u = m_d$. However, both terms in Eq. (11) break chiral symmetry. Since the up and down quark masses (Eqs. (4) and (5)) are small as compared to the typical hadronic mass scale of $\sim$1 GeV, the explicit chiral symmetry breaking due to nonvanishing quark masses is very small.

3.1.3. *Spontaneous symmetry breaking*

A (continuous) symmetry is said to be *spontaneously broken* if a symmetry of the Lagrangian is not realized in the ground state of the system. There is evidence that the (approximate) chiral symmetry of the QCD Lagrangian is spontaneously broken — for dynamical reasons of nonperturbative origin which are not fully understood at this time. The most plausible evidence comes from the hadron spectrum.

From chiral symmetry, one naively expects the existence of degenerate hadron multiplets of opposite parity, i.e., for any hadron of positive parity one would expect a degenerate hadron state of negative parity and vice versa. However, these "parity doublets" are not observed in nature. For example, take the ρ-meson which is a vector meson of negative parity ($J^P = 1^-$) and mass 776 MeV. There does exist a 1^+ meson, the a_1, but it has a mass of 1230 MeV and, therefore, cannot be perceived as degenerate with the ρ. On the other hand, the ρ meson comes in three charge states (equivalent to three isospin states), the $\rho^\pm$ and the ρ^0, with masses that differ by at most a few MeV. Thus, in the hadron spectrum, $SU(2)_V$ (isospin) symmetry is well observed, while axial symmetry is broken: $SU(2)_R \times SU(2)_L$ is broken down to $SU(2)_V$.

A spontaneously broken global symmetry implies the existence of (massless) Goldstone bosons. The Goldstone bosons are identified with the isospin triplet of the (pseudoscalar) pions, which explains why pions are so light. The pion masses are not exactly zero because the up and down quark masses are not exactly zero either (explicit symmetry breaking). Thus, pions are a truly remarkable species: they reflect spontaneous as well as explicit symmetry breaking. Goldstone bosons interact weakly at low energy. They are degenerate with the vacuum and, therefore, interactions between them must vanish at zero momentum and in the chiral limit ($m_\pi \to 0$).

3.2. *Chiral effective Lagrangians*

The next step in our EFT program is to build the most general Lagrangian consistent with the (broken) symmetries discussed above. An elegant formalism for the construction of such Lagrangians was developed by Callan, Coleman, Wess and Zumino (CCWZ)[33] who worked out the group-theoretical foundations of nonlinear realizations of chiral symmetry.[e] It is characteristic for these nonlinear realizations that, whenever functions of the Goldstone bosons appear in the Langrangian, they are always accompanied with at least one space-time derivative. The Lagrangians given below are built upon the CCWZ formalism.

As discussed, the relevant degrees of freedom are pions (Goldstone bosons) and nucleons. Since the interactions of Goldstone bosons must vanish at zero momentum transfer and in the chiral limit ($m_\pi \to 0$), the low-energy expansion of the Lagrangian is arranged in powers of derivatives and pion masses. The hard scale is the chiral-symmetry breaking scale, $\Lambda_\chi \approx 1$ GeV. Thus, the expansion is in terms of powers of Q/Λ_χ where Q is a (small) momentum or pion mass. This is ChPT.

[e] An accessible introduction into the rather involved CCWZ formalism can be found in Ref. 31.

The effective Lagrangian can formally be written as,

$$\mathcal{L} = \mathcal{L}_{\pi\pi} + \mathcal{L}_{\pi N} + \mathcal{L}_{NN} + \cdots , \tag{12}$$

where $\mathcal{L}_{\pi\pi}$ deals with the dynamics among pions, $\mathcal{L}_{\pi N}$ describes the interaction between pions and a nucleon and $\mathcal{L}_{NN}$ contains two-nucleon contact interactions which consist of four nucleon-fields (four nucleon legs) and no meson fields. The ellipsis stands for terms that involve two nucleons plus pions and three or more nucleons with or without pions, relevant for nuclear many-body forces (an example for this in lowest order are the last two terms of Eq. (18)). The individual Lagrangians are organized order by order:

$$\mathcal{L}_{\pi\pi} = \mathcal{L}^{(2)}_{\pi\pi} + \mathcal{L}^{(4)}_{\pi\pi} + \cdots , \tag{13}$$

$$\mathcal{L}_{\pi N} = \mathcal{L}^{(1)}_{\pi N} + \mathcal{L}^{(2)}_{\pi N} + \mathcal{L}^{(3)}_{\pi N} + \mathcal{L}^{(4)}_{\pi N} + \mathcal{L}^{(5)}_{\pi N} + \cdots , \tag{14}$$

$$\mathcal{L}_{NN} = \mathcal{L}^{(0)}_{NN} + \mathcal{L}^{(2)}_{NN} + \mathcal{L}^{(4)}_{NN} + \mathcal{L}^{(6)}_{NN} + \cdots , \tag{15}$$

where the superscript refers to the number of derivatives or pion mass insertions (chiral dimension) and the ellipsis stands for terms of higher dimensions.

Above, we have organized the Lagrangians by the number of derivatives or pion-mass insertions. This is the standard way, appropriate particularly for considerations of π–π and π–N scattering. As it turns out (cf. Sec. 4.1), for interactions among nucleons, it is sometimes useful to also consider the so-called index of the interaction,

$$\Delta \equiv d + \frac{n}{2} - 2, \tag{16}$$

where d is the number of derivatives or pion-mass insertions and n the number of nucleon field operators (nucleon legs). We will now write down the Lagrangian in terms of increasing values of the parameter Δ and we will do so using the so-called heavy-baryon formalism which we indicate by a "hat".[34]

The leading-order Lagrangian reads

$$\begin{aligned}
\widehat{\mathcal{L}}^{\Delta=0} &= \frac{1}{2}\partial_\mu \boldsymbol{\pi} \cdot \partial^\mu \boldsymbol{\pi} - \frac{1}{2} m_\pi^2 \boldsymbol{\pi}^2 + \frac{1-4\alpha}{2f_\pi^2}(\boldsymbol{\pi}\cdot\partial_\mu\boldsymbol{\pi})(\boldsymbol{\pi}\cdot\partial^\mu\boldsymbol{\pi}) - \frac{\alpha}{f_\pi^2}\boldsymbol{\pi}^2\partial_\mu\boldsymbol{\pi}\cdot\partial^\mu\boldsymbol{\pi} \\
&\quad + \frac{8\alpha-1}{8f_\pi^2} m_\pi^2 \boldsymbol{\pi}^4 + \bar{N}\left[i\partial_0 - \frac{g_A}{2f_\pi}\boldsymbol{\tau}\cdot(\boldsymbol{\sigma}\cdot\vec{\nabla})\boldsymbol{\pi} - \frac{1}{4f_\pi^2}\boldsymbol{\tau}\cdot(\boldsymbol{\pi}\times\partial_0\boldsymbol{\pi})\right] N \\
&\quad + \bar{N}\left\{ \frac{g_A(4\alpha-1)}{4f_\pi^3}(\boldsymbol{\tau}\cdot\boldsymbol{\pi})[\boldsymbol{\pi}\cdot(\boldsymbol{\sigma}\cdot\vec{\nabla})\boldsymbol{\pi}] + \frac{g_A\alpha}{2f_\pi^3}\boldsymbol{\pi}^2[\boldsymbol{\tau}\cdot(\boldsymbol{\sigma}\cdot\vec{\nabla})\boldsymbol{\pi}]\right\} N \\
&\quad - \frac{1}{2} C_S \bar{N}N\bar{N}N - \frac{1}{2} C_T (\bar{N}\boldsymbol{\sigma}N)\cdot(\bar{N}\boldsymbol{\sigma}N) + \cdots ,
\end{aligned} \tag{17}$$

and subleading Lagrangians are

$$\widehat{\mathcal{L}}^{\Delta=1} = \bar{N}\left\{\frac{\vec{\nabla}^2}{2M_N} - \frac{ig_A}{4M_N f_\pi}\boldsymbol{\tau}\cdot\left[\boldsymbol{\sigma}\cdot(\overleftarrow{\nabla}\,\partial_0\boldsymbol{\pi} - \partial_0\boldsymbol{\pi}\,\vec{\nabla})\right]\right.$$

$$\left. - \frac{i}{8M_N f_\pi^2}\boldsymbol{\tau}\cdot\left[\overleftarrow{\nabla}\cdot(\boldsymbol{\pi}\times\vec{\nabla}\boldsymbol{\pi}) - (\boldsymbol{\pi}\times\vec{\nabla}\boldsymbol{\pi})\cdot\vec{\nabla}\right]\right\}N$$

$$+\bar{N}\left[4c_1 m_\pi^2 - \frac{2c_1}{f_\pi^2}m_\pi^2\boldsymbol{\pi}^2 + \left(c_2 - \frac{g_A^2}{8M_N}\right)\frac{1}{f_\pi^2}(\partial_0\boldsymbol{\pi}\cdot\partial_0\boldsymbol{\pi})\right.$$

$$\left. + \frac{c_3}{f_\pi^2}(\partial_\mu\boldsymbol{\pi}\cdot\partial^\mu\boldsymbol{\pi}) - \left(c_4 + \frac{1}{4M_N}\right)\frac{1}{2f_\pi^2}\epsilon^{ijk}\epsilon^{abc}\sigma^i\tau^a(\partial^j\pi^b)(\partial^k\pi^c)\right]N$$

$$- \frac{D}{4f_\pi}(\bar{N}N)\bar{N}[\boldsymbol{\tau}\cdot(\boldsymbol{\sigma}\cdot\vec{\nabla})\boldsymbol{\pi}]N - \frac{1}{2}E(\bar{N}N)(\bar{N}\boldsymbol{\tau}N)\cdot(\bar{N}\boldsymbol{\tau}N) + \cdots, \tag{18}$$

$$\widehat{\mathcal{L}}^{\Delta=2} = \mathcal{L}_{\pi\pi}^{(4)} + \widehat{\mathcal{L}}_{\pi N}^{(3)} + \widehat{\mathcal{L}}_{NN}^{(2)} + \cdots, \tag{19}$$

$$\widehat{\mathcal{L}}^{\Delta=3} = \widehat{\mathcal{L}}_{\pi N}^{(4)} + \cdots, \tag{20}$$

$$\widehat{\mathcal{L}}^{\Delta=4} = \widehat{\mathcal{L}}_{\pi N}^{(5)} + \widehat{\mathcal{L}}_{NN}^{(4)} + \cdots, \tag{21}$$

$$\widehat{\mathcal{L}}^{\Delta=6} = \widehat{\mathcal{L}}_{NN}^{(6)} + \cdots, \tag{22}$$

where we included terms relevant for a calculation of the 2NF up to sixth-order. The Lagrangians $\widehat{\mathcal{L}}_{\pi N}^{(3)}$ and $\widehat{\mathcal{L}}_{\pi N}^{(4)}$ can be found in Ref. 35 and *NN* contact Lagrangians are given below. The pion fields are denoted by $\boldsymbol{\pi}$ and the heavy baryon nucleon field by N ($\bar{N} = N^\dagger$). Furthermore, g_A, f_π, m_π and M_N are the axial-vector coupling constant, pion decay constant, pion mass and nucleon mass, respectively. Numerical values for these quantities will be given later. The c_i are low-energy constants (LECs) from the dimension 2 πN Lagrangian and α is a parameter that appears in the expansion of the pion fields, see Ref. 30 for more details. Results are independent of α.

The lowest order (or leading order) *NN* Lagrangian has no derivatives and reads[29]

$$\widehat{\mathcal{L}}_{NN}^{(0)} = -\frac{1}{2}C_S\bar{N}N\bar{N}N - \frac{1}{2}C_T(\bar{N}\boldsymbol{\sigma}N)\cdot(\bar{N}\boldsymbol{\sigma}N), \tag{23}$$

where C_S and C_T are unknown constants which are determined by a fit to the *NN* data.

The second-order *NN* Lagrangian can be found in Ref. 36. The *NN* contact potentials that emerge from the *NN* Lagrangians are given in Sec. 5.1.

4. Nuclear Forces from EFT: Overview

In the beginning of Sec. 3, we listed the steps we have to take for carrying out the EFT program of a derivation of the nuclear potential. So far, we discussed

Steps (1)–(3). What is left are Steps (4) (low-momentum expansion) and (5) (Feynman diagrams). In this section, we will say more about the expansion we are using and give an overview of the Feynman diagrams that arise order by order.

4.1. *Chiral perturbation theory and power counting*

Effective Langrangians have infinitely many terms, and an unlimited number of Feynman graphs can be calculated from them. Therefore, we need a scheme that makes the theory manageable and calculable. This scheme which tells us how to distinguish between large (important) and small (unimportant) contributions is ChPT.

In ChPT, graphs are analyzed in terms of powers of small external momenta over the large scale: $(Q/\Lambda_\chi)^\nu$, where Q is generic for a momentum (nucleon three-momentum or pion four-momentum) or a pion mass and $\Lambda_\chi \sim 1\,\text{GeV}$ is the chiral symmetry breaking scale (hadronic scale, hard scale). Determining the power ν has become known as power counting.

For the moment, we will consider only so-called irreducible graphs. By definition, an irreducible graph is a diagram that cannot be separated into two by cutting only nucleon lines. Following the Feynman rules of covariant perturbation theory, a nucleon propagator is Q^{-1}, a pion propagator Q^{-2}, each derivative in any interaction is Q, and each four-momentum integration Q^4. This is also known as naive dimensional analysis. Applying then some topological identities, one obtains for the power of an irreducible diagram involving A nucleons[30]

$$\nu = -2 + 2A - 2C + 2L + \sum_i \Delta_i, \tag{24}$$

with

$$\Delta_i \equiv d_i + \frac{n_i}{2} - 2, \tag{25}$$

where C denotes the number of separately connected pieces and L the number of loops in the diagram; d_i is the number of derivatives or pion-mass insertions and n_i the number of nucleon fields (nucleon legs) involved in vertex i; the sum runs over all vertices i contained in the diagram under consideration. Note that $\Delta_i \geq 0$ for all interactions allowed by chiral symmetry. Purely pionic interactions have at least two derivatives ($d_i \geq 2$, $n_i = 0$); interactions of pions with a nucleon have at least one derivative ($d_i \geq 1$, $n_i = 2$); and *NN* contact terms ($n_i = 4$) have $d_i \geq 0$. This demonstrates how chiral symmetry guarantees a low-energy expansion.

An important observation from power counting is that the powers are bounded from below and, specifically, $\nu \geq 0$. This fact is crucial for the convergence of the low-momentum expansion.

Moreover, the power formula Eq. (24) allows to predict the leading orders of connected multi-nucleon forces. Consider a m-nucleon irreducibly connected diagram (m-nucleon force) in an A-nucleon system ($m \leq A$). The number of separately connected pieces is $C = A - m + 1$. Inserting this into Eq. (24) together with $L = 0$

and $\sum_i \Delta_i = 0$ yields $\nu = 2m-4$. Thus, 2NFs ($m = 2$) start at $\nu = 0$, three-nucleon forces ($m = 3$) at $\nu = 2$ (but they happen to cancel at that order) and four-nucleon forces (4NFs) at $\nu = 4$ (they do not cancel). More about this in the next subsection.

For later purposes, we note that for an irreducible NN diagram ($A = 2$, $C = 1$), the power formula collapses to the very simple expression

$$\nu = 2L + \sum_i \Delta_i . \tag{26}$$

In summary, the chief point of the ChPT expansion is that, at a given order ν, there exists only a finite number of graphs. This is what makes the theory calculable. The expression $(Q/\Lambda_\chi)^{\nu+1}$ provides a rough estimate of the relative size of the contributions left out and, thus, of the accuracy at order ν. In this sense, the theory can be calculated to any desired accuracy and has predictive power.

4.2. *The hierarchy of nuclear forces*

ChPT and power counting imply that nuclear forces emerge as a hierarchy controlled by the power ν, Fig. 7.

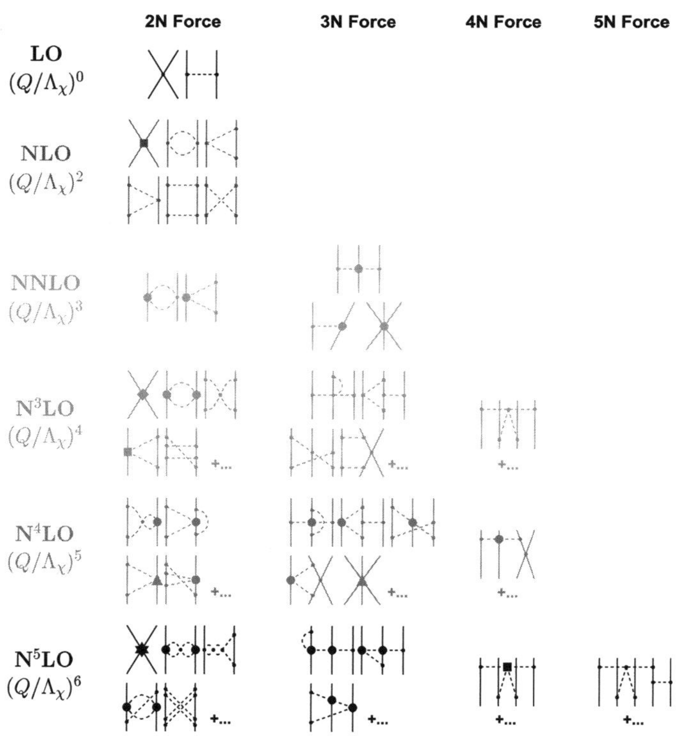

Fig. 7. Hierarchy of nuclear forces in ChPT. Solid lines represent nucleons and dashed lines pions. Small dots, large solid dots, solid squares, triangles, diamonds and stars denote vertices of index $\Delta = 0, 1, 2, 3, 4$ and 6, respectively. Further explanations are given in the text.

In lowest order, better known as leading order (LO, $\nu = 0$), the NN amplitude is made up by two momentum-independent contact terms ($\sim Q^0$), represented by the four-nucleon-leg graph with a small-dot vertex shown in the first row of Fig. 7, and static one-pion exchange (1PE), second diagram in the first row of the figure. This is, of course, a rather rough approximation to the 2NF, but accounts already for some important features. The 1PE provides the tensor force, necessary to describe the deuteron, and it explains NN scattering in peripheral partial waves of very high orbital angular momentum. At this order, the two contacts which contribute only in S-waves provide the short- and intermediate-range interaction which is somewhat crude.

In the next order, $\nu = 1$, all contributions vanish due to parity and time-reversal invariance.

Therefore, the next-to-leading order (NLO) is $\nu = 2$. 2PE occurs for the first time ("leading 2PE") and, thus, the creation of a more sophisticated description of the intermediate-range interaction is starting here. Since the loop involved in each pion-diagram implies already $\nu = 2$ (cf. Eq. (26)), the vertices must have $\Delta_i = 0$. Therefore, at this order, only the lowest order πNN and $\pi\pi NN$ vertices are allowed which is why the leading 2PE is rather weak. Furthermore, there are seven contact terms of $\mathcal{O}(Q^2)$, shown by the four-nucleon-leg graph with a solid square, which contribute in S- and P-waves. The operator structure of these contacts includes a spin-orbit term besides central, spin–spin and tensor terms. Thus, essentially all spin–isospin structures necessary to describe the 2NF phenomenologically have been generated at this order. The main deficiency at this stage of development is an insufficient intermediate-range attraction.

This problem is finally fixed at order three ($\nu = 3$), next-to-next-to-leading order (NNLO or N^{2}LO). The 2PE involves now the two-derivative $\pi\pi NN$ seagull vertices (proportional to the c_i LECs) denoted by a large solid dot in Fig. 7.[37] These vertices represent correlated 2PE as well as intermediate $\Delta(1232)$-isobar contributions.[38] As discussed in the context of the meson theory of nuclear forces[11,14] (cf. Sec. 2.2), these two contributions are crucial for a realistic and quantitative 2PE model. Consequently, the 2PE now assumes a realistic size and describes the intermediate-range attraction of the nuclear force about right. There are no new contacts.

The reason why we talk of a hierarchy of nuclear forces is that two- and many-nucleon forces are created on an equal footing and emerge in increasing number as we go to higher and higher orders. At NNLO, the first set of nonvanishing three-nucleon forces (3NF) occur,[39,40] cf. column "3N Force" of Fig. 7. In fact, at the previous order, NLO, irreducible 3N graphs appear already, however, it has been shown by Weinberg[41] that these diagrams all cancel. Since nonvanishing 3NF contributions happen first at order $(Q/\Lambda_\chi)^3$, they are very weak as compared to the 2NF which starts at $(Q/\Lambda_\chi)^0$.

More 2PE is produced at $\nu = 4$, next-to-next-to-next-to-leading order (N^{3}LO), of which we show only a few symbolic diagrams in Fig. 7. There is a large attractive one-loop 2PE contribution (the bubble diagram with two large solid dots $\sim c_i^2$), which is slightly over-doing the intermediate-range attraction of the 2NF. Two-loop 2PE graphs show up for the first time and so does three-pion exchange (3PE) which necessarily involves two loops. 3PE was found to be negligible at this order.[42,43] Most importantly, 15 new contact terms $\sim Q^4$ arise and are represented by the four-nucleon-leg graph with a solid diamond. They include a quadratic spin-orbit term and contribute up to D-waves. Mainly due to the increased number of contact terms, a quantitative description of the two-nucleon interaction up to about 300 MeV lab. energy is possible at N^{3}LO.[30,44] Besides further 3NF, 4NF start at this order. Since the leading 4NF come into existence one order higher than the leading 3NF, 4NF is weaker than 3NF. Thus, ChPT provides a straightforward explanation for the empirically known fact that 2NF $\gg$ 3NF $\gg$ 4NF $\cdots$.

Further 2PE and 3PE occur at N^{4}LO (fifth-order). The contribution to the 2NF at this order has been first calculated by Entem *et al.*[45] It turns out to be moderately repulsive, thus, compensating for the attractive surplus generated at N^{3}LO by the bubble diagram with two solid dots. The long- and intermediate-range 3NF contributions at this order have been evaluated,[35,46] but not applied yet in nuclear structure calculations. They are expected to be sizeable. Moreover, a new set of 3NF contact terms appear.[47] The N^{4}LO 4NF has not been derived yet. Due to the subleading $\pi\pi NN$ seagull vertex (large solid dot $\sim c_i$), this 4NF could be sizeable.

Finally turning to N^{5}LO (sixth-order): The dominant 2PE and 3PE contributions to the 2NF have been derived by Entem *et al.*,[48] which represents the most sophisticated investigation ever conducted in chiral EFT for the *NN* system. The effects are small indicating the desired trend toward convergence of the chiral expansion for the 2NF. Moreover, a new set of 26 *NN* contact terms $\sim Q^6$ that contribute up to F-waves occurs (represented by the *NN* diagram with a star in Fig. 7) bringing the total number of *NN* contacts to 50.[49] The three-, four- and five-nucleon forces of this order have not yet been derived.

To summarize, we show in Fig. 8 the contributions to the phase shifts of peripheral *NN* scattering through all orders from LO to N^{5}LO. Note that the difference between the LO prediction (1PE, dotted line) and the data (filled and open circles) is to be provided by two- and 3PEs, i.e., the intermediate-range part of the nuclear force. How well that is accomplished is a crucial test for any theory of nuclear forces. NLO produces only a small contribution, but N^{2}LO creates substantial intermediate-range attraction (most clearly seen in $^1G_4$, $^3G_5$). In fact, N^{2}LO is the largest contribution among all orders. This is due to the one-loop 2π-exchange triangle diagram which involves one $\pi\pi NN$-contact vertex proportional to c_i. As discussed, the one-loop 2π-exchange at N^{2}LO is attractive and describes the intermediate-range attraction of the nuclear force about right. At N^{3}LO, more one-loop 2PE is added by the bubble diagram with two c_i-vertices, a contribution that seemingly is overestimating the attraction. This attractive surplus is then compensated by the prevailingly repulsive two-loop 2π- and 3π-exchanges that occur at N^{4}LO and N^{5}LO.

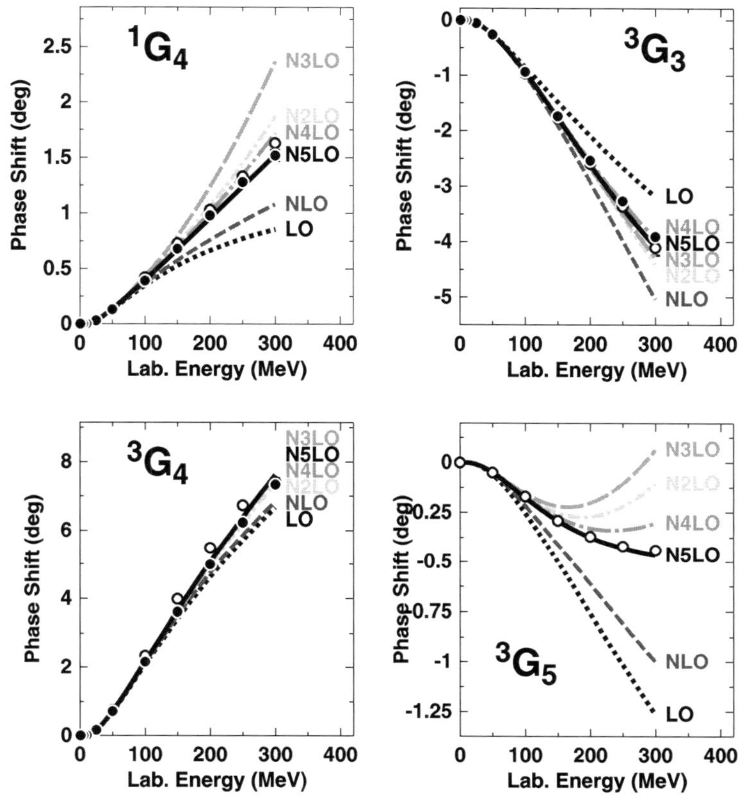

Fig. 8. Phase-shifts of neutron–proton scattering in G waves at all orders from LO to N^5LO. The filled and open circles represent the results from the Nijmegen multi-energy np phase-shift analysis[50] and the GWU np analysis SP07,[51] respectively.

In this context, it is worth noting that also in conventional meson theory[11] (Sec. 2.2) the one-loop models for the 2PE contribution always show some excess of attraction (cf. Fig. 10 of Ref. 30). The same is true for the dispersion theoretic approach pursued by the Paris group (see, e.g., the predictions for $^1D_2$, $^3D_2$ and $^3D_3$ in Fig. 8 of Ref. 14 which are all too attractive). In conventional meson theory, this attraction is reduced by heavy-meson exchanges (ρ-, ω- and $\pi\rho$-exchange) which, however, has no place in chiral EFT (as a finite-range contribution). Instead, in the latter approach, two-loop 2π- and 3π-exchanges provide the corrective action.

5. Constructing Chiral *NN* Potentials

In the previous section, we mainly discussed the pion-exchange contributions to the NN interaction. They describe the long- and intermediate-range parts of the nuclear force, which are governed by chiral symmetry and rule the peripheral partial waves (cf. Fig. 8). However, for a "complete" nuclear force, we have to describe correctly all partial waves, including the lower ones. In fact, in calculations of NN

observables at low energies (cross-sections, analyzing powers, etc.), the partial waves with $L \leq 2$ are the most important ones, generating the largest contributions. The same is true for microscopic nuclear structure calculations. The lower partial waves are dominated by the dynamics at short distances. Therefore, we need to look now more closely into the short-range part of the *NN* potential.

5.1. *NN contact terms*

In conventional meson theory, the short-range nuclear force is described by the exchange of heavy mesons, notably the $\omega(782)$. Qualitatively, the short-distance behavior of the *NN* potential is obtained by Fourier transform of the propagator of a heavy meson,

$$\int d^3q \frac{e^{i\mathbf{q}\cdot\mathbf{r}}}{m_\omega^2 + \mathbf{q}^2} \sim \frac{e^{-m_\omega r}}{r}. \tag{27}$$

ChPT is an expansion in small momenta Q, too small to resolve structures like a $\rho(770)$ or $\omega(782)$ meson, because $Q \ll \Lambda_\chi \approx m_{\rho,\omega}$. But the latter relation allows us to expand the propagator of a heavy meson into a power series,

$$\frac{1}{m_\omega^2 + Q^2} \approx \frac{1}{m_\omega^2}\left(1 - \frac{Q^2}{m_\omega^2} + \frac{Q^4}{m_\omega^4} - \cdots\right), \tag{28}$$

where the ω is representative for any heavy meson of interest. The above expansion suggests that it should be possible to describe the short distance part of the nuclear force simply in terms of powers of Q/m_ω, which fits in well with our over-all power expansion since $Q/m_\omega \approx Q/\Lambda_\chi$. Since such terms act directly between nucleons, they are dubbed contact terms.

Contact terms play an important role in renormalization. Regularization of the loop integrals that occur in multi-pion exchange diagrams typically generates polynomial terms with coefficients that are, in part, infinite or scale dependent (cf. Appendix B of Ref. 30). Contact terms absorb infinities and remove scale dependences, which is why they are also known as counter terms.

Due to parity, only even powers of Q are allowed. Thus, the expansion of the contact potential is formally given by

$$V_{\rm ct} = V_{\rm ct}^{(0)} + V_{\rm ct}^{(2)} + V_{\rm ct}^{(4)} + V_{\rm ct}^{(6)} + \cdots, \tag{29}$$

where the superscript denotes the power or order.

We will now present, one by one, the various orders of *NN* contact terms.

5.1.1. *Zeroth-order (LO)*

The contact Lagrangian $\widehat{\mathcal{L}}_{NN}^{(0)}$, Eq. (23), which is part of $\widehat{\mathcal{L}}^{\Delta=0}$, Eq. (17), leads to the following *NN* contact potential,

$$V_{\rm ct}^{(0)}(\mathbf{p}', \mathbf{p}) = C_S + C_T \boldsymbol{\sigma}_1 \cdot \boldsymbol{\sigma}_2, \tag{30}$$

and, in terms of partial waves, we have

$$\begin{aligned} V_{\rm ct}^{(0)}(^1S_0) &= \widetilde{C}_{^1S_0} = 4\pi(C_S - 3C_T), \\ V_{\rm ct}^{(0)}(^3S_1) &= \widetilde{C}_{^3S_1} = 4\pi(C_S + C_T). \end{aligned} \tag{31}$$

5.1.2. *Second-order (NLO)*

The contact Lagrangian $\widehat{\mathcal{L}}^{(2)}_{NN}$, which is part of $\widehat{\mathcal{L}}^{\Delta=2}$, Eq. (19), generates the following NN contact potential

$$\begin{aligned} V_{\rm ct}^{(2)}(\mathbf{p}',\mathbf{p}) = {} & C_1 q^2 + C_2 k^2 + (C_3 q^2 + C_4 k^2)\boldsymbol{\sigma}_1 \cdot \boldsymbol{\sigma}_2 + C_5(-i\mathbf{S}\cdot(\mathbf{q}\times\mathbf{k})) \\ & + C_6(\boldsymbol{\sigma}_1\cdot\mathbf{q})(\boldsymbol{\sigma}_2\cdot\mathbf{q}) + C_7(\boldsymbol{\sigma}_1\cdot\mathbf{k})(\boldsymbol{\sigma}_2\cdot\mathbf{k}). \end{aligned} \tag{32}$$

There are many ways to perform the partial-wave decomposition of the above potential. We perceive the method presented by Erkelenz, Alzetta and Holinde[52] as the most elegant one. Thus, one obtains

$$\begin{aligned} V_{\rm ct}^{(2)}(^1S_0) &= C_{^1S_0}(p^2 + p'^2), \\ V_{\rm ct}^{(2)}(^3P_0) &= C_{^3P_0}pp', \\ V_{\rm ct}^{(2)}(^1P_1) &= C_{^1P_1}pp', \\ V_{\rm ct}^{(2)}(^3P_1) &= C_{^3P_1}pp', \\ V_{\rm ct}^{(2)}(^3S_1) &= C_{^3S_1}(p^2 + p'^2), \\ V_{\rm ct}^{(2)}(^3S_1 - {}^3D_1) &= C_{^3S_1 - {}^3D_1}p^2, \\ V_{\rm ct}^{(2)}(^3D_1 - {}^3S_1) &= C_{^3S_1 - {}^3D_1}p'^2, \\ V_{\rm ct}^{(2)}(^3P_2) &= C_{^3P_2}pp' \end{aligned} \tag{33}$$

with

$$\begin{aligned} C_{^1S_0} &= 4\pi\left(C_1 + \frac{1}{4}C_2 - 3C_3 - \frac{3}{4}C_4 - C_6 - \frac{1}{4}C_7\right), \\ C_{^3P_0} &= 4\pi\left(-\frac{2}{3}C_1 + \frac{1}{6}C_2 - \frac{2}{3}C_3 + \frac{1}{6}C_4 - \frac{2}{3}C_5 + 2C_6 - \frac{1}{2}C_7\right), \\ C_{^1P_1} &= 4\pi\left(-\frac{2}{3}C_1 + \frac{1}{6}C_2 + 2C_3 - \frac{1}{2}C_4 + \frac{2}{3}C_6 - \frac{1}{6}C_7\right), \\ C_{^3P_1} &= 4\pi\left(-\frac{2}{3}C_1 + \frac{1}{6}C_2 - \frac{2}{3}C_3 + \frac{1}{6}C_4 - \frac{1}{3}C_5 - \frac{4}{3}C_6 + \frac{1}{3}C_7\right), \\ C_{^3S_1} &= 4\pi\left(C_1 + \frac{1}{4}C_2 + C_3 + \frac{1}{4}C_4 + \frac{1}{3}C_6 + \frac{1}{12}C_7\right), \end{aligned}$$

$$C_{^3S_1-^3D_1} = 4\pi\left(-\frac{2\sqrt{2}}{3}C_6 - \frac{\sqrt{2}}{6}C_7\right),$$

$$C_{^3P_2} = 4\pi\left(-\frac{2}{3}C_1 + \frac{1}{6}C_2 - \frac{2}{3}C_3 + \frac{1}{6}C_4 + \frac{1}{3}C_5\right). \tag{34}$$

5.1.3. *Fourth-order (N^3LO)*

The contact potential of order four reads

$$\begin{aligned} V_{\text{ct}}^{(4)}(\mathbf{p}',\mathbf{p}) &= D_1q^4 + D_2k^4 + D_3q^2k^2 + D_4(\mathbf{q}\times\mathbf{k})^2 + (D_5q^4 + D_6k^4 + D_7q^2k^2 \\ &\quad + D_8(\mathbf{q}\times\mathbf{k})^2)\boldsymbol{\sigma}_1\cdot\boldsymbol{\sigma}_2 + (D_9q^2 + D_{10}k^2)(-i\mathbf{S}\cdot(\mathbf{q}\times\mathbf{k})) \\ &\quad + (D_{11}q^2 + D_{12}k^2)(\boldsymbol{\sigma}_1\cdot\mathbf{q})(\boldsymbol{\sigma}_2\cdot\mathbf{q}) + (D_{13}q^2 + D_{14}k^2)(\boldsymbol{\sigma}_1\cdot\mathbf{k})(\boldsymbol{\sigma}_2\cdot\mathbf{k}) \\ &\quad + D_{15}(\boldsymbol{\sigma}_1\cdot(\mathbf{q}\times\mathbf{k})\boldsymbol{\sigma}_2\cdot(\mathbf{q}\times\mathbf{k})). \end{aligned} \tag{35}$$

The rather lengthy partial-wave expressions of this order are given in Appendix E of Ref. 30.

5.1.4. *Sixth-order (N^5LO)*

At sixth-order, 26 new contact terms appear, bringing the total number to 50. These terms as well as their partial-wave decomposition have been worked out in Ref. 49. So far, these terms have not been used in the construction of *NN* potentials.

5.2. *Definition of NN potential*

We have now rounded up everything needed for a realistic nuclear force — long, intermediate and short ranged components — and so we can finally proceed to the lower partial waves. However, here we encounter another problem. The two-nucleon system at low angular momentum, particularly in S-waves, is characterized by the presence of a shallow bound state (the deuteron) and large scattering lengths. Thus, perturbation theory does not apply. In contrast to π–π and π–N, the interaction between nucleons is not suppressed in the chiral limit ($Q \to 0$). Weinberg[29] showed that the strong enhancement of the scattering amplitude arises from purely nucleonic intermediate states ("infrared enhancement"). He therefore suggested to use perturbation theory to calculate the *NN* potential (i.e., the irreducible graphs) and to apply this potential in a scattering equation to obtain the *NN* amplitude. We will follow this prescription.

The potential V as discussed in previous sections is, in principal, an invariant amplitude and, thus, satisfies a relativistic scattering equation, for which we choose the BbS equation,[30] which reads explicitly

$$T(\mathbf{p}',\mathbf{p}) = V(\mathbf{p}',\mathbf{p}) + \int \frac{d^3p''}{(2\pi)^3} V(\mathbf{p}',\mathbf{p}'') \frac{M_N^2}{E_{p''}} \frac{1}{p^2 - p''^2 + i\epsilon} T(\mathbf{p}'',\mathbf{p}) \tag{36}$$

with $E_{p''} \equiv \sqrt{M_N^2 + p''^2}$. The advantage of using a relativistic scattering equation is that it automatically includes relativistic corrections to all orders. Thus, in the scattering equation, no propagator modifications are necessary when raising the order to which the calculation is conducted.

Defining

$$\widehat{V}(\mathbf{p}', \mathbf{p}) \equiv \frac{1}{(2\pi)^3} \sqrt{\frac{M_N}{E_{p'}}} V(\mathbf{p}', \mathbf{p}) \sqrt{\frac{M_N}{E_p}} \tag{37}$$

and

$$\widehat{T}(\mathbf{p}', \mathbf{p}) \equiv \frac{1}{(2\pi)^3} \sqrt{\frac{M_N}{E_{p'}}} T(\mathbf{p}', \mathbf{p}) \sqrt{\frac{M_N}{E_p}}, \tag{38}$$

where the factor $1/(2\pi)^3$ is added for convenience, the BbS equation collapses into the usual, nonrelativistic LS equation

$$\widehat{T}(\mathbf{p}', \mathbf{p}) = \widehat{V}(\mathbf{p}', \mathbf{p}) + \int d^3p'' \widehat{V}(\mathbf{p}', \mathbf{p}'') \frac{M_N}{p^2 - p''^2 + i\epsilon} \widehat{T}(\mathbf{p}'', \mathbf{p}). \tag{39}$$

Since $\widehat{V}$ satisfies Eq. (39), it can be used like a nonrelativistic potential, and $\widehat{T}$ may be perceived as the conventional nonrelativistic T-matrix.

5.3. *Regularization and nonperturbative renormalization*

Iteration of $\widehat{V}$ in the LS equation, Eq. (39), requires cutting $\widehat{V}$ off for high momenta to avoid infinities. This is consistent with the fact that ChPT is a low-momentum expansion which is valid only for momenta $Q \ll \Lambda_\chi \approx 1\,\text{GeV}$. Therefore, the potential $\widehat{V}$ is multiplied with the regulator function $f(p', p)$,

$$\widehat{V}(\mathbf{p}', \mathbf{p}) \mapsto \widehat{V}(\mathbf{p}', \mathbf{p}) f(p', p) \tag{40}$$

with

$$f(p', p) = \exp[-(p'/\Lambda)^{2n} - (p/\Lambda)^{2n}], \tag{41}$$

such that

$$\widehat{V}(\mathbf{p}', \mathbf{p}) f(p', p) \approx \widehat{V}(\mathbf{p}', \mathbf{p}) \left\{ 1 - \left[\left(\frac{p'}{\Lambda} \right)^{2n} + \left(\frac{p}{\Lambda} \right)^{2n} \right] + \cdots \right\}. \tag{42}$$

Typical choices for the cutoff parameter Λ that appears in the regulator are $\Lambda \approx 0.5\,\text{GeV} < \Lambda_\chi \approx 1\,\text{GeV}$. At N^{3}LO and N^{4}LO, an appropriate choice for n is three.

Equation (42) provides an indication of the fact that the exponential cutoff does not necessarily affect the given order at which the calculation is conducted. For sufficiently large n, the regulator introduces contributions that are beyond the given order. Assuming a good rate of convergence of the chiral expansion, such orders are small as compared to the given order and, thus, do not affect the accuracy at the given order. In calculations, one uses, of course, the exponential form, Eq. (41), and not the expansion Eq. (42).

It is pretty obvious that results for the T-matrix may depend sensitively on the regulator and its cutoff parameter. The removal of such regulator dependence is known as renormalization. Because of lack of space, I cannot discuss this issue here and like to refer the interested reader to Sec. 4.5 of Ref. 30.

5.4. *NN potentials order by order*

As discussed, NN potentials can be calculated at various orders and the accuracy improves as the order increases. How well the chiral expansion converges in the important lower partial waves is demonstrated in Figs. 9 and 10, where we show the $J \leq 2$ phase parameters for potentials constructed through all orders from LO to N^{3}LO. These figures clearly reveal substantial improvements in the reproduction of the empirical phase shifts with increasing order.

There is an even better way to confront theory with experiment. One calculates observables of NN scattering and compares this directly to the experimental data. It is customary to state the result of such a comparison in terms of the χ^2/datum where a value around unity would signify a perfect fit.

Let us start with potentials developed to NLO and NNLO. In Table 1, we show the χ^2/datum for the fit of the world np data below 290 MeV for families of np potentials at NLO and NNLO constructed by the Bochum group.[53] The NLO potentials produce the very large χ^2/datum between 67 and 105, and the NNLO are between 12 and 27. The rate of improvement from one order to the other is very encouraging, but the quality of the reproduction of the np data at NLO and NNLO is obviously insufficient for reliable predictions.[f]

Based upon these facts, it has been pointed out in 2002 by Entem and Machleidt[57,58] that one has to proceed to N^{3}LO. Consequently, the first N^{3}LO potential was published in 2003.[44]

At N^{3}LO (Q^4), there are a total of 24 contact terms (24 parameters) which contribute to the partial waves with $L \leq 2$, while at NLO and NNLO there are only nine contacts with $L \leq 1$ (cf. Sec. 5.1 and Table 2). These LECs are essentially free constants which parametrize the short-ranged phenomenological part of the interaction. Table 2 shows how these terms are distributed over partial waves. One reason for the improved reproduction of the NN phase shifts (and NN observables) at N^{3}LO is the fact that, at that order, contacts appear for the first time in D-waves. D-waves are not truly peripheral and, therefore, 1PE plus 2PE alone do not describe them well (too attractive, cf. the NNLO predictions for $^1D_2$ and $^3D_2$ in Fig. 10). The D-wave contacts provide the necessary short-range corrections to predict the D-phases right. Furthermore, at N^{3}LO, another contact is added to each P-wave, which leads to substantial improvements, particularly, in $^3P_0$ and $^3P_1$ above 100 MeV (cf. Fig. 9).

[f]For an optimized NNLO potential see Ref. 55 and for local NLO and NNLO potentials see Ref. 56.

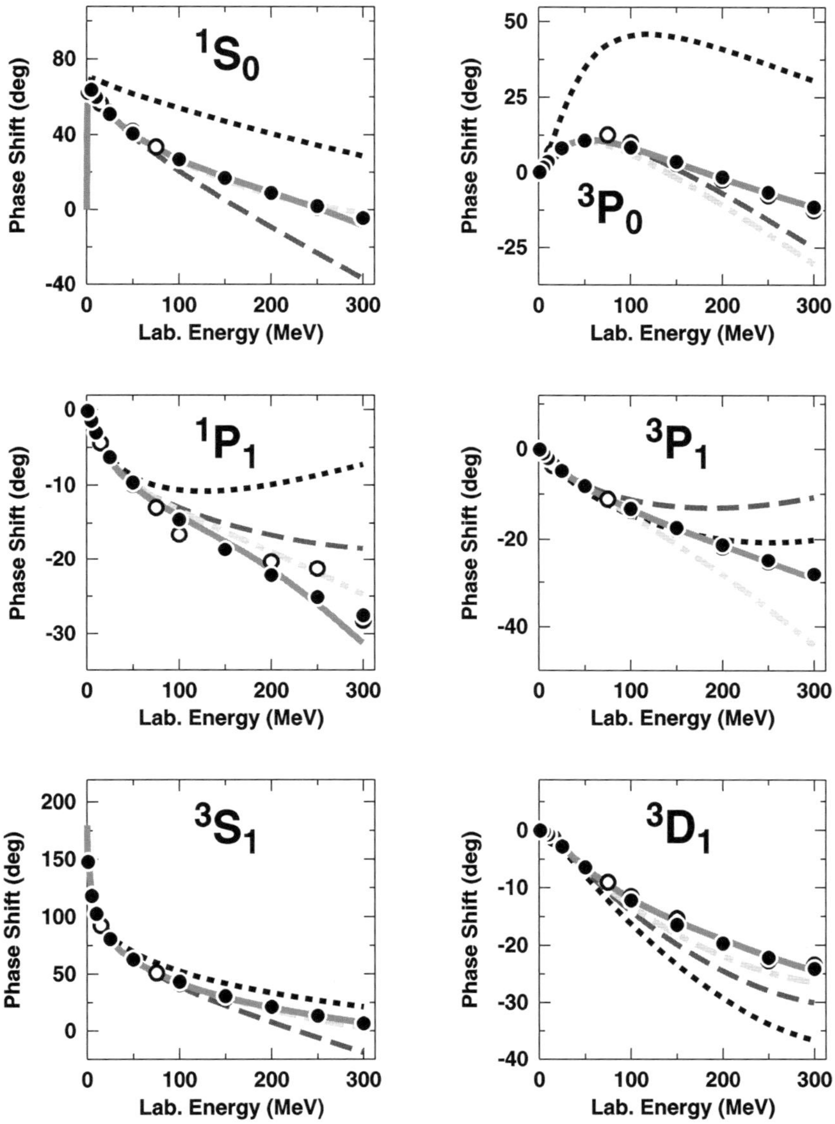

Fig. 9. (Color online) Phase shifts of np scattering as calculated from NN potentials at different orders of ChPT. The black dotted line is LO (500), the blue dashed line is NLO (550/700),[53] the green dash-dotted line is NNLO (600/700),[53] and the red solid line is N^{3}LO (500),[44] where the numbers in parentheses denote the cutoffs in MeV. Phase parameters with total angular momentum $J \leq 1$ are displayed. Filled circles as in Fig. 8 while open circles represent here the GWU np analysis SM99.[54]

In Table 2, we also show the number of parameters used in the Nijmegen partial wave analysis (PWA93)[50] and in the high-precision CD-Bonn potential.[23] The table reveals that, for S- and P-waves, the number of parameters used in high-precision phenomenology and in EFT at N^{3}LO (Q^4) are about the same. *Thus, the EFT*

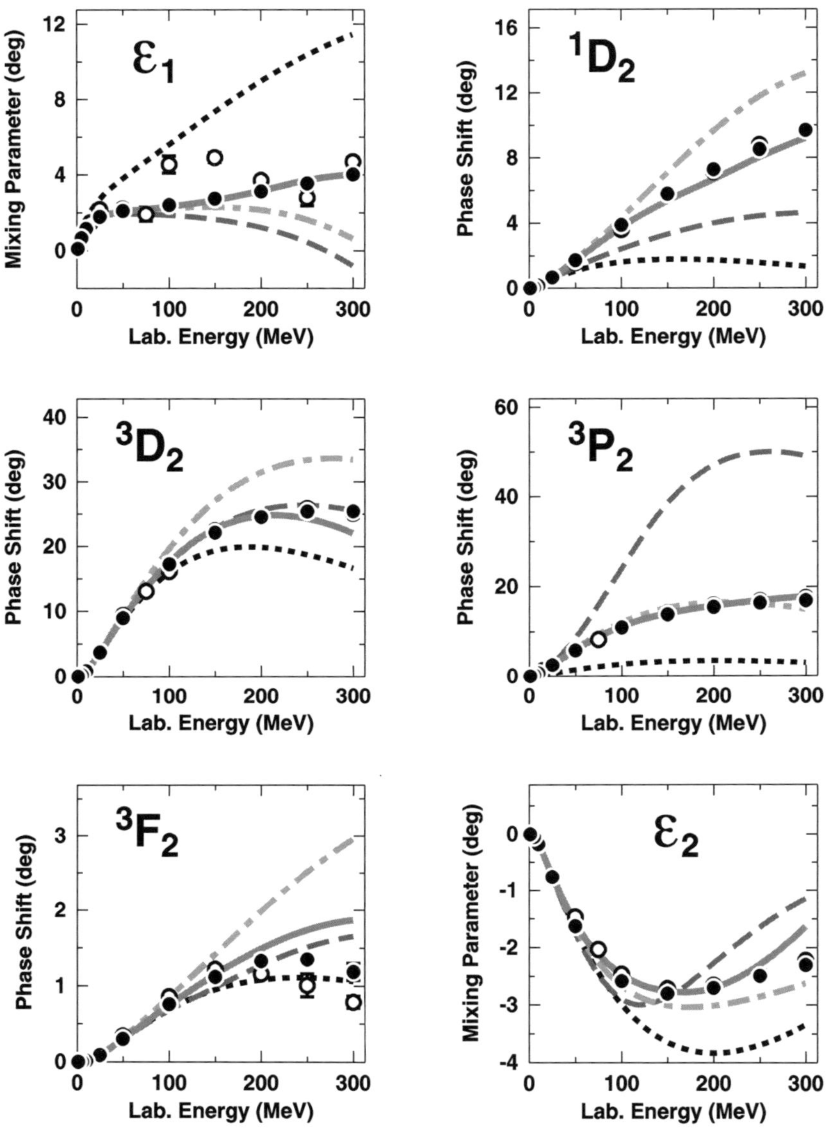

Fig. 10. Same as Fig. 9, but $J = 2$ phase shifts and $J \leq 2$ mixing parameters.

approach provides retroactively a justification for the phenomenology used in the 1990s to obtain high-precision fits.

The 24 parameters of N^3LO are close to the 30+ used in PWA93 and high precision potentials. Consequently, at N^3LO, potentials can be constructed which are of about the same quality as the high-precision *NN* potentials of the 1990s.[23,59,60] This fact is clearly revealed in the χ^2/datum for the fit of the *np* and *pp* data below 290 MeV shown in Tables 3 and 4, respectively. The Idaho N^3LO potential[44] with $\Lambda = 500$ MeV produces a χ^2/datum $= 1.1$ for the world *np* data below 290 MeV

Table 1. Columns three and four show the χ^2/datum for the reproduction of the 1999 *np* database (defined in Ref. 23) by families of *np* potentials at NLO and NNLO constructed by the Bochum group.[53] The χ^2/datum is stated in terms of ranges which result from a variation of the cutoff parameters used in the regulator functions. The values of these cutoff parameters in units of MeV are given in parentheses. $T_{\rm lab}$ denotes the kinetic energy of the incident neutron in the laboratory system.

$T_{\rm lab}$ (MeV)		Bochum *np* potentials	
energy bin	# of *np* data	NLO (550/700–400/500)	NNLO (600/700–450/500)
0–100	1058	4–5	1.4–1.9
100–190	501	77–121	12–32
190–290	843	140–220	25–69
0–290	2402	67–105	12–27

Table 2. Number of parameters needed for fitting the *np* data in the Nijmegen phase-shift analysis and by the high-precision CD-Bonn potential versus the total number of *NN* contact terms of EFT-based potentials to different orders.

			EFT contact potentials[49]			
	Nijmegen PWA93[50]	CD-Bonn pot.[23]	Q^0	Q^2	Q^4	Q^6
$^1S_0$	3	4	1	2	4	6
$^3S_1$	3	4	1	2	4	6
$^3S_1$-$^3D_1$	2	2	0	1	3	6
$^1P_1$	3	3	0	1	2	4
$^3P_0$	3	2	0	1	2	4
$^3P_1$	2	2	0	1	2	4
$^3P_2$	3	3	0	1	2	4
$^3P_2$-$^3F_2$	2	1	0	0	1	3
$^1D_2$	2	3	0	0	1	2
$^3D_1$	2	1	0	0	1	2
$^3D_2$	2	2	0	0	1	2
$^3D_3$	1	2	0	0	1	2
$^3D_3$-$^3G_3$	1	0	0	0	0	1
$^1F_3$	1	1	0	0	0	1
$^3F_2$	1	2	0	0	0	1
$^3F_3$	1	2	0	0	0	1
$^3F_4$	2	1	0	0	0	1
$^3F_4$-$^3H_4$	0	0	0	0	0	0
$^1G_4$	1	0	0	0	0	0
$^3G_3$	0	1	0	0	0	0
$^3G_4$	0	1	0	0	0	0
$^3G_5$	0	1	0	0	0	0
Total	35	38	2	9	24	50

Table 3. Columns three to five display the χ^2/datum for the reproduction of the 1999 *np* database (defined in Ref. 23) by various ***np*** potentials. For the chiral potentials, the χ^2/datum is stated in terms of ranges which result from a variation of the cutoff parameters used in the regulator functions. The values of these cutoff parameters in units of MeV are given in parentheses. $T_{\rm lab}$ denotes the kinetic energy of the incident nucleon in the laboratory system.

$T_{\rm lab}$ (MeV) energy bin	# of ***np*** data	Idaho N^{3}LO (500–600)[44]	Bochum N^{3}LO (600/700–450/500)[61]	Argonne V_{18} (Ref. 59)
0–100	1058	1.0–1.1	1.0–1.1	0.95
100–190	501	1.1–1.2	1.3–1.8	1.10
190–290	843	1.2–1.4	2.8–20.0	1.11
0–290	2402	1.1–1.3	1.7–7.9	1.04

Table 4. Same as Table 3 but for ***pp***.

$T_{\rm lab}$ (MeV) energy bin	# of ***pp*** data	Idaho N^{3}LO (500–600)[44]	Bochum N^{3}LO (600/700–450/500)[61]	Argonne V_{18} (Ref. 59)
0–100	795	1.0–1.7	1.0–3.8	1.0
100–190	411	1.5–1.9	3.5–11.6	1.3
190–290	851	1.9–2.7	4.3–44.4	1.8
0–290	2057	1.5–2.1	2.9–22.3	1.4

which compares well with the χ^2/datum = 1.04 by the Argonne potential. In 2005, also the Bochum group produced several N^{3}LO *NN* potentials,[61] the best of which fits the *np* data with a χ^2/datum = 1.7 and the worse with 7.9 (Table 3).

As we turn now to *pp*, note first that the χ^2 for *pp* data are typically larger than for *np* because of the higher precision of *pp* data (Table 4). Thus, the Argonne V_{18} produces a χ^2/datum = 1.4 for the world *pp* data below 290 MeV and the best Idaho N^{3}LO *pp* potential obtains 1.5. The fit by the best Bochum N^{3}LO *pp* potential results in a χ^2/datum = 2.9 and the worst produces 22.3. In view of these poor χ^2, the Bochum group has recently launched an attempt toward improving their chiral potentials.[62,63] However, as in their previous work,[61] they have fitted their new potentials only to *NN* phase shifts and not the *NN* data. The χ^2 for the reproduction of the *NN* data by the new Bochum potentials is not available and, thus, no reliable statement about the quality of the new potentials can be made. In the 1990s, the Nijmegen group has pointed out repeatedly that for high quality potentials it is insufficient to fit phase shifts only. A seemingly "good" fit of phase shifts can be misleading and can result in a poor χ^2 for the reproduction of the data.

Concerning alternative N^{3}LO potentials, we note that a minimally nonlocal *NN* potential of this kind has been constructed in Ref. 64 which produces a χ^2/datum of about 1.3 for the *pp* plus *np* data.

Turning now to N^{4}LO: Based upon the derivation of the 2PE and 3PE contributions to the *NN* interaction at N^{4}LO by Entem *et al.*,[45] *NN* potential at N^{4}LO has recently been developed.[45,63] Note that the lower partial waves, which are crucial for a quantitative reproduction of the NN data, are ruled by the contact terms. The number of contacts at N^{4}LO (Q^5) is the same as at N^{3}LO (Q^4). Thus, the N^{4}LO potentials are not very different from the N^{3}LO ones. Note also that the high quality of some of the N^{3}LO potentials[30,44] leaves little room for improvements.

A further increase in accuracy (if needed) could be achieved at N^{5}LO (Q^6), where the number of contact terms advances to 50 (Table 2).[49] The dominant 2PE and 3PE contributions at N^{5}LO have been derived.[48] Thus, all the mathematical materials for the construction of N^{5}LO potentials are available. However, it is debatable if there is a need for them.

6. Comparison: Meson Theory Versus Chiral EFT

In the memory of Gerry Brown, we have reviewed the two major theoretical approaches that can describe the NN interaction quantitatively. Both approaches go back to early suggestions and initiatives by Gerry.[5,6] It is now natural to ask, what the main differences are between the two approaches and how to value each of them.

In chiral EFT, the nuclear potential is expanded in terms of increasing powers of small momenta, $(Q/\Lambda_\chi)^\nu$. In meson theory, the expansion is in terms of increasing mesons masses corresponding to decreasing ranges.

Two- and many-body forces emerge from both approaches. Because of lack of space, we did not discuss the issue of many-body forces and, therefore, we like to refer the interested reader to the review paper of Ref. 30.

Since both theories describe the same complicated object quantitatively, they should be equivalent to a large extent. First, there is a 1PE in both cases, which is trivial. The 2PE contributions may look diagrammatically quite different (cf. Fig. 3 and the 2π diagrams contained in Fig. 7). The main disparity is that, in chiral EFT, the 2PE is build up order by order, while in conventional meson theory it comes as one set. Finally, the short-range contributions appear graphically very different with heavy boson exchange (like ω-exchange) in meson theory and contacts in chiral EFT. However, if $Q < m_\omega$, the propagator of a heavy-meson can be expanded into a power series generating contacts of increasing order, as demonstrated in Eq. (28). Moreover, there is repulsive two-loop 2π exchange and 3π exchange in chiral EFT.

In spite of these equivalences, there are arguments why chiral EFT may be perceived as superior, at least, for conventional low-energy nuclear physics. Chiral EFT

- is more closely connected to QCD;
- comes with an organizational scheme (power counting) that allows to estimate the accuracy of the predictions (at a given order);
- generates two- and many-body forces on an equal footing.

However, there is also a caveat in place. Since chiral EFT is a low-momentum expansion, we have to expect limitations concerning its applicability. This is

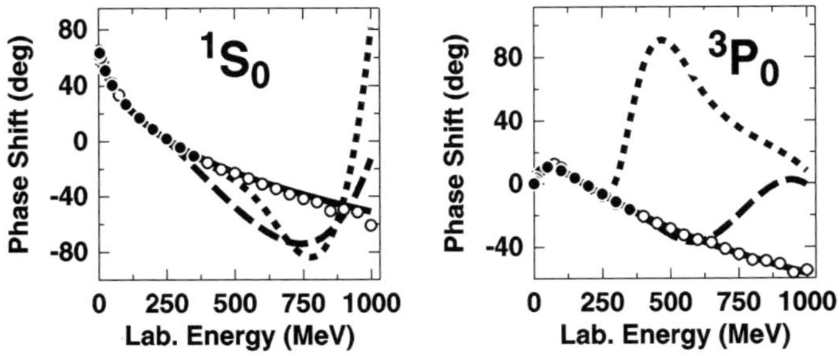

Fig. 11. *np* phase shifts of the $^1S_0$ and $^3P_0$ partial waves for lab. energies up to 1000 MeV. The solid curve shows the phase shifts predicted by the CD-Bonn potential, which is based upon meson theory. Note that this curve is hardly visible because it agrees with the data up to 1000 MeV and, thus, is buried under the symbols representing the data. The dashed and the dotted lines are the predictions by the N^{3}LO chiral potentials constructed by the Idaho[44] and the Bochum[61] groups, respectively. Solid dots represent the Nijmegen multi-energy *np* phase shift analysis[50] and open circles the GWU/VPI single-energy *np* analysis SM99.[54]

demonstrated in Fig. 11, where phase shift predictions by various *NN* potentials are shown up to 1000 MeV lab. energy for the incident nucleon and compared to the results from phase shift analyses. The figure includes one representative of meson theory namely, the CD-Bonn potential[23] (solid line), which obviously predicts the phase shifts correctly up to the highest energies shown, even though it was adjusted only up to 350 MeV. On the other hand, the chiral *NN* potentials at order N^{3}LO (dashed line[44] and dotted line[61]) do not make any reasonable predictions beyond about 300 MeV lab. energy. This is, of course, not unexpected since chiral EFT applies only for momenta $Q < \Lambda_\chi \approx 1\,\text{GeV}$, which is enforced by the regulator Eq. (41) where a typical choice for the cutoff parameter is $\Lambda \approx 500\,\text{MeV}$. Thus, chiral potentials are reliable only for CMS momenta $p, p' \lesssim 2.2\,\text{fm}^{-1}$. A Fermi momentum $k_F \approx 2.2\,\text{fm}^{-1}$ is equivalent to a nuclear matter density $\rho \approx 4\rho_0$ where ρ_0 denotes nuclear matter saturation density. Thus, nuclear matter calculations in which chiral potentials are applied may be trusted only up to densities around $4\rho_0$. In contrast, relativistic meson theory can be trusted to very high momenta (cf. the CD-Bonn curve in Fig. 11) and densities equivalent to those high momenta.

Acknowledgment

This work was supported in part by the U.S. Department of Energy under Contract No. DE-FG02-03ER41270.

References

1. G. E. Brown, A. M. Green and W. J. Gerace, *Nucl. Phys. A* **115** (1968) 435.
2. G. E. Brown, *Comments Nucl. Part. Phys.* **4** (1970) 140.
3. G. E. Brown, *Comments Nucl. Part. Phys.* **5** (1972) 6.
4. S. Barshay and G. E. Brown, *Phys. Rev. Lett.* **34** (1975) 1106.

5. G. E. Brown and A. D. Jackson, *The Nucleon–Nucleon Interaction* (North-Holland, Amsterdam, 1976).
6. G. E. Brown, in *Mesons in Nuclei*, Vol. 1, eds. M. Rho and D. H. Wilkinson (North-Holland, Amsterdam, 1979), pp. 330–356.
7. S. O. Băckman, G. E. Brown and J. A. Niskanen, *Phys. Rep.* **124** (1985) 1.
8. K. Erkelenz, *Phys. Rep. C* **13** (1974) 191.
9. K. Holinde and R. Machleidt, *Nucl. Phys. A* **247** (1975) 495; **256** (1976) 479.
10. K. Holinde and R. Machleidt, *Nucl. Phys. A* **280** (1977) 429.
11. R. Machleidt, K. Holinde and Ch. Elster, *Phys. Rep.* **149** (1987) 1.
12. M. Chemtob, J. W. Durso and D. O. Riska, *Nucl. Phys. B* **38** (1972) 141; A. D. Jackson, D. O. Riska and B. Verwest, *Nucl. Phys. A* **249** (1975) 397.
13. R. Vinh Mau, J. M. Richard, B. Loiseau, M. Lacombe and W. N. Cottingham, *Phys. Lett. B* **44** (1973) 1.
14. R. Vinh Mau, in *Mesons in Nuclei*, Vol. 1, eds. M. Rho and D. H. Wilkinson, The Paris Nucleon-Nucleon Potential (North-Holland, Amsterdam, 1979), pp. 151–196.
15. M. Lacombe, B. Loiseau, J. M. Richard, R. Vinh Mau, J. Côté, P. Pirès and R. de Tourreil, *Phys. Rev. C* **21** (1980) 861.
16. H. Yukawa, *Proc. Phys. Math. Soc. Japan* **17** (1935) 48.
17. Particle Data Group (K. A. Olive *et al.*), *Chin. Phys. C* **38** (2014) 0900001.
18. R. Machleidt, *Adv. Nucl. Phys.* **19** (1989) 189.
19. R. Bryan and B. L. Scott, *Phys. Rev.* **177** (1969) 1435.
20. M. M. Nagels, T. A. Rijken and J. J. de Swart, *Phys. Rev. D* **17** (1978) 768.
21. R. Machleidt and G. Q. Li, *Phys. Rep.* **242** (1994) 5.
22. A. Gersten, R. Thompson and A. E. S. Green, *Phys. Rev. D* **3** (1971) 2076; G. Schierholz, *Nucl. Phys. B* **40** (1972) 335; J. Fleischer and J. A. Tjon, *Nucl. Phys. B* **84** (1975) 375; *Phys. Rev. D* **24** (1980) 87.
23. R. Machleidt, *Phys. Rev. C* **63** (2001) 024001.
24. S. R. Beane, P. F. Bedaque, K. Orginos and M. J. Savage, *Phys. Rev. Lett.* **97** (2006) 012001.
25. N. Ishii, S. Aoki and T. Hatsuda, *Phys. Rev. Lett.* **99** (2007) 022001.
26. S. Weinberg, *Physica A* **96** (1979) 327.
27. J. Gasser and H. Leutwyler, *Ann. Phys.* **158** (1984) 142; *Nucl. Phys. B* **250** (1985) 465.
28. J. Gasser, M. E. Sainio and A. Švarc, *Nucl. Phys. B* **307** (1986) 779.
29. S. Weinberg, *Nucl. Phys. B* **363** (1991) 3.
30. R. Machleidt and D. R. Entem, *Phys Rep.* **503** (2011) 1.
31. S. Scherer, *Adv. Nucl. Phys.* **27** (2003) 277.
32. E. Epelbaum, H.-W. Hammer and U.-G. Meißner, *Rev. Mod. Phys.* **81** (2009) 1773.
33. S. Coleman, J. Wess and B. Zumino, *Phys. Rev.* **177** (1969) 2239; C. G. Callan, S. Coleman, J. Wess and B. Zumino, *Phys. Rev.* **177** (1969) 2247.
34. N. Fettes, U.-G. Meißner, M. Mojžiš and S. Steininger, *Ann. Phys. (N.Y.)* **283** (2000) 273; **288** (2001) 249.
35. H. Krebs, A. Gasparyan and E. Epelbaum, *Phys. Rev. C* **85** (2012) 054006.
36. C. Ordóñez, L. Ray and U. van Kolck, *Phys. Rev. C* **53** (1996) 2086.
37. N. Kaiser, R. Brockmann and W. Weise, *Nucl. Phys. A* **625** (1997) 758.
38. N. Kaiser, S. Gerstendörfer and W. Weise, *Nucl. Phys. A* **637** (1998) 395.
39. U. van Kolck, *Phys. Rev. C* **49** (1994) 2932.
40. E. Epelbaum, A. Nogga, W. Glöckle, H. Kamada, U.-G. Meißner and H. Witala, *Phys. Rev. C* **66** (2002) 064001.
41. S. Weinberg, *Phys. Lett. B* **295** (1992) 114.

42. N. Kaiser, *Phys. Rev. C* **61** (2000) 014003.
43. N. Kaiser, *Phys. Rev. C* **62** (2000) 024001.
44. D. R. Entem and R. Machleidt, *Phys. Rev. C* **68** (2003) 041001.
45. D. R. Entem, N. Kaiser, R. Machleidt and Y. Nosyk, *Phys. Rev. C* **91** (2015) 014002.
46. H. Krebs, A. Gasparyan and E. Epelbaum, *Phys. Rev. C* **87** (2013) 054007.
47. L. Girlanda, A. Kievsky and M. Viviani, *Phys. Rev. C* **84** (2011) 014001.
48. D. R. Entem, N. Kaiser, R. Machleidt and Y. Nosyk, *Phys. Rev. C* **92** (2015) 064001.
49. D. R. Entem and R. Machleidt, unpublished.
50. V. G. J. Stoks, R. A. M. Klomp, M. C. M. Rentmeester and J. J. de Swart, *Phys. Rev. C* **48** (1993) 792.
51. W. J. Briscoe, I. I. Strakovsky and R. L. Workman, SAID Partial-Wave Analysis Facility, Data Analysis Center, The George Washington University, solution SP07 (2007).
52. K. Erkelenz, R. Alzetta and K. Holinde, *Nucl. Phys. A* **176** (1971) 413.
53. E. Epelbaum, W. Glöckle and U.-G. Meißner, *Eur. Phys. J. A* **19** (2004) 401.
54. R. A. Arndt, I. I. Strakovsky and R. L. Workman, SAID, Scattering Analysis Interactive Dial-in computer facility, George Washington University (formerly Virginia Polytechnic Institute), solution SM99 (1999).
55. A. Ekström *et al.*, *Phys. Rev. Lett.* **110** (2013) 192502.
56. A. Gezerlis *et al.*, *Phys. Rev. C* **90** (2014) 054323.
57. D. R. Entem and R. Machleidt, *Phys. Lett. B* **524** (2002) 93.
58. D. R. Entem and R. Machleidt, *Phys. Rev. C* **66** (2002) 014002.
59. R. B. Wiringa, V. G. J. Stoks and R. Schiavilla, *Phys. Rev. C* **51** (1995) 38.
60. V. G. J. Stoks, R. A. M. Klomp, C. P. F. Terheggen and J. J. de Swart, *Phys. Rev. C* **49** (1994) 2950.
61. E. Epelbaum, W. Glöckle and U.-G. Meißner, *Nucl. Phys. A* **747** (2005) 362.
62. E. Epelbaum, H. Krebs and U.-G. Meißner, *Eur. Phys. J. A* **51** (2015) 53.
63. E. Epelbaum, H. Krebs and U.-G. Meißner, *Phys. Rev. Lett.* **115** (2015) 122301.
64. M. Piarulli, L. Girlanda, R. Schiavilla, R. Navarro Perez, J. E. Amaro and E. Ruiz Arriola, *Phys. Rev. C* **91** (2015) 024003.

Foundations of strangeness nuclear physics derived from chiral effective field theory

Ulf-G. Meißner
HISKP and BCTP, Universität Bonn, D-53115 Bonn, Germany
IKP-3, IAS-4, JARA-HPC and JARA-FAME, Forschungszentrum Jülich, D-52425 Jülich, Germany
meissner@hiskp.uni-bonn.de

Johann Haidenbauer
IKP-3 and IAS-4, Forschungszentrum Jülich, D-52425 Jülich, Germany
j.haidenbauer@fz-juelich.de

Dense compact objects like neutron stars or black holes have always been one of Gerry Brown's favorite research topics. This is closely related to the effects of strangeness in nuclear physics. Here, we review the chiral Effective Field Theory approach to interactions involving nucleons and hyperons, the possible existence of strange dibaryons, the fate of hyperons in nuclear matter and the present status of three-body forces involving hyperons and nucleons.

Keywords: Baryon-baryon interactions; strangeness nuclear physics; chiral effective field theory.

1. Introduction

Gerry Brown was a fascinating scholar and human being with a broad range of physics interests. One of his prime foci was the physics of dense and compact objects like neutron stars and black holes, see e.g., Refs. 1– 8. With the observation of two solar mass neutron stars, this physics has taken a new twist, commonly referred to as the "hyperon puzzle". This puzzle relates to the fact that the equation of state in the presence of hyperons is generally too soft to allow for such heavy neutron stars. Some repulsive mechanism, may be related to the hyperon–nucleon or hyperon–hyperon interactions, or even to more exotic three-body forces involving hyperons, is thus required to reconcile the presence of hyperons within such dense and compact objects with their global properties, see e.g., Refs. 9 and 10 (and references therein). This is where some of our recent research comes in, namely the chiral effective field theory (EFT) based description of the hyperon–nucleon (YN) in Sec. 2 and the hyperon–hyperon (YY) interactions in Sec. 3. Another related fascinating topic is the possible existence of exotic states with baryon number two, formed from two hyperons, which we will also discuss in Sec. 4. Next, we discuss the behavior of hyperons in nuclear matter in Sec. 5. Last but not least, we consider the final frontier, namely the three-body forces involving hyperons in Sec. 6. We are quite certain that Gerry would have loved these developments and we therefore dedicate this paper to his memory.

2. Hyperon–Nucleon Interactions

The basic ingredient in hyper-nuclear physics is the hyperon–nucleon interaction. Conventionally, it has been studied based on meson-exchange models, however, with the advent of successful EFT methods for the nucleon–nucleon (NN) interaction (see Refs. 11 and 12 for reviews), one now has a better and more systematic handle on these fundamental interactions, firmly rooted in the symmetries of QCD. In contrast to the NN interaction, the amount of data on YN ($Y = \Lambda, \Sigma$) scattering is scarce, about 35 data points supplemented by the binding energies of a few light hyper-nuclei. Therefore, any EFT description of these data has a more exploratory character than it is the case for the NN problem. In fact, the first work on the YN interaction in EFT was due to Korpa *et al.*,[13] who made use of the so-called KSW power counting.[14] In the Bonn-Jülich group, we already had made good experiences with the so-called Weinberg power counting for the NN interactions. In that scheme, chiral perturbation theory is utilized for calculating the potential between two or more baryons in a systematic manner. This potential is then used within a regularized Lippmann–Schwinger (LS) equation to generate the bound and the scattering states, $T = V + VG_0T$, with T the T-matrix and G_0 the two-baryon propagator. The leading order (LO) $YN \to YN$ calculations were performed by Polinder and the authors[15] and later, the next-to-leading order (NLO) contributions were analyzed in collaboration with the Munich group of Norbert Kaiser, Stephan Petschauer and Wolfram Weise.[16] For a comparison of the EFT and more conventional approaches, see Ref. 17. In what follows, we eschew models.

The EFT is constructed from the asymptotically observed particles, the hadrons, here from the baryon octet (for short, baryons) and the Goldstone-boson octet (for short, mesons). The basic ingredient in the EFT approach to the YN interaction is the effective potential, with its various terms ordered according to the power counting. At LO, $\mathcal{O}(Q^0)$, where Q denotes an external momentum or a Goldstone boson mass, there are two types of contributions. First, one has one-meson-exchange graphs, with all couplings constants expressed in terms of the pion–nucleon coupling $g_{\pi NN}$ and the ratio of the SU(3) axial-couplings $F/(F+D)$. Second, based on group theory arguments, there are six four-baryon contact terms without derivatives. From those, only five contribute to the scattering process $YN \to YN$ while all six occur in the YY sector. Extending to NLO, one has now also contributions from two-meson-exchange diagrams and further contact interactions with two derivatives. The pertinent Feynamn diagrams are depicted in Fig. 1. All these have been calculated and included in the potential presented in Ref. 16. However, it should be stressed that some two-meson-exchanges like $K\bar{K}$ or ηK are so short-ranged

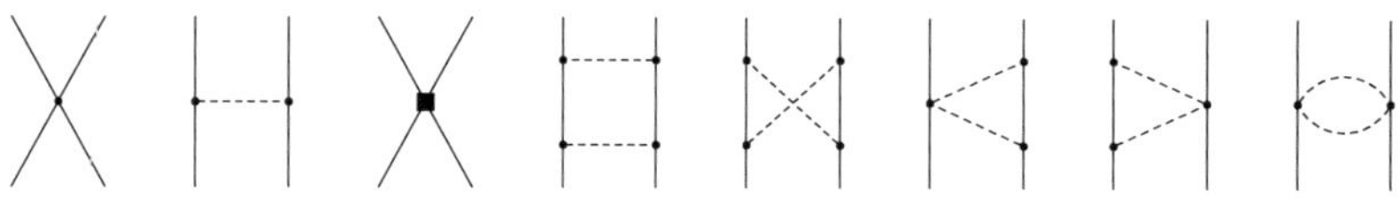

Fig. 1. Relevant diagrams for the effective potential up-to-and-including NLO. Solid and dashed lines denote baryons (N, Λ, Σ, Ξ) and mesons (π, K, η), respectively. The square symbolizes a contact vertex with two derivatives. From left to right: LO contact term, one-meson exchange, NLO contact term, planar box, crossed box, left triangle, right triangle, football diagram. From the planar box graph, only the irreducible part contributes to the potential.

that they could be also effectively absorbed in the contact terms. The number of additional contact interactions is large, so in general we resort to account only for SU(3) symmetric terms and fit only the low-energy constants (LECs) related to the S-wave terms and use P-wave NN phase shifts as constraints. However, physical masses of the mesons and the baryons are used throughout. This is needed to account properly for the fact that the pion mass is much smaller than that of the other Goldstone bosons and also to have the correct threshold energies for the various baryon–baryon channels, and it introduces a certain amount of SU(3) breaking.

In Fig. 2, we show some assorted results for the reactions $\Lambda p \to \Lambda p$ and $\Sigma^- p \to \Lambda n$. The middle panel zooms on the region around the $\Sigma^+ n$ and the $\Sigma^0 p$ thresholds for Λp scattering. In both cases, the bands at the given order are obtained by variations of the cutoff in the LS equation, that is applied in terms of the regulator function $f_R(\Lambda) = \exp[-(p'^4 + p^4)/\Lambda^4]$, with Λ the cutoff. This is only a very rough estimate of the theoretical uncertainty and it should be revisited in the future upon

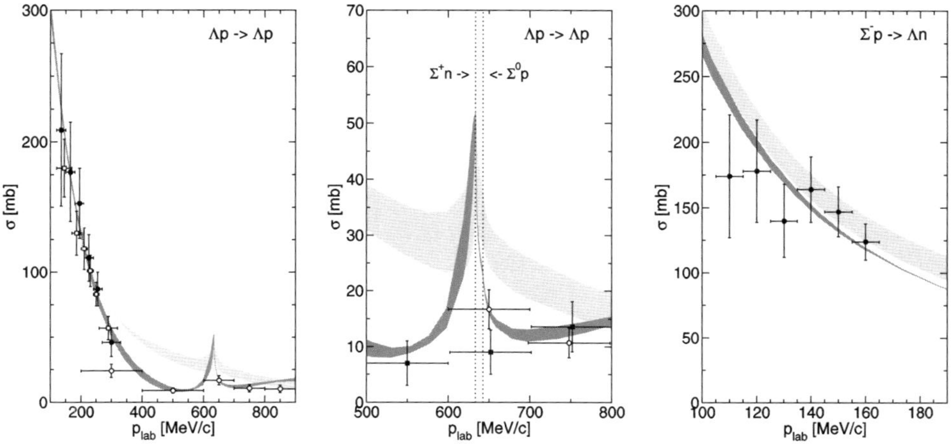

Fig. 2. (Color online) Total cross-section σ for $\Lambda p \to \Lambda p$ (left and middle panel) and for $\Sigma^- p \to \Lambda n$ (right panel) as a function of p_{lab}. The red/dark band shows the chiral EFT results to NLO for variations of the cutoff in the range $\Lambda = 500, \ldots, 650\,\text{MeV}$, while the cyan/light band are results to LO for $\Lambda = 550, \ldots, 700\,\text{MeV}$. The data can be traced back from Ref. 16.

the lines suggested e.g., in Refs. 18 and 19. However, we remark that the NLO contributions improve the description of the data and that the uncertainty bands shrink with increasing order.

3. Strangeness $S = -2, -3, -4$ Baryon–Baryon Interactions

The experimental situation in the sector with strangeness $S = -2$ (YY and ΞN interactions), with $S = -3$ (ΞY interactions) and $S = -4$ ($\Xi\Xi$ interactions) is even poorer than for the hyperon–nucleon interaction. There are some constraints on the $\Lambda\Lambda$ interaction from double-Λ hyper-nuclei and from final-state interactions in production reactions. Similarly, there are bounds on the $\Xi^- p$ elastic cross-section and for $\Xi^- p \to \Lambda\Lambda$ in a broad momentum range. There are some further results on in-medium cross-sections and a few constraints from production experiments. For a detailed discussion, we refer the reader to Ref. 20.

It is therefore interesting to perform exploratory studies of the baryon–baryon interaction with $S = -2$ and with $S = -3, -4$ at LO, see Refs. 21 and 22. As already noted, at LO one has one more contact interaction in the $S = -2$ system compared to $S = -1$. Denoting the corresponding LEC by C_1 and varying its value within the natural range, given by[23] $|C_1| = 4\pi/F_\pi^2$, with F_π the pion decay constant, one observes that the chiral EFT predictions are consistent with the experimental information on the $S = -2$ sector. Using the LECs fixed in the fit to the ΛN and ΣN data, one can make predictions for the $\Xi\Lambda$, $\Xi\Sigma$ and $\Xi\Xi$ interactions. Strong attraction is found in some of the $S = -3$ and $S = -4$ channels, suggesting the possible existence of bound states. However, as we will argue below, these seem to be artifacts of the SU(3) symmetry imposed on the contact interactions.

In the approximation of taking only the SU(3) symmetric contact interactions, a consistent description of the NN and YN data is not possible. However, at NLO SU(3) breaking contact terms arise in the employed Weinberg power counting scheme.[24] These allow one to account for the leading SU(3) breaking terms in the $^1S_0$ wave for the baryon–baryon channels with maximal isospin, that contributes e.g., to nucleon–nucleon scattering via the potential $V_{NN}^{I=1} = C_1^\chi (M_K^2 - M_\pi^2)/2$, see Ref. 25. From a simultaneous description of the pp and the $\Sigma^+ p$ phase shifts one can pin down the LEC C_1^χ and then predict the $\Sigma\Sigma$ interaction, see Fig. 3. In the left panel of Fig. 4 we show the LO and NLO results for $\Xi^- p \to \Lambda\Lambda$ in comparison to the data from Ahn *et al.*[27] and Kim *et al.*.[28] Both are consistent with these data. Note that the LO results are solely determined by the underlying SU(3) symmetry whereas in the NLO case the aforementioned SU(3) symmetry breaking in the $^1S_0$ wave is taken into account. The results for $\Xi^- p \to \Xi^- p$ shown in the right panel of Fig. 4 indicate that some degree of SU(3) symmetry breaking is also required in the $^3S_1$ partial wave. Specifically, we see that the inclusion of the leading SU(3) symmetry breaking term considerably lowers the prediction (red/dark band), giving a better agreement with the upper bound from Ref. 27. Relying strictly on SU(3) symmetry would yield cross-sections that are apparently too large, cf. the hatched

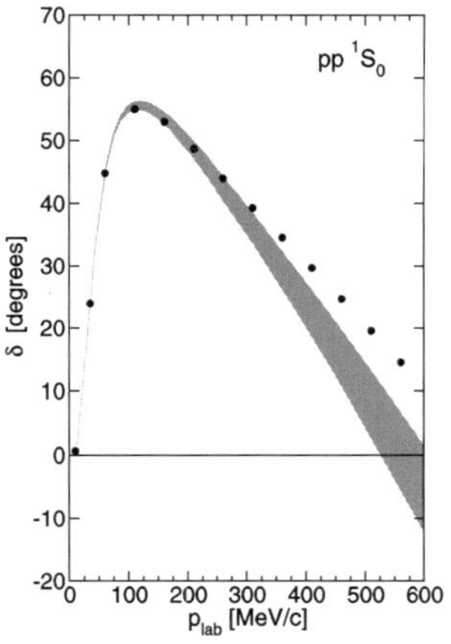

Fig. 3. pp, Σ^+p and $\Sigma^+\Sigma^+$ phase shifts in the $^1S_0$ partial wave. The filled band represents the NLO result. The pp phase shifts of the GWU analysis[26] are shown by circles. In case of Σ^+p the circles indicate upper limits for the phase shifts, deduced from the Σ^+p cross-section.

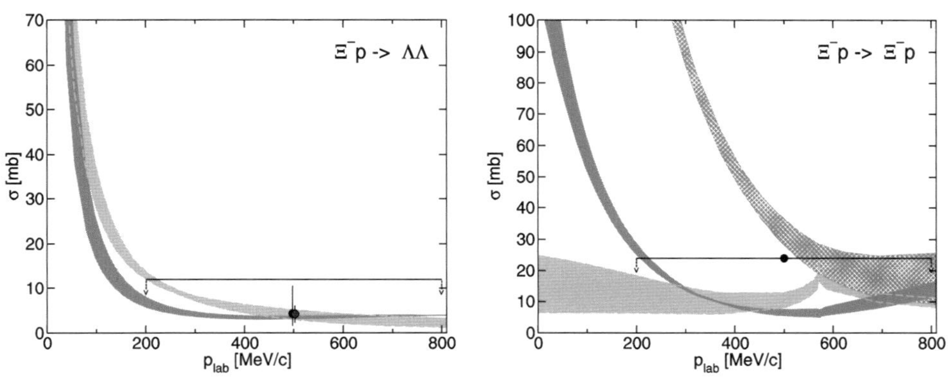

Fig. 4. (Color online) Total cross-section σ for $\Xi^-p \to \Lambda\Lambda$ (left panel) and for $\Xi^-p \to \Xi^-p$ (right panel) as a function of $p_{\rm lab}$. The red/dark band shows the chiral EFT results to NLO for variations of the cutoff in the range $\Lambda = 500, \ldots, 650$ MeV, while the cyan/light band are results to LO for $\Lambda = 550, \ldots, 700$ MeV. The hatched area in the right panel refers to a calculation where all LECs are taken over from the YN fit. The data can be traced back from Ref. 20.

band. A more detailed account of these topics, including also discussions of the pertinent baryon–baryon scattering lengths and effective ranges can be found in Ref. 20. It is fairly obvious that to make further progress in this field, one needs more data, that are expected to come from the J-PARC facility in Japan and from FAIR at Darmstadt (Germany).

4. Exotic Bound States?

Another fascinating topic related to the baryon–baryon interactions is the possible appearance of (exotic) bound states. Historically, the so-called dibaryons have been conceived either as tightly bound six quark objects or as shallow bound states of two baryons. For the first category, the recently observed dibaryon $d^*(2380)$ at COSY (Forschungszentrum Jülich)[29,30] qualifies, whereas an established loosely bound state of a proton and the neutron, the deuteron, is one of the best studied nuclei. The deuteron is a bound state in the $^3S_1$-$^3D_1$ channel, whereas the binding of the proton–neutron system in the $^1S_0$ channel is just not strong enough to produce a bound state and only a virtual state is created. May be the most famous, — or should we say: infamous? — dibaryon is the H-particle predicted by Jaffe within the bag model as a compact $|uuddss\rangle$ state.[31] After decades of failed experiments to establish its presence, it regained popularity a few years ago from lattice sightings reported in Refs. 32 and 33, though at unphysical quark masses. The quantum numbers of the H are these of the $\Lambda\Lambda$ system in an S-wave state, namely strangeness $S = -2$, isospin $I = 0$ and $J^P = 0^+$. Other predicted dibaryons involving strangeness are cousins of the $d^*(2380)$[34] or generated from the attractive interaction between certain baryons, e.g., in the $\Xi\Xi$ but also for the $\Xi\Sigma$ and $\Xi\Lambda$ systems.[22,35] Note that the NPLQCD Collaboration reported evidence for a $\Xi^-\Xi^-$ bound state.[36] This list is by far not exhaustive but should merely serve as an illustration that there might be very rich physics in systems involving strangeness–strange exotics.

The chiral EFT discussed so far is very well suited to shed light on the H-dibaryon, should it indeed exist. In particular, one can study the implications of the imposed (approximate) SU(3) symmetry and further explore the dependence of its properties on the involved meson and baryon masses. The latter aspect is important as the existing lattice QCD calculations were not performed at the physical quark and thus hadron masses. In particular, one can use the flavor singlet LEC C_1 as a dial to generate a bound state with a given binding energy.[37,38]

The first issue to be discussed in this context is the effective range expansion in the $^1S_0$ channel of the $\Lambda\Lambda$ interaction. Let us assume that the H is a loosely bound two-baryon state much like the deuteron. For illustration, let us fix the value of the LEC C_1 such that the binding momentum of the H is the same as for the deuteron, $\gamma = 45.7\,\mathrm{MeV}$, related to the binding energy via $E = -\gamma^2/m_B$, with m_B either m_N or m_Λ. As it is well known, the effective range expansion of Bethe and Schwinger relates the binding momentum to the scattering length a and effective range r, $1/a \simeq \gamma - r\gamma^2/2$. While this is very well fulfilled for the deuteron, the corresponding results for the $\Lambda\Lambda$-system in the $^1S_0$ wave are very different, $a \simeq -0.65\,\mathrm{fm}$ and $r \simeq 6\,\mathrm{fm}$.[20] Thus, the properties of the H are very different from the ones of the deuteron, despite the fact that both are close-to-threshold bound states. This can also be understood from the effective potentials in the $I = 0$ channel, the SU(3) flavor singlet $\sim C_1$ contributes with a much larger strength to ΞN than to $\Lambda\Lambda$. This means that the H should predominantly be a ΞN bound state. This can be sharpened by looking at the corresponding phase shifts, the one in the ΞN channel is rather similar to the NN $^3S_1$ phase shift, see Refs. 37 and 38, see also Fig. 5.

Second, one can use the chiral EFT to vary the quark/pion mass to make contact to the lattice QCD results. If one sticks to the SU(3) symmetric case, one finds that

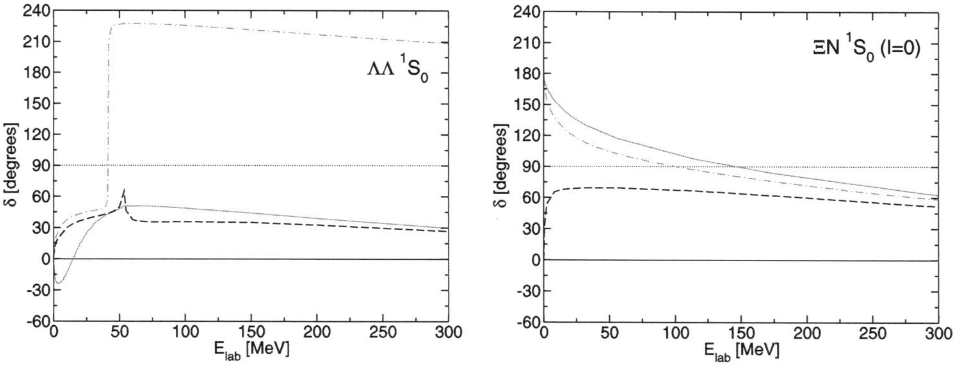

Fig. 5. Phase shifts for $\Lambda\Lambda$ ($^1S_0$) (left) and ΞN ($^1S_0$) (right) as a function of the pertinent laboratory energies. The solid line is the result for our reference BB interaction that produces a bound H at $E_H = -1.87\,\text{MeV}$. The other curves are results for interactions that are fine-tuned to the H binding energies found in the lattice QCD calculations of the HAL QCD (dashed) and NPLQCD (dash-dotted) collaborations, respectively, for the pertinent meson (pion) and baryon masses as given by the lattice collaborations.

for pion masses below 400 MeV, the dependence of the binding energy of the H on M_π is linear, see also Ref. 39. In particular, if one adjusts the H binding energy to the value found by the NPLQCD collaboration at $M_\pi = 389\,\text{MeV}$, it is reduced by 7 MeV when going to the physical point. For larger pion masses, this dependence is weakened but one should be aware that the EFT cannot be trusted anymore at too large pion masses.

A third, and much more drastic effect, is caused by the SU(3) breaking related to the three thresholds for $\Lambda\Lambda$, $\Sigma\Sigma$ and ΞN, which are located at 2231.2, 2257.7 and 2385.0 MeV, in order. For physical values, the binding energy of the H is reduced by as much as 60 MeV as compared to an SU(3) symmetric interaction with degenerate two-baryon thresholds. For the lattice QCD results of the HAL QCD collaboration, this means that the bound state has disappeared at the physical point where as for the NPLQCD case, a resonance in the $\Lambda\Lambda$ system might survive, cf. also Fig. 5. The intricacies of coupled channel systems at unphysical quark masses are further discussed in the appendix.

As noted above, there have been speculations about bound states in two-baryon systems with $S = -3$ and $S = -4$, as e.g., given by the LO and NLO SU(3) symmetric interactions in our EFT. However, we have reexamined the role of SU(3) breaking in Ref. 25. As worked out by Kaiser and Petschauer,[24] there are 12 LO symmetry breaking terms with corresponding LECs, from which 6 appear in the $^1S_0$ and 6 in the $^3S_1$ partial wave. At present, there is simply not enough information to determine these all. However, if we restrict ourselves to the BB systems in the $^1S_0$ wave with maximal isospin, only two such LECs survive, called C_1^χ and C_2^χ. We had already discussed in Sec. 3 how to fix C_1^χ. The other LEC cannot be pinned down reliably at present, but must be varied within a reasonable range. Including the symmetry breaking $\sim C_1^\chi$ in the $\Sigma\Sigma$ channel, the scattering length and the corresponding attraction is much reduced so that practically any bound state in this system can be ruled out. It is reasonable to assume that C_2^χ is of the same size as C_1^χ and that the increase in repulsion when going from NN to ΣN to $\Sigma\Sigma$ is not reversed for the $S = -3$ and $S = -4$ systems. In that case, bound states in these systems are also rather unlikely. For a more detailed discussion, we refer the reader to Ref. 25.

5. Hyperons in Nuclear Matter

Apart from the quest to find repulsion in the YN interaction, there are other interesting aspects of the behavior of hyperons in nuclei or nuclear matter. In particular, the repulsive nature of the Σ-nucleus potential[40] and the weak Λ-nucleus spin–orbit interaction[41–43] are long-standing issues, on which our calculations of hyperons in nuclear matter have shed some light.[44,45]

First, we consider the antisymmetric spin–orbit force. It is generated from the NLO potential $V \sim (\boldsymbol{\sigma}_1 - \boldsymbol{\sigma}_2) \cdot (\mathbf{q} \times \mathbf{k})$, in terms of the Pauli spin matrices, the exchange momentum $\mathbf{q}$ and the total momentum $\mathbf{k}$. This term gives rise to (spin) singlet-triplet ($^1P_1$-$^3P_1$) transitions. In our study of the YN interaction in free space,[16] this antisymmetric spin–orbit force was set to zero because it cannot be determined from YN scattering data. Matters are different in nuclei. Here, the strength of the spin–orbit interaction is frequently parameterized in terms of the so-called Scheerbaum factor S_Y,[46] which is related to the hyper-nuclear spin–orbit potential via $U_Y^{\ell s}(r) = -(\pi/2)S_Y(1/r)(d\rho(r)/dr)\boldsymbol{\ell}\cdot\boldsymbol{\sigma}$, with $\rho(r)$ the nucleon density distribution and $\boldsymbol{\ell}$ the single-particle orbital angular momentum operator. The EFT approach allows us to tune the LEC related to the $^1P_1$-$^3P_1$ transition to achieve a value of[44] $S_\Lambda = -3.7\,\text{MeV fm}^5$, in accordance with phenomenological determinations that give S_Λ in the range -4.6 to $-3.0\,\text{MeV fm}^5$.[47,48]

The standard method to calculate the properties of hyperons in a nuclear medium is Brueckner theory. The Brueckner reaction matrix (G-matrix) is determined from a solution of the Bethe–Goldstone equation, $G(\omega) = V + V[Q/(e(\omega) + i\epsilon)]G(\omega)$, with V the pertinent free-space potential, $e(\omega)$ the energy denominator depending on the starting energy ω, and Q is the (angle-averaged) Pauli operator. Medium effects are thus generated from the Pauli operator as well as the density-dependent single-particle potential $U(\omega)$ in the energy denominator $e(\omega)$. This single-particle potential is obtained self-consistently from the G-matrix. There are two commonly used methods to calculate the G-matrix. In the so-called gap choice, only the free particle energies of the intermediate states appear in the energy denominator of the Bethe–Goldstone equation. This method was e.g., used in Ref. 44. In the so-called continuous choice, the dependence of the energy denominator on the full single-particle energies is retained. This method is computationally more difficult but allows to reliably access the imaginary parts of the single-particle potentials, which is not possible when the gap

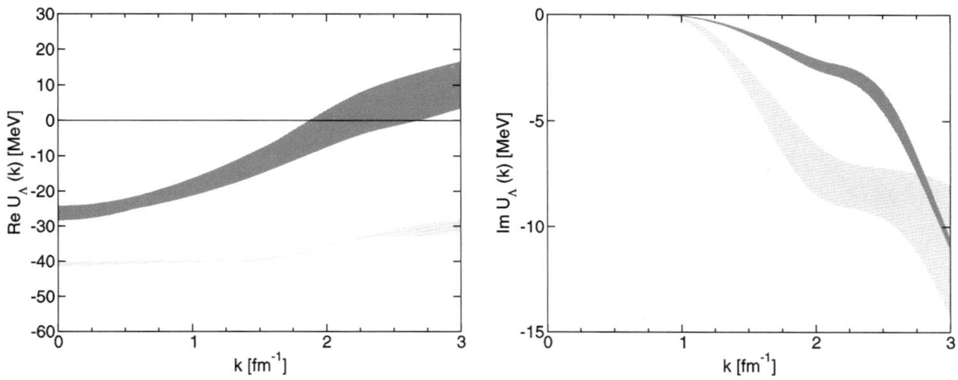

Fig. 6. Momentum dependence of the real and imaginary parts of the single-particle potential of a Λ hyperon in isospin symmetric nuclear matter at saturation density. The bands represent the variation of our results with the cutoff, see text.

choice is employed. The results presented in Ref. 45 are based on the continuous choice.

In Fig. 6, we show the momentum dependence of the real and the imaginary part of the Λ single-particle potential in isospin-symmetric nuclear matter. The bands refer to the usual cutoff variations, using the continuous choice. We find $U_\Lambda(k = 0) = -28, \ldots, -24\,\text{MeV}$, consistent with the empirical value of about $-28\,\text{MeV}$ as deduced from binding energies of Λ hyper-nuclei.[49,50] The corresponding results for the neutral Σ in isospin-symmetric matter at saturation density are shown in Fig. 7, with small differences for the charged Σ hyperons given by the small inter-multiplet mass splittings. In pure neutron matter, matters are very different. Due to the maximal asymmetry between protons and neutrons, the single-particle potentials for the $\Sigma^+, \Sigma^0, \Sigma^-$ hyperons are rather different. These are discussed in

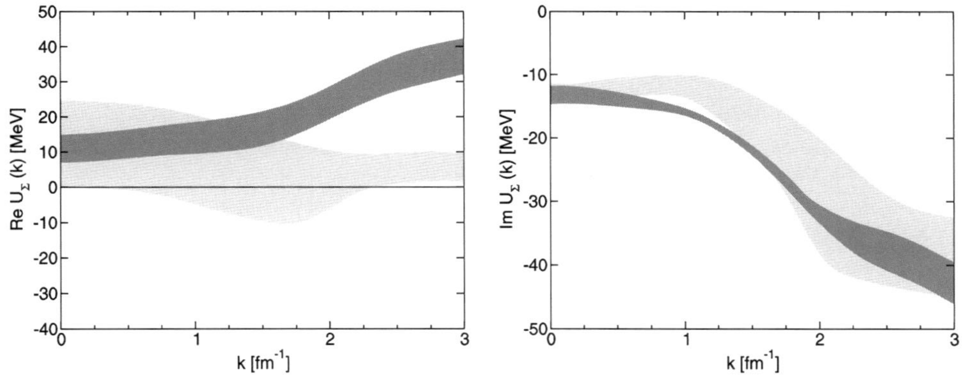

Fig. 7. Momentum dependence of the real and imaginary parts of the single-particle potential of a Σ^0 hyperon in isospin symmetric nuclear matter at saturation density. The bands represent the variation of our results with the cutoff, see text.

detail in Ref. 45. As already said above, the resulting Λ-nuclear spin–orbit potential is small, which is partly due to cancellations between contributions from the symmetric and the anti-symmetric spin–orbit forces, and partly due to the repulsive interactions in some of the *P*-waves. The corresponding Scheerbaum factors for the Σ hyperons are much bigger than for the Λ. Specifically, for the neutral Σ we find $S_\Sigma = -15, \ldots, -18\,\text{MeV fm}^5$ at NLO. For more details, especially also a comparison between the results obtained in the gap and the continuous choice, see Refs. 44 and 45. Note that the in-medium hyperon spin–orbit interaction was also investigated in Refs. 47, 51 and 52.

6. Three-Baryon Interactions

The need for (repulsive) three-baryon forces was already alluded to in the introduction. Another indication that points toward the necessity of such forces are the calculations of the binding energies of light hyper-nuclei at NLO by Nogga.[53] Here, the remaining cutoff dependence hints at missing three-body forces. More generally, it is well established that three-body forces are required in nuclei and nuclear matter, see e.g., Refs. 54 and 55.

In the Weinberg power counting, three-baryon forces (3BFs) appear at next-to-next-to-leading order and are given in terms of the three topologies shown in Fig. 8. The first and second topology require the meson-baryon Lagrangian at NLO. Its minimal form is given in Ref. 56 (see also Refs. 57 and 58). The six-baryon contact terms given by the topology (3) require the general SU(3) Lagrangian that was constructed by Petschauer and Kaiser.[24] Armed with these ingredients, the leading three-baryon forces were derived in Ref. 59 within SU(3) chiral EFT. For that, the chiral Lagrangian in the nonrelativistic limit with the minimal number of terms in the full SU(3) sector was derived. One finds that there are in total 18 different structures corresponding to topology (3) accompanied by 18 LECs. However, only a limited number of combinations of these LECs are involved in a given process. In general, one can write this local contribution to the 3BFs as

$$V^{\text{ct}} = N_1 \mathbb{1} + N_2 \boldsymbol{\sigma}_1 \cdot \boldsymbol{\sigma}_2 + N_3 \boldsymbol{\sigma}_1 \cdot \boldsymbol{\sigma}_3 + N_4 \boldsymbol{\sigma}_2 \cdot \boldsymbol{\sigma}_3 + N_5 i \boldsymbol{\sigma}_1 \times \boldsymbol{\sigma}_2 \cdot \boldsymbol{\sigma}_3, \tag{1}$$

in terms of the spin matrices of the three baryons and the N_i are prefactors containing LECs and kinematical factors, and we have suppressed the corresponding

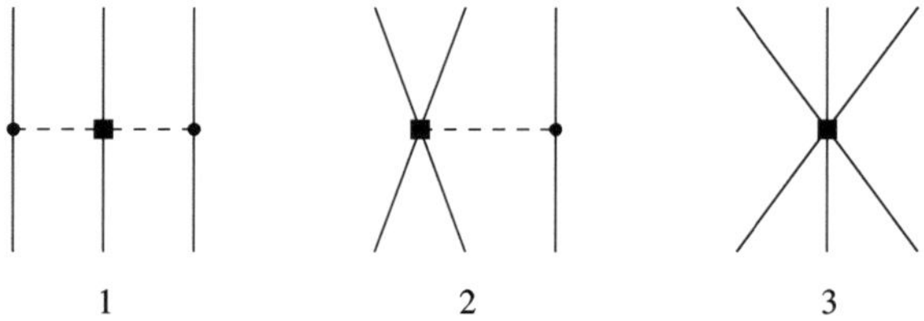

Fig. 8. Topologies of the leading three-baryon forces. (1) Two-pion exchange terms, (2) one-pion exchange terms and (3) six-baryon contact terms.

isospin factors, as detailed in Ref. 59. It can be shown that in the SU(2) limit one recovers the purely nucleonic contact term that is accompanied by the LEC E.[60] The graphs corresponding to the one-meson exchange topology (2) require the knowledge of combinations of the 14 four-baryon-one-meson vertices listed in Ref. 24. In spin-momentum space, this translates into the following structures:

$$V^{1\phi} = -\frac{1}{2F_\pi^2}\frac{\boldsymbol{\sigma}_1\cdot\mathbf{q}_1}{\mathbf{q}_1^2+M_\phi^2}\{N_6\boldsymbol{\sigma}_2\cdot\mathbf{q}_1 + N_7\boldsymbol{\sigma}_3\cdot\mathbf{q}_1 + N_8(\boldsymbol{\sigma}_2\times\boldsymbol{\sigma}_3)\cdot\mathbf{q}_1\}, \tag{2}$$

where the N_i are again composed of LECs and kinematical factors. Further, $\mathbf{q}_i = \mathbf{p}_i' - \mathbf{p}_i$, with p_i and p_i' the initial and final momenta of baryon i, respectively. Again, in the SU(2) limit, one recovers the well-known term that is parameterized in terms of the LEC D, that can e.g., be determined from single pion production.[61] Finally, the diagrams corresponding to topology (1) are the generalizations of the well-known Fujita–Miyazawa three-nucleon force, when the decuplet is considered as very heavy. The spin-momentum structure of these terms takes the form

$$\begin{aligned} V^{2\phi} &= -\frac{1}{4F_\pi^2}\frac{\boldsymbol{\sigma}_1\cdot\mathbf{q}_1\boldsymbol{\sigma}_3\cdot\mathbf{q}_3}{(\mathbf{q}_1^2+M_{\phi_1}^2)(\mathbf{q}_3^2+M_{\phi_3}^2)} \\ &\quad\times\{N_9M_\pi^2 + N_{10}M_K^2 + N_{11}\mathbf{q}_1\cdot\mathbf{q}_3 + N_{12}\boldsymbol{\sigma}_2\cdot(\mathbf{q}_1\times\mathbf{q}_3)\}, \end{aligned} \tag{3}$$

in the same notations as before. Again, in the SU(2) limit this expression recovers the well-known result in terms of the pion-nucleon LECs $c_{1,3,4}$, cf. e.g. Refs. 62 and 60. The complete expressions for the much thought after ΛNN three-body force can be found in Ref. 59.

At first sight, the large number of LECs (or combinations thereof) seems to be discomforting. It certainly is not possible to fix all these from data at present. However, one can use decuplet saturation for estimating these LECs. This is based on the successful resonance saturation hypothesis in the pion–nucleon sector[63] that has already been utilized to estimate nucleonic three-body forces in Refs. 64, 62 and 65. For that, consider the theory with explicit decuplet degrees of freedom. Here, the three-body forces appear already at NLO, and the pertinent diagrams are shown in Fig. 9. These diagrams involve two kinds of vertices. First, there is the

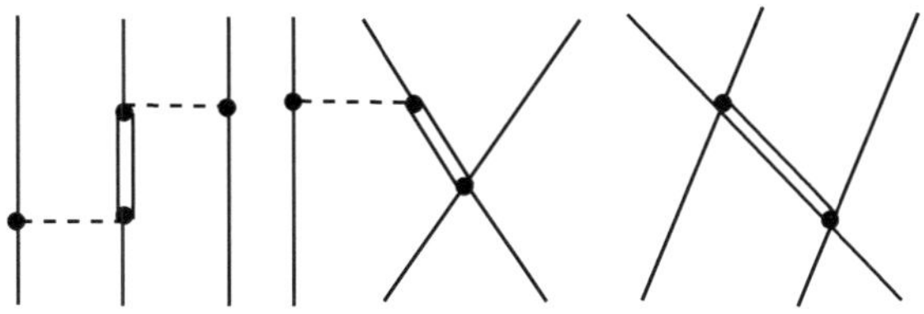

(a) (b) (c)

Fig. 9. Leading contributions to the three-baryon force at NLO in the theory with decuplet baryons. Solid, double and dashed lines denote octet baryons, decuplet baryons and Goldstone bosons, in order. The circles represent leading vertices. The diagram of type b) with a decuplet baryon in the initial state is not shown.

leading meson–baryon–baryon vertex with an octet baryon and a decuplet baryon (graphs (a) and (b)). The corresponding LEC $\mathcal{C}$ (sometimes also called h_A) can be estimated from the decay width $\Delta \to N\pi$, and one finds $\mathcal{C} = 3g_A/4 \simeq 1$, with g_A the nucleon axial-vector coupling constant. SU(3) symmetry can then be used to fix the values for all other possible combinations of octet and decuplet baryons. The other kind of vertices are the ones with three octet baryons and one decuplet baryon (appearing in graphs (b) and (c)). Those vertices involve two new constants. In this case the pertinent constants cannot be deduced from considering the $3N$-Δ vertex simply because in LO the latter is Pauli forbidden. That there are exactly two such terms can be easily understood from group theoretical considerations, for details see Ref. 66. Let us call the corresponding LECs G_1 and G_2 and consider the various topologies in detail and their behavior in the limit of infinite decuplet masses, keeping the ratio of coupling constants over decuplet mass fixed. The graph (a) is nothing but the dominant contribution to the famous Fujita–Miyazawa force. Its emergence in chiral EFT is discussed in detail in Ref. 67. In the heavy decuplet limit, this type of graphs generates the topology (1) of Fig. 8, with the LECs in the theory without decuplet given in very symbolic notation by $C^{2\phi} \sim \alpha_{2\phi}\mathcal{C}^2/\Delta$, where $\alpha_{2\phi}$ is some numerical factor and Δ denotes the decuplet-octet mass splitting. Similarly, the one-pion exchange topology (2) in the theory without the decuplet is generated from the terms of the type (b) in Fig. 9, leading to LECs with their strengths given by $C^{1\phi} \sim \alpha_{1\phi}\mathcal{C}(G_1 + \beta_{1\phi}G_2)/\Delta$, again with $\alpha_{1\phi}$ and $\beta_{1\phi}$ numerical factors. Finally, the graphs of the type (c) lead to local six-baryon contact interactions with their LECs given by $C^{\rm ct} \sim \alpha_{\rm ct}(G_1 + \beta_{\rm ct}G_2)^2/\Delta$. Explicit expressions for the ΛNN force can be found in Ref. 66. Eventually, these LECs might also be computed directly from lattice QCD, say along the lines of Ref. 68. We are looking forward to such calculations. To end this section, we stress that it will be of utmost interest to work out the consequences of these three-baryon forces within hyper-nuclei and compact dense objects like neutron stars.

Acknowledgments

We thank Tom Kuo and Ismail Zahed for inviting us to write this paper. We acknowledge fruitful collaborations with Norbert Kaiser, Andreas Nogga, Stefan Petschauer and Wolfram Weise. This work was supported in part by DFG and NSFC (CRC 110), by the HGF Virtual Institute NAVI (Grant No. VH-VI-417) and by the Chinese Academy of Sciences (CAS) President's International Fellowship Initiative (PIFI) (Grant No. 2015VMA076).

Appendix A.1. Coupled Channel Dynamics on the Lattice: Intricacies

Here, we briefly summarize the work of Ref. 69 that nicely exhibits the intricacies one faces when one is dealing with a coupled channel system at unphysical quark masses in a finite volume. The starting point is the observation, first made by the

Munich group,[70] that the baryon resonance $S_{11}(1535)$ can be generated through coupled channel dynamics in the πN, ηN, $K\Lambda$ and $K\Sigma$ systems with total isospin $I = 1/2$ in the odd-parity S_{11} partial wave of pion–nucleon scattering. Including also NLO terms in the interaction kernel and using a field theoretical regularization method, this calculation was sharpened in Ref. 71, where it was shown that also the next resonance, the $S_{11}(1650)$, is generated dynamically, see also Ref. 72 for a similar conclusion.

In Ref. 69, we have considered two lattice set-ups that allow to investigate the rich phenomenology in the odd-parity S_{11} partial wave for varying quark masses. Set A is related to the work of the European Twisted Mass collaboration (ETMC). In this set-up the meson masses and pion decay constant are taken from the recent calculation in $N_f = 2 + 1 + 1$ twisted mass lattice QCD, i.e., ensemble $B25.32$ of Ref. 73. For the lattice size of $L/a = 32$ and spacing $a = 0.078\,\text{fm}$, the pion mass is fixed there to $M_\pi = 269\,\text{MeV}$, whereas the strange quark mass is held approximately at the physical value. As the kaon and eta decay constants are not available in this calculation at the moment, we decided to relate them to F_π with typical ratios of 1.15 and 1.3, respectively. The baryon masses are also taken from a calculation by the ETMC, however, with only two dynamical quarks and an older lattice action, see Ref. 74. Nevertheless, the strange quark mass is held again approximately at the physical value and $M_\pi = 269\,\text{MeV}$ for the identical lattice size and comparable lattice spacing, i.e., $a = 0.0855\,\text{fm}$. The S_{11} amplitude, with the masses and decay constants of the ETMC, is shown in the left upper panel of Fig. A.1. Comparing to the physical situation, all thresholds have moved to higher energies. The cusp at the ηN threshold has become more pronounced, but no clear resonance shapes are visible. The structure of the amplitude becomes clearer by inspecting the complex

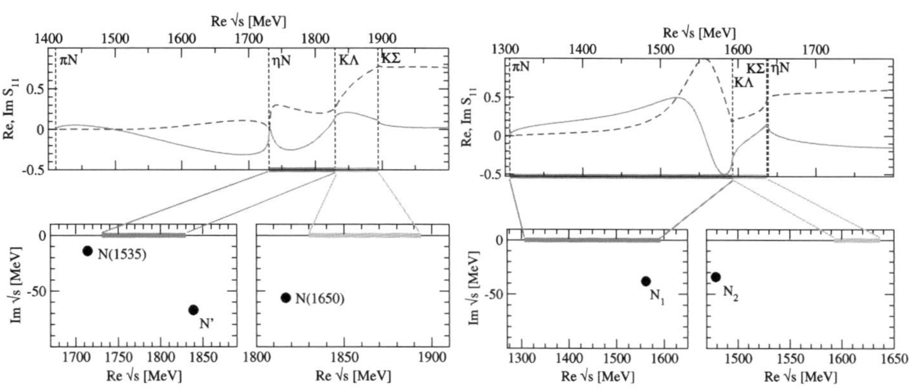

Fig. A.1. Left upper panel: Real (solid line) and imaginary part (dashed line) of the S_{11} amplitude, chirally extrapolated using masses and decay constants of the ETM collaboration. Left lower panels: two of the Riemann sheets with poles (left: Riemann sheet connected to the physical axis between the ηN and the $K\Lambda$ threshold. right: sheet connected to the physical axis between $K\Lambda$ and $K\Sigma$ threshold.). Right upper panel: Real (solid line) and imaginary part (dashed line) of the S_{11} amplitude, chirally extrapolated using masses and decay constants of the QCDSF collaboration.[75] Left lower panels: two of the Riemann sheets with poles (same as for ETMC).

energy plane on different Riemann sheets. This is visualized in the lower left panels of Fig. A.1. Compared to the physical point, the imaginary parts of the pole positions became much smaller due to the reduced phase space. Both the thresholds and the real parts of the pole positions have moved to higher energies. However, the thresholds have moved farther than the pole positions, such that the $N(1535)$ and $N(1650)$ poles are no longer situated below the part of the respective sheet, that is connected to the physical axis (thick horizontal lines). The poles are thus hidden and no clear resonance signals are visible in the physical amplitude. Instead, the amplitude is dominated by cusp effects.

Matters are very different for set B, that refers to calculation from QCDSF.[75] Here, baryon and meson masses are determined from an alternative approach to tune the quark masses, namely to start with the SU(3) symmetry limit and work at a fixed sum of the quark masses. Most importantly, while the lattice size and spacing are comparable to those of the ETMC, i.e., $L/a = 32$ and $a = 0.075\,\text{fm}$, the strange quark mass differs significantly from the physical value. The latter results in a different ordering of the masses of the ground-state octet mesons and, consequently, in a different ordering of meson–baryon thresholds. For further details we refer the interested reader to Ref. 75. The amplitude using the QCDSF parameter set is shown in the left upper panel of Fig. A.1. In contrast to the ETMC case, a clear resonance signal is visible below the $K\Lambda$ threshold, that is the first inelastic channel in this parameter setup. Indeed, we find a pole N_1 on the corresponding Riemann sheet, as indicated in the right lower first panel. Unlike in the ETMC case, it is not hidden behind a threshold. Between the $K\Lambda$ and the $K\Sigma$ threshold, there is only the hidden pole N_2 (right lower second panel). The $K\Sigma$ and ηN thresholds are almost degenerate and on sheets corresponding to these higher-lying thresholds we only find hidden poles.

This shows that the extraction of the scattering amplitude from lattice QCD data is a major challenge as we demonstrate by extrapolating the physical S_{11} amplitude of pion–nucleon scattering to the finite volume and unphysical quark masses, using a unitarized chiral framework including all NLO contact terms. As shown, the pole movement of the resonances $N(1535)1/2^-$ and $N(1650)1/2^-$ with varying quark masses is nontrivial. In addition, one can also calculate the finite volume energy levels. One finds that there are several strongly coupled S-wave thresholds that induce a similar avoided level crossing as narrow resonances. Consequently, one has to be extremely careful in comparing lattice results at unphysical quark masses when a strong coupled channel dynamics is present. For more details, we refer the reader to Ref. 69. Finally, we note that pion–nucleon scattering in the negative parity channel in lattice QCD was considered by Lang and Verduci.[76]

References

1. O. Maxwell, G. E. Brown, D. K. Campbell, R. F. Dashen and J. T. Manassah, *Astrophys. J.* **216** (1977) 77.

2. G. E. Brown, K. Kubodera, D. Page and P. Pizzochero, *Phys. Rev. D* **37** (1988) 2042.
3. G. E. Brown and H. Bethe, *Astrophys. J.* **423** (1994) 659.
4. G. E. Brown, *Nucl. Phys. A* **574** (1994) 217C.
5. H. A. Bethe and G. E. Brown, *Astrophys. J.* **445** (1995) L129.
6. C. H. Lee, G. E. Brown and R. A. M. J. Wijers, *Astrophys. J.* **575** (2002) 996, arXiv:astro-ph/0109538.
7. A. Schwenk, B. Friman and G. E. Brown, *Nucl. Phys. A* **713** (2003) 191, arXiv:nucl-th/0207004.
8. C. H. Lee and G. E. Brown, *J. Korean Phys. Soc.* **49** (2006) S803, arXiv:astro-ph/0510380.
9. D. Chatterjee and I. Vidana, *Eur. Phys. J. A* **52** (2016) 29, arXiv:1510.06306 [nucl-th].
10. I. Bombaci, arXiv:1601.05339 [nucl-th].
11. E. Epelbaum, H. W. Hammer and U.-G. Meißner, *Rev. Mod. Phys.* **81** (2009) 1773, arXiv:0811.1338 [nucl-th].
12. R. Machleidt and D. R. Entem, *Phys. Rep.* **503** (2011) 1, arXiv:1105.2919 [nucl-th].
13. C. L. Korpa, A. E. L. Dieperink and R. G. E. Timmermans, *Phys. Rev. C* **65** (2002) 015208, arXiv:nucl-th/0109072.
14. D. B. Kaplan, M. J. Savage and M. B. Wise, *Phys. Lett. B* **424** (1998) 390, arXiv:nucl-th/9801034.
15. H. Polinder, J. Haidenbauer and U.-G. Meißner, *Nucl. Phys. A* **779** (2006) 244, arXiv:nucl-th/0605050.
16. J. Haidenbauer, S. Petschauer, N. Kaiser, U.-G. Meißner, A. Nogga and W. Weise, *Nucl. Phys. A* **915** (2013) 24, arXiv:1304.5339 [nucl-th].
17. J. Haidenbauer, U.-G. Meißner, A. Nogga and H. Polinder, *Lect. Notes Phys.* **724** (2007) 113, arXiv:nucl-th/0702015.
18. E. Epelbaum, H. Krebs and U.-G. Meißner, *Eur. Phys. J. A* **51** (2015) 53, arXiv:1412.0142 [nucl-th].
19. R. J. Furnstahl, N. Klco, D. R. Phillips and S. Wesolowski, *Phys. Rev. C* **92** (2015) 024005, arXiv:1506.01343 [nucl-th].
20. J. Haidenbauer, U.-G. Meißner and S. Petschauer, *Nucl. Phys. A* **954** (2016) 273, arXiv:1511.05859 [nucl-th].
21. H. Polinder, J. Haidenbauer and U.-G. Meißner, *Phys. Lett. B* **653** (2007) 29, arXiv:0705.3753 [nucl-th].
22. J. Haidenbauer and U.-G. Meißner, *Phys. Lett. B* **684** (2010) 275, arXiv:0907.1395 [nucl-th].
23. E. Epelbaum, *Prog. Part. Nucl. Phys.* **57** (2006) 654, arXiv:nucl-th/0509032.
24. S. Petschauer and N. Kaiser, *Nucl. Phys. A* **916** (2013) 1, arXiv:1305.3427 [nucl-th].
25. J. Haidenbauer, U.-G. Meißner and S. Petschauer, *Eur. Phys. J. A* **51** (2015) 17, arXiv:1412.2991 [nucl-th].
26. R. A. Arndt, W. J. Briscoe, I. I. Strakovsky and R. L. Workman, *Phys. Rev. C* **76** (2007) 025209; http://gwdac.phys.gwu.edu/.
27. J. K. Ahn *et al.*, *Phys. Lett. B* **633** (2006) 214.
28. S. J. Kim, *12th Int. Conf. Measurement of the* $\Xi^- p \to \Lambda\Lambda$ *Cross Sections at* $p_\Xi = 0.2 - 0.8$ *GeV/c with a scintillating fiber target, Hypernuclear and Strange Particle Physics*, Sendai, Japan, 2015, http://lambda.phys.tohoku.ac.jp/hyp2015/.
29. WASA-at-COSY Collab. (P. Adlarson *et al.*), *Phys. Rev. Lett.* **106** (2011) 242302, arXiv:1104.0123 [nucl-ex].
30. WASA-at-COSY Collab. (P. Adlarson *et al.*), *Phys. Rev. Lett.* **112** (2014) 202301, arXiv:1402.6844 [nucl-ex].

31. R. L. Jaffe, *Phys. Rev. Lett.* **38** (1977) 195 [*Phys. Rev. Lett.* **38** (1977) 617].
32. NPLQCD Collab. (S. R. Beane *et al.*), *Phys. Rev. Lett.* **106** (2011) 162001, arXiv:1012.3812 [hep-lat].
33. HAL QCD Collab. (T. Inoue *et al.*), *Phys. Rev. Lett.* **106** (2011) 162002, arXiv:1012.5928 [hep-lat].
34. F. Dyson and N. H. Xuong, *Phys. Rev. Lett.* **13** (1964) 815.
35. V. G. J. Stoks and T. A. Rijken, *Phys. Rev. C* **59** (1999) 3009, arXiv:nucl-th/9901028.
36. NPLQCD Collab. (S. R. Beane *et al.*), *Phys. Rev. D* **85** (2012) 054511, arXiv:1109.2889 [hep-lat].
37. J. Haidenbauer and U.-G. Meißner, *Phys. Lett. B* **706** (2011) 100, arXiv:1109.3590 [hep-ph].
38. J. Haidenbauer and U.-G. Meißner, *Nucl. Phys. A* **881** (2012) 44, arXiv:1111.4069 [nucl-th].
39. S. R. Beane *et al.*, *Mod. Phys. Lett. A* **26** (2011) 2587, arXiv:1103.2821 [hep-lat].
40. E. Friedman and A. Gal, *Phys. Rep.* **452** (2007) 89.
41. O. Hashimoto and H. Tamura, *Prog. Part. Nucl. Phys.* **57** (2006) 564.
42. A. Gal, *Prog. Theor. Phys. Suppl.* **186** (2010) 270.
43. E. Botta, T. Bressani and G. Garbarino, *Eur. Phys. J. A* **48** (2012) 41.
44. J. Haidenbauer and U.-G. Meißner, *Nucl. Phys. A* **936** (2015) 29, arXiv:1411.3114 [nucl-th].
45. S. Petschauer, J. Haidenbauer, N. Kaiser, U.-G. Meißner and W. Weise, *Eur. Phys. J. A* **52** (2016) 15, arXiv:1507.08808 [nucl-th].
46. R. R. Scheerbaum, *Nucl. Phys. A* **257** (1976) 77.
47. M. Kohno, *Phys. Rev. C* **81** (2010) 014003, arXiv:0912.4330 [nucl-th].
48. M. Kohno, private communication.
49. D. J. Millener, C. B. Dover and A. Gal, *Phys. Rev. C* **38** (1988) 2700.
50. Y. Yamamoto, H. Bandō and J. Zofka, *Prog. Theor. Phys.* **80** (1988) 757.
51. N. Kaiser and W. Weise, *Nucl. Phys. A* **804** (2008) 60, arXiv:0802.1190 [nucl-th].
52. HAL QCD Collab. (N. Ishii *et al.*), *PoS Lattice* **2013** (2014) 234; *Eur. Phys. J. A* **48** (2012) 41.
53. A. Nogga, *Few Body Syst.* **55** (2014) 757.
54. N. Kalantar-Nayestanaki, E. Epelbaum, J. G. Messchendorp and A. Nogga, *Rep. Prog. Phys.* **75** (2012) 016301, arXiv:1108.1227 [nucl-th].
55. H. W. Hammer, A. Nogga and A. Schwenk, *Rev. Mod. Phys.* **85** (2013) 197, arXiv:1210.4273 [nucl-th].
56. M. Frink and U.-G. Meißner, *Eur. Phys. J. A* **29** (2006) 255, arXiv:hep-ph/0609256.
57. A. Krause, *Helv. Phys. Acta* **63** (1990) 3.
58. J. A. Oller, M. Verbeni and J. Prades, arXiv:hep-ph/0701096.
59. S. Petschauer, N. Kaiser, J. Haidenbauer, U.-G. Meißner and W. Weise, *Phys. Rev. C* **93** (2016) 014001, arXiv:1511.02095 [nucl-th].
60. E. Epelbaum, A. Nogga, W. Gloeckle, H. Kamada, U.-G. Meißner and H. Witala, *Phys. Rev. C* **66** (2002) 064001, arXiv:nucl-th/0208023.
61. C. Hanhart, U. van Kolck and G. A. Miller, *Phys. Rev. Lett.* **85** (2000) 2905, arXiv:nucl-th/0004033.
62. J. L. Friar, D. Huber and U. van Kolck, *Phys. Rev. C* **59** (1999) 53, arXiv:nucl-th/9809065.
63. V. Bernard, N. Kaiser and U.-G. Meißner, *Nucl. Phys. A* **615** (1997) 483, arXiv:hep-ph/9611253.
64. U. van Kolck, *Phys. Rev. C* **49** (1994) 2932.

65. E. Epelbaum, H. Krebs and U.-G. Meißner, *Nucl. Phys. A* **806** (2008) 65, arXiv:0712.1969 [nucl-th].
66. S. Petschauer, Thesis, Technische Universität München (2015).
67. U.-G. Meißner, *AIP Conf. Proc.* **1011** (2008) 49.
68. S. R. Beane, W. Detmold, T. C. Luu, K. Orginos, M. J. Savage and A. Torok, *Phys. Rev. Lett.* **100** (2008) 082004, arXiv:0710.1827 [hep-lat].
69. M. Döring, M. Mai and U.-G. Meißner, *Phys. Lett. B* **722** (2013) 185, arXiv:1302.4065 [hep-lat].
70. N. Kaiser, P. B. Siegel and W. Weise, *Phys. Lett. B* **362** (1995) 23, arXiv:nucl-th/9507036.
71. P. C. Bruns, M. Mai and U.-G. Meißner, *Phys. Lett. B* **697** (2011) 254, arXiv:1012.2233 [nucl-th].
72. J. Nieves and E. Ruiz Arriola, *Phys. Rev. D* **64** (2001) 116008, arXiv:hep-ph/0104307.
73. ETM Collab. (K. Ottnad *et al.*), *J. High Energy Phys.* **1211** (2012) 048, arXiv:1206.6719 [hep-lat].
74. ETM Collab. (C. Alexandrou *et al.*), *Phys. Rev. D* **80** (2009) 114503, arXiv:0910.2419 [hep-lat].
75. W. Bietenholz *et al.*, *Phys. Rev. D* **84** (2011) 054509, arXiv:1102.5300 [hep-lat].
76. C. B. Lang and V. Verduci, *Phys. Rev. D* **87** (2013) 054502, arXiv:1212.5055.

Nuclear matrix elements for neutrinoless double-beta decay in covariant density functional theory

J. Meng*,†,‡,‖, L. S. Song* and J. M. Yao§,¶

*State Key Laboratory of Nuclear Physics and Technology, School of Physics, Peking University, Beijing 100871, P. R. China

†School of Physics and Nuclear Energy Engineering, Beihang University, Beijing 100191, P. R. China

‡Department of Physics, University of Stellenbosch, Stellenbosch 7602, South Africa

§Department of Physics and Astronomy, University of North Carolina, Chapel Hill, NC 27516-3255, USA

¶School of Physical Science and Technology, Southwest University, Chongqing 400715, P. R. China

‖mengj@pku.edu.cn

We review the first fully relativistic description for the neutrinoless double-beta decay by the beyond-mean-field covariant density functional theory. The calculations of the nuclear transition matrix elements are based on a full relativistic transition operator and nuclear wave functions in which the dynamic effects of particle-number, angular-momentum, and parity conservations as well as shape fluctuations are incorporated. We find that the commonly used nonrelativistic approximation for the transition operator is justified within the mechanism of light-Majorana-neutrino exchange. We stress that reducing the discrepancies in the predicted nuclear matrix elements by different nuclear models is a major challenge in nuclear physics.

Keywords: Double beta decay; neutrino mass; covariant density functional theory; beyond mean field approaches.

1. Introduction

1.1. *Nuclear weak process*

As one of the typical radioactive decays in an atomic nucleus, beta decay is a weak process in which a neutron (proton) is transformed into a proton (neutron), producing an electron (positron) and an antineutrino (neutrino). This process plays an important role in understanding many fundamental questions in physics. It allows the nucleus to move closer to the β-stability line in the nuclear chart and plays a key role in the nucleosynthesis.

‖Corresponding author.

Historically, the study of beta decay provided the first physical evidence of the neutrino. Following the pioneering works in 1911 by Meitner and Hahn and in 1913 by Danysz, Chadwick established the continuous feature of the kinetic energy spectrum of the beta-decay electron in 1914, via an accurate measurement performed by a magnetic spectrometer and one of Geiger's new counters.[1,2] This finding, confirmed by experiments during 1920–1927, was in an apparent contradiction with the conservation law of energy. Because if beta decay was known at that time as simply an electron emission, the energy of the emitted electron should equal the energy difference between the initial and final nuclear states and lead to a narrow energy distribution. Moreover, considering the fact that the spin is integer for nuclei with even mass number and half-integer for nuclei with odd mass number, the change of nuclear spin must be an integer in beta decay since the mass number is the same before and after. The angular momentum would not be conserved either if beta decay was simply an emission of a spin-1/2 electron.

In 1930, Pauli suggested, in an open letter to a physics meeting in Tübingen which he was unable to attend, that there exists a spin-1/2, weakly-interacting neutral particle with small mass produced together with the electron in beta decay, thus accounting for the missing energy and angular momentum.[3] In 1934, Fermi published his famous four-fermion theory of beta decay, based on Pauli's neutrino hypothesis and the discovery of the proton–neutron structure of nuclei.[4,5] In analogy with electromagnetic interactions, he proposed the following Hamiltonian of beta decay:

$$\mathcal{H}_I = G_F \bar{p}\gamma^\alpha n \bar{e}\gamma_\alpha \nu + \text{h.c.},$$

where G_F is the Fermi's constant. Fermi's theory was a great success, especially after Cowan and Reines detected the neutrino,[6] 25 years after Pauli's suggestion.

Nuclear double-beta decay is a type of radioactive decay in an atomic nucleus with two neutrons (protons) simultaneously transformed into two protons (neutrons) accompanied with the emission of two electrons (positrons) and possibly two antineutrinos (neutrinos). Double-beta decay cannot be simply regarded as two single-beta decays. It is a second-order weak process and thus much slower than the single-beta decay. There are 35 naturally occurring isotopes capable of double-beta decay. The decay can be observed in practice when the single-beta decay is energetically forbidden. Therefore, a typical candidate of double-beta decay is an even–even nucleus which is more bound than its neighboring odd–odd nucleus because of pairing correlations but with energy higher than the neighboring even–even nucleus.

On the basis of Fermi's theory, Goeppert-Mayer proposed in 1935 the first consideration of the simultaneous emission of two electrons and two antineutrinos in a nucleus, i.e., the ordinary double-beta ($2\nu\beta\beta$) decay.[7] Two years later, Majorana formulated a theory of neutrinos in which there was no distinction between neutrinos and antineutrinos,[8] based on which Furry introduced a novel scheme of double-beta decay, where the decay process occurs without the emission of antineutrinos.[9] This process is now called neutrinoless double-beta ($0\nu\beta\beta$) decay. This kind of neutrino is named the Majorana particle. At that time nothing was known about the parity violation in weak processes and the chirality suppression of the neutrinoless decay mode. It was believed that due to a considerable phase-space advantage, the neutrinoless mode dominated the double-beta-decay rate. The first experiment on search of double-beta decay was done in 1948 using Geiger counters, establishing a half-life limit of $T_{1/2} \geq 3 \cdot 10^{15}$ years for ^{124}Sn.[10] The first actual detection of double-beta decay was realized in a geochemical experiment in 1950 for the isotope ^{130}Te.[11] In 1952, Primakoff calculated the electron–electron angular correlations and electron energy spectra for both the two-neutrino decay and neutrinoless decay, producing a useful tool for distinguishing between the two processes in experiments.[12] This in principle enables verifying Majorana's theory through the observation of the neutrinoless decay, as this process occurs only if the neutrino is a Majorana particle. However, the distinction was not achievable at that time due to the limitation of sensitivity in the direct counting experiments.

In the mid-1950s, Lee and Yang questioned a hypothetical law of elementary particle physics, the conservation law of parity. Lee and Yang's research into existing experimental results convinced them that parity was conserved for the electromagnetic or strong interaction, but not for the weak interaction.[13] In 1957, Wu and coworkers proved in the Wu experiment that the parity conservation is clearly violated in the beta decay of ^{60}Co.[14] The discovery of parity violation prompted Salam,[15] Landau,[16] and Lee and Yang[17] to propose the theory of a massless two-component neutrino, in which the neutrino is a particle with definite helicity (projection of the spin on the direction of momentum) and parity is violated maximally. In 1958, Feynman and Gell-Mann,[18] Sudarshan and Marshak[19] proposed the vector minus axial vector or left-handed theory (V–A theory) of the weak interaction with the Hamiltonian

$$\mathcal{H}_I = \frac{G_F}{\sqrt{2}} j_L^\alpha j_{L\alpha}^\dagger,$$

where j_L^α is the weak charged current in which enter only the left-handed components of the fields. It was soon after the publication of this phenomenological V–A theory, in the 1960s, that the Glashow–Weinberg–Salam unified theory of the weak and electromagnetic interactions was established,[20–22] known as the Standard Model.

The study of nuclear weak processes has played a revolutionary role in the development of the Standard Model of electroweak interactions. If neutrinos were massless Dirac particles, the $0\nu\beta\beta$ decay could never happen. In the past decades, the discovery of massive neutrinos implies that the helicity is not a label for neutrinos, providing a source for the lepton number violation. The study of $0\nu\beta\beta$ decay has therefore become a renewed research highlight in nuclear and particle physics. Detection and study of this process may clarify several fundamental questions associated with physics beyond the Standard Model, including questions regarding the properties of neutrinos, such as the Majorana or Dirac nature of a neutrino, the absolute neutrino mass scale and the type of neutrino mass hierarchy.[23] The discovery of the

$0\nu\beta\beta$ process would also demonstrate the violation of total lepton number in nature, which could provide an explanation for the observed baryon–antibaryon asymmetry of our Universe with the mechanism of baryogenesis through leptogenesis.[24]

1.2. *Neutrinoless double-beta decay and neutrino masses*

Since 1998, nonvanishing neutrino mass has been observed in the form of neutrino oscillations in experiments including the atmospheric neutrino experiment Super-Kamiokande,[25] the solar neutrino experiment SNO,[26] the long-baseline accelerator neutrino experiments MINOS,[27] K2K[28] and T2K,[29] as well as the long-baseline reactor neutrino experiments KamLAND,[30] Double Chooz,[31] Daya Bay[32] and RENO.[33] However, the minimal Standard Model was built based on the assumption of massless neutrinos. It is also highly unlikely that the standard Higgs mechanism,[34–36] which generates masses of the quarks, the charged leptons and the $W^{\pm}$ and Z^0 bosons, is responsible for the generation of very small neutrino masses. Thus, a new mechanism of mass generation beyond the Standard Model is required and new physics is to be explored.

There are two fundamentally different possibilities for the massive neutrinos ν_i: they can be Dirac particles, if the total lepton number L is conserved; otherwise they can be Majorana particles. To reveal the mysterious nature of massive neutrinos, it is necessary to study processes in which the total lepton number L is violated. The $0\nu\beta\beta$ decay:

$$(A, Z) \to (A, Z+2) + e^- + e^-, \tag{1}$$

where the total lepton number is changed by two ($\Delta L = 2$), provides a sensitive probe to explore the Dirac or Majorana nature of the neutrino with definite mass. The measurement of the $0\nu\beta\beta$ decay is also regarded as an effective tool to distinguish the hierarchy of the neutrino mass spectrum, identify the unknown Majorana phases in the mixing matrix and determine the upper bound of the absolute value of the small neutrino mass.[23]

In the assumption of mass mechanism, which means that the $0\nu\beta\beta$-decay process is mediated by the exchange of light virtual Majorana neutrinos, the decay rate is determined by

$$\Gamma^{0\nu} = G^{0\nu}|M^{0\nu}|^2|\langle m_\nu\rangle|^2, \tag{2}$$

where $G^{0\nu}$ is a kinematic phase-space factor that can be accurately calculated from the Q value of the process and the nuclear charge Z of mother nucleus, $M^{0\nu}$ is the nuclear matrix element that is determined by the complex structures of initial and final nuclei and $\langle m_\nu\rangle$ is the effective Majorana mass given by the relation $\langle m_\nu\rangle = \sum_i |U_{ei}|^2 m_i$, with the 3×3 unitary mixing matrix $U = U^D S(\alpha)$. Here, the diagonal phase matrix $S(\alpha) = \mathrm{diag}(e^{i\alpha_1}, e^{i\alpha_2}, 1)$ is characterized by two Majorana phases α_1 and α_2, which only show up in lepton number violating processes, and U^D is the standard Dirac mixing matrix of neutrinos.[23]

The effective Majorana mass $\langle m_\nu \rangle$ depends on seven physical parameters, including two mixing angles $(\theta_{12}, \theta_{13})$, two Majorana phases (α_1, α_2) and three neutrino masses (m_1, m_2, m_3). Of these parameters, currently we do not know the Majorana phases, the lightest neutrino mass and the hierarchy of the neutrino mass spectrum (normal ordering $m_3 > m_2 > m_1$ or inverted ordering $m_2 > m_1 > m_3$). Therefore, the physical constraints on $\langle m_\nu \rangle$ drawn from the $0\nu\beta\beta$-decay measurement are expected to shed light on these fundamental questions, especially when combining with the cosmology data on total neutrino mass $\sum_i m_i$ and the data from tritium beta-decay experiments on $m_\beta = \sqrt{\sum_i |U_{ei}|^2 m_i^2}$.[37]

1.3. *Status on search of the* $0\nu\beta\beta$ *decay*

In the 1980s, the first observation of $2\nu\beta\beta$ decay in direct counting experiment was achieved by Moe's group, and in this experiment a half-life of $1.1 \cdot 10^{20}$ years for ^{82}Se was established.[38] To date, the $2\nu\beta\beta$-decay process has been reported in 11 isotopes, with the half-lives from 10^{18} years to 10^{24} years. Moreover, the $2\nu\beta\beta$ decay has been observed for ^{100}Mo and ^{150}Nd to the first 0^+ excited states of daughter nuclei and the 2ν ECEC process (double-electron capture with the emission of two neutrinos) has been observed in ^{130}Ba. Table 2 in Ref. 39 displays the present-day averaged and recommended values of half-life $T_{1/2}^{2\nu}$.

For the search of $0\nu\beta\beta$ decay, although numerous experimental data have been released, no actual signal is observed so far (for comprehensive and recent reviews, refer to Refs. 40–43). Only the lower bounds of the half-lives can be derived from those data. According to Eq. (2), one can estimate the upper bounds on the effective neutrino mass $\langle m_\nu \rangle$ from those limits on half-lives, given the calculated phase-space factor $G^{0\nu}$ and the nuclear matrix elements $M^{0\nu}$. The best available limits (90%) confidence level (C.L.) on the half-lives of $0\nu\beta\beta$ decay in various candidate isotopes are listed in Table 1, together with the corresponding Q values and phase-space factors.

Table 1. Experimental results to date on $0\nu\beta\beta$ decay (limits at 90% C.L.), together with the corresponding Q values $Q_{\beta\beta}$ and phase-space factors $G^{0\nu}$ adopted from Ref. 44.

Isotope	$Q_{\beta\beta}$ (MeV)	$G_{0\nu}$ (10^{-15} year $^{-1}$)	$T_{1/2}^{0\nu}$ (year)	Reference
^{48}Ca	4.27226(404)	24.81	$> 5.8 \cdot 10^{22}$	ELEGANT VI,[45] 2008
^{76}Ge	2.03904(16)	2.363	$> 3.0 \cdot 10^{25}$	GERDA,[46] 2013
^{82}Se	2.99512(201)	10.16	$> 3.6 \cdot 10^{23}$	NEMO-3,[47] 2011
^{96}Zr	3.35037(289)	20.58	$> 9.2 \cdot 10^{21}$	NEMO-3,[48] 2010
^{100}Mo	3.03440(17)	15.92	$> 1.1 \cdot 10^{24}$	NEMO-3,[49] 2014
^{116}Cd	2.81350(13)	16.70	$> 1.7 \cdot 10^{23}$	Solotvina,[50] 2003
^{124}Sn	2.28697(153)	9.04	$> 5.0 \cdot 10^{19}$	KIMS,[51] 2009
^{128}Te	0.86587(131)	0.5878	$> 1.6 \cdot 10^{24}$	Geochemistry,[52] 2011
^{130}Te	2.52697(23)	14.22	$> 2.8 \cdot 10^{24}$	CUORICINO,[53] 2011
^{136}Xe	2.45783(37)	14.58	$> 1.1 \cdot 10^{26}$	KamLAND-Zen,[54] 2016
^{150}Nd	3.37138(20)	63.03	$> 2.0 \cdot 10^{22}$	NEMO-3,[55] 2016

Previously, the best limit was set by the Heidelberg–Moscow experiment.[56] Part of the Heidelberg–Moscow collaboration claimed to have evidence for the $0\nu\beta\beta$ decay in ^{76}Ge with $T^{0\nu}_{1/2} = (1.19^{+0.37}_{-0.23}) \cdot 10^{25}$ years at a 4σ level,[57] which, however, has been overruled by the constraints from the cosmology observations[58,59] and by the latest $0\nu\beta\beta$ experiments.

Recent sensitive experiments are being carried out in ^{76}Ge (GERDA[46]) and ^{136}Xe (KamLAND-Zen[54] and EXO-200[60]), respectively. The limit of the ^{76}Ge half-life imposed by the available results of GERDA is $T_{1/2} > 2.1 \cdot 10^{25}$ years (90% C.L.) or, $T_{1/2} > 3.0 \cdot 10^{25}$ years (90% C.L.) if combined with earlier Germanium experiments.[56,61] A similarly strong limit is obtained by the recent results from KamLAND-Zen, which is $T_{1/2} > 1.1 \cdot 10^{26}$ years (90% C.L.). So far, the experiments of both isotopes set essentially the same limit on the effective neutrino mass, which is $|\langle m_\nu \rangle| \lesssim 0.3\,\text{eV}$, taking into account the uncertainty of the nuclear matrix elements calculated by various nuclear structure models.[37]

Several future experiments on the search of $0\nu\beta\beta$ decay are in preparation at present, including the next-phase experiments of the currently running projects (GERDA, KamLAND-Zen, EXO-200), the projects under construction (COURE, Majorana, CANDLES, NEXT, SuperNEMO, SNO+), as well as those in proposal (Cobra, Lucifer, DCBA/MTD, MOON, AMoRE).[41] In the future experiments, much larger detectors will be utilized, and immense efforts will be made to significantly decrease the background and improve the energy resolution. The anticipated sensitivity to the effective Majorana mass will be $|\langle m_\nu \rangle| \simeq$ a few 10^{-2} eV, which is in the range of the possible values $(1.8 \cdot 10^{-2}\,\text{eV} \leq |\langle m_\nu \rangle| \leq 4.9 \cdot 10^{-2}\,\text{eV})$ given by the inverted hierarchy of the neutrino masses. Thus, the next-generation experiments will allow for testing the Majorana nature of the neutrino in the case of inverted hierarchy for the neutrino mass spectrum.

1.4. *Status on computing the nuclear matrix elements for the $0\nu\beta\beta$ decay*

According to Eq. (2), the effective neutrino mass can be determined from the half-life of the $0\nu\beta\beta$ decay provided that the nuclear matrix element (NME) $M^{0\nu}$ is known. The NME depends on the transition operator and the wave functions of the initial and final nuclear states. The transition operator is derived based on the underlying mechanism, while the nuclear wave functions rely on complex nuclear many-body calculations. The nuclear models frequently adopted for computing the NMEs are the configuration-interacting shell model (CISM)[62–77] and the quasiparticle random phase approximation (QRPA).[78–97] In recent years, some other methods have also been adopted for this purpose, including the projected Hartree–Fock–Bogoliubov (PHFB) model,[98–103] the interacting boson model (IBM),[104–106] and the energy density functional (EDF) method based on a Gogny interaction.[107–109] The predicted NMEs by different models deviate from each other by a factor of 2–3, resulting in about one order-of-magnitude uncertainty for the predicted half-lives. Reducing these discrepancies is a major challenge in nuclear physics. Tracing the sources responsible for the discrepancy by comparing in detail different model calculations are underway.[110,111]

The decay operator constitutes another essential ingredient of the NME in $0\nu\beta\beta$ decay. It is derived from the second-order weak Hamiltonian constructed by charged nucleonic and leptonic currents. However, within the nuclear models descried above, which are all based on nonrelativistic frameworks, one has to adopt a nonrelativistic reduced form for the derived operator in the calculation of NMEs. Thus, it is desirable to examine this nonrelativistic approximation via a fully relativistic study. Covariant density functional theory (CDFT) provides such a relativistic framework in which comparative studies can be conducted using both the relativistic and the nonrelativistic reduced decay operators.

In the past decades, CDFT has been proven to be a very powerful tool in nuclear physics, which, on the mean-field level, provides a satisfying description for the static ground state properties of finite nuclei.[112–116] The covariant version of energy density functional takes into account Lorentz invariance, which puts stringent restrictions on the number of parameters. The spin-orbit potential is included naturally, as well as the time-odd components of the nuclear mean field. With the merits inherited, this method has also been generalized beyond the static mean-field level by RPA[117,118] and QRPA,[119–122] or by generator coordinate method (GCM),[123–129] so that it could be applied for the description of the excited states, the electromagnetic properties and the weak transitions including the beta and double-beta decays.[130]

The first fully relativistic description for the NMEs of $0\nu\beta\beta$ decay has been constructed within the beyond-mean-field GCM scheme based on CDFT.[130–133] This approach is able to give a unified description for all $0\nu\beta\beta$ candidates including the heavy deformed nuclei. For both initial and final nuclear states, the dynamic effects of particle-number, angular-momentum and parity conservations as well as shape fluctuations are taken into account in the wave functions via the GCM. Moreover, as the wave functions are Dirac spinors, the full relativistic decay operator, which is a 4×4 matrix, can be directly sandwiched between the initial and final states to calculate the NME.

1.5. *Aim of this paper*

The aim of this paper is to review the NMEs of $0\nu\beta\beta$ decay based on CDFT, including the formalism of constructing the nuclear many-body wave functions and deriving the $0\nu\beta\beta$-decay operator by assuming the mechanism of light-Majorana-neutrino exchange. Results of a systematic calculation for some popular $0\nu\beta\beta$-decay candidates are summarized. Detailed derivations and a proof-of-principle calculation for the $0\nu\beta\beta$ decay in ^{150}Nd can be found in Ref. 130.

2. Beyond-Mean-Field CDFT for Nuclear Wave Functions

We start from a universal relativistic energy functional PC-PK1[134] that is based on the concept that the nucleons interact strongly with each other in a contact coupling form. The nuclear many-body reference states $|q\rangle$, labeled by different deformation parameters q, are generated by the deformation-constrained self-consistent mean-field calculations. The BCS approximation is applied in constructing these reference states to account for pairing correlations that are very important in nuclei with open shells. A zero-range force $V_\tau^{pp}(\boldsymbol{r}_1, \boldsymbol{r}_2) = V_0^{pp}\delta(\boldsymbol{r}_1 - \boldsymbol{r}_2)$ with a smooth cutoff[135] is implemented in the particle–particle channel. The pairing strength parameters V_0^{pp} are determined by fitting the neutron and proton average pairing gaps in ^{150}Nd provided by the separable finite-range pairing force[136] (see Fig. 1 of Ref. 130), and kept constant in the calculations of all $0\nu\beta\beta$ nuclei.

The self-consistent CDFT has been extended to include dynamical correlations related to symmetry restoration and nuclear shape fluctuation via the GCM combined with quantum-number (parity, particle-number and angular-momentum) projections. The symmetry-conserved wave function is thus constructed by superposing a set of quantum-number-projected nonorthogonal states around the equilibrium shape

$$|J^\pi NZ;\alpha\rangle = \sum_{\kappa\in\{q,K\}} f_\kappa^{J\pi\alpha}\hat{P}_{MK}^J\hat{P}^N\hat{P}^Z\hat{P}^\pi|q\rangle, \tag{3}$$

where the discretized deformation parameters $q \equiv \{\beta_2, \gamma, \beta_3, \ldots,\}$ of the reference states are chosen as the generator coordinates. The $\hat{P}^G$s $(G \equiv J, \pi, N, Z)$ are projection operators onto spin-parity (J^π) and numbers (N, Z) of neutrons and protons.[137] The weight function $f_\kappa^{J\pi\alpha}$ is determined by solving the Hill–Wheeler–Griffin equation

$$\sum_{\kappa_b}[\mathscr{H}_{\kappa_a,\kappa_b}^{J\pi} - E_\alpha^{J\pi}\mathscr{N}_{\kappa_a,\kappa_b}^{J\pi}]f_{\kappa_b}^{J\pi\alpha} = 0, \tag{4}$$

where the Hamiltonian kernel $\mathscr{H}_{\kappa_a,\kappa_b}^{J\pi}$ and norm kernel $\mathscr{N}_{\kappa_a,\kappa_b}^{J\pi}$ are given by $\mathscr{O}_{\kappa_a,\kappa_b}^{J\pi} = \langle q_a|\hat{O}\hat{P}_{MK_b}^J\hat{P}^N\hat{P}^Z\hat{P}^\pi|q_b\rangle$, with the operator $\hat{O}$ representing $\hat{H}$ and 1, respectively. More details about the beyond-mean-field calculation have been discussed in Refs. 126, 127, 129 and 138.

In recent years, the symmetry-conserved GCM approach has been developed rapidly.[139] In particular, based on CDFT, this approach has been applied to study different aspects in the nuclear low-spin states, including the triaxiality in ^{76}Kr,[129] the stabilization of octupole deformation with the spin in ^{224}Ra (Ref. 138) and the diatomic-like cluster structure (^{16}O+α) in ^{20}Ne.[140] These studies demonstrate partially that the symmetry-conserved GCM approach is capable of describing the main features of the nuclear low-lying states provided that the generator coordinates are chosen properly.

3. Relativistic Description for the $0\nu\beta\beta$ Decay

3.1. *The $0\nu\beta\beta$ decay with light-neutrino exchange*

The starting point is the weak interaction Hamiltonian with the left-handed fields only:

$$\mathcal{H}_\beta(x) = \frac{G_F \cos\theta_C}{\sqrt{2}} 2(\bar{e}_L \gamma^\mu \nu_{eL}) j^\dagger_\mu + \text{h.c.}, \tag{5}$$

where G_F is the Fermi constant, θ_C is the Cabbibo angle and $j^\dagger_\mu$ is the strangeness-conserving charged hadronic current. In terms of nucleon field $\psi(x)$, the current has the form:

$$j^\dagger_\mu(x) = \bar{\psi}(x)\left[g_V(q^2)\gamma_\mu + ig_M(q^2)\frac{\sigma_{\mu\nu}}{2m_p}q^\nu - g_A(q^2)\gamma_\mu\gamma_5 - g_P(q^2)q_\mu\gamma_5\right]\tau_-\psi(x), \tag{6}$$

where m_p is the nucleon mass, q^μ is the momentum transferred from leptons to hadrons, $\tau_- \equiv (\tau_1 - i\tau_2)/2$ is the isospin lowering operator, and $\sigma_{\mu\nu} = \frac{i}{2}[\gamma_\mu, \gamma_\nu]$. The form factors $g_V(q^2)$, $g_A(q^2)$, $g_M(q^2)$ and $g_P(q^2)$, in which the effects of the finite nucleon size are incorporated, represent, respectively, the vector, axial-vector, weak-magnetism and induced pseudoscalar coupling constants in the zero-momentum-transfer limit.

Here, the underlying assumption is that the $0\nu\beta\beta$ decay, which is a second-order weak process, is mediated by the exchange of virtual Majorana neutrinos with small masses. The scattering matrix element is given by

$$\begin{aligned}\langle f|S^{(2)}|i\rangle = {}& 4\frac{(-i)^2}{2!}\left(\frac{G_F\cos\theta_C}{\sqrt{2}}\right)^2 N_{p_1}N_{p_2}\iint d^4x_1 d^4x_2 \\ & \times \bar{u}_L(p_1)e^{ip_1x_1}\gamma^\mu\langle 0|T(\nu_{eL}(x_1)\nu^T_{eL}(x_2))|0\rangle\gamma^{\nu T}\bar{u}^T_L(p_2)e^{ip_2x_2} \\ & \times \langle\Psi_F|T(\mathcal{J}^\dagger_\mu(x_1)\mathcal{J}^\dagger_\nu(x_2))|\Psi_I\rangle - (p_1 \leftrightarrow p_2),\end{aligned} \tag{7}$$

where p_1 and p_2 are the four-momenta of outgoing electrons, $|\Psi_I\rangle$ and $|\Psi_F\rangle$ are the initial and final nuclear states with the respective four-momenta $P_I = (E_I, \boldsymbol{p}_I)$ and $P_F = (E_F, \boldsymbol{p}_F)$ and $N_p = ((2\pi)^{3/2}\sqrt{2p^0})^{-1}$ is the standard normalization factor. Notably, $\mathcal{J}^\dagger_{\mu(\nu)}(x)$ is the hadronic current in Heisenberg representation so that the strong interactions are taken into consideration.

The left-handed electron neutrino $\nu_{eL}(x)$ is assumed to be a mixing of the field of light neutrino with mass m_i ($m_i \ll 1\,\text{keV}$): $\nu_{eL}(x) = \sum_i U_{ei}\nu_{iL}(x)$. The field ν_i satisfies the Majorana condition, i.e., $\nu_i = (\nu_i)^c = C\bar{\nu}^T_i$, where C denotes the charge conjugation. Thus, the neutrino propagator $\langle 0|T(\nu_{eL}(x_1)\nu^T_{eL}(x_2))|0\rangle$ takes the following form:

$$-\sum_i |U_{ei}|^2 m_i \frac{i}{(2\pi)^4}\int \frac{e^{-iq(x_1-x_2)}}{q^2 - m_i^2} d^4q \frac{1-\gamma_5}{2}C, \tag{8}$$

and it is straightforward to derive the scattering matrix[23]

$$\langle f|S^{(2)}|i\rangle = i\left(\frac{G_F\cos\theta_C}{\sqrt{2}}\right)^2 N_{p_1}N_{p_2}\bar{u}(p_1)\gamma^\mu(1-\gamma_5)\gamma^\nu C\bar{u}^T(p_2)$$
$$\times \iint d^3x_1 d^3x_2 e^{-i\boldsymbol{p}_1\cdot\boldsymbol{x}_1 - i\boldsymbol{p}_2\cdot\boldsymbol{x}_2}\sum_i (U_{ei})^2 m_i$$
$$\times \int \frac{d^3q}{(2\pi)^3}\frac{e^{i\boldsymbol{q}\cdot(\boldsymbol{x}_1-\boldsymbol{x}_2)}}{\omega_i}\left[\sum_n \frac{\langle\Psi_F|\mathcal{J}^\dagger_\mu(\boldsymbol{x}_1)|n\rangle\langle n|\mathcal{J}^\dagger_\nu(\boldsymbol{x}_2)|\Psi_I\rangle}{\omega_i+p^0_2+E_n-E_I-i\varepsilon}\right.$$
$$\left.+\sum_n \frac{\langle\Psi_F|\mathcal{J}^\dagger_\nu(\boldsymbol{x}_2)|n\rangle\langle n|\mathcal{J}^\dagger_\mu(\boldsymbol{x}_1)|\Psi_I\rangle}{\omega_i+p^0_1+E_n-E_I-i\varepsilon}\right] 2\pi\delta(E_F-E_I+p^0_1+p^0_2), \quad (9)$$

where the time components x^0_1, x^0_2 and q^0 have been integrated out, and $\omega_i = \sqrt{|\boldsymbol{q}|^2+m_i^2}$. The following approximations are usually considered to simplify the calculation.

(i) Light neutrino mass: Because the average momentum carried by the virtual neutrino $|\boldsymbol{q}|$ is estimated to be of the order of $1/\bar{r}_{NN} \sim 100\,\text{MeV}$ with $\bar{r}_{NN} \sim 2\,\text{fm}$ (mean inter-nucleon distance), the small masses for light neutrinos ($m_i < 2.2\,\text{eV}$) can be safely neglected, and we have $\omega_i = \sqrt{|\boldsymbol{q}|^2+m_i^2} \simeq |\boldsymbol{q}| \equiv q$. Now one can define the effective Majorana mass through

$$\langle m_\nu\rangle \equiv \sum_i |U_{ei}|^2 m_i, \quad (10)$$

since the rest of Eq. (9) no longer depends on m_i.

(ii) Long-wave approximation: From the condition $|\boldsymbol{p}_1|R, |\boldsymbol{p}_2|R \ll 1$ and $R \simeq 1.2A^{1/3}\,\text{fm}$, we know the average momenta of outgoing electrons $|\boldsymbol{p}_1|, |\boldsymbol{p}_2| \lesssim 1\,\text{MeV}$. Thus, the approximation $e^{-i\boldsymbol{p}_1\cdot\boldsymbol{x}_1 - i\boldsymbol{p}_2\cdot\boldsymbol{x}_2} \simeq 1$ can be made. In other words, the two electrons are emitted in the S-states.

(iii) Closure approximation: Momentum of the virtual neutrino $q \simeq 100\,\text{MeV}$ is much larger than the excitation energy $(E_n - E_I)$. Thus, the state-dependent energy can be replaced with an average one: $E_n \to \bar{E}$, and the intermediate states can be eliminated with closure relation ($\sum_n |n\rangle\langle n| = 1$), so that

$$\sum_n \frac{\langle\Psi_F|\mathcal{J}^\dagger_\mu(\boldsymbol{x}_1)|n\rangle\langle n|\mathcal{J}^\dagger_\nu(\boldsymbol{x}_2)|\Psi_I\rangle}{q+p^0_{1,2}+E_n-E_I} \to \frac{\langle\Psi_F|\mathcal{J}^\dagger_\mu(\boldsymbol{x}_1)\mathcal{J}^\dagger_\nu(\boldsymbol{x}_2)|\Psi_I\rangle}{q+\bar{E}-(E_I+E_F)/2}, \quad (11)$$

where the approximation $q+p^0_{1,2}+\bar{E}-E_I \simeq q+\bar{E}-(E_I+E_F)/2$ has been applied for the energy denominator.

Then the scattering matrix element becomes

$$\langle f|S^{(2)}|i\rangle = i\left(\frac{G_F\cos\theta_C}{\sqrt{2}}\right)^2 \frac{1}{(2\pi)^3}\frac{1}{\sqrt{p^0_1p^0_2}}\frac{1}{R}\bar{u}(p_1)(1+\gamma_5)C\bar{u}^T(p_2)$$
$$\times g^2_A(0)\langle m_\nu\rangle M^{0\nu}\delta(E_F-E_I+p^0_1+p^0_2), \quad (12)$$

where the nuclear matrix element is defined as $M^{0\nu} \equiv \langle \Psi_F | \hat{\mathcal{O}}^{0\nu} | \Psi_I \rangle$. The $0\nu\beta\beta$-decay operator reads

$$\hat{\mathcal{O}}^{0\nu} = \frac{4\pi R}{g_A^2(0)} \iint d^3x_1 d^3x_2 \int \frac{d^3q}{(2\pi)^3} h(q) e^{i\boldsymbol{q}\cdot(\boldsymbol{x}_1 - \boldsymbol{x}_2)} \mathcal{J}_\mu^\dagger(\boldsymbol{x}_1) \mathcal{J}^{\mu\dagger}(\boldsymbol{x}_2), \tag{13}$$

with $h(q) = (q(q + E_d))^{-1}$ and $E_d \equiv \bar{E} - (E_I + E_F)/2$.

Finally the $0\nu\beta\beta$-decay rate is deduced from the above scattering matrix element, by integrating over the phase-space of final states,[23]

$$[T_{1/2}^{0\nu}]^{-1} = G_{0\nu} g_A^4(0) \left| \frac{\langle m_\nu \rangle}{m_e} \right|^2 |M^{0\nu}|^2, \tag{14}$$

which is essentially the same as Eq. (2). In this expression of decay rate, it is obvious that the axial-vector coupling constant $g_A(0)$ and the electron mass m_e are constants. The kinematic phase-space factor $G_{0\nu}$ can be determined precisely.[44,141] Hence, accurate calculations for the NMEs $M^{0\nu}$ are crucial for deducing the effective mass $|\langle m_\nu \rangle|$ from possible $0\nu\beta\beta$ measurements.

3.2. *Relativistic description for the nuclear matrix elements*

Now we deal with the NME of the major $0\nu\beta\beta$ transition between the $J^\pi = 0^+$ nuclear states. In the current framework of CDFT, the shape fluctuations of quadrupole and octupole characters have been taken into consideration within the GCM scheme by mixing the deformed mean-field wave functions. The axial symmetry is assumed throughout this work for the sake of simplicity. The reflection and rotation symmetries and the particle-number conservation that are violated at the mean-field level are restored by parity, angular-momentum and particle-number projections (PNPs), respectively, prior to the mixing.

With the $J^\pi = 0^+$ GCM wave functions the NME can be expressed as

$$M^{0\nu} = \langle 0_F^+ | \hat{\mathcal{O}}^{0\nu} | 0_I^+ \rangle = \sum_{q_I, q_F} f_{0_F^+}^*(q_F) f_{0_I^+}(q_I) M^{0\nu}(q_I, q_F), \tag{15}$$

which is the weighted superposition of the projected matrix elements of different deformations:

$$\begin{aligned} M^{0\nu}(q_I, q_F) &= \langle q_F | \hat{\mathcal{O}}^{0\nu} \hat{P}^{J=0} \hat{P}^{N_I} \hat{P}^{Z_I} \hat{P}^{\pi=+} | q_I \rangle \\ &= \int \frac{d\Omega}{8\pi} \int_0^{2\pi} \frac{d\varphi_n}{2\pi} e^{-i\varphi_n N_I} \int_0^{2\pi} \frac{d\varphi_p}{2\pi} e^{-i\varphi_p Z_I} \delta_{\pi+} \\ &\quad \times \sum_{abcd} \langle ab | \hat{O} | cd \rangle \, \langle q_F | c_a^{(\pi)\dagger} c_b^{(\pi)\dagger} c_d^{(\nu)} c_c^{(\nu)} | \tilde{q}_I \rangle, \end{aligned} \tag{16}$$

with the shorthand notation $|\tilde{q}_I\rangle \equiv \frac{1}{2}(1 + \pi\hat{\mathcal{P}}) e^{i\varphi_n \hat{N}} e^{i\varphi_p \hat{Z}} \hat{R}(\Omega) |q_I\rangle$. Note that $\hat{R}(\Omega)$, $\hat{\mathcal{P}}$ and $\hat{N}$ or $\hat{Z}$ are rotation, parity and neutron or proton number operators, respectively.

The coefficients $f_{0_I^+}(q_I)$ and $f_{0_F^+}(q_F)$ in Eq. (15) are obtained by solving the Hill–Wheeler–Griffin equation for the 0^+ ground states of initial and final nuclei. The collective coordinate q stands for the intrinsic deformation parameters (β_2, β_3) of the mean-field states. The projection operators are responsible for projecting the nuclear wave function onto state with $J^\pi = 0^+$ and with correct neutron and proton numbers. Note that by performing the projections for the initial wave function $|q_I\rangle$, the symmetries of the final state $|q_F\rangle$ are restored automatically.[130] In the second line of Eq. (16), the projection operators are written explicitly and the decay operator (13) is expressed in its second-quantized form. Operators $c_d^{(\nu)}, c_c^{(\nu)}$ are neutron annihilation and $c_a^{(\pi)\dagger}, c_b^{(\pi)\dagger}$ are proton creation operators. The indices c, d run over a complete set of single neutron states and a, b over a complete set of single proton states.

The crucial part that contains the nuclear structural information in the NME (16) is the two-body transition density, $\langle q_F|c_a^{(\pi)\dagger}c_b^{(\pi)\dagger}c_d^{(\nu)}c_c^{(\nu)}|\tilde{q}_I\rangle$. Provided that the states $|q_F\rangle$ and $|\tilde{q}_I\rangle$ are nonorthogonal, one can use the extended Wick's theorem[142,143] to express it as a product of two one-body transition densities (abnormal) and a norm overlap:

$$\frac{\langle q_F|c_a^{(\pi)\dagger}c_b^{(\pi)\dagger}|\tilde{q}_I\rangle^{(\pi)}}{\langle q_F|\tilde{q}_I\rangle^{(\pi)}} \times \frac{\langle q_F|c_d^{(\nu)}c_c^{(\nu)}|\tilde{q}_I\rangle^{(\nu)}}{\langle q_F|\tilde{q}_I\rangle^{(\nu)}} \times \langle q_F|\tilde{q}_I\rangle. \tag{17}$$

The two-body matrix element $\langle ab|\hat{O}|cd\rangle$ in the NME (16) is

$$\frac{4\pi R}{g_A^2(0)}\int \frac{d^3q}{(2\pi)^3}\, h(q)\langle a|\gamma^0\Gamma_\mu(q)e^{i\boldsymbol{q}\cdot\boldsymbol{r}_1}|c\rangle\langle b|\gamma^0\Gamma^\mu(q)e^{-i\boldsymbol{q}\cdot\boldsymbol{r}_2}|d\rangle, \tag{18}$$

where $\Gamma_\mu(q) \equiv g_V(q^2)\gamma_\mu + ig_M(q^2)\frac{\sigma_{\mu\nu}}{2m_p}q^\nu - g_A(q^2)\gamma_\mu\gamma_5 - g_P(q^2)q_\mu\gamma_5$, and the following decomposition for $(\gamma^0\Gamma_\mu)^{(1)}(\gamma^0\Gamma^\mu)^{(2)}$ is considered:

$$VV : g_V^2(\boldsymbol{q}^2)(\gamma^0\gamma_\mu)^{(1)}(\gamma^0\gamma^\mu)^{(2)}, \tag{19a}$$

$$AA : g_A^2(\boldsymbol{q}^2)(\gamma^0\gamma_\mu\gamma_5)^{(1)}(\gamma^0\gamma^\mu\gamma_5)^{(2)}, \tag{19b}$$

$$AP : 2g_A(\boldsymbol{q}^2)g_P(\boldsymbol{q}^2)(\gamma^0\boldsymbol{\gamma}\gamma_5)^{(1)}(\gamma^0\boldsymbol{q}\gamma_5)^{(2)}, \tag{19c}$$

$$PP : g_P^2(\boldsymbol{q}^2)(\gamma^0\boldsymbol{q}\gamma_5)^{(1)}(\gamma^0\boldsymbol{q}\gamma_5)^{(2)}, \tag{19d}$$

$$MM : g_M^2(\boldsymbol{q}^2)\left(\gamma^0\frac{\sigma_{\mu i}}{2m_p}q^i\right)^{(1)}\left(\gamma^0\frac{\sigma^{\mu j}}{2m_p}q_j\right)^{(2)}. \tag{19e}$$

Here, the dependence on the time component of the four-momemtum transfer q^0 has been neglected. The vector and axial vector form factors have the usual dipole form: $g_V(\boldsymbol{q}^2) = g_V(0)/(1+\boldsymbol{q}^2/M_V^2)^2$, $g_A(\boldsymbol{q}^2) = g_A(0)/(1+\boldsymbol{q}^2/M_A^2)^2$, and the weak magnetism and induced pseudoscalar form factors are given, respectively, by $g_M(\boldsymbol{q}^2) = (\mu_p - \mu_n)g_V(\boldsymbol{q}^2)$ and $g_P(\boldsymbol{q}^2) = 2m_p g_A(\boldsymbol{q}^2)/(\boldsymbol{q}^2 + m_\pi^2)\cdot(1 - m_\pi^2/M_A^2)$,[144] where $g_V(0) = 1$, $g_A(0) = 1.254$, $(\mu_p - \mu_n) = 3.70$, $M_V = 842\,\text{MeV}$,[145] $M_A = 1090\,\text{MeV}$ (Ref. 144) and m_π is the mass of pion. Details of the evaluation of the two-body transition density and the two-body matrix element can be found in Ref. 130.

4. Nuclear Matrix Elements for the $0\nu\beta\beta$ Decay

4.1. *Relativistic effects*

The calculation of NME within CDFT is fully relativistic and, therefore, it allows us to examine the nonrelativistic approximation applied to the transition operator in most calculations.

Figure 1 compares the NMEs for the $0\nu\beta\beta$ decay in ^{150}Nd calculated with the full operator (13) and its nonrelativistic reduced form[90,130] using the GCM+PN1DAMP method. In this comparison, the relativistic effect $\Delta_{\text{Rel.}} \equiv (M^{0\nu} - M^{0\nu}_{\text{NR}})/M^{0\nu}$ is roughly 5% in the ground state to ground state ($0^+_1 \to 0^+_1$) transition and 24% in the ground state to excited state ($0^+_1 \to 0^+_2$) transition.

Within the same GCM+PN1DAMP framework, the relativistic effects in the $0\nu\beta\beta$ decay have been investigated in Ref. 131 over a series of candidate nuclei, where the normalized NMEs $\tilde{M}^{0\nu}(q_I, q_F)$, defined as

$$\tilde{M}^{0\nu}(q_I, q_F) = \mathcal{N}_F \mathcal{N}_I \langle q_F | \hat{\mathcal{O}}^{0\nu} \hat{P}^{J=0} \hat{P}^{N_I} \hat{P}^{Z_I} | q_I \rangle, \tag{20}$$

are calculated at the spherical configurations ($q_I = q_F = 0$). The normalization factors are given by the norms of projected states, i.e., $\mathcal{N}_a^{-2} = \langle q_a | \hat{P}^{J=0} \hat{P}^{N_a} \hat{P}^{Z_a} | q_a \rangle$ with $a = I, F$. As the angular-momentum projection is not needed for the states with spherical symmetry, the matrix element reduces to

$$M^{0\nu}_{\text{sph}} = \frac{\langle q_F = 0 | \hat{\mathcal{O}}^{0\nu} \hat{P}^{N_I} \hat{P}^{Z_I} | q_I = 0 \rangle}{\prod\limits_{a=I,F} \sqrt{\langle q_a = 0 | \hat{P}^{N_a} \hat{P}^{Z_a} | q_a = 0 \rangle}}. \tag{21}$$

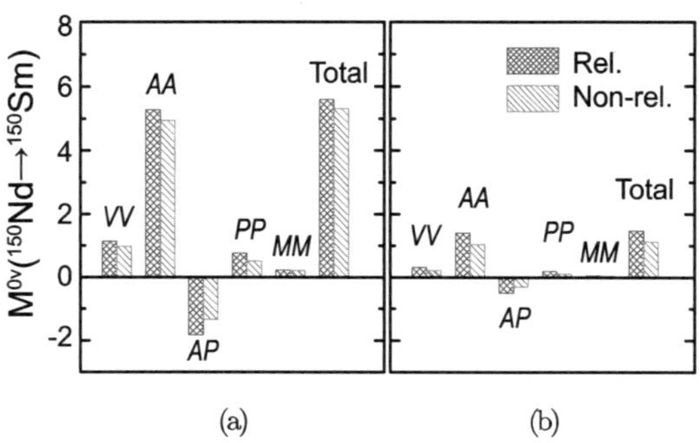

Fig. 1. (Color online) NMEs of the $0\nu\beta\beta$ decay from ^{150}Nd to ^{150}Sm for both the (a) $0^+_1 \to 0^+_1$ and (b) $0^+_1 \to 0^+_2$ transitions. Values of the matrix elements evaluated using the full relativistic operator (Rel.) are compared with those obtained with the nonrelativistic reduced operator (Non-rel.). The results are calculated with the GCM wave functions with particle-number and one-dimensional angular-momentum projection (GCM+PN1DAMP) method without considering the triaxial and octupole deformations of nuclei.

Source: Reproduced from Ref. 130.

From Table I of Ref. 131 one finds that, in the spherical case, the error arisen from the nonrelativistic reduction is marginal, which can either increase or decrease the matrix elements within 2%.

These results have provided evidence that the nonrelativistic decay operator is a good approximation to the full operator within the assumption of light-neutrino exchange, although the conclusion is to be scrutinized by further systematic calculations that utilize more general nuclear wave functions.

4.2. *Particle-number effects*

The nuclear states from the mean-field-based models usually do not conserve proton and neutron numbers. In this subsection, we will analyze the influence of particle-number conservation in nuclear ground states on the calculated NMEs. To this end, we perform a comparative calculation using the spherical mean-field wave functions with and without projection of neutron and proton numbers. In this simplest case, the NME of the $0\nu\beta\beta$ decay is given by Eq. (21).

The effect of particle-number projection usually reduces both the nominator and the denominator in Eq. (21), giving rise to marginal changes in the NMEs, such as for ^{150}Nd $\to$ ^{150}Sm. However, there are exceptions as shown in Table 2. This effect increases significantly the NMEs for ^{48}Ca $\to$ ^{48}Ti, ^{116}Cd $\to$ ^{116}Sn, ^{124}Sn $\to$ ^{124}Te, and ^{136}Xe $\to$ ^{136}Ba. In these four cases, pairing correlation of neutrons or protons collapses artificially either in the initial nucleus or in the final nucleus. Taking the decay ^{48}Ca $\to$ ^{48}Ti as an example, pairing collapse is found for both neutrons and protons in ^{48}Ca. While the nominator $\langle q_F = 0(^{48}\text{Ti})|\hat{O}^{0\nu}\hat{P}^{N_I}\hat{P}^{Z_I}|q_I = 0(^{48}\text{Ca})\rangle$ does not change by the implementation of particle-number projection because the initial state $|q_I = 0(^{48}\text{Ca})\rangle$ is a (particle-number conserving) Hartree–Fock state, the denominator of Eq. (21) decreases by the projection due to the decrease of ^{48}Ti normalization factor, so that an increase of the normalized NME is expected.

4.3. *Quadrupole deformation correlations*

Nuclear collective correlations associated with quadrupole deformation degree of freedom are very important in the nuclear low-energy structures. The low-lying spectra of spherical nuclei usually exhibit the characters of quadrupole surface vibrational motions, while those of open-shell nuclei are generally interpreted as rotational excitations of quadrupole deformed rotors. The nuclei in between exhibit transitional characters with a rather flat energy surface around the equilibrium

Table 2. NMEs of the $0\nu\beta\beta$ decay in several candidate nuclei obtained with the spherical wave functions with and without particle-number projection.

	^{48}Ca$\to$^{48}Ti	^{116}Cd$\to$^{116}Sn	^{124}Sn$\to$^{124}Te	^{136}Xe$\to$^{136}Ba	^{150}Nd$\to$^{150}Sm
Without PNP	1.61	4.14	4.29	4.18	13.32
With PNP	3.66	6.18	6.66	6.59	13.25

shape with respect to the quadrupole deformation, in which case, the mixing of a set of quadrupole deformed configurations is expected in nuclear low-lying states. The previous study based on deformed QRPA[146] showed that the quadrupole deformations have a great influence on the $2\nu\beta\beta$ decay. In particular, the transition matrix elements are increasingly quenched with respect to the spherical case when the differences in quadrupole deformations of mother and daughter nuclei increase. It is natural to ask if the quadrupole deformation also affects the NMEs of the $0\nu\beta\beta$ decay. In this subsection, we discuss this topic in detail.

The influence of quadrupole deformation on the NMEs of $0\nu\beta\beta$ decay has been studied by the phenomenological pseudo SU(3) model,[147] the PHFB approach,[98,100] the IBM,[104] the deformed QRPA,[92,95] and the beyond-mean-field density functional theory based on nonrelativistic[107,109] and relativistic[130,131] energy functionals.

In the approaches based on nuclear single-reference state, the deformation parameters of candidate nuclei play an important role in the determination of the NMEs. Therefore, an accurate knowledge on the deformation of both nuclei is very important in these approaches. The ground state quadrupole deformations of even–even nuclei in the intrinsic frame can be learnt from the measured electric quadrupole transition strengths between the ground state and the first collective 2^+ state based on the picture of a rigid rotor: $\beta_2 = (4\pi/3ZR^2)\sqrt{B(E2: 0_1^+ \to 2_1^+)}$, with $R = 1.2A^{1/3}$. This formula is valid for well-deformed nuclei, but might not be applied for weakly deformed and transitional nuclei, where the quadrupole deformations of nuclear ground states rely solely on model calculations.

Figure 2(a) shows the uncertainties in the predicted quadrupole deformation parameters for the ground states of the $0\nu\beta\beta$-transition pairs from different self-consistent mean-field calculations. It is seen that these three density functional calculations predicted similar quadrupole deformation parameters for all the $0\nu\beta\beta$-transition pairs, except for ^{96}Zr–Mo, ^{116}Cd-Sn and ^{124}Sn–Te. Since the deformation

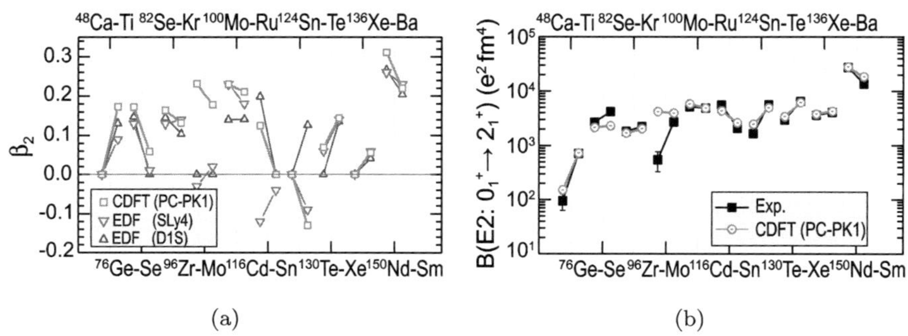

Fig. 2. (Color online) (a) The quadrupole deformation parameter β_2 for the ground states of the $0\nu\beta\beta$-decay candidate nuclei given by the mean-field calculations based on the relativistic PC-PK1, and the nonrelativistic Skyrme SLy4 and Gogny D1S functionals. The results by the Skyrme SLy4 and Gogny D1S functionals are taken from Refs. 148 and 149, respectively. (b) The experimental $E2$ transition strengths $B(E2: 0_1^+ \to 2_1^+)$ in these nuclei, in comparison with the GCM calculations based on the PC-PK1 functional.

parameters cannot be measured directly, the justification of these three calculations is not clear. However, one can compare the predicted $B(E2 : 0_1^+ \to 2_1^+)$ value from the beyond-mean-field calculation to the data. Figure 2(b) displays the measured $B(E2 : 0_1^+ \to 2_1^+)$ in these nuclei in comparison with the GCM calculations[127] based on the PC-PK1 functional. It is shown that all the $B(E2 : 0_1^+ \to 2_1^+)$ values are reproduced rather well by PC-PK1, except for ^{96}Zr. The overestimation of the collectivity in ^{96}Zr is a common problem of most density functional-based GCM or collective Hamiltonian calculations.[149,150] The lowest excited states in ^{96}Zr are of particle–hole type, the description of which requires the inclusion of noncollective configurations. The inclusion of these noncollective configurations is expected to decrease the $M^{0\nu}$ of ^{96}Zr.

To clarify the deformation-dependence of the NMEs, Fig. 3 shows the normalized NME $\tilde{M}^{0\nu}$ as a function of the intrinsic quadrupole deformations of the mother and daughter nuclei. The normalized NME $\tilde{M}^{0\nu}$ is concentrated rather symmetrically along the diagonal line $\beta_2^I = \beta_2^F$, implying that the decay between nuclei with different deformations is strongly hindered. In other words, when the deformations of parent and daughter nuclei are similar the NMEs have a maximum, and when the absolute value of the difference in deformations increases the NMEs are strongly reduced. Moreover, the $\tilde{M}^{0\nu}$ has the largest value at the spherical configuration for most candidate nuclei except for ^{48}Ca–Ti, ^{96}Zr–Mo and ^{136}Xe–Ba. It implies that generally the $0\nu\beta\beta$ decay is favored if both nuclei are spherical. However, this

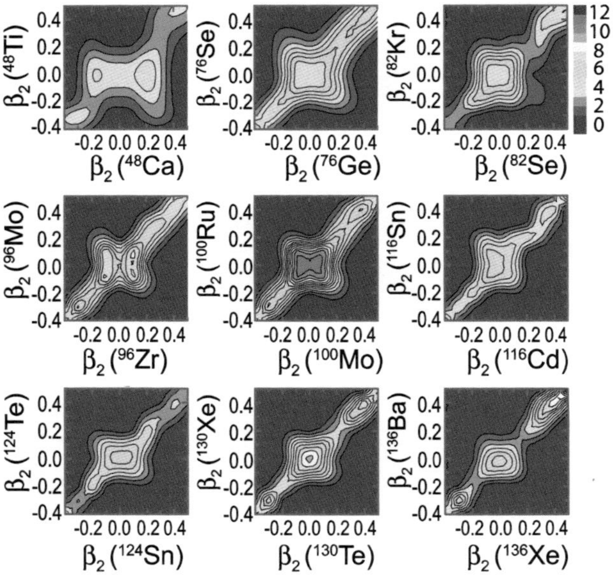

Fig. 3. (Color online) Normalized NME $\tilde{M}^{0\nu}$ as a function of the intrinsic deformation parameters β_2 of the initial ^AZ and final $^A(Z+2)$ nuclei.

Source: Taken from Ref. 131.

conclusion might be artificial because of the missing of higher-seniority components in the spherical mean-field state. A comparison study of the results by the density functional method and the shell model[110] has shown that the spherical mean-field wave function gives a similar result as that by the shell model calculation restricted to the model space with only zero seniority. The extension of the model space with higher seniority in the shell model continually decreases the $M^{0\nu}$. It indicates that the higher-seniority components are essential to be included in the future density functional studies.

The total $M^{0\nu}$ (15) by the beyond-mean-field calculation is a superposition of the nonnormalized transition matrix elements with various deformations multiplied by specific weights. It is evident that configuration mixing occurs and that the regions of maximal overlap between the three quantities $f^*_{0^+_F}(q_F)$, $f_{0^+_I}(q_I)$ and $M^{0\nu}(q_I, q_F)$ contribute mostly to the total matrix element $M^{0\nu}$. A similar deformation of the initial and final states is favored by the decay process. A large overlap between the collective wave functions of the initial and final states is important. In Fig. 4 we show the probabilities $|g^{J=0}_\alpha(\beta_2)|^2$ in the ground states of the candidate nuclei as a function of the deformation. It is clearly seen, for instance, that these probability distributions are peaked at $\beta_2 \simeq 0.3$ for the nucleus ^{150}Nd and at $\beta_2 \simeq 0.2$ for the nucleus ^{150}Sm. However, the distributions show a relatively large width and therefore there is an overlapping region of considerable size in between. It is clear

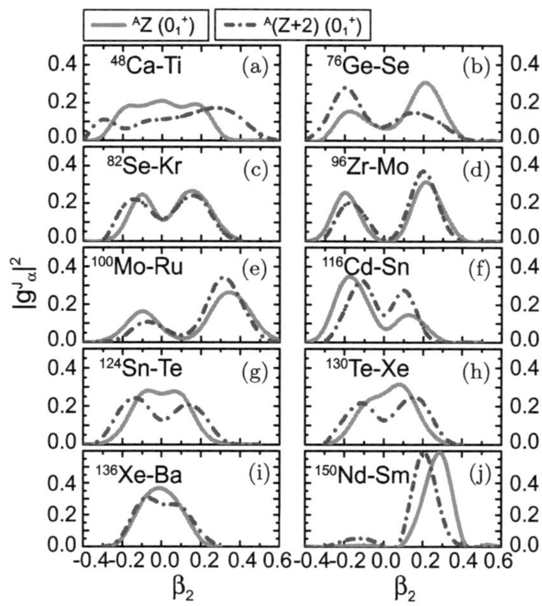

Fig. 4. (Color online) Distribution of collective wave functions $|g^J_\alpha(\beta_2)|^2$ as a function of deformation parameters β_2 for the ground states of initial ^AZ and final $^A(Z+2)$ nuclei in the $0\nu\beta\beta$ decay.

Source: Taken from Ref. 131.

that deformation fluctuation plays an essential role in the description of the decay matrix element.

4.4. *Octupole deformation correlations*

In this subsection, we discuss the influence of octupole deformation on the NME of the $0\nu\beta\beta$ decay of ^{150}Nd $\to$ ^{150}Sm. Strong octupole correlations have been found in their ground states.

Figure 5(a) displays the off-diagonal elements of $\tilde{M}^{0\nu}(q_I, q_F)$ in the β_3^I–β_3^F plane cut at $\beta_2^I = \beta_2^F = 0.2$ where the dominate configuration of the ground state is located. Figure 5(b) shows that the $\tilde{M}^{0\nu}$ with $\beta_3^I = \beta_3^F \equiv \beta_3$ and $\beta_2^I = \beta_2^F = 0.2$ is decreasing smoothly from 6.35 to 2.17 as β_3 increases from 0 to 0.3. The calculated $\tilde{M}^{0\nu}$ is only 1.38 by the configurations with deformation parameters $q_I = (0.285, 0.113)$ and $q_F = (0.193, 0.145)$ chosen according to the data of $B(E2 : 0_1^+ \to 2_1^+)$ and $B(E3 : 0_1^+ \to 3_1^-)$.

The GCM calculation with quadrupole–octupole shape mixing gives the $M^{0\nu}(0_1^+ \to 0_1^+) = 5.2$, which is just smaller by 7% than the previous result 5.6 by mixing only axially symmetric configurations.[130] It means that the shape-mixing effect significantly moderates the quenching effect on $M^{0\nu}$. Moreover, the transition matrix element $M^{0\nu}(0_1^+ \to 0_2^+)$ turns out to be 0.72.

4.5. *Uncertainties in the nuclear matrix elements*

The predicted NMEs by various nuclear structure models (CISM,[67] (renormalized) QRPA,[151] PHFB,[101] EDF,[107] CDFT[130,131] and the IBM[104]) deviate from each other by a factor of 2–3, as shown in Fig. 6. There are also other calculations that are

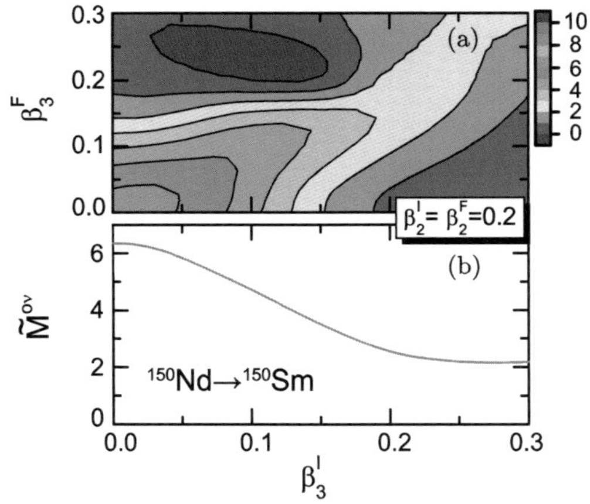

Fig. 5. (Color online) Normalized NMEs $\tilde{M}^{0\nu}(q_I, q_F)$ for the $0\nu\beta\beta$ decay of ^{150}Nd, where $q \equiv (\beta_2, \beta_3)$. (a) $\tilde{M}^{0\nu}$ as a function of the octupole deformation parameters of the initial and final nuclei, with the quadruple deformation parameters β_2^I and β_2^F fixed at 0.2. (b) The same quantity with the restriction $\beta_3^I = \beta_3^F$.

Source: Taken from Ref. 133.

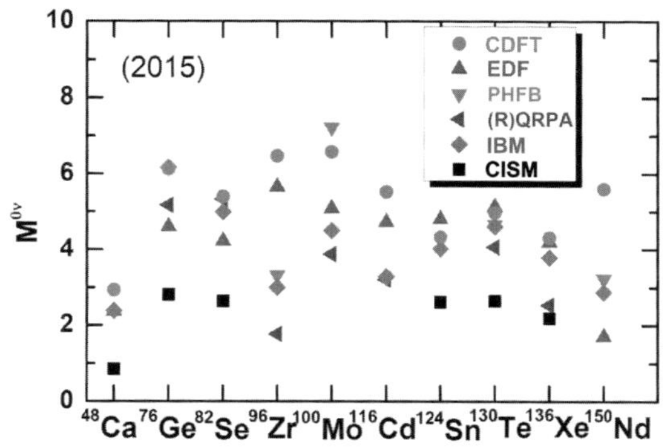

Fig. 6. (Color online) Comparison of the NMEs of $0\nu\beta\beta$ decay from different model calculations.

not included here for comparison. It is shown that the shell-model and the density functional calculations provide the lower and upper boundaries, respectively. The relativistic (CDFT) and the nonrelativistic (EDF) density functional calculations give basically similar results for the NMEs, except for the case of ^{150}Nd.

The large systematic uncertainty in the predicted NMEs is mainly originated from the different ways of modeling the wave functions.[152] Since one cannot carry out a full configuration-interacting calculation by diagonalizing a many-body Hamiltonian in the full model space for (medium) heavy nuclei with the present computer resource, various kinds of approximations are employed, which usually fall into two categories.

- In the shell-model calculation, instead of using the full single-particle basis, one diagonalizes an effective many-body Hamiltonian in a model space composed of several valence single-particle states (usually within one shell). The NME of $0\nu\beta\beta$ decay is calculated with the two-body transition densities defined in this model space. The advantage of the shell model is that all possible correlations within the space are included. However, the configuration space is significantly smaller than other models.
- In contrast with the shell model, the mean-field-based models are formulated in a full single-particle basis, but based on the mean-field approximation. Even though some static correlations can be taken into account by breaking symmetries in the mean fields, other correlations, such as the dynamical ones, necessitate a beyond-mean-field calculation. Along this direction, both QRPA and symmetry-restored GCM have been implemented to take into account the correlations beyond the mean-field approximation. Within this framework, it turns out that generally the more correlations are taken into account, a smaller NME is obtained. For example, in the GCM calculation, the inclusion of dynamical deformation and neutron–proton pairing significantly quenches the NME. An inclusion of noncollective configurations with nonzero seniority number is expected to reduce the NME further.[110]

The IBM locates roughly in between the shell model and the mean-field-based approaches, where the nuclear wave functions are modeled in terms of different ranked bosons. This method can be regarded as an approximation of the large-space shell model by truncating the model space to states built from pairs of nucleons with $J = 0$ and 2. Therefore, compared with the conventional shell model within one shell, the single-particle basis could be larger, but at the cost of missing some correlations. Therefore, the predicted NMEs are systematically larger than those by the shell model.

Apart from the difference in nuclear models, numerous discussions are also found on the uncertainties stemming from the treatment of intermediate nuclear states and short-range correlations between nucleons in the NME calculation.

(i) Intermediate nuclear states:
The calculations with exact treatment of the energies of the intermediate nucleus are achieved within the QRPA methods. In other calculations, the NMEs of the $0\nu\beta\beta$ decay are usually obtained using the closure approximation, which avoids exact treatment of the intermediate states and simplifies the numerical calculation drastically. The operator expansion method,[153–156] which aims at a step beyond closure approximation, takes account of effects of the intermediate states implicitly. The effects of both the closure approximation and the operator expansion methods have been studied in detail in Ref. 157 and the differences in NMEs were found to be within 10%. It is suggested that a closure calculation with an average energy $\bar{E}$ of a few MeV simulates the exact calculation excellently well.

(ii) Short-range correlations:
In most of previous calculations, the two-nucleon short-range correlations have been treated in a conventional way via a Jastrow-type correlation function in the parametrization of Miller and Spencer.[63,158] A suppression of 15–25% has been found for the NME of $0\nu\beta\beta$ decay.[67,73,104,159] Alternatively, the two-nucleon short-range correlations have been implemented using the unitary correlation operator method,[160] which leads to a much smoother correlation, of the order of 5% compared to the 15–25% of the Jastrow-type correlator.[67] Recently, the short-range correlations have also been studied within the coupled clusters method.[161] Differences between the results calculated with Miller–Spencer Jastrow-type and coupled-cluster short-range correlations are in the range of about 20%–30%,[73,106,159] and the results with Miller–Spencer treatment are considered to overestimate the quenching effect due to short-range correlations. The best way to treat the short-range correlations remains an open question today, leading to an uncertainty in the NME.

5. Conclusion

Since the discovery of neutrino oscillations, the search for the $0\nu\beta\beta$ decay has attracted lots of renewed interests in nuclear and particle physics. The worldwide set of double-beta-decay measurements running, under construction and planned for the near future represents "current-generation experiments," approaching the half-lives of $0\nu\beta\beta$ decay of about 10^{26} years. In particular, it is expected that these will lead to one or more "next-generation" experiments which should have sufficient sensitivity to, with high confidence level, resolve the issue of Majorana versus. Dirac nature of neutrinos for the so-called "inverted hierarchy" of neutrino mass values. However, the large uncertainties coming from the NMEs will translate into the effective neutrino mass and thus complicate the situation. Therefore, lots of theoretical efforts are being devoted into reducing these uncertainties to be as small as possible.

In this paper, we have briefly overviewed the current status of experimental measurements and the computation of the NMEs for $0\nu\beta\beta$ decay. In particular, we have discussed in detail the first relativistic description of the NMEs within the framework of the beyond-mean-field CDFT based on a relativistic point-coupling functional PC-PK1, where the dynamic correlations related to the restoration of broken symmetries and to the fluctuations of collective coordinates are incorporated in the nuclear wave functions. The NMEs have been calculated with the full relativistic transition operators derived from one-body charge-changing nuclear current. The reliability of the nuclear wave functions has been examined by comparing the calculated low-energy structural properties with the corresponding data.

We point out that there are still large systematic uncertainties in the NMEs by different models. On one hand, more correlations are to be taken into account in the density functional calculations. The important ones include the neutron–proton isoscalar pairing fluctuations and the mixing of the noncollective configurations with high seniority. On the other hand, the effect outside the shell-model space has to be studied. One of the promising ways to this end is by introducing the techniques of in-medium similarity renormalization group.[162] Hopefully, the gap between the density functional-predicted and shell-model predicted NMEs will be reduced to an acceptable value after the efforts from both sides. Moreover, the contribution from the exchange of heavy neutrino to the decay process and the quenching mechanism for the coupling constant g_A needs to be investigated further in the near future.

Acknowledgments

This work is dedicated to the memory of Prof. Gerry Brown who had made important contributions not only in nuclear physics but also in promoting nuclear physics research worldwide. In the 1980s, many Chinese visitors from the Chinese Academy of Sciences, China Institute of Atomic Energy, Fudan University, Jilin University and other institutes were hosted by Gerry and were doing research at his Stony Brook Nuclear Theory Group. Gerry had contributed a great deal in promoting nuclear physics in China as well as in bringing up generations of theoretical nuclear physicists there. He is well remembered and appreciated in China.

We thank J. Engel, K. Hagino, Z. P. Li, P. Ring and E. F. Zhou for discussions and contributions to the work reviewed in this paper.

This work was supported in part by the Major State 973 Program of China (Grant No. 2013CB834400), the National Natural Science Foundation of China (Grant Nos. 11175002, 11335002, 11461141002, 11575148, 11475140 and 11305134) and the Scientific Discovery through Advanced Computing (SciDAC) Program funded by US Department of Energy, Office of Science, Advanced Scientific Computing Research and Nuclear Physics, under the Contract No. DE-SC0008641, ER41896.

References

1. J. Chadwick, *Verh. Dtsch. Phys. Ges.* **16** (1914) 383.
2. C. Jensen, *Controversy and Consensus: Nuclear Beta Decay* 1911–1934, 1st edn., Vol. 24 (Birkhäuser, Basel, 2000).
3. L. M. Brown, *Phys. Today* **31** (1978) 23.
4. E. Fermi, *Nuovo Cimento* **11** (1934) 1.
5. E. Fermi, *Z. Phys.* **88** (1934) 161.
6. C. L. Cowan, Jr., F. Reines, F. B. Harrison, H. W. Kruse and A. D. McGuire, *Science* **124** (1956) 103.
7. M. Goeppert-Mayer, *Phys. Rev.* **48** (1935) 512.
8. E. Majorana, *Nuovo Cimento* **14** (1937) 171.
9. W. H. Furry, *Phys. Rev.* **56** (1939) 1184.
10. E. L. Fireman, *Phys. Rev.* **75** (1949) 323.
11. M. G. Inghram and J. H. Reynolds, *Phys. Rev.* **78** (1950) 822.
12. H. Primakoff, *Phys. Rev.* **85** (1952) 888.
13. T. D. Lee and C. N. Yang, *Phys. Rev.* **104** (1956) 254.
14. C. S. Wu, E. Ambler, R. W. Hayward, D. D. Hoppes and R. P. Hudson, *Phys. Rev.* **105** (1957) 1413.
15. A. Salam, *Nuovo Cimento* **5** (1957) 299.
16. L. Landau, *Nucl. Phys.* **3** (1957) 127.
17. T. D. Lee and C. N. Yang, *Phys. Rev.* **105** (1957) 1671.
18. R. P. Feynman and M. Gell-Mann, *Phys. Rev.* **109** (1958) 193.
19. E. C. G. Sudarshan and R. E. Marshak, *Phys. Rev.* **109** (1958) 1860.
20. S. L. Glashow, *Nucl. Phys.* **22** (1961) 579.
21. S. Weinberg, *Phys. Rev. Lett.* **19** (1967) 1264.
22. A. Salam, in *Proc. Eighth Nobel Symp.: Elementory Particle Physics: Groups and Analyticity*, ed. N. Svartholm (Almquvist and Wiksell, Stockholm, 1968), p. 367.
23. S. Bilenky, Introduction to the Physics of Massive and Mixed Neutrions, 1st edn., Lecture Notes in Physics, Vol. 817 (Springer-Verlag, Berlin, 2010).
24. A. Strumia, in *Proc. 84th Session Summer School on Theoretical Phyiscs: Particle Physics Beyond the Standard Model*, Les Houches, France, 1–26 August, 2005, eds. D. Kazakov, S. Lavignac and J. Dalibard (Elsevier Science, Amsterdom, 2006), pp. 655–680.
25. Y. Fukuda *et al.*, *Phys. Rev. Lett.* **81** (1998) 1562.
26. Q. R. Ahmad *et al.*, *Phys. Rev. Lett.* **87** (2001) 071301.
27. D. G. Michael *et al.*, *Phys. Rev. Lett.* **97** (2006) 191801.
28. M. H. Ahn *et al. Phys. Rev. D* **74** (2006) 072003.
29. K. Abe *et al.*, *Phys. Rev. Lett.* **107** (2011) 041801.

30. S. Abe *et al.*, *Phys. Rev. Lett.* **100** (2008) 221803.
31. Y. Abe *et al.*, *Phys. Rev. Lett.* **108** (2012) 131801.
32. F. P. An *et al.*, *Phys. Rev. Lett.* **108** (2012) 171803.
33. J. K. Ahn *et al.*, *Phys. Rev. Lett.* **108** (2012) 191802.
34. F. Englert and R. Brout, *Phys. Rev. Lett.* **13** (1964) 321.
35. P. W. Higgs, *Phys. Rev. Lett.* **13** (1964) 508.
36. G. S. Guralnik, C. R. Hagen and T. W. Kibble, *Phys. Rev. Lett.* **13** (1964) 585.
37. H. Päs and W. Rodejohann, *New J. Phys.* **17** (2015) 115010.
38. S. R. Elliott, A. A. Hahn and M. K. Moe, *Phys. Rev. Lett.* **59** (1987) 2020.
39. A. Barabash, *Nucl. Phys. A* **935** (2015) 52.
40. O. Cremonesi and M. Pavan, *Adv. High Energy Phys.* **2014** (2014) 951432.
41. B. Schwingenheuer, *Ann. Phys.* **525** (2013) 269.
42. A. Barabash, *Phys. Proc.* **74** (2015) 416.
43. W. Maneschg, *Nucl. Part. Phys. Proc.* **260** (2015) 188.
44. J. Kotila and F. Iachello, *Phys. Rev. C* **85** (2012) 034316.
45. S. Umehara *et al.*, *Phys. Rev. C* **78** (2008) 058501.
46. M. Agostini *et al.*, *Phys. Rev. Lett.* **111** (2013) 122503.
47. A. S. Barabash, V. B. Brudanin and NEMO Collab., *Phys. At. Nucl.* **74** (2011) 312.
48. J. Argyriades *et al.*, *Nucl. Phys. A* **847** (2010) 168.
49. R. Arnold *et al.*, *Phys. Rev. D* **89** (2014) 111101.
50. F. A. Danevich *et al.*, *Phys. Rev. C* **68** (2003) 035501.
51. M. J. Hwang *et al.*, *Astropart. Phys.* **31** (2009) 412.
52. A. S. Barabash, *Phys. At. Nucl.* **74** (2011) 603.
53. E. Andreotti *et al.*, *Astropart. Phys.* **34** (2011) 822.
54. A. Gando *et al.* (KamLAND-Zen Collaboration), *Phys. Rev. Lett.* **117** (2016) 082503.
55. R. Arnold *et al.* (NEMO-3 Collaboration), *Phys. Rev. D* **94**, 072003 (2016).
56. H. V. Klapdor-Kleingrothaus *et al.*, *Eur. Phys. J. A* **12** (2001) 147.
57. H. V. Klapdor-Kleingrothaus, I. V. Krivosheina, A. Dietz and O. Chkvorets, *Phys. Lett. B* **586** (2004) 198.
58. S. Hannestad, A. Mirizzi, G. G. Raffelt and Y. Y. Wong, *J. Cosmol. Astropart. Phys.* **2010** (2010) 001.
59. E. Komatsu *et al.*, *Astrophys. J. Suppl.* **192** (2011) 18.
60. M. Auger *et al.*, *Phys. Rev. Lett.* **109** (2012) 032505.
61. C. E. Aalseth *et al.*, *Phys. Rev. D* **65** (2002) 092007.
62. W. Haxton and G. Stephenson, *Prog. Part. Nucl. Phys.* **12** (1984) 409.
63. H. Wu, H. Song, T. Kuo, W. Cheng and D. Strottman, *Phys. Lett. B* **162** (1985) 227.
64. J. Retamosa, E. Caurier and F. Nowacki, *Phys. Rev. C* **51** (1995) 371.
65. E. Caurier, F. Nowacki, A. Poves and J. Retamosa, *Phys. Rev. Lett.* **77** (1996) 1954.
66. E. Caurier, F. Nowacki and A. Poves, *Eur. Phys. J. A* **36** (2008) 195.
67. J. Menéndez, A. Poves, E. Caurier and F. Nowacki, *Nucl. Phys. A* **818** (2009) 139.
68. J. Menéndez, A. Poves, E. Caurier and F. Nowacki, *Phys. Rev. C* **80** (2009) 048501.
69. J. Menéndez, D. Gazit and A. Schwenk, *Phys. Rev. Lett.* **107** (2011) 062501.
70. R. A. Sen'kov and M. Horoi, *Phys. Rev. C* **88** (2013) 064312.
71. R. A. Sen'kov, M. Horoi and B. A. Brown, *Phys. Rev. C* **89** (2014) 054304.
72. R. A. Sen'kov and M. Horoi, *Phys. Rev. C* **90** (2014) 051301 (R).
73. M. Horoi and S. Stoica *Phys. Rev. C* **81** (2010) 024321.
74. M. Horoi, *Phys. Rev. C* **87** (2013) 014320.
75. A. Neacsu and M. Horoi, *Phys. Rev. C* **91** (2015) 024309.
76. M. Horoi and A. Neacsu, *Phys. Rev. C* **93** (2016) 024308.

77. Y. Iwata, N. Shimizu, T. Otsuka, Y. Utsuno, J. Menéndez, M. Honma and T. Abe, *Phys. Rev. Lett.* **116** (2016) 112502.
78. T. Tomoda and A. Faessler, *Phys. Lett. B* **199** (1987) 475.
79. K. Muto, E. Bender and H. Klapdor, *Z. Phys. A* **334** (1989) 187.
80. A. Staudt, T. Kuo and H. Klapdor-Kleingrothaus, *Phys. Lett. B* **242** (1990) 17.
81. D. Stout and T. Kuo, *Phys. Rev. Lett.* **69** (1992) 1900.
82. A. Staudt, T. Kuo and H. Klapdor-Kleingrothaus, *Phys. Rev. C* **46** (1992) 871.
83. G. Pantis, A. Faessler, W. Kaminski and J. Vergados, *J. Phys. G, Nucl. Part. Phys.* **18** (1992) 605.
84. M. Kortelainen and J. Suhonen, *Phys. Rev. C* **76** (2007) 024315.
85. M. Kortelainen and J. Suhonen, *Phys. Rev. C* **75** (2007) 051303.
86. G. Pantis, F. Šimkovic, J. Vergados and A. Faessler, *Phys. Rev. C* **53** (1996) 695.
87. F. Šimkovic, J. Schwieger, M. Veselský, G. Pantis and A. Faessler, *Phys. Lett. B* **393** (1997) 267.
88. F. Šimkovic, G. Pantis, J. Vergados and A. Faessler, *Phys. Rev. C* **60** (1999) 055502.
89. V. Rodin, A. Faessler, F. Šimkovic and P. Vogel, *Nucl. Phys. A* **766** (2006) 107.
90. F. Šimkovic, A. Faessler, V. Rodin, P. Vogel and J. Engel, *Phys. Rev. C* **77** (2008) 045503.
91. F. Šimkovic, A. Faessler and P. Vogel, *Phys. Rev. C* **79** (2009) 015502.
92. D.-L. Fang, A. Faessler, V. Rodin and F. Šimkovic, *Phys. Rev. C* **82** (2010) 051301.
93. D.-L. Fang, A. Faessler, V. Rodin and F. Šimkovic, *Phys. Rev. C* **83** (2011) 034320.
94. M. T. Mustonen and J. Engel, *Phys. Rev. C* **87** (2013) 064302.
95. J. Terasaki, *Phys. Rev. C* **91** (2015) 034318.
96. F. Šimkovic, V. Rodin, A. Faessler and P. Vogel, *Phys. Rev. C* **87** (2013) 045501.
97. D.-L. Fang, A. Faessler and F. Šimkovic, *Phys. Rev. C* **92** (2015) 044301.
98. K. Chaturvedi, R. Chandra, P. K. Rath, P. K. Raina and J. G. Hirsch, *Phys. Rev. C* **78** (2008) 054302.
99. R. Chandra, K. Chaturvedi, P. Rath, P. Raina and J. Hirsch, *Europhys. Lett.* **86** (2009) 32001.
100. P. K. Rath, R. Chandra, K. Chaturvedi, P. K. Raina and J. G. Hirsch, *Phys. Rev. C* **80** (2009) 044303.
101. P. K. Rath, R. Chandra, K. Chaturvedi, P. K. Raina and J. G. Hirsch, *Phys. Rev. C* **82** (2010) 064310.
102. P. K. Rath, R. Chandra, P. K. Raina, K. Chaturvedi and J. G. Hirsch, *Phys. Rev. C* **85** (2012) 014308.
103. P. K. Rath, R. Chandra, K. Chaturvedi, P. Lohani, P. K. Raina and J. G. Hirsch, *Phys. Rev. C* **88** (2013) 064322.
104. J. Barea and F. Iachello, *Phys. Rev. C* **79** (2009) 044301.
105. J. Barea, J. Kotila and F. Iachello, *Phys. Rev. Lett.* **109** (2012) 042501.
106. J. Barea, J. Kotila and F. Iachello, *Phys. Rev. C* **87** (2013) 014315.
107. T. R. Rodríguez and G. Martínez-Pinedo, *Phys. Rev. Lett.* **105** (2010) 252503.
108. T. R. Rodríguez and G. Martínez-Pinedo, *Prog. Part. Nucl. Phys.* **66** (2011) 436.
109. N. L. Vaquero, T. R. Rodríguez and J. L. Egido, *Phys. Rev. Lett.* **111** (2013) 142501.
110. J. Menéndez, T. R. Rodríguez, G. Martínez-Pinedo and A. Poves, *Phys. Rev. C* **90** (2014) 024311.
111. J. Menéndez, N. Hinohara, J. Engel, G. Martínez-Pinedo and T. R. Rodríguez, *Phys. Rev. C* **93** (2016) 014305.
112. B. D. Serot and J. D. Walecka, *Adv. Nucl. Phys.* **16** (1986) 1.
113. P.-G. Reinhard, *Rep. Prog. Phys.* **52** (1989) 439.
114. P. Ring, *Prog. Part. Nucl. Phys.* **37** (1996) 193.

115. D. Vretenar, A. Afanasjev, G. Lalazissis and P. Ring, *Phys. Rep.* **409** (2005) 101.
116. J. Meng, H. Toki, S. Zhou, S. Zhang, W. Long and L. Geng, *Prog. Part. Nucl. Phys.* **57** (2006) 470.
117. P. Ring, Z.-Y. Ma, N. Van Giai, D. Vretenar, A. Wandelt and L.-G. Cao, *Nucl. Phys. A* **694** (2001) 249.
118. H. Liang, N. V. Giai and J. Meng, *Phys. Rev. Lett.* **101** (2008) 122502.
119. N. Paar, D. Vretenar, E. Khan and G. Colò, *Rep. Prog. Phys.* **70** (2007) 691.
120. N. Paar, Y. F. Niu, D. Vretenar and J. Meng, *Phys. Rev. Lett.* **103** (2009) 032502.
121. Y. Niu, N. Paar, D. Vretenar and J. Meng, *Phys. Lett. B* **681** (2009) 315.
122. Z. Niu, Y. Niu, H. Liang, W. Long, T. Nikšić, D. Vretenar and J. Meng, *Phys. Lett. B* **723** (2013) 172.
123. T. Nikšić, D. Vretenar and P. Ring, *Phys. Rev. C* **73** (2006) 034308.
124. T. Nikšić, D. Vretenar and P. Ring, *Phys. Rev. C* **74** (2006) 064309.
125. J. M. Yao, J. Meng, D. P. Arteaga and P. Ring, *Chin. Phys. Lett.* **25** (2008) 3609.
126. J. M. Yao, J. Meng, P. Ring and D. P. Arteaga, *Phys. Rev. C* **79** (2009) 044312.
127. J. M. Yao, J. Meng, P. Ring and D. Vretenar, *Phys. Rev. C* **81** (2010) 044311.
128. J. M. Yao, H. Mei, H. Chen, J. Meng, P. Ring and D. Vretenar, *Phys. Rev. C* **83** (2011) 014308.
129. J. M. Yao, K. Hagino, Z. P. Li, J. Meng and P. Ring, *Phys. Rev. C* **89** (2014) 054306.
130. L. S. Song, J. M. Yao, P. Ring and J. Meng, *Phys. Rev. C* **90** (2014) 054309.
131. J. M. Yao, L. S. Song, K. Hagino, P. Ring and J. Meng, *Phys. Rev. C* **91** (2015) 024316.
132. P. Ring, J. M. Yao, L. S. Song, K. Hagino and J. Meng, *AIP Conf. Proc.* **1681** (2015) 050008.
133. J. M. Yao and J. Engel, *Phys. Rev. C* **94** (2016) 014306.
134. P. W. Zhao, Z. P. Li, J. M. Yao and J. Meng, *Phys. Rev. C* **82** (2010) 054319.
135. Y. Gambhir, P. Ring and A. Thimet, *Ann. Phys.* **198** (1990) 132.
136. Y. Tian, Z. Ma and P. Ring, *Phys. Lett. B* **676** (2009) 44.
137. P. Ring and P. Schuck, *The Nuclear Many-Body Problem* (Springer-Verlag, Berlin, 1980).
138. J. M. Yao, E. F. Zhou and Z. P. Li, *Phys. Rev. C* **92** (2015) 041304.
139. J. L. Egido, *Phys. Scr.* **91** (2016) 073003.
140. E. Zhou, J. Yao, Z. Li, J. Meng and P. Ring, *Phys. Lett. B* **753** (2016) 227.
141. M. Mirea, T. Pahomi and S. Stoica, arXiv:1411.5506V3 [nucl-th].
142. N. Onishi and S. Yoshida, *Nucl. Phys.* **80** (1966) 367.
143. R. Balian and E. Brezin, *Nuovo Cimento* **64** (1969) 37.
144. I. S. Towner and J. C. Hardy, in *Symmetries and Fundamental Interactions in Nuclei* eds. W. C. Haxton and E. M. Henley (World Scientific, 1995), pp. 183–249.
145. O. Dumbrajs, R. Koch, H. Pilkuhn, G. Oades, H. Behrens, J. de Swart and P. Kroll, *Nucl. Phys. B* **216** (1983) 277.
146. R. Álvarez-Rodríguez, P. Sarriguren, E. M. de Guerra, L. Pacearescu, A. Faessler and F. Šimkovic, *Phys. Rev. C* **70** (2004) 064309.
147. J. G. Hirsch, O. Castanos and P. Hess, *Nucl. Phys. A* **582** (1995) 124.
148. M. Bender, G. F. Bertsch and P.-H. Heenen, *Phys. Rev. C* **73** (2006) 034322.
149. J. P. Delaroche, M. Girod, J. Libert, H. Goutte, S. Hilaire, S. Péru, N. Pillet and G. F. Bertsch, *Phys. Rev. C* **81** (2010) 014303.
150. H. Mei, J. Xiang, J. M. Yao, Z. P. Li and J. Meng, *Phys. Rev. C* **85** (2012) 034321.
151. A. Faessler, V. Rodin and F. Šimkovic, *J. Phys. G, Nucl. Part. Phys.* **39** (2012) 124006.
152. J. Engel, *J. Phys. G, Nucl. Part. Phys.* **42** (2015) 034017.

153. C.-R. Ching and T.-H. Ho, *Commun. Theor. Phys.* **10** (1988) 45.
154. C.-R. Ching and T.-H. Ho, *Commun. Theor. Phys.* **11** (1989) 433.
155. C.-R. Ching, T.-H. Ho and X.-R. Wu, *Commun. Theor. Phys.* **12** (1989) 167.
156. X. Wu, A. Staudt, T. Kuo and H. Klapdor-Kleingrothaus, *Phys. Lett. B* **276** (1992) 274.
157. K. Muto, *Nucl. Phys. A* **577** (1994) 415c.
158. G. A. Miller and J. E. Spencer, *Ann. Phys.* **100** (1976) 562.
159. F. Šimkovic, A. Faessler, H. Müther, V. Rodin and M. Stauf, *Phys. Rev. C* **79** (2009) 055501.
160. H. Feldmeier, T. Neff, R. Roth and J. Schnack, *Nucl. Phys. A* **632** (1998) 61.
161. C. Giusti, H. Müther, F. D. Pacati and M. Stauf, *Phys. Rev. C* **60** (1999) 054608.
162. H. Hergert, S. K. Bogner, T. Morris, A. Schwenk and K. Tsukiyama, *Phys. Rep.* **621** (2016) 165.
163. J. Meng (ed.), *Relativistic Density Functional for Nuclear Structure* (International Review of Nuclear Physics, Vol. 10) (World Scientific, 2016).

Partial conservation law in a schematic single j shell model

Wesley Pereira, Ricardo Garcia, Larry Zamick* and Alberto Escuderos
Department of Physics and Astronomy,
Rutgers University, Piscataway, New Jersey 08854, USA
**lzamick@gmail.com; lzamick@physics.rutgers.edu*

Kai Neergård

Fjordtoften 17, 4700 Næstved, Denmark

We report the discovery of a partial conservation law obeyed by a schematic Hamiltonian of two protons and two neutrons in a j shell. In our Hamiltonian, the interaction matrix element of two nucleons with combined angular momentum J is linear in J for even J and constant for odd J. It turns out that in some stationary states, the sum of the angular momenta J_p and J_n of the proton and neutron pairs is conserved. The energies of these states are given by a linear function of $J_p + J_n$. The systematics of their occurrence is described and explained.

Keywords: Schematic interactions; shell model.

1. Introduction

Among the many contributions of Gerry Brown to Nuclear Physics, one of the first that comes to the minds of many is his development with Tom Kuo of realistic nuclear matrix elements.[1] These involve the very complicated nucleon–nucleon interaction and the added complication of handling the hard core by obtaining a G matrix which a researcher could easily handle. However, our present work is inspired by another aspect of Gerry Brown's contributions — his use of simple schematic models to bring out the physics of the more complex calculations. One example is his early article with Marc Bolsterli in Physical Review Letter on dipole states in nuclei.[2] Their simple model employs a delta interaction with radial integrals set to a constant. One state gets elevated to a high energy and contains all the dipole strength. Gerry and Marc compared their results with a more detailed calculation of Elliott and Flowers.[3] These authors obtained two collective states, and Gerry and Marc noted that a defect of their model was the neglect of the spin–orbit interaction. However, they expected that it could work better for heavier nuclei. A quote from the end of their paper: "The schematic model is of course no substitute for detailed calculations but indicates the possibility of these coherent features in a simple way."

*Corresponding author.

In Gerry's first book *Unified Theory of Nuclear Models*[4] he discusses besides more elaborate schemes of calculation such schematic models as Elliott's SU(3) model to describe nuclear rotation[5] and Racah's seniority scheme displaying the physics of pairing in nuclei.[6]

Below, we consider a simple model with only one j shell, where we put both protons and neutrons. Such a model was applied in the early days to the description of nuclear spectra, magnetic moments, beta delay, etc. in the $1f_{7/2}$ shell.[7–11] The interaction matrix elements were taken from the spectra of ^{42}Ca and ^{42}Sc. The ^{42}Sc, $T = 0$ spectrum was poorly known at that time and some of the assignments were wrong. Revised matrix elements were later extracted from the correct ^{42}Sc spectrum by Zamick and Robinson,[12] and these matrix elements were employed by Escuderos *et al.* in complete calculations for the $1f_{7/2}$ shell.[13] Despite large differences between the original and revised matrix elements, especially a lowering of those for two nucleon angular momentum $J = 1$, 3 and 5 by about half an MeV, no red flags were raised. This indicates a certain insensitivity to the $T = 0$ matrix, a theme that will pervade this work.

In our present investigation j is arbitrary, and we adopt a schematic interaction. The nuclei considered are such which have two protons and two neutrons in the given shell. It is well known that such a model also applies to the case of two proton holes and two neutron holes. Our choice of schematic interaction is motivated by the gross structure of the matrix elements of Ref. 12, which are displayed in Fig. 1. Shown there are the interaction matrix elements $E^J = \langle (jj)J|v|(jj)J\rangle$, where $j = 7/2$. It is seen that while the even J matrix element rises steeply with J, the odd J matrix element varies much less and its average slope as a function of J is approximately zero. This suggests to approximate the even J matrix elements by a function linear in J and the odd J matrix elements by a constant c. The only effect of this constant

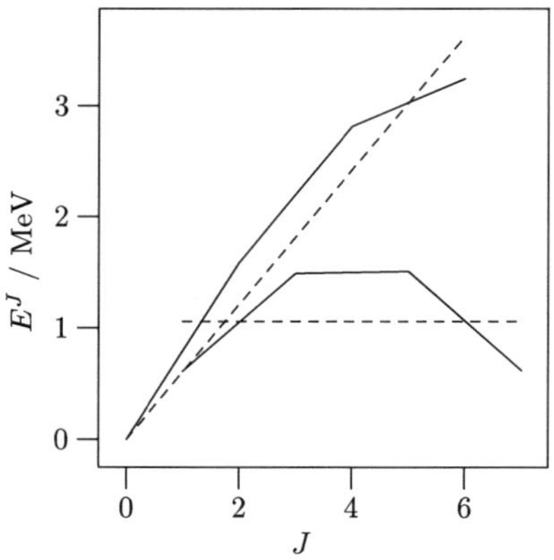

Fig. 1. Empirical interaction of two nucleons in the $1f_{7/2}$ shell derived from the spectrum of ^{42}Sc. The matrix elements E^J are connected by broken lines separately for even and odd J. The dashed lines suggest an approximation of the even J matrix elements by a function linear in J and the odd J matrix elements by a constant.

is to add $(3-\frac{1}{2}T(T+1))c$ to all energies, where T is the total isospin. The stationary wave functions are not affected. As we consider only states with $T = 0$, we can therefore choose $c = 0$ just as well. The interaction then depends only on an energy scale factor. Choosing this scale factor in the simplest possible way, we arrive at the following schematic interaction to be studied in the subsequent part of this paper.

$$E^J = \begin{cases} J, & \text{even } J, \\ 0, & \text{odd } J. \end{cases} \tag{1}$$

The next section shows examples of results derived numerically from this interaction. We illustrate, in particular, the occurrence for certain values of j and the total angular momentum I, of stationary states where the sum $J_p + J_n$ of the angular momenta of the proton and neutron pairs is conserved. We also illustrate that these states, which we call *special* states, always have absolute energies (that is, energies before the ground state energy is subtracted to give an excitation energy) equal to $3(J_p + J_n)/2$. To finish the section, we report a systematic search of special states for all $j \leq 15/2$ and give empiric rules for their occurrence. In Sec. 3, we then explain these observations, and the present paper is summarized in Sec. 4.

2. Numeric Results

Figure 2 shows the even I yrast bands calculated for $j = 7/2$ and $9/2$. The top half of each band is seen to be strictly linear. In fact, the absolute energies equal $3I/2$. The wave functions, shown in Table 1 for $j = 9/2$, have a very simple structure.

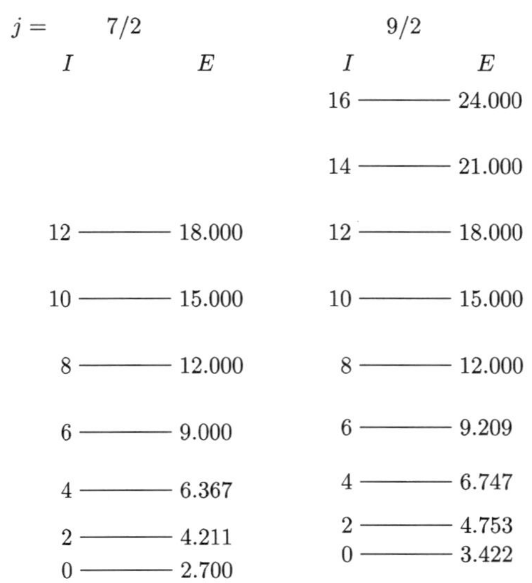

Fig. 2. Calculated even I yrast bands for $j = 7/2$ and $9/2$. The absolute energy E of each level is indicated.

Table 1. Wave functions in the calculated even I yrast band for $j = 9/2$ and $I \geq 8$. Shown are the coefficients of the states $|J_pJ_n\rangle_e$ defined by Eq. (3).

J_p	J_n	$I =$	8	10	12	14	16
4	4		0.595				
6	2		0.700				
6	4		0.000	0.885			
6	6		0.000	0.000	0.745		
8	0		0.395				
8	2		0.000	0.466			
8	4		0.000	0.000	0.667		
8	6		0.000	0.000	0.000	1.000	
8	8		0.000	0.000	0.000	0.000	1.000

As all these states have $T = 0$, which implies that the coefficient of a basic state

$$|J_pJ_n\rangle = |((jj)J_p(jj)J_n)IM\rangle, \tag{2}$$

acquires a sign factor $(-)^I$ when J_p and J_n are interchanged, we show in the table the coefficients of the basic states

$$|J_pJ_n\rangle_e = 2^{-\frac{1+\delta_{J_pJ_n}}{2}}(|J_pJ_n\rangle + (-)^I|J_nJ_p\rangle). \tag{3}$$

All the states listed in Table 1 are seen to have only components with $J_p + J_n = I$. In Eq. (2), the first two angular momenta j are those of the individual protons and the last two those of the neutrons. The total magnetic quantum number M is arbitrary. In Eq. (3), the angular momenta J_p and J_n are even, $J_p \geq J_n$ for even I and $J_p > J_n$ for odd I. The subscript 'e' stands for 'even' to indicate that these states span the space where T is even for the given j, I and M. This is used in Sec. 3.

Several other states are degenerated with these even I yrast states. They are listed in Table 2. All these states have $T = 0$. As this holds for all the states discussed in this paper, we do not mention it any more. Most of the states in Table 2 have odd I. The lowest state for each of $I = 9$, 11 and 13 is an yrast state and degenerate with the yrast state with one unit higher angular momentum. (The only state with

Table 2. Energies E and wave functions of $j = 9/2$ special states not belonging to the even I yrast band. The wave functions are shown as coefficients of the states $|J_pJ_n\rangle_e$.

		$E =$	15	15	18	18	21	21	24
J_p	J_n	$I =$	7	9	10	11	11	13	14
6	2		0.000						
6	4		0.872	0.459	0.000				
6	6				0.689				
8	0		0.000						
8	2		−0.489	0.888	0.000				
8	4		0.000	0.000	−0.725	1.000	0.000		
8	6		0.000	0.000	0.000	0.000	1.000	1.000	0.000
8	8				0.000				1.000

Table 3. All special states occurring for $j \leq 15/2$.

j	$J_p + J_n$	I	j	$J_p + J_n$	I
1/2	0	0	13/2	12	12
3/2	2	2		14	13, 14
	4	2, 4		16	15, 16
5/2	4	4		18	15, 17, 18
	6	3, 5, 6		20	18, 19, 20
	8	6, 8		22	19, 21, 22
7/2	6	3, 6		24	22, 24
	8	6, 7, 8	15/2	14	14
	10	3, 7, 9, 10		16	15, 16
	12	10, 12		18	17, 18
9/2	8	8		20	19, 20
	10	7, 9, 10		22	19, 21, 22
	12	10, 11, 12		24	22, 23, 24
	14	11, 13, 14		26	23, 25, 26
	16	14, 16		28	26, 28
11/2	10	10			
	12	11, 12			
	14	11, 13, 14			
	16	14, 15, 16			
	18	15, 17, 18			
	20	18, 20			

$I = 15$, which as such is necessarily the yrast state for this angular momentum, has $T = 1$.) Inspecting the wave functions, one notices again a conservation of $J_p + J_n$. Furthermore, the energy is always $3(J_p + J_n)/2$.

An analogous situation emerges for any j we have examined. Table 3 shows the result of a complete search of special states for $j \leq 15/2$. Always the absolute energy is $3(J_p + J_n)/2$. The following systematics is inferred from Table 3.

Rule 1: For a given j, there is a special state for any I from $2j - 1$ to $4j - 2$ except $4j - 3$ (which is impossible for $j = 1/2$ and accommodates for $j \geq 3/2$ just a single $T = 1$ state). These states have $J_p + J_n = I$ for even I and $J_p + J_n = I + 1$ for odd I and are yrast states.

Rule 2: Besides, there are special states with $(J_p + J_n, I) = (4j - 8, 4j - 11)$, $(4j - 6, 4j - 8)$, $(4j - 4, 4j - 7)$ and $(4j - 2, 4j - 4)$ provided this I is not negative.

These rules have only two exceptions, both of which occur for fairly low j: First, there is no $(J_p + J_n, I) = (4j - 6, 4j - 8) = (4, 2)$ special state for $j = 5/2$. Second, there is an additional $(J_p + J_n, I) = (10, 3) = (4j - 4, 4j - 11)$ special state for $j = 7/2$.

The four degenerate levels with $J_p + J_n = 10$ and $I = 3, 7, 9$ and 10 occurring for $j = 7/2$ are familiar from studies by Robinson and Zamick.[14,15] These authors consider an interaction in the $1f_{7/2}$ shell with $E^J = 0$ for odd J and arbitrary E^J is for even J. (As noted in the introduction, their results then apply essentially unaltered to the case when E^J is constant for odd J.) From properties of 9-j symbols, they derive in Ref. 14 that for these I, there is a stationary state whose wave function is just $|64\rangle_\mathrm{e}$. Because for all these I, this is the only $|J_pJ_n\rangle_\mathrm{e}$ with $J_p + J_n = 10$, these are the same states as considered presently. A slight extension of the arguments in Ref. 14 shows that for the more general interaction considered there they have energies $3(E^6 + E^4)/2$, so they are degenerate. In Ref. 15, the properties of 9-j symbols employed in Ref. 14 are derived from the fact that none of the four angular momenta accommodate $T = 2$. It is shown in Sec. 3 that when this happens and $E^J = 0$ for odd J, then quite generally any $|J_pJ_n\rangle_\mathrm{e}$ is a stationary state. Its energy is $3(E^{J_p} + E^{J_n})/2$.

3. Explanation

How is it possible that J_p+J_n is conserved in some stationary states of our schematic Hamiltonian, and why do these states always have energy $3(J_p + J_n)/2$? In order to see how this comes about notice that for given (j, I and M), this Hamiltonian H has matrix elements

$$\langle J_pJ_n|H|J_p'J_n'\rangle = \delta_{J_pJ_p'}\delta_{J_nJ_n'}(E^{J_p} + E^{J_n}) + 4\sum_{J_1J_2}\langle J_pJ_n\|J_1J_2\rangle E^{J_1}\langle J_1J_2\|J_p'J_n'\rangle, \tag{4}$$

where $\langle J_1J_2\|J_1'J_2'\rangle$ is shorthand for a unitary 9-j symbol,

$$\langle J_1J_2\|J_1'J_2'\rangle = \langle((j_1j_2)J_1(j_3j_4)J_2)IM|((j_1j_3)J_1'(j_2j_4)J_2'))IM\rangle, \tag{5}$$

where all j's equal j. While the angular momenta J_p, J_n, J_p' and J_n' are even, J_1 and J_2 take all values allowed by the triangle inequalities. It is convenient to define an operator X such that

$$\langle J_1J_2|X|J_1'J_2'\rangle = \langle J_1J_2\|J_1'J_2'\rangle. \tag{6}$$

The space with even T is spanned by the states $|J_pJ_n\rangle_\mathrm{e}$. By the symmetry of $\langle J_pJ_n\|J_1J_2\rangle$, the matrix element $\langle J_pJ_n|_\mathrm{e}X|J_1J_2\rangle$ vanishes unless J_1 and J_2 have equal parities. Therefore, in the even T space, when $E^J = 0$ for odd J, only even J_1 and J_2 contribute to the sum in (4), and we have

$$H = \Omega + 2W\Omega W \tag{7}$$

with operators Ω and W acting within the even T space and defined by

$$\langle J_pJ_n|\Omega|J_p'J_n'\rangle_\mathrm{e} = \delta_{J_pJ_p'}\delta_{J_nJ_n'}(E^{J_p} + E^{J_n}), \tag{8}$$

$$\langle J_pJ_n|W|J_p'J_n'\rangle_\mathrm{e} = \langle J_pJ_n|X|J_p'J_n'\rangle_\mathrm{e}. \tag{9}$$

The subscript 'e' indicates that the matrix element is taken between states $|J_pJ_n\rangle_\mathrm{e}$.

We denote by (ik) the interchange of the states of the ith and kth nucleons, where the nucleons are numbered in the order of appearance of their angular momenta in

Eq. (2). Due to $(12)|J_pJ_n\rangle = (34)|J_pJ_n\rangle = -|J_pJ_n\rangle$ one can make in Eq. (9) the substitution

$$4X = (13) + (14) + (23) + (24). \tag{10}$$

By Eq. (4) of Ref. 16, we have

$$\sum_{i<k}(ik) = 4 - 4^2/4 - T(T+1) = -T(T+1). \tag{11}$$

As a result, the matrix W has the eigenvalue $(-T(T+1) - 2 \times (-1))/4 = 1/2$ for $T = 0$. In particular, if some $T = 0$ state is an eigenstate of Ω it is an eigenstate of H with eigenvalue $1 + 2 \times (1/2)^2 = 3/2$ times that of Ω.

This explains the finding of Robinson and Zamick in Ref. 15. If $T = 2$ is not accommodated for the given j and I then the states $|J_pJ_n\rangle_\mathrm{e}$ have $T = 0$. They are also eigenstates of Ω with eigenvalue $E^{J_p} + E^{J_n}$. Therefore, they are eigenstates of H with eigenvalue $3(E^{J_p} + E^{J_n})/2$.

For the Hamiltonian presently considered any linear combination of states $|J_pJ_n\rangle_\mathrm{e}$ with $J_p + J_n = k$, where k is a constant, is an eigenstate of Ω with eigenvalue k. What then remains to be explained is that for the combinations of j, k and I obeying the above rules 1 and 2 with the two exceptions mentioned, there exist such linear combinations which have $T = 0$. The rest of this section is devoted to a proof of this. The proof is divided into separate parts for the two rules. Notice that the second exception is explained already. The special state with $(j, k, I) = (7/2, 10, 3)$ is one of the states discussed by Robinson and Zamick in Refs. 14 and 15. An explanation of the first exception is deferred to Sec. 3.2.

3.1. *Rule* 1

We discuss the cases of even and odd I separately.

Even *I* Let

$$|\psi\rangle = \sum_{J_p+J_n=k} c_{J_p}|(J_pJ_n)kk\rangle \tag{12}$$

with some set of coefficients c_J, where we have included explicitly I and M on the left-hand side of Eq. (2). This state evidently has $I = k$. We assume $k \geq 2j - 1$, so the range $\mathcal{S}$ of J_p in the summation is the set of even integers J with $k - 2j + 1 \leq J \leq 2j - 1$. From formulas for vector coupling coefficients[17] one gets

$$\begin{aligned}
&\langle m_1m_2m_3m_4|\psi\rangle \\
&\quad := (\langle jm_1| \times \langle jm_2| \times \langle jm_3| \times \langle jm_4|)|\psi\rangle \\
&\quad = \delta_{m_1+m_2+m_3+m_4,k}(-)^{m_1-m_3}a(m_1)a(m_2)a(m_3)a(m_4)f(m_1+m_2),
\end{aligned} \tag{13}$$

where the m's are single nucleon magnetic quantum numbers, and

$$a(m) = \sqrt{\frac{(j+m)!}{(j-m)!}}, \tag{14}$$

$$f(\mu) = \begin{cases} b(\mu)b(k-\mu)c_\mu, & \mu \in \mathcal{S}, \\ 0, & \text{otherwise}, \end{cases} \tag{15}$$

$$b(J) = \frac{1}{J!}\sqrt{\frac{(2j-J)!(2J+1)!}{(2j+J+1)!}}. \tag{16}$$

By Eq. (11), the state $|\psi\rangle$ has $T=0$ when it belongs to the kernel of

$$K = (13) + (14) + (23) + (24) - 2. \tag{17}$$

This is seen to be equivalent to

$$\begin{aligned}(-)^{m_3-m_1}f(m_3+m_2) + (-)^{m_4-m_3}f(m_4+m_2) + (-)^{m_1-m_2}f(m_1+m_3) \\ + (-)^{m_1-m_3}f(m_1+m_4) - 2(-)^{m_1-m_3}f(m_1+m_2) = 0\end{aligned} \tag{18}$$

for $m_1+m_2+m_3+m_4=k$. Equation (18) holds when $f(\mu)$ is constant for $\mu \in \mathcal{S}$. Indeed, when $m_1+m_2+m_3+m_4=k$, no sum μ of two of the m's is greater that $2j$ or less than $k-2j$, so $\mu \in \mathcal{S}$ if μ is even. First assume that m_1+m_2 is even. If m_4+m_2 is even then the sign factor in the second term in Eq. (18) becomes $(-)^{m_1-m_3}$. If it is odd, the term vanishes. If m_1+m_3 is even, the sign factor in the third term becomes $(-)^{m_1-m_3}$. If it is odd, the term vanishes. All sign factors are thus effectively equal to $(-)^{m_1-m_3}$. Because with even m_1+m_2 the sum m_3+m_4 is also even and the m's are half-integral, the numbers m_3+m_2 and m_4+m_2 have opposite parities. So do the numbers m_1+m_3 and m_1+m_4. Therefore, the equation holds. If m_1+m_2 is odd, because m_3+m_4 is also odd, all of m_3+m_2, m_4+m_2, m_1+m_3 and m_1+m_4 have the same parities. If all of them are even, m_1+m_4, in particular, is even, so $(-)^{m_3-m_1}(-)^{m_4-m_3} = (-)^{m_4-m_1} = -$. Similarly, because m_2+m_3 is even, $(-)^{m_3-m_1}(-)^{m_1-m_2} = (-)^{m_3-m_2} = -$. So again the equation holds.

Thus, $|\psi\rangle$ is special when

$$c_J \propto \frac{1}{b(J)b(k-J)}. \tag{19}$$

Odd *I* We now consider a state

$$|\psi\rangle = \sum_{J_p+J_n=k} c_{J_n}|(J_pJ_n)(k-1)(k-1)\rangle, \tag{20}$$

which has $I=k-1$, and we assume so far again $k \geq 2j-1$. This limit is going to be sharpened. For $k-1$ to be non-negative necessarily $k \geq 2$. We also assume

$$c_J = -c_{k-J} \tag{21}$$

as required for T to be even. This rules out $k=4j-2$ because in that case $\mathcal{S}$ has only one element $J=2j-1$, whose c_J would then vanish. (It was noted already, indeed, that $I=(4j-2)-1=4j-3$ accommodates only a single $T=1$ state.)

Using again formulas from Ref. 17, we then get

$$\begin{aligned}\langle m_1 m_2 m_3 m_4|\psi\rangle &= \delta_{m_1+m_2+m_3+m_4,k-1}(-)^{m_1-m_3}\sqrt{\frac{1}{2k}}\\ &\quad\times a(m_1)a(m_2)a(m_3)a(m_4)g(m_1+m_2)\\ &\quad\times\begin{cases} m_3-m_4, & \text{even } m_1+m_2,\\ m_1-m_2, & \text{odd } m_1+m_2\end{cases}\end{aligned} \tag{22}$$

with

$$g(\mu) = \begin{cases} d(\mu)d(k-\mu)c_\mu, & \mu\in\mathcal{S},\\ g(k-1-\mu), & k-1-\mu\in\mathcal{S},\\ 0, & \text{otherwise},\end{cases} \tag{23}$$

$$d(J) = \sqrt{J}b(J). \tag{24}$$

As $\langle m_1 m_2 m_3 m_4|\psi\rangle$ vanishes unless $m_1+m_2+m_3+m_4 = k-1$, this is understood in the following. The state $|\psi\rangle$ is even under the permutation (13) (24) and odd under (12), both of which commute with K. (That $|\psi\rangle$ is even under (13) (24) is seen explicitly from Eqs. (22) and (23). Quite generally a state with definite T of equally many protons and neutrons has the parity $(-)^T$ under the interchange of the entire states of the proton and neutron subsystems.) Because the m's are half-integral, we can therefore, assume without loss of generality that m_1+m_2 and m_1+m_3 are even. Then m_3+m_4, m_2+m_4 and m_2+m_3 are odd and m_1+m_4 is even. A sufficient condition for $|\psi\rangle$ to belong to the kernel of K is then

$$\begin{aligned}&(-)^{m_3-m_1}(m_3-m_2)g(m_3+m_2)+(-)^{m_4-m_3}(m_4-m_2)g(m_4+m_2)\\ &\quad+(-)^{m_1-m_2}(m_2-m_4)g(m_1+m_3)+(-)^{m_1-m_3}(m_3-m_2)g(m_1+m_4)\\ &\quad-2(-)^{m_1-m_3}(m_3-m_4)g(m_1+m_2)=0.\end{aligned} \tag{25}$$

By $(-)^{m_3-m_1}(-)^{m_4-m_3} = (-)^{m_4-m_1} = -$, $(-)^{m_3-m_1}(-)^{m_1-m_2} = (-)^{m_3-m_2} = +$ and $g(\mu) = g(k-1-\mu)$ this is reduced to

$$(m_3-m_2)g(m_3+m_2)+(m_2-m_4)g(m_2+m_4)+(m_4-m_3)g(m_4+m_3)=0. \tag{26}$$

An odd sum μ of two m's cannot be greater than $2j$ or less than $k-1-(2j-1) = k-2j$. If $\mu = 2j$ both m's equal j, which eliminates the term with this $g(\mu)$ from Eq. (26). For $k-2j \le \mu \le 2j-2$, the number $k-1-\mu$ belongs to $\mathcal{S}$. Equation (26) holds if $g(\mu)$ is a polynomial of first degree in μ for odd sums μ of two m's, and it is by the preceding remark sufficient that $k-1-\mu\in\mathcal{S}$. This is by $g(\mu) = g(k-1-\mu)$ equivalent to $g(\mu)$ being a polynomial of first degree in μ for $\mu\in\mathcal{S}$.

Choosing

$$c_J \propto \frac{k-2J}{d(J)d(k-J)}, \tag{27}$$

gives the polynomial $g(\mu) \propto k - 2\mu$ of first degree, which satisfies Eq. (21). The state $|\psi\rangle$ is then special. As the denominator in Eq. (27) vanishes for $J = 0$, this must not be allowed. Then $k = 2j - 1$ is ruled out and the final scope of the proof is $2j + 1 \leq k \leq 4j - 4$, corresponding to odd I with $2j \leq I \leq 4j - 5$.

3.2. *Rule 2*

We already introduced the notion of the *even T space*, which is the space of states with given j, I and M and even T. The condition $J_p + J_n = k$ defines a subspace, which we call the *k space*. Its dimension is called the *k dimension*. The condition $T = 0$ similarly defines a subspace. This we call the *$T = 0$ space* and its dimension the *$T = 0$ dimension*. A *$T = 2$ space* and a *$T = 2$ dimension* are defined analogously. If the k dimension is greater than the $T = 2$ dimension, then at least one state in the k space is perpendicular to the $T = 2$ space and thus belongs to the $T = 0$ space. It is then a special state. A special state thus exist for given j, k and I whenever the k dimension exceeds the $T = 2$ dimension. Note that this is a sufficient but not a necessary condition. As we shall see, it is not satisfied in some cases covered by rule 1.

In particular, if the $T = 2$ space is zero-dimensional, then each entire k space consists of special states. It turns out, as discussed below, that the k dimension never exceeds the $T = 2$ dimension by more that one, so in that case any positive k dimension is just one. That is, the k space is spanned by a single $|J_pJ_n\rangle_e$. These are the states discussed by Robinson and Zamick in Refs. 14 and 15.

The k and $T = 2$ dimensions are determined by combinatorics. In particular, because a state $|\psi\rangle$ has $T = 2$ if and only if $\langle m_1m_2m_3m_4|\psi\rangle$ is antisymmetric in the m's, the $T = 2$ dimension is given as the number of combinations of $m_1 > m_2 > m_3 > m_4$ such that $\sum m = I$. The counts are simplified if one assumes $I \geq 2j - 1$ because then, in counting the combinations of $J_p \geq J_n$ that give $J_p + J_n = k$ and the combinations of $m_1 > m_2 > m_3 > m_4$ that give $\sum m = I$, one can neglect the lower limits $J_n \geq 0$ and $m_4 \geq -j$. The condition $I \geq 2j - 1$ also secures the triangle inequality $J_p \leq J_n + I$. The triangle inequality $I \leq J_n + J_p$ is secured by $k \geq I$. Therefore, if $I \geq 2j - 1$ the k dimension is a function of $x = 4j - k$ and the $T = 2$ dimension a function of $y = 4j - I$.

The following tables show the result of this combinatoric analysis.

x	2	4	6	8	10	12	14	16
k dim., even y	1	1	2	2	3	3	4	4
k dim., odd y	0	1	1	2	2	3	3	4

y	6	7	8	9	10	11	12	13	14	15	16	17
$T = 2$ dim.	1	0	1	1	2	1	3	2	4	3	5	4

Because $J_p, J_n \leq 2j - 1$ both x and y are at least 2. The $T = 2$ dimension vanishes for $y < 6$ because no combination of four different m's have a sum greater than $4j - 6$. The condition $k \geq I$ translates to $x \leq y$. It is evident that the $T = 2$

dimension rises more rapidly than the k dimension with increasing y so that the values of x and y included in the tables suffice to determine the cases when the latter dimension exceeds the former.

This is seen to happen when $y = x \le 10$ or $4 \le x = y - 1 \le 12$, which corresponds to rule 1 with an additional upper limit on x. As rule 1 does not have this upper limit, it is thus more general than can be inferred from this dimensional analysis. The only other cases when the k dimension exceeds the $T = 2$ dimension are $(x, y) = (2, 4)$, $(4, 7)$, $(6, 8)$ and $(8, 11)$, which correspond exactly to rule 2.

It was assumed that $I \ge 2j - 1$, and all the cases of the k dimension exceeding the $T = 2$ dimension that were identified have $y \le 13$. When $y \le 13$, the condition $I \ge 2j - 1$ is satisfied for $j \ge 13/2$. The dimensional analysis is thus exhaustive for these j. The combinations of j, k and I that occur for $j \le 11/2$ are finite in number, so they, can be examined individually. This was done in the search of special states with $j \le 15/2$ reported in Sec. 2. It turns out that all the special states with $j \le 11/2$ appear when the k dimension is greater than the $T = 2$ dimension with one exception: For $j = 11/2$ and $I = k = 10$ both dimensions equal 3. This state is covered by rule 1, and the equality of the two dimensions is, in fact, consistent with the combinatoric analysis, which is valid for $j \ge 9/2$ when I is even and then requires $I \ge 4j - 10$ for the k dimension to exceed the $T = 2$ dimension. For $(j, I) = (5/2, 2)$ both the $T = 2$ dimension and the maximal k dimension equal 1, and the $k = 4$ special state anticipated by rule 2 indeed does not appear.

The tables above show that for $j \ge 13/2$, the k dimension never exceeds the $T = 2$ dimension by more than one. This is found to hold also for $j \le 11/2$. It can be turned around to say that the dimension of the configuration space of four identical fermions with given j, I and M is never less than the maximal k dimension minus one. As an empiric rule, a special state is always unique to the given j, k, I and M, that is, any k space has at most a one-dimensional intersection with the $T = 0$ space.

3.3. *Wave functions*

The wave functions of special states occurring by rule 1 are given by Eqs. (19) and (27). For the special states occurring by rule 2 with its two exceptions, the k dimension never exceeds two. This follows for $I \ge 2j - 1$ from the first table in Sec. 3.2, and it holds, as well, in the four cases (see Table 3) with $I < 2j - 1$. If the k dimension is one, the special state $|\psi\rangle$ is a single $|J_p J_n\rangle_e$. If the k dimension is two, $|\psi\rangle$ is a linear combination

$$|\psi\rangle = \alpha|a\rangle + \beta|b\rangle, \tag{28}$$

where $|a\rangle$ and $|b\rangle$ are states $|J_p J_n\rangle_e$. The ratio of the coefficients α and β is determined by the fact that, having $T = 0$, the state $|\psi\rangle$ is an eigenstate with eigenvalue 1/2 of the operator W defined by Eq. (9). Explicitly

$$\frac{\alpha}{\beta} = \frac{\langle a|W|b\rangle}{\frac{1}{2} - \langle a|W|a\rangle} = \frac{\frac{1}{2} - \langle b|W|b\rangle}{\langle a|W|b\rangle}. \tag{29}$$

By Eqs. (9), (6) and (5) the matrix elements of W are[17] 9-j symbols multiplied by factors $\sqrt{2q+1}$ for some angular momenta q and possibly factors $\sqrt{2}$.

For any special state occurring by either rule 1 or rule 2 with its two exceptions, if the k dimension is one the $T = 2$ dimension vanishes. For $I \geq 2j - 1$ this can be inferred again from the tables in Sec. 3.2, and again it holds, as well, in the four cases with $I < 2j - 1$. It implies that for these j and I, the entire even T space has $T = 0$ so that within this space, W is equal to the constant 1/2 and every $|J_pJ_n\rangle_e$ is special. As shown in Ref. 15, that W is equal to the constant 1/2 within the even T space can be inferred also directly from the fact that these j and I do not accommodate $T = 2$. It implies that the 9-j symbols in the matrix elements of W between different $|J_pJ_n\rangle_e$ vanish.

When a special state belongs to a two-dimensional (2D) k space, the requirement that both expressions in Eq. (29) give the same result entails relations between the 9-j symbols involved. So does the requirement that the matrix element of W between the state (28) and any $|J_pJ_n\rangle_e$ with a different k vanishes. The expressions (19) and (27) give rise to similar relations involving several 9-j symbols.

4. Summary

We studied the system of two protons and two neutrons in a j shell with the two nucleon interaction matrix element equal to the two nucleon angular momentum J for even J and zero for odd J. This model has a straightforward generalization to the case when the matrix element is linear in J for even J and constant for odd J. It was found to exhibit for any $j \geq 3/2$ several stationary states where the sum $J_p + J_n$ of the angular momenta of the proton and neutron pairs is conserved. The absolute energies of these states, which we call *special*, that is, their energies before the ground state energy is subtracted to give excitation energies, are $3(J_p + J_n)/2$. Special states in particular form the even and odd I yrast bands from $I = 2j - 1$ to the maximal $I = 4j - 2$ except $I = 4j - 3$, where I is the total angular momentum. Other, nonyrast states are also special.

It was shown that any state which conserves $J_p + J_n$ in this model is a stationary state with absolute energy $3(J_p + J_n)/2$ provided it has isospin $T = 0$. Using explicit expressions for vector coupling coefficients, we then demonstrated that such states exist for all the yrast total angular momenta I specified above. The nonyrast special states could be explained by a combinatoric analysis of the dimensions of various subspaces of the configuration space. Explicit expressions for the wave functions of all special states are provided by our study.

Acknowledgments

Wesley Pereira is a student at Essex College, Newark, New Jersey, 07102. His research at Rutgers is funded by a Garden State Stokes Alliance for Minorities Participation (G.S.L.S.A.M.P.) internship.

Ricardo Garcia has two institutional affiliations: Rutgers University, and the University of Puerto Rico, Rio Piedras Campus. The permanent address associated with the UPR-RP is University of Puerto Rico, San Juan, Puerto Rico 00931. He acknowledges that to carry out this work, he has received support via the Research Undergraduate Experience program (REU) from the U.S. National Science Foundation through grant PHY-1263280 and thanks the REU Physics program at Rutgers University for their support.

References

1. T. T. S. Kuo and G. E. Brown, *Nucl. Phys.* **85** (1966) 40.
2. G. E. Brown and M. Bolsterli, *Phys. Rev. Lett.* **3** (1959) 477.
3. J. P. Elliott and B. F. Flowers, *Proc. Roy. Soc. (London) A* **242** (1957) 57.
4. G. E. Brown, *Unified Theory of Nuclear Models* (North-Holland Publishing Company, Amsterdam-London, 1964).
5. J. P. Elliott, *Proc. Roy. Soc. (London) A* **245** (1958) 128.
6. G. Racah, *Phys. Rev.* **63** (1943) 367.
7. B. F. Bayman, J. D. McCullen and L. Zamick, *Phys. Rev. Lett.* **11** (1963) 215.
8. J. D. McCullen, B. F. Bayman and L. Zamick, *Phys. Rev.* **134** (1964) B515, Technical Report NYO-9891.
9. J. N. Ginocchio and J. B. French, *Phys. Lett.* **7** (1963) 137.
10. J. N. Ginocchio, *Nucl. Phys.* **63** (1965) 449.
11. J. N. Ginocchio, *Phys. Rev.* **144** (1966) 952.
12. L. Zamick and J. Q. Robinson, *Yad. Fiz.* **65** (2002) 773; *Phys. At. Nucl.* **65** (2002) 740.
13. A. Escuderos, L. Zamick and B. F. Bayman, arXiv:nucl-th/0506050.
14. J. Q. Robinson and L. Zamick, *Phys. Rev. C* **63** (2001) 064316.
15. J. Q. Robinson and L. Zamick, *Phys. Rev. C* **64** (2001) 057302.
16. K. Neergård, *Phys. Rev. C* **90** (2014) 014318.
17. A. R. Edmonds, *Angular Momentum in Quantum Mechanics* (Princeton University Press, Princeton, 1957).

Chiral electroweak currents in nuclei

D. O. Riska*
Finnish Society of Science and Letters, Helsinki, Finland
The Cyprus Institute, Nicosia, Cyprus
**riska@scientiarum.fi; riska@promail.tsv.fi*

R. Schiavilla
Theory Center, Jefferson Lab, Newport News, Virginia, USA
Physics Department, Old Dominion University, Norfolk, Virginia, USA
schiavilla@jlab.org

The development of the chiral dynamics based description of nuclear electroweak currents is reviewed. Gerald E. (Gerry) Brown's role in basing theoretical nuclear physics on chiral Lagrangians is emphasized. Illustrative examples of the successful description of electroweak observables of light nuclei obtained from chiral effective field theory are presented.

Keyword: Nuclear currents.

1. Introduction

The phenomenological success of the systematic application of chiral effective field theory (χEFT) to the electromagnetic and weak observables of light nuclei in the mass range $A = 2$–10 has been remarkable. Here the background and early application of chiral Lagrangians to nuclear current operators, and Gerry Brown's role as an initiator of this approach are reviewed in the next section. The third and fourth sections contain a summary of the present stage of the χEFT approach and a set of illustrative examples of its application to nuclear electroweak observables.

2. Historical Perspective

2.1. *Gerry Brown's early work with chiral Lagrangians*

Gerry Brown was one of the first physicists to appreciate the utility of effective chiral Lagrangians in theoretical nuclear physics. In the late 1960s, he planned a major effort to derive a realistic nucleon–nucleon (NN) interaction model from meson exchange, as he saw that the pair suppression built into Weinberg's nonlinear chiral Lagrangian for the pion–nucleon interaction[1] might solve the over-binding

*Corresponding author.

problem, which the two-pion exchange interaction described with the conventional pseudoscalar pion–nucleon coupling model inevitably led to. That a realistic description of the nucleon–nucleon amplitude could be constructed in this way was then demonstrated by Brown and Durso[2] and Chemtob *et al.*.[3]

In parallel with this development Chemtob and Rho derived expressions for the exchange current contributions to the electromagnetic and axial two-nucleon current operators that arise from the effective chiral Lagrangians for pion and vector meson exchange.[4] The ρ-meson exchange interaction complements the pion exchange one by counteracting the strong tensor component of the latter and improving the interaction models of Refs. 2 and 3. Concurrently, it enhances the effects of the long range electromagnetic pion exchange current.

Gerry Brown's interest in the role of exchange currents in nuclei was stimulated by Chemtob and Rho's estimate of the axial exchange current enhancement of the Gamow–Teller matrix element in Tritium β-decay.[5] He suggested that the D-state components in the trinucleon wave functions, even if small, could enhance the calculated value. This was illustrated with a schematic model for those components.[6]

The axial exchange current is related to the NN interaction indirectly through the partially conserved axial current (PCAC) relation to the pion-production operator. The form of the electromagnetic exchange current is in contrast directly constrained by the NN interaction through the continuity equation.[7,8] The first demonstrations that electromagnetic exchange currents could play a significant role in nuclear observables were in fact related to those. In 1971 Gerry Brown's attention was drawn to the fact the well measured total cross-section for capture of thermal neutrons on protons — the process $^1\mathrm{H}(n,\gamma)^2\mathrm{H}$ — could not be fully explained by the sum of the neutron and proton magnetic moments. He asked one of us (DOR) to take into account the pion exchange current operators derived by Chemtob and Rho.[4] The result was that the exchange current contribution can account for the $\sim 10\%$ difference between the calculated and the experimental value for the cross-section.[9] The key part of the pion exchange current operator was related to the chiral Lagrangian for the pion–nucleon interaction. A smaller, nevertheless significant, effect was due to the pion exchange operator, which involved intermediate Δ resonances.[10]

2.2. *The sequel*

It was soon afterward shown that the experimental values of ^{3}H and ^{3}He magnetic moments could also be almost fully accounted for in the same way with realistic wave functions.[11] Moreover it was found that the pion exchange current contribution could provide about one half of the cross-section for capture of thermal neutrons on ^{2}H.[12] Later it was shown that the exchange current contribution to the calculated cross-section for thermal neutron capture on ^{3}He is about 5 times larger than that from the sum of the three nucleon magnetic moments.[13] Finally, it was shown that the cross-section for radiative neutron capture on ^{3}He is almost totally due to the exchange current contribution.[14]

While the strength of the NN interaction scales with the mass of the exchanged system, the meson exchange magnetization operator scales with the inverse mass of the exchanged system.[15] This makes the relative contribution of short-range mechanisms to the matrix elements of the exchange current operators much weaker than to the matrix elements of the interaction. The calculated cross-section for backward electro-disintegration of the deuteron, which is very strongly dependent on the electromagnetic exchange currents, illustrates this well.[16] In this reaction the matrix element of the single-nucleon current operators changes sign at fairly low momentum transfer and therefore the cross-section near that zero is built up entirely from the exchange current contribution. In this case the cross-section obtained with the pion-exchange current alone is quite similar to that calculated later with wave functions that are consistent with a realistic phenomenological interaction model.[17] In the case of the magnetic form factors of the trinucleons there is a similar destructive interference between the matrix elements of the single-nucleon current operators for the S- and D-state components of the wave functions[18] with the consequence that the exchange current contribution is very large.[19] With only the single-nucleon current operator, the calculated magnetic form factors of the trinucleons would have zeroes at fairly low values of momentum transfer, in disagreement with experiment. This has later been demonstrated with improved wave functions and exchange current operators that satisfy the continuity equation with realistic interactions.[20] In larger nuclei the main features in elastic and transition electromagnetic form factors are due to the shell structure. Even so, it has been shown that in the case of Li, the inclusion of the exchange current contribution does markedly improve the agreement with experiment for these form factors, once the interaction currents are consistent with wave functions corresponding to realistic interaction models.[21] The effect is smaller in ^{12}C [22] and in ^{7}Li, ^{9}Be and ^{10}B.[23]

In the examples above, it is the isovector part of the pion exchange current operator, which is numerically most important. In the case of the magnetic form factor of the deuteron, only the isoscalar part of the pion exchange current contributes, and the most important term in that operator involves a $\rho\pi\gamma$ transition. Without this exchange current operator the calculated magnetic moment of the deuteron would have a node close to momentum transfer values $\sim 6\,\mathrm{GeV/c}$, in disagreement with experiment.[17,24] Since the $\rho\pi\gamma$ exchange current is transverse, its form is not constrained by the NN interaction by the continuity equation. Its longer range form can however be determined in the Skyrme model approach by the chiral anomaly.[25,26]

While the two-nucleon exchange current operators give large contributions to nuclear electromagnetic observables, it has been demonstrated that the three-nucleon exchange current operators that are associated with two-pion exchange with pion scattering off an intermediate nucleon, give but very small contributions to the magnetic form factors of the trinucleons.[27] The remarkably successful nuclear phenomenology based on the simple pion exchange operators that are related to the lowest order chiral Lagrangian for the pion–nucleon interaction has later been explained within the context of χEFT.[28]

2.3. *The axial exchange current*

The role of the pion exchange axial exchange current was first considered for the case of the Gamow–Teller transition in the β-decay of Tritium. Those results were reaffirmed with more accurate wave functions.[29] Gari and Huffman noted that this axial exchange current also contributes a small enhancement of the cross-section for the basic solar burning reaction ${}^1\mathrm{H}(p, e^+ \nu_e)^2\mathrm{H}$.[30] This was confirmed by Dautry *et al.* in a study of muon absorption in the deuteron ${}^2\mathrm{H}(\mu^-, \nu_\mu)nn$ with an improved version of the axial exchange current operator, which was checked for consistency against the P-wave piece of the cross-section for the reaction ${}^1\mathrm{H}(p, \pi^+)^2\mathrm{H}$ near threshold.[31] The same Hamiltonian was then used to calculate the matrix elements for the solar neutrino processes ${}^1\mathrm{H}(p, e^+\nu_e)^2\mathrm{H}$ and ${}^1\mathrm{H}(pe^-, \nu_e)^2\mathrm{H}$ as well. A later calculation, with wave functions obtained from the realistic Argonne v_{14} NN interaction[32] of the weak proton capture reactions ${}^1\mathrm{H}(p, e^+\nu_e)^2\mathrm{H}$ and ${}^3\mathrm{He}(p, e^+\nu_e)^4\mathrm{He}$ normalized the axial exchange current operator against the known rate of the Gamow–Teller component in Tritium.[33] The result was that the exchange current increased the cross-section of the reaction ${}^1\mathrm{H}(p, e^+\nu_e)^2\mathrm{H}$ by 1.5% and that of the reaction ${}^3\mathrm{He}(p, e^+\nu_e)^4\mathrm{He}$ by almost a factor 5. The axial exchange current contributions to the solar burning reaction ${}^1\mathrm{H}(p, e^+\nu_e)^2\mathrm{H}$ have later been recalculated on the basis of χEFT, and have been found to enhance the cross-section obtained with single-nucleon currents by $\sim 4\%$[34] — however, see below for a more recent assessment. Finally, parameter free calculations of the cross-section for this reaction and the associated ${}^3\mathrm{He}(p, e^+\nu_e)^4\mathrm{He}$ reaction have been carried out in Ref. 35.

2.4. *Nuclear charge form factors*

The phenomenological success of the exchange current operators described above was mainly due to the chiral Lagrangians for the pion–nucleon couplings. The corresponding contributions to nuclear charge operators involve terms of higher power in the momentum transfer. The observation by Kloet and Tjon that there is a significant pion exchange contribution to the charge form factors of the trinucleons therefore came as a surprise.[36] This pion exchange operator brings the first diffraction minimum to lower values of momentum transfer and therefore closer to the experimental data. This observation was subsequently confirmed by a calculation of the charge form factor of the α-particle.[37] Later calculations of the charge form factors of ${}^3\mathrm{H}$, ${}^3\mathrm{He}$ and ${}^4\mathrm{He}$ with realistic variational wave functions reaffirm the need for a substantial exchange charge effect for agreement with the empirical values.[38] This exchange current effect is most prominent in the lightest nuclei, and less so in the case of heavier nuclei as ${}^{16}\mathrm{O}$ and ${}^{40}\mathrm{Ca}$, where the shell structure is most prominent.[39,40] The pion exchange effect nevertheless does improve slightly the agreement between the calculated and empirically extracted charge distributions throughout the periodic table.[41] The best indication of the role of the pion exchange charge operator was finally provided by the measurement of the charge form factor of the deuteron.[42] Inclusion of the exchange charge operator clearly improved the agreement with the experimental values. While the exchange charge operators that involve two nucleons were found to give substantial contributions to nuclear charge form factors, the corresponding exchange charge operators that involve three nucleons were found to give only minor contributions, because of cancellations between the pion and rho-meson exchange operators. Those involving 4 nucleons were found to be insignificant.[43]

2.5. *The axial charge operator*

That there is a significant nuclear enhancement of the axial charge of the nucleon was demonstrated by Kirchbach *et al.*, who extended the Adler–Weisberger sum rule to light nuclei.[44] A subsequent explicit calculation of the nuclear enhancement of the axial charge based on meson exchange indicated that pion-exchange mechanisms yield enhancements of the order 45–60%. In combination with short range mechanisms the total enhancement in heavy nuclei is of the order 85–100%, depending on the interaction model.[45,46] This is sufficient to explain almost all of the empirically observed $\sim$ 100% enhancement of first forbidden β-decay transitions in the lead region.[47] The large nuclear enhancement of the axial charge operator has later been explained within χEFT, which provides a dynamical basis for the utility of early chiral charge algebra.[48] Finally, Lee and Riska employed the PCAC relation between the axial current and pion-production operators to show that the axial exchange current could explain the large difference between the empirical cross-section for the reaction $^1\mathrm{H}(p,\pi^0)pp$ and the value given by the single-nucleon pion-production operators alone.[49]

2.6. *Gerry Brown and the Skyrmion*

In the late 1970's Gerry Brown engaged in a project to develop a chiral dynamics basis for quark bag models of baryons.[50] In the limit of a small bag radius this led to a connection to Skyrmes topological soliton model for the nucleon,[51] which had been shown by Witten to give results that are consistent with quantum chromodynamics (QCD) in the large color limit.[52] Skyrme's topological soliton model is based on a chiral Lagrangian for pions yielding finite size solutions, which may be interpreted as baryons. The Noether currents of this Lagrangian can be viewed as models for the electromagnetic and axial currents of nucleons and nuclei. When the meson field is described by the common product ansatz for the two-nucleon system, the current operators separate into single nucleon and exchange current operators. Indeed, Nyman and Riska[25,53] showed that if the chiral profile of the Skyrmion field for a nucleon is fitted to its electric form factor, then the deuteron form factors can be calculated with good qualitative agreement with data. In the case of the magnetic form factor the exchange current contribution is large, and essential for agreement with the empirical form factor at large values of moment transfer. In the long-range limit there is a simple relation between the isoscalar exchange current and the phenomenological exchange current that is associated with the $\rho\pi\gamma$ vertex.[26] The long-range component of the isovector magnetic moment operator too is similar to the corresponding conventional pion-exchange magnetic-moment operator.[54]

3. The Chiral Effective Field Theory Approach

The last two decades have witnessed remarkable developments in nuclear χEFT, originally proposed by Weinberg.[55–57] Chiral symmetry is an approximate symmetry of QCD, which becomes exact in the limit of vanishing quark masses. Nuclear χEFT is the theoretical framework that permits the derivation of nuclear interactions and electroweak currents with hadronic degrees of freedom, while preserving the symmetries of QCD — the exact Lorentz, parity, and time-reversal symmetries, and the approximate chiral symmetry. The latter requires the pion couplings to hadrons to be proportional to powers of its momentum Q and, as a consequence, the Lagrangian for these interactions may be expanded in powers of Q/Λ_χ, where $\Lambda_\chi \sim 1$ GeV is the chiral symmetry breaking scale. The Lagrangians may be ordered into classes defined by the power of Q/Λ_χ, or equivalently by order of the gradients of the pion field and/or pion mass factors. Each of these contain a certain number of parameters or "low-energy constants" (LECs), which in practice are fixed by comparison with experimental data. These LECs could in principle be calculated from the underlying QCD theory of quarks and gluons, but the nonperturbative nature of the theory at low energies makes this task extremely difficult. Thus, nuclear χEFT provides a direct connection between QCD and the strong and electroweak interactions in nuclei, and at the same time a practical calculational scheme which, at least in principle, may be improved systematically. In this sense it provides a fundamental basis for low-energy nuclear physics.

The nuclear χEFT approach has been applied in a number of studies to derive the two- and three-nucleon potentials[58–66] including isospin-symmetry-breaking corrections.[67–70] In the electroweak sector, there have been derivations of parity-violating two-nucleon potentials induced by hadronic weak interactions,[71–74] constructions of nuclear electroweak currents[48] and studies of Compton scattering on nucleons and nuclei with the explicit inclusion of Δ-resonance degrees of freedom.[75,76] Here the focus is on nuclear electroweak current operators. These were originally derived up to one loop level in the heavy-baryon formulation of covariant perturbation theory by Park *et al.*.[35,48,77] More recently two independent derivations, based on time-ordered perturbation theory (TOPT), have been published — one by the present authors (RS)[78–81] and the other by Kölling *et al.*,[82,83] although these latter works only deal with electromagnetic currents. In the following, we outline the derivation of these electroweak operators, referring to the original papers[78–81] for the more technical aspects.

3.1. *Interaction Hamiltonians*

In the simplest implementation, χEFT Lagrangians are constructed in terms of nucleon and pion degrees of freedom. This has been described in a number of papers,[84,85] and $\pi\pi$ and πN Lagrangians, denoted respectively as $\mathcal{L}_{\pi\pi}^{(m)}$ and $\mathcal{L}_{\pi N}^{(n)}$, have been derived up to high order in the chiral expansion. Contributions that arise from additional degrees of freedom, such as Δ-resonances and heavier mesons, are subsumed in the LECs of $\mathcal{L}_{\pi N}^{(n)}$ and $\mathcal{L}_{\pi\pi}^{(m)}$. In principle these Lagrangians contain an infinite number of interactions compatible with the QCD symmetries, but as the transition amplitudes obtained from them may be expanded in powers of Q/Λ_χ, the number of terms that contribute to the amplitude at any given order of the expansion is finite.[55–57] The Hamiltonians are constructed from the chiral Lagrangians by the canonical formalism.

The leading interaction terms in $\mathcal{L}_{\pi N}^{(1)}$, $\mathcal{L}_{\pi N}^{(2)}$ and $\mathcal{L}_{\pi N}^{(3)}$ in the πN sector, and $\mathcal{L}_{\pi\pi}^{(2)}$ and $\mathcal{L}_{\pi\pi}^{(4)}$ in the $\pi\pi$ sector, which are relevant to the derivation of nuclear potentials and electroweak operators at one loop level lead to the following Hamiltonians:

$$H_{\pi N} = \int \mathrm{d}\mathbf{x} N^\dagger \left[\frac{g_A}{2f_\pi}\tau_a \boldsymbol{\sigma}\cdot\boldsymbol{\nabla}\pi_a + \frac{1}{4f_\pi^2}\boldsymbol{\tau}\cdot(\boldsymbol{\pi}\times\boldsymbol{\Pi}) + \cdots\right] N, \tag{1}$$

$$\begin{aligned} H_{\gamma N} = e\int \mathrm{d}\mathbf{x} N^\dagger \Big[& e_N V^0 + i\frac{e_N}{2m}\left(-\overleftarrow{\boldsymbol{\nabla}}\cdot\mathbf{V} + \mathbf{V}\cdot\overrightarrow{\boldsymbol{\nabla}}\right) - \frac{\mu_N}{2m}\boldsymbol{\sigma}\cdot\boldsymbol{\nabla}\times\mathbf{V} \\ & - \frac{2\mu_N - e_N}{8m^2}(\boldsymbol{\nabla}^2 V^0 + \boldsymbol{\sigma}\times\boldsymbol{\nabla}V^0\cdot\overrightarrow{\boldsymbol{\nabla}} - \overleftarrow{\boldsymbol{\nabla}}\cdot\boldsymbol{\sigma}\times\boldsymbol{\nabla}V^0) + \cdots\Big] N, \end{aligned} \tag{2}$$

$$H_{\gamma\pi} = e\int \mathrm{d}\mathbf{x}[V^0(\boldsymbol{\pi}\times\boldsymbol{\Pi})_z + \epsilon_{zab}\pi_a(\boldsymbol{\nabla}\pi_b)\cdot\mathbf{V} + \cdots], \tag{3}$$

$$\begin{aligned} H_{\gamma\pi N} = \frac{e}{2f_\pi}\int \mathrm{d}\mathbf{x} N^\dagger \Big[& \frac{g_A}{2m}(\boldsymbol{\tau}\cdot\boldsymbol{\pi} + \pi_z)\boldsymbol{\sigma}\cdot\boldsymbol{\nabla}V^0 - [8d_8\boldsymbol{\nabla}\pi_z \\ & + 8d_9\tau_a\boldsymbol{\nabla}\pi_a - (2d_{21} - d_{22})\epsilon_{zab}\tau_b\boldsymbol{\sigma}\times\boldsymbol{\nabla}\pi_a]\cdot\boldsymbol{\nabla}\times\boldsymbol{V} + \cdots\Big] N, \end{aligned} \tag{4}$$

$$H_{AN} = \frac{g_A}{2}\int \mathrm{d}\mathbf{x} N^\dagger(\tau_a\boldsymbol{\sigma}\cdot\mathbf{A}_a + \cdots)N, \tag{5}$$

$$H_{A\pi} = f_\pi\int \mathrm{d}\mathbf{x}(\mathbf{A}_a\cdot\boldsymbol{\nabla}\pi_a + A_a^0\Pi_a + \cdots), \tag{6}$$

$$\begin{aligned} H_{A\pi N} = \int \mathrm{d}\mathbf{x} N^\dagger \Big[& -\frac{1}{4f_\pi}A_a^0(\boldsymbol{\tau}\times\boldsymbol{\pi})_a + \frac{2c_3}{f_\pi}\mathbf{A}_a\cdot\boldsymbol{\nabla}\pi_a - \frac{c_4}{f_\pi}\epsilon_{abc}\tau_a(\mathbf{A}_b\times\boldsymbol{\nabla}\pi_c)\cdot\boldsymbol{\sigma} \\ & - \frac{c_6}{4mf_\pi}(\boldsymbol{\tau}\times\boldsymbol{\pi})_a(\boldsymbol{\nabla}\times\mathbf{A}_a)\cdot\boldsymbol{\sigma} + \cdots\Big] N, \end{aligned} \tag{7}$$

where g_A, f_π, e and m are, respectively, the nucleon axial coupling constant, pion decay amplitude, proton electric charge and nucleon mass, and the parameters c_i and d_i are LECs in the $\mathcal{L}_{\pi N}^{(2)}$ and $\mathcal{L}_{\pi N}^{(3)}$ Lagrangians.

The isospin doublet of (nonrelativistic) nucleon fields, isospin triplet of pion fields and conjugate fields, electromagnetic vector field and weak axial field are denoted by N, $\boldsymbol{\pi}$ and $\boldsymbol{\Pi}$, V^μ and A^μ respectively, and $\boldsymbol{\sigma}$ and $\boldsymbol{\tau}$ are spin and isospin Pauli matrices. The arrow over the gradient specifies whether it acts on the left or right nucleon field. The isospin operators e_N and μ_N are defined as

$$e_N = (1+\tau_z)/2, \quad \kappa_N = (\kappa_S + \kappa_V \tau_z)/2, \quad \mu_N = e_N + \kappa_N, \tag{8}$$

where κ_S and κ_V are the isoscalar and isovector combinations of the anomalous magnetic moments of the proton and neutron. The power counting of the resulting vertices follows by noting that each gradient brings in a factor of Q, so, for example, the two terms in $H_{\pi N}$ are both of order Q, (ignoring the counting Q assumed for the external fields V^μ and A^μ) the first term in $H_{\gamma\pi N}$ ($H_{A\pi N}$) is of order Q (Q^0), while the remaining ones are of order Q^2 (Q^1).

In addition to the chiral Hamiltonians above, up to and including order Q^2 there are 14 contact interaction terms allowed by the symmetries of the strong interactions, each one multiplied by a LEC. Two of these contact terms (proportional to the LECs C_S and C_T in standard notation) are of a nonderivative type, and therefore are of order Q^0, while the remaining 12 (proportional to the LECs C_i') of order Q^2 involve two gradients acting on the nucleon fields (these are listed in Ref. 86). The contact potential at order Q^2, derived from them in the two-nucleon center-of-mass system, in fact depends on C_S and C_T, and seven linear combinations of the C_i', which are customarily denoted as $C_1, \ldots, C_7$. The remaining five linear combinations of C_i' have been shown to be related to C_S and C_T by requiring that the Poincaré covariance of the theory be satisfied to order Q^2.[86] The Q^2 potential therefore involves nine independent LECs. (As a side remark, the contact potential at order Q^4 requires an additional 15 independent LECs.) These LECs are determined by fits to two-nucleon elastic scattering data.

Minimal substitution in the gradient terms leads to a (two-nucleon) electromagnetic contact current denoted as $\mathbf{j}^{(1)}_{\gamma,\mathrm{min}}$ in Refs. 78 and 80, where the superscript (n) specifies the power counting Q^n. Nonminimal couplings through the electromagnetic field tensor $F_{\mu\nu}$ are also allowed. It may be shown[78] that only two independent operator structures enter at order Q^1, which lead to the contact term given by

$$\mathbf{j}^{(1)}_{\gamma,\mathrm{nm}} = -ie[C_{15}'\boldsymbol{\sigma}_1 + C_{16}' \times (\tau_{1,z} - \tau_{2,z})\boldsymbol{\sigma}_1] \times \mathbf{q} + (1 \rightleftharpoons 2), \tag{9}$$

where $\mathbf{q}$ is the external field momentum, and the isoscalar C_{15}' and isovector C_{16}' LECs (as well as the d_i's multiplying the higher order terms in the $\gamma\pi N$ Hamiltonian) can be determined by fitting photo-nuclear data in the few-nucleon systems.[80]

In the weak axial sector, there is a single contact term at order Q^0,

$$\mathbf{j}^{(0)}_{5,a} = z_0(\boldsymbol{\tau}_1 \times \boldsymbol{\tau}_2)_a \left[\boldsymbol{\sigma}_1 \times \boldsymbol{\sigma}_2 - \frac{\mathbf{q}}{q^2+m_\pi^2}\mathbf{q}\cdot(\boldsymbol{\sigma}_1 \times \boldsymbol{\sigma}_2)\right] \tag{10}$$

(here the second term of Eq. (10) is the pion-pole contribution), and none at order Q^1. This term is due to an interaction of the type $(\overline{N}\gamma^\mu\gamma_5 N)(\overline{N}u_\mu N)$ and, as first pointed out in Ref. 87, the LEC z_0 is related to the LEC c_D (in standard notation), which enters the three-nucleon potential at leading order. The two LECs c_D and c_E which fully characterize this potential have recently been constrained by reproducing the empirical value of the Gamow–Teller matrix element in tritium β decay and the binding energies of the trinucleons[88,89] (see below). Lastly, in the limit of small momentum transfers, there are two independent two-nucleon contact terms in the axial charge at order Q^{181}:

$$\rho^{(1)}_{5,a} = i z_1 (\boldsymbol{\tau}_1 \times \boldsymbol{\tau}_2)_a (\boldsymbol{\sigma}_1 \cdot \mathbf{k}_1 - \boldsymbol{\sigma}_2 \cdot \mathbf{k}_2) + i z_2 (\boldsymbol{\sigma}_1 \times \boldsymbol{\sigma}_2) \cdot (\tau_{1,a} \mathbf{k}_2 - \tau_{2,a} \mathbf{k}_1). \tag{11}$$

The LECs z_1 and z_2 have, however yet to be determined.

3.2. *From amplitudes to potentials and currents*

Application of χEFT to nuclear structure and bound states requires going beyond perturbation theory. As suggested by Weinberg,[55–57] the formalism briefly described below for constructing nuclear potentials and currents is based on TOPT although it differs from Weinberg's in the way reducible contributions are dealt with. This has been employed in Refs. 78–81 and 90 for constructing nuclear potentials and currents.

The terms in the TOPT expansion are conveniently represented by diagrams. Here a distinction is made between reducible diagrams, which involve at least one pure nucleonic intermediate state, and irreducible diagrams, which include pionic and nucleonic intermediate states. The contributions of the former are enhanced with respect to those of corresponding irreducible ones by a factor of Q for each pure nucleonic intermediate state. In the static limit, in which $m \to \infty$ or, equivalently, nucleon kinetic energy terms are dropped, the reducible contributions are infrared-divergent. The prescription proposed by Weinberg[55–57] to treat these is to define the nuclear potential (and currents) as given by the irreducible contributions only. The reducible contributions are generated by solution of the Lippmann–Schwinger (or Schrödinger) equation iteratively, with the nuclear potential (and currents) given by the irreducible amplitudes.

The formalism originally developed in Ref. 90 is based on this approach. The omission of the reducible contributions from the definition of the interaction operators requires care, when the irreducible amplitudes are evaluated in the static approximation, which is commonly used. The iterative process will in that limit generate only part of the reducible amplitude. The reducible part of the amplitude beyond the static approximation has then to be incorporated order by order — along with the irreducible amplitude — in the definition of nuclear operators. This scheme in combination with TOPT, which is best suited to separate the reducible content from the irreducible one, has been implemented in Refs. 78–81 and is described below. The method does however lead to nuclear operators, which are not uniquely defined because of the nonuniqueness of the transition amplitude off-the-energy shell. This lack of uniqueness is immaterial, however, because the resulting operators are unitarily equivalent, and therefore the description of physical observables is not affected by this ambiguity.[79]

Another approach for overcoming the difficulties posed by the reducible amplitudes, has been introduced by Epelbaum and collaborators.[59] That method is usually referred to as the unitary transformation method and is based on TOPT. It exploits the Okubo (unitary) transformation[91] to decouple the Fock space of pions and nucleons into two subspaces, one that has pure nucleonic states and the other with states which retain at least one pion. In this decoupled space, the amplitude does not involve enhanced contributions associated with the reducible diagrams. The subspaces are not uniquely defined, as it is always possible to perform additional unitary transformations on them, with a consequent change in the formal definition of the resulting nuclear operators. This, of course, does not affect the calculated physical observables.

The two TOPT-based methods outlined above lead to formally equivalent operator structures for the nuclear potential and electromagnetic currents up to loop-corrections included.[80] It is natural to conjecture that the two methods are closely related, although this remains to be proved. Below we briefly outline the methods developed in Refs. 78–81 and sketch how nuclear operators are derived from transition amplitudes.

We start from the conventional perturbative expansion of the NN scattering amplitude T:

$$\langle f \mid T \mid i\rangle = \langle f \mid H_1 \sum_{n=1}^{\infty} \left(\frac{1}{E_i - H_0 + i\eta} H_1 \right)^{n-1} \mid i\rangle. \tag{12}$$

Here $\mid i\rangle$ and $\mid f\rangle$ represent the initial and final NN states with energy $E_i = E_f$, H_0 is the Hamiltonian describing free pions and nucleons, and H_1 is the Hamiltonian describing interactions between them (see Sec. 3.1). The evaluation of this amplitude is in practice carried out by inserting complete sets of H_0 eigenstates between successive terms of H_1. Power counting is then used to organize the expansion.

In the perturbation expansion of Eq. (12), a generic (reducible or irreducible) contribution is characterized by a certain number, say M, of vertices, each scaling as $Q^{\alpha_i} \times Q^{-\beta_i/2}(i = 1, \ldots, M)$, where α_i is the power counting implied by the relevant interaction Hamiltonian and β_i is the number of pions in and/or out of the vertex, a corresponding $M - 1$ number of energy denominators, and possibly L loops. Out of these $M - 1$ energy denominators, M_K will involve only nucleon kinetic energies, which scale as Q^2, and the remaining $M - M_K - 1$ will involve, in addition, pion energies, which are of order Q. Loops, on the other hand, contribute a factor Q^3 each, since they imply integrations over intermediate three momenta. Hence the power counting associated with such a contribution is

$$\left(\prod_{i=1}^{M} Q^{\alpha_i - \beta_i/2} \right) \times [Q^{-(M-M_K-1)} Q^{-2M_K}] \times Q^{3L}. \tag{13}$$

Clearly, each of the $M - M_K - 1$ energy denominators can be further expanded as

$$\frac{1}{E_i - E_I - \omega_\pi} = -\frac{1}{\omega_\pi}\left[1 + \frac{E_i - E_I}{\omega_\pi} + \frac{(E_i - E_I)^2}{\omega_\pi^2} + \cdots\right], \tag{14}$$

where E_I denotes the kinetic energy of the intermediate two-nucleon state, ω_π the pion energy (or energies, as the case may be), and the ratio $(E_i - E_I)/\omega_\pi$ is of order Q. The terms proportional to powers of $(E_i - E_I)/\omega_\pi$ lead to nonstatic corrections.

The Q-scaling of the interaction vertices and the considerations above show that the amplitude T admits the following expansion:

$$T = T^{(\nu)} + T^{(\nu+1)} + T^{(\nu+2)} + \cdots, \tag{15}$$

where $T^{(n)} \sim Q^n$, and chiral symmetry ensures that ν is finite. In the case of the two-nucleon potential $\nu = 0$, a two-nucleon potential v can then be derived, which when iterated in the Lippmann–Schwinger (LS) equation,

$$v + vG_0v + vG_0vG_0v + \cdots, \tag{16}$$

leads to the on-the-energy-shell ($E_i = E_f$) T-matrix in Eq. (15), order by order in the power counting. In practice, this requirement can only be satisfied up to a given order n^*, and the resulting potential, when inserted into the LS equation, will generate contributions of order $n > n^*$, which do not match $T^{(n)}$. In Eq. (16), G_0 denotes the free two-nucleon propagator, $G_0 = 1/(E_i - E_I + i\eta)$, and we assume that

$$v = v^{(0)} + v^{(1)} + v^{(2)} + \cdots, \tag{17}$$

where the still to be determined term $v^{(n)}$ is of order Q^n. We also note that, generally, a term like $v^{(m)}G_0v^{(n)}$ is of order Q^{m+n+1}, since G_0 is of order Q^{-2} and the implicit loop integration brings in a factor Q^3.

Having established the above power counting, we obtain

$$v^{(0)} = T^{(0)}, \tag{18}$$

$$v^{(1)} = T^{(1)} - [v^{(0)}G_0v^{(0)}], \tag{19}$$

$$v^{(2)} = T^{(2)} - [v^{(0)}G_0v^{(0)}G_0v^{(0)}] - [v^{(1)}G_0v^{(0)} + v^{(0)}G_0v^{(1)}]. \tag{20}$$

The leading-order (LO) Q^0 term, $v^{(0)}$, consists of (static) one-pion exchange (OPE) and two (non-derivative) contact interactions, while the next-to-leading (NLO) Q^1 term, $v^{(1)}$, is easily seen to vanish,[79] since the leading nonstatic corrections $T^{(1)}$ to the (static) OPE amplitude add up to zero on the energy shell, while the remaining diagrams in $T^{(1)}$ represent iterations of $v^{(0)}$, whose contributions are exactly canceled by $[v^{(0)}G_0v^{(0)}]$ (complete or partial cancellations of this type persist at higher $n \geq 2$ orders). The next-to-next-to-leading (N2LO) Q^2 term, which follows from Eq. (20), contains two-pion-exchange (TPE) and contact (involving two gradients of the nucleon fields) interactions.

The inclusion (in first order) of electroweak interactions in the perturbative expansion of Eq. (12) is in principle straightforward. The transition operator can be expanded as[79,81]:

$$T_{\rm ext} = T^{(\nu_e)}_{\rm ext} + T^{(\nu_e+1)}_{\rm ext} + T^{(\nu_e+2)}_{\rm ext} + \cdots \,, \tag{21}$$

where $T^{(n)}_{\rm ext}$ is of order Q^n and $\nu_e = -3$ in this case. The nuclear electromagnetic (weak axial) charge, ρ_γ ($\rho_{5,a}$), and current, $\mathbf{j}_\gamma$ ($\mathbf{j}_{5,a}$), operators follow from $v_\gamma = V^0\rho_\gamma - \mathbf{V}\cdot\mathbf{j}_\gamma$ ($v_5 = A^0_a\rho_{5,a} - \mathbf{A}_a\cdot\mathbf{j}_{5,a}$), where $V^\mu = (V^0, \mathbf{V})$ $[A^\mu_a = (A^0_a, \mathbf{A}_a)]$ is the electromagnetic vector (weak axial) field, and it is assumed that $v_{\rm ext}$ has a similar expansion as $T_{\rm ext}$. The requirement that, in the context of the LS equation, $v_{\rm ext}$ matches $T_{\rm ext}$ order by order in the power counting implies relations for the $v^{(n)}_\gamma = V^0\rho^{(n)}_\gamma - \mathbf{V}\cdot\mathbf{j}^{(n)}_\gamma$ and $v^{(n)}_5 = A^0_a\rho^{(n)}_{5,a} - \mathbf{A}_a\cdot\mathbf{j}^{(n)}_{5,a}$, which can be found in Refs. 79 and 81, similar to those derived above for $v^{(n)}$, the strong-interaction potential.

The lowest order terms that contribute to the electromagnetic charge and axial current operators have $\nu_e = -3$,

$$\rho^{(-3)}_\gamma = e\,\frac{1+\tau_{1,z}}{2} + (1 \rightleftharpoons 2), \tag{22}$$

$$\mathbf{j}^{(-3)}_{5,a} = -\frac{g_A}{2}\tau_{1,a}\left(\boldsymbol{\sigma}_1 - \frac{\mathbf{q}}{q^2+m_\pi^2}\boldsymbol{\sigma}_1\cdot\mathbf{q}\right) + (1 \rightleftharpoons 2). \tag{23}$$

There are no Q^{-3} contributions to $\mathbf{j}$ and $\rho_{5,a}$, and the lowest order ($\nu_e = -2$) consists of electromagnetic current and axial charge operators, given by

$$\mathbf{j}^{(-2)}_\gamma = \frac{e}{2m}\left(2\mathbf{K}_1\frac{1+\tau_{1,z}}{2} + i\boldsymbol{\sigma}_1\times\mathbf{q}\,\frac{\mu^S+\mu^V\tau_{1,z}}{2}\right) + (1 \rightleftharpoons 2), \tag{24}$$

$$\rho^{(-2)}_{5,a} = -\frac{g_A}{2m}\tau_{1,a}\boldsymbol{\sigma}_1\cdot\mathbf{K}_1 + (1 \rightleftharpoons 2), \tag{25}$$

where $\mathbf{k}_i$ and $\mathbf{K}_i$ denote hereafter the combinations of initial and final nucleon momenta

$$\mathbf{k}_i = \mathbf{p}'_i - \mathbf{p}_i, \quad \mathbf{K}_i = (\mathbf{p}'_i + \mathbf{p}_i)/2. \tag{26}$$

The counting Q^{-3} (Q^{-2}) in the electromagnetic charge and axial current (electromagnetic current and axial charge) operators follows from the product of the power counting associated with the γNN, ANN, $A\pi$ and πNN vertices, and the Q^{-3} factor due to the momentum-conserving δ-function implicit in disconnected terms of this type.

The contributions up to one loop to the electromagnetic current and charge operators are illustrated diagrammatically in Figs. 1 and 2, while those to the weak axial current and charge operators in Figs. 3 and 4. As already noted, the LO starts at $\nu_e = -2$ for the electromagnetic current and axial charge and at $\nu_e = -3$ for the electromagnetic charge and axial current; NnLO corrections to these are labeled as $Q^n \times$ LO. We begin by discussing the electromagnetic operators.

The electromagnetic currents from LO, NLO and N2LO terms and from N3LO loop corrections depend only on the known parameters g_A and f_π (NLO and N3LO),

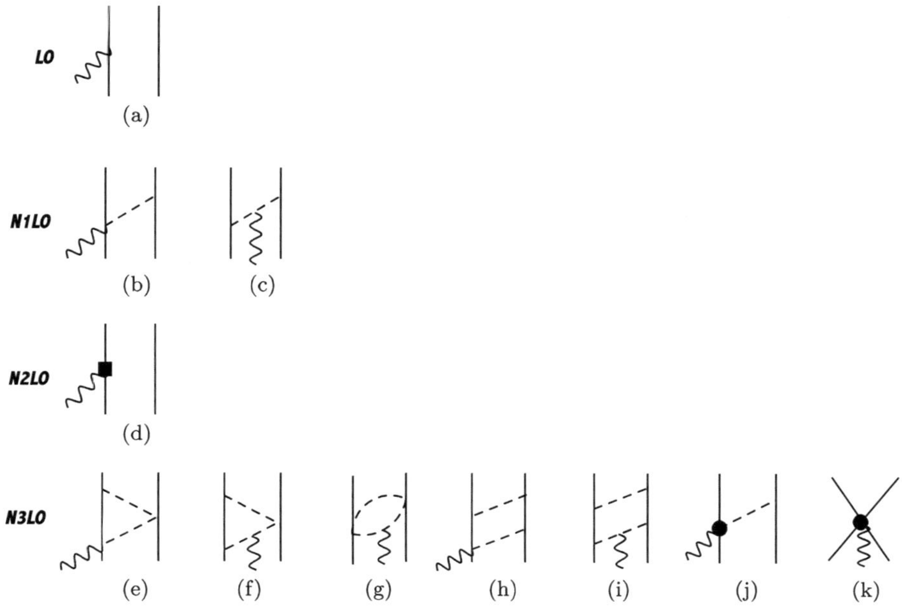

Fig. 1. Diagrams illustrating one- and two-body electromagnetic currents entering at Q^{-2} (LO), Q^{-1} (NLO), Q^0 (N2LO) and Q^1 (N3LO). Nucleons, pions and photons are denoted by solid, dashed and wavy lines, respectively. The square in panel (d) represents the $(Q/m)^2$ relativistic correction to the LO one-body current; the solid circle in panel (j) is associated with the $\gamma\pi N$ current coupling of order Q, involving the LECs d_8, d_9 and $2d_{21} - d_{22}$; the solid circle in panel (k) denotes two-body contact terms of minimal and nonminimal nature, the latter involving the LECs C'_{15} and C'_{16}. Only one among all possible time orderings is shown for the NLO and N3LO currents, so that all direct- and crossed-box contributions are accounted for.

and the nucleon magnetic moments (LO and N2LO). Unknown LECs enter the N3LO OPE contribution involving the $\gamma\pi N$ vertex of order Q^2 from $H_{\gamma\pi N}$, the term proportional to the d_i in Eq. (4), as well as the contact currents implied by nonminimal couplings, Eq. (9), discussed in the next subsection. On the other hand, in the charge operator there are no unknown LECs up to one loop level, and OPE contributions, illustrated in panels (c)–(e) of Fig. 2, only appear at N3LO. The contributions in panels (d) and (e) involve nonstatic corrections,[79] while the contribution in panel (c) is associated with the $\gamma\pi N$ coupling of order Q originating from the first term in Eq. (4). It leads to a two-body charge operator:

$$\rho_\gamma^{(0)}(\text{OPE}) = \frac{e}{8m}\frac{g_A^2}{f_\pi^2}(\boldsymbol{\tau}_1 \cdot \boldsymbol{\tau}_2 + \tau_{2z})\frac{\boldsymbol{\sigma}_1 \cdot \mathbf{q}\boldsymbol{\sigma}_2 \cdot \mathbf{k}_2}{k_2^2 + m_\pi^2} + (1 \rightleftharpoons 2). \tag{27}$$

In the present χEFT context, $\rho_\pi^{(0)}$ was derived first by Phillips in 2003.[92] However, the operator of Eq. (27) is the same as the π-exchange contribution derived within the conventional approach (see Ref. 93 and references therein). This operator plays an important role in yielding predictions for the $A = 2$–4 charge form factors that are in very good agreement with the experimental data at low and moderate values

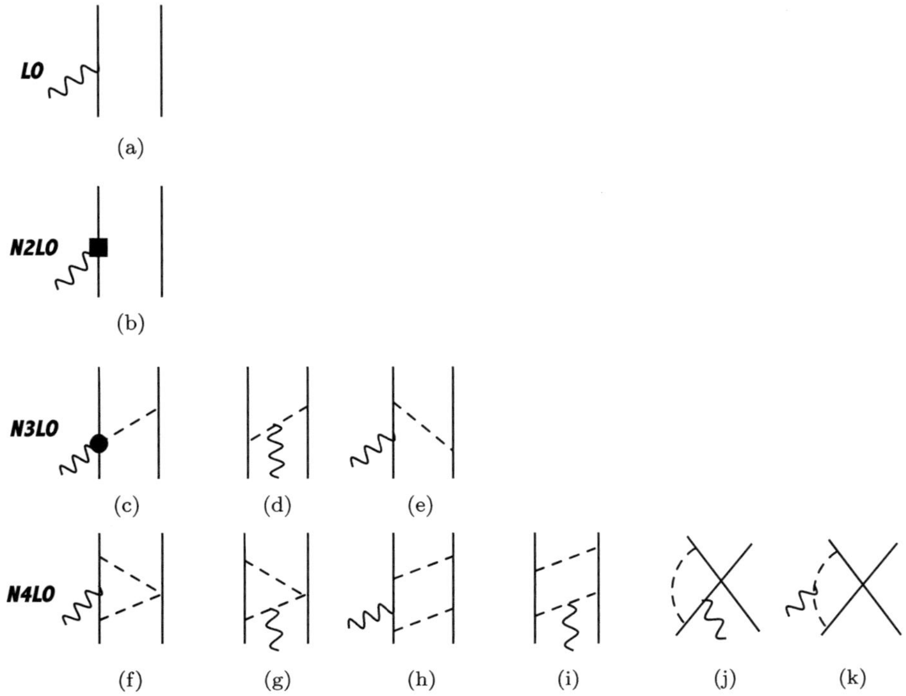

Fig. 2. Diagrams illustrating one- and two-body electromagnetic charge operators entering at Q^{-3} (LO), Q^{-1} (N2LO), Q^0 (N3LO), Q^1 (N4LO). The square in panel (b) represents the $(Q/m)^2$ relativistic correction to the LO one-body charge operator, whereas the solid circle in panel (c) is associated with a $\gamma\pi N$ charge coupling of order Q. As in Fig. 1, only a single time ordering is shown for the N3LO and N4LO charge operators.

of the momentum transfer ($q \lesssim 5\,\mathrm{fm}^{-1}$).[80,94] The calculations in Ref. 80 also showed that the OPE contributions from panels (d) and (e) of Fig. 2 are typically an order of magnitude smaller than those generated by panel (c).

The axial current and charge operators illustrated in Figs. 3 and 4 include pion-pole contributions, which are crucial for the current to be conserved in the chiral limit[81] (these contributions were ignored in the earlier studies of Park *et al.*[35,48]; obviously, they are suppressed in low momentum transfer processes). It is also interesting to note that there are no direct couplings of A_a^0 to the nucleon, see panel (a) in Fig. 4. In the axial current pion-range contributions enter at N3LO, panels (e) and (f) of Fig. 3, and involve vertices form the sub-leading $\mathcal{L}_{\pi N}^{(2)}$ Lagrangian, proportional to the LECs c_3, c_4 and c_6. It is given by (the complete operator, including pion-pole contributions, is listed in Ref. 81)

$$\mathbf{j}_{5,a}^{(0)}(\mathrm{OPE}) = \frac{g_A}{2f_\pi^2}\left\{2c_3\tau_{2,a}\mathbf{k}_2 + (\boldsymbol{\tau}_1 \times \boldsymbol{\tau}_2)_a\left[\frac{i}{2m}\mathbf{K}_1 - \frac{c_6+1}{4m}\boldsymbol{\sigma}_1 \times \mathbf{q} + \left(c_4 + \frac{1}{4m}\right)\boldsymbol{\sigma}_1 \times \mathbf{k}_2\right]\right\}\frac{\boldsymbol{\sigma}_2 \cdot \mathbf{k}_2}{k_2^2 + m_\pi^2} + (1 \rightleftharpoons 2). \tag{28}$$

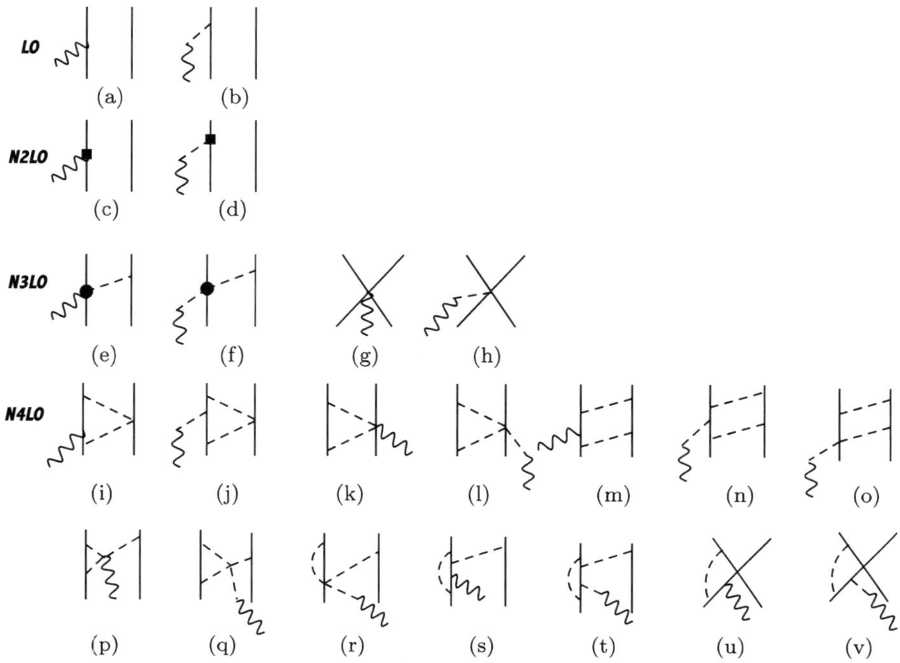

Fig. 3. Diagrams illustrating one- and two-body axial currents entering at Q^{-3} (LO), Q^{-1} (N2LO), Q^0 (N3LO) and Q^1 (N4LO). Nucleons, pions and axial fields are denoted by solid, dashed and wavy lines, respectively. The squares in panels (c) and (d) denote relativistic corrections to the one-body axial current, while the circles in panels (e) and (f) represent vertices implied by the $\mathcal{L}_{\pi N}^{(2)}$ chiral Lagrangian, involving the LECs c_i (see Ref. 81 for additional explanations). As in Fig. 1, only a single time ordering is shown.

In contrast, the axial charge has a OPE contribution at NLO, illustrated in panels (b) and (c) of Fig. 4, which reads

$$\rho_{5,a}^{(-1)}(\mathrm{OPE}) = i\frac{g_A}{4f_\pi^2}(\boldsymbol{\tau}_1 \times \boldsymbol{\tau}_2)_a \frac{\boldsymbol{\sigma}_2 \cdot \mathbf{k}_2}{k_2^2 + m_\pi^2} + (1 \rightleftharpoons 2). \tag{29}$$

In fact, an operator of precisely this form was derived by Kubodera *et al.*[95] in the late seventies, long before the systematic approach based on chiral Lagrangians now in use was established. Corrections to the axial current at N4LO in panels (i)–(v) of Fig. 3 and those to the axial charge at N3LO in panels (d)–(n) of Fig. 4 have yet to be included in actual calculations of weak transitions in nuclei. It is worthwhile noting that vertices involving three or four pions, such as those, for example, occurring in panels (l), (p), (q) and (r) of Fig. 3, depend on the pion field parametrization. This dependence must cancel out after summing the individual contributions associated with these diagrams, as indeed it does[81] (this and the requirement, remarked below, that the axial current to be conserved in the chiral limit provide useful checks of the calculation).

The loop integrals in the diagrams of Figs. 1–4 are ultraviolet divergent and are regularized in dimensional regularization.[78,79,81–83] In the electromagnetic current,

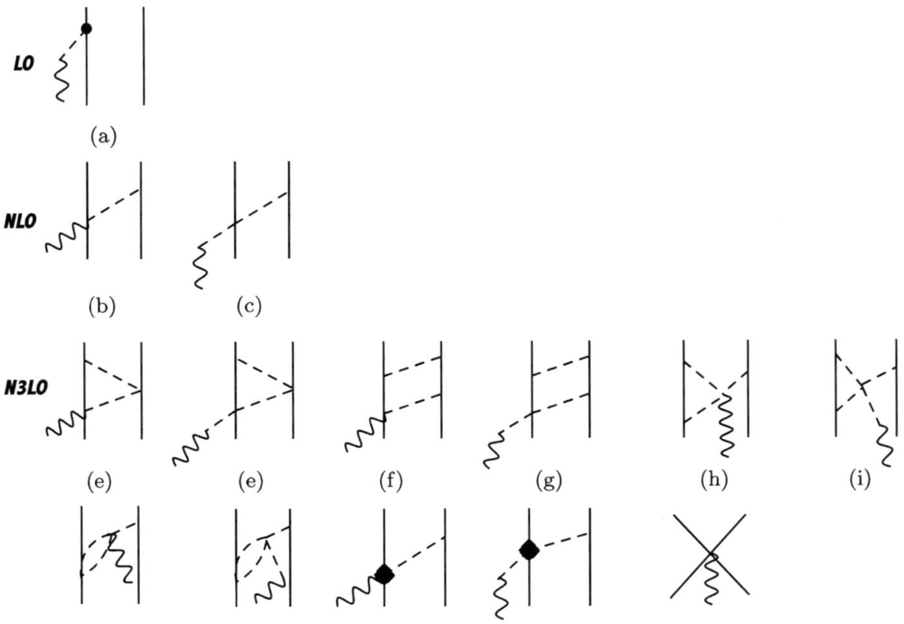

Fig. 4. Diagrams illustrating one- and two-body axial charge operators entering at Q^{-2} (LO), Q^{-1} (NLO), and Q^{1} (N3LO). Nucleons, pions, and axial fields are denoted by solid, dashed, and wavy lines, respectively. The diamonds in panels (l) and (m) indicate higher order $A\pi N$ vertices implied by the $\mathcal{L}^{(3)}_{\pi N}$ chiral Lagrangian, involving the LECs d_i (see Ref. 81 for additional explanations). As in Fig. 1, only a single time ordering is shown.

the divergent parts of these loop integrals are reabsorbed by the LECs C_i',[78,83] while those in the electromagnetic charge cancel out, in line with fact that there are no counter-terms at N4LO.[79,83] In the case of the axial operators,[48,81] there are no divergencies in the current, while those in the charge lead to renormalization of the LECs multiplying contact-type contributions. In particular, the infinities in loop corrections to the OPE axial charge (not shown in Fig. 4) are re-absorbed by renormalization of the LECs d_i in the $\mathcal{L}^{(3)}_{\pi N}$ Lagrangian. For a discussion of these issues we refer to Ref..[81]

We conclude this subsection by pointing out that at the present time two-nucleon potentials have been derived, and widely used, up to order $(Q/\Lambda_\chi)^4$ (or $v^{(4)}$, requiring two-loop contributions). Very recently, a new derivation up to order $(Q/\Lambda_\chi)^5$ has appeared.[96] Some of these high-order potentials have been used, in conjunction with the one-loop operators presented here, in calculations of electroweak observables of light nuclei, as reported below. Conservation of the electromagnetic current $\mathbf{q}\cdot\mathbf{j}_\gamma = [H, \rho_\gamma]$ with the two-nucleon Hamiltonian given by $H = T^{(-1)} + v^{(0)} + v^{(2)} + \cdots$ and where the (two-nucleon) kinetic energy $T^{(-1)}$ is counted as Q^{-1}, implies,[78] order by order in the power counting, a set of nontrivial relations between the $\mathbf{j}_\gamma^{(n)}$ and the $T^{(-1)}$, $v^{(n)}$ and $\rho_\gamma^{(n)}$ (note that commutators implicitly bring in factors of Q^3) — incidentally, similar considerations also apply to the conservation of the axial current in the chiral limit.[81] These relations couple different orders in the power counting of the operators, making it impossible to carry out a calculation, which at a given n for $\mathbf{j}_\gamma^{(n)}$, $v^{(n)}$ and $\rho_\gamma^{(n)}$ (and hence "consistent" from a power-counting perspective) also leads to a conserved current.

4. Results

In this section, we provide a sample of results obtained with χEFT electroweak currents for light systems, including $A = 2$–4 nuclei and s- and p-shell nuclei in the mass range $A = 6$–10, in the last five years or so. The few-nucleon calculations are based on the chiral two-nucleon potentials developed by Entem and Machleidt[60,61] at order Q^4 in the power counting including up to two-loop corrections, and chiral three-nucleon potentials at leading order.[97] (Below, this combination of two and three-nucleon potentials will be referred to as N3LO/N2LO, as is customarily done in the literature, even though such a classification does not conform to the power-counting notation adopted in the present chapter.) As noted earlier, the two LECs c_D — related to the LEC z_0 in the contact axial current of Eq. (10) — and c_E entering the three-nucleon potential have been constrained by fitting the Gamow–Teller matrix element in tritium β-decay and the binding energies of the trinucleons.[88,89]

The electroweak operators in Figs. 1–4 have power law behavior for large momenta, and need to be regularized before they can be sandwiched between nuclear wave functions. The regulator is taken of the form $C_\Lambda(k) = \exp[-(k/\Lambda)^n]$ with $n = 4$ and Λ in the range (500–600) MeV. For processes involving low momentum and energy transfers one would expect predictions to be fairly insensitive to variations of Λ. As shown below, this expectation is borne out in actual calculations, at least in the case of processes which are not inhibited at leading order, such as the $n\,^3$He radiative capture or $p\,^3$He weak fusion.

There are 5 unknown LECs in $\mathbf{j}$ — see Fig. 5 or panels (j) and (k) of Fig. 1 — and none in ρ.[78,80,82,83] Two (three) of these LECs multiply isoscalar (isovector)

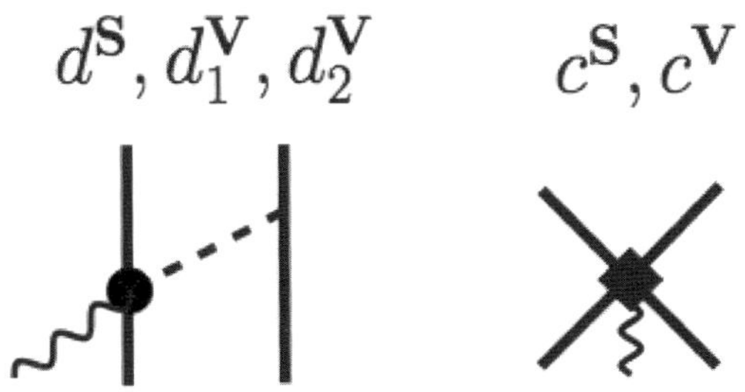

Fig. 5. The isoscalar d^S and c^S, and isovector d_1^V, d_2^V and c^V LECs characterizing the electromagnetic current at N3LO.

Table 1. Values for the LECs in units $1/\Lambda^2$ for d^S and $1/\Lambda^4$ for c^S and c^V; see text for further explanations.

Λ MeV	c^S	$d^S \times 10$	$c^V(\sigma_{np})$	$c^V(\mu^V)$
500	4.1	2.2	−13	−8.0
600	11	3.2	−22	−12

operators. For each Λ the two isoscalar LECs are fixed by reproducing the deuteron and isoscalar trinucleon magnetic moments. Two of the isovector LECs are then constrained by assuming Δ-resonance saturation,[80] while the remaining LEC is determined by reproducing (again for each Λ) either the np radiative capture cross-section σ_{np} at thermal neutron energies or the isovector trinucleon magnetic moment μ^V.[80] There are no three-body currents entering at the order of interest,[98] and so it is possible to use three-nucleon observables to fix some of these LECs. Their values are listed in Table 1. They are generally rather large, particularly when c^V is determined by the np radiative capture cross-section. The exception is the isoscalar LEC d^S multiplying the OPE current involving a sub-subleading $\gamma\pi N$ vertex from the chiral Lagrangian $\mathcal{L}^{(3)}_{\pi N}$, which in a resonance-saturation picture reduces to the $\rho\pi\gamma$ transition current.

The calculations of $A = 6$–10 nuclei are carried out in the hybrid approach, in which χEFT electroweak currents are used in combination with he Argonne v_{18} two-nucleon (AV18) and Illinois-7 three-nucleon (IL7) potentials. The AV18 consists of a long-range component induced by OPE and intermediate-to-short range components modeled phenomenologically and constrained to fit the NN database beyond the pion-production threshold ($E_{\text{lab}} = 350\,\text{MeV}$). The IL7 includes a central (albeit isospin dependent) short-range repulsive term and two- and three-pion-exchange mechanisms involving excitation of intermediate Δ resonances. Its strength is determined by four parameters which are fixed by a best fit to the energies of 17 low-lying states of nuclei in the mass range $A \leq 10$, obtained in combination with the AV18 potential. The AV18/IL7 Hamiltonian then leads to predictions of ~ 100 ground- and excited-state energies up to $A = 12$, including the ^{12}C ground- and Hoyle-state energies, in good agreement with the corresponding empirical values (for a recent review of these as well as results obtained for nuclear and neutron matter, see Ref. 99).

4.1. *Electromagnetic observables of $A = 2$–4 nuclei*

The deuteron magnetic form factor, calculated in Ref. 80, is shown in Fig. 6. The bands reflect the sensitivity to cutoff variations in the range $\Lambda = (500$–$600)\,\text{MeV}$. The black bands include all corrections up to N3LO in the (isoscalar) electromagnetic (EM) current. The NLO OPE and N3LO TPE currents are isovector and therefore give no contributions to this observable. The right panel of Fig. 6 contains a comparison of the results of Ref. 80 with those of a calculation based on a lower

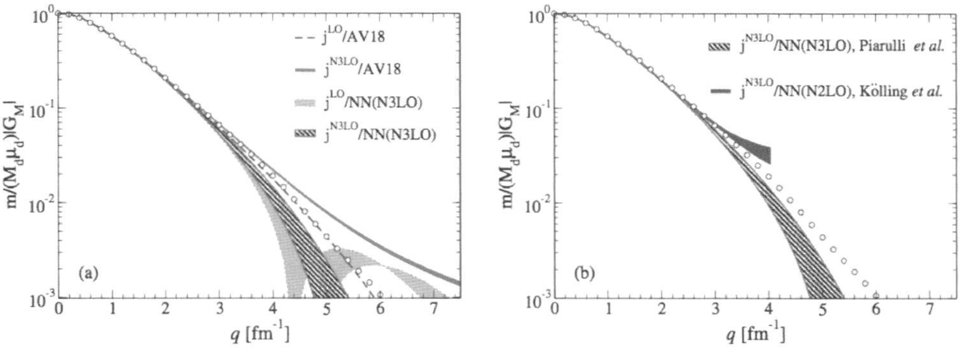

Fig. 6. Magnetic form factor of the deuteron: the left panel shows the results obtained with LO and N3LO currents and either the chiral N3LO or conventional AV18 potential; the right panel shows results obtained with N3LO currents and either the chiral N3LO (same as in left panel) or a chiral N2LO potential by Kölling *et al.* The bands reflect cutoff variation. Experimental data are the empty circles.

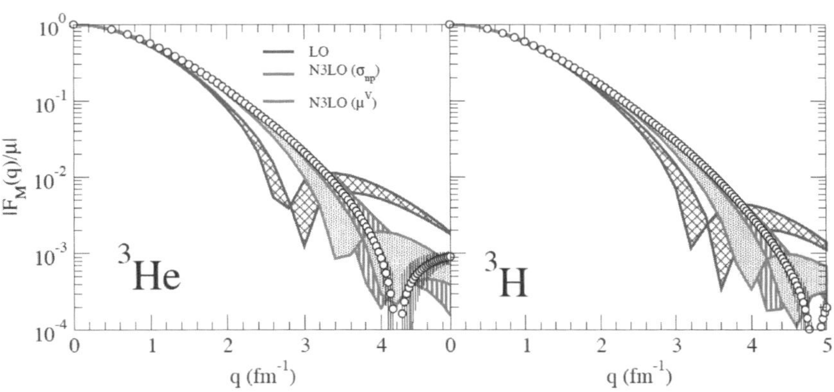

Fig. 7. Magnetic form factors of ^{3}He (left panel) and ^{3}H (right panel); see text for further explanations.

order potential and in which a different strategy was adopted for constraining the LEC's d^S and c^S in the N3LO EM current.[100] This figure and the following Fig. 2 are from the recent review paper by Bacca and Pastore.[101]

The predicted magnetic form factors of the ^{3}He and ^{3}H ground states are compared to experimental data in Fig. 7.[80] Isovector OPE and TPE two-body terms in the EM current play an important role in these observables, confirming previous results obtained in the conventional meson-exchange framework. We show the N3LO results corresponding to the two different ways used to constrain the LEC c^V in the isovector contact current (recall the LECs d_1^V and d_2^V are assumed to be saturated by the Δ resonance), namely by reproducing (i) the empirical value for the np cross-section — curve labeled N3LO(σ_{np}) — or (ii) the isovector magnetic moment of ^{3}He/^{3}H — curve labeled N3LO(μ^V). The bands display the cutoff sensitivity, which becomes rather large for momentum transfers $q \gtrsim 3$ fm^{-1}. The N3LO(σ_{np})

results are in better agreement with the data at higher momentum transfers; however, they overestimate μ^V by $\sim 2\%$. On the other hand, the N3LO(μ^V) results, while reproducing μ^V by construction, under-predict σ_{np} by $\sim 1\%$.

Moving on to the EM charge operator, we show in Figs. 8 and 9 very recent calculations of the deuteron monopole and quadrupole form factors[80] and ^{4}He (charge) form factor.[102] There are no unknown LECs beyond g_A, f_π and the nucleon magnetic

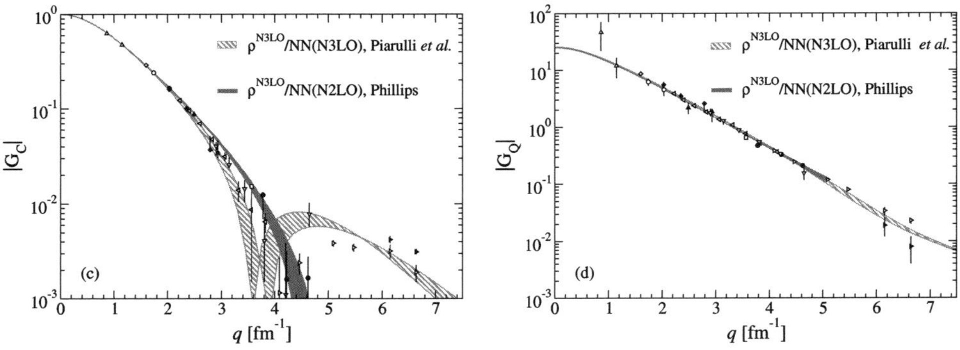

Fig. 8. The deuteron monopole and quadrupole form factors obtained from measurements of the A structure function and tensor polarization are compared to predictions based on N2LO and N3LO chiral potentials.

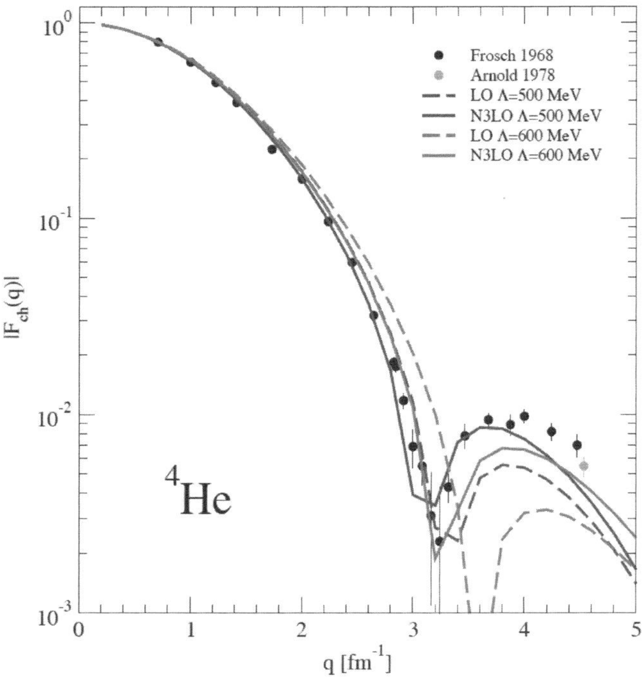

Fig. 9. The ^{4}He charge form factor obtained from elastic electron scattering data is compared to results obtained with the LO and N3LO charge operator.

moments — the latter enter a relativistic correction, suppressed by Q^2 relative to the LO charge operator, i.e., the well-known spin-orbit term. The loop contributions (at N4LO) from two-pion exchange are isovector and hence vanish for these observables.

The deuteron monopole and quadrupole form factor data are obtained from measurements of the A structure function and tensor polarization observable in electron–deuteron scattering. In Fig. 8, the two bands correspond to two different calculations, one of which, labeled as NN(N2LO), is based on a lower order chiral potential.[92,103] There is good agreement between theory and experiment. Differences between the two sets of theory predictions merely reflect differences in the deuteron wave functions obtained with the N3LO and N2LO potentials. These differences are amplified in the diffraction region of the monopole form factor.

The ^{4}He charge form factor is obtained from elastic electron scattering cross-section data. These data now extend up to momentum transfers $q \lesssim 10\,\text{fm}^{-1}$,[104] well beyond the range of applicability of χEFT. In Fig. 9, only data up to $q \lesssim 5\,\text{fm}^{-1}$ are shown. They are in excellent agreement with theory.

Predictions for the charge radii of the deuteron and helium isotopes and for the deuteron quadrupole moment (Q_d) are listed in Table 2.[80] They are within 1% of experimental values. It is worth noting that until recently calculations based on the conventional meson-exchange framework were used to consistently underestimate Q_d. However, this situation has now changed, and a relativistic calculation in the covariant spectator theory based on a one-boson exchange model of the NN interaction has led to a value for the quadrupole moment[105] which is in agreement with experiment.

As a last example we show in Fig. 10 predictions for the nd and $n\,^3$He radiative capture cross-sections at thermal neutron energies.[98] It is well known that these $M1$ transitions are suppressed when the magnetic dipole operator is taken to consist only of proton and neutron contributions, i.e., in the impulse approximation (IA). The results shown in Fig. 10 have been obtained from highly accurate (essentially exact) solutions of the bound and continuum states of the $A = 3$ and 4 systems with the hyperspherical-harmonics technique,[106] based on both chiral and conventional two- and three-nucleon potentials, the N3LO/N2LO and AV18/UIX, and chiral electromagnetic currents up to N3LO. However, in this earlier study, the procedure adopted to fix the three isovector LECs is different from that utilized in the calculations discussed so far. Here Δ-resonance saturation is exploited only to fix the

Table 2. The charge radii of the ^{2}H, ^{3}He and ^{4}He nuclei, and ^{2}H quadrupole moment. The numbers in parentheses at the side of the χEFT predictions give the cutoff dependence of the results.

	$r_c(^2\text{H})$ (fm)	Q_d (fm^2)	$r_c(^3\text{He})$ (fm)	$r_c(^4\text{He})$ (fm)
χEFT	2.126(4)	0.2836(16)	1.962(4)	1.663(11)
EXP	2.130(10)	0.2859(6)	1.973(14)	1.681(4)

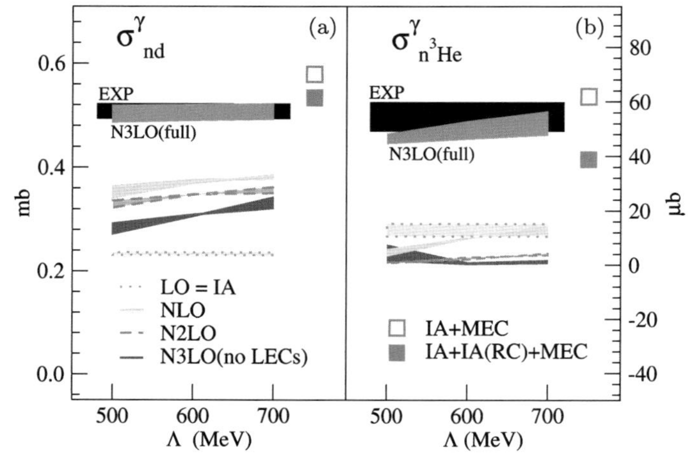

Fig. 10. Results for the nd and $n\,^3$He radiative capture cross-sections, obtained by including cumulatively the LO, NLO, N2LO, N3LO(no LECs) and N3LO(full) contributions from the χEFT electromagnetic current. Also shown are predictions obtained in the conventional approach based on the AV18/UIX Hamiltonian and accompanying effective-meson exchange currents, the square labeled IA+MEC and IA+IA(RC)+MEC, the latter including relativistic correction to the IA operator. The black band represents the experimental data, see text for further explanations.

ratio of the two LECs d_1^V and d_2^V in the tree-level contribution of Fig. 5, and the remaining two (isovector) LECs are then determined by a simultaneous fit to σ_{np} and the isovector combination of the trinucleon magnetic moment. Furthermore, the LECs in the contact current originating from minimal couplings have been taken from a lower order (NLO) chiral potential[78] rather than from the N3LO potential of Refs. 60 and 61.

In Fig. 10, the experimental data are from Ref. 107 for nd and Refs. 108 and 109 for $n\,^3$He, the band thickness denoting the error. Results obtained with the complete N3LO χEFT operator are shown by the orange band labeled N3LO(full): those corresponding to the N3LO/N2LO (AV18/UIX) model delimit the lower (upper) end of the band in the case of nd, and its upper (lower) end in the case of $n\,^3$He. There is considerable cutoff dependence, particularly for the four-body capture. In this connection, it is interesting to note the crucial role played by the N3LO currents in Fig. 5: indeed, retaining only the minimal contact currents and the currents from TPE loop corrections — bands labeled N3LO(no LECs) — would severely under-estimate the measured cross-sections. It is clear that the convergence of the chiral expansion for these processes is problematic. The LO (or IA) is unnaturally small, since the associated operator cannot connect the dominant S-states in the hydrogen and helium bound states. This leads to an enhancement of the NLO contribution, which, however, in the case of $n\,^3$He, is offset by the destructive interference between it and the LO contribution. Thus, a satisfactory description of these processes remains particularly challenging for nuclear theory and nuclear χEFT in particular.

4.2. *Electromagnetic transitions in $A = 6$–10 nuclei*

Heavier systems offer new challenges and opportunities for applications of nuclear χEFT, and *ab initio* studies of electroweak processes based on this approach in systems with $A > 4$ have only just begun. In the mass range $A = 6$–12 variational Monte Carlo (VMC) and Green's function Monte Carlo (GFMC) methods allow us to carry out accurate, in fact exact in the case of GFMC, first-principles calculations of many nuclear properties (see Ref. 99 for a recent review). However, since these methods are formulated in configuration space, it has not been possible to use them in conjunction with the chiral two-and three-nucleon potentials above, which are given in momentum-space (and are strongly nonlocal in configuration space). A first step in this direction is the very recent development of a class of configuration-space, minimally nonlocal two-nucleon chiral potentials that fit the *np* and *pp* database up to the pion-production threshold with a χ^2 per datum close to 1.3,[110] i.e., of the same quality as the well established N3LO models of Refs. 60 and 61. Use of these potentials in VMC and, especially, GFMC calculations of light s- and p-shell nuclei will expand the scope of the nuclear χEFT approach, and in particular test its validity beyond the realm of few-nucleon systems, to which it has primarily been limited so far.

As mentioned earlier, the results presented in this subsection for the magnetic moments and transition widths of $A = 6$–10 nuclei[111] have been obtained in the hybrid approach, which combines conventional potentials (AV18 and IL7) with chiral electromagnetic currents, see Figs. 11 and 12. Figure 11 makes it plain that the inclusion of corrections beyond LO is necessary in order to have a satisfactory description of the experimental data: their effect is particularly pronounced in the

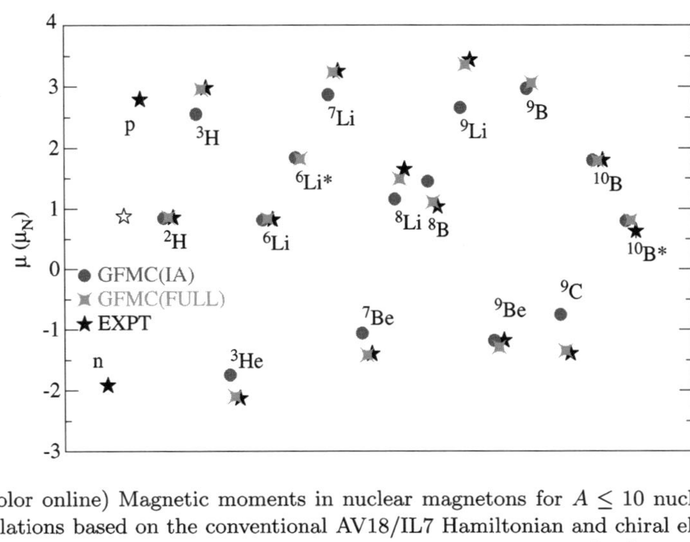

Fig. 11. (Color online) Magnetic moments in nuclear magnetons for $A \leq 10$ nuclei obtained in GFMC calculations based on the conventional AV18/IL7 Hamiltonian and chiral electromagnetic currents. Black stars indicate the experimental values, while the blue (red) dots represent the results of calculations including the LO (N3LO) chiral electromagnetic current.

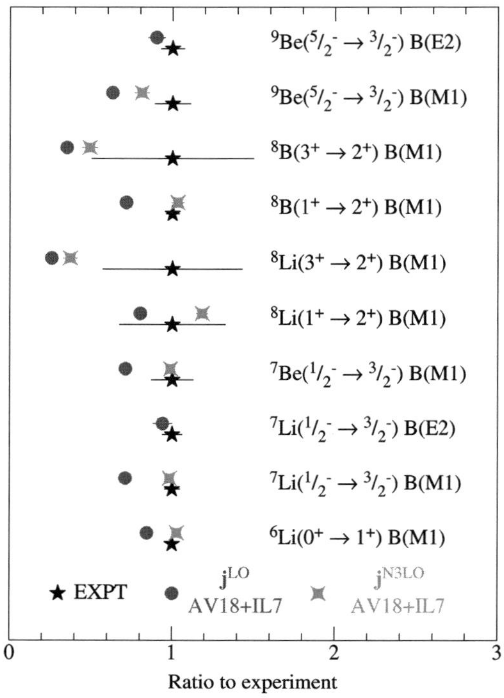

Fig. 12. (Color online) Ratio to the experimental $M1$ and $E2$ transition widths in $A \leq 9$. Black stars with error bars indicate the experimental values, while the blue dots (red diamonds) represent GFMC AV18/IL7 calculations including chiral electromagnetic currents at LO (up to N3LO).

$A = 9$ and isospin $T = 3/2$ systems, in which they provide up to $\sim 20\%$ (40%) of the total predicted value for the ^{9}Li (^{9}C) magnetic moments.

In Fig. 12, the calculated $M1$ and $E2$ transition widths for $A = 6$–9 nuclei are compared to experimental data. Overall, there is good agreement between theory and experiment, particularly when one considers the fact that for systems like ^{8}Li and ^{8}B the errors bars are so large to prevent any robust conclusions to be drawn from the apparent under-prediction by theory of the associated widths. It should be noted that the $E2$ widths have been obtained with the LO EM charge operator. Higher order corrections are expected to be very small, in particular the OPE charge operator in Eq. (27) vanishes in the static limit.

4.3. *Weak transitions in few-nucleon systems*

Most calculations of nuclear axial current matrix elements, such as those discussed below for the *pp* weak fusion and for muon capture on ^{2}H and ^{3}He, have used axial current operators up to N3LO or Q^0 (one exception is Ref. 112, which included effective one-body reductions, for use in a shell-model study, of some of the TPE corrections to the axial current derived in Ref. 35). A recent application of these N3LO transition operators is the calculation of the rates for μ^- capture on deuteron and ^{3}He.[89] These rates have been predicted with $\sim 1\%$ accuracy,

$$\Gamma(^2\mathrm{H}) = (399 \pm 3)\,\mathrm{s}^{-1}, \quad \Gamma(^3\mathrm{He}) = (1494 \pm 21)\,\mathrm{s}^{-1}.$$

At this level of precision, it is necessary to also account for electroweak radiative corrections, which have been evaluated for these processes in Ref. 113. The error quoted on the predictions above results from a combination of (i) the experimental error on the ^{3}H GT matrix element used to fix the LEC in the contact axial current, (ii) uncertainties in the electroweak radiative corrections — overall, these corrections increase the rates by 3% — and (iii) the cutoff dependence.

There is a very accurate and precise measurement of the rate on ^{3}He: $\Gamma^{\mathrm{EXP}}(^3\mathrm{He}) = (1496{\pm}4)\,\mathrm{s}^{-1}$.[114] It can be used to constrain the induced pseudo-scalar form factor of the nucleon. It gives $G_{PS}(q_0^2 = -0.95\,m_\mu^2) = 8.2 \pm 0.7$, which should be compared to a direct measurement on hydrogen at PSI, $G_{PS}^{\mathrm{EXP}}(q_0^2 = -0.88\,m_\mu^2) = 8.06 \pm 0.55$,[115] and a chiral perturbation theory prediction of 7.99 ± 0.20.[116,117]

The situation for μ^- capture on ^{2}H remains, to this day, somewhat confused: there is a number of measurements that have been carried out, but they all have rather large error bars. However, this unsatisfactory state of affairs should be cleared by an upcoming measurement of this rate by the MuSun collaboration at PSI with a projected 1% error.

Another recent example is the proton weak capture on protons.[35,118] This process is important in solar physics: it is the largest source of energy and neutrinos in the Sun. The astrophysical S-factor for this weak fusion reaction is one of the inputs in the standard model of solar (and stellar) evolution.[119] A recent calculation

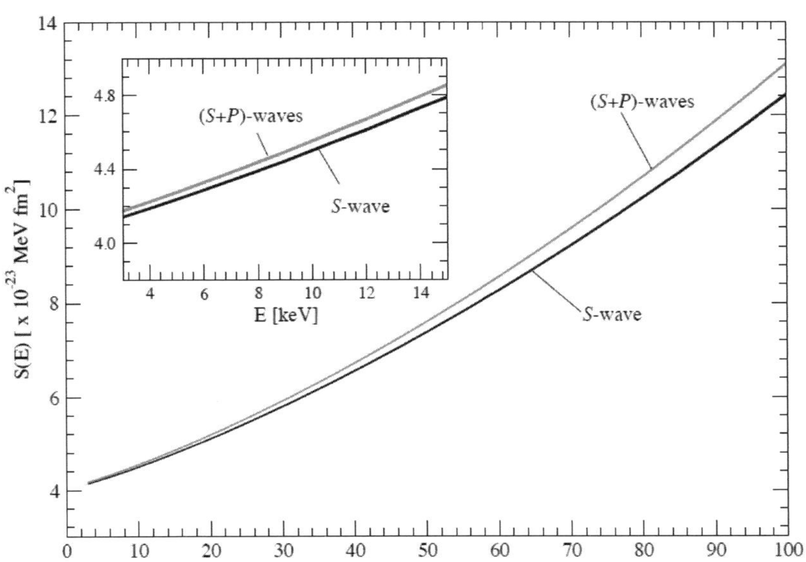

Fig. 13. The S-factor for pp weak fusion due to S- and (S+P)-wave capture as function of energy.

based on N3LO chiral potentials including a full treatment of EM interactions up to order α^2 (α is the fine structure constant), shows that it is now predicted with an accuracy of much less than 1%: $S(0) = (4.030 \pm 0.006) \times 10^{-23}$ MeV-fm^2.

This calculation also included the (small) effects from capture of the two protons in relative P-wave, see Fig. 13.[118] The increase due to P-wave capture offsets the decrease from higher order EM effects, in particular vacuum polarization.

5. Conclusion

The presentation above illustrates the remarkable progress of the development and application of the chiral Lagrangian-based description of nuclear electroweak current operators beginning with the first steps taken by Gerry Brown, Mannque Rho and their colleagues and students around 1970. The early work in the 1970s was, however, based on the lowest order terms in the chiral Lagrangians and, to a large extent, phenomenological wave functions. The advent of systematic chiral effective field theory has brought the theoretical work to a quantitative level and has provided a basis for it in the fundamental theory of the strong interactions.

Acknowledgments

We wish to thank our collaborators A. Baroni, J. Carlson, L. Girlanda, A. Kievsky, L. E. Marcucci, S. Pastore, M. Piarulli, S. C. Pieper, M. Viviani and R. B. Wiringa for their many contributions to the work presented here. The support of the U.S. Department of Energy under contract DE-AC05-06OR23177 is also gratefully acknowledged.

References

1. S. Weinberg, *Phys. Rev. Lett.* **17** (1966) 616.
2. G. Brown and J. Durso, *Phys. Lett. B* **35** (1971) 120.
3. M. Chemtob, J. Durso and D. O. Riska, *Nucl. Phys. B* **38** (1972) 141.
4. M. Chemtob and M. Rho, *Nucl. Phys. A* **163** (1971) 1.
5. M. Chemtob and M. Rho, *Phys. Lett. B* **29** (1969) 540.
6. D. O. Riska and G. Brown, *Phys. Lett. B* **32** (1970) 662.
7. D. O. Riska, *Phys. Scr.* **31** (1985) 471.
8. F. Gross and D. O. Riska, *Phys. Rev. C* **36** (1987) 1928.
9. D. O. Riska and G. Brown, *Phys. Lett. B* **38** (1972) 193.
10. G. Stranahan, *Phys. Rev.* **135** (1964) 953.
11. E. Harper, Y. Kim, A. Tubis and M. Rho, *Phys. Lett. B* **40** (1972) 533.
12. E. Hadjimichael, *Phys. Rev. Lett.* **31** (1973) 183.
13. I. Towner and F. Khanna, *Nucl. Phys. A* **356** (1981) 445.
14. J. Carlson, D. O. Riska, R. Schiavilla and R. B. Wiringa, *Phys. Rev. C* **42** (1990) 830.
15. D. O. Riska, *Nucl. Phys. A* **606** (1996) 251.
16. J. Hockert, D. O. Riska, M. Gari and A. Huffman, *Nucl. Phys. A* **217** (1973) 14.
17. R. Schiavilla and D. O. Riska, *Phys. Rev. C* **43** (1991) 437.
18. R. A. Brandenburg, Y. E. Kim and A. Tubis, *Phys. Rev. Lett.* **32** (1974) 1325.

19. A. Barroso and E. Hadjimichael, *Nucl. Phys. A* **238** (1975) 422.
20. R. Schiavilla, V. R. Pandharipande and D. O. Riska, *Phys. Rev. C* **40** (1989) 2294.
21. R. B. Wiringa and R. Schiavilla, *Phys. Rev. Lett.* **81** (1998) 4317.
22. M. Chemtob and A. Lumbroso, *Nucl. Phys. B* **17** (1970) 401.
23. J. Dubach, J. Koch and T. W. Donnelly, *Nucl. Phys. A* **271** (1976) 279.
24. M. Gari and H. Hyuga, *Phys. Rev. Lett.* **36** (1976) 345.
25. E. M. Nyman and D. O. Riska, *Phys. Rev. Lett.* **57** (1986) 3007.
26. M. Wakamatsu and W. Weise, *Nucl. Phys. A* **477** (1988) 559.
27. L. E. Marcucci, D. O. Riska and R. Schiavilla, *Phys. Rev. C* **58** (1998) 3069.
28. T.-S. Park, D.-P. Min and M. Rho, *Phys. Rev. Lett.* **74** (1995) 4153.
29. E. Fischbach, E. Harper, Y. Kim, A. Tubis and W. Cheng, *Phys. Lett. B* **38** (1972) 8.
30. M. Gari and A. Huffman, *Astrophys. J.* **174** (1972) L153.
31. F. Dautry, M. Rho and D. O. Riska, *Nucl. Phys. A* **264** (1976) 507.
32. R. B. Wiringa, V. G. J. Stoks and R. Schiavilla, *Phys. Rev. C* **51** (1995) 38.
33. J. Carlson, D. O. Riska, R. Schiavilla and R. B. Wiringa, *Phys. Rev. C* **44** (1991) 619.
34. T.-S. Park, K. Kubodera, D.-P. Min and M. Rho, *Astrophys. J.* **507** (1998) 443.
35. T.-S. Park, L. E. Marcucci, R. Schiavilla, M. Viviani, A. Kievsky, S. Rosati, K. Kubodera, D.-P. Min and M. Rho, *Phys. Rev. C* **67** (2003) 055206.
36. W. Kloet and J. Tjon, *Phys. Lett. B* **49** (1974) 419.
37. J. Borysowicz and D. O. Riska, *Nucl. Phys. A* **254** (1975) 301.
38. R. Schiavilla, V. R. Pandharipande and D. O. Riska, *Phys. Rev. C* **41** (1990) 309.
39. M. Radomski and D. O. Riska, *Nucl. Phys. A* **274** (1976) 428.
40. A. Lovato, S. Gandolfi, R. Butler, J. Carlson, E. Lusk, S. C. Pieper and R. Schiavilla, *Phys. Rev. Lett.* **111** (2013) 092501.
41. J. W. Negele and D. O. Riska, *Phys. Rev. Lett.* **40** (1978) 1005.
42. D. Abbott *et al.*, *Eur. Phys. J. A* **7** (2000) 421.
43. D. O. Riska and M. Radomski, *Phys. Rev. C* **16** (1977) 2105.
44. M. Kirchbach, R. Mach and D. O. Riska, *Nucl. Phys. A* **511** (1990) 592.
45. M. Kirchbach, D. O. Riska and K. Tsushima, *Nucl. Phys. A* **542** (1992) 616.
46. I. Towner, *Nucl. Phys. A* **542** (1992) 631.
47. E. K. Warburton, *Phys. Rev. Lett.* **66** (1991) 1823.
48. T.-S. Park, D.-P. Min and M. Rho, *Phys. Rep.* **233** (1993) 341.
49. T.-S. H. Lee and D. O. Riska, *Phys. Rev. Lett.* **70** (1993) 2237.
50. G. Brown and M. Rho, *Phys. Lett. B* **82** (1979) 177.
51. G. Brown, A. Jackson, M. Rho and V. Vento, *Phys. Lett. B* **140** (1984) 285.
52. E. Witten, *Nucl. Phys. B* **160** (1979) 57.
53. E. M. Nyman and D. O. Riska, *Nucl. Phys. A* **468** (1987) 473.
54. U. Blom and D. O. Riska, *Nucl. Phys. A* **476** (1988) 603.
55. S. Weinberg, *Phys. Lett. B* **251** (1990) 288.
56. S. Weinberg, *Nucl. Phys. B* **363** (1991) 3.
57. S. Weinberg, *Phys. Lett. B* **295** (1992) 114.
58. C. Ordonez, L. Ray and U. van Kolck, *Phys. Rev. C* **53** (1996) 2086.
59. E. Epelbaum, W. Gloeckle and U. G. Meissner, *Nucl. Phys. A* **637** (1998) 107.
60. D. Entem and R. Machleidt, *Phys. Rev. C* **68** (2003) 041001.
61. R. Machleidt and D. Entem, *Phys. Rep.* **503** (2011) 1.
62. P. Navratil, *Few-Body Syst.* **41** (2007) 117.
63. E. Epelbaum *et al.*, *Phys. Rev. C* **66** (2002) 064001.
64. U. van Kolck, *Phys. Rev. C* **49** (1994) 2932.

65. V. Bernard, E. Epelbaum, H. Krebs and U.-G. Meissner, *Phys. Rev. C* **84** (2011) 054001.
66. L. Girlanda, A. Kievsky and M. Viviani, *Phys. Rev. C* **84** (2011) 014001.
67. J. L. Friar and U. van Kolck, *Phys. Rev. C* **60** (1999) 034006.
68. E. Epelbaum and U. G. Meissner, *Phys. Lett. B* **461** (1999) 287.
69. J. L. Friar, U. van Kolck, M. C. M. Rentmeester and R. G. E. Timmermans, *Phys. Rev. C* **70** (2004) 044001.
70. J. L. Friar, G. L. Payne and U. van Kolck, *Phys. Rev. C* **71** (2005) 024003.
71. W. C. Haxton and B. R. Holstein, *Prog. Part. Nucl. Phys.* **71** (2013) 185.
72. S. L. Zhu, C. M. Maekawa, B. R. Holstein, M. J. Ramsey-Musolf and U. van Kolck, *Nucl. Phys. A* **748** (2005) 435.
73. L. Girlanda, *Phys. Rev. C* **77** (2008) 067001.
74. M. Viviani, A. Baroni, L. Girlanda, A. Kievsky, L. E. Marcucci and R. Schiavilla, *Phys. Rev. C* **89** (2014) 064004.
75. V. Pascalutsa and D. R. Phillips, *Phys. Rev. C* **67** (2003) 055202.
76. H. Griesshammer, J. McGovern, D. R. Phillips and G. Feldman, *Prog. Part. Nucl. Phys.* **67** (2012) 841.
77. T.-S. Park, D.-P. Min and M. Rho, *Nucl. Phys. A* **596** (1996) 515.
78. S. Pastore, L. Girlanda, R. Schiavilla, M. Viviani and R. Wiringa, *Phys. Rev. C* **80** (2009) 034004.
79. S. Pastore, L. Girlanda, R. Schiavilla and M. Viviani, *Phys. Rev. C* **84** (2011) 024001.
80. M. Piarulli, L. Girlanda, L. Marcucci, S. Pastore, R. Schiavilla and M. Viviani, *Phys. Rev. C* **87** (2013) 014006.
81. A. Baroni, L. Girlanda, S. Pastore, R. Schiavilla and M. Viviani, *Phys. Rev. C* **93** (2016) 015501.
82. S. Kölling, E. Epelbaum, H. Krebs and U.-G. Meissner, *Phys. Rev. C* **80** (2009) 045502.
83. S. Kölling, E. Epelbaum, H. Krebs and U.-G. Meissner, *Phys. Rev. C* **84** (2011) 054008.
84. J. Gasser and H. Leutwyler, *Ann. Phys.* **158** (1984) 142.
85. N. Fettes, U.-G. Meissner, M. Mojzis and S. Steininger, *Ann. Phys. (N.Y.)* **283** (2000) 273.
86. L. Girlanda, S. Pastore, R. Schiavilla and M. Viviani, *Phys. Rev. C* **81** (2010) 034005.
87. A. Gårdestig and D. R. Phillips, *Phys. Rev. Lett.* **96** (2006) 232301.
88. D. Gazit, S. Quaglioni and P. Navrátil, *Phys. Rev. Lett.* **103** (2009) 102502.
89. L. E. Marcucci, A. Kievsky, S. Rosati, R. Schiavilla and M. Viviani, *Phys. Rev. Lett.* **108** (2012) 052502.
90. S. Pastore, R. Schiavilla and J. L. Goity, *Phys. Rev. C* **78** (2008) 064002.
91. S. Okubo, *Prog. Theor. Phys.* **12** (1954) 603.
92. D. R. Phillips, *Phys. Lett. B* **567** (2003) 12.
93. D. O. Riska, *Phys. Rep.* **181** (1989) 207.
94. J. Carlson and R. Schiavilla, *Rev. Mod. Phys.* **70** (1998) 743.
95. K. Kubodera, J. Delorme and M. Rho, *Phys. Rev. Lett.* **40** (1978) 755.
96. E. Epelbaum, H. Krebs and U. G. Meissner, *Phys. Rev. Lett.* **115** (2015) 122301.
97. E. Epelbaum, A. Nogga, W. Glöckle, H. Kamada, U.-G. Meißner and H. Witała, *Phys. Rev. C* **66** (2002) 064001.
98. L. Girlanda, A. Kievsky, L. Marcucci, S. Pastore, R. Schiavilla and M. Viviani, *Phys. Rev. Lett.* **105** (2010) 232502.
99. J. Carlson, S. Gandolfi, F. Pederiva, S. C. Pieper, R. Schiavilla, K. Schmidt and R. Wiringa, *Rev. Mod. Phys.* **87** (2015) 1067.

100. S. Kölling, E. Epelbaum and D. R. Phillips, *Phys. Rev. C* **86** (2012) 047001.
101. S. Bacca and S. Pastore, *J. Phys. G: Nucl. Part. Phys.* **41** (2014) 123002.
102. L. E. Marcucci, F. Gross, M. T. Pea, M. Piarulli, R. Schiavilla, I. Sick, A. Stadler, J. W. V. Orden and M. Viviani, *J. Phys. G: Nucl. Part. Phys.* **43** (2016) 023002.
103. D. R. Phillips, *J. Phys. G: Nucl. Part. Phys.* **34** (2007) 365.
104. A. Camsonne *et al.*, *Phys. Rev. Lett.* **112** (2014) 132503.
105. F. Gross, *Phys. Rev. C* **91** (2015) 014005.
106. A. Kievsky, S. Rosati, M. Viviani, L. E. Marcucci and L. Girlanda, *J. Phys. G: Nucl. Part. Phys.* **35** (2008) 063101.
107. E. T. Jurney, P. J. Bendt and J. C. Browne, *Phys. Rev. C* **25** (1982) 2810.
108. F. L. H. Wolfs, S. J. Freedman, J. E. Nelson, M. S. Dewey and G. L. Greene, *Phys. Rev. Lett.* **63** (1989) 2721.
109. R. Wervelman, K. Abrahams, H. Postma, J. Booten and A. V. Hees, *Nucl. Phys. A* **526** (1991) 265.
110. M. Piarulli, L. Girlanda, R. Schiavilla, R. N. Pérez, J. E. Amaro and E. R. Arriola, *Phys. Rev. C* **91** (2015) 024003.
111. S. Pastore, S. C. Pieper, R. Schiavilla and R. B. Wiringa, *Phys. Rev. C* **87** (2013) 035503.
112. P. Klos, J. Menéndez, D. Gazit and A. Schwenk, *Phys. Rev. D* **88** (2013) 083516.
113. A. Czarnecki, W. J. Marciano and A. Sirlin, *Phys. Rev. Lett.* **99** (2007) 032003.
114. P. Ackerbauer *et al.*, *Phys. Lett. B* **417** (1998) 224.
115. V. A. Andreev *et al.*, *Phys. Rev. Lett.* **110** (2013) 012504.
116. V. Bernard, N. Kaiser and U.-G. Meissner, *Phys. Rev. D* **50** (1994) 6899.
117. N. Kaiser, *Phys. Rev. C* **67** (2003) 027002.
118. L. E. Marcucci, R. Schiavilla and M. Viviani, *Phys. Rev. Lett.* **110** (2013) 192503.
119. J. N. Bahcall and M. H. Pinsonneault, *Phys. Rev. Lett.* **92** (2004) 121301.

In search of a pristine signal for (scale-)chiral symmetry in nuclei

Mannque Rho
Institut de Physique Théorique,
CEA Saclay, 91191 Gif-sur-Yvette, France
mannque.rho@cea.fr

I describe the long-standing search for a "smoking-gun" signal for the manifestation of (scale-)chiral symmetry in nuclear interactions. It is prompted by Gerry Brown's last unpublished note, reproduced verbatim below, on the preeminent role of pions and vector (ρ, ω) mesons in providing a simple and elegant description of strongly correlated nuclear interactions. In this note written in tribute to Gerry Brown, I first describe a case of an unambiguous signal in axial-charge transitions in nuclei and then combine his ideas with the more recent development on the role of hidden symmetries in nuclear physics. What transpires is the surprising conclusion that the Landau–Migdal fixed point interaction G'_0, the nuclear tensor forces and Brown–Rho scaling, all encoded in scale-invariant hidden local symmetry, as Gerry put, "run the show and make all forces equal."

Keywords: Proton mass; BR scaling; intrinsic density dependence (IDD); scale-chiral symmetry; hidden symmetries; tensor forces.

1. Introduction

The currently active theoretical nuclear physics research is to calculate, "*ab initio*," nuclear properties in an effective field theory starting from chiral symmetry of QCD associated with the light-mass quarks relevant in nuclear interactions. This approach consists of calculating m-nucleon potentials for $m \geq 2$ with "irreducible diagrams" to high orders $\kappa \gg 1$ in N$^\kappa$LO in the standard chiral counting and computing many-body nuclear correlations summing "reducible diagrams" in a variety of sophisticated many-body techniques. Here the term "*ab initio*" refers then to the putative contact with QCD via effective field theory in the spirit of Weinberg's "folk theorem".[1] In consistency with the folk theorem, the higher one can go up in κ, while preserving the required conditions such as symmetries, etc., the better the calculation will fare in confronting Nature. Of course, given the nature of effective field theory, one is currently limited in scope by rapidly increasing number of parameters as κ is increased, but unless the effective field theory in question breaks down — which could happen under certain extreme conditions — with more refined experimental information and increasing computer power, it is reasonable to expect that our understanding of what goes on in nuclear systems will be greatly improved in the years to come. One could say this is a nuclear physics proof of the "folk theorem."

In this note, I would like to describe what Gerry and I undertook, initially in 1970's, then more intensively in 1980's and 1990's, to uncover what we considered as "preeminent features" of chiral symmetry, combined with a presumed scale symmetry, in QCD. Looked at from the present state of art in high-order chiral effective field theory, most, if not all, of those features could very well be captured in high κ calculations. Thus, one might say, "what's the big deal?"

Our philosophy has been this: Whatever beautiful phenomena there may be in the processes accessed — and however well they are described — by high-order and consistent chiral perturbation theory, they are likely buried and difficult to single out in full-fledged high-κ computations. By identifying the prominent features by relying on simplicity and intuition, one can reveal in what elegant way Nature works and make certain predictions that are hidden in potentially accurate "*ab initio*" approaches.

2. Low-Energy Theorems

What one might identify as the first "signal" for chiral symmetry was "seen" in the 1972 calculation by Riska and Brown for the thermal np capture $n + p \to d + \gamma$ (see Ref. 2) where the dominant source for $\sim$ 10% meson-exchange effects in the cross-section was identified. This calculation was prompted by the observation that the soft-pion theorems given by the current algebra relations, fairly well established then, could give an important contribution to the M1 matrix element figuring in low-energy processes of the np-capture type.[3] Soon afterwards, it was realized that the soft-pion theorems must play an even more important role in weak processes in nuclei. Indeed it was predicted that the exchange of a nearly zero-mass pion could give a lot bigger contribution to the exchange axial-charge matrix elements in nuclear beta decay than in the M1 process.[4] This prediction was confirmed, convincingly, by several experiments of first-forbidden beta decay transitions as I will describe below (see for an early discussion, Ref. 5). The corollary to these two (confirmed) observations was that both the EM charge operators and the weak Gamow–Teller operators could not receive contributions from soft-pion exchanges and hence must be subject to higher-order corrections. These observations were made before Weinberg's 1979 paper[6] established that these current algebra terms are the first terms in the chiral perturbation expansion, which is the source of the subsequent developments in EFT for QCD and the current lively activity in the nuclear physics community. In the modern language, therefore, the soft-pion terms operative in the M1 and axial-charge operators are the leading exchange-current operators with next-order corrections suppressed by one or two (in the latter case) chiral orders. This was already understood in 1981 when we started[5,7] to formulate the chiral counting rule in nuclear interactions motivated by Weinberg's 1979 paper. It is perhaps fair to say that our work foresaw the arrival of nuclear

EFT largely triggered by Weinberg's influential paper on nuclear chiral effective field theory.[8,a]

As I shall stress below, the enhanced soft-pion exchange currents offer a *clear signal* of how chiral symmetry (more precisely scale-chiral symmetry, specified below) manifests in nuclei. This point, reinforced later in the context of what is known as "BR scaling" as mentioned below, was underlined in our 1981 note[5] where it was stated "Meson exchange currents, therefore, probe the structure of the "vacuum" inside the nucleus." I will come back to this matter to argue for a "pristine" signal for scale-chiral symmetry.

3. Scale-Chiral Symmetry

3.1. *Vector mesons and scalar meson*

When Gerry Brown received the first draft of Weinberg's 1991 article with a request for comments, his first reaction was that he preferred that the vector-meson degrees of freedom, in particular, that of the ρ meson, be explicit in the Lagrangian, instead of being generated at higher chiral orders as in the chiral Lagrangian with pions only, adopted by Weinberg and in almost all of the current applications. The reason for this was Gerry's conviction that certain properties of nuclear forces could be most economically and efficiently captured if the vector mesons were treated explicitly. He was persuaded on this by a variety of nuclear observables connected to the nuclear tensor force,[11] which is one of the most important component of nuclear forces, in particular, spin–isospin response functions, the vector dominance and most significantly the effect of vacuum change in dense nuclear medium.

Although Gerry relied mostly on intuitive reasoning at the early stage, the most rigorous way presently available to address the problems involved is now recognized to be to resort to flavor gauge symmetry for the vector mesons $V = (\rho, \omega)$ supplemented by scale symmetry for a scalar degree of freedom, with the vector and scalar degrees of freedom making up the crucial ingredients of the argument.

It is now pretty convincingly established (as is summarized for example in the review of Ref. 10) that up to nuclear matter density, most, if not all, of nuclear properties are well described by chiral EFT.[b] This means that relevant fluctuations with the quantum numbers of vector and scalar mesons, if important, could be properly captured in higher-loop terms in chiral EFT. However, such an EFT must break down when the energy scale probed becomes comparable to the mass of heavier mesons, possibly at some density above normal. This could happen if the "effective" vector meson mass in medium went down as suggested in the structure of the tensor forces,[11] and more seriously at high density if the mass went to near

[a]The formulation that we initiated in early 1980's was interrupted by the rediscovery in the context of QCD of the Skyrme soliton model for nucleons in 1983 which took us away from our activity on chiral EFT, and was made more complete after Weinberg's paper. See Ref. 9.

[b]By chiral EFT, I mean EFT based on chiral Lagrangian with pions only (with or without baryon fields).

zero according to the "vector manifestation" (VM).[12] One way the dropping vector meson mass can be handled is to treat the vector meson as a local gauge boson. In fact, it is only in gauge symmetry that one can take the ρ meson as light and gets, at the leading order, the KSRF relations and vector dominance that agree well with experiments. In addition, one can set up a chiral perturbative scheme, if the vector meson mass is formally considered as light as the pion mass (although it is ~ 6 times the pion mass in the vacuum), with a systematic chiral expansion[12] that works fairly well in the vacuum. In medium it would work even better, the lighter the ρ mass dropped as predicted in HLS at high density. There are only two cases known in gauge theory where the notion of "light" (strong-interaction) vector meson makes sense; one is the case of hidden gauge symmetry we are dealing with and the other is a supersymmetric QCD in some special parameter space.[13]

An equally important degree of freedom in nuclear physics is a scalar meson of mass $\sim 600\,\text{MeV}$ that *effectively* provides the attraction that binds nuclei. In the particle data booklet, there is a scalar of comparable mass, namely, $f_0(500)$, with a large width. In the view Gerry and I have advocated since 1991, the scalar is a dilaton resulting from spontaneously broken scale symmetry. We consider it as "light" in the same sense as the ρ mass ($m_\rho \approx 770\,\text{MeV}$) is "light." This is an assumption that of course needs still to be confirmed by higher-order calculations (in the scheme mentioned below). In fact it is a long-standing controversy, with no clear consensus, whether such a scalar — that we will denote as ϕ — can be associated with scale invariance. On one hand, lattice calculations indicate there can be an infrared (IR) fixed point within the conformal window but at large number of flavors $N_f \sim 8$. This is the case for which the Higgs may be identified as a dilaton.[14] An active work on this issue is in progress in anticipation of further discoveries at LHC.[14,15] But so far there is no firm nonperturbative evidence, lattice or otherwise, for an IR fixed point for $N_f \leq 3$ that we are concerned with in QCD. There is therefore a school — call it "no-go school" — that dismisses the notion of a dilaton scalar for $f_0(500)$. On the other hand, there is a conjecture that $f_0(500)$ could be interpreted as a dilaton with an IR fixed point with the β function for the QCD gauge coupling $\alpha_s(= g^2_{\text{QCD}}/4\pi)$ vanishing and hence the trace of the energy-momentum tensor θ^μ_μ vanishing in the chiral limit.[16] In this scheme, scale symmetry and chiral symmetry merge into what is called "scale-chiral symmetry"[16] with their scales locked to each other, $4\pi f_\chi \approx 4\pi f_\pi$ where f_χ is the dilaton decay constant. At present, there is no rigorous no-go theorem against scale-chiral symmetry either, so that possibility cannot be ruled out on theoretical ground. Furthermore this scheme has a great advantage not only for particle physics (such as, among others, giving a simple explanation for the famous $\Delta I = 1/2$ rule[c]) but also for nuclear physics where the dilaton scalar ϕ can provide a systematic scale-chiral expansion including a scalar meson, generalizing the standard chiral expansion. It can provide

[c] This is somewhat like the status of the KSRF relations before the notion of hidden local symmetry was introduced.

justification to the long-standing use — with success — of a local scalar field for nuclear potentials (e.g., Bonn potential), Walecka-type mean-field models, etc. In addition, it offers an additional procedure to calculate its mass, width, etc. at low loop-orders both in and out of medium. Perhaps more importantly, it would provide a more efficient method to do calculations where strange hadrons, such as hyperons and kaons, relevant for compact-star matter — as suggested in the counting rule in Ref. 16 — are involved. We will see below that this scalar as a dilaton plays a key role in what Gerry and I have been doing all along.

The question remains, however, as to whether the failure for the lattice calculations to "see" the putative IR fixed point will not invalidate what I will be discussing below. I have no clear answer to this. In my opinion, one way to address this issue is to view the scale symmetry we are exploiting is an emergent symmetry in a way analogous to hidden local symmetry in baryonic medium. The $U_A(1)$ anomaly offers another analogy.

The "no-go school" argument against a possible IR fixed point in QCD is anchored on the trace anomaly which cannot be turned off in the vacuum. The trace anomaly is due to the regularization required for the quantum theory, or put differently, the dimensional transmutation, and is renormalization-group (RG) invariant.[d] Similarly, the $U_A(1)$ anomaly cannot be turned off. It can be tuned to zero if the number of colors N_c is tuned to ∞. However, in Nature, $N_c \ll \infty$, so the axial anomaly is there to stay. Nonetheless it has been argued that the $U_A(1)$ symmetry could be restored at high temperature.[17,e] In a similar vein, it is possible that the trace anomaly could be turned off, in the chiral limit, by density with the symmetry exposed at some high density. This possibility is being explored.[19] Since I am not concerned here with densities much higher than that of normal nuclear matter, I will not go into it.

3.2. *Hidden symmetries*

It should be recognized that *both* the local symmetry for the ρ and the scale symmetry for the ϕ are "hidden" symmetries: They are not visible or may even be absent in QCD proper. The hidden local symmetry for the ρ becomes manifest only when the ρ mass is driven toward — but not exactly onto — zero.[12,f] Since this flavor local symmetry is not present in QCD proper, the $m_\rho = 0$ limit may not be accessible in QCD. I will however suggest that it can emerge via strong nuclear correlations in dense medium. As for scale symmetry, it can be shown that the familiar linear sigma model has the scale symmetry *hidden* in it. It has been shown[15] that by dialing one parameter λ in a potential term in the standard linear sigma model (which is equivalent to the standard Higgs model) from ∞ to 0, the Lagrangian

[d] I am grateful to Koichi Yamawaki for his point of view on this matter.

[e] Up to date lattice calculations fail to see the phase transition up to ~ 1.5 times the chiral transition temperature for $N_f = 2$.[18]

[f] Unless otherwise noted, I will be working with the chiral limit.

can go from the nonlinear sigma model with no conformal (scale) symmetry to a conformal-invariant nonlinear sigma model. With hidden local gauge fields suitably incorporated, the latter turns into scale-invariant hidden local symmetry (sHLS for short). Below I will use sHLS with the scale symmetry spontaneously broken by a potential $V(\chi)$ where χ is what is referred to as "conformal compensator field" connected to the dilaton ϕ (defined below). What the baryon density does is to drive the parameter between $\lambda = \infty$ and $\lambda = 0$ and expose the hidden symmetries somewhere along the way.

4. Scale-Invariant Hidden Local Symmetric Nuclear EFT

Let me start with the mesonic Lagrangian denoted as sHLS that combines scale symmetry and hidden local symmetry that can be written in a schematic form:

$$\mathcal{L}_{\rm sHLS} = \mathcal{L}_0(U, \chi, V_\mu) + \mathcal{L}_{SB}(U, \chi), \tag{1}$$

where the conformal compensator field χ is related to the dilaton field ϕ as

$$\chi = f_\chi e^{\phi/f_\chi} \tag{2}$$

and the chiral field is given by the familiar form $U = e^{i\frac{2\pi}{f_\pi}}$. In this paper, I will deal with flavor $SU_f(2)$ and assume $U(2)$ symmetry for the vector mesons $V = (\rho, \omega)$.[g] The mass dimension-one χ transforms linearly under scale transformation while ϕ transforms nonlinearly as the pion field π does under chiral transformation, that is, as a Nambu–Goldstone. The first term (1) is of scale dimension 4, so gives scale-invariant action and also HLS (chiral) invariant. The second term contains the pseudo-scalar meson mass term, hence breaking explicitly the chiral symmetry, and a potential that breaks scale symmetry both explicitly (due to the trace anomaly) and spontaneously. Although most of my discussions in application to nuclei can be done in the chiral limit, the chiral symmetry breaking pion term will be necessary to fix the property of the pion decay constant which sets the density-scaling behavior in nuclear matter.

Expanded in fields, the Lagrangian (1) in unitary gauge in hidden gauge symmetry is, to $\mathcal{O}(p^2)$ chiral order, of the form

$$\begin{aligned}\mathcal{L}_{\chi_s} &= \frac{f_\pi^2}{4}\kappa^2\,\mathrm{Tr}(\partial_\mu U^\dagger \partial^\mu U) + \kappa^3 v^3\,\mathrm{Tr}\mathcal{M}(U + U^\dagger)\\ &\quad - \frac{f_\pi^2}{4} a\kappa^2\,\mathrm{Tr}[\ell_\mu + r_\mu + i(g/2)(\boldsymbol{\tau}\cdot\boldsymbol{\rho}_\mu + \omega_\mu)]^2 - \frac{1}{4}\boldsymbol{\rho}_{\mu\nu}\cdot\boldsymbol{\rho}^{\mu\nu}\\ &\quad - \frac{1}{4}\omega_{\mu\nu}\omega^{\mu\nu} + \frac{1}{2}\partial_\mu\chi\partial^\mu\chi + V(\chi),\end{aligned} \tag{3}$$

where a is related to the ratio f_π/f_σ (where f_π is the pion decay constant and f_σ is the decay constant of the would-be Goldstone boson Higgsed to give the vector

[g]There is a strong indication that this symmetry is badly broken in sHLS at a density denoted $n_{1/2} \sim (2-3)n_0$ where n_0 stands for normal nuclear matter density. I will not deal with this high density which is relevant for compact stars.[19]

meson mass), $\kappa = \chi/f_\chi$ with $f_\chi = \langle 0|\chi|0\rangle$, $l_\mu = \partial_\mu \xi \xi^\dagger$, $r_\mu = \partial_\mu \xi^\dagger \xi$ with $\xi = \sqrt{U}$ and v is a constant of mass dimension 1.

4.1. *Intrinsic density dependence*

In order to apply (1) to baryonic matter, baryon degrees of freedom are needed. There are two ways to bring in baryon fields. One way, perhaps most consistent with QCD, is to generate baryons as skyrmions of the mesonic Lagrangian. There is a work along this line with some progress. At present, however, it is not developed well enough to quantitatively describe nuclear processes. The alternative is to put baryon fields explicitly in consistency with the folk theorem, staying as faithful as possible to the symmetries involved. I will follow this approach below. Let me call the baryon-implemented effective Lagrangian bsHLS.

Since the ultimate aim is to probe the density regime where the vector mesons, i.e., ρ and ω, and the scalar are relevant, perhaps overlapping the regime where the explicit QCD degrees of freedom may intervene, the strategy to take is to have the EFT matched to QCD at the scale where the cutoff for the EFT is set. In Ref. 12, the matching was done by means of the vector and axial-vector correlators, tree order in HLS (i.e., "bare" HLS) and OPE in QCD, at $\Lambda_M = \Lambda_\chi \sim 4\pi f_\pi$. With sHLS, the energy-momentum tensor needs also to be matched. What the matching does is then to endow the "bare" parameters of the EFT Lagrangian with dependence on the condensates, i.e., the quark condensate $\langle\bar{q}q\rangle$, the gluon condensate $\langle G^2\rangle$, mixed forms, etc. Since those condensates depend on the vacuum, if the vacuum is modified by density, then they will necessarily depend on density. This dependence will then render the parameters of the Lagrangian *intrinsically* density-dependent. This density dependence, of QCD origin, is called "intrinsic density dependence" (IDD for short).

Given the Lagrangian so matched to QCD, then one has an effective Lagrangian, the "bare" parameters of which are density-dependent, with which one can do quantum theory. The IDD so defined is related — but not identical — to what is known in the literature as "Brown-Rho scaling" (BR scaling for short).[h]

[h]The BR scaling as applied to certain nuclear processes may contain other density dependence than IDD. For example, when short-range contact three-body forces are integrated out, the resulting two-body force can inherit the three-body force effect in the form of BR scaling. One can understand this by that the zero-range three-body forces figuring in chiral EFT, involving ω and heavier meson exchanges of bsHLS, are of the same or higher scale than the cutoff scale $\tilde{\Lambda}$, hence their effects get captured in BR scaling when integrated out to arrive at chiral EFT. This will be the case with the C14 dating problem mentioned below. This means that the BR scaling used there has a contribution from the three-body force effect in addition to IDD. Another example: The axial current coupling g_A is different for Gamow–Teller transitions (space component of the current) from axial charge transitions, also discussed below. The former is BR and the latter is IDD. Note that IDD in EFT Lagrangian is a Lorentz-invariant object while BR in physical observables may contain Lorentz-breaking contributions. A most prominent example where BR scaling in practice can contain more than IDD is the anomalous orbital gyromagnetic ratio δg_l which can be well

The most efficient and flexible approach, presently available, to treat many-body nuclear dynamics with bsHLS is the RG approach employing $V_{\rm lowk}$. It involves "double decimations".[21] For nuclear processes, one should be able to do the decimation from a cutoff $\tilde{\Lambda}$ somewhat lower than ρ mass. From the vector meson mass scale, the "bare" parameters of the EFT Lagrangian — except for f_π[i] — do not flow to the scale picked, $\tilde{\Lambda}$, from which the decimation is to be done, so it should be justified to lower the cutoff to $\tilde{\Lambda}$ without modifying the "bare" Lagrangian. In practice, the first decimation is made from $\sim (2-3)$ fm^{-1} to obtain the $V_{\rm lowk}$ and then the 2nd decimation consists of doing Fermi-liquid calculations with this $V_{\rm lowk}$.[22]

In HLS taken to $\mathcal{O}(p^2)$ in the chiral (derivative) counting, there are only three parameters g, f_π and a. In the skyrmion description, nucleon properties including couplings to the vector mesons involved do not require additional parameters. With the scalar field included, there is of course an additional parameter, namely, f_χ. However the locking of scale symmetry and chiral symmetry makes f_χ equal to f_π, so it does not require additional IDD.

Now the question is how these parameters vary as a function of density and how their dependence affects hadron masses and coupling constants of the "bare" Lagrangian?

The answer to this question requires knowing whether there is any phase change in the matter structure as density increases. It is obvious that the parameters will not necessarily continue moving smoothly in density. For example, at some density, QCD degrees of freedom could enter. In the skyrmion description of baryonic matter, there is a robust topological transition from a skyrmion matter to a half-skyrmion matter at a density around $n_{1/2} \sim (2-3)n_0$. In fact this transition is in a sense equivalent to what is called "quarkyonic" in which quark degrees of freedom figure at about the same density.[19,23] In terms of the "bare" Lagrangian, such a transition would imply changes in the density dependence of the bare parameters. I would not go into what happens after such transition which matters for compact star structure — which has been studied,[19] so let me focus on the density regime $n < n_{1/2}$.

First consider the ρ mass. The "bare" mass at the matching scale $\Lambda_M \sim \Lambda_\chi$ is given by

$$m_\rho^2 = a f_\pi^2 g^2. \tag{4}$$

What is remarkable about this relation, known as KSRF formula, is that it holds to all orders of loop corrections with the HLS Lagrangian taken to $\mathcal{O}(p^2)$ — and believed to be valid at higher chiral (derivative) orders — with corrections of $\mathcal{O}(m_\rho^2/\Lambda_\chi^2)$.[12,24] This expression therefore becomes more accurate, the lighter the

described by a BR scaling expressed in terms of certain Fermi-liquid parameters. It is in Fermi-liquid parameters that the IDD is lodged.[20] Here the link between BR scaling and IDD is indirect and complicated.

[i]The pion decay constant does, however, flow by pion loops below $\tilde{\Lambda}$.

vector mass becomes as is predicted in dense medium.[12] This means that the "bare" ρ mass in the Lagrangian will always be of the form of (4), regardless of the cutoff for decimation, with the IDD reflecting *entirely* the effect of density. The explicit calculation of the EFT-QCD matching formulas shows that both g and a depend quite weakly on the quark and gluon condensates[12] and hence the density dependence will be mainly in the pion decay constant, hence in the dilaton condensate since $f_\pi \approx f_\chi$. Therefore, the only scaling factor in the density regime $n \leq n_{1/2}$ is

$$f_\pi^*/f_\pi \approx f_\chi^*/f_\chi \equiv \Phi(n). \tag{5}$$

Thus via (4)

$$m_\rho^*/m_\rho \approx \Phi. \tag{6}$$

It follows from the bare Lagrangian (3) with the expansion $\chi = \langle 0^*|\chi|0^*\rangle + \chi'$ (where 0^* is the in-medium vacuum) that

$$m_H^*/m_H \approx \Phi(n) \tag{7}$$

for $H = N, \rho, \omega, \phi$ assuming $U(2)$ symmetry for (ρ, ω) for $n < n_{1/2}$. The pion mass, with broken chiral symmetry — and hence broken scale symmetry, scales differently,

$$m_\pi^*/m_\pi \approx \sqrt{\Phi}. \tag{8}$$

Equations (7) and (8) are the same as the expressions derived in 1991 using the skyrmion model. Note that they follow not directly from chiral symmetry but from scale symmetry locked to chiral symmetry. In other words, *it is the dilaton condensate that "runs" the show.* This means that the symmetry involved is the "scale-chiral symmetry" as defined precisely in Ref. 16.

4.2. *Soft-pion signal for scale-chiral symmetry*

The first "pristine" signal for scale-chiral symmetry in nuclei is in the axial-charge beta decay process in nuclei. It is in the first-forbidden beta transition of the form

$$I(0^-) \to F(0^+) + e^- + \bar{\nu}_e \quad \Delta T = 1. \tag{9}$$

This beta decay process from the initial nucleus I to the final nucleus F goes via the axial change operator A_0^a.

As first recognized in 1978[4] from current algebras and later confirmed in chiral perturbation theory,[9] the exchange axial-charge two-body operator receives a large contribution from a soft-pion exchange term, with the next contribution suppressed by two chiral order. Furthermore, the leading one-body operator, being first-forbidden, is kinematically suppressed. Therefore, the two-body "correction" term is expected to contribute to the decay at an order comparable to or bigger than the "leading" single-particle operator. Written in effective one-body operator, the corresponding Feynman diagram is of the form Fig. 1. The sum of the one-body

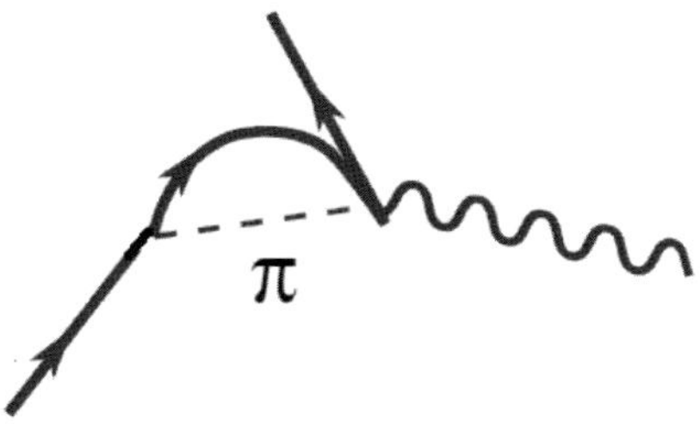

Fig. 1. Effective single-particle soft-pion-exchange axial charge operator. The solid line is the nucleon and the wiggly line the external weak field. The right vertex $A^a_\mu \pi$NN is large for the time component A^a_0 and is suppressed for the space component $\mathbf{A}^a$.

and two-body axial charge operators with the IDD incorporated into the constants of the EFT Lagrangian has the extremely simple form[25]

$$A^a_0 = g_A \frac{\tau^a}{2} \frac{\sigma \cdot \mathbf{p}}{m_N \Phi} \left(1 + \frac{R}{\Phi}\right), \tag{10}$$

where $\mathbf{p}$ is the nucleon momentum, R is the ratio of the matrix element of the two-body operator (M_2) to that of one-body operator (M_1). The factor Φ corresponds to the IDD dependence in the EFT Lagrangian. It may be that the numerical values of M_1 and M_2 depend on how nuclear wavefunctions are calculated. However, the ratio R is highly insensitive to it. Thus, one can take either Fermi-liquid model or Fermi-gas model in place of more sophisticated wavefunctions. One gets essentially the same value. What is significant is that the R is big $R \sim \mathcal{O}(1)$, and varies slowly in density, reflecting the robust nature of the soft-pion exchange.

One can readily make a simple estimate of what comes out. Let us look at the quantity defined and measured experimentally by Warburton[26]

$$\epsilon_{\rm MEC} = \frac{1}{\Phi}\left(1 + \frac{R}{\Phi}\right). \tag{11}$$

This represents the enhancement factor due to both the exchange-current contribution in the transition matrix element relative to the single-particle operator contribution *and* the IDD. Take the lead $A = 205 - 212$ nuclei for which data are available which have densities comparable to nuclear matter density. Calculating R in Fermi-gas model, it is found at nuclear matter density n_0 that $R(n_0) \approx 0.5$. Now from pionic atom data, one has $\Phi(n_0) \approx 0.8$.[27] Thus, (11) gives $\epsilon_{\rm MEC} \approx 2.0$. This agrees very well with the measured enhancement factor $\epsilon^{\rm exp}_{\rm MEC} = 2.01 \pm 0.05$. Taking into account the density dependence of Φ and R, one can also reproduce the observed enhancements in $A = 12$ and $A = 16$ systems, $\epsilon^{\rm exp}_{\rm MEC}(A = 12) = 1.64 \pm 0.05$ (see Ref. 28) and $\epsilon^{\rm exp}_{\rm MEC}(A = 16) \approx 1.7$.[29] Although the theoretical estimate as well as the experimental values for $\epsilon_{\rm MEC}$ are rough, this is a clear evidence for both the soft-pion and IDD effects: The density dependence of $\epsilon_{\rm MEC}$ is found to be consistent with what is predicted of Φ and R.

Two remarks are in order here.

One is the crucial role of soft pions. It zeroes in on the Nambu–Golddstone-boson nature of the pion with a mass nearly vanishing (on the strong interaction scale). For very low-energy processes, $E \ll m_\pi \approx 140\,\text{MeV}$, according to the standard lore of EFT, one may be justified to integrate out the pions leaving only the nucleons as relevant degrees of freedom. One then gets what is called "pionless effective field theory" ($\not\pi$EFT), which is generally thought to be consistent with the "folk theorem." With the pions absent, the resulting Lagrangian is blind to chiral symmetry but it does not mean chiral symmetry is violated. So does it always work? The question is: Is $\not\pi$EFT applicable to the axial-charge transition which receives big contributions from soft pions? If the pion mass were strictly zero for which the soft-pion theorems hold, clearly the pion could not be integrated out from the chiral Lagrangian. Thus, it seems inevitable that the $\not\pi$EFT Lagrangian with the pions got rid of would miss the soft-pion effect (at the $\kappa = 1$ order) and hence fail even if it were applicable to Gamow–Teller transitions.

The other remark is on the possibility to do precision calculations and test the combined enhancement by soft pions and IDD. The operators are well defined to the leading chiral order with higher-order terms strongly suppressed, so given accurate wavefunctions, one could then do a precision test of the scaling parameter Φ. Recent developments on "*ab initio*" approaches with sophisticated many-body techniques could be exploited to calculate ϵ_{MEC} with accurate error estimates for both theory and experiment.

5. What Runs the Show in Nuclear Interactions?

Let me now come to the startling, if not puzzling, observation — the main thrust of Gerry's note — that the π, ρ, ω and ϕ, the principal degrees of freedom of bsHLS at mean-field, play the dominant and even clear-cut role in nuclear dynamics. For this part, reading Gerry's note (added below) will be helpful.

Given the bsHLS Lagrangian, one could perhaps perform a (covariant) density functional analysis for the nuclear ground-state properties along the line set up by the Hohenberg–Kohn theorem for atomic/molecular physics and chemical physics. The currently popular covariant energy-density functional approaches employed in nuclear theory typically have six or more free parameters. In contrast, up to at least nuclear matter density, the bsHLS Lagrangian was found to have basically only one parameter[j] governing the d-scaling factor Φ associated with the dilaton condensate. It would be extremely interesting to see how an "*ab initio*" covariant density functional given by bsHLS compares with the standard approach with many more parameters. However, one is ultimately interested in the equation of state relevant to compact stars. For this the density functional approach does not seem appropriate.

[j] Or at most three if fine-tuning is needed to obtain a precision fit to data. Since the only parameter of the theory, Φ, can be fixed by experiments at least up to nuclear matter density, there is no real free parameter in the theory.

Instead the strategy followed below is the double-decimation RG procedure in terms of $V_{\rm lowk}$,[21] which is to start with the first RG decimation to go from the effective cutoff $\tilde{\Lambda}$ down to the scale at which the $V_{\rm lowk}$ is obtained. For this, in principle, the "irreducible graphs" are to be summed to high orders in scale-chiral expansion to give the potential with which to do the decimation. In doing this, the IDDs enter. In practice, BonnS-type potentials are used and the cutoff is put at (2–3) fm^{-1}, somewhat lower than $\tilde{\Lambda}$. Performing the second decimation corresponds to doing Landau Fermi-liquid theory with $V_{\rm lowk}$ as formulated in Refs. 30 and 22. The result is then the set of Fermi-liquid fixed point parameters, i.e., the effective masses and interactions. I will focus on the Fermi-liquid parameters and the fixed point quasiparticle interactions in which the pion and the ρ play the main role, in particular, the Landau parameter G'_0 and the tensor forces.

5.1. *The EELL effect or G'_0*

In his note, Gerry Brown presents strong intuitive arguments, drawing from the previous works,[22,30] to show that the Landau parameter G'_0, coming from the pion and ρ exchanges, is by far the largest among the Fermi-liquid interactions and dominates the Kuo–Brown effective interactions at mean-field level, with higher order terms suppressed.[31] How and to what extent the suppression takes place more generally is yet to be worked out. However, in the Wilsonian RG approach, the beta function for the quasiparticle interactions should tend to zero in the large N limit where N is related to the Fermi momentum k_F, as beautifully explained in Ref. 32. Gerry relies on the double-decimation $V_{\rm lowk}$ approach with BR scaling implemented to argue that the Kuo–Brown interaction, with just one bubble, which is classical in nature, has the most of the physics in it, giving an extremely simple interpretation of why and how the high-order core polarization contributions are suppressed as found in Ref. 31. Both the scalar ϕ and the vector ω figure in bringing the interaction $\sim G'_0(\tau_1 \cdot \tau_2)(\sigma_1 \cdot \sigma_2)$ to the fixed point, the former "holding the ball of pions together" and the latter providing the short-range repulsion giving rise to the Ericson–Ericson–Lorentz–Lorenz effect.

5.2. *Nuclear tensor forces*

The exchange of π and ρ with the parameters endowed with IDD of the EFT Lagrangian gives the effective in-medium nuclear tensor forces. While the pion tensor is more or less unaffected by density, an effect which could be attributed to protection by chiral symmetry, the ρ tensor, with the dropping mass, increases in magnitude with a sign opposite to that of the pion tensor as density increases. Because of the cancellation between the two, the net tensor force strength gets weaker due to the BR scaling at increasing density. The attraction in the tensor channel goes nearly to zero when density reaches $2n_0$. This has been well known. In fact precisely this effect has been exploited with an impressive success to explain the long C14 life-time.[33] What is involved in this process is a delicate density-dependent cancelation in the Gamow–Teller matrix element, which nearly vanishes in the density regime involved. This is a spectacular signal of working of the BR scaling, although it is not as clear-cut an evidence as in the case of the first-forbidden beta decay process described above. As noted by Gerry, in the context of double decimation, the effect of BR scaling here is equivalent to the effect of contact three-body forces in the sense described in footnote h.

I will now propose that there is a possibility of "seeing" the IDD scaling — not just BR scaling — in nuclei via the tensor forces. I would like to describe this using the $V_{\rm lowk}$ formalism.

In a series of beautiful papers,[34] Takaharu Otsuka showed that the tensor forces played a remarkable role in the "monopole" matrix element of the two-body interaction between two single-particle states labeled j and j' and total two-particle isospin T

$$V_{j,j'}^{T} = \frac{\sum_J (2J+1)\langle jj'|V|jj'\rangle_{JT}}{\sum_J (2J+1)}. \tag{12}$$

What is special with this matrix element is that it affects the evolution of single-particle energy;

$$\Delta\epsilon_p(j) = \frac{1}{2}(V_{jj'}^{T=0} + V_{jj'}^{T=1})n_n(j'), \tag{13}$$

where $\Delta\epsilon_p(j)$ represents the change of the single-particle energy of protons in the state j when $n_n(j')$ neutrons occupy the state j'. It turns out that the matrix elements $V_{jj'}$ and $V_{j'j}$ have opposite sign for the tensor forces if j and j' are spin–orbit partners.

Let me summarize the salient features of the shell evolution connected to the tensor force found by Otsuka:

Otsuka works with the phenomenological potential Av18' and the one given by ChPT at N^{3}LO, both treated à la $V_{\rm lowk}$. Other realistic potentials are found to give the same results. He varies the cutoff $\tilde\Lambda$ and finds $\tilde\Lambda$ independence around 2.1 fm^{-1}.

Otsuka calculated the shell evolution in the pf and sd regions by including high-order correlations using the Q-box formalism to 3rd order. While he finds the central part of the potential strongly renormalized by high-order terms, the tensor forces are left unrenormalized, leaving the "bare" tensors more or less intact. Shown in Fig. 2 is Otsuka's result (copied from his paper) in the pf shell region. The result shows that the sum of the short-range correlation and medium effects as taken into account by the 3rd order Q-box leaves the bare tensor force unchanged. Otsuka looks at a variety of other realistic potentials, both phenomenological as well as ChPT at N^{3}LO, and finds, remarkably, that they all give the same result, showing that the effect is robust. It implies

$$\frac{d}{d\tilde\Lambda}V_{\rm lowk}^{\rm tensor} = \beta([V_{\rm lowk}^{\rm tensor}], \tilde\Lambda) \approx 0. \tag{14}$$

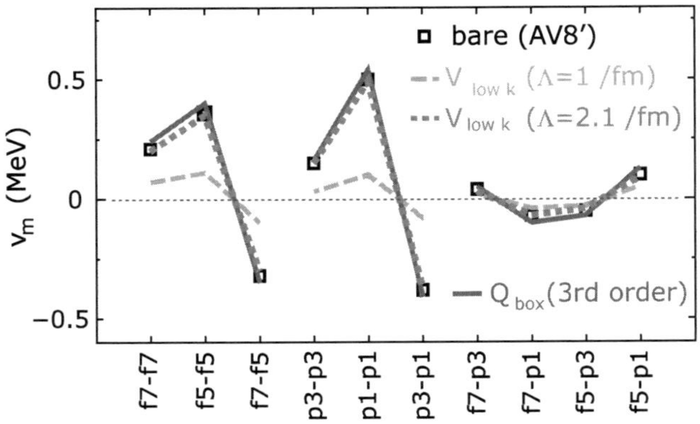

(a) $T = 0$ tensor force

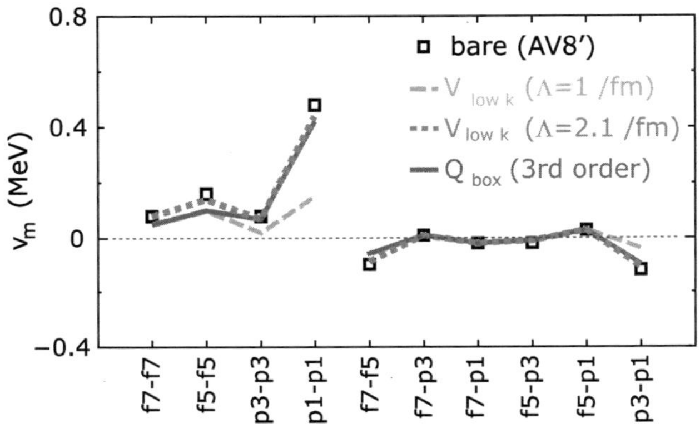

(b) $T = 1$ tensor force

Fig. 2. Tensor forces in AV8' interaction, in low-momentum interactions in the pf shell obtained from AV8', and in the 3rd-order Q_{box} interaction for (a) $T = 0$ and (b) $T = 1$.

Some experimental data are available, e.g., Jahn–Teller effect,[34] that verify the tensor implemented calculations to be in agreement with experimental data quite well. Forthcoming experiments in RIB accelerators promise to reveal more surprising results.

The result (14) says that the beta function is zero *both* in the first decimation and in the second. The latter could perhaps be understood as the net tensor force at a given density being at the Landau fixed point with all correlation effects suppressed. However, the former is surprising since it implies that the tensor force does not RG-flow in the vacuum. Why the net tensor force is free of all strong interaction effects, in and out of medium, is mysterious. Learning of Otsuka's results, Tom Kuo kindly performed a $V_{\rm lowk}$ analysis in the vacuum. Shown in Fig. 3 are his results of the tensor potentials in momentum space $V^{\rm tensor}(k_1, k_2)$ for $k_1 = k_2$ and $k_1 \neq k_2$. The $V^{\rm tensor}_{\rm lowk}$ potential is identical to the "bare" BonnS tensor potential, independent of $\tilde{\Lambda}$.

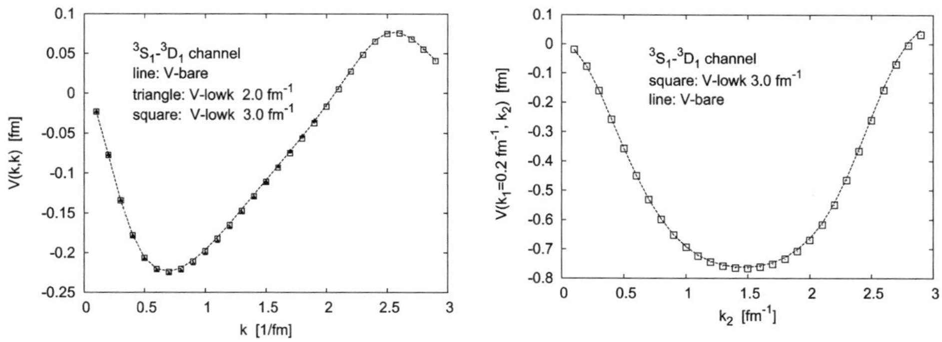

Fig. 3. "Bare" BonnS and V_{lowk} tensor forces in matter-free space. Courtesy of Tom Kuo.

6. Seeing "Pristine" Signals

Let me close this note with what "pristine" signal could mean for chiral symmetry (or more precisely scale-chiral symmetry). At the early stage of dilepton production in relativistic heavy-ion collisions, Gerry imagined that it would mean the ρ mass going to zero at the critical temperature T_c (in the chiral limit). Gerry's idea was that in medium the ρ mass, rather than the quark condensate, is the relevant order parameter to measure for chiral restoration, and in HLS which is a natural framework for addressing the issue, this meant going toward the VM fixed point as T_c is approached. However, the VM fixed point with an enhanced symmetry turns out not to be in QCD in the vacuum.[12,35] It thus could make sense — if at all — only as an emergent phenomenon. Furthermore, HLS predicts that in approaching the VM fixed point, while the ρ mass could go to zero with a vanishing width, the photon tends to largely decouple from the dropping-mass ρ. This is quite unlike seeing the ρ in the vacuum by measuring the dileptons of the given invariant mass of ρ with a detector in the vacuum. In our view, dileptons in heavy-ion processes are not at all a suitable probe for scale-chiral symmetry: the ρ meson produced in the process is so strongly distorted by background nuclear correlations in the medium so that the signal for the ρ meson carrying the order parameter subject to the VM, when measured with a detector outside of the medium, will be like a "needle in the haystack" as Gerry and I, with our collaborators, have argued.

So how does one go about "seeing" the signal?

I have argued in this note that "seeing" (scale-)chiral symmetry in action in nuclei is much like "seeing" the meson-exchange currents in nuclei. Three such signals for scale-chiral symmetry in nuclei are described.

The first is the combined effect of soft pions and IDD, giving a whopping factor of ~4 effect in the decay rate. It reveals *both* the presence of meson-exchange currents *and* the influence of scale-chiral symmetry manifested by the excitation of pions. It further shows that pions cannot be integrated out for certain processes — such as first-forbidden transitions — that single out soft-pion dominated effects in near zero-energy processes.

The second is the Landau parameter G_0' dominated by the π and ρ-meson exchanges in strong correlations with the scalar and ω mesons (binding and short-range) which in Gerry's words, runs the "show" in nuclear dynamics.

The third is the nonrenormalization of the net tensor force in and out of medium, offering the possibility of a pristine probing for IDD by zeroing in on those processes that are controlled by the (net) tensor force and by dialing the density of the system. Of course how to dial the density is a big open issue.

Finally, all these are simple and elegant aspects of nuclear interactions which could be sharpened in "*ab initio*" precision calculations — in progress and to come.

Acknowledgments

I am grateful for fruitful discussions and collaborations with Masa Harada, Tom Kuo, Hyun Kyu Lee, Yong-Liang Ma and Won-Gi Paeng and would like to acknowledge extensive comments and tutorials from Rod Crewther, Lewis Tunstall and Koichi Yamawaki on scale symmetry in gauge theory. I would particularly like to thank Tom Kuo for his help on the nuclear tensor force at its putative fixed point.

References

1. S. Weinberg, arXiv:hep-th/9702027.
2. D. O. Riska and G. E. Brown, *Phys. Lett. B* **38** (1972) 193.
3. M. Chemtob and M. Rho, *Nucl. Phys. A* **163** (1971) 1.
4. K. Kubodera, J. Delorme and M. Rho, *Phys. Rev. Lett.* **40** (1978) 755.
5. M. Rho and G. E. Brown, *Comments Nucl. Part. Phys.* **10** (1981) 201.
6. S. Weinberg, *Physica A* **96** (1979) 327.
7. M. Rho, *Prog. Part. Nucl. Phys.* **8** (1982) 103.
8. S. Weinberg, *Nucl. Phys. B* **363** (1991) 3.
9. M. Rho, *Phys. Rev. Lett.* **66** (1991) 1275.
10. J. W. Holt, M. Rho and W. Weise, *Phys. Rep.* **621** (2016) 2.
11. G. E. Brown and M. Rho, *Phys. Lett. B* **237** (1990) 3.
12. M. Harada and K. Yamawaki, *Phys. Rep.* **381** (2003) 1.
13. Z. Komargodski, *J. High Energy Phys.* **1102** (2011) 019.
14. S. Matsuzaki and K. Yamawaki, *J. High Energy Phys.* **1512** (2015) 053.
15. H. S. Fukano, S. Matsuzaki, K. Terashi and K. Yamawaki, *Nucl. Phys. B* **904** (2016) 400.
16. R. J. Crewther and L. C. Tunstall, *Phys. Rev. D* **91**(3) (2015) 034016.
17. R. D. Pisarski and F. Wilczek, *Phys. Rev. D* **29** (1984) 338.
18. S. Sharma, V. Dick, F. Karsch, E. Laermann and S. Mukherjee, arXiv:1602.02197 [hep-lat].
19. H. K. Lee, W. G. Paeng and M. Rho, *Phys. Rev. D* **92**(12) (2015) 125033; W. G. Paeng, T. T. S. Kuo, H. K. Lee and M. Rho, arXiv:1508.05210 [hep-ph].
20. B. Friman and M. Rho, *Nucl. Phys. A* **606** (1996) 303.
21. G. E. Brown and M. Rho, *Phys. Rep.* **396** (2004) 1.
22. J. W. Holt, G. E. Brown, J. D. Holt and T. T. S. Kuo, *Nucl. Phys. A* **785** (2007) 322.
23. K. Fukushima and T. Kojo, *Astrophys. J.* **817**(2) (2016) 180.
24. M. Harada, T. Kugo and K. Yamawaki, *Phys. Rev. Lett.* **71** (1993) 1299.
25. K. Kubodera and M. Rho, *Phys. Rev. Lett.* **67** (1991) 3479.

26. E. K. Warburton, *Phys. Rev. Lett.* **66** (1991) 1823.
27. P. Kienle and T. Yamazaki, *Prog. Part. Nucl. Phys.* **52** (2004) 85.
28. T. Miniamisono, K. Matsuta, T. Yamaguchi, K. Minamisono, M. Fukuda, A. Kitagawa and K. Koshigiri, *Phys. Rev. Lett.* **82** (1999) 1644.
29. C. A. Gagliardi, G. T. Garvey and J. R. Wrobel, *Phys. Rev. Lett.* **48** (1982) 914.
30. A. Schwenk, G. E. Brown and B. Friman, *Nucl. Phys. A* **703** (2002) 745.
31. J. D. Holt, J. W. Holt, T. T. S. Kuo, G. E. Brown and S. K. Bogner, *Phys. Rev. C* **72** (2005) 041304.
32. R. Shankar, *Rev. Mod. Phys.* **66** (1994) 129.
33. J. W. Holt, G. E. Brown, T. T. S. Kuo, J. D. Holt and R. Machleidt, *Phys. Rev. Lett.* **100** (2008) 062501.
34. N. Tsunoda, T. Otsuka, K. Tsukiyama and M. Hjorth-Jensen, *Phys. Rev. C* **84** (2011) 044322 and references given therein; Y. Utsuno, T. Otsuka, B. A. Brown, M. Honma, T. Mizusaki and N. Shimizu, *Phys. Rev. C* **86** (2012) 051301.
35. H. Georgi, *Phys. Rev. Lett.* **63** (1989) 1917.

Note added: Unpublished paper by G. E. Brown "the dominant role of the EELL interaction in nuclear structure"

Forewords

This is Gerry's last manuscript, hand-written and as he used to do, faxed to me on 4 April 2007, the cover letter and the first page of which are scanned and put in Fig. 4. Below, the article is transcribed verbatim, totally unedited and unrevised. It contained no abstract, but a short statement on the cover letter conveyed the essence of the idea. Referring to the Ericson–Ericson–Lorentz–Lorenz (EELL or E^2L^2 for short) interaction, which captures the RG fixed point interaction g_0' in Landau–Migdal Fermi liquid theory, Gerry stated "I believe this is as universal as Brown–Rho scaling. It's magical how E^2L^2 runs the show and makes, together with B-R scaling, all forces equal." In the contribution to this volume given above, I describe how Gerry's ideas can be synthesized into the more recently developed notion of scale-invariant hidden local symmetry in nuclear interactions.

1. Introduction

We develop the renormalization group description of Schwenk *et al.*[1] which introduced the effective interaction V_{lowK}. The effective pion energies in terms of low

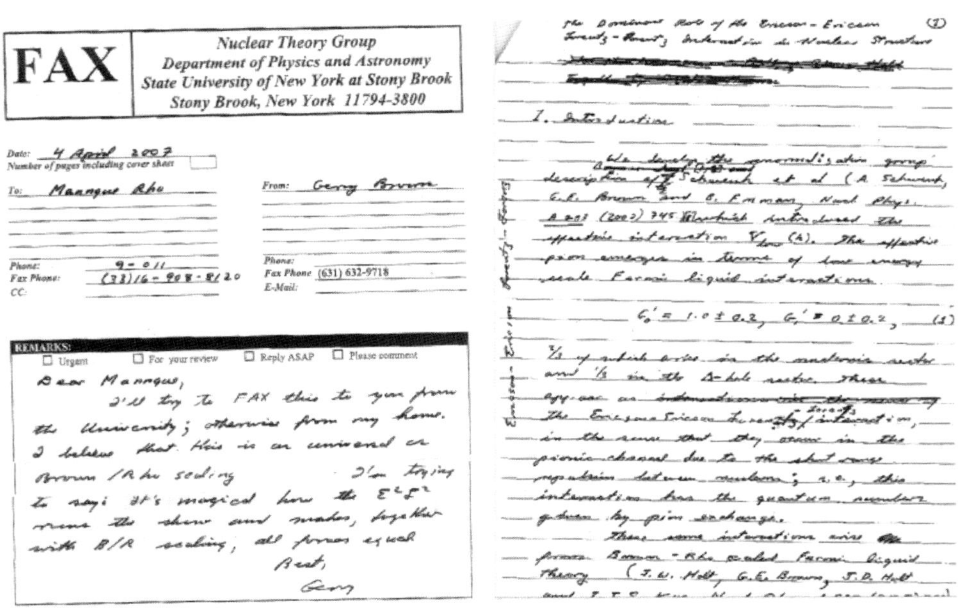
FAX

Nuclear Theory Group
Department of Physics and Astronomy
State University of New York at Stony Brook
Stony Brook, New York 11794-3800

Date: 4 April 2007
Number of pages including cover sheet
To: Mannque Rho
From: Gerry Brown
Phone: 9-011
Fax Phone: (33)16-908-8120
CC:
Phone:
Fax Phone (631) 632-9718
E-Mail:

REMARKS: ☐ Urgent ☐ For your review ☐ Reply ASAP ☐ Please comment

Dear Mannque,
I'll try to FAX this to you from the University; otherwise from my home. I believe that this is as universal as Brown/Rho scaling. I'm trying to say: It's magical how the E^2L^2 runs the show and makes, together with B/R scaling, all forces equal.
Best,
Gerry

The Dominant Role of the Ericson-Ericson Lorentz-Lorentz Interaction in Nuclear Structure (1)

1. Introduction

We develop the renormalization group description of Schwenk et al (A. Schwenk, G.E. Brown and B. Friman, Nucl. Phys. A 703 (2002) 745) which introduced the effective interaction $V_{low\,k}(\Lambda)$. The effective pion energies in terms of low energy scale Fermi liquid interactions.

$G_0' = 1.0 \pm 0.2, \quad G_1' = 0 \pm 0.2, \quad (1)$

Fig. 4. Cover letter (left) and title page (right) of the faxed paper.

energy scale Fermi-liquid interactions

$$G_0' = 1.0 \pm 0.2, \quad G_1' = 0 \pm 0.2. \tag{1}$$

2/3 of which arise in the nucleonic sector and 1/3 in the Δ-hole sector. These appear as the Ericson–Ericson Lorentz–Lorenz interaction, in the sense that they occur in the pionic channel due to the short-range repulsion between nucleons; i.e., this interaction has the quantum numbers given by pion exchange.

These same interactions arise from Brown–Rho scaled Fermi-liquid theory[2] as repulsive interactions between nucleons, replacing the attractive piece of the zero range pion–nucleon interaction which is removed by the short-range repulsion between nucleons. The effective $\boldsymbol{\sigma}_1 \cdot \boldsymbol{\sigma}_2 \tau_1 \cdot \tau_2$ component of the interaction found by Holt *et al.*[2] is not only the strongest component of the effective interaction, but is exactly found by Schwenk *et al.*.[1] The long-range scalar exchange through the σ-meson and coupled two-pion exchange tends to hold the ball of pions together.

As far as we can see, were we to include the Δ with coupling constant $g'_{N\Delta} = 0.3$ and the double Δ with $g'_{\Delta\Delta} = 0.3$ then we would have a description in the double-decimation decimation of Brown and Rho[3] dual to that of Schwenk *et al.*[1] of the screening by the Δ-hole channel of the screening of the interaction in the pionic interaction. We return to this later.

2. Nuclear Matter with Brown–Rho Scaled Fermi-Liquid Interactions

We believe that it is useful to review the work of Holt *et al.*[2] quantitatively in comparison with that of Schwenk *et al.*[1] because the double decimation is essentially the same as the latter with introduction of three-body forces. In any case, the substantial improvement in going over to the double decimation from the usual Fermi-liquid approach in the long wavelength limit should serve a stimulus for pursuing this direction. We reproduce Table 4 of Ref.2 as Table 1 below.

In Table 1, we show the effective mass, compression modulus, symmetry energy and anomalous orbital gyromagnetic ratio for the Nijmegen I ($V_{\rm NI}$) and II ($V_{\rm NII}$) and CD-Bonn ($V_{\rm CDB}$) potentials with the in-medium modification à la Brown–Rho. We also show for comparison the results from the Nijmegen 93 (V_{N93}) one-boson exchange potential, which has only 15 parameters and is not fine-tuned separately in each partial wave. The iterative solution is in better agreement with all nuclear

Table 1. Nuclear observables from the self-consistent solution obtained by iterating the Babu–Brown equations.

	$V_{\rm NI}$	$V_{\rm NII}$	V_{N93}	$V_{\rm CDB}$
$m^\star/m$	0.721	0.763	0.696	0.682
K MeV	218	142	190	495
β MeV	20.4	25.5	23.7	19.2
δg_l	0.296	0.181	0.283	0.267

Table 2. Fermi liquid coefficients for the self-consistent solution to the Babu–Brown equations.

l	F_l	G_l	F'_l	G'_l
0	-0.20 ± 0.39	0.04 ± 0.11	0.24 ± 0.16	0.53 ± 0.09
1	-0.86 ± 0.10	0.19 ± 0.06	0.18 ± 0.05	0.17 ± 0.01
2	-0.21 ± 0.01	0.12 ± 0.01	0.10 ± 0.02	0.01 ± 0.02
3	-0.09 ± 0.01	0.05 ± 0.01	0.05 ± 0.01	0.01 ± 0.01

variables. The anomalously large compression modulus in the CD-Bonn potential results almost completely from the presence of an ω coupling $g^2_{\omega NN}/4\pi = 20$ as discussed in Ref. 2. Otherwise we do not see much difference between the generally good fit to observables.

Since the double decimation, although crudely done, generally gives observables close to the empirical ones we shall not try to distinguish between them and following Ref. 2 we take the average of them and quote a deviation, reproducing Table 5 of Ref. 2 as Table 2 below.

Aside from the very large compression modulus $K = 495\,\text{MeV}$ in the CDB, the various observables are not significantly different from each other.

The results of Holt *et al.* differ from those of Schwenk *et al.* in that they do not include the Δ isobar. They later renormalized G'_0 to take effects from the Δ into account. Without their effects, Kawahigashi *et al.*[4] found $g'_{NN} = 0.6$ to compare with $G'_0 = 0.53 \pm 0.09$ from our Table 2. Generalization by Schwenk *et al.* to a model corrected for the screening due to Δ-hole excitations to all orders with $NN \to N\Delta$ and $N\Delta \to N\Delta$ interaction strengths of $g'_{N\Delta} = 0.3$ and $g'_{\Delta\Delta} = 0.3$ by Körfgen *et al.*[5] gives $G'_0 = 1.0$. Thus, we believe that adding the Δ to our description will have the same consequence and that we are starting with a g'_{NN} close enough so that we have essentially the same screening by a small change in our Δ coupling.

3. G'_0 as the Main Interaction Strength

Where there is no two-body correlation function keeping the two nucleons apart, there would be no pionic interaction with them in the long wavelength limit since it is derivative in nature

$$\delta H = \frac{f}{4\pi}\bar{\psi}\sigma \cdot \nabla \boldsymbol{\tau} \cdot \boldsymbol{\pi}\psi(r) \tag{2}$$

and the momentum $\mathbf{p}$

$$\nabla \boldsymbol{\pi} = \mathbf{p}\boldsymbol{\pi} \tag{3}$$

goes to zero as the volume goes to ∞. The total interaction by way of pion exchange is

$$V(r) = \frac{f^2}{r}\frac{1}{3}\sigma_1 \cdot \sigma_2 \boldsymbol{\tau}_1 \cdot \boldsymbol{\tau}_2 \left(\frac{e^{-m_\pi r}}{m_\pi r} - \frac{4\pi}{m_\pi^3}\delta(r)\right) \tag{4}$$

plus a tensor interaction which averages to zero over angles. The integration of the above $V(r)$

$$\int d^3 r V(r) = 0 \tag{5}$$

does go to zero. The Ericsons[6] included short-range correlations by multiplying $V(r)$ by a short-range correlation function $g(r_{12})$ which had the property that

$$g(r_{12}) = 0, \quad r_{12} = 0 \tag{6}$$

but otherwise it was of sufficiently short range that it did not affect anything.

If we leave the $\delta(r)$ out we have a minimalist description of the effect of short-range correlations. We are not finished because it is well-known that the exchange of ρ-mesons with tensor coupling between two nucleons contributes to the Lorentz–Lorenz effect. Brown[7] found that inclusion of this contribution increases the pionic one by a factor of 1.8, increasing the 1/3 in Eq. (4) to 0.6. We note that this is close to the G_0', the average $\sigma\tau$ Fermi liquid interaction in the $l = 0$ state of the interactions shown in Table 2. Furthermore, the 0.6 is precisely what Kawahigashi *et al.*[4] found for the contribution to the E^2L^2 interaction from the nucleon channel alone.

We see that the $G_0' = 1$, $G_1' = 0$ is the largest interaction by far. (It will be clear below why we group them.) In Table 2, the next largest ones are $F_l = -0.20$ and $F_l' = 0.24$. We divide F_l by 3 because of its angle dependence.

We see that $G_0' = 1$, $G_1' = 0$ implies a δ-function interaction, the potential

$$V(r) = \frac{\pi^2}{2k_f/m^\star} \sigma_1 \cdot \sigma_2 \boldsymbol{\tau}_1 \cdot \boldsymbol{\tau}_2 \delta(r) \tag{7}$$

representing the potential energy necessary to pull the nucleon and antinucleon apart from each other to overcome the pionic attraction. Note that this potential is always attractive, because the two fermions are at the same point and since they are antisymmetrical, if they are spin triplet they must be isospin singlet and vice versa.

Note that $V_{\rm lowk}$ gives quite a good description of the ^{18}O and ^{18}F spectra.[8] These calculations do not have the Brown–Rho-scaled masses in them, however, and would be expected to change as have results of Tables 1 and 2.

4. Conclusion

We believe that we can offer a qualitative understanding of one of the main points of nuclear structure physics without detailed calculations; namely, why the one-bubble correction to the mean-field (shell-model) spectrum is very important as is well known in the Kuo–Brown interactions, whereas higher order corrections, as found in Holt *et al.*[8] do not change the pattern qualitatively. This is because the E^2L^2 interaction is by far the strongest one and it is largely spent in the single bubble. In preliminary estimates, we find that the direct and exchange effects largely cancel each other in two-bubble corrections and suggest that one bubble which is classical in nature has most of the physics in it. The last paragraph is highly speculative.

References

1. A. Schwenk, G. E. Brown and B. Friman, *Nucl. Phys. A* **703** (2002) 745.
2. J. W. Holt, G. E. Brown, J. D. Holt and T. T. S. Kuo, *Nucl. Phys. A* **785** (2007) 322.
3. G. E. Brown and M. Rho, *Phys. Rep.* **396** (2004) 1.
4. K. Kawahigashi, K. Nishida, A. Itabashi and M. Ichimura, *Phys. Rev. C* **63** (2001) 044609.
5. B. Koerfgen, F. Osterfeld and T. Udagawa, *Phys. Rev. C* **50** (1994) 1637; B. Korfgen, P. Oltmanns, F. Osterfeld and T. Udagawa, *Phys. Rev. C* **55** (1997) 1819.
6. M. Ericson and T. E. O. Ericson, *Ann. Phys.* **36** (1966) 323.
7. G. E. Brown, *Nucl. Phys. A* **518** (1990) 99.
8. J. D. Holt, J. W. Holt, T. T. S. Kuo, G. E. Brown and S. K. Bogner, *Phys. Rev. C* **72** (2005) 041304.

Confinement, chiral symmetry breaking and the effective quark mass

Edward Shuryak
Department of Physics and Astronomy,
Stony Brook University, Stony Brook NY 11794, USA
Edward.Shuryak@stonybrook.edu

Paradigm shift in gauge topology, from instantons to their constituents — *instanton-dyons* — has recently lead to very significant advances. Unlike instantons, these objects have three different set of charges, therefore allowing to explain and to tie together several important phenomena: confinement, chiral symmetry breaking and magnetic screening.

Keywords: QCD; confinement mechanism; chiral symmetry breaking; instantons.

Preface

While Gerry Brown had a profound influence on me as a person, and obviously on my worldline, in this book I will not write about that. The excuse, if needed, is that I was a Co-organizer of the memorial conference at Stony Brook and in its proceedings[1] one can find a lot of recallections: there is no need to get repetitive. The best tribute to Gerry is to continue working on issues which were so dear to him, and some of these results and discussions are presented below.

1. Topology, Confinement and Chiral Symmetry Breaking

1.1. *Brief introduction*

Confinement is the most famous nonperturbative feature of the gauge theories. Its most intuitive explanation from 1970's is the *dual superconductor model* by Nambu, 't Hooft and Mandelstam.[2] Lattice studies did reveal the monopoles (defined in certain gauges) and prove that they indeed form Bose–Einstein condensate at $T < T_c$, see e.g., Ref. 4.

Original discussion of the chiral symmetry and its breaking predate QCD and even quarks. In a classic paper, Nambu and Jona-Lasinio[3] showed that an attraction in the scalar $\bar{q}q$ channel, if strong enough, can dynamically "gap" the surface of the Dirac sea. The origin of this interaction — claimed to be the origin of the mass of the "constinuent quarks", the nucleons and thus ourselves — was of course in 1961 completely unknown. Two decades later, in my paper,[5] it was suggested that NJL attraction is nothing else but 't Hooft effective $2N_f$ interaction induced by instantons. Two parameters of the NJL model has been substituted by another two: the total instanton–antiinstanton density $n \sim 1/fm^4$ and their typical size $\rho \sim 1/3\,fm$. Of course, 't Hooft vertex does more than the NJL operator, it in particular explicitly breaks the $U(1)_A$ symmetry.

A decade later instantons with such properties were found on the lattice. Statistical mechanics of instanton ensemble, including 't Hooft interaction to all orders, known as the Interacting Instanton Liquid Model, has been solved numerically, for a review see Ref. 6. Among other things, it introduced the notion of "collectivized zero mode zone", or ZMZ for short. Lattice practitioners struggle with it till now, since most of numerical fluctuations in simulations come from it.

And yet, some nasty questions remained, such as e.g., *Is there any connection between confinement and chiral symmetry breaking? Why is it that the corresponding finite temperature transitions happen close, $T_c \approx T_\chi$ for fundamental quarks, but not for adjoint ones*[25]*? Apart of color representations, are there other parameters one can introduce, affecting these transitions?*

1.2. *Nonzero holonomy, instanton-dyons and confinement*

The "holonomy" issue refers to the observation that a (gauge invariant) Polyakov loop

$$P = P\exp\left(i\int_0^\beta A_0^a T^a d\tau\right) \tag{1}$$

has a nonzero vacuum expectation value (VEV). Note that there is no trace in this formula, so it is a color matrix. One can always gauge rotate it to be a diagonal one, to Cartan sub-algebra with $N_c - 1$ parameters. These different phases appear in propagation amplitudes of gluons and quarks, different for different colors. For the simplest $SU(2)$ color group there is only one diagonal generator T^3 and only one parameter. The notations to be used below are

$$\langle P\rangle = \mathrm{diag}(e^{i\pi\nu}, e^{-i\pi\nu}). \tag{2}$$

The color averaged Polyakov line is then

$$\left\langle \frac{1}{2}\mathrm{Tr}\,P \right\rangle = \cos(\pi\nu). \tag{3}$$

At high T this phenomenon is nearly absent as $\langle P\rangle \approx 1$. However in the temperature interval $(2..1)T_c$ it changes to zero (or small value in QCD with quarks). Accounting for this phenomenon leads Pisarski and collaborators to "semi-QGP" paradigm[15] and construction of the so-called PNJL model.

A nontrivial average value of the Polyakov line $\langle P\rangle \neq 1$, indicating that an expectation value of the gauge potential is nonzero . This calls for re-defining the boundary condition of A_4 at infinity, for any solitons made of gauge fields, including the instantons. When such solutions were found in 1998[7,8] it has been realized that nonzero VEV splits instantons into N_c (number of colors) constituents, the self-dual *instanton-dyons*.[a] Since these objects have nonzero electric and magnetic charges and source Abelian (diagonal) massless gluons, the corresponding ensemble is an "instanton-dyon plasma", with long-range Coulomb-like forces between constituents. For the SU(2) gauge group the self-dual ones are called M with electric and magnetic charges $(e,m) = (+,+)$ and L with charges $(e,m) = (-,-)$, the anti-selfdual antidyons are called $\bar{M}$, $(e,m) = (+,-)$ and $\bar{L}$, $(e,m) = (-,+)$.

Diakonov and collaborators (for review see Ref. 9) emphasized that, unlike the (topologically protected) instantons, the dyons interact directly with the holonomy field. They suggested that since such dyon (anti-dyon) density increases at low temperature, their back reaction may overcome perturbative holonomy potential and drive it to its confining value, leading to vanishing of the mean Polyakov line, or confinement.

Specifically, Diakonov and collaborators focused on the self-dual sector L, M and studied the one-loop contribution to the partition function.[10] The volume element of the moduli space was written in terms of dyons coordinates as a determinant of certain matrix G, to be referred to as Diakonov determinant. In a dilute limit it leads to Coulomb interactions between the dyons, but in the dense region it becomes strongly repulsive, till at certain density the moduli volume vanishes.

A semi-classical confining regime has been defined by Poppitz *et al.*[11,12] in a carefully devised setting of softly broken supersymmetric models. While the setting includes a compactification on a small circle, with weak coupling and an *exponentially small* density of dyons, the minimum at the confining holonomy value is induced by the repulsive interaction in the dyon–antidyon molecules (called *bions* by these authors). The crucial role of the supersymmetry is the cancellation of the perturbative Gross–Pisarski–Yaffe (GPY)[13] holonomy potential: as a result, in this setting there is no deconfined phase with trivial holonomy at all, unless supersymmetry is softly broken. Sulejmanpasic and myself[14] proposed a simple analytic model for the dyon ensemble with dyon–antidyon "repulsive cores", and have shown how they may naturally induce confinement in dense enough dyonic ensemble.

Two recent papers took into account a back reaction of the instanton–dyons on the holonomy value. Important ingredient of both of them is classical dyon–antidyon interaction, determined Larsen and myself in Ref. 16: its discussion we skip for space reasons.

The first one is by Liu, Zahed and myself.[19] Its main idea is that when the ensemble of the dyons is dense enough, it leads to strong electric screening, which in turn allows to use analytic mean field approximation. Here it is not the place to present technical details of this work, and I just summarize the results. It is shown, that dense enough dyon ensemble does shift the minimum of the holonomy potential to the confining value, $\nu = 1/2$ for the SU(2) gauge theory considered.

[a]They are called "instanton-monopoles" by Unsal *et al.*, and in some ways similar to "instanton quarks" discussed by Zhitnitsky *et al.*

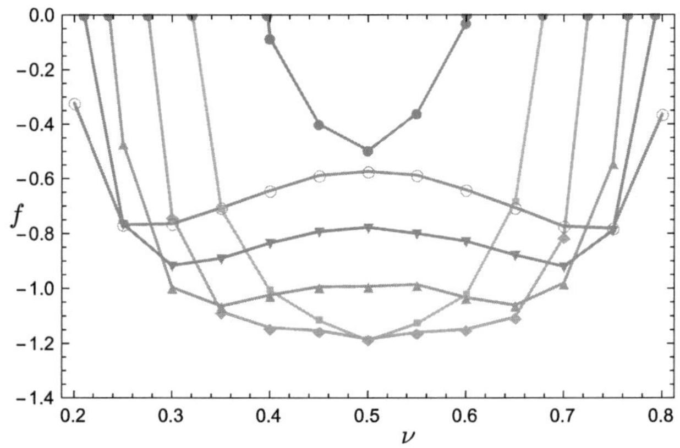

Fig. 1. (Color online) Free Energy density f as a function of holonomy ν at $S = 6$, $M_D = 2$ and $N_M = N_L = 16$. The different curves corresponds to different densities. • $n = 0.53$, ■ $n = 0.37$, ♦ $n = 0.27$, ▲ $n = 0.20$, ▼ $n = 0.15$, ○ $n = 0.12$. Not all densities are shown.

The second one, by Larsen and myself[18] uses direct numerical simulation of the instanton–dyon ensemble, both in the high-T dilute and low-T dense regimes. The holonomy potential as a function of all parameters of the model is determined and minimized. In Fig. 1 from this work, we show the dependence of the total free energy on holonomy value, for different ensemble densities. As one can see, at high density of the dyons their back reaction shifts the minimum to $\nu = 1/2$, which is the confining value for SU(2) ($\cos(\pi\nu) = 0$): the confinement transition is thus generated. The self-consistent parameters of the ensemble, minimizing the free energy, is determined for each density.

1.3. *Instanton–dyon ensemble and chiral symmetry breaking*

The first direct simulation of the instanton–dyon ensemble with dynamical fermions has been made by Faccioli and myself in Ref. 21. The general setting follows the example of the "instanton liquid", it included the determinant of the so called "hopping matrix", a part of the Dirac operator in the quasizero-mode sector. It has been done for the $SU(2)$ color group and the number of fermions flavors $N_f = 1, 2, 4$. Except in the last case, chiral symmetry breaking has been clearly observed, for dense enough dyon ensemble.

Recent work[20] is a continuation of the analytic mean-field approximation, now applied to the confining ensemble with $\nu = 1/2$. It is shown that the gap equation for quark condensate has nonzero solution for $N_f < 2N_c$. A number of other quantities is calculated in this paper.

Larsen and myself[17] addressed the issue of chiral symmetry breaking using numerical simulations, with the fermionic determinant included. Using two sizes of the system, with 64 and 128 dyons, we identify the finite-size effects in the eigenvalue distribution, and extrapolate to infinite size system. The location of the chiral

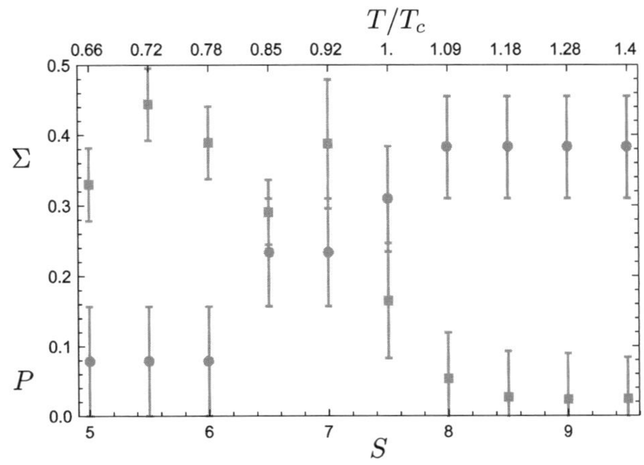

Fig. 2. (Color online) The Polyakov loop P (blue circles) and the chiral condensate Σ (red squares) as a function of action $S = 8\pi^2/g^2(T) = 6\log(T/\Lambda_{\rm QCD})$ of the normalized temperature T/Tc (upper scale).

transition temperature is defined both by extrapolation of the quark condensate, from below, and the so called "gaps" in the Dirac spectra, from above.

We do indeed observe see Fig. 2, that for SU(2) gauge theory with two flavors of light fundamental quarks both the confinement–deconfinement transition and chiral symmetry breaking, as the density of dyons goes up at lower temperature.

Determination of the transition point by vanishing of $\langle P\rangle$ or $\langle\bar{\psi}\psi\rangle$ is difficult for technical reasons, since both transitions appear to be in this case just a smooth crossover ones. It is by now well-established procedure to define the transition points via maxima of corresponding susceptibilities. Those should correspond to *inflection points* (change of curvature) on the plots to be shown. Looking from this perspective at Fig. 2, one would locate the inflection points of both curves, for $\langle P\rangle$ or $\langle\bar{\psi}\psi\rangle$, at the same location, namely $S = 7 - 7.5$. Thus, within the accuracy our simulations, we conclude that both phase transitions happen at the same conditions.

2. Quark Effective Mass

The "fundamental quark masses" are the coefficients in the QCD Lagrangian $\sum_f m_f\bar{q}_f q_f$. Their magnitude and dependence on scale is described in Particle Data Tables. They do violate the chiral symmetry, but for our discussion one can put them to zero, going into the so-called exact chiral limit.

The subject of our discussion are "effective" quark masses, originated from their interaction with the gauge fields. We will split them into two categories, the "nonchiral" $M_{{\rm non}\,\chi}$ violating the chiral symmetry at temperatures below T_c, and the chiral M_χ.

By definition, the former ones are induced by spontaneous chiral symmetry breaking and thus are only present below the chiral phase transition $T < T_\chi$. It is defined in terms of certain models and thus is somewhat ambigous by definition. In particular, the NJL model — built soon after BCS theory of superconductivity and resembling it — provided a dynamically generated gap in the fermion spectrum, known also as "constituent quark mass". The value comes from a "gap equation", and with parameters used by NJL practitioners take it is about

$$M_{\mathrm{non}\,\chi} \approx M_N/3 \sim 320\,\mathrm{MeV}.$$

The interacting instanton liquid model does not use mean field and has no gap equation *per se*, but from fermion propagators one finds the value for the constituent quark mass in this model. It is larger, of about $M_{\mathrm{non}\,\chi} \approx 420\,\mathrm{MeV}$ which is close to $m_\rho/2$ or $m_\Delta/3$, so-to-say "more normal" meson and baryon. (The nucleon should not be used, since it contains a scalar diquark, subject to additional strong instanton-induced pairing, which violates the notion of "quark additivity")

The famous Brown and Rho[22] idea was that since the chiral condensate is "melting" at T_c, the quark effective mass should melt there as well. This idea leads to their famous suggestion, that hadronic masses should *decrease* with T and density, according to certain scaling relation.

In the opposite limit, at very high T well above T_c, one should find a weakly coupled QGP and be able to use perturbative QCD. The gluonic polarization operator calculated in Ref. 24 and predicted normal screening with the Debye mass m_D and the on-shell gluon mass

$$m_g^2 = \frac{m_D^2}{2} = \left(\frac{g^2T^2}{2}\right)\left(\frac{N_c}{3} + \frac{N_f}{6}\right). \tag{4}$$

A new mode of oscillations appear, called "plasmon". Quark quasiparticles in QGP form two brunches of excitation — known as quark quasiparticle and plasmino — which share the same on-shell mass, known as the Klimov–Weldon mass[27,28]

$$m_{KW}^2 = \left(\frac{N_c^2-1}{2N_c}\right)\left(\frac{g^2T^2}{8}\right). \tag{5}$$

The quark-to-gluon mass ratio is in this order just a number, for $N_c = N_f = 3$ it gives

$$m_g/m_{KW} = \sqrt{2}/3 \approx 0.47.$$

This mass comes from Compton-like quark scattering on gluons, obviously respecting the chiral symmetry.

In Fig. 3, we plotted it by a straight line, even at low T. However, a number of corrections should in principle be included. First of all, the T^2 in it comes from the thermal density of massless gluons. This density should be strongly modified near and below T_c, in a spirit of semi-QGP one should perhaps multiply it by a factor

$$\frac{N_c^2-N_c}{N_c^2-1}\left\langle\frac{1}{2}TrP(T)\right\rangle^2 + \frac{N_c-1}{N_c^2-1} \tag{6}$$

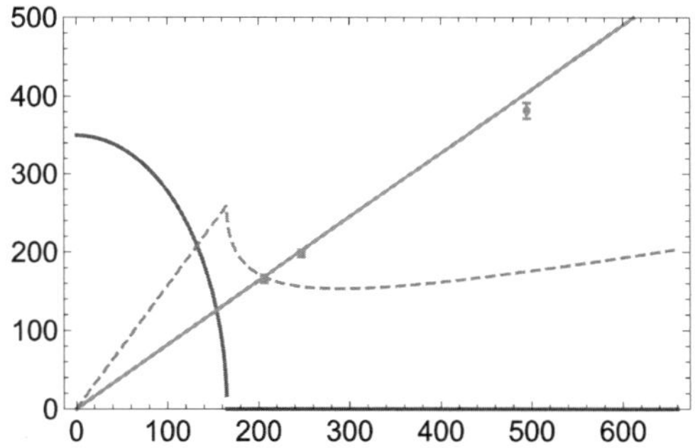

Fig. 3. (Color online) Quark effective mass (MeV) versus temperature T (MeV). Black solid line is the nonchiral mass related to chiral symmetry breaking; Brown dashed line is for "Polyakov-induced mass" (7); Red solid line is a perturbative mass due to quark–gluon scattering (4); Data points with error bars are from lattice simulations Ref. [26].

in which the first term stands for color-nondiagonal gluons and the second for the remaining diagonal gluons which are not "higgsed" by the holonomy. On the other hand, some nonperturbative contribution to the mass appear there, e.g., from forward scattering on monopoles or instanton–dyons, which may more than compensate that. This uncertainty remains the reason why we are not sure how strong the Brown–Rho suppression of mass may actually be.

There are direct lattice studies of the gluon and quark quasiparticles on the lattice, at $T > T_c$. Three points plotted in Fig. 3 are from Karsch and Kitazawa[26] work, in which quenched simulation with Wilson fermions were used. The masses come from numerical Landau gauge propagators fitted by the two-pole ansatz. One can see that they correspond to perturbative expectations rather well.

Because of the Euclidean time used on the lattice, there are no decays there and thus no imaginary part of the mass: but the real part has been be calculated. On the other hand, as T decreases toward T_c, one enters the domain studied experimentally, in RHIC/LHC heavy ion collisions. The main conclusion from those is that here matter becomes "strongly coupled" sQGP, with surprisingly large rescattering (and jet quenching) and small viscosity. In terms of effective quark mass this implies existence of a large imaginary part of nonperturbative origin. Its possible magnitude remains a subject of a debate.

2.1. *The holonomy and the quark phases*

On top of effective masses we discussed, the Dirac equation for a quark in QCD vacuum (heat bath) has other terms coming from the interaction with the gauge field. It of course has highly fluctuating quantum (thermal) components, but now I want to emphasize the role of the constant terms coming from $\langle \mathrm{Tr}\, P \rangle$. In some calculations — such as total contribution of quarks to thermodynamics — the colors are summed by a trace. It is the resulting $\langle \mathrm{Tr}\, P \rangle$ which is usually measured on the lattice, which shows how this quantity goes to zero in the confining phase

$T < T_c$ due to destructive interference of different colors. Thus, by multiplying quark contribution by $\langle \mathrm{Tr}\, P \rangle$ one effectively switches it off: this is the essence of "semi-QGP" approach by Pisarski and the PNJL model we already mentioned.

Some idea of the magnitude of the total effect in terms of extra "P-induced mass"

$$m_P = T \times \log\left(\frac{N_c}{\langle \mathrm{Tr}\, P \rangle} \right) \tag{7}$$

plotted with masses, by a dashed line in Fig. 3. One can see that in the phase transition region this effect is comparable to other components of the "effective mass". (For definiteness, we use data on $\langle \mathrm{Tr}\, P(T) \rangle$, in $2+1$ flavor SU(3) QCD, from.[23] For the purpose of plotting we fitted its results by a Fermi-like function

$$P(T) = \frac{1}{\exp[(T_1 - T)/T_2] + 1} \tag{8}$$

with $T_1 = 233\,\mathrm{MeV}$, $T_2 = 33\,\mathrm{MeV}$).

However, the actual point I want to make here is that usage of such color-averaged (traced) P may be misleading, because the effects of holonomy for individual colors are all different. In particular, the instanton–dyons we discussed above have certain fixed color arrangements, related to phases of the matrix $\langle P \rangle$. As a consequence, fermionic zero modes of these dyons are also restricted to particular eigenvalues of P and thus to particular colors.

The issue of quark phases depending on their color can be complemented by additional phases related with the *periodicity condition* of quarks on thermal Matsubara circle. Standard fermions are antiperiodic, $\psi(\beta) = \exp(i\pi)\psi(0)$, bosons are periodic, but one can introduce an arbitrary phase

$$\psi(\beta) = e^{i2\pi z}\psi(0) \tag{9}$$

If one wishes, one can imagine certain Aharonov–Bohm solenoid in extra dimensions providing it. Also one may say that by doing that one changes quarks from being fermions to "*anyons*". Furthermore, these phases can be set to be *flavor* dependent, by considering z a flavor matrix $\hat{z}$.

So, in general one finds many interesting models, easily implementable on the lattice, in which quark motion is affected *both* by color-dependence phases from $\langle P \rangle$ and flavor dependent ones from $\hat{z}$. While the former ones are generated dynamically, the latter can be changed, and the result of their interplay will tell us about color phases. Therefore, the quark phases is not a toy, but a useful tool which can tell us a lot about quark motion at finite T and the mechanism of chiral symmetry breaking.

As an example, let me discuss one concrete model suggested by Kouna *et al.*.[29–33] They consider especially a symmetric case $N_f = N_c$. In particular, for the SU(3) color, the fermionic angles for u, d, s quarks are taken of the form $(0, \theta, -\theta)$, respectively. The parameter θ can be varied from $\theta = 0$ (the usual QCD) to $\theta = 2\pi/3$, at which point the theory becomes "center symmetric" , so they called it *$Z(N_c)$-symmetric QCD*. While the original authors studied this model in PNJL model context, subsequently lattice studies of it has been made by Misumi *et al.*[34] They observe that, in contrast to ordinary QCD, in this case the deconfinement transition is significantly strengthened to the first-order phase transition, with clear hysteresis, etc. The chiral transition for $Z(3)$-symmetric theory has indeed moved upward: the exact location is hard to tell from the paper, since small mass extrapolation is not yet performed. It is more important that chiral condensates for different flavors become clearly different, so T_χ should get split for different flavors.

These results can be qualitatively understood in the instanton–dyon framework rather easily. The key to it is the arrangement of the zero fermionic modes, related to chiral symmetry breaking. Remember, according to van Baal *et al.* the instanton is an assembly of N_c dyons, but it has only one unit of the topological charge and thus only one zero mode. For the usual antiperiodic boundary condition for a fermion, this zero mode goes to "twisted" L dyons only. That means that in ordinary QCD all quarks "see" only $L, \bar{L}$-dyon subcomponent, leaving $M_1, \ldots, M_{N_c-1}$ without zero modes. The $Z(N_c)$-*symmetric* QCD changes this assignment to much more symmetric one: u, d, s quarks have zero modes with L, M_1, M_2 dyons, respectively. No doubt the deconfinement transition is much stregthened, and the chiral ones split in flavors. For recent developments see Refs. [35–37].

3. Summary

Paradigm shift in gauge topology, from instantons to their constituents — *instanton–dyons* — has lead to very significant advances. Unlike instantons, these objects have three different set of charges, therefore allowing to tie several important phenomena together.

Like the instantons, dyons have topological charges and are subject to topological index theorems. It means that for appropriate fermionic boundary conditions they must have fermionic zero modes. Collectivization of those into a nonzero quark condensate follows, provided the ensemble is dense enough.

Unlike instantons, the dyons possess (Euclidean) electric charges, and interact directly with the holonomy. Therefore, they back-react and are able to modify the holonomy potential. Furthermore, as the calculations showed, the potential's minimum *shifts to confining value of the holonomy*, at which all types of dyons obtain equal actions.

Unlike instantons, the dyons possess magnetic charges, and thus their ensemble generates *the magnetic screening mass.* Recall that perturbative polarization tensor does not generate it[24]: but, according to lattice data, in the near-T_c region it even surpasses the electric mass. While we have not discussed it above, let me just mention that it clearly indicates a transition from electric (QGP) to magnetic plasma, as the coupling grows with decreasing temperature.

In section 2, we discussed various "known" contributions to the effective quark mass, both chiral symmetry violating and preserving ones. Taken literally, they support existence of the Brown–Rho dip in the effective quark mass near T_c. However, effects due to nonzero holonomy — interaction with mean $\langle A_0 \rangle$ — has comparable magnitude and is peaked near T_c. Furthermore, the story is much more complicated since different phases appear for different color fermions, which make up topological zero modes and thus the quark condensate. The colors of the holonomy and quarks in the condensate are in correlation, their independent averaging was not justified. So, much more work is still needed to understand these correlations and perhaps to get a final judgement on the BR scaling.

My final point starts in an obvious way: the theory of instanton-dyons and its multiple predictions need to be extensively checked on the lattice. An excellent way to do so is via studies of QCD-like theories with quarks possessing extra periodicity phases. Needless to say, those should also be supplemented by a transition from occasional "hunt" for instanton–dyons, to systematic studies of their density and parameters.

Acknowledgments

I am greatly indebted to Gerry Brown: 20 plus years of daily discussions with him was an invaluable experience. The progress reported would not be possible without breakthroughs by Pierre van Baal and Mitya Diakonov, the consequences of whose legacy remains to be worked out. The particular results reported here were obtained with my collaborators on the instanton–dyon projects, P. Faccioli, T. Sulejmanpasic, I. Zahed and R. Larsen.

References

1. E. Shuryak, *Nucl. Phys. A* **928** (2014) 138, doi:10.1016/j.nuclphysa.2014.03.006, arXiv:1401.2032 [nucl-th].
2. Y. Nambu, Gauge Theory for String Interactions, *Phys. Rev. D* **10** (1974) 4262; S. Mandelstam, *Phys. Rep.* **23C** (1976) 145; G. 't Hooft, in *Proceedings of the European Physics Society 1975*, ed. A. Zichichi (Editrice Compositori, Bologna, 1976), p. 1225.
3. Y. Nambu, *Phys. Rev. Lett.* **4** (1960) 380; *Phys. Rev.* **117** (1960) 648; Y. Nambu and G. Jona-Lasinio, *Phys. Rev.* **122** (1961) 345.
4. A. D'Alessandro, M. D'Elia and E. V. Shuryak, *Phys. Rev. D* **81** (2010) 094501, doi:10.1103/PhysRevD.81.094501, arXiv:1002.4161 [hep-lat].
5. E. V. Shuryak, *Nucl. Phys. B* **203** (1982) 93.
6. T. Schafer and E. V. Shuryak, *Rev. Mod. Phys.* **70** (1998) 323, arXiv:hep-ph/9610451.
7. T. C. Kraan and P. van Baal, *Phys. Lett. B* **435** (1998) 389, arXiv:hep-th/9806034.
8. K.-M. Lee and C.-H. Lu, *Phys. Rev. D* **58** (1998) 025011, arXiv:hep-th/9802108.
9. D. Diakonov, *Topology and Confinement*, arXiv:0906.2456v1 (2009).
10. D. Diakonov, N. Gromov, V. Petrov and S. Slizovskiy, *Phys. Rev. D* **70** (2004) 036003, arXiv:hep-th/0404042.
11. E. Poppitz, T. Schaefer and M. Unsal, *J. High Energy Phys.* **1210** (2012) 115, arXiv:1205.0290 [hep-th].
12. E. Poppitz and M. Unsal, *J. High Energy Phys.* **1107** (2011) 082, arXiv:1105.3969 [hep-th].
13. D. J. Gross, R. D. Pisarski and L. G. Yaffe, *Rev. Mod. Phys.* **53** (1981) 43.

14. E. Shuryak and T. Sulejmanpasic, *Phys. Lett. B* **726** (2013) 257, arXiv:1305.0796 [hep-ph].
15. R. D. Pisarski, *Nucl. Phys. Proc. Suppl.* **195** (2009) 157.
16. R. Larsen and E. Shuryak, *Nucl. Phys. A* **950** (2016) 110, arXiv:1408.6563 [hep-ph].
17. R. Larsen and E. Shuryak, *Phys. Rev. D* **93** (2016) 054029, arXiv:1511.02237 [hep-ph].
18. R. Larsen and E. Shuryak, *Phys. Rev. D* **92** (2015) 094022, arXiv:1504.03341 [hep-ph].
19. Y. Liu, E. Shuryak and I. Zahed, *Phys. Rev. D* **92** (2015) 085006, arXiv:1503.03058 [hep-ph].
20. Y. Liu, E. Shuryak and I. Zahed, *Phys. Rev. D* **92** (2015) 085007, arXiv:1503.09148 [hep-ph].
21. P. Faccioli and E. Shuryak, *Phys. Rev. D* **87** (2013) 074009, arXiv:1301.2523 [hep-ph].
22. G. E. Brown and M. Rho, *Phys. Rev. Lett.* **66** (1991) 2720.
23. A. Bazavov and P. Petreczky, *Phys. Rev. D* **87** (2013) 094505, arXiv:1301.3943 [hep-lat].
24. E. V. Shuryak, *Sov. Phys. JETP* **47** (1978) 212 [*Zh. Eksp. Teor. Fiz.* **74** (1978) 408].
25. F. Karsch and M. Lutgemeier, *Nucl. Phys. B* **550** (1999) 449, arXiv:hep-lat/9812023.
26. F. Karsch and M. Kitazawa, *Phys. Rev. D* **80** (2009) 056001, arXiv:0906.3941 [hep-lat].
27. V. V. Klimov, *Sov. J. Nucl. Phys.* **33** (1981) 934 [*Yad. Fiz.* **33** (1981) 1734].
28. H. A. Weldon, *Phys. Rev. D* **61** (2000) 036003, arXiv:hep-ph/9908204.
29. H. Kouno, Y. Sakai, T. Makiyama, K. Tokunaga, T. Sasaki and M. Yahiro, *J. Phys. G* **39** (2012) 085010.
30. Y. Sakai, H. Kouno, T. Sasaki and M. Yahiro, *Phys. Lett. B* **718** (2012) 130, arXiv:1204.0228 [hep-ph].
31. H. Kouno, T. Makiyama, T. Sasaki, Y. Sakai and M. Yahiro, *J. Phys. G* **40** (2013) 095003, arXiv:1301.4013 [hep-ph].
32. H. Kouno, T. Misumi, K. Kashiwa, T. Makiyama, T. Sasaki and M. Yahiro, *Phys. Rev. D* **88** (2013) 016002, arXiv:1304.3274 [hep-ph].
33. H. Kouno, K. Kashiwa, J. Takahashi, T. Misumi and M. Yahiro, arXiv:1504.07585 [hep-ph].
34. T. Misumi, T. Iritani and E. Itou, Finite-temperature phase transition of $N_f = 3$ QCD with exact center symmetry, arXiv:1510.07227 [hep-lat].
35. Y. Liu, E. Shuryak and I. Zahed, *Phys. Rev. D* **94**(10) 105013 (2016) doi: 1001103/ PhysRevD.94.105013.
36. R. Larsen and E. Shuryak, *Phys. Rev. D* **94**(9) 014009 (2016) doi: 10.1103/PhysRevD.94.094009 [arXiv: 1605.07474 [hep-ph]].
37. A. Cherman, T. Schäfer and M. Ünsal, *Phys. Rev. Lett.* **171**(8) 081601 (2016) doi: 10.1103/PhysRevLett.117.081601 [arXiv: 1604.06168 [hep-th]].

Many-body Green functions in nuclear physics

J. Speth*
Institut für Kernphysik, Forschungszentrum Jülich,
D-52425 Jülich, Germany
j.speth@fz-juelich.de

N. Lyutorovich
Saint Petersburg State University, 7/9 Universitetskaya nab.,
Saint Petersburg, 198504, Russia
n_lyutor@mail.ru

Our contribution is dedicated to the memory of G.E. Brown on the occasion of his 90th birthday. JS: He was my friend and teacher. His seminal article[1] on Landau, Bruechner-Bethe, and Migdal theories of Fermi Systems had a strong influence on my research.

Many-body Green functions are a very efficient formulation of the many-body problem. We review the application of this method to nuclear physics problems. The formulas which can be derived are of general applicability, e.g., in self-consistent as well as in nonself-consistent calculations. With the help of the *Landau renormalization*, one obtains relations without any approximations. This allows to apply conservation laws which lead to important general relations. We investigate the one-body and two-body Green functions as well as the three-body Green function and discuss their connection to nuclear observables. The generalization to systems with pair correlations are also presented. Numerical examples are compared with experimental data.

Keywords: Many-body Green functions; nuclear structure.

1. Introduction

The application of many-body Green functions to concrete physical problems goes back to Landau who derived his originally purely phenomenological *Fermi liquid theory*[2] within the frame work of many-body Green functions. His approach is based on the response function which is the correlated part of the two-body Green function. The response function can be derived from an integral equation which includes one-body Green functions and a very complex two-body interaction. The key assumption of the Fermi liquid theory is the separation of the one-body Green function into a singular quasiparticle pole part and a regular rest. With this ansatz (*Landau renormalization*), Landau transformed the full many-body problem into a system of

*Corresponding author.

interacting quasiparticles in the vicinity of the Fermi surface. He parametrized the quasiparticle–quasihole (ph) interaction in the corresponding equation and introduced the Landau parameters which are adjusted to experimental data. With this equation, Landau predicted the famous *zero sound* in ^{3}He.

Migdal applied Landaus approach to the nuclear many-body problem. The *Theory of finite Fermi system* (FFS) by Migdal[3] is therefore a generalization of Landau's theory of Fermi liquid. In nuclear physics, the Landau quasiparticle are the single-particle excitations and the corresponding ph interaction is parametrized in the same way as in the Fermi-liquid theory, only the parameters of the nuclear ph-interaction, the Landau–Migdal parameters, are density dependent. The renormalization procedure is much more involved. A detailed presentation is given in Refs. 4 and 5.

Originally, the theory was applied to spherical nuclei only with the shell model as basis for the quasiparticles. In Ref. 6, the range of applications was extended to strongly deformed nuclei, where the energies and wave functions of the Nilsson model were considered as Landau quasiparticle. Here also pair correlations had to be taken into account by introducing *anomalous Green functions.* The basis of all these investigations is the response function which is, as mentioned above, the correlated part of the two-body Green function.

A further step to a generalization of the FFS theory was the investigation of the three-body Green functions. Of special importance was here the response function in second-order (with pair correlations) which allowed to calculate isomer shifts of rotational states in deformed nuclei.[7] This phenomena was of great interest 40 years ago as the change of the radius of *rotational states* was in some nuclei positive, as expected, but in some nuclei it was also negative which was unexpected. The theoretical explanation[7] was quite surprising. An other generalization is the extension beyond the one-particle one-hole ($1p1h$) approximation. Here, the fragmentation of the single-particle strength due to the coupling to phonons is taken into consideration. This allows to determine, e.g., the spreading width of giant resonances.

In our contribution, we review the application of many-body Green functions on nuclear structure problem. The equations which can be derived from the response function are formally identical with the well known RPA equation and the linear response. The great advantage of the Green function approach is that one derives these equations without any approximations. For that reason, one can apply conservation law and obtains important relations like the Ward identities. We present applications of these equations within the FFS approach but also in a self-consistent model.

The paper is organized as follows. In Sec. 2.1, we define the many-body Green functions, the equation of motion for the one-particle Green function and the response function and we discuss the physical content of the response function. In Sec. 2.2, we discuss the renormalization of the response function and present the final equations for the various nuclear quantities which can be calculated with the help of the response function. In Sec. 2.3, the response function with pair-correlations is introduced and we outline the second-order response functions. In Sec. 3, we discuss the basics of the FFS theory and present two problems of strongly deformed nuclei in the rare earth region and show in a third example the predictive power of the FFS theory. In Sec. 4, a self-consistent approach is discussed and two examples are presented. Finally, we give in Sec. 5 a short summary.

2. The Method

2.1. *Many-body Green functions*

The one-particle, two-particle and three-body Green functions are defined in a single-particle basis as:

$$g_{\nu_1\nu_2}(t_1;t_2) = (-i)\langle A0|T\{a_{\nu_1}(t_1)a^+_{\nu_2}(t_2)\}|A0\rangle, \tag{1}$$

$$\begin{aligned} &g_{\nu_1\nu_3\nu_2\nu_4}(t_1t_3;t_2t_4) \\ &\quad = (-i)^2\langle A0|T\{a_{\nu_1}(t_1)a_{\nu_3}(t_3)a^+_{\nu_4}(t_4)a^+_{\nu_2}(t_2)\}|A0\rangle, \end{aligned} \tag{2}$$

$$\begin{aligned} &g_{\nu_1\nu_3\nu_5\nu_2\nu_4\nu_6}(t_1t_3t_5;t_2t_4t_6) \\ &\quad = (-i)^3\langle A0|T\{a_{\nu_1}(t_1)a_{\nu_3}(t_3)a_{\nu_5}(t_5)a^+_{\nu_6}(t_6)a^+_{\nu_4}(t_4)a^+_{\nu_2}(t_2)\}|A0\rangle, \end{aligned} \tag{3}$$

with the time-dependent creation and annihilation operators

$$a_\nu(t) = e^{iHt}a_\nu e^{-iHt}, \quad a^+_\nu(t) = e^{iHt}a^+_\nu e^{-iHt}, \tag{4}$$

where T denotes the time-ordering operator of Wick. For further investigation, it is convenient to introduce Green functions with a source field q:

$$g_q(1,2) = \frac{(-i)\langle A0|T\{Ua_{\nu_1}(t_1)a^+_{\nu_2}(t_2)\}|A0\rangle}{\langle A0|T\{U\}|A0\rangle} \tag{5}$$

with

$$U = e^{i\int d5d6q(5,6)a^+_{\nu_5}(t_5)a_{\nu_6}(t_6)} \tag{6}$$

and the short hand notation:

$$\sum_{\nu_1}\int dt_1 = \int d1. \tag{7}$$

The equation of motion for the one-particle Green function, *the Dyson equation*, has the form:

$$\begin{aligned} &\frac{i}{2}\int d3\{G_0^{-1}(1,3) + q(1,3) + \Sigma_q(1,3)\}g_q(3,2) = \delta(1,2), \\ &G_0^{-1}(1,3) \equiv \delta(t_1-t_3)\left\{\delta_{\nu_1\nu_3}i\frac{\delta}{\delta t_1} - h_0(\nu_1\nu_3)\right\}. \end{aligned} \tag{8}$$

Here, h_0 is a single-particle Hamiltonian. The *mass operator* Σ_q contains the complete many-body information. It is in a complex way connected with bare two-body interaction.[4] The higher Green functions can now be obtained by functional derivation of the one-body Green function with respect to the external field q. In this way, one gets the response function L

$$\frac{\delta g_q(1,2)}{\delta q(4,3)} = L(13,24) = g(13,24) - g(1,2)g(3,4) \tag{9}$$

(*Note:* after the functional derivation we always put $q = 0$). The functional derivative of the Dyson equation gives an integral equation for the response function L:

$$L(13,24) = -g(1,4)g(3,2) - i\int d5d6d7d8g(1,5)K(57,68)L(83,74)g(6,2), \tag{10}$$

the kernel K is the functional derivative of the mass operator Σ

$$\frac{\delta\Sigma(1,2)}{\delta g(3,4)} = iK(13,24). \tag{11}$$

The second-order response functions $L^{(2)}$ follows from the second derivative of the one-body Green function and represents the correlated part of the three-body Green function[8]:

$$\frac{\delta^2 g_q(1,2)}{\delta q(6,5)\delta q(4,3)} = L(135,246). \tag{12}$$

This function has an important application as we will see in Sec. 3.

The linear response function L is connected with the following nuclear quantities:

(i) From Eq. (9), we obtain in linear approximation the change of the one-particle Green function due to an external field δq:

$$\delta g_q(1,2) = \sum_{3,4} L(13,24)\delta q(4,3). \tag{13}$$

The Fourier transformation gives $\delta g_q(\omega,\Omega)$ and, after integration over ω, one obtains the change of the density $\delta\rho_q(\Omega)$, where Ω is the energy transferred by the external field $\delta q(\Omega)$.

$$\delta\rho_{\alpha\beta}(\Omega) = \sum_{\kappa,\lambda} L(\Omega)_{\alpha\beta\lambda\kappa}\delta q_{\kappa\lambda}(\Omega). \tag{14}$$

(ii) With the time order $t_3, t_4 > t_1, t_2$, we insert into the two-particle Green function a complete system of eigenfunctions of the A-particle system between the two particle-hole pairs

$$g(13,24) = \sum_{n=0}^{\infty} g^{0n}(34)g^{n0}(12) \tag{15}$$

with

$$g^{0n}(34) = -i\langle A0|T\{a(3)a^+(4)\}|An\rangle \tag{16}$$

and

$$g^{n0}(12) = -i\langle An|T\{a(1)a^+(2)\}|A0\rangle. \tag{17}$$

As the response function is defined as

$$L(13,24) = g(13,24) - g(34)g(12), \tag{18}$$

its spectral representation is also given by Eq. (15) but the sum begins at $n = 1$. Thus, one obtains the transition amplitudes between the ground state and the excited states of the A-particle system.

(iii) With the time order $t_3 \to \pm\infty$ and $t_4 \to \mp\infty$ in the two-particle Green function one can move the corresponding operators to left and right side of the time ordering and may introduce two complete sets of eigenstates of the $(A+1)$- system or $(A-1)$-system, respectively. From this, one obtains another generalized Green function

$$g^{\alpha,\beta}(12) = -i\langle A\pm 1,\alpha|T\{a(1)a^+(2)\}|A\pm 1,\beta\rangle, \tag{19}$$

where α and β are two given states in the odd mass system. As we know only the response function, one has to subtract the one-particle Green function of the A-particle system and obtains finally the change of the density between two given states in the $(A\pm 1)$-system and the ground state of the A-particle system. With this quantity, one can calculate moments and transitions in odd-mass nuclei. Details are given in Ref. 4.

2.2. *Landau renormalization and the basic equations*

The crucial step in the *Landau–Migdal* theory is the assumption, that the one-body (Fourier transformed) Green function can be separated into a pole part and into a regular rest:

$$g_{\nu_1\nu_2}(\omega) = \frac{\delta_{\nu_1\nu_2} z_{\nu_1}}{\omega - \epsilon_{\nu_1} + i\eta\ \mathrm{sign}(\epsilon_{\nu_1} - \mu)} + g^{(r)}_{\nu_1\nu_2}(\omega) \tag{20}$$

with the single-particle energies ϵ_ν and the single-particle strength z_ν. With help of the renormalization procedure by Landau, the single-particle strengths and the regular part of the one-body Green functions do no longer appear explicitly in the equations. They give rise to a *renormalized* kernel K which is now the effective particle–hole (ph)-interaction F^{ph} and effective single-particle operators Q^{eff}. In the following, we present the renormalized equations which follow from the linear response equation. The detail of the derivation are given in Ref. 5.

(i) We start with the expression for the transition strength as a function of the excitation energy which one needs to calculate reaction cross-sections.

$$S(\Omega,\Delta) = -\frac{i}{\pi}\mathrm{Im}\sum_{\nu_1\nu_2} Q^{\mathrm{eff}}_{\nu_1\nu_2}(\Omega)\rho_{\nu_2\nu_1}(\Omega + i\Delta). \tag{21}$$

Here, Δ is a smearing parameter that simulates the finite experimental resolution and the more complex configurations not included in the present approach, Q^{eff} is the renormalized one-particle operator mentioned above, ρ is the quasiparticle density:

$$\rho_{\nu_1\nu_2}(\Omega) = \frac{n_{\nu_1} - n_{\nu_2}}{\epsilon_{\nu_1} - \epsilon_{\nu_2} - \Omega} Q^{\text{eff}}_{\nu_1\nu_2}(\Omega) - \frac{n_{\nu_1} - n_{\nu_2}}{\epsilon_{\nu_1} - \epsilon_{\nu_2} - \Omega} \sum_{\nu_3\nu_4} F^{ph}_{\nu_1\nu_3\nu_2\nu_4} \rho_{\nu_4\nu_3}(\Omega) \tag{22}$$

with the single-particle energies ϵ_ν and the occupation probabilities n_ν which are one or zero, respectively, for quasiparticles and quasiholes. These formulas refer to even–even nuclei because only in that case one can define the Landau quasiparticles as the single-particle excitation in the neighboring odd mass nuclei. This holds for spherical as well as for deformed nuclei.[8]

(ii) The equation for the excitation energies Ω_m and transition probabilities χ^m in even–even nuclei has the same form in the conventional RPA. The derivation of the equation and the definition of F^{ph}, however, is much more general. In addition an effective operators Q^{eff} appear. One obtains:

$$(\epsilon_{\nu_1} - \epsilon_{\nu_2} - \Omega_m)\chi^m_{\nu_1\nu_2} = (n_{\nu_1} - n_{\nu_2}) \sum_{\nu_3\nu_4} F^{ph}_{\nu_1\nu_4\nu_2\nu_3} \chi^m_{\nu_3\nu_4}, \tag{23}$$

where the transition matrix elements of a one-body operator Q are given by

$$\langle Am|Q|A0\rangle = \sum_{\nu_1\nu_2} Q^{\text{eff}}_{\nu_1\nu_2} \chi^m_{\nu_1\nu_2}. \tag{24}$$

(iii) The expression for moments and transitions in odd mass nuclei has the form:

$$\langle A \pm 1, \alpha|Q|\beta, A \pm 1\rangle - \langle A, 0|Q|0, A\rangle \delta_{\alpha\beta} = \sum_{\nu_1\nu_2} Q^{\text{eff}}_{\nu_1\nu_2} \rho^{\alpha\beta}_{\nu_1\nu_2} \tag{25}$$

with

$$\rho^{\alpha\beta}_{\nu_1\nu_2} = \delta_{\nu_1\alpha}\delta_{\nu_2\beta} + \frac{n_{\nu_1} - n_{\nu_2}}{\epsilon_{\nu_1} - \epsilon_{\nu_2} - \Omega_{\alpha\beta}} \sum_{\nu_3\nu_4} F^{ph}_{\nu_1\nu_3,\nu_2\nu_4} \rho^{\alpha\beta}_{\nu_3\nu_4}. \tag{26}$$

Note: From Eq. (25), one obtains the difference between the expectation value of the even–even and odd nucleus. This is important for the charge distribution where we can only calculate the change of the charge distribution. But this is exactly what one needs for the calculations of isotope shifts.[9] All relations have been derived without any approximations. Therefore, conservation laws can be applied, e.g., the effective electric operators reduces to the bare ones due to Ward identities in the long-wave length limit.

2.3. *Green functions including pair correlations*

In most of the nuclei, pair correlations are important and has to be considered in the corresponding calculations. Such an extension has been first obtained by Larkin and Migdal.[10] Here, we follow the derivation by Ref. 6 which is based on the matrix formulation by Ref. 11. One introduces a one-body Green function matrix of the form:

$$G^{kl}_{\nu_1\nu_2}(t_1t_2) = \begin{pmatrix} G^{11}_{\nu_1\nu_2}(t_1t_2) & G^{12}_{\nu_1\nu_2}(t_1t_2) \\ G^{21}_{\nu_1\nu_2}(t_1t_2) & G^{22}_{\nu_1\nu_2}(t_1t_2) \end{pmatrix}. \tag{27}$$

The normal and anomalous Green functions are defined in the following way:

$$G^{kl}_{\nu_1\nu_2}(t_1t_2) = \frac{i}{\langle A|T\{U\}|A\rangle} \times \begin{pmatrix} \langle A|T\{Ua_{\nu_1}(t_1)a^+_{\nu_2}(t_2)\}|A\rangle & -\langle A|T\{Ua_{\nu_1}(t_1)a_{\overline{\nu_2}}(t_2)\}|A+2\rangle \\ \langle A+2|T\{Ua^+_{\overline{\nu_1}}(t_1)a^+_{\nu_2}(t_2)\}|A\rangle & -\langle A+2|T\{Ua^+_{\overline{\nu_1}}(t_1)a_{\overline{\nu_2}}(t_2)\}|A+2\rangle \end{pmatrix}. \tag{28}$$

The bar on the indices denote the time-reversed state. The higher Green functions are obtained by functional derivative like before. To begin with, we restrict ourselves first on the two-body Green function and obtain as in the previous case three equations, (i) the linear response under the influence of an external field, (ii) the equations for moments and transitions in odd mass nuclei and (iii) the equation for the excitation energies and transition probabilities in even–even nuclei. This equation has the identical form of the quasiparticle RPA which has been previously derived in different ways by Bogoliubov,[12] Baranger[13] and Belyaev.[14] Birbrair[15] and Kamerdzhiev[16] who also used the GF formalism making explicit the difference between *pp*- and *ph*-interaction. The advantage of the derivation within the Green function formalism is that one has to make much less restrictive assumptions.[6] Here, we present the formula for moments and transitions in odd mass nuclei[17]:

$$\delta\rho^{\alpha\alpha}_{\alpha\alpha} = u^2{}_\alpha - v^2{}_\alpha, \quad \delta\kappa^{\alpha\alpha}_{\alpha\alpha} = 2u_\alpha v_\alpha, \tag{29}$$

$$\begin{aligned}\delta\rho^{\alpha\alpha}_{\nu_1\nu_2} &= -\frac{\eta^2_{\nu_1\nu_2}}{E_{\nu_1}+E_{\nu_2}}\left(\Delta\mu^P N^P_{\nu_1\nu_2} + \Delta\mu^N N^N_{\nu_1\nu_2} + \sum_{\nu_1\nu_2} F^{ph}_{\nu 1\nu 4,\nu 2\nu 3}\delta\rho^{\alpha\alpha}_{\nu_3\nu_4}\right) \\ &\quad - \frac{\eta_{\nu_1\nu_2}\xi_{nu_1\nu_2}}{E\nu_1+E\nu_2}F^{pp}_{\nu 1\nu 4,\nu 2\nu 3}\delta\kappa^{\alpha\alpha}_{\nu_3\nu_4},\end{aligned} \tag{30}$$

$$\begin{aligned}\delta\kappa^{\alpha\alpha}_{\nu_1\nu_2} &= -\frac{\eta_{\nu_1\nu_2}\xi_{\nu_1\nu_2}}{E\nu_1+E\nu_2}\left(\Delta\mu^P N^P_{\nu_1\nu_2} + \Delta\mu^N N^N_{\nu_1\nu_2} + \sum_{\nu_1\nu_2} F^{ph}_{\nu 1\nu 4,\nu 2\nu 3}\delta\rho^{\alpha\alpha}_{\nu_3\nu_4}\right) \\ &\quad - \frac{\xi^2{}_{\nu_1\nu_2}}{E\nu_1+E\nu_2}F^{pp}_{\nu_1\nu_4,\nu_2\nu_3}\delta\kappa^{\alpha\alpha}_{\nu_3\nu_4}.\end{aligned} \tag{31}$$

In addition to the *ph*-interaction F^{ph} one has to introduce also a particle–particle interaction F^{pp}. The quasiparticle energy E_ν and the gap Δ_ν are given as:

$$(\nu_1,\nu_2) \neq (\alpha,\alpha), \quad E = \sqrt{\epsilon^2{}_\nu + \Delta^2{}_\nu}, \quad \Delta_\nu = \sum_{\nu_2} F^{pp}_{\overline{\nu_1}\nu_1,\overline{\nu_2}\nu_2}\frac{\Delta_{\nu_2}}{2E_{\nu_2}}. \tag{32}$$

The other BCS quantities are defined in the usual way: e.g., Ref. 17. The quantities $\Delta\mu^P$ and $\Delta\mu^N$ denote the change of the Fermi energy of protons and neutrons and are determined by the condition (if one adds a proton):

$$\sum_{\text{protons}} \delta\rho_{\nu\nu}^{\alpha\alpha} = 1, \qquad \sum_{\text{neutrons}} \delta\rho_{\nu\nu}^{\alpha\alpha} = 0. \tag{33}$$

All quantities which we have discuss so far are derived from the two-body Green function, i.e., the *linear response function* which is, as mentioned before, its correlated part. An important extension of this approach is the *second-order response function* which has been derived in Ref. 7 as second derivative of the Green function matrix in Eq. (27):

$$\frac{\delta^2 G^{kl}_{\nu_1\nu_2}(U(t_1t_2))}{\delta q_{\nu_3\nu_4}(t_3t_4)\delta q_{\nu_5\nu_6}(t_5t_6)} = L^{kl}_{\nu_1\nu_3\nu_5,\nu_2\nu_4\nu_6}(t_1t_3t_5, t_2t_4t_6). \tag{34}$$

After a lengthy calculation[7] one arrives at the change of the density $\rho^{(2)}$ which depends (in a complicated way) quadratically on the external field:

$$\rho^{(2)} \propto \delta q(1)\delta q(2). \tag{35}$$

3. Application to Nuclear Properties

From the huge variety of existing results, we discuss three topics which have been investigated within the FFS theory and a fourth one which has been obtained within in a self-consistent approach. The first example concerns the isomer-shifts in the strongly deformed nucleus ^{153}Eu which has been measured with help of the Mössbauer technique and muonic atoms. In the second example, we demonstrate the predictive power of the FFS-theory where we discuss the close correlation between isotope shifts and breathing mode which led to the correct prediction of its energy. The third example concerns isomer shifts of rotational states in deformed nuclei where one was able to explain and predict extremely small effects correctly. In the fourth case, we present results of the giant electric resonances in ^{40}Ca and ^{208}Pb within a self-consistent approach.

3.1. *Landau–Migdal approach*

The input data in all cases are single-particle energies and wave functions and the *ph* interaction. In the FFS theory, a phenomenological single-particle model is used which is adjusted to experiments and as far as possible, the experimental energies are directly used. The *ph* interaction is expanded in Legendre polynomials on the Fermi surface. Its spin-independent part reads:

$$F^{ph}\left(\frac{\mathbf{p}\cdot\mathbf{p}'}{p_F^2}\right) = C_0 \sum_{l=0}^{\infty} [f_l + f_l' \hat{\tau}_1 \cdot \hat{\tau}_2] P_l \left(\frac{\mathbf{p}\cdot\mathbf{p}'}{p_F^2}\right), \tag{36}$$

$$C_0 = \frac{\pi^2\hbar^3}{2m^* p_F}, \quad p_F = \hbar\left(\frac{3\pi^2\rho_0}{2}\right)^{1/3}, \tag{37}$$

where $P_l(x)$ is the Legendre polynomial of order l and p_F is the Fermi momentum. In lowest order, one obtains a zero range interaction where Migdal introduced density dependent parameters of the form[3]:

$$f(\rho) = f^{(\mathrm{ex})} + (f^{(\mathrm{in})} - f^{(\mathrm{ex})})\frac{\rho_0(r)}{\rho_0(0)}, \tag{38}$$

where $f^{(\mathrm{ex})}$ stands for the exterior region of the nucleus and $f^{(\mathrm{in})}$ for the interior. The corresponding parameters, the famous *Landau–Migdal parameters*, have been adjusted to experimental data and applied to all nuclei except the lightest ones.

3.1.1. *Isomer shifts in deformed nuclei*

Isomer shifts allows to extract the difference of the charge radius $\delta\langle r^2\rangle$ between the ground state and an excited states of a nucleus. This extremely small effect can be measured with help of the Mössbauer technique as well as with muonic atoms. The isotope ^{153}Eu has been investigated experimentally by both methodes. It possesses two low lying levels around 100 keV as shown in Fig. 1 which are especially suitable for Mössbauer experiments.

As we have seen in the previous section, one obtains from the linear response function the change of the density between an even–even nucleus and the neighboring odd mass nuclei which seems on the first sight a restriction turns out to be of great advantage for the present nuclear structure problem. The change of the density $\Delta(\rho)$ between the ground state of the even–even nucleus and the ground states of the neighboring odd mass nuclei allows to calculate, as mentioned before, the isotope shifts of the corresponding nuclei.[9] But we can also calculate the change between the ground state of the even–even nucleus and excited states of the odd mass nuclei. These changes of the densities $\delta(\rho)$ between the excited states and the ground state of the odd mass nuclei are related to the isomer shifts. As pairing correlations are important for nuclei in the rare earth region, the authors of Ref. 17 had to solve Eqs. (29)–(31). All details of the numerical calculations can be found in Ref. 17. Here, we only mention the most essential points. It turned out that

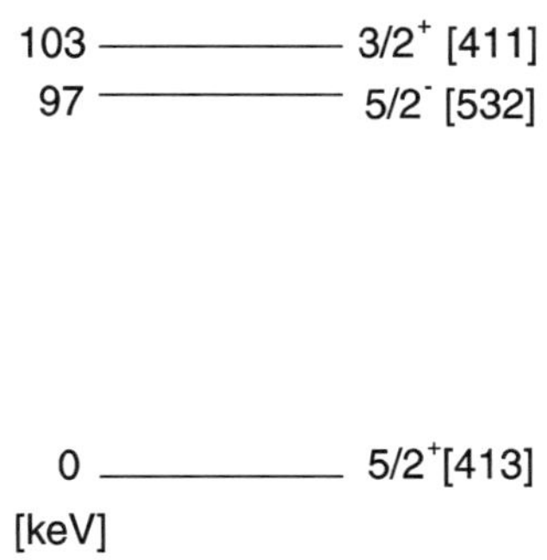

Fig. 1. Low lying level scheme of ^{153}Eu. On the right side spin, parity and the Nilsson quantum numbers are given.

the results depend very sensitively on the single-particle energies. For this reason, the energies of Ref. 18 were used which have been deduced from experiments. The ph-interaction was restriced to lowest order, so only $f_0(\rho)$ and $f'_0(\rho)$ enter which introduces four parameters: the scalar parameters f_0^{in}, f_0^{ex}, and the isovector parameters $f_0'^{(\mathrm{in})}$, $f_0'^{(\mathrm{ex})}$. The parameters by Krainov[19] where used as closely as possible. The muonic isomer shift were used to adjust essentially the deformation parameter δ; the changes of the radii and the corresponding ratio are then independent results. In Table 1, the theoretical results for the muonic isomer shifts in ^{153}Eu are compared with the experimental data. The agreement is very good, but one has to consider that parameters have been adjusted. The changes of the radii which are shown in Table 2 are now parameter free results and also agree nicely with the data. So one may conclude that the theory is able to reproduce extremely tiny nuclear structure effects.

3.1.2. *Breathing mode and incompressibility*

After the discovery of the giant quadrupole resonance, there were great experimental efforts to find also the giant monopole resonance (GMR), the so called breathing mode. This resonance is of special interest for nuclear physics and astrophysics as it is directly connected with the incompressibility of nuclei. At that time, nearly all many-body theories have been applied to calculate this important quantity. Unfortunately, the predicted energies of the various models in ^{208}Pb range from about 8 MeV to nearly 22 MeV. It turned out that the results were strongly dependent on the interaction used. Self-consistent calculations with Skyrme forces were considered as a very reliable method. Therefore, the results of Ref. 21 with an incompressibility $K = 342$ MeV and $K = 370$ MeV were regarded as the most probable values. This corresponds to an energy of the breathing mode between 16 MeV and 17 MeV which

Table 1. Muonic isomer shifts in ^{153}Er. The data are taken from Ref. 20.

	Theory	Exp.
$\delta E_{103}^{\mathrm{is}}$ (keV)	-6.00	-6.02 ± 0.13
$\delta E_{97}^{\mathrm{is}}$ (keV)	-4.70	-4.72 ± 0.17
$\delta E_{103}^{\mathrm{is}}/\delta E_{97}^{\mathrm{is}}$	1.28	1.27 ± 0.05

Table 2. Change of the radii in ^{153}Eu. The data are taken from Ref. 20.

	Theory	Exp.
$\delta\langle r^2\rangle_{103}$ (10^{-3} fm^2)	-112.5	-101 ± 12
$\delta\langle r^2\rangle_{97}$ (10^{-3} fm^2)	-86.9	-93 ± 11
$\delta\langle r^2\rangle_{103}/\delta\langle r^2\rangle_{97}$	1.29	1.09 ± 0.06

is far away of the later detected resonance at around 14 MeV. After the energy of the GMR was confirmed, one realized that the number of parameters in the Skyrme force used so far, was too restricted and introduced an exponent for the density ρ^α with $\alpha < 1$, Ref. 22.

The FFS theory is not self-consistent but much closer to experiments. As the breathing mode energy depends so sensitively on the interaction, isotope and isomer shifts are the best tool to adjust the interaction because these quantities depend on the virtual 0^+-excitations (see Refs. 19 and 17). In the FFS theory exists a relation between the incompressibility and the parameter $f_0^{\rm in}$ of the ph interaction:

$$K = 3\frac{\hbar^2 c^2}{mc^2} p_{\rm F}^2 (1 + 2f_0^{\rm in}). \tag{39}$$

From the investigation of various nuclear properties, it turned out that $f_0^{\rm in}$ is close to zero. As one can see from Table 3, the GMR predicted within the FFS theory is close to the well known GDR. There existed strong arguments against the GMR in the vicinity of the isovector dipole resonance (GDR) from the analyses of inelastic electron scattering. One would expect a nearly equally large contribution to the (e, e') cross-section from the GMR as from the GDR if it exhaust the energy-weighted sum rule (EWSR). This effect was obviously not needed as all (e, e') experiments reported until the end of 1977 including the GDR only gave good agreement with the data. However, the authors of Ref. 26 pointed out that the GDR analyses with phenomenological models are strongly model dependent. With the use of the same B(E1) value, the transition density derived from the Steinwedel–Jensen (SJ) model gives a much larger (e, e') than that of the Goldhaber–Teller (GT) model which is concentrated on the surface of the nuclei. In Ref. 27 cross-sections of inelastic scattering experiments with various probes were calculated with microscopic form factors based on RPA calculations within the FFS theory. In Fig. 2, we present the (e, e') cross-section. Here, the most important question concerns the transition density of the GDR. The authors of Ref. 27 found that: (a) the GDR is fragmented in several discrete states with a centroid energy at $E_c^{1^-}$=13.2 MeV and a B(E1) strength which exhaust nearly 80% of EWSR, (b) most of dipole states possess transition densities which are surface peaked like in the GT model, only less than 20% of the EWSR are SJ-like. For this reason, the (e, e') is somewhat too small if only the GDR is included, therefore allowing strength from the GMR. This theoretical investigation confirmed that the isoscalar resonance at around 14 MeV which was

Table 3. Incompressibility and breathing mode. The lower value of energy of the GMR was published in Ref. 23 and the higher value in Ref. 9. The Skyrme(SIII) results are taken from Ref. 24 and Skyrme(SkM^*) from Ref. 22.

	FFS	Skyrme (SIII)	Skyrme (SkM*)
K (MeV)	230 ± 20	355	216
GMR (MeV)	13.97; 14.76	17.3	13.5
$\mathrm{GMR_{exp}}$ (MeV)	13.96 ± 0.2		

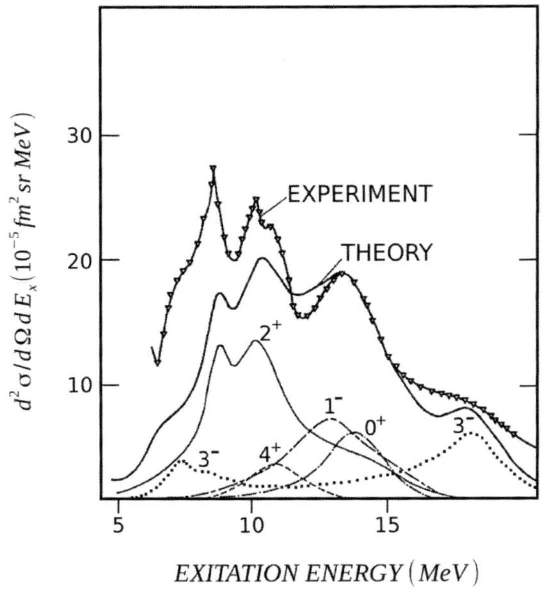

Fig. 2. The electron spectrum for 90 MeV (e, e') at 75° calculated using RPA $0^+, 1^-, 2^+, 3^-, 4^+$ form factors. Data are from Ref. 25.

detected in (α, α') scattering[28] is indeed the GMR. The final confirmation came from an (α, α') experiment in forwards direction.[29] These investigations show that the FFS theory has a strong predictive power as it allows to combine related experimental quantities with each other: here the isotope and isomer shifts with the GMR.

3.1.3. *Change of nuclear radii due to rotation*

In this section, we present again an investigation of isomer shifts measured as before with help of the Mössbauer technique and muonic atoms. From both experiments, one can derive the change of the charge radius. In Sec. 3.1.1, we presented the results of single-particle transitions. In the present section $\delta\langle r^2\rangle$ are due to collective rotation of deformed nuclei. These effects are one order of magnitude smaller than the effects in ^{153}Eu nonetheless one is able to reproduce these extremely small numbers. Rotational states are described within the cranking model (see, e.g., Ref. 30) which is given by the Hamiltonian:

$$H' = H - \Omega J_x. \tag{40}$$

H denotes the Hamiltonian of the many-body system, J_x is a component of the total angular momentum perpendicular to the intrinsic symmetry axis and the parameter Ω is determined from the condition:

$$\mathrm{Tr}(J_x \rho^{(\Omega)}) = \hbar\sqrt{I(I+1)}. \tag{41}$$

For the present case, one has to solve Eq. (22) (respectively, the corresponding equation with pairing) for the change of the density due to an external field, where the external field is the cranking field ΩJ_x. As the external field is a pseudo vector and the expectation value which has to be calculated is a scalar, one has to go to second-order. For this reason, the second-order response theory has been developed in Ref. 17 where all details are given. The crucial input into the equations turned out to be the single-particle energies. Here, the level scheme of Refs. 18 and 31 were used.

As one has rotational nuclei, one expected that the radii in the excited 2^+ states is larger than in the ground state. This was the case in the majority of the nuclei. However, surprisingly, one observed also negative values which causes some discussion, e.g., that antipairing might be important.

The calculations in Ref. 17 yield positive and negative $\delta\langle r^2\rangle$ in agreement with the data. The explanation of the effect was very surprising. The effect is not a collective one but only a few single-particle states near the diffuse Fermi edge contribute. In this mass region "Nilsson levels" with the principle quantum number $N = 4$ and $N = 5$ are mixed around the Fermi edge for the single-particle states. The radii of the single-particle states increase with increasing quantum number N. The rotational field gives rise to a redistribution of the occupation probabilities of nucleons. This occurs only for a few levels within the range Δ around the diffuse Fermi

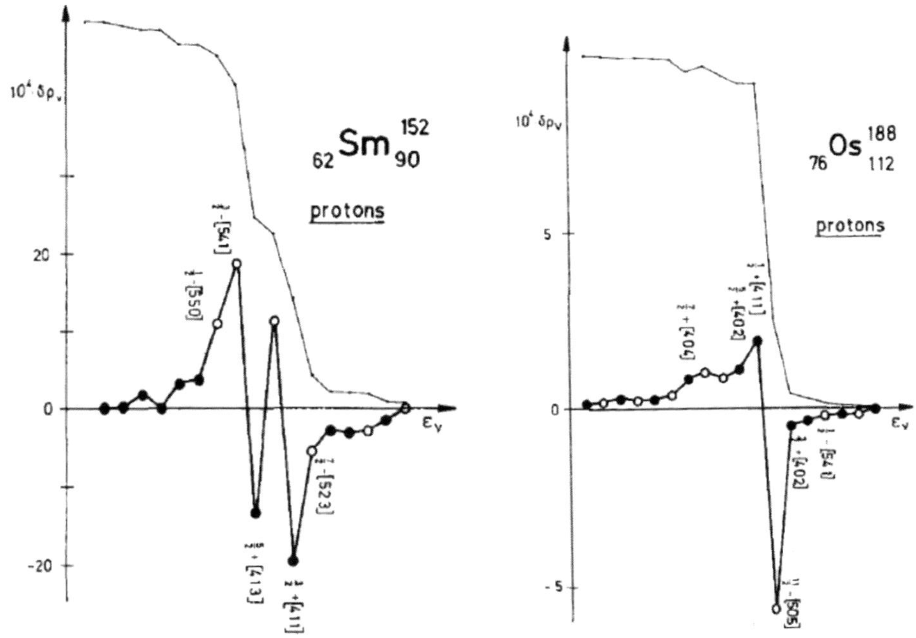

Fig. 3. The calculated change $\delta\rho_\nu$ of the occupation probabilities of the Nilsson levels near the Fermi energy the for protons of ^{152}Sm on the left side and ^{188}Os on the right side. The levels are drawn in a schematic way, equally spaced in the order of increasing energies ε_ν. Open circles refer to $N = 5$ and the full circles to $N = 4$ proton levels, respectively. The thin lines give the occupation probabilities in the ground state (scale different from $\delta\rho_\nu$!) showing the diffuse Fermi edge.

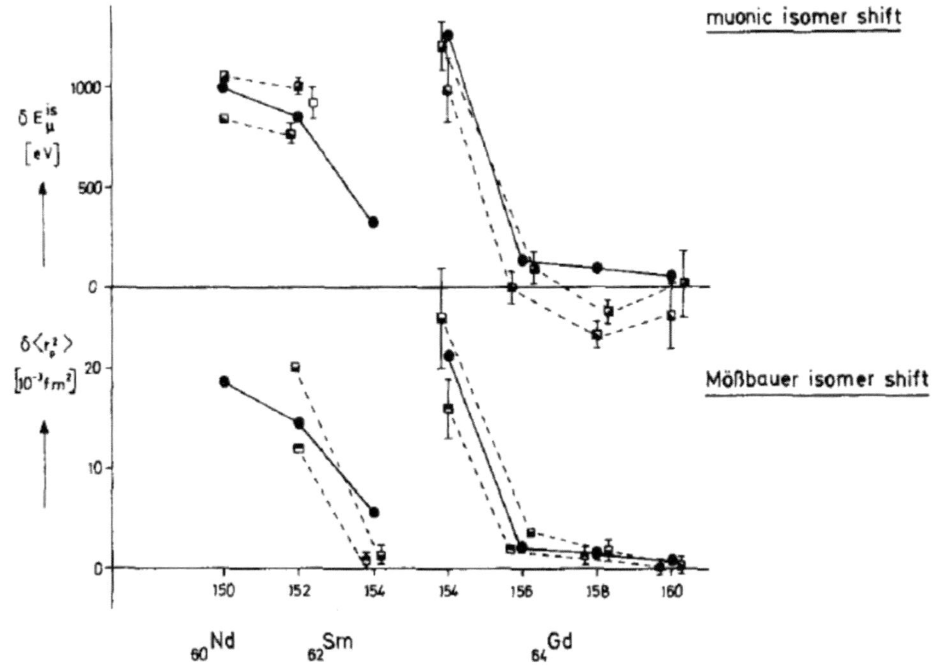

Fig. 4. Comparison of experimental and theoretical isomer shifts. The experimental data are taken from Refs. 32 and 33.

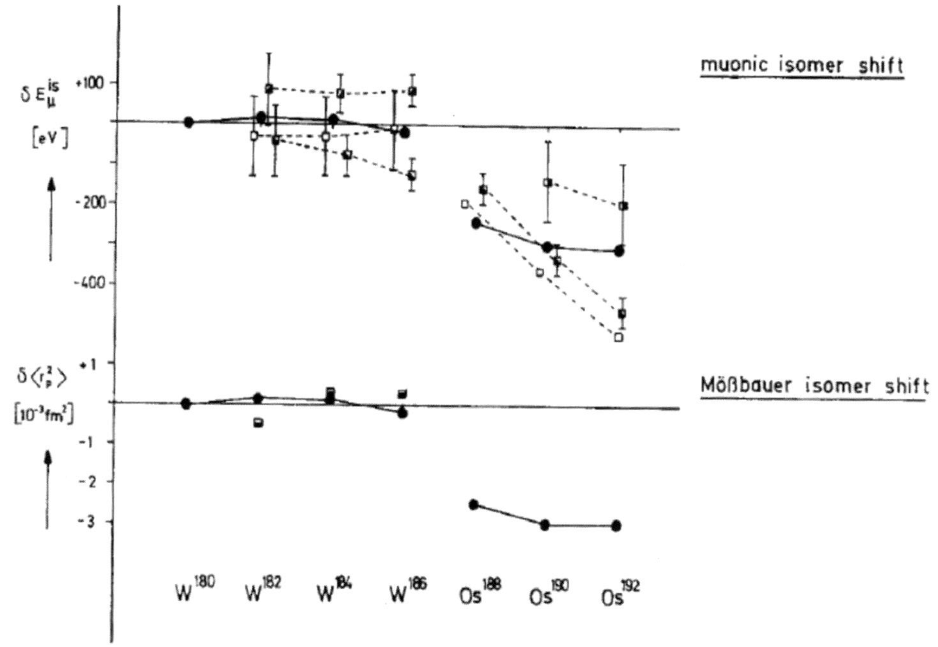

Fig. 5. Comparison of experimental and theoretical isomer shifts. The experimental data are taken from Refs. 32 and 33.

edge. This is demonstrated on the left side of Fig. 3 for the protons of ^{152}Sm. The *antipairing characteristic* tending to depopulate levels just above the Fermi energy in favor of levels below clearly shows up in the present case. Here, the increase of the occupation probabilities of the $N = 5$ states is the dominant effect and therefore the $\delta\langle r^2\rangle$ in this region are positive in agreement with the data as shown in Fig. 4. The situation changes at the upper end of the rare earth region where the isomer shifts in the W isotopes are close to zero and negative in the Os isotopes. This is shown in Fig. 5. The explanation for the negative isomer shifts is given on the right side of Fig. 3. The dominate depletion comes from an $N = 5$ level and therefore the theoretical isomer shift is negative in agreement with the muonic data. The theory agreed well with the positive and negative shifts and as the physical explanation was convincing but not very exciting, the interest of the experimental colleagues vanished.

4. Extended Theory of Finite Fermi Systems

4.1. *Phonon coupling*

Up to now all theoretical formulations are based on the explicit inclusion of $1p1h$ configurations and the effective ph interaction is corresponding irreducible with respect to the ph propagator. If we considers, e.g., giant resonances, one is able to calculate the mean energy and the escape width if the continuum is correctly included. This contribution is in heavier mass nuclei small compared to the spreading width which is due to the fragmentation of the single-particle strength. One possibility to include this effect is to couple phonons to the single-particle states. Many authors have suggested this method, e.g., Refs. 16, 34 and 35, but the most complete formulation has been published by Tselyaev[36,37] which is the basis of the numerical results presented in the following.

The contributions in lowest order are shown in Fig. 6. The coupling of the phonons gives rise to a fragmentation of the single-particle strength Fig. 6(b) and

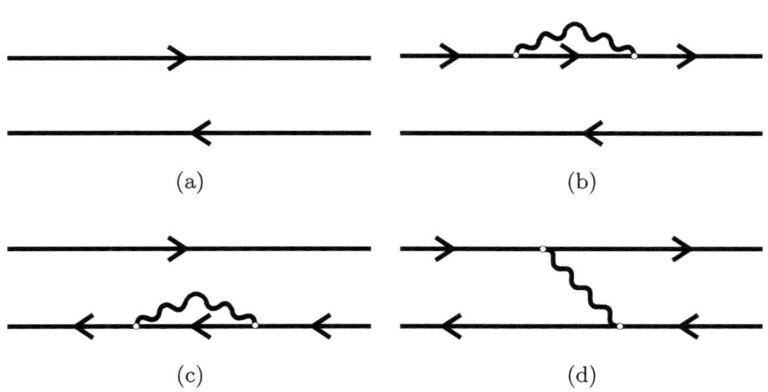

Fig. 6. (a) is the uncorrelated ph propagator, (b) and (c) give rise to the splitting of the single-particle strengths and (d) contribute to the ph-interaction.

Fig. 6(c) and to a contribution to the ph-interaction Fig. 6(d). It is important to mention that no new parameters have to be introduced as the phonon vertex γ which determines the coupling of the phonons to the propagators is given by:

$$\gamma_{\nu}^{\mu;i} = \sum_{\alpha,\beta} F^{ph}_{\nu\alpha;\mu\beta} \chi^{i}_{\alpha\beta}, \tag{42}$$

where χ^i is the RPA wave function of the phonon considered. All details can be found in the papers by Tselyaev who called his approach, *Time Blocking Approximation* (TBA). The basic equations are only slightly modified compared to the conventional RPA. One obtains for the response function in TBA:

$$R^{\text{TBA}}(\omega) = R^{(0)}(\omega) - R^{(0)}(\omega)(F^{ph} + \tilde{W}(\omega))R^{\text{TBA}}(\omega), \tag{43}$$

$$\tilde{W}(\omega) = W(\omega) - W(0), \tag{44}$$

where $R^0(\omega)$ is the uncorrelated ph propagator and the induced interaction $\tilde{W}(\omega)$ serves to include contributions of $1p1h\otimes$ phonon configurations.

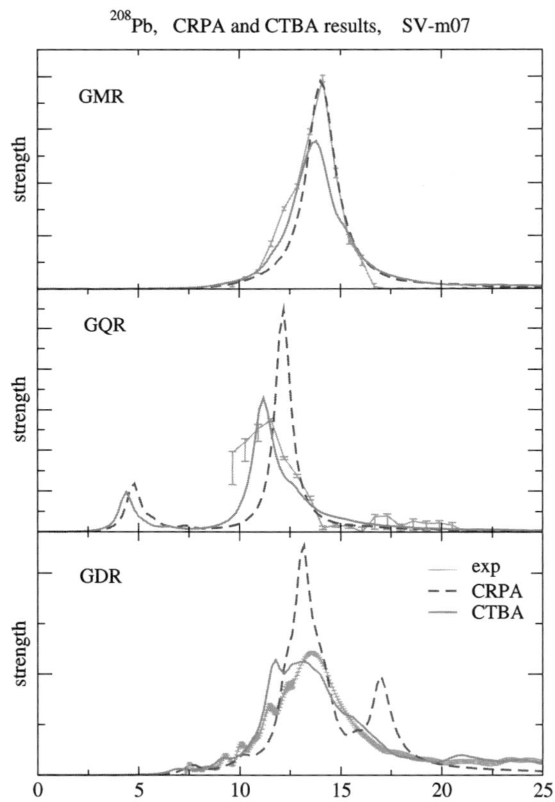

Fig. 7. Strength distribution of the GMR, GQR and GDR calculated self-consistently within the continuum RPA (dashed lines) and continuum TBA (full line). The theoretical results are compared with data.

In theories going beyond the RPA, the problem of double counting and of ground-state stability arises.[38] To avoid this problem in the TBA, in Ref. 37 a subtraction method was introduced. It consists in the replacement of the amplitude $W(\omega)$ by the quantity $\widetilde{W}(\omega) = W(\omega) - W(0)$ as it is given in Eq. (43). This procedure eliminates double counting and ensures stability of solutions of the TBA eigenvalue equations.

4.2. *Self-consistent results*

We present two examples for the TBA. The calculations were performed self-consistently on the basis of an energy functional $E(\rho)$ of the Skyrme type.[39] Here, the mean-field Hamiltonian is derived from the energy functional by functional derivative

$$h_{12} = \frac{\delta E[\rho]}{\delta \rho_{21}} \tag{45}$$

and the residual ph interaction is the second functional derivative

$$V_{12,34} = \frac{\delta^2 E[\rho]}{\delta \rho_{21} \delta \rho_{34}}. \tag{46}$$

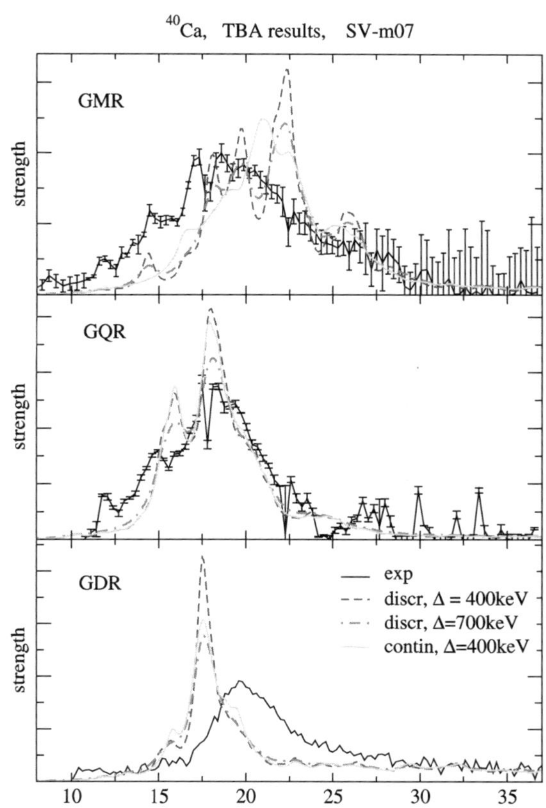

Fig. 8. Discrete TBA calculations and continuum TBA calculations are compared with experimental data.

In the numerical approach the continuum was treated in an exact way, so the escape width as well as the spreading widths are included. In the following examples the Skyrme parameter set SV-mas07 of Ref. 40 was used, which give rise to an effective mass of $m^*/m = 0.7$ and an incompressibility of $K = 234\,\text{MeV}$. In Fig. 7, the three major giant resonances in ^{208}Pb are presented. This nucleus is the ideal example for this kind of calculations. The *ph*-space is large and the nucleus possesses many collective low-lying phonons which contribute to the fragmentation of the single-particle states and correspondingly to the spreading width of the giant resonances. In the figure, the continuum RPA (CRPA) results and the continuum TBA (CTBA) results are shown and compared with the data. The agreement for the GQR[41] and GDR[42] with the TBA results is very good. It is some what surprising that in the case of the GMR,[41] the RPA results is nearly identical with the experiment. In Fig. 8, the influence of the continuum for ^{40}Ca is investigated where discrete TBA calculations are compared with continuum TBA calculations. In the discrete calculations, two different smearing factors Δ were used. Here, it is show that, for light nuclei, that if one increases the smearing parameter the artificial fine structure of the discrete approach is damped. But, at the same time, it wipes out important physical features. Hence, it is impossible to reproduce CTBA results for strength functions of light nuclei by using the DTBA, both with small and large smearing parameters

5. Summary

In our contribution, we have reviewed the application of many-body Green function on nuclear structure problems. We discuss the two-body and three-body Green functions with and without pairing correlations. The very complex integral equations for the two- and three-body Green functions can be transformed into solvable equations with help of the quasiparticle pole ansatz and the well known Landau renormalization. The form of the equations are identical with the RPA equation and the linear response equations. However, the equations have been derived without any approximations, therefore one can apply conservation law which lead to important relations. The equations are very general and we present two different numerical schemes: the FFS approach, and a self-consistent example based on a Skyrme energy density functional. Within the FFS approach, we present investigations of the isomer shifts in deformed nuclei which can be nicely reproduced. As an example of the predictive power of the FFS approach, we discussed the close connection between isomer shifts and the giant monopole resonance. We also presented two examples within a self-consistent approach, namely giant resonances in ^{208}Pb and ^{40}Ca. In all cases, theory and experiment are in fair agreement.

Acknowledgment

This work has been supported by contract Re322-13/1 from the DFG. We thank S. Krewald, P.-G. Reinhard and V. Tselyaev for discussions and comments.

N.L. acknowledged financial support from the Russian Science foundation (project No. 16-12-10155).

References

1. G. E. Brown, *Rev. Mod. Phys.* **43** (1971) 1.
2. L. Landau, *J. Exp. Theor. Phys.* **8** (1959) 70.
3. A. B. Migdal, *Theory of Finite Fermi Systems and Application to Atomic Nuclei* (Wiley, New York, 1967).
4. J. Speth, E. Werner and W. Wild, *Phys. Rep. C* **33** (1977) 127.
5. F. Grümmer and J. Speth, *J. Phys. G, Nucl. Part. Phys.* **32** (2006) R193.
6. J. Speth, *Nucl. Phys. A* **135** (1969) 445.
7. J. Meyer and J. Speth, *Nucl. Phys. A* **203** (1973) 17.
8. J. Speth, *Z. Phys.* **239** (1970) 249.
9. J. Speth, P. Ring and L. Zamick, *Nucl. Phys. A* **232** (1974) 1.
10. A. L. Larkin and A. B. Migdal, *J. Exp. Theor. Phys.* **17** (1963) 1146.
11. W. Brenig and H. Wagner, *Z. Phys.* **173** (1963) 484.
12. N. N. Bogolyubov, *Usp. Fiz. Nauk* **67** (1959) 549.
13. M. Baranger, *Phys. Rev.* **120** (1960) 957.
14. S. T. Belyaev, *Nucl. Phys.* **64** (1965) 17.
15. B. L. Birbrair, *Nucl. Phys. A* **108** (1968) 449.
16. S. P. Kamerdzhiev, *Sov. J. Nucl. Phys.* **9** (1969) 190.
17. J. Meyer and J. Speth, *Phys. Lett. B* **39** (1972) 330.
18. I. L. Gustafson, I. L. Lamm, B. Nilsson and S. G. Nilsson, *Ark. Fys.* **36** (1967) 69.
19. V. P. Krainov, *Phys. Lett. B* **27** (1968) 341.
20. H. Walter, H. Backe, R. Engfer, K. Kankeleit, C. Petijean, H. Schneuwly and W. Schröder, *Phys. Lett. B* **38** (1972) 64.
21. D. Vautherin and D. Brink, *Phys. Rev. C* **5** (1972) 626.
22. J. Bartel, P. Quentin, M. Brack, C. Guet and H.-B. Hakansson, *Nucl. Phys. A* **386** (1982) 79.
23. P. Ring and J. Speth, *Nucl. Phys. A* **235** (1974) 315.
24. M. Beiner, H. Vlocard, N. Van Giai and P. Quentin, *Nucl. Phys. A* **361** (1975) 29.
25. R. Pitthan *et al.*, *Phys. Rev. Lett.* **33** (1974) 849.
26. M. Sasao and Y. Torizuka, *Phys. Rev. C* **15** (1977) 217.
27. J. Wambach, V. M. Madson, G. A. Rinker and J. Speth, *Phys. Rev. Lett.* **39** (1977) 1443.
28. M. N. Harakeh *et al.*, *Phys. Rev. Lett.* **38** (1977) 676.
29. D. H. Youngblood *et al.*, *Phys. Rev. Lett.* **39** (1977) 1188.
30. J. Meyer, *Nucl. Phys. A* **137** (1969) 193.
31. W. Ogle, S. Wahlborn, R. Piepenbring and S. Frederiksson, *Rev. Mod. Phys.* **43** (1971) 424.
32. G. Goldring and R. Kalish, *Hyperfine Interaction in Excited Nuclei* (Gordon and Breach, New York, 1971).
33. C. S. Wu and L. Wilets, *Ann. Rev. Nucl. Sci.* **19** (1969) 527.
34. V. G. Soloviev, *Theory of Complex Nuclei* (Pergamon Press, Oxford, 1976).
35. J. Dehesa, S. Krewald, J. Speth and A. Faessler, *Phys. Rev. C* **15** (1977) 1858.
36. V. I. Tselyaev, *Sov. J. Nucl. Phys.* **50** (1989) 1252.
37. V. I. Tselyaev, *Phys. Rev. C* **88** (2013) 054301.
38. C. Toepffer and P.-G. Reinhard, *Ann. Phys. (N.Y.)* **181** (1988) 1.
39. N. Lyutorovich, V. Tselyaev, J. Speth, S. Krewald, F. Grümmer and P.-G. Reinhard, *Phys. Lett. B* **749** (2015) 292.

40. N. Lyutorovich, V. I. Tselyaev, J. Speth, S. Krewald, F. Grümmer and P. G. Reinhard, *Phys. Rev. Lett.* **109** (2012) 092502.
41. D. H. Youngblood, Y.-W. Lui, H. L. Clark, B. John, Y. Tokimoto and X. Chen, *Phys. Rev. C* **69** (2004) 034315.
42. S. N. Belyaev, O. V. Vasiliev, V. V. Voronov, A. A. Nechkin, V. Y. Ponomarev and V. A. Semenov, *Phys. Atom. Nucl.* **58** (1995) 1883.

Validity of SU(3) in the $0p1f$ shell

D. D. Strottman
Theoretical Division
Los Alamos National Laboratory,
Los Alamos, New Mexico, USA
Frankfurt Institute for Advanced Studies (FIAS),
Johann Wolfgang Goethe University
Frankfurt, Germany

Complexities of using an SU(4) and SU(3) basis in the $1p0f$ shell are discussed. SU(3) representations are detailed for 2–5 particles. Earlier calculations by McGrory are discussed with the view of using the SU(3) model to describe multi-particle multi-hole states.

Keywords: Nuclear shell model; SU(3) and SU(4) symmetries; $1p0f$ shells.

1. Introduction

In the early 1960s, the structure of ^{16}O was under debate, particularly the mechanism whereby the first excited 0^+ could appear at the low energy of 6.06 MeV. At the Paris Nuclear Physics conference, Brown[1] introduced his — now widely accepted — explanation that the level was a four-particle four-hole state and its energy was due to the role of essential deformation. His explanation was further elucidated by the work of Brown and Green[2] and Engeland[3] and subsequently by Ellis and Engeland[4] and others.

Gerry's model was loosely based on the Elliott SU(3)[5] model. Subsequent calculations by Millener[6] using a computer code based on SU(3) helped affirm the earlier work. These calculations as well as a number of other calculations[7–11] throughout the $1s0d$ shell and its usefulness in describing cluster-transfer reactions[12,13] and alpha decay[14] demonstrated the usefulness of the Elliott model in shell model calculations. Additionally, spurious states[15] that will invariably arise when the model space involves more than one shell can be eliminated exactly[16] when using SU(3) basis states. It is primarily for this reason that we are developing an SU(3)-based code to study multi-particle multi-hole states around mass 40.

From the earliest times, it was recognized that the one-body spin-orbit interaction as well as noncentral two-body forces would tend to destroy the SU(3) symmetry in nuclei. However, in the early part of the $1s0d$ shell, the leading SU(3) state in states of low isospin usually accounted for 80%–90% of the wave functions of the lowest states, e.g., for ^{20}Ne, the leading or dominant SU(3) representation of $(8\,0)$ has amplitude 0.882 for the ground state wave function[17] using Kuo–Brown matrix elements.[18,19] It was assumed that in the $1p0f$ shell wherein the lowest states are dominated by $0f^n_{\frac{7}{2}}$ configurations, SU(3) would cease to be a reasonable approximation of the wave functions. This seemed to be confirmed by calculations of McGrory[20] in which the leading SU(3) representations accounted for around 20% of the wave function of ^{44}Ti. In this paper, these conclusions will be reexamined.

One must remember that today's computers are many millions of times faster than those of the late 1960s; then diagonalizing a matrix of dimension 100 was a considerable challenge. The increased computer speed and the introduction by Whitehead[21] of the Lanczos method into nuclear structure has transformed the predictive capability of the shell model, see e.g., the review by Dean *et al.*[22] Nevertheless, the role of SU(4) and SU(3) models remain useful to better understand some of the physical processes such as E1, M1 and E2 transitions as well as alpha decay and transfer reactions. In addition, although computer codes using the m-scheme and variants can attack model spaces of huge dimension, the importance of removing the influence of spurious states on E1 transitions means that for calculations of states with many-particle many-hole states, one may still need to rely on SU(3)-based codes for the near future.

2. The SU(3) Model

In this section, we briefly describe the SU(3) model of Elliott,[5,24–26] primarily to introduce the notation that will be used.

The Elliott model assumes the use of harmonic oscillator — the symmetry group for which is SU(3) — basis states. All states are products of three components: space, spin and isospin. Following the work of Wigner,[23] the spin and isospin is combined in a wave function described by SU(4) and labeled as $[\tilde{f}]$. The N-particle spatial wave functions — being three-dimensional harmonic oscillator functions — will be labeled by the representations of the symmetric group, $S(N)$, $[\tilde{f}]$, and by the representations of SU(n), $[f]$ and its sub-group SU(3), $(\lambda\mu)$; here, $n = (\frac{1}{2}(q{+}1)(q{+}2)$ where q is the number of quanta of the harmonic oscillator shell being studied, e.g., $q = 2$ or 3 for the $1s0d$ or $1p0f$ shells, respectively. The total N particle wave function is necessarily antisymmetric which imposes the restriction that $[\tilde{f}]$ must be the contragradient of $[f]$. The $[f]$ are Young tableux and this restriction requires $[f]$ to have fewer than n rows or four columns.

$$|q^N[f](\lambda\mu)LTSJ\rangle. \tag{1}$$

The SU(3) representations are labeled by $(\lambda\mu)$, the dimension of which is $\frac{1}{2}(\lambda+1)(\mu+1)(\lambda+\mu+2)$. The possible $(\lambda\mu)$ for the $1s0d$ shell were enumerated by Peres and Flores[27] and by the author[28] for up to nine $1p0f$ particles and up to six $2s1d0g$ particles.

To calculate the two-body matrix elements, the N particle wave function is expanded as usual using the techniques introduced by Racah[29–31] and generalized[32–34] to the case of the unitary groups involved:

$$
\begin{aligned}
&|q^N\,[f](\lambda\mu)\,S\,T\,J\rangle \\
&= \sum_{f'',(\lambda'',\mu''),I'',[f_2],I_2} \sqrt{\frac{n_{f''}}{n_{[f]}}} \left\langle \begin{matrix} [f''] & [f_2] \\ (\lambda''\mu'')\,S'' & (\lambda_2\mu_2)\,S_2 \end{matrix} \middle| \begin{matrix} [f] \\ (\lambda\mu)\,S \end{matrix} \right\rangle \\
&\quad\times \left\langle \begin{matrix} [f''] & [f_2] \\ S''T'' & S_2T_2 \end{matrix} \middle| \begin{matrix} [f] \\ ST \end{matrix} \right\rangle \left\langle \begin{matrix} (\lambda''\mu'') & (\lambda_2\,\mu_2) \\ L'' & L_2 \end{matrix} \middle| \begin{matrix} (\lambda\mu) \\ L \end{matrix} \right\rangle \\
&\quad\times [|q^{N-2}\,[f''](\lambda''\mu'')L''S''T''\rangle \times |q^2[f_2](\lambda_2\,\mu_2)L_2S_2T_2\rangle]^{(L,S,T,J)}. \qquad (2)
\end{aligned}
$$

The angular momenta and SU(3) representations can then be recoupled using the nine-j and nine-$(\lambda\mu)$ symbols, the details of which we spare the reader. The first factor is the ratio of the dimensions of the representation $[f'']$ and $[f]$ of the symmetric group. The next three factors on the right-hand side of Eq. (2) are the Clebsch–Gordan coefficients for the groups SU(N), SU(4) and SU(3), respectively. Also, the $\tilde{f}$ indicates the representation $[\tilde{f}]$ is contragradient to that of $[f]$. The SU(N) coefficients are usually referred to as fractional-parentage coefficients following the work of Racah.

For the 1s0d shell, the fractional coefficients up to sd^6 were calculated by Akiyama[36]; for the remainder of the shell they were calculated by the current author.[37] Any necessary SU(3) recoupling coefficients and Wigner coefficients were calculated in closed form were heroically calculated by Hecht[38] and Vergados.[35] Subsequently, computer codes for the recoupling coefficients and the SU(3) $\supset$ SU(2) and SU(3) $\supset$ R(3) Wigner coefficients were published by Akiyama and Draayer.[39,40] Nine-$(\lambda\mu)$ coefficients were defined by Millener.[41]

Fractional parentage coefficients needed for 0p-shell calculations for SU(4) $\supset$ SU(4) $\times$ SU(4) were calculated by Jahn and coworkers.[32–34] Closed forms for many other cases are given by Hecht and Pang.[42] Those SU(4) coefficients that are needed but not available from the above references have been calculated.[28]

The formulae to calculate matrix elements for the 1p0f shell are virtually identical to those of the lower shells although allowance must be made for a greater multiplicity in the representations, the number of possible SU(3) representations is much greater. The Akiyama–Draayer codes are sufficiently general for the 1p0f shell, so only the SU(10) $\supset$ SU(3) fractional parentage coefficients are needed. These have been calculated[28] by the author.

3. SU(3) Representations in the 1p0f Shell

As noted above, it is necessary to know which SU(3) representations occur for a specified pf^N SU(10) state. These are a byproduct of the calculation of the fractional parentage coefficients. The technique to calculate the coefficients was inspired by the work of Bayman and Lande.[43] It is necessary to insure the phases are correct; the methods developed by Jahn and co-workers[32–34] were employed. The method

Table 1. The SU(3) multiplicities for specified SU(4) representations for two through five particles for allowed $(\lambda\mu)$ representations in the $1p0f$ shell.

	λ	0	0	0	1	1	1	2	2	3	3	3	4	4	5	5	6	6	7	7	8	9	9	10	11	12
$[f]$	μ	0	3	6	1	4	7	2	5	0	3	6	1	4	2	5	0	3	1	4	2	0	3	1	2	0
[2]		0	0	0	0	0	0	1	0	0	0	0	0	0	0	0	1	0	0	0	0	0	0	0	0	0
[11]		0	1	0	0	0	0	0	0	0	0	0	1	0	0	0	0	0	0	0	0	0	0	0	0	0
[3]		0	1	0	0	0	0	0	0	1	1	0	0	0	1	0	0	0	0	0	0	1	0	0	0	0
[21]		0	0	0	1	1	0	1	0	0	1	0	1	0	1	0	0	0	1	0	0	0	0	0	0	0
[111]		1	0	0	0	0	0	1	0	0	1	0	0	0	0	0	1	0	0	0	0	0	0	0	0	0
[4]		1	0	1	0	0	0	1	0	0	1	0	1	1	0	0	1	1	0	0	1	0	0	0	0	1
[31]		0	1	0	1	1	0	2	2	1	2	0	2	1	2	0	1	2	1	0	1	0	0	1	0	0
[22]		0	0	1	1	1	0	2	0	0	1	0	1	2	1	0	1	0	1	0	1	0	0	0	0	0
[211]		0	1	0	1	2	0	1	1	2	2	0	2	1	2	0	0	1	1	0	0	1	0	0	0	0
[1111]		0	0	1	0	0	0	1	0	1	1	0	0	0	1	0	0	0	0	0	0	0	0	0	0	0
[41]		0	1	1	1	3	1	3	2	2	4	2	3	3	4	2	2	3	3	2	2	1	2	1	1	0
[32]		0	2	1	2	3	1	3	3	2	5	2	4	3	5	2	1	3	3	2	2	2	1	1	1	0
[311]		1	1	2	2	3	1	6	3	1	5	1	4	5	4	2	4	3	3	1	3	0	1	1	0	1
[221]		0	2	0	2	3	1	3	3	2	4	1	4	3	4	1	1	3	2	1	1	1	0	1	0	0
[2111]		0	2	1	1	2	0	3	2	1	3	1	3	2	2	0	2	2	1	0	1	0	0	0	0	0
[11111]		0	1	0	0	1	0	0	1	1	0	0	1	0	1	0	0	0	0	0	0	0	0	0	0	0

used is described in Refs. 44 and 45 although the calculations described therein were for the group SU(6) $\supset$ SU(3)$\times$SU(2).[a] The formulae for the current group SU(10)$\supset$SU(3) and SU(4)$\supset$SU(4)$\times$SU(4) is straightforward.

The SU(3) multiplicities for two, three, four and five particles in the $1p0f$ shell are shown in Table 1.

The maximum multiplicity for three particles in the pf shell is one, but for pf^4, the maximum multiplicity is two which occurs for several $[f](\lambda\mu)$ as listed in Table 1. For the case of pf^5, the maximum multiplicity is six for [311](2 2); for the cases of pf^6, pf^7 and pf^8, the maximum multiplicity increases rapidly to 15: [321](2 2), 37: [421](4 4) and 77: [4211](4 4), respectively.

The number of distinct $[f](\lambda\mu)$ is 4, 16, 62, 239 and 854 for 2–6 $1p0f$ particles, respectively. For comparison, the analogous number for sd^6 is 90 and reaches a maximum number of 625 for sd^{12}.

4. Validity of SU(3) in the $0f1p$ Shell

It has been long known that the $0f1p$ shell is not the optimal place to find the SU(3) model to be a wonderful tool for truncation of the shell model basis. There

[a]This provides another example of Gerry's breadth of interests. He suggested I should apply my techniques to the calculation of multi quark hadrons[46,47] using the Jaffee model[48,49] which is a variant of the MIT bag model.[50] In the Jaffe model, the color-magnetic interaction is $E_g = -\frac{\alpha_c}{R}\sum_{i<j}\sum_a M(m_iR, m_jR)\boldsymbol{\sigma}_i\boldsymbol{\sigma}_j\lambda_i^a\lambda_j^a$ which is handled analogously to the nucleon–nucleon interaction of the Wigner supermultiplet model $E_{NN} = V_0\sum_{i<j}\sum_a \boldsymbol{\sigma}_i\boldsymbol{\sigma}_j\tau_i^a\tau_j^a$ and the spin–isospin interaction is replaced by a spin–color interaction.

are at least two reasons for this. First, and the usual argument given, is the presence of strong spin–orbit coupling. The spin–orbit interaction will tend to admix states of different SU(4) symmetry as well as different SU(3) states. One reason the SU(3) model works in the $1d0s$ shell is that the different SU(4) representations are relatively well separated in energy which implies that the different SU(3) representations belonging to different SU(4) representations will also be well separated, at least at the beginning of the shell.

A second, less appreciated reason, is the relative $0s$ content of the SU(3) states. The nucleon–nucleon interaction is most attractive for two nucleons in a relative $0s$ state, and those states that maximize the amount of relative $0s$ content will lie lowest in energy absent the spin–orbit interaction. One can show that for fewer than five particles, the SU(3) state $(\lambda 0)$ has *all* the $0s$ content — this is one of the reasons why the SU(3) model works — it is also the reason it is useful for alpha decay and transfer reactions. One can demonstrate from the SU(3) coefficients calculated by Hecht and Vergados that the $(\lambda 0)$ states are dominated by the low angular momentum states, i.e., the $1s$ and $1p$ states. In the $1s0d$ shell, the centroid of the $1s$ state lies 1.15 MeV lower than the $0d$ centroid whereas in the $1p0f$ shell, the $1p$ and the $0f$ centroids are essentially degenerate. Thus, the $(\lambda 0)$ states do not have the advantages in the $1p0f$ shell that are exhibited in the $1s0d$ shell.

The real question is how good is SU(3) in the $1p0f$ shell? McGrory[20] made an initial attempt to answer this question. Using the Oak Ridge jj-coupling code, he diagonalized the Kuo–Brown interaction for $(1p0f)^{3,4}$ and the quadrupole–quadrupole interaction for the same basis. Then using the wave functions from the latter calculation, he was able to transform the wave function of $(1p0f)^3$ and $(1p0f)^4$ into a basis that exhibited the contents of the SU(3) representations. The results of his calculations for $(1p0f)^3 T = \frac{1}{2}$ states are that the intensities of the $(9\,0)$ states for the lowest levels of each angular momentum varied from 15% to 32%. For the $(1p0f)^4$ states, the intensity of the $(12\,0)$ states varied from 15% to 55%. These intensities are appreciably less than found in the $1s0d$ shell.[17]

There are two potential issues with these calculations. First, although the intensities of each SU(3) representation can in principle be extracted in this manner, the relative phases of each representation is more problematic. If one wishes to use these results to predict transfer reactions, one must ensure the relevant phases are correct. Second, there are two possible phase conventions for single-particle wave functions, for when there are two or more in a major shell: the wave functions can be positive valued as they approach the origin or they can have the same phase at the surface of the nucleus. The Kuo–Brown matrix elements use the first convention.

However, the SU(3) model requires the wave functions to have the same phase at the surface of the nucleus. This implies that in calculations, the two-body matrix elements of Kuo–Brown must have a sign change for each 1p particle appearing in a two-body matrix element. In LS coupling, this implies that the matrix elements

$$\langle 1p\,0f\,LS\,TJ\,|\,V\,|\,1p^2\,L'S' : TJ\rangle$$

and

$$\langle 1p\,0f\, LSTJ \,|\, V \,|\, 0f^2\, L'S' : TJ\rangle$$

will be affected. How will this affect multi-particle calculations is not obvious. Since one can envision the change of phase as a unitary transformation, one can conclude that the eigenvalues will be unaffected. However, the off-diagonal matrix elements will change, and it is *a priori* unclear if the resulting mixing will increase or decrease.

A detailed comparison must await more investigation, but some preliminary results are available. Changing the phase of the $1p$ single-particle states in accord with Elliott increases the intensity of the leading SU(3) states for low-lying states; the amount varies depending upon the total angular momentum, e.g., for levels of maximum angular momentum, $J = \frac{19}{2}$ for ^{43}Ti and $J = 12$ for ^{44}Ti, there is no change; these states are entirely composed of $f_{\frac{7}{2}}$ and $f_{\frac{5}{2}}$ single-particle states. For the ground state of ^{43}Ti, the intensity of (9 0) increases from 18% to 23%. Other states see a similar, small change. It should be emphasized that none of these new results, however, invalidate the main conclusions of McGrory, namely, SU(3) is not as good a symmetry in the $1p0f$ shell as in lighter nuclei.

However, an examination of the wave functions does reveal some interesting aspects, e.g., the lowest $J = \frac{7}{2}, T = \frac{1}{2}$ level of $(1p0f)^3$ is 94% $L = 3$, 84% $S = \frac{1}{2}$ and has 67% spatial symmetry.[3] Other low-lying states exhibit similar numbers. This suggests that in future multi-particle multi-hole calculations one can restrict the $(1p0f)^n$ states to a relatively few (perhaps 10–50) SU(3) and SU(4) states and obtain multi-particle multihole states that are free of spurious contamination and yet reflect the physical characteristics of the calculation.

Acknowledgments

I was motivated to investigate the structure of light nuclei using SU(3) by Gerry Brown when I was his thesis student at Stony Brook. In subsequent years, he introduced me to many other topics such as the bag model, pion-induced charge exchange, and relativistic heavy ion reactions. I am deeply grateful for his many suggestions and his guidance over the years.

Some of these calculations were performed in Frankfurt while the author was a Fellow of the Humboldt Stiftung; their support and the hospitality of Walter Greiner is greatly appreciated. I also thank John Millener for his comments on the SU(3) and SU(4) coefficients used in this work.

Appendix. The SDG Shell

There may be occasions where there is a need to eliminate spurious states or to study more exotic excitations for which one has excitations to the $2s1d0g$ shell. For completeness, we briefly mention some results. The allowed SU(3)states for two particles are given in Table A. 1. The maximum multiplicity for three particles in the *sdg* shell is two as listed in Table A.2 and for sdg^4, the maximum multiplicity

Table A.1. The SU(3) multiplicities for specified SU(4) representations for two and three particles for allowed $(\lambda\mu)$ representations in the sdg shell.

	λ	0	2	4	6	8
$[f]$	μ	4	3	2	1	0
[200000]		1	0	1	0	1
[110000]		0	1	0	1	0

Table A.2.

	λ	0	0	0	1	1	2	2	3	3	4	4	5	6	6	7	8	9	10	12
$[f]$	μ	0	3	6	1	4	2	5	0	3	1	4	2	0	3	1	2	0	1	0
[300000]		1	0	1	0	0	1	0	0	1	0	1	0	1	1	0	1	0	0	1
[210000]		0	0	0	1	1	1	1	0	1	1	2	1	0	1	1	1	0	1	0
[111000]		0	1	0	0	0	0	1	1	1	0	0	1	0	1	0	0	1	0	0

is four. For the case of sdg^5, the maximum multiplicity is 14 for [311](3 4); for the case of sdg^6, [321](4 4) the maximum multiplicity increases to 64.

The number of distinct $[f](\lambda\mu)$ is 5, 29, 179, 971 and 5,187 2–6 *sdg* particles, respectively. For details on more than three particles, the reader is referred to Ref. 28.

References

1. G. E. Brown, *Proc. Int. Congress for Nuclear Physics*, ed. P. Gugenberger (CNRS, Paris, France, 1964), p. 129.
2. G. E. Brown and A. M. Green, *Nucl. Phys.* **75** (1966) 401; **85** (1966) 87.
3. T. Engeland, Nucl. Phys. **72** (1965) 68.
4. P. J. Ellis and T. Engeland, *Nucl. Phys. A* **144** (1970) 161; **181** (1972) 368.
5. J. P. Elliott, *Proc. Roy. Soc. Lond. A* **245** (1958) 562.
6. D. J. Millener, D. Phil. Thesis, Oxford University (1972) (unpublished); *Proc. of the Int. Nuclear Physics Conference* (Munich, 1974).
7. T. Inoue, T. Sebe, H. Hagiwara and A. Arima, *Nucl. Phys.* **59** (1964) 1; **85** (1966) 184.
8. Y. Akiyama, A. Arima and T. Sebe, *Nucl. Phys. A* **138** (1969) 273.
9. D. Strottman, *Phys. Lett. B* **39** (1972) 457.
10. P. Manakos, *Z. Phys.* **247** (1971) 127; H. Feldmeier, P. Manakos and T. Wolff, *Z. Phys.* **258** (1973) 81.
11. A. Arima and D. Strottman, *Nucl. Phys. A* **162** (1971) 423; D. Strottman and A. Arima, *Nucl. Phys. A* **162** (1971) 605.
12. D. Strottman *et al.*, *Phys. Lett. B* **47** (1973) 16.
13. N. Anyas-Weiss, J. C. Cornell, P. S. Fischer, P. N. Hudson, A. Menchaca-Rochka, A. D. Panagiotou, D. K. Scott, D. J. Millener, D. Strottman, D. M. Brink, B. Buck, P. J. Ellis and T. Engeland, *Phys. Rep. C* **12** (1974) 202.
14. M. Ichimura, A. Arima, E. Halbert and T. Terasawa, *Nucl. Phys. A* **204** (1973) 225.
15. J. P. Elliott and T. H. R. Skyrme, *Proc. R. Soc. Lond. A* **232** (1955) 561.
16. K. T. Hecht, *Nucl. Phys. A* **170** (1971) 34.

17. D. Strottman, Oxford Nuclear Physics Laboratory preprint 14/73.
18. T. T. S. Kuo and G. E. Brown, *Nucl. Phys.* **85** (1966) 40.
19. T. T. S. Kuo, *Nucl. Phys.* **103** (1967) 71.
20. J. B. McGrory, *Phys. Lett. B* **47** (1973) 481.
21. R. Whitehead, *Nucl. Phys. A* **182** (1972) 290; R. R. Whitehead, A. Watt, B. J. Cole and I. Morrison, *Adv. Nucl. Phys.* **9** (1977) 123.
22. D. J. Dean, T. Engeland, M. Hjorth-Jensen, M. P. Kartamyshev and E. Osnes, *Prog. Part. Nucl. Phys.* (2004).
23. E. P. Wigner, *Phys. Rev.* **51** (1937) 106; **51** (1937) 947; **56** (1939) 519.
24. J. P. Elliott and M. Harvey, *R. Soc. A* **272** (1963) 557.
25. K. T. Hecht, *Selected Topics in Nuclear Spectroscopy*, ed. B. J. Verhaar (North Holland and John Wiley, 1964).
26. M. Harvey, *Advances in Nuclear Physics*, Vol. 1, eds. M. Baranger and E. Vogt (Plenum, New York, 1968), p. 67.
27. R. Perez and J. Flores, *Nucl. Data Sheets A* **4** (1968) 265.
28. D. Strottman, 1974, unpublished; D. Strottman, LANL Technical Report, in preparation.
29. G. Racah, *Phys. Rev.* **61** (1942) 186; **62** (1942) 438; **63** (1943) 367; **76** (1949) 1352.
30. D. Littlewood, *The Theory of Group Characters* (Clarendon Press, Oxford, 1940).
31. A. de-Shalit and I. Talmi, *Nuclear Shell Theory* (Academic Press, US, 1963).
32. H. A. Jahn and H. van Wieringen, *Proc. R. Soc. A* **209** (1951) 502.
33. J. P. Elliott, J. Hope and H. A. Jahn, *Phil. Tran. R. Soc. A* **246** (1953) 241.
34. J. P. Elliott and B. F. Flowers, *Proc. R. Soc. Lond. A* **242** (1957) 57.
35. J. D. Vergados, *Nucl. Phys.* **111** (1968) 681.
36. Y. Akiyama, *Nucl. Data A* **2** (1966) 403.
37. D. Strottman, Oxford Nuclear Physics Laboratory Technical Report (1973).
38. K. T. Hecht, *Nucl. Phys.* **62** (1965) 1.
39. J. P. Draayer and Y. Akiyama, *J. Math. Phys.* **14** (1973) 1904.
40. Y. Akiyama and J. P. Draayer, *Comput. Phys. Commun.* **5** (1973) 405.
41. D. J. Millener, *J. Math. Phys.* **19** (1978) 1513.
42. K. T. Hecht and S. C. Pang, *J. Math. Phys.* **10** (1969) 1571.
43. B. F. Bayman and A. Lande, *Nucl. Phys.* **77** (1966) 1.
44. S. I. So and D. Strottman, *J. Math. Phys.* **20** (1979) 153.
45. D. Strottman, *J. Math. Phys.* **20** (1979) 1643.
46. D. Strottman, *Phys. Rev. D* **18** (1978) 2716.
47. D. Strottman, *Phys. Rev. D* **20** (1979) 748.
48. R. L. Jaffe, *Phys. Rev. Lett.* **38** (1977) 195.
49. R. L. Jaffe, *Phys. Rev. D* **15** (1977) 281.
50. A. Chodos, R. L. Jaffe, K. Johnson and C. Thorn, *Phys. Rev. D* **10** (1974) 2599.

Multi-regulator functional renormalization group for many-fermion systems

Yuya Tanizaki*,§ and Tetsuo Hatsuda†,‡,¶
*RIKEN BNL Research Center,
Brookhaven National Laboratory,
Upton, NY 11973-5000, USA
†iTHES Research Group, RIKEN, Wako,
Saitama 351-0198, Japan
‡Nishina Center, RIKEN, Wako,
Saitama 351-0198, Japan
§yuya.tanizaki@riken.jp
¶thatsuda@riken.jp

We propose a method of multi-regulator functional renormalization group (MR-FRG) which is a novel formulation of functional renormalization group with multiple infrared (IR) regulators. It is applied to a two-component fermionic system with an attractive contact interaction to study crossover phenomena between the Bardeen–Cooper–Schrieffer (BCS) phase and the Bose–Einstein condensation (BEC) phase. To control both the fermionic one-particle excitations and the bosonic collective excitations, IR regulators are introduced, one for the fermionic two-point function and another for the four-fermion vertex. It is shown that the Nozières–Schmitt-Rink (NSR) theory, which is successful to capture qualitative features of the BCS–BEC crossover, can be derived from MR–FRG. Some aspects of MR-FRG to go beyond the NSR theory are also discussed.

Keywords: Fermionic functional renormalization group; BCS-BEC crossover.

1. Introduction

Functional renormalization group (FRG)[1–3] is a pragmatic realization of Wilson's idea of renormalization group (RG).[4] Functional implementation of coarse-graining realizes an exact evolution of an effective action involving quantum and thermal fluctuations, and enables us to calculate observables of various quantum systems. Especially for interacting many fermions, the fermionic FRG provides a systematic way to analyze the system without *a priori* knowledge on their ground state properties[5–7] and is complementary to bosonized FRG where auxiliary bosonic fields are introduced.[8–10]

We have previously studied the Bardeen–Cooper–Schrieffer (BCS) to Bose–Einstein condensation (BEC) crossover of the two-component fermionic system with an attractive contact interaction[11,12] by treating the BCS regime and the BEC

§Corresponding author.

regime separately on the basis of fermionic FRG: In the BCS side, the Gorkov and Melik-Barkhudarov correction[13] to the BCS theory has been derived from fermionic FRG by introducing an infrared (IR) regulator for the two-point function,[14,15] while, in the BEC side, fermionic FRG with an IR regulator for the four-fermion vertex has been introduced.[16]

The purpose of this paper is to propose a new fermionic FRG with multiple IR regulators, which we call multi-regulator functional renormalization group (MR-FRG), to control the low-energy fermionic and bosonic excitations simultaneously, so that it can treat the whole region of the BCS–BEC crossover. To show how our formalism works, we discuss its relation to the Nozières–Schmitt-Rink (NSR) theory,[17] the conventional formalism to treat the BCS–BEC crossover with pairing fluctuations.

This paper is organized as follows. In Sec. 2, we consider a model for the two-component fermionic system, and introduce a new formulation of FRG with two different IR regulators and two-parameter flow equations. In Sec. 3, formal properties of the flow equations are discussed. After carrying out the ultraviolet (UV) renormalization of the flow equations, we optimize the RG flow by choosing appropriate IR regulators under the vertex expansion of the flow equations. In Sec. 4, we derive the NSR theory for the BCS–BEC crossover based on MR-FRG. We first study the RG flow of the four-fermion vertex and observe that one-particle fermion excitations decouple from the flow equation. Using this property, we can solve the flow equation of the self-energy, whose solution correctly reproduces the number equation of the NSR theory. Section 5 is devoted to summary and perspectives.

2. Fermionic FRG with Multiple IR Regulators

Let us consider the following action with nonrelativistic two-component fermions $\Psi = \binom{\Psi_\uparrow}{\Psi_\downarrow}$:

$$S[\overline{\Psi}, \Psi] = \int_0^\beta d\tau \int d^3\boldsymbol{x} \left[\overline{\Psi} \left(\partial_\tau - \frac{\nabla^2}{2m} - \mu \right) \Psi(x) + g \overline{\Psi}_\uparrow \overline{\Psi}_\downarrow \Psi_\downarrow \Psi_\uparrow(x) \right], \tag{1}$$

with $\beta(= 1/T)$ the inverse temperature, μ the chemical potential, m the fermion mass, and g the bare coupling constant of the contact interaction. The UV regularized form of (1) in the momentum space is given by

$$S[\overline{\Psi}, \Psi] = \int_p^{(T)} \overline{\Psi}_p G^{-1}(p) \Psi_p + g(\Lambda) \int_p^{(T)} e^{-ip^0 0^+} \int_{q,q' \le \Lambda}^{(T)} \overline{\Psi}_{\uparrow, \frac{p}{2}+q} \overline{\Psi}_{\downarrow, \frac{p}{2}-q} \Psi_{\downarrow, \frac{p}{2}-q'} \Psi_{\uparrow, \frac{p}{2}+q'}, \tag{2}$$

where $G^{-1}(p) = ip^0 + \boldsymbol{p}^2/2m - \mu$ with $p = (p^0, \boldsymbol{p})$, $\psi_{\sigma,p}$ denotes the Fourier expansion coefficient of $\psi_\sigma(x)$, $\int_p^{(T)} = \int \frac{d^3\boldsymbol{p}}{(2\pi)^3} \frac{1}{\beta} \sum_{p^0}$, and Λ denotes a UV cutoff for the spatial momenta, $\boldsymbol{q}$ and $\boldsymbol{q}'$.

The fermionic field theory (2) contains UV divergence only in the loop of particle–particle scattering,

$$\Pi(p) = \int_{l\leq\Lambda}^{(T)} \frac{1}{G^{-1}\left(\frac{p}{2}+l\right)G^{-1}\left(\frac{p}{2}-l\right)} = \frac{m\Lambda}{2\pi^2} + \mathcal{O}(1). \tag{3}$$

Such a divergence can be absorbed into the bare coupling $g(\Lambda)$ and renormalized into the physical scattering length a_s:

$$\frac{m}{4\pi a_s} = g^{-1}(\Lambda) + \frac{m\Lambda}{2\pi^2}. \tag{4}$$

Observables are written in terms of m, a_s, T, and μ by absorbing all the UV divergences through Eq. (4) and taking $\Lambda \to \infty$.

In order to describe the BCS–BEC crossover by fermionic FRG, we need to control both fermionic one-particle excitations and bosonic collective excitations. Since fermionic excitations are described by the two-point Green's function, we introduce an additional action

$$\delta S_{k_1}^{(f)}[\overline{\Psi}, \Psi] = \int_p^{(T)} \overline{\Psi}_p R_{k_1}^{(f)}(\boldsymbol{p})\Psi_p, \tag{5}$$

where $R_{k_1}^{(f)}$ regulates fermionic one-particle excitations with energy smaller than $k_1^2/2m$. On the other hand, the collective bosonic excitation is described by the pole of the four-point Green's function. In order to suppress bosonic excitations with energy less than $k_2^2/4m$, we introduce the following action[16]:

$$\delta S_{k_2}^{(b)}[\overline{\Psi}, \Psi] = \int_p^{(T)} \frac{g^2 R_{k_2}^{(b)}(\boldsymbol{p})e^{-ip^0 0^+}}{1 - gR_{k_2}^{(b)}(\boldsymbol{p})} \int_{q,q'\leq\Lambda}^{(T)} \overline{\Psi}_{\uparrow,\frac{p}{2}+q}\overline{\Psi}_{\downarrow,\frac{p}{2}-q}\Psi_{\downarrow,\frac{p}{2}-q'}\Psi_{\uparrow,\frac{p}{2}+q'}, \tag{6}$$

with $R_{k_2}^{(b)}$ being a vertex IR regulator. For later convenience, we define

$$g_{k_2}(p) \equiv \frac{g^2 R_{k_2}^{(b)}(\boldsymbol{p})}{1 - gR_{k_2}^{(b)}(\boldsymbol{p})}. \tag{7}$$

Note that adding this vertex IR regulator introduces an effective shift of the inverse coupling, $g^{-1} \to g^{-1} - R_{k_2}^{(b)}(\boldsymbol{p})$.

With an abbreviated notation $\phi \equiv (\overline{\Psi}, \Psi)$, the Schwinger functional $W_{k_1,k_2}[J]$ depending on two scales (k_1, k_2) can be defined as

$$W_{k_1,k_2}[J] = \ln \int \mathcal{D}\phi \exp[-(S[\phi] + \delta S_{k_1}^{(f)}[\phi] + \delta S_{k_2}^{(b)}[\phi]) + J^\alpha \phi_\alpha]. \tag{8}$$

In this expression, α runs over all the arguments of ϕ (space-time coordinates x, internal degrees of freedom σ, and the global $U(1)$ charges ±1 for $\overline{\Psi}$ and Ψ). The (k_1, k_2)-dependent one-particle-irreducible (1PI) effective action $\Gamma_{k_1,k_2}[\varphi]$ with

$\varphi \equiv (\overline{\psi}, \psi)$ is defined as the Legendre transform of the Schwinger functional:

$$\Gamma_{k_1,k_2}[\varphi] + \delta S^{(f)}_{k_1}[\varphi] = J^{\alpha}[\varphi]\varphi_{\alpha} - W_{k_1,k_2}[J[\varphi]], \tag{9}$$

where $J[\varphi]$ is the solution of $\delta_L W_{k_1,k_2}[J]/\delta J = \varphi$ with $\delta_L/\delta J$ being the left functional derivative. This is the generating functional of the 1PI effective vertices at the scale (k_1, k_2).

In terms of the parameter k_1, the 1PI effective action $\Gamma_{k_1,k_2}[\varphi]$ obeys the flow equation[1–3]:

$$\partial_{k_1}\Gamma_{k_1,k_2}[\varphi] = \frac{1}{2}\mathrm{STr}\left[\frac{\partial_{k_1}R^{(f)}_{k_1}}{\Gamma^{(2)}_{k_1,k_2}[\varphi] + R^{(f)}_{k_1}}\right], \tag{10}$$

where $\Gamma^{(2)}_{k_1,k_2}[\varphi] = \frac{\delta_L}{\delta\varphi}\frac{\delta_R}{\delta\varphi}\Gamma_{k_1,k_2}[\varphi]$ is the field-dependent propagator, and STr denotes the supertrace in the functional space. For fixed and large UV cutoff Λ, the initial condition of the flow equation is given by $\Gamma_{k_1=\Lambda\to\infty,k_2}[\varphi] = S[\varphi] + \delta S^{(b)}_{k_2}[\varphi]$ with a possible field-independent part. For this boundary condition to be satisfied, we need to choose $R^{(f)}_{k_1=\Lambda\to\infty} = \infty$ and $R^{(f)}_{k_1=0} = 0$.

The k_2-dependence of Γ_{k_1,k_2} is obtained from the following flow equation[16]:

$$\begin{aligned}\partial_{k_2}\Gamma_{k_1,k_2}[\varphi] = \frac{1}{4!}\partial_{k_2}g^{\alpha_1\alpha_2\alpha_3\alpha_4}_{k_2}(&\varphi_{\alpha_1}\varphi_{\alpha_2}\varphi_{\alpha_3}\varphi_{\alpha_4} + 6\varphi_{\alpha_1}\varphi_{\alpha_2}G_{k_1k_2,\alpha_3\alpha_4}\\ &+ 3G_{k_1k_2,\alpha_1\alpha_2}G_{k_1k_2,\alpha_3\alpha_4} + 4\varphi_{\alpha_1}G_{k_1k_2,\alpha_2\beta_2}G_{k,\alpha_3\beta_3}G_{k_1k_2,\alpha_4\beta_4}\\ &\times\Gamma^{(3),\beta_2\beta_3\beta_4}_{k_1,k_2} + G_{k_1k_2,\alpha_1\beta_1}G_{k_1k_2,\alpha_2\beta_2}G_{k_1k_2,\alpha_3\beta_3}G_{k_1k_2,\alpha_4\beta_4}\\ &\times\Gamma^{(4),\beta_1\beta_2\beta_3\beta_4}_{k_1,k_2} + 3G_{k_1k_2,\alpha_1\beta_1}G_{k_1k_2,\alpha_2\beta_2}G_{k_1k_2,\alpha_3\beta_3}G_{k_1k_2,\alpha_4\beta_4}\\ &\times G_{k_1k_2,\gamma_1\gamma_2}\Gamma^{(3),\beta_1\beta_2\gamma_1}_{k_1,k_2}\Gamma^{(3),\gamma_2\beta_3\beta_4}_{k_1,k_2}),\end{aligned} \tag{11}$$

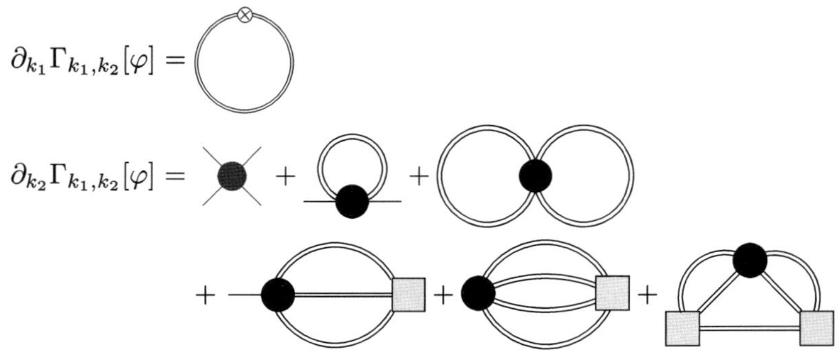

Fig. 1. The flow equations for the effective action, $\Gamma_{k_1,k_2}[\varphi]$, in terms of k_1 and k_2. The thin lines represent the field φ, the double lines represent the field-dependent propagator, $G_{k_1,k_2}[\varphi]$, the circle with cross represents $\partial_{k_1}R^{(f)}_{k_1}$, the filled circles represent $-\partial_{k_2}g_{k_2}$, and the square boxes represent the field-dependent 1PI effective vertice, $\Gamma^{(3,4)}_{k_1,k_2}[\varphi]$.

with $G_{k_1,k_2}[\varphi] = (\Gamma^{(2)}_{k_1,k_2}[\varphi] + R^{(f)}_{k_1})^{-1}$ being the field-dependent propagator, and $\Gamma^{(n)}_{k_1,k_2}[\varphi]$ being the nth functional derivative of $\Gamma_{k_1,k_2}[\varphi]$. The boundary condition of the RG flow is specified by requiring $R^{(b)}_{k_2=\Lambda\to\infty} \to \infty$ and $R^{(b)}_{k_2=0} = 0$. For large UV cutoff Λ, the 1PI effective action $\Gamma_{k_1,k_2}[\varphi]$ converges to the free action in the limit $k_2 = \Lambda \to \infty$, because $g + g_{k_2=\Lambda\to\infty} = g - g = 0$.

Diagrammatic expressions of the flow equations in MR-FRG, (10) and (11), are shown in Fig. 1, where the double lines represent the field-dependent propagator, $G_{k_1,k_2}[\varphi]$.

3. Formal Aspects of the Flow Equations in MR-FRG

Equations (10) and (11) constitute the basic equations of MR-FRG. In this section, we study formal aspects of the flow equations in MR-FRG, and consider optimization of the flow with respect to the IR regulators.

3.1. *UV renormalization of the flow equations*

In Eq. (10), the derivative of the fermion IR regulator $\partial_{k_1} R^{(f)}_{k_1}$ introduces a UV cutoff at the energy scale $k_1^2/2m$ in the loop integration, and thus it is free from UV divergences in the limit $\Lambda \to \infty$. Let us now show that Eq. (11) does not have UV divergences too.

We first note that the bare coupling appears in the flow equation as $\partial_{k_2} g_{k_2} = g^2 \partial_{k_2} R^{(b)}_{k_2}/(1 - g R^{(b)}_{k_2})^2$. Therefore, together with the renormalization condition, Eq. (4), we have

$$\partial_{k_2} g_{k_2} \xrightarrow[\Lambda\to\infty]{} \left(\frac{2\pi^2}{m\Lambda}\right)^2 \partial_{k_2} R^{(b)}_{k_2} \sim O(1/\Lambda^2). \tag{12}$$

Now, the UV-finite subdiagrams originate only from particle–particle scattering contribution (see Fig. 2)[a] in the present nonrelativistic system:

$$\lim_{\Lambda\to\infty} [\Pi_{k_1,k_2}(p)]^2 [-\partial_{k_2} g_{k_2}(p)] = -\partial_{k_2} R^{(b)}_{k_2}(\boldsymbol{p}), \tag{13}$$

where $\Pi_{k_1,k_2}(p)$ is a particle–particle one-loop amplitude of the field-independent propagator, $G_{k_1,k_2}[0] = (\Gamma^{(2)}_{k_1,k_2}[0] + R^{(f)}_{k_1})^{-1}$. Since all the other subdiagrams

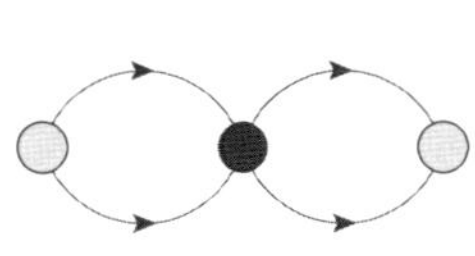

Fig. 2. The UV-finite diagram which appears in the flow equation. The grey circles denote the bare vertex g, while the filled circle represents $-\partial_{k_2} g_{k_2}$. The internal lines with arrows are the field-independent propagator, $G_{k_1,k_2}[\varphi = 0]$.

[a]From now on, we will attach arrows to fermion lines in order to identify whether the fermion is emitted or absorbed at each vertex.

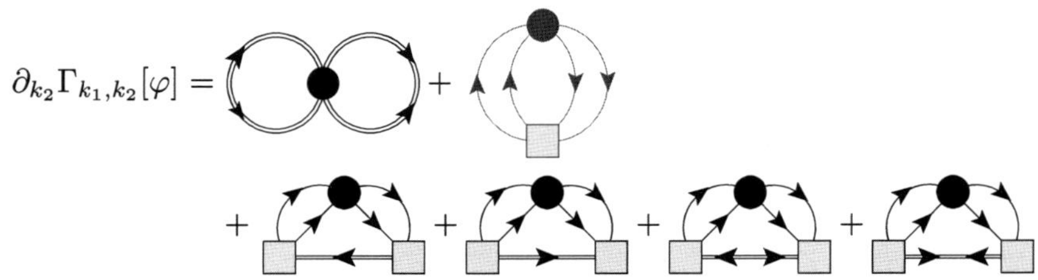

Fig. 3. UV-renormalized k_2-flow equation of the effective action. The field-independent propagator ($G_{k_1,k_2}[0]$) and the field-independent propagator ($G_{k_1,k_2}[\varphi]$) are represented by the single line and the double line, respectively.

are suppressed by the powers of $1/\Lambda$, the k_2-flow equation in Fig. 1 can be much simplified in the UV limit as shown in Fig. 3. To make clear distinction between the field-independent propagator $G_{k_1,k_2}[0]$ and the field-dependent propagator $G_{k_1,k_2}[\varphi]$, we represent the former (latter) by a single (double) line in this figure.

3.2. *Vertex expansion of the flow equations*

In order to solve the flow equations (10) and (11) for each 1PI vertex function, we need to perform the vertex expansion of $\Gamma_{k_1,k_2}[\overline{\psi},\psi]$. For simplicity, we assume the symmetry-unbroken phase in the following, so that each vertex couples to the same numbers of ψ and $\overline{\psi}$. Then we consider the following vertex expansion of Γ_{k_1,k_2} up to the fourth-order:

$$\Gamma_{k_1,k_2}[\overline{\psi},\psi] = \int_p^{(T)} \overline{\psi}_p[G^{-1} - \Sigma_{k_1,k_2}](p)\psi_p + \int_{p,q,q'}^{(T)} \Gamma^{(4)}_{k_1,k_2}(p,q,q')\overline{\psi}_{\uparrow,\frac{p}{2}+q}\overline{\psi}_{\downarrow,\frac{p}{2}-q}\psi_{\downarrow,\frac{p}{2}-q'}\psi_{\uparrow,\frac{p}{2}+q'}, \tag{14}$$

where Σ_{k_1,k_2} and $\Gamma^{(4)}_{k_1,k_2}$ are the self-energy and the four-point vertex, respectively. The flow equations in terms of k_1 with this truncation are shown in Fig. 4 with $\widetilde{\partial}_{k_1}$, the k_1-derivative, acting only on the explicit k_1-dependence of the fermionic

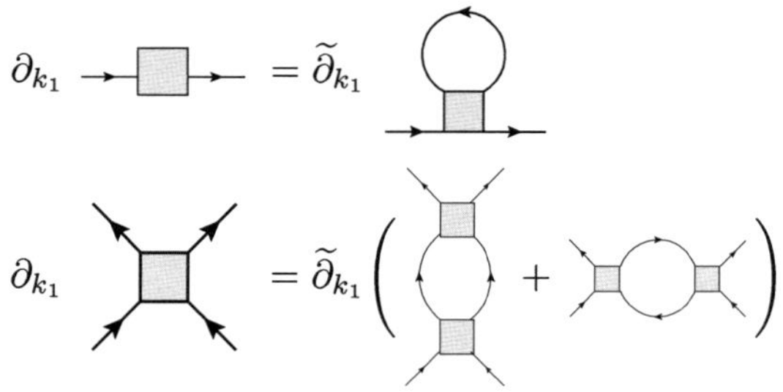

Fig. 4. The flow equations for the self-energy Σ_{k_1,k_2} and the four-point vertex $\Gamma^{(4)}_{k_1,k_2}$ in terms of k_1.

IR regulator $R_{k_1}^{(f)}$. Their analytic expressions are obtained by substituting (14) into (10):

$$\partial_{k_1}\Sigma_{k_1,k_2}(p) = \widetilde{\partial}_{k_1}\int_l^{(T)} e^{-il^0 0^+}\frac{\Gamma^{(4)}_{k_1,k_2}(p+l)}{[G^{-1}-\Sigma_{k_1,k_2}+R^{(f)}_{k_1}](l)}, \tag{15}$$

$$-\partial_{k_1}\Gamma^{(4)}_{k_1,k_2}(p) = \widetilde{\partial}_{k_1}\left[\int_l^{(T)}\frac{\Gamma^{(4)}_{k_1,k_2}(p)\Gamma^{(4)}_{k_1,k_2}(p)}{\prod_{\pm}\left([G^{-1}-\Sigma_{k_1,k_2}+R^{(f)}_{k_1}]\left(\frac{p}{2}\pm l\right)\right)} + \frac{1}{2}\sum_{\pm}\int_l^{(T)}\right.$$
$$\left.\times\frac{\Gamma^{(4)}_{k_1,k_2}\left(\frac{p}{2}+q+l\right)\Gamma^{(4)}_{k_1,k_2}\left(\frac{p}{2}\pm q'+l\right)}{[G^{-1}-\Sigma_{k_1,k_2}+R^{(f)}_{k_1}](l)[G^{-1}-\Sigma_{k_1,k_2}+R^{(f)}_{k_1}](q\pm q'+l)}\right]. \tag{16}$$

Here, relative momenta dependence of the 1PI four-point vertex is not shown explicitly just for simplicity.

The flow equations for Σ_{k_1,k_2} and $\Gamma^{(4)}_{k_1,k_2}$ in terms of k_2 can be obtained by substituting the vertex expansion (14) into the flow equation in Fig. 3. Their diagrammatic expressions are shown in Fig. 5, and the analytic forms are given by

$$\partial_{k_2}\Sigma_{k_1,k_2}(p) = \int_l^{(T)} e^{-il^0 0^+}\frac{(\Gamma^{(4)}_{k_1,k_2}(p+l))^2\partial_{k_2}R^{(b)}_{k_2}(\boldsymbol{p}+\boldsymbol{l})}{[G^{-1}-\Sigma_{k_1,k_2}+R^{(f)}_{k_1}](l)}, \tag{17}$$

$$-\partial_{k_2}\Gamma^{(4)}_{k_1,k_2}(p) = -(\Gamma^{(4)}_{k_1,k_2}(p))^2\partial_{k_2}R^{(b)}_{k_2}(\boldsymbol{p}) + \sum_{\pm}\int_l^{(T)}$$
$$\times\frac{\Gamma^{(4)}_{k_1,k_2}\left(\frac{p}{2}+q+l\right)\left(\Gamma^{(4)}_{k_1,k_2}\left(\frac{p}{2}\pm q'+l\right)\right)^2\partial_{k_2}R^{(b)}_{k_2}\left(\frac{\boldsymbol{p}}{2}\pm\boldsymbol{q}'+\boldsymbol{l}\right)}{[G^{-1}-\Sigma_{k_1,k_2}+R^{(f)}_{k_1}](l)[G^{-1}-\Sigma_{k_1,k_2}+R^{(f)}_{k_1}](q\pm q'+l)}. \tag{18}$$

When we write down the full momentum dependence of this flow equation, we should notice that only the relative momentum-independent part of each 1PI vertex

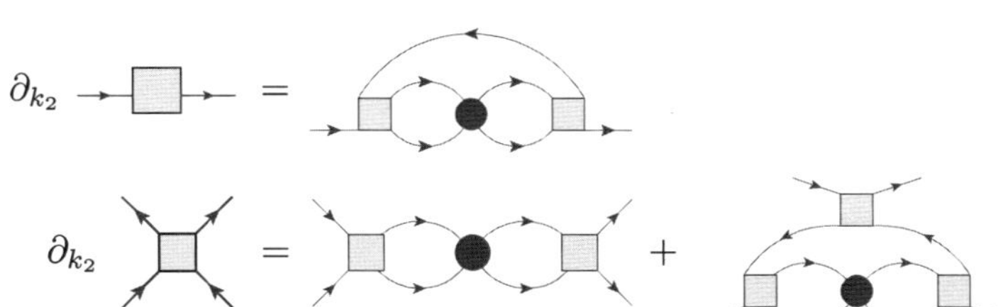

Fig. 5. Flow equation of the self-energy Σ_{k_1,k_2} and the four-point vertex $\Gamma^{(4)}_{k_1,k_2}$ in terms of k_2.

can couple to $\partial_{k_2} g_{k_2}$ in the limit $\Lambda \to \infty$. Such momentum-independent parts can be extracted by putting the corresponding relative momentum to be infinity.

After solving the flow equations (15)–(17), we apply the Thouless criterion for the superfluid phase transition[18]:

$$\frac{1}{\Gamma^{(4)}_{0,0}(p=0)} = 0, \tag{19}$$

which corresponds to the gap equation at $T = T_c$. On the other hand, the number density n is obtained by

$$n = -2 \int_p^{(T)} \frac{e^{-ip^0 0^+}}{G^{-1}(p) - \Sigma_{0,0}(p)}. \tag{20}$$

By combining (19) and (20), one eventually obtains T_c/ε_F as a function of $(k_F a_s)^{-1}$ with $k_F = (3\pi n)^{1/3}$ and $\varepsilon_F = k_F^2/2m$.

3.3. *Optimized flow and IR regulators*

There are many degrees of freedom in the choice of IR regulators. The idea of optimizing the FRG flow is to use such degrees of freedom to construct a systematic approximation scheme. The optimized IR regulator depends naturally on the approximation scheme and the order of truncation. For bosonic theories, regulator dependence of the approximate FRG flow has been studied in details within the local potential approximation (LPA).[19–21] According to these studies, the optimized IR regulators have been found to share the following properties besides the boundary condition at $k = \infty$:

(i) IR regulator R_k makes the system gapped by a typical energy scale $k^2/2m$ with k being the scale parameter.
(ii) High-energy excitations with the energy larger than $k^2/2m$ decouple from the flow equation.

With these properties, low-energy excitations are equally gapped and high-energy excitations do not affect the low-energy dynamics after renormalization. This helps us to establish a series of systematic truncations of the flow equations.

For the system treated in this paper, there are two different degrees of freedom to be gapped. The first one is the one-particle fermion excitation. The above two conditions for such excitations can be satisfied by taking the Litim-type optimized regulator[19,20]:

$$R^{(f)}_{k_1}(\boldsymbol{p}) \equiv Z_f \left[\frac{k_1^2}{2m}\mathrm{sgn}(\xi(\boldsymbol{p})) - \xi(\boldsymbol{p})\right] \theta\left(\frac{k_1^2}{2m} - |\xi(\boldsymbol{p})|\right), \tag{21}$$

where $\xi(\boldsymbol{p})$ denotes the excitation energy relative to the Fermi level, with Z_f being the wave function renormalization of fermionic fields. For the simplest case, $Z_f = 1$ and $\xi(\boldsymbol{p}) = \frac{\boldsymbol{p}^2}{2m} - \mu$, it is easy to check that this regulator opens an energy gap by the amount $\pm\frac{k_1^2}{2m}$ near the Fermi surface, $|\frac{\boldsymbol{p}^2}{2m} - \mu| \leq \frac{k_1^2}{2m}$.

The second one is the bosonic collective excitation given by the pole of $\Gamma^{(4)}_{k_1,k_2}(p)$. According to the flow equation in Fig. 5, $\partial_{k_2} R^{(b)}_{k_2}$ can make high-energy collective excitations decouple from the flow equation. Therefore, the Litim-type regulator is again expected to satisfy the above two conditions[19,20]:

$$R^{(b)}_{k_2} \equiv Z_b \left(\frac{k_2^2}{4m} - \frac{\boldsymbol{p}^2}{4m} \right) \theta \left(\frac{k_2^2}{4m} - \frac{\boldsymbol{p}^2}{4m} \right), \tag{22}$$

where Z_b is the wave function renormalization of the effective boson propagator. For the simplest case, $Z_b = 1$, this introduces a gap by $\frac{k_2^2}{4m}$ at low momentum, $\frac{\boldsymbol{p}^2}{4m} \leq \frac{k_2^2}{4m}$.

We can convert the two-parameter flow equations (15)–(17) into a single-parameter flow equation by relating k_1 and k_2. In this paper, we simply take $k \equiv k_1 = k_2$, and adopt a simplified notation $\Gamma_k[\overline{\psi}, \psi] \equiv \Gamma_{k,k}[\overline{\psi}, \psi]$. One may alternatively choose $k_1 = ck_2$ with a constant c which can be determined, e.g., by applying the principle of minimal sensitivity[22] to some observable such as T_c/ε_F.

4. NSR Theory from MR-FRG

In this section, we derive the NSR theory[17] on the basis of MR-FRG. In the NSR theory, the particle–particle random phase approximation (pp-RPA) is adopted to take into account the effect of collective fermion pairs in the free energy. Then the critical temperature in the BCS–BEC crossover can be calculated as a function of $(k_F a_s)^{-1}$ by solving the following two equations:

$$\frac{1}{a_s} = -\frac{2}{\pi} \int_0^\infty \sqrt{2m\varepsilon} d\varepsilon \left[\frac{\tanh \frac{\beta}{2}(\varepsilon - \mu)}{2(\varepsilon - \mu)} - \frac{1}{2\varepsilon} \right], \tag{23}$$

$$n = -2 \int_p^{(T)} e^{-ip^0 0^+} G(p) - \frac{\partial}{\partial \mu} \int_p^{(T)} e^{-ip^0 0^+} \ln \left[1 + \frac{4\pi a_s}{m} \left(\Pi(p) - \frac{m\Lambda}{2\pi^2} \right) \right]. \tag{24}$$

4.1. *Pairing approximation on the four-point vertex*

We first consider the flow equation of the four-point 1PI vertex function, and identify an approximation corresponding to pp-RPA. Flow equations of the four-point 1PI vertex $\Gamma^{(4)}_{k_1,k_2}(p; q, q')$ are given in Figs. 4 and 5. As we have announced at the end of the previous section, we set $k = k_1 = k_2$ and analyze the flow equation for $\Gamma^{(4)}_k = \Gamma^{(4)}_{k,k}$. Let us consider only the particle–particle correlation in the flow equation of the four-point vertex by neglecting the self-energy correction:

$$- \partial_k \Gamma^{(4)}_k(p) = \int_l^{(T)} \frac{-2(\Gamma^{(4)}_k(p))^2 \partial_k R^{(f)}_k \left(\frac{\boldsymbol{p}}{2} - \boldsymbol{l} \right)}{[G^{-1} + R^{(f)}_k] \left(\frac{p}{2} + l \right) [G^{-1} + R^{(f)}_k]^2 \left(\frac{p}{2} - l \right)} + (\Gamma^{(4)}_k(p))^2 \partial_k R^{(b)}_k(\boldsymbol{p}). \tag{25}$$

Under this approximation, the solution of the flow equation becomes the four-point vertex in pp-RPA.[14,16] Since the fermion dispersion relation is given by the bare propagator, parameters of the IR regulator $R_k^{(f)}$ in (21) can be specified to be $Z_f = 1$ and $\xi(\boldsymbol{p}) = \boldsymbol{p}^2/2m - \mu$. Therefore, the analytic expression for the four-point vertex becomes

$$\frac{1}{\Gamma_k^{(4)}(p)} = \frac{m}{4\pi a_s} - R_k^{(b)}(\boldsymbol{p}) + \int \frac{d^3\boldsymbol{\ell}}{(2\pi)^3}\left[\frac{1 - \sum_{\pm} n_F\left(\frac{(\boldsymbol{p}/2 \pm \boldsymbol{\ell})^2}{2m} - \mu + R_k^{(f)}\left(\frac{\boldsymbol{p}}{2} \pm \boldsymbol{\ell}\right)\right)}{\frac{\boldsymbol{\ell}^2}{m} + \left(ip^0 + \frac{\boldsymbol{p}^2}{4m} - 2\mu\right) + \sum_{\pm} R_k^{(f)}\left(\frac{\boldsymbol{p}}{2} \pm \boldsymbol{\ell}\right)} - \frac{1}{\frac{\boldsymbol{\ell}^2}{m}}\right]. \tag{26}$$

Then the Thouless criterion (19) leads to

$$0 = \frac{1}{\Gamma_0^{(4)}(p=0)} = \frac{m}{4\pi a_s} + \int \frac{d^3\boldsymbol{\ell}}{(2\pi)^3}\left[\frac{1 - 2n_F\left(\frac{\boldsymbol{\ell}^2}{2m} - \mu\right)}{\boldsymbol{\ell}^2/m - 2\mu} - \frac{1}{\boldsymbol{\ell}^2/m}\right], \tag{27}$$

with n_F being the Fermi distribution function. Equation (27) is identical to Eq. (23) in NSR theory. From (27), the chemical potential μ must be positive in the BCS region, $(k_F a_s)^{-1} \lesssim -1$, and negative in the BEC region, $(k_F a_s)^{-1} \gtrsim 1$. There exist low-energy fermions in the BCS side, while all the fermions are gapped due to the binding energy $1/2ma_s^2$ in the BEC side.

Now, the nontrivial part in the derivation of NSR theory comes from the pairing fluctuation in the number equation (24). In Sec. 4.2, we first study the k-dependence of the four-point 1PI vertex function. Then we derive the number equation by analyzing the structure of the flow equation for the self-energy in Sec. 4.3.

4.2. *Structure of the k-dependent four-point* 1*PI vertex*

Since the following analysis depends on the sign of the chemical potential, we consider the BCS side and BEC side separately.

In the BCS side where the chemical potential μ is positive, the condition $\xi(\boldsymbol{q}) = 0$ defines a Fermi sphere $\{|\boldsymbol{q}| = \sqrt{2m\mu}\}$. Using the Thouless criterion (27), Eq. (26) can be rewritten as

$$\frac{-1}{\Gamma_k^{(4)}(p)} = -\int \frac{d^3\boldsymbol{\ell}}{(2\pi)^3}\left[\frac{1 - \sum_{\pm} n_F\left(\xi\left(\frac{\boldsymbol{p}}{2} \pm \boldsymbol{\ell}\right) + R_k^{(f)}\left(\frac{\boldsymbol{p}}{2} \pm \boldsymbol{\ell}\right)\right)}{2\xi(\boldsymbol{\ell}) + \left(ip^0 + \frac{\boldsymbol{p}^2}{4m}\right) + \sum_{\pm} R_k^{(f)}\left(\frac{\boldsymbol{p}}{2} \pm \boldsymbol{\ell}\right)} - \frac{1 - 2n_F(\xi(\boldsymbol{\ell}))}{2\xi(\boldsymbol{\ell})}\right] + R_k^{(b)}(\boldsymbol{p}). \tag{28}$$

Since it suffices to calculate the coefficient of p^0 to determine Z_b, let us put $\boldsymbol{p} = 0$ in (28). In order to find the properties of low-energy collective excitations, we assume that $k^2/2m$ is much smaller than other energy scales such as the temperature and the Fermi energy. Then, Eq. (28) becomes

$$\frac{-1}{\Gamma_k^{(4)}(p^0, \mathbf{0})} = Z_k i p^0 + R_k^{(b)}(\mathbf{0}) + \mathcal{O}(\sqrt{\mu} \cdot k^6/T^3), \tag{29}$$

where Z_k is given by

$$Z_k = \int \frac{d^3\boldsymbol{\ell}}{(2\pi)^3} \frac{\tanh \frac{\beta}{2}(\xi(\boldsymbol{\ell}) + R_k^{(f)}(\boldsymbol{\ell}))}{[2(\xi(\boldsymbol{\ell}) + R_k^{(f)}(\boldsymbol{\ell}))]^2}. \tag{30}$$

Since Z_k converges to a finite value in the limit $k \to 0$, we can set $Z_b = Z_0$ in (22) with

$$Z_0 = \text{p.v.} \int \frac{d^3\boldsymbol{\ell}}{(2\pi)^3} \frac{\tanh\left(\frac{\beta}{2}\xi(\boldsymbol{\ell})\right)}{[2\xi(\boldsymbol{\ell})]^2}. \tag{31}$$

Here, p.v. denotes the principal-value integral. Due to the vertex IR regulator, the collective excitation is also gapped by $k^2/4m$ even when $k^2 \ll 2mT$. The effect of $R_k^{(f)}$ vanishes of the order of k^6 as $k \to 0$.

In the BEC side where the chemical potential μ is negative, $\xi(\boldsymbol{q})$ is always positive, so that $R_k^{(f)} = 0$ for $k < \sqrt{2m|\mu|} \sim 1/a_s$. When k is smaller than $\sqrt{2m|\mu|}$, the inverse propagator of the collective excitation is given by

$$\frac{-1}{\Gamma_k^{(4)}(p)} = R_k^{(b)}(\boldsymbol{p}) - \int \frac{d^3\boldsymbol{\ell}}{(2\pi)^3} \left[\frac{1 - \sum_{\pm} n_F\left(\xi\left(\frac{\boldsymbol{p}}{2} \pm \boldsymbol{\ell}\right)\right)}{2\xi(\boldsymbol{\ell}) + \left(ip^0 + \frac{\boldsymbol{p}^2}{4m}\right)} - \frac{1 - 2n_F(\xi(\boldsymbol{\ell}))}{2\xi(\boldsymbol{\ell})} \right]. \tag{32}$$

Setting $Z_b = Z_0$ with Z_0 given in (31), the bosonic excitation becomes gapped by $k^2/4m$. Since the regulator inside the fermion propagator vanishes for small k, the vertex IR regulator plays a key role in the BEC side.

For small k, the k-dependence of $\Gamma_k^{(4)}$ mainly comes from that of the vertex IR regulator so that we find

$$\partial_k \Gamma_k^{(4)}(p) \simeq \partial_{k_2} \Gamma_{k_1,k_2}^{(4)}(p)|_{k_1=k_2=k} = (\Gamma_k^{(4)}(p))^2 \partial_k R_k^{(b)}(p). \tag{33}$$

The neglected term vanishes in the BEC side and is also smaller by a factor $\mathcal{O}(\sqrt{\mu} k^4/Z_0 T^3)$ even for positive $\mu > 0$ in the BCS side. This approximation will play an important role in Sec. 4.3 to calculate the momentum dependence of the self-energy.

4.3. *Self-energy flow and number density*

According to (15) and (18), the flow equation for the self-energy $\Sigma_k(p) = \Sigma_{k,k}(p)$ is given by

$$\partial_k \Sigma_k(p) = -\int_l^{(T)} e^{-il^0 0^+} \left[\frac{\Gamma_k^{(4)}(p+l)\partial_k R_k^{(f)}(\boldsymbol{l})}{[G^{-1}+R_k^{(f)}]^2(l)} - \frac{(\Gamma_k^{(4)}(p+l))^2 \partial_k R_k^{(b)}(\boldsymbol{p}+\boldsymbol{l})}{[G^{-1}+R_k^{(f)}](l)} \right], \tag{34}$$

where the self-energy corrections in the loop integrals are neglected, while $\Gamma_k^{(4)}$ is given in Sec. 4.1.

Let us first analyze the self-energy correction for large k. Since all kinds of excitations are gapped if k is large, the momentum-dependent part of Σ_k must be small. In the BCS side, there are gapped fermions inside the Fermi sphere even when k is large, so that the self-energy itself can be comparable with the chemical potential: It only shifts the chemical potential and does not affect the low-energy dynamics. As a result, its effect on T_c/ε_F is not significant since T_c and ε_F are shifted in the same way.[14,15] In the BEC side, since the number density becomes exponentially small for large k, so is the self-energy.[16]

When k becomes small, we may use the approximation given in (33), which leads to the flow equation of the self-energy (34):

$$\partial_k \Sigma_k(p) = \int_l^{(T)} e^{-il^0 0^+} \left[-\frac{\Gamma_k^{(4)}(p+l)\partial_k R_k^{(f)}(\boldsymbol{l})}{[G^{-1}+R_k^{(f)}]^2(l)} + \frac{\partial_k \Gamma_k^{(4)}(p+l)}{[G^{-1}+R_k^{(f)}](l)} \right]. \tag{35}$$

Since the right-hand side of (35) is a total derivative in terms of k, the solution is found to be

$$\Sigma_k(p) = \int_l^{(T)} e^{-il^0 0^+} \frac{\Gamma_k^{(4)}(p+l)}{[G^{-1}+R_k^{(f)}](l)}. \tag{36}$$

The number density of fermions n can now be evaluated by using the formula (20) with the self-energy in (36). By taking into account the effect of Σ_0 up to first-order in (20), the number density n is given by

$$\begin{aligned} n &= -2\int_p^{(T)} G(p) - 2\int_p^{(T)} G(p)^2 \Sigma_0(p) \\ &= -2\int_p^{(T)} G(p) + 2\int_{p,l}^{(T)} e^{-il^0 0^+} G(p)^2 G(l-p) \Gamma_0^{(4)}(l) \\ &= -2\int_p^{(T)} G(p) + \int_l^{(T)} e^{-il^0 0^+} \frac{\partial}{\partial \mu} \ln \Gamma_0^{(4)}(l). \end{aligned} \tag{37}$$

To obtain the last line of (37), we used the following equality given by taking the μ-derivative of both sides of (26):

$$2\int_l^{(T)} G(l)^2 G(p-l) = (\Gamma_0^{(4)}(p))^{-2} \frac{\partial}{\partial \mu} \Gamma_0^{(4)}(p). \tag{38}$$

Equation (37) is nothing but the number equation of the NSR theory, (24), since pp-RPA gives

$$\frac{1}{\Gamma_0^{(4)}(p)} = \frac{1}{g} + \Pi(p) = \frac{m}{4\pi a_s} + \left(\Pi(p) - \frac{m\Lambda}{2\pi^2}\right). \tag{39}$$

Thus we could derive the NSR theory from the MR-FRG flow equations: The Thouless criterion (27) and the number equation (37) give the BCS gap equation (23) at $T = T_c$ and the number equation (24) of the NSR theory, respectively. Accordingly, Eqs. (15)–(18) turn out to be a minimal setup of MR-FRG to describe the entire region of the BCS–BEC crossover.

4.4. *Number equation from the free energy*

The same expression as (37) for the number density can be also derived from the free energy evaluated in MR-FRG:

$$n = \frac{-1}{\beta V}\frac{\partial}{\partial\mu}\Gamma_0(\beta,\mu), \tag{40}$$

with V being the volume of the system, and $\Gamma_0(\beta,\mu)/\beta$ being the free energy obtained as the field-independent part of the 1PI effective action $\Gamma_0[\overline{\psi},\psi]$.

Equation (10) and the flow equation in Fig. 3 with the vertex IR regulator give

$$\partial_k\Gamma_k(\beta,\mu)/\beta V = \int_l^{(T)} \frac{-2\partial_k R_k^{(f)}(\boldsymbol{l})}{[G^{-1} - \Sigma_k + R_k^{(f)}](l)} - \int_p^{(T)} \Gamma_k^{(4)}(p)\partial_k R_k^{(b)}(\boldsymbol{p}). \tag{41}$$

By taking into account the effect of the self-energy correction in the fermion-loop up to the first-order, the flow equation (41) becomes

$$\partial_k\Gamma_k(\beta,\mu)/\beta V = \int_l^{(T)} \frac{-2\partial_k R_k^{(f)}(\boldsymbol{l})}{[G^{-1} + R_k^{(f)}](l)} + \int_l^{(T)} \frac{-2\Sigma_k(l)\partial_k R_k^{(f)}(\boldsymbol{l})}{[G^{-1} + R_k^{(f)}]^2(l)}$$
$$- \int_p^{(T)} \Gamma_k^{(4)}(p)\partial_k R_k^{(b)}(\boldsymbol{p}). \tag{42}$$

Substitution of the explicit form (36) into (42) gives

$$\partial_k\Gamma_k(\beta,\mu)/\beta V = \int_l^{(T)} \frac{-2\partial_k R_k^{(f)}(\boldsymbol{l})}{[G^{-1} + R_k^{(f)}](l)}$$
$$- \int_p^{(T)} \Gamma_k^{(4)}(p)\left[\int_l^{(T)} \frac{2\partial_k R_k^{(f)}\left(\frac{\boldsymbol{p}}{2} - \boldsymbol{l}\right)}{\prod_{\pm}\left([G^{-1} + R_k^{(f)}]\left(\frac{p}{2} \pm l\right)\right)} + \partial_k R_k^{(b)}(\boldsymbol{p})\right]. \tag{43}$$

Combining (25) and (43), we obtain

$$\partial_k \Gamma_k(\beta,\mu)/\beta V = -2\int_l^{(T)} \partial_k \ln[G^{-1}(l) + R_k^{(f)}(\boldsymbol{l})] - \int_p^{(T)} \partial_k \ln \Gamma_k^{(4)}(p). \quad (44)$$

Since both sides of (44) are total derivatives in terms of k, we obtain the following expression for $\Gamma_k(\beta,\mu)$ up to a μ-independent constant:

$$\Gamma_k(\beta,\mu)/\beta V = -2\int_l^{(T)} \ln[G^{-1}(l) + R_k^{(f)}(\boldsymbol{l})] + \int_p^{(T)} \ln\left[\frac{4\pi a_s}{m}(\Gamma_k^{(4)}(p))^{-1}\right]. \quad (45)$$

By combining (40) and (45), we obtain the same expression as (37).

5. Summary and Perspectives

In this paper, we proposed a new formulation of fermionic FRG with multiple infrared regulators, MR-FRG, and applied the method to the two-component fermionic system which exhibits the BCS–BEC crossover. To control both the fermionic one-particle excitations and the bosonic collective excitations, two IR regulators were introduced; one for the fermionic two-point function and another for the four-fermion vertex. The NSR theory, which is successful to capture the qualitative features of the BCS–BEC crossover, can be derived from MR-FRG.

There are various possibilities to improve the approximations adopted in MR-FRG to go beyond the NSR theory. Solving the coupled flow equations for the four-point vertex and the self-energy is necessary when the magnitude of the self-energy becomes comparable with that of the chemical potential. The frequency dependence of the four-point function and the higher-point vertex functions would play significant roles in such a situation.[14,15] (Importance of the frequency dependences is also suggested from the study of unitary Fermi gas in the spin-imbalanced case.[23])

In the BCS regime ($(k_F a_s)^{-1} \ll -1$), each closed fermion loop for Feynman diagram in the flow equation produces a factor $n_k \sim k_F^3/3\pi^2$, so that the number of closed loops is a good expansion parameter. This is similar to the hole-line expansion of the Bethe–Brueckner–Goldstone (BBG) theory[24] for degenerate fermion system: A formal connection between fermionic FRG and the BBG theory will be reported elsewhere. In the BCS regime, the particle–hole loops in the flow of the four-point coupling are also important to generate the Gorkov and Melik-Barkhudarov correction which reduces the critical temperature by a factor 2.2 from the BCS critical temperature.[13–15] The six- or higher-point vertices must be also important to consider the dynamics of collective bosonic excitations in the BCS regime.

In the BEC regime ($(k_F a_s)^{-1} \gg +1$), the real part of the fermion propagator is always positive and there are no hole excitations. Nevertheless, the interaction between composite bosons takes place through the fermion closed loops, so that the critical temperature and the number density are affected through the self-energy and the four-point vertex function.

Since the present formulation provides a good starting point to make systematic analysis of the interacting fermion systems, we expect it to be useful not only for cold atomic systems but also for strongly interacting nuclear/neutron matter and for quark matter.

Acknowledgments

The authors are grateful for useful comments by Gergely Fejős. T. H. thanks late Gerry Brown for his stimulating discussions on various aspects of quantum many-body problems in nuclear, hadron, and particle physics. Y. T. was supported by JSPS Research Fellowships for Young Scientists. This work was partially supported by RIKEN iTHES project and by the Program for Leading Graduate Schools, MEXT, Japan. This work was completed at the Aspen Center for Physics, which is supported by National Science Foundation Grant PHY-1066293.

References

1. C. Wetterich, *Phys. Lett. B* **301** (1993) 90.
2. T. R. Morris, *Int. J. Mod. Phys. A* **9** (1994) 2411.
3. U. Ellwanger, *Z. Phys. C, Part. Fields* **62** (1994) 503.
4. K. G. Wilson and J. Kogut, *Phys. Rep.* **12** (1974) 75.
5. R. Shankar, *Rev. Mod. Phys.* **66** (1994) 129.
6. M. Salmhofer and C. Honerkamp, *Prog. Theor. Phys.* **105** (2001) 1.
7. W. Metzner, M. Salmhofer, C. Honerkamp, V. Meden and K. Schönhammer, *Rev. Mod. Phys.* **84** (2012) 299.
8. M. C. Birse, B. Krippa, J. A. McGovern and N. R. Walet, *Phys. Lett. B* **605** (2005) 287.
9. S. Floerchinger, M. Scherer, S. Diehl and C. Wetterich, *Phys. Rev. B* **78** (2008) 174528.
10. S. Floerchinger, M. M. Scherer and C. Wetterich, *Phys. Rev. A* **81** (2010) 063619.
11. D. M. Eagles, *Phys. Rev.* **186** (1969) 456.
12. A. Leggett, Diatomic molecules and Cooper pairs, in *Modern Trends in the Theory of Condensed Matter*, Lecture Notes in Physics, Vol. 115 (Springer, Berlin, 1980), pp. 13–27.
13. L. Gorkov and T. Melik-Barkhudarov, *Sov. Phys.-JETP* **13** (1961) 1018.
14. Y. Tanizaki, G. Fejős and T. Hatsuda, *Prog. Theor. Exp. Phys.* **2014** (2014) 043I01.
15. Y. Tanizaki, G. Fejős and T. Hatsuda, *J. Phys., Conf. Ser.* **529** (2014) 012027.
16. Y. Tanizaki, *Prog. Theor. Exp. Phys.* **2014** (2014) 023A04.
17. P. Nozières and S. Schmitt-Rink, *J. Low Temp. Phys.* **59** (1985) 195.
18. D. Thouless, *Ann. Phys.* **10** (1960) 553.
19. D. Litim, *Phys. Lett. B* **486** (2000) 92.
20. D. Litim, *Phys. Rev. D* **64** (2001) 105007.
21. J. M. Pawlowski, *Ann. Phys.* **322** (2007) 2831.
22. P. Stevenson, *Phys. Rev. D* **23** (1981) 2916.
23. R. Schmidt and T. Enss, *Phys. Rev. A* **83** (2011) 063620.
24. B. D. Day, *Rev. Mod. Phys.* **39** (1967) 719.

Spectral functions and in-medium properties of hadrons

R.-A. Tripolt
European Centre for Theoretical Studies in Nuclear Physics and Related Areas (ECT), Strada delle Tabarelle 286, 38123 Villazzano (TN), Italy*
tripolt@ectstar.eu

L. von Smekal
Institut für theoretische Physik, Justus-Liebig-Universität Giessen, Heinrich-Buff-Ring 16, 35392 Giessen, Germany
lorenz.smekal@theo.physik.uni-giessen.de

J. Wambach
European Centre for Theoretical Studies in Nuclear Physics and Related Areas (ECT), Strada delle Tabarelle 286, 38123 Villazzano (TN), Italy*
jwambach@ectstar.eu

The in-medium modifications of hadron properties such as masses and decay widths have been a major focus of the scientific work of Gerry Brown and the insights gained by him and his collaborators made them major drivers of this field for several decades. Their prediction of experimental signals in di-lepton pair production in relativistic heavy-ion collisions was instrumental in initiating large experimental campaigns which continue till today. In this chapter, we review recent results which elucidate the relation of hadronic spectral properties at finite temperature and density to the restoration of spontaneously broken chiral symmetry.

Keywords: Spectral functions; functional renormalization group; analytic continuation; QCD phase diagram; chiral phase transition.

1. Introduction

The spectral properties of hadrons for matter under extreme conditions in temperature and density (QCD matter, i.e. matter that is governed by Quantum Chromodynamics) as encountered in the early universe and the core of neutron stars are of fundamental importance for the understanding of the relevant degrees of freedom in the equation of state and the transport properties. It was realized early on by Gerry and his collaborators that spontaneously broken chiral symmetry and its restoration play an important role in this context. Spontaneously broken chiral symmetry endows the up and down quarks with almost all of their mass as confirmed by QCD lattice studies of the quark propagator as well as by Dyson–Schwinger results.[1,2] It seems therefore rather natural to conclude that all hadrons composed of light quarks (except for the pion as a Goldstone boson) essentially loose most of their mass as chiral symmetry gets gradually restored. This has led to the conjecture of a scaling law of hadron masses with the in-medium chiral condensate.[3,4] The discussions are far from over and they are at the very heart of the question of mass generation in QCD. In this context, chiral effective quark Lagrangians have played an important role, as they are able to describe the quark mass generation through the chiral condensate. At the same time, hadrons (mesons and baryons) emerge as dressed quark–antiquark or three-quark composites.

In the vacuum-spectra of hadrons containing light quarks, broken chiral symmetry manifests itself through the absence of doublets of opposite parity. As chiral symmetry gets restored with increasing temperature and (net) quark density, the parity partners approach degeneracy and their spectral properties eventually become identical. For vector and axial-vector mesons, this is encoded in the Weinberg sum rules[5,6] for instance. How this degeneracy happens and whether it is accompanied by significant drops in (pole) masses is the subject of the present study. It will be found that this is a rather subtle issue, which we will comment on further in the end. We employ the Quark-Meson (QM) model (a variant of the Gell–Mann–Levy linear sigma model), which shares the most important chiral aspects with QCD. The aim is to compute the chiral restoration transition and the resulting medium modification of pions and their chiral partner (the sigma meson) as a function of temperature T and quark chemical potential μ in a thermodynamically consistent manner.

As the chiral transition may become a true phase transition for specific values of T and μ, a method is needed that is able to deal with phase transitions in a reliable manner. Such a method is the functional renormalization group (FRG), which is widely used in quantum field theory and statistical physics.[7–14] It goes beyond mean-field theory (MFT) by incorporating thermal and quantum fluctuations through scale-dependent evolution equations for the effective action of a given theory (in fact MFT is the starting point at the ultraviolet scale from which the evolution is initiated). In MFT, the thermodynamically consistent excitation spectrum is obtained from the 'random-phase-approximation (RPA)' (an infinite sum of one-loop diagrams) first introduced for the electron gas by Bohm and Pines in 1953.[15] Gerry together with Bolsterli has made extensive use of the RPA in the context of nuclear collective excitations with devising an analytic model (the 'schematic model'[16]), which has provided valuable insight into the role of residual nuclear interactions in the collectivity of giant resonances and low-lying vibrational modes of spherical nuclei.

Although MFT often captures the gross features of the equilibrium properties of a given system, quantitative predictions and correct descriptions of phase transitions and critical phenomena require the proper inclusion of thermal and quantum fluctuations, in particular those of the order parameters due to collective excitations.

However, a thermodynamically consistent calculation of spectral properties beyond MFT poses an inherently difficult problem as it involves real-time correlation functions. Within the FRG, the thermal equilibrium state is usually obtained in an Euclidean approach, making use of the Matsubara formalism. To compute correlation functions in Minkowski space-time, an analytic continuation is required. The analytic continuation problem is common to all Euclidean approaches to quantum field theory and describes the need to analytically continue from imaginary to real energies. This technical difficulty, for example, also arises in lattice QCD, where one has to use numerical methods to reconstruct real-time correlation functions from the Euclidean data, cf. Refs. 17–20. The method[21–24] applied in the following avoids this exponentially hard inverse problem as it deals with the analytic continuation on the level of the FRG flow equations, which can then be solved for real frequencies without the need for any numerical reconstruction.

2. The Functional Renormalization Group

The FRG represents a powerful nonperturbative continuum framework to study quantum field theories as statistical systems near criticality in Euclidean space-time. The basic idea is to implement Wilson's coarse-graining procedure[25,26] within a functional formulation. One usually starts with an ansatz for the classical or mean-field action S to describe the microphysics at some high energy–momentum scale and then successively lowers this scale to include fluctuations with lower and lower momenta in order to arrive at the full quantum effective action Γ. The connection between the ultraviolet (UV) and the infrared (IR) regimes is established by an FRG flow equation for a scale-dependent effective action, the so-called effective average action.

We briefly recall that the effective action $\Gamma[\phi]$ is the generating functional for the one-particle-irreducible (1PI) correlation functions of a theory. It is obtained from the generating functional $Z[J]$ for the full Green functions, which we formally write as an Euclidean path integral over some generic field φ,

$$Z[J] = \int \mathcal{D}\varphi \exp\Big(-S[\varphi] + \int d^4x J(x)\varphi(x)\Big), \tag{1}$$

where $S[\varphi]$ is the classical action. The generating functional for the connected n-point Green functions is then given by

$$W[J] = \log Z[J] \tag{2}$$

and the effective action $\Gamma[\phi]$ is obtained as the Legendre transform of $W[J]$ with respect to the expectation value $\phi(x) = \langle\varphi(x)\rangle_J$ of the field $\varphi(x)$ in presence of the source $J(x)$,

$$\Gamma[\phi] = \sup_J \Big(\int d^4x J(x)\phi(x) - W[J]\Big). \tag{3}$$

When evaluated for a homogeneous equilibrium state with (spatially) constant $\phi = \phi_0$ which satisfies

$$\left.\frac{\delta\Gamma[\phi]}{\delta\phi}\right|_{\phi=\phi_0} = 0, \tag{4}$$

the effective action is related to the Grand Canonical potential by

$$\Omega(T,\mu) = \frac{T}{V}\Gamma[\phi_0]. \tag{5}$$

Wilson's coarse-graining idea is implemented by introducing a regulator function R_k, which suppresses fluctuations of modes with momenta lower than the RG scale k. Successively lowering this scale from the UV to the IR, thermal and quantum fluctuations are taken into account momentum shell by momentum shell. The scale-dependent analogues of the generating functionals $W[J]$ and $\Gamma[\phi]$ are then given by

$$W_k[J] = \log \int \mathcal{D}\varphi \exp\left(-S[\varphi] - \Delta S_k[\varphi] + \int d^4x J(x)\varphi(x)\right), \tag{6}$$

$$\Gamma_k[\phi] = \sup_J \left(\int d^4x J(x)\phi(x) - W_k[J]\right) - \Delta S_k[\phi], \tag{7}$$

where the regulator insertion $\Delta S_k[\phi]$ is given by

$$\Delta S_k[\phi] = \frac{1}{2}\int \frac{d^4q}{(2\pi)^4}\phi(-q)R_k(q)\phi(q), \tag{8}$$

which acts as a scale-dependent mass term. The effective average action Γ_k interpolates between the classical action S at the UV scale Λ and the effective action $\Gamma[\phi]$ at $k = 0$, i.e.,

$$\Gamma_\Lambda \approx S, \quad \lim_{k\to 0}\Gamma_k = \Gamma. \tag{9}$$

The scale dependence of Γ_k is given by the following exact flow equation, also known as the 'Wetterich equation'[27,28]:

$$\partial_k\Gamma_k[\phi] = \frac{1}{2}\mathrm{Tr}\{\partial_k R_k(\Gamma_k^{(2)}[\phi] + R_k)^{-1}\}, \tag{10}$$

where the 2-point functions $\Gamma_k^{(2)}[\phi]$ represents the second functional derivative with respect to the field ϕ, and the trace includes integration over the loop momentum as well as summation over all internal indices. The Wetterich equation has the simple diagrammatic representation as shown in Fig. 1.

Despite its simple one-loop structure, due to the full field dependence in the 2-point functions this functional integro–differential equation can usually not be solved without truncations. In order to describe phase transitions and critical phenomena, it has been proven to be successful to use the so-called 'derivative

$$\partial_k \Gamma_k = \frac{1}{2}$$

Fig. 1. Diagrammatic representation of the flow equation for the effective average action, Eq. (10). The dashed line represents the full field and scale-dependent propagator and the circle is the regulator insertion $\partial_k R_k$.

expansion' as an ansatz for Γ_k, i.e.,

$$\Gamma_k[\phi] = \int d^4x \left\{ U_k(\phi) + \frac{1}{2} Z_k(\phi)(\partial_\mu \phi)^2 + \frac{1}{8} Y_k(\phi)(\partial_\mu \phi^2)^2 + \mathcal{O}(\partial^4) \right\}, \tag{11}$$

where $U_k(\phi)$, $Z_k(\phi)$ and $Y_k(\phi)$ are the field and scale-dependent potential and derivative terms.[7,29] The latter includes wave-function renormalization factors. In many cases, already the leading order of this derivative expansion, at which $Z_k(\phi) = 1$ and $Y_k(\phi) = 0$, can provide a reasonably good quantitative description of thermodynamics and critical phenomena, see, e.g., Refs. 30 and 31. This truncation is also known as the 'local potential approximation' (LPA) and will be employed in the following.

3. Flow Equations for the QM Model

As mentioned in the introduction, the aim of the present study is to elucidate the role of broken chiral symmetry and it restoration at high temperature and density in the spectral properties of light hadrons. As this remains very difficult in QCD proper, we resort to the QM model which represents a low-energy effective theory of QCD and is based on chiral symmetry.[32,33] The elementary degrees of freedom are given by the two lightest quarks, i.e., the up and down quark, as well as the three pions and the sigma meson. The QM model is renormalizable and falls into the same $O(4)$ universality class as expected for QCD with two flavors.

3.1. *Flow of the effective action*

Applying the LPA yields the following ansatz for the scale-dependent effective action of the QM model,

$$\Gamma_k = \int d^4x \left\{ \bar{\psi}(\gamma_\mu \partial^\mu + h(\sigma + i\boldsymbol{\tau}\boldsymbol{\pi}\gamma^5) - \mu\gamma_0)\psi + \frac{1}{2}(\partial_\mu \phi)^2 + U_k(\phi^2) - c\sigma \right\}. \tag{12}$$

where the field $\phi = (\sigma, \boldsymbol{\pi})$ combines the mesonic fields, h is the Yukawa coupling, μ the quark chemical potential and $-c\sigma$ an explicit chiral symmetry breaking term. The mesonic potential $U_k(\phi^2)$ as a function of the $O(4)$ invariant $\phi^2 = \sigma^2 + \boldsymbol{\pi}^2$ can in principle assume an arbitrary form and encodes the mesonic interactions. Inserting this ansatz into the Wetterich equation gives the flow equation for Γ_k, depicted diagrammatically in Fig. 2. It immediately reduces to a flow equation for the effective potential, because U_k is the only scale-dependent quantity within the LPA.

Fig. 2. Diagrammatic representation of the flow equation for the effective average action of the QM model. The dashed (solid) line represents the full scale-dependent mesonic (quark) propagator and the circles the regulator insertions $\partial_k R_k$.

In order to solve the flow equation for Γ_k, we need to specify the shape of U_k at the UV scale $k = \Lambda$. We choose a UV potential with a chirally symmetric minimum at $\phi_0^2 = 0$ of the form

$$U_\Lambda(\phi^2) = \frac{1}{2} m_\Lambda^2 \phi^2 + \frac{1}{4} \lambda_\Lambda (\phi^2)^2. \tag{13}$$

The explicit values for the parameters are listed in Table 1. The fermionic minus sign in the flow equation for the effective potential drives the minimum away from zero analogous to a negative index of refraction in a self-focusing nonlinear medium and leads to the physical parameters listed in the table in the IR.

Moreover, we have to choose suitable bosonic and fermionic regulator functions R_k.[34] In the following, we adopt three-dimensional regulator functions that only regulate the spatial momenta but leave the energy component unaffected, cf. Ref. 35 for an alternative approach involving 4D regulators. The choice of 3D regulators will allow us to perform the analytic continuation from imaginary to real energies on the level of the flow equations. For explicit expression of the regulator functions and the flow equations we refer to Refs. 23 and 24.

The flow equation for $U_k(\phi^2)$ is solved numerically on an one-dimensional grid in field space, i.e., the field ϕ^2 is discretized in the direction of the σ-field and the flow equation, which turns into a set of coupled ordinary differential equations, is solved on each grid point ϕ_i^2. For numerical reasons, these flow equations are not integrated down to $k = 0$ but only to $k_{\mathrm{IR}} = 40\,\mathrm{MeV}$, which does, however, not affect our results since the minimum of the potential as well as the values of the particle masses are basically fixed as soon as the lightest degree of freedom decouples from the flow.

In the IR, the explicit symmetry breaking term $c\sigma$ is subtracted from the effective potential and the resulting global minimum defines the pion decay constant as well as the curvature masses of the particles, cf. Table 1. Moreover, the IR potential can be identified with the Grand Canonical potential $\Omega(T, \mu)$ (Eq. (5)), which allows

Table 1. Parameter set for the QM model and obtained vacuum values for the particle masses and the pion decay constant in the IR. The UV scale is chosen to be $\Lambda = 1\,\mathrm{GeV}$ and the IR scale is $k_{\mathrm{IR}} = 40\,\mathrm{MeV}$.

m_Λ/Λ	λ_Λ	c/Λ^3	h	$\sigma_0 \equiv f_\pi$	m_π	m_σ	m_ψ
0.794	2.00	0.00175	3.2	93.5 MeV	138 MeV	509 MeV	299 MeV

access to the various thermodynamic quantities such as pressure, entropy density and quark number density,

$$p(T,\mu) = -\Omega(T,\mu) + \Omega(0,0), \tag{14}$$

$$s(T,\mu) = \frac{\partial p(T,\mu)}{\partial T}, \tag{15}$$

$$n_\psi(T,\mu) = \frac{\partial p(T,\mu)}{\partial \mu}. \tag{16}$$

3.2. *Flow equations for spectral functions*

We now turn to the calculation of spectral functions, which are derived from the retarded 2-point functions $\Gamma^{(2),R}$. The corresponding flow equations are obtained by taking two functional derivatives of the flow equation for the effective action, cf. Fig. 2, with respect to the appropriate fields. The structure of the resulting equations for the sigma, pion and quark 2-point functions is displayed in Fig. 3. The mesonic three- and four-point vertices are generated from derivatives of the scale-dependent effective potential U_k and the QM vertices are basically given by the Yukawa coupling. We note that this truncation is thermodynamically consistent in the sense that the (Euclidean) 2-point functions, evaluated at vanishing external momenta, agree with the curvature masses as extracted from the scale-dependent effective potential.[23,36] The truncation is also symmetry conserving in that the pion is massless in the chiral limit in the spontaneously broken phase as required by the Goldstone theorem.

To begin with, we are working in Euclidean space-time involving imaginary Matsubara frequencies. In order to be able to calculate real-time 2-point functions and spectral functions, we have to perform an analytic continuation from imaginary to real energies. This analytic continuation is performed on the level of the flow equations and is achieved by the following two-step procedure. First, the periodicity of the bosonic and fermionic occupation numbers with respect to the Euclidean

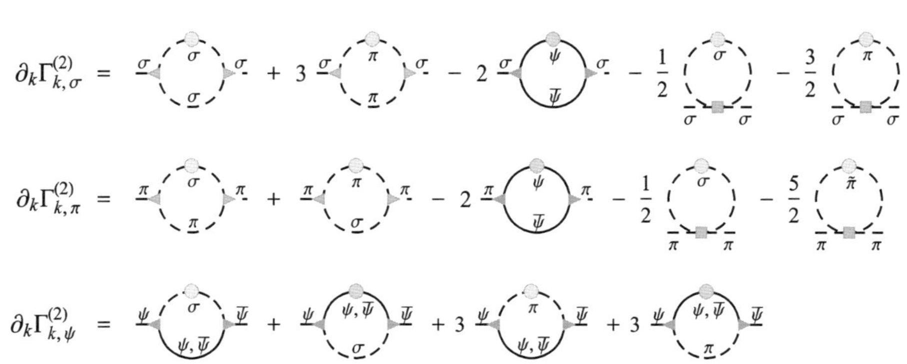

Fig. 3. (Color online) Diagrammatic representation of the flow equations for the sigma, pion and quark 2-point functions. Blue (red) triangles indicate the mesonic (QM) three-point vertices, blue squares denote mesonic four-point vertices.

Matsubara frequencies $p_0 = i\,2\pi nT$ is exploited,

$$n_{B,F}(E + ip_0) \to n_{B,F}(E). \tag{17}$$

In a second step, p_0 is replaced by a continuous real energy ω,

$$\partial_k \Gamma_k^{(2),R}(\omega, \mathbf{p}) = -\lim_{\epsilon \to 0} \partial_k \Gamma_k^{(2),E}(p_0 = -\mathrm{i}(\omega + \mathrm{i}\epsilon), \mathbf{p}), \tag{18}$$

where the limit $\epsilon \to 0$ can be taken analytically for the imaginary part of the 2-point functions. This analytic continuation yields the physical Baym–Mermin boundary conditions.[37,38] The resulting flow equations for the real and the imaginary part of the retarded 2-point functions are then solved numerically and the spectral functions are obtained as

$$\rho(\omega, \mathbf{p}) = -\frac{1}{\pi} \mathrm{Im} \frac{1}{\Gamma^{(2),R}(\omega, \mathbf{p})}. \tag{19}$$

4. Vacuum Potential and Spectral Functions

To elucidate the relevant physical effects generated by the flow with resolution scale k, we first present results for the flow of the effective potential U_k and the pion- and sigma spectral functions in the vacuum, see also Refs. 23 and 36. As shown in Fig. 4, the UV shape of the effective potential is essentially symmetric, up to a small effect due to the explicit symmetry breaking term $c\sigma$. When solving the corresponding flow equation, cf. Fig. 2, with decreasing k, the quark fluctuations push the minimum of the effective potential to larger values of the sigma field, thus establishing the well-known Mexican-hat shape of the chirally broken potential.

In Fig. 4, we also show the vacuum sigma and pion spectral functions at the UV and the IR scale. As a result of the quasi-symmetric UV potential, the UV spectral functions are almost degenerate with masses of the order of 800 MeV. Both the sigma meson and the pion are stable in the UV since all available decay channels are essentially boosted to very high energies due to the large scale k which can be loosely interpreted as the three-momentum $|\mathbf{p}|$. In the IR, however, the sigma meson becomes unstable with a large width, since it can decay into two pions. The pion

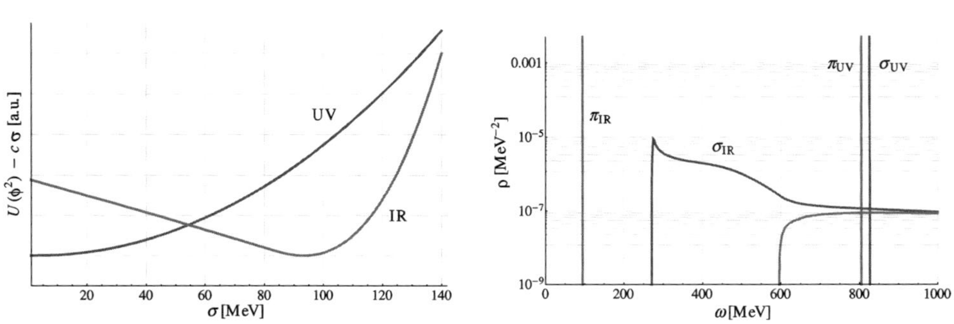

Fig. 4. Left: The effective potential $U_k(\phi^2) - c\sigma_k$ is shown in the UV (k_Λ) and the IR ($k = 0$). Right: The sigma- and pion spectral functions are shown in the UV and the IR.

as a Goldstone particle remains stable. These results nicely illustrate the transition from a linear realization of chiral symmetry to a nonlinear one through the RG-flow.

5. Phase Diagram and Thermodynamics

In this section, we present results for the phase diagram of the QM model and other thermodynamical quantities. The phase diagram, as obtained from the location of the global minimum of the effective potential, $\sigma_{k=0}(\mu, T)$, is shown in Fig. 5.

It exhibits a chiral crossover transition, which is located at $T \approx 170\,\text{MeV}$ for $\mu = 0$, a chiral critical endpoint (CEP) at $T_c \approx 9\,\text{MeV}$ and $\mu \approx 292\,\text{MeV}$ and a first-order phase transition for $T < T_c$.

To study the medium effects, we will focus on two regions in the phase diagram — the temperature axis at $\mu = 0$ and a horizontal line at $T \approx 9\,\text{MeV}$ near the CEP. In Fig. 6, we show the behavior of the chiral order parameter and the curvature (screening) masses within these two regions of the phase diagram. Near the crossover transition, the sigma mass decreases and then becomes degenerate with the pion mass as chiral symmetry gets restored at high temperatures. At the same time, the quark mass and the chiral order parameter (identified with the weak pion decay constant) decrease. Near the CEP, which represents a second-order phase transition, the sigma meson is expected to become massless, which is reproduced by our numerical calculations to a very good degree. At higher chemical potentials, we again observe a progressing degeneration of the sigma meson and the pion.

For later discussions, we show in Fig. 7 the entropy density of the mesons and of the complete QM system. The entropy density of the pions agrees well with the chiral perturbation theory (χPT) result obtained in Ref. 39 at low temperatures. The total entropy density is dominated by the quarks and approaches the corresponding

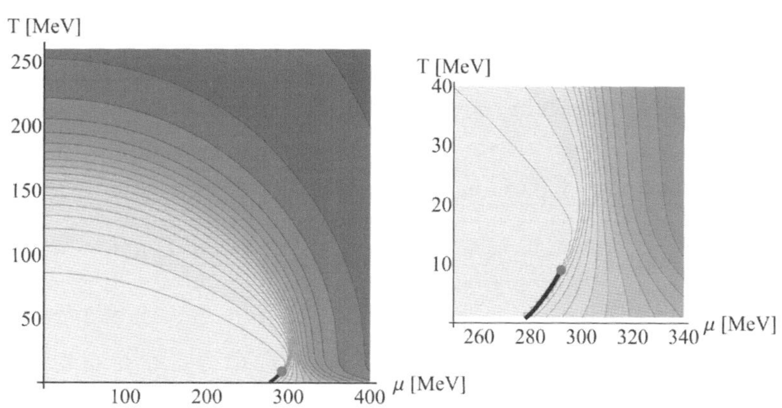

Fig. 5. (Color online) The phase diagram of the QM model is shown as a contour plot of the IR chiral order parameter $\sigma_{k=0}$, which decreases towards higher temperatures and larger quark chemical potentials, cf. Ref. 23. The red dot represents the CEP and the black line the first-order phase transition. Right: Enlargement of the regime around the CEP.

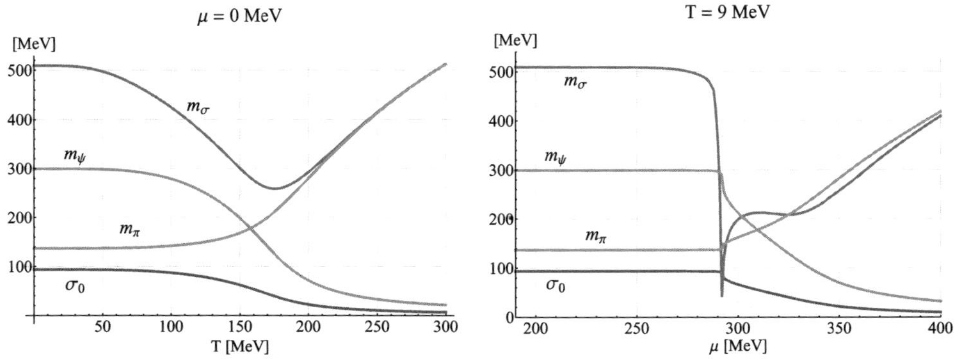

Fig. 6. The IR curvature (screening) masses are shown together with the chiral order parameter σ_0 versus temperature at $\mu = 0$ (left) and versus chemical potential at $T = 9\,\mathrm{MeV}$ (right), cf. Ref. 23.

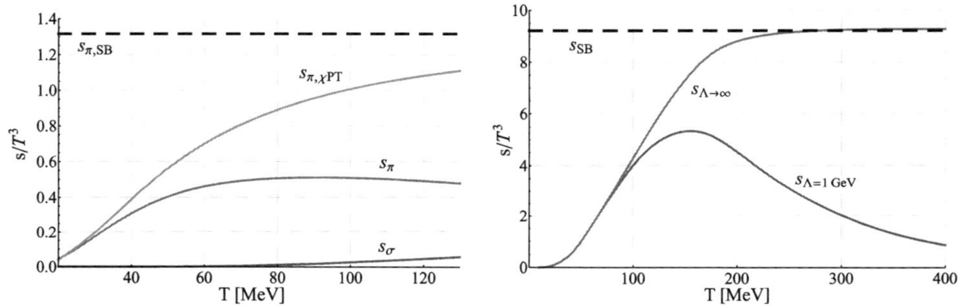

Fig. 7. The entropy density is shown for the pions and the sigma mesons versus temperature at $\mu = 0$ in comparison to the result from χPT, [see Ref. 39] and the SB value for three massless noninteracting pions of $s_{\pi,\mathrm{SB}}/T^3 = 2\pi^2/15$ (dashed line). Right: The total entropy density of the QM model is shown versus temperature at $\mu = 0$ without UV corrections, $s_{\Lambda=1\,\mathrm{GeV}}$, and with UV corrections, $s_{\Lambda\to\infty}$, in comparison to the corresponding SB value of $s_{\mathrm{SB}}/T^3 = 14\pi^2/15$ (dashed line).

Stefan–Boltzmann (SB) value at high temperatures, once effects from scales beyond the UV cutoff Λ are taken into account, see Refs. 40–43 for details.

6. In-Medium Spectral Functions

In this section, we present results for the mesonic spectral functions of the QM model. The structure of these spectral functions is determined by the different physical processes that can occur within our truncation, cf. Fig. 8. We divide these processes into two groups: time-like processes, where the energy ω of the external particle is larger than its spatial momentum $|\mathbf{p}|$, and space-like processes with $\omega < |\mathbf{p}|$. The kinematic thresholds of these processes are determined by energy- and momentum-conservation, see also Refs. 24 and 43.

In Fig. 9, we show the spectral functions for pions and sigma mesons which are at rest relative to the heat bath, i.e., $|\mathbf{p}| = 0$, for different combinations of

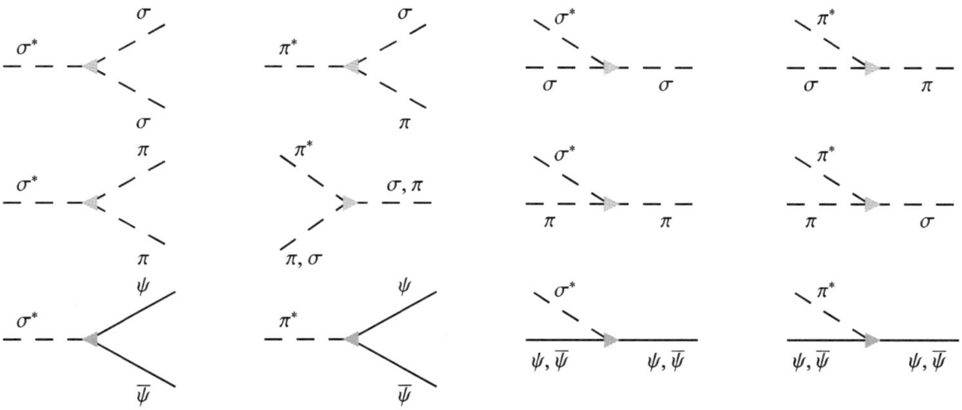

Fig. 8. Graphical representation of the possible time-like, $p^2 = \omega^2 - \mathbf{p}^2 > 0$, and space-like, $p^2 < 0$, processes for an off-shell sigma, σ^*, or pion, π^*. First column: time-like decay channels for a sigma meson. Second column: time-like decay channels for a pion. Third column: absorption or emission of a space-like sigma excitation. Fourth column: absorption or emission of a space-like pion excitation. The figure is taken from Ref. 24.

temperature and chemical potential. The left panel shows the evolution along the temperature axis in the phase diagram and the right panel the evolution along the $T = 9\,\text{MeV}$-line towards the CEP.

As shown in Sec. 4, the stable vacuum pion is represented by a Dirac δ-function whose location defines the pole mass. The sigma meson in the vacuum is unstable since it can decay into two pions, $\sigma^* \to \pi+\pi$. At higher energies, other processes such as the decay into a quark–antiquark pair give rise to modifications of the spectral functions. At $T = 110\,\text{MeV}$, the pion has become unstable since it can capture a pion from the heat bath and turn into a sigma meson, $\pi^*+\pi \to \sigma$. At $T = 150\,\text{MeV}$, i.e., near the crossover transition, also the sigma meson becomes stable since it can no longer decay into two pions. At higher temperatures beyond the crossover, the mesons become unstable again, since they can decay into light quark–antiquark pairs, and the spectral functions eventually become identical. It should be noted that complete degeneracy is only reached at temperatures significantly higher than the (pseudo) critical temperature derived from the evolution of the chiral condensate. This implies that chiral correlations persist well into the restored phase, as can also be inferred from the corresponding evolution of the curvature masses in Fig. 6.

When looking at the dependence of the spectral functions on the chemical potential at $T = 9\,\text{MeV}$, we observe that they remain unchanged over a wide range of μ, as expected from the Silver Blaze property.[44] Near the CEP, however, the sigma spectral function undergoes significant changes, while the pion spectral function remains almost unchanged. In particular, we observe that the sigma meson becomes stable and almost massless near the CEP as it should be.

We finally turn to the full energy- and momentum-dependence of the spectral functions, which contains information on all physical processes that can happen in

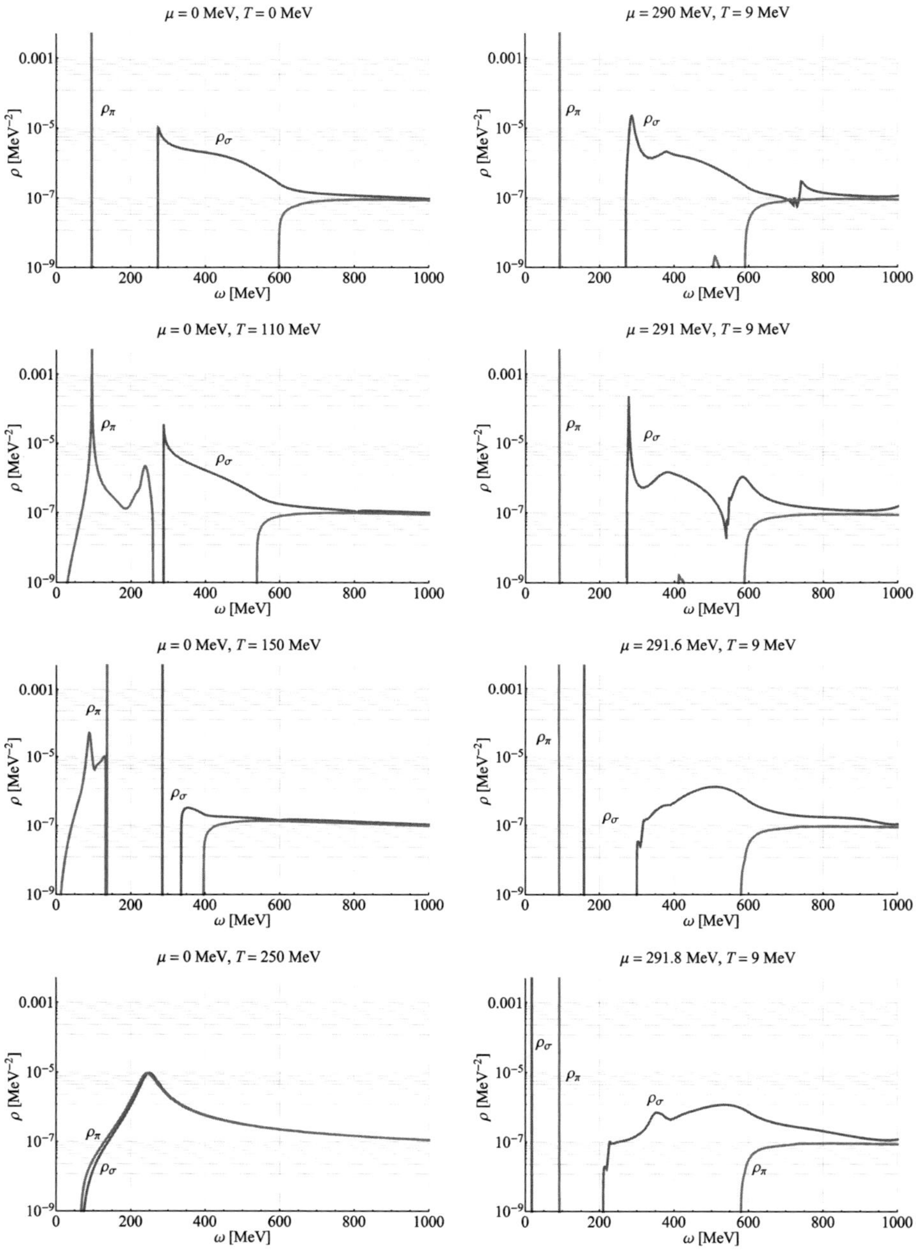

Fig. 9. The in-medium sigma and pion spectral functions, ρ_σ and ρ_π, are shown versus energy ω at $|\mathbf{p}| = 0$ for different temperatures and chemical potentials. Left panel: $\mu = 0$ and T increases from top to bottom. Right panel: $T = 9\,\text{MeV}$ and μ increases from top to bottom, thus approaching the CEP.

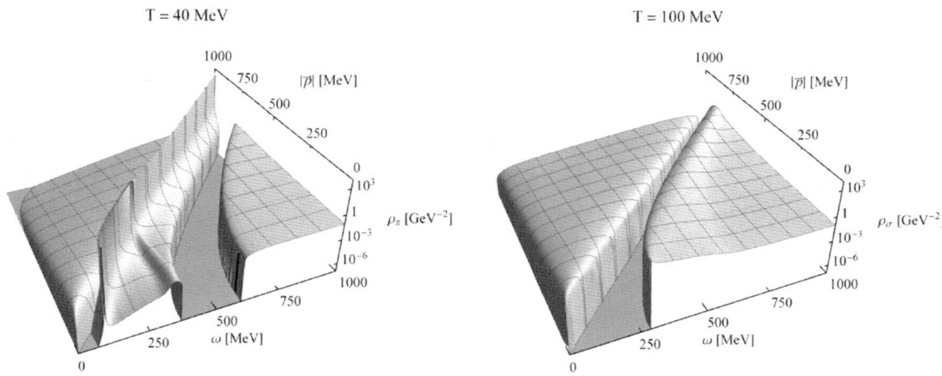

Fig. 10. Left: The pion spectral function is shown versus energy ω and spatial momentum $|\mathbf{p}|$ at $T = 40\,\text{MeV}$ and $\mu = 0$. Right: The sigma spectral function is shown versus ω and $|\mathbf{p}|$ at $T = 100\,\text{MeV}$ and $\mu = 0$.

the medium. They are also the essential input for computing the transport properties of the matter such as the shear viscosity. In Fig. 10, the energy- and momentum-dependence of the pion and sigma spectral function is shown for finite temperatures and $\mu = 0$. The space-like region ($|\mathbf{p}| > \omega$) is nonvanishing at finite temperature due to the various space-like processes, cf. Fig. 8, while the decay thresholds and the pion-peak within the time-like region ($|\mathbf{p}| < \omega$) are Lorentz-boosted to higher energies as the spatial momentum increases.

7. Shear Viscosity and η/s

As mentioned above, the spectral functions in Minkowski space-time contain the information of all physical scattering processes and they are thus suitable to determine the transport properties of the matter. When evaluated in the FRG, it is then possible to treat the influence of phase transitions reliably.

The suitable framework is the Green–Kubo formalism[45,46] of linear response theory. It is more general than kinetic theory since nowhere the quasiparticle picture has to be invoked. For the shear viscosity η, which is of great interest for the space-time evolution of relativistic heavy-ion collisions, one has

$$\eta = \frac{1}{24} \lim_{q_0 \to 0} \lim_{|\mathbf{q}| \to 0} \frac{1}{q_0} \int d^4x e^{iqx} \langle [T_{ij}(x), T^{ij}(0)] \rangle, \tag{20}$$

where T_{ij} are the spatial components of the (local) energy–momentum tensor. Inserting the energy–momentum tensor of the QM model, which can be straightforwardly evaluated gives

$$\eta_{\sigma,\pi} \propto \int \frac{d\omega}{2\pi} \int \frac{d^3p}{(2\pi)^3} p_x^2 p_y^2 n_B'(\omega) \rho_{\sigma,\pi}^2(\omega, \mathbf{p}), \tag{21}$$

as the main contribution to the mesonic shear viscosities, where n_B' denotes the energy derivative of the Bose occupation factor. In Fig. 11, we show results for the

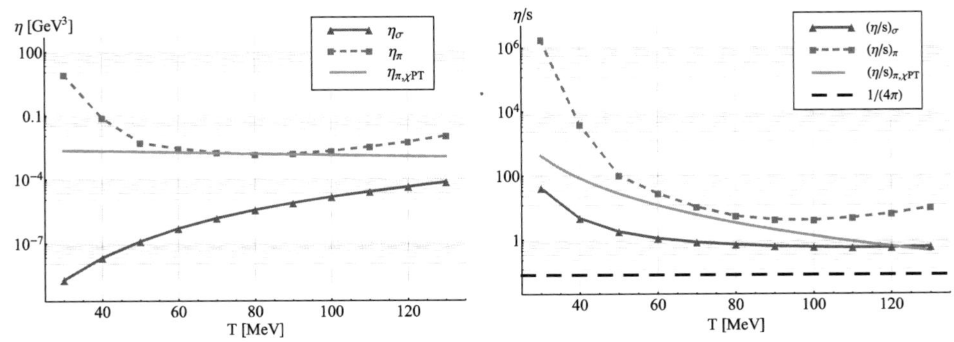

Fig. 11. The shear viscosity (left) and the shear viscosity over entropy density ratio (right) of the sigma mesons and the pions are shown versus temperature at $\mu = 0$ in comparison to results based on χPT.[39]

shear viscosity as well as for η/s of the mesons in comparison to a result based on χPT.[39] We find good agreement for η and η/s of the pions at intermediate temperatures, while towards lower temperatures, i.e., $T \lesssim 50\,\mathrm{MeV}$, and higher temperatures, i.e., $T \gtrsim 100\,\mathrm{MeV}$, our results for η and η/s are larger than expected from χPT. These differences partly stem from the fact that two-pion scattering processes are not yet included in our truncation. The decrease of η_σ towards lower temperatures is simply a temperature effect, which, in the case of the pions, is overcompensated by a sharpening of the pion peak, cf. Fig. 10.

8. Summary and Outlook

Spectral functions of physical systems in equilibrium contain the information on the measurable elementary excitations and, taking suitable limits, determine the transport properties. A well-known self-consistent method for evaluating spectral functions is the RPA, first introduced in the electron gas. In nuclear physics, Gerry has applied this method extensively for collective modes like the giant dipole resonance. However, the RPA is based on a mean-field picture, it has difficulties in properly dealing with phase transitions. Since the QCD phase diagram may contain such transitions as chiral symmetry gets restored with increasing temperature and density, it is therefore imperative to link the spectral properties of QCD matter to the modifications of the chiral order parameter in a framework that goes beyond the mean-field approximation. Such a method is the FRG in which the equilibrium thermodynamics is obtained from suitably truncated momentum-flow equations of the Grand Potential in Euclidean space-time.

To compute spectral functions in this framework, flow equations for the 2-point functions have to be derived with suitable truncations for the pertinent vertex functions. These are then analytically continued to Minkowski space-time. We have recently proposed a truncation scheme that is thermodynamically consistent and symmetry conserving. The latter is important to properly account for the Goldstone nature of spontaneously broken symmetries that is chiral symmetry in QCD.

Within the QM model as an effective theory for QCD, we have applied our truncation scheme to the scalar parity partners, the pion and the sigma meson. Starting from the unbroken phase at the ultraviolet cutoff, for which both mesons have large and practically degenerate masses, the momentum flow generates spontaneous chiral symmetry breaking and the pion and sigma spectral functions split. In the vacuum, the decay of the sigma mesons into two pions as the dominant channel is obtained naturally. As a consequence of chiral symmetry restoration in the medium, both the pion and sigma meson spectral functions are modified in a complicated fashion and eventually become identical when the symmetry is fully restored.

Let us come back to Gerry. He and his collaborators became interested early on in the question of how spontaneous chiral symmetry breaking and its restoration with temperature and density influences the spectral properties of vector mesons. The ρ-meson was of special interest since it strongly couples to the e^+e^- channel and hence its in-medium modification should be observable in the di-lepton rates of relativistic heavy-ion collisions. This is a long story (see Ref. 47 for a recent review) but the question persists to which extent the observed effects are indicators for the restoration of chiral symmetry. From the discussions above, it is obvious that in a consistent chiral theory, the parity partner of the ρ meson has to be treated on equal footing.[48] Chiral symmetry restoration then manifests itself in the complete degeneracy of the spectral functions of the parity partners. It should be clear that the FRG approach outlined above is suitable to address this question. Based on a gauged QM model,[49] we have started to tackle this question and first promising results have been obtained.[50] We are sure that Gerry would have appreciated them.

Acknowledgments

We dedicate this paper to Gerry Brown, an inspiring 'father figure' for generations of theoretical nuclear physicists and a great human being. This work was supported by the BMBF, project number 05P12RDGHD, and the Helmholtz International Center for FAIR within the LOEWE initiative of the state of Hesse. R.-A. T. was furthermore supported by the Helmholtz Research School for Quark Matter Studies, H-QM.

References

1. M. B. Parappilly, P. O. Bowman, U. M. Heller, D. B. Leinweber, A. G. Williams and J. B. Zhang, *Phys. Rev. D* **73** (2006) 054504, doi: 10.1103/PhysRevD.73.054504.
2. R. Williams, C. S. Fischer and W. Heupel, *Phys. Rev. D* **93** (2016) 034026, doi: 10.1103/PhysRevD.93.034026.
3. G. Brown and M. Rho, *Phys. Rev. Lett.* **66** (1991) 2720, doi: 10.1103/PhysRevLett.66.2720.
4. G. Brown and M. Rho, *Phys. Rep.* **396** (2004) 1, doi, 10.1016/j.physrep.2004.02.002.
5. S. Weinberg, *Phys. Rev. Lett.* **18** (1967) 507, doi: 10.1103/PhysRevLett.18.507.
6. J. I. Kapusta and E. V. Shuryak, *Phys. Rev. D* **49** (1994) 4694, doi: 10.1103/PhysRevD.49.4694.

7. J. Berges, N. Tetradis and C. Wetterich, *Phys. Rept.* **363** (2002) 223.
8. J. Polonyi, *Cent. Eur. J. Phys.* **1** (2003) 1, doi: 10.2478/BF02475552.
9. J. M. Pawlowski, *Ann. Phys.* **322** (2007) 2831, doi: 10.1016/j.aop.2007.01.007.
10. B.-J. Schaefer and J. Wambach, *Phys. Part. Nucl.* **39** (2008) 1025, doi: 10.1134/S1063779608070083.
11. P. Kopietz, L. Bartosch and F. Schutz, *Lect. Notes Phys.* **798** (2010) 1, doi: 10.1007/978-3-642-05094-7.
12. J. Braun, *J. Phys. G* **39** (2012) 033001, doi: 10.1088/0954-3899/39/3/033001.
13. B. Friman *et al.*, *Lect. Notes Phys.* **814** (2011) 1, doi: 10.1007/978-3-642-13293-3.
14. H. Gies, *Lect. Notes Phys.* **852** (2012) 287.
15. D. Bohm and D. Pines, *Phys. Rev.* **92** (1953) 609. doi: 10.1103/PhysRev.92.609.
16. G. E. Brown and M. Bolsterli, *Phys. Rev. Lett.* **3** (1959) 472, doi: 10.1103/PhysRevLett.3.472.
17. H. J. Vidberg and J. W. Serene, *J. Low Temp. Phys.* **29** (1977) 179, doi: 10.1007/BF00655090.
18. M. Jarrell and J. Gubernatis, *Phys. Rep.* **269** (1996) 133, ISSN 0370-1573, doi: 10.1016/0370-1573(95)00074-7.
19. M. Asakawa, T. Hatsuda and Y. Nakahara, *Prog. Part. Nucl. Phys.* **46** 459 (2001), doi: 10.1016/S0146-6410(01)00150-8.
20. D. Dudal, O. Oliveira and P. J. Silva, *Phys. Rev. D* **89** (2014) 014010, doi: 10.1103/PhysRevD.89.014010.
21. N. Strodthoff, B.-J. Schaefer and L. von Smekal, *Phys. Rev. D* **85**, 074007 (2012), doi: 10.1103/PhysRevD.85.074007.
22. K. Kamikado, N. Strodthoff, L. von Smekal and J. Wambach, *Phys. Lett. B* **718** (2013) 1044, doi: 10.1016/j.physletb.2012.11.055.
23. R.-A. Tripolt, N. Strodthoff, L. von Smekal and J. Wambach, *Phys. Rev. D* **89** (2014) 034010, doi: 10.1103/PhysRevD.89.034010.
24. R.-A. Tripolt, L. von Smekal and J. Wambach, *Phys. Rev. D* **90** (2014) 074031, doi: 10.1103/PhysRevD.90.074031.
25. K. G. Wilson, *Phys. Rev. B* **4** (1971) 3184, doi: 10.1103/PhysRevB.4.3184.
26. K. Wilson and J. B. Kogut, *Phys. Rept.* **12** (1974) 75, doi: 10.1016/0370-1573(74)90023-4.
27. C. Wetterich, *Phys. Lett. B* **301** (1993) 90, doi: 10.1016/0370-2693(93)90726-X.
28. T. R. Morris, *Int. J. Mod. Phys. A* **9** (1994) 2411, doi: 10.1142/S0217751X94000972.
29. N. Tetradis and C. Wetterich, *Nucl. Phys. B* **422** (1994) 541, doi: 10.1016/0550-3213(94)90446-4.
30. D. F. Litim, *J. High Energy Phys.* **0111** (2001) 059, doi: 10.1088/1126-6708/2001/11/059.
31. J. Braun, *Phys. Rev. D* **81** (2010) 016008, doi: 10.1103/PhysRevD.81.016008.
32. D. Jungnickel and C. Wetterich, *Phys. Rev. D* **53** (1996) 5142, doi: 10.1103/PhysRevD.53.5142.
33. B.-J. Schaefer and J. Wambach, *Nucl. Phys. A* **757** 479 (2005), doi: 10.1016/j.nuclphysa.2005.04.012.
34. J. M. Pawlowski, M. M. Scherer, R. Schmidt and S. J. Wetzel, arXiv: 1512.03598 (2015).
35. J. M. Pawlowski and N. Strodthoff, *Phys. Rev. D* **92** (2015) 094009, doi: 10.1103/PhysRevD.92.094009.
36. K. Kamikado, N. Strodthoff, L. von Smekal and J. Wambach, *Eur. Phys. J. C* **74** (2014) 2806, doi: 10.1140/epjc/s10052-014-2806-6.

37. G. Baym and N. D. Mermin, *J. Math. Phys.* **2** (1961) 232, doi: http://dx.doi.org/10.1063/1.1703704.
38. N. Landsman and C. van Weert, *Phys. Rept.* **145** (1987) 141, doi: 10.1016/0370-1573(87)90121-9.
39. R. Lang, N. Kaiser and W. Weise, *Eur. Phys. J. A* **48** (2012) 109, doi: 10.1140/epja/i2012-12109-3.
40. J. Braun, K. Schwenzer and H.-J. Pirner, *Phys. Rev. D* **70** (2004) 085016, doi: 10.1103/PhysRevD.70.085016.
41. T. K. Herbst, J. M. Pawlowski and B.-J. Schaefer, *Phys. Lett. B* **696** (2011) 58, doi: 10.1016/j.physletb.2010.12.003.
42. V. Skokov, B. Stokic, B. Friman and K. Redlich, *Phys. Rev. C* **82** (2010) 015206, doi: 10.1103/PhysRevC.82.015206.
43. R.-A. Tripolt, Spectral Functions and Transport Coefficients from the Functional Renormalization Group, Ph.D. thesis, TU Darmstadt (2015).
44. T. D. Cohen, *Phys. Rev. Lett.* **91** (2003) 222001, doi: 10.1103/PhysRevLett.91.222001.
45. M. S. Green, *J. Chem. Phys.* **22** (1954) 398.
46. R. Kubo, *J. Phys. Soc. Jpn.* **12** (1957) 570.
47. R. Rapp, J. Wambach and H. van Hees, *Landolt-Bornstein* **23** (2010) 134, doi: 10.1007/978-3-642-01539-7_6.
48. P. M. Hohler and R. Rapp, *Ann. Phys.* **368** (2016) 70, doi: 10.1016/j.aop.2016.01.018.
49. M. Urban, M. Buballa and J. Wambach, *Nucl. Phys. A* **697** (2002) 338, doi: 10.1016/S0375-9474(01)01248-9.
50. C. Jung, R.-A. Tripolt, L. von Smekal and J. Wambach, 2015, arXiv:1610.08754.

Skyrmions at high density

Vicente Vento
Departamento de Física Teórica,
Universidad de Valencia-CSIC, C/Dr. Moliner, 50,
E-46100 Burjassot, Valencia, Spain
Instituto de Física Corpuscular, Universidad de Valencia-CSIC,
C/Catedrático José Beltrán 2, E-46980 Paterna,
Valencia, Spain
vicente.vento@uv.es

The phase diagram of quantum chromodynamics is conjectured to have a rich structure containing at least three forms of matter: hadronic nuclear matter, quarkyonic matter and quark–gluon plasma. We justify the origin of the quarkyonic phase transition in a chiral-quark model and describe its formulation in terms of Skyrme crystals.

Keywords: Quark; pion; Skyrmion; dilation.

1. Introduction

The phase diagram of quantum chromodynamics (QCD) is closely related to the history of the universe and can be probed by heavy ion collisions and compact stars studies. Heavy ion collision experiments aim at discovering the deconfinement and chiral transitions which determine the QCD phase diagram. The phase diagram is conjectured to have a rich phase structure. At low temperature QCD has the conventional hadron nuclear matter phase. At high temperatures and/or densities QCD matter develops a qualitatively different phase where quarks are liberated from confinement which is called the quark–gluon plasma (QGP) phase. This has been shown by data taken at RHIC and LHC at high temperature (T) and almost zero baryon chemical potential (μ_b). There is strong evidence from lattice QCD for a crossover transition from the hadronic phase to QGP for small μ_b (Fig. 1).[1] The crossover nature of the transition justifies the standard scenario of the homogeneous Big Bang nucleosynthesis.

In heavy ion collisions at lower energies nucleons get into the mid-rapidity region and the nuclear stopping power gives rise to the formation of hot matter at finite μ_b. Lattice studies aimed to extend the simulation to finite chemical potential are limited to small μ_b due to the so-called sign problem. One expects the existence of a critical point at the end of the crossover region. This scenario is supported by the phase diagram of lattice QCD in the strong coupling limit which extends to the full $\mu_b - T$ plane featuring in the chiral limit a tricritical point which may be related to the critical point expected in the QCD phase diagram.[2]

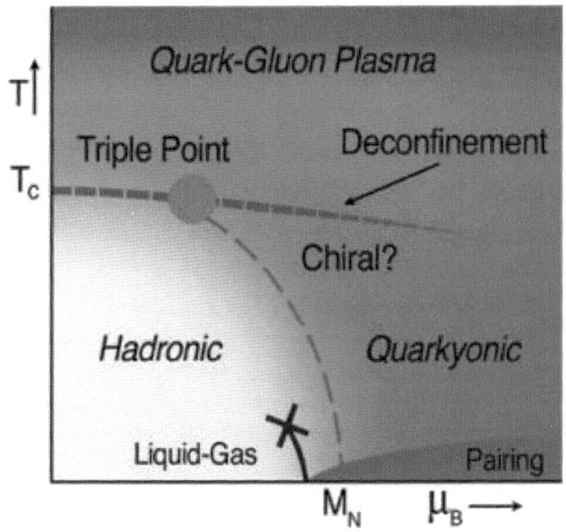

Fig. 1. The QCD phase diagram.

It has been argued that some features of hadroproduction in relativistic heavy ion collisions may be explained by the existence of three forms of matter: hadronic nuclear matter, quarkyonic matter and QGP.[3] These forms of matter meet at a triple point in the QCD phase diagram. Quarkyonic matter is an exotic form of matter which can be understood as a quark Fermi sea with baryonic Fermi energy. This picture can be enriched at low temperature and very high density. Before all condensates melt away and QGP is realized, condensation of quark–quark pairs can take place leading to color superconductivity. Color superconductivity allows for a very rich phase structure in that region depending on the various flavor–color symmetry structures, the so-called color–flavor locking (CFL) scenarios.[4]

Our aim is to describe the transition at high density between ordinary matter and exotic matter in terms of effective theories where chiral symmetry plays a crucial role. Chiral symmetry is a flavor symmetry of QCD which is essential to understand hadronic physics. At low temperatures and densities it is spontaneously broken leading to the existence of the pion, its Goldstone boson. These models do not have explicit color degrees of freedom and therefore no gluons. One can therefore not investigate the confinement/deconfinement transition directly, but may study the chiral symmetry restoration transition. Lattice studies seem to indicate that chiral symmetry is restored in the high temperature and/or high baryon density phases and that it may go hand-in-hand with the confinement/deconfinement transition.[2]

Another important ingredient of effective theories is scale invariance. The classical QCD action in the chiral limit is invariant under the scale transformation. The quark mass term breaks the scale invariance. At the quantum level, scale invariance is also broken by dimensional transmutation even for massless quarks, as signaled by the nonvanishing of the trace of the energy–momentum tensor. This phenomenon can be formulated by the nonvanishing divergence of the dilatation current, the so-called trace anomaly, which can be hadronized and implemented into effective Lagrangians.

Our study of the chiral phase transition will proceed by describing the quarkyonic phase in terms of a quark chiral model in Sec. 2. This description leads directly to the so-called skyrmion–half-skyrmion phase transition which has been studied in great detail in the Skyrme model and which we address in Sec. 3. We end by drawing some conclusions.

2. A Model for the Quarkyonic Phase Transition

The gap between the underlying QCD theory and phenomenology in the nonperturbative regime is occupied by effective theories, whose Lagrangians contain the relevant degrees of freedom at that energy scale implementing the fundamental symmetries of QCD. Moreover, the intrinsic numerical difficulties in the evaluation of observables nonperturbatively leads to the lack of any reliable expansion scheme and one is even forced to resort to model Lagrangians.[5–9] As a result the high baryon density region allows for various scenarios associated with the restoration of symmetries such as chiral symmetry and scale invariance. For example, it has been suggested, using the limit of large number of colors N_c,[3] that before deconfinement, the so-called quarkyonic transition will take place, where several symmetries are eventually restored leading to a rich set of phases.

We decided to understand this phase transition in terms of the chiral dilaton model (CDM) a model with only flavor degrees of freedom where baryons are made of quarks interacting with the sigma and pion mesons.[10] The model has been already studied in detail in Ref. 11 and the corresponding simplified density Lagrangian[12–14] reads

$$\mathcal{L} = \bar{\psi}[i\gamma^\mu\partial_\mu - g_\pi(\sigma + i\boldsymbol{\pi}\cdot\boldsymbol{\tau}\gamma_5)]\psi + \frac{1}{2}(\partial_\mu\sigma\partial^\mu\sigma + \partial_\mu\boldsymbol{\pi}\cdot\partial^\mu\boldsymbol{\pi}) - V(\sigma,\pi). \quad (1)$$

The potential is given by

$$V(\sigma,\boldsymbol{\pi}) = \lambda_1^2(\sigma^2+\boldsymbol{\pi}^2) - \lambda_2^2\ln(\sigma^2+\boldsymbol{\pi}^2) - \sigma_0 m_\pi^2\sigma, \quad (2)$$

where

$$\lambda_1^2 = \frac{1}{2}\frac{B\delta\phi_0^4 + m_\pi^2\sigma_0^2}{\sigma_0^2} = \frac{1}{4}(m_\sigma^2 + m_\pi^2),$$

$$\lambda_2^2 = \frac{1}{2}B\delta\phi_0^4 = \frac{\sigma_0^2}{4}(m_\sigma^2 - m_\pi^2).$$

Here σ is the scalar–isoscalar field, $\boldsymbol{\pi}$ is the pseudoscalar-isotriplet meson field, ϕ is the dilaton field and ψ describes the isodoublet quark fields. This Lagrangian density, besides being invariant under chiral symmetry, is also invariant under scale invariance, which is also spontaneously broken, and has been implemented above following Refs. 15–17.

The dilaton field will be frozen in the present calculation at its vacuum value ϕ_0. This approximation is based on the results obtained in Refs. 14 and 18, which showed that at low temperatures the dilaton remains close to its vacuum value even at large densities.

The constants B and ϕ_0 in the potential are fixed by choosing a value for the mass of the glueball and for the vacuum energy, while $\delta = 4/33$ is given by the QCD beta function and it represents the relative weight of the fermionic and gluonic degrees of freedom. The vacuum state is chosen, as usually, at $\sigma_0 = f_\pi$ and $\boldsymbol{\pi} = 0$.

To describe the CDM at large densities we have adopted the Wigner–Seitz approximation, in which solitonic matter is described by a lattice of spherical cells of radius $R_{\rm WS}$ and finite density effects are obtained by imposing boundary conditions on fields at the boundaries of the cells.[20,21] The presence of a periodical potential generated by the sigma and the pion fields leads to a band structure, in which quarks move. As the cell radius decreases, the baryon density increases, we have shown that the development of the quark band gives rise to two separate phenomena. As the system approaches densities around ρ_0, the band starts to get wider and the quarks, still in the lowest positive parity state, $G = 0^+$, where $G = S + T$ denotes grandspin, are free to move inside the cell. As the quarks get squeezed inside the cell, they get excited and as the first excited state $G = 1^+$ plunges into the lower band, some quarks will move into it (see Fig. 2). In this moment the baryon density profile inside the cell, which was initially peaking at the center as corresponding to a well-localized soliton, begins to change. The appearance of one excited quark within the cell leads to a modification of the baryon density distribution, which develops a minimum at the center of the cell and two bumps on each side (see Fig. 3). This

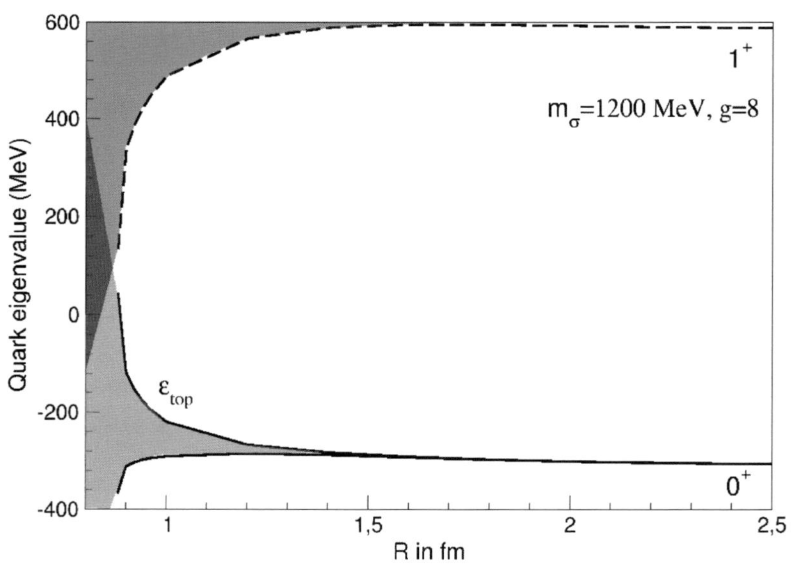

Fig. 2. Quark eigenvalue as a function of the cell radius R in CDM. The shaded area represents the band estimated following Ref. 19. The first excited state 1^+ and the corresponding lower part of the band are also shown.

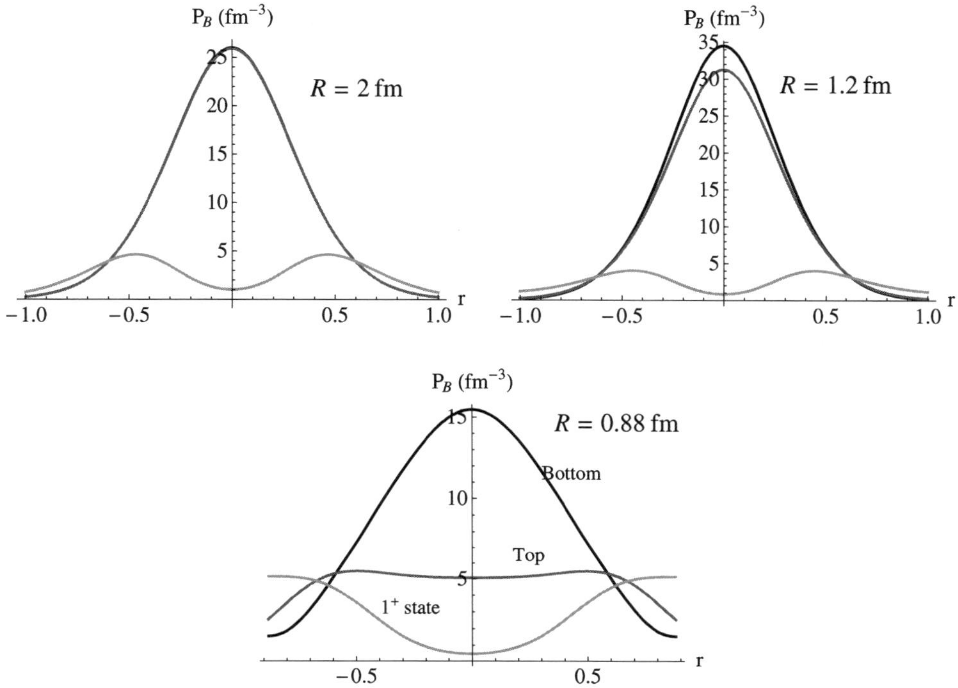

Fig. 3. (Color online) Baryon density profiles inside the Wigner–Seitz cell, for several values of $R_{\rm WS}$. The profiles are shown for the ground state $G = 0^+$ (black line), the top of the lower band (blue line) and the first excited state $G = 1^+$ (red line). The three panels are chosen in order to show, going from the left to right, the confined $B = 1$ phase, the delocalized $B = \frac{1}{2}$ phase and the deconfined phase.

mechanism suggests a transition to the $B = \frac{1}{2}$ quarkyonic phase at intermediate densities.[10] This relation is in line with the close connection found between the discrete half-skyrmion symmetry and the continuous chiral symmetry.[22]

If we look with more detail Fig. 2, we see two distinct phenomena both correlated to the presence of a band structure. On one hand, the exchange of quarks between the lower band $G = 0^+$ and the first excited one $G = 1^+$ occurring at densities larger than $3\rho_0$ can be interpreted as a signal of *deconfinement*, since the crystal becomes a color conductor.[23] On the other hand, the plunging of the excited state into the lower band and the sharing between neighboring cells is the origin of a *delocalization* of the baryon number carried here only by quarks, leading to a $B = \frac{1}{2}$ phase. The delocalization mechanism is the seed for the quarkyonic interpretation of the half-skyrmion phase.[24]

At very low densities (for $R = 2\,{\rm fm}$, $\rho_B \approx 0.2\rho_0$) the solitons are well localized inside the cell, all the quarks occupy the lower state, as can be seen in the left panel of Fig. 3. The baryon density at the top of the band almost coincides with that at the bottom, and is clearly peaked at the center of the cell vanishing at the edges. As the radius shrinks, the band gets wider (middle panel of Fig. 3) as the $G = 1^+$ state plunges into the lower band (see Fig. 2). Now, the quarks are free to move to the excited state whose baryon density profile shows a valley at the center of the cell and two bumps close to the edges. We can imagine a $B = 1$ soliton with one quark in the $G = 1^+$ state and the other two lying in the lower band. This mechanism provides a delocalization of the baryon number from the center of the cell to the edges and leads to a $B = \frac{1}{2}$ scenario. In our calculation, scale invariance cannot be restored due to the frozen dilaton's dynamics, and therefore chiral symmetry restoration cannot take place. The bosonized half-skyrmion phase, from the point of view of a fermionic description, is characterized by the fact that there are two types of quarks according to their spatial distribution, those which are in the $G = 0^+$ band and those which are in the $G = 1^+$ band.[10]

At densities larger than ρ_0, the band keeps broadening and before the breakdown of the solution takes place, the upper and the lower bands merge. The quarks are now free to populate also the excited states and therefore all are free to move everywhere in the lattice, this being the behavior of the deconfined phase as discussed in Ref. 23.

The main result of our discussion is that inside the quarkyonic phase scenario, within the structural changes that nuclear matter could undergo before deconfinement, the half-skyrmion phase arising from bosonized QCD might take place along with the restoration of QCD symmetries.[10] A detailed description of the transition appears in Skyrme matter.

3. Skyrme Matter

We proceed to describe next this phase transition in a fully bosonized model defined therefore only in terms of mesonic fields.

3.1. *The Skyrme model*

The most essential ingredients of the Skyrme model are the pions, Goldstone bosons associated with the spontaneous breakdown of chiral symmetry.[6–8] Baryons arise as topological solitons of the meson Lagrangian. The pion Lagrangian can be realized nonlinearly as $U = \exp(i\boldsymbol{\tau}\cdot\boldsymbol{\pi}/f_\pi)$, which transforms as $U \to g_L U g_R^\dagger$ under the global chiral transformations $SU_L(N_f) \times SU_L(N_f)$; $g_L \in SU_L(N_f)$ and $g_R \in SU_R(N_f)$. Hereafter, we will restrict our consideration to $N_f = 2$. In the case of $N_f = 2$, the meson field π represents three pions as

$$\boldsymbol{\tau}\cdot\boldsymbol{\pi} = \begin{pmatrix} \pi^0 & \sqrt{2}\pi^+ \\ \sqrt{2}\pi^- & -\pi^0 \end{pmatrix}. \tag{3}$$

The Lagrangian for their dynamics can be expanded in powers of the right and left invariant currents $R_\mu = U\partial_\mu U^\dagger$ and $L_\mu = U^\dagger\partial_\mu U$, which transforms as $R_\mu \to g_L R_\mu g_L^\dagger$ and $L_\mu \to g_R L_\mu g_R^\dagger$. The lowest order term is

$$\mathcal{L}_\sigma = \frac{f_\pi^2}{4}\mathrm{Tr}(\partial_\mu U^\dagger \partial^\mu U). \tag{4}$$

Here, $f_\pi = 93\,\mathrm{MeV}$ is the pion decay constant.[8]

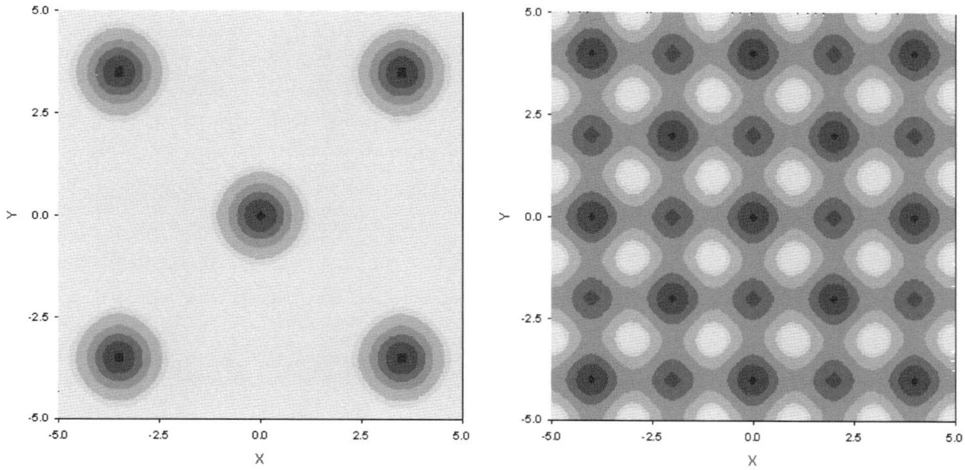

Fig. 4. Local baryon number densities at low density (left) and beyond the critical density (right) with massive pions. For high density the system is (almost) a half-skyrmion in a CC crystal configuration.

The classical nature of the topological solitons hereafter called skyrmions enables us to construct a dense system quite conveniently by putting more and more skyrmions into a given volume. Skyrmions shape and arrange themselves to minimize the energy of the system. The ground state configuration of skyrmion matter are crystals.[24] At low density it is made of well localized single skyrmions. At a critical density, the system undergoes a structural phase transition to a new kind of crystal. It is made of 'half-skyrmions' which are still well localized but carry only half winding number (see Fig. 4). In the half-skyrmion phase, the system develops an additional symmetry which leads to a vanishing average value of $\sigma = \frac{1}{2}\mathrm{Tr}(U)$, the normalized trace of the U field.[24] In the studies of the late 80s,[25] the vanishing of this average value $\langle\sigma\rangle$ was interpreted as chiral symmetry restoration by assuming that $\langle\sigma\rangle$ is related to the QCD order parameter $\langle\bar{q}q\rangle$. However, in Ref. 26, it was shown that the vanishing of $\langle\sigma\rangle$ cannot be an indication of a genuine chiral symmetry restoration, because the decay constant of the pion fluctuating in such a half-skyrmion matter does not vanish. This was interpreted as a signal of the appearance of a pseudogap phase similar to what happens in high T_c superconductors.[27]

The puzzle was solved in Refs. 24 and 28 by incorporating a suitable degree of freedom, the dilaton field χ, associated to the scale anomaly of QCD.

3.2. *Dilaton dynamics*

The dynamics introduced in Sec. 3.1 as an effective theory for the hadronic interactions is incomplete. In fact, it is not clear that the intrinsic density dependence required by the matching to QCD is fully implemented in the model. One puzzling feature is that the Wigner phase represented by the half-skyrmion matter with

$\langle\sigma\rangle = 0$ supports a nonvanishing pion decay constant. This may be interpreted as a possible signal for a pseudogap phase. However, at some point, the chiral symmetry should be restored and there the pion decay constant should vanish.

This difficulty can be circumvented in our framework by incorporating in the standard skyrmion dynamics the trace anomaly of QCD in an effective manner.[16] The end result is the skyrmion Lagrangian introduced by Ellis and Lanik[29] and employed by Brown and Rho[30] for nuclear physics which contains an additional scalar field, the so-called scale dilaton. Broken scale invariance can be implemented into large N_c physics by modifying the standard skyrmion Lagrangian[24]:

$$\mathcal{L} = \frac{f_\pi^2}{4}\left(\frac{\chi}{f_\chi}\right)^2 \mathrm{Tr}(\partial_\mu U^\dagger \partial^\mu U) + \frac{1}{32e^2}\mathrm{Tr}([U^\dagger\partial_\mu U, U^\dagger \partial_\nu U])^2$$
$$+ \frac{f_\pi^2 m_\pi^2}{4}\left(\frac{\chi}{f_\chi}\right)^3 \mathrm{Tr}(U + U^\dagger - 2) + \frac{1}{2}\partial_\mu\chi\partial^\mu\chi$$
$$- \frac{1}{4}\frac{m_\chi^2}{f_\chi^2}\left[\chi^4\left(\ln(\chi/f_\chi) - \frac{1}{4}\right) + \frac{1}{4}\right]. \tag{5}$$

We have denoted the nonvanishing vacuum expectation value of χ as f_χ, a constant which describes the decay of the scalar into pions. The gluonic component of the trace anomaly can be reproduced by the potential energy $V(\chi)$, which is adjusted in the Lagrangian (5) so that $V = dV/d\chi = 0$ and $d^2V/d\chi^2 = m_\chi^2$ at $\chi = f_\chi$.[16]

The vacuum state of the Lagrangian at zero baryon number density is defined by $U = 1$ and $\chi = f_\chi$. The fluctuations of the pion and the scalar fields about this vacuum, defined through

$$U = \exp(i\boldsymbol{\tau}\cdot\boldsymbol{\phi}/f_\pi) \quad \text{and} \quad \chi = f_\chi + \tilde{\chi}, \tag{6}$$

give physical meaning to the model parameters: f_π as the pion decay constant, m_π as the pion mass, f_χ as the scalar decay constant and m_χ as the scalar mass. For the pions, we use their empirical values as $f_\pi = 93\,\mathrm{MeV}$ and $m_\pi = 140\,\mathrm{MeV}$. We fix the Skyrme parameter e to 4.75 from the axial-vector coupling constant g_A as in Ref. 31. However, for the scalar field χ, no experimental values for the corresponding parameters are available.

In Ref. 32, the scalar field is incorporated into a relativistic hadronic model for nuclear matter not only to account for the anomalous scaling behavior but also to provide the mid-range nucleon–nucleon attraction. Then, the parameters f_χ and m_χ are adjusted so that the model fits finite nuclei. One of the parameter sets is $m_\chi = 550\,\mathrm{MeV}$ and $f_\chi = 240\,\mathrm{MeV}$ (set A). On the other hand, Song *et al.*[33] obtain the "best" values for the parameters of the effective chiral Lagrangian with the "soft" scalar fields so that the results are consistent with "Brown–Rho" scaling,[30] explicitly, $m_\chi = 720\,\mathrm{MeV}$ and $f_\chi = 240\,\mathrm{MeV}$ (set B). For completeness, we consider also a parameter set of $m_\chi = 1\,\mathrm{GeV}$ and $f_\chi = 240\,\mathrm{MeV}$ (set C) corresponding to a mass scale comparable to that of chiral symmetry $\Lambda_\chi \sim 4\pi f_\pi$.

3.3. *Dynamics of the single skyrmion*

One has to follow the procedure as in Refs. 24 and 28. The first step is to find the solution for the single skyrmion which includes the dilaton dynamics. The skyrmion with the baryon number $B = 1$ can be found by generalizing the spherical hedgehog ansatz of the original Skyrme model as

$$U_0(\mathbf{r}) = \exp(i\boldsymbol{\tau} \cdot \hat{r}F(r)), \quad \text{and} \quad \chi_0(\mathbf{r}) = f_\chi C(r), \tag{7}$$

with two radial functions $F(r)$ and $C(r)$. Minimization of the mass equation leads to a coupled set of equations of motion for these functions. In order for the solution to carry a baryon number, U_0 has the value -1 at the origin, i.e., $F(x = 0) = \pi$, while there is no such topological constraint for $C(x = 0)$. All that is required is that it be a positive number below 1. At infinity, the fields $U_0(\mathbf{r})$ and $\chi_0(\mathbf{r})$ should reach their vacuum values.

Shown in Fig. 5 are profile functions as a function of $x(= ef_\pi r)$. $F(r)$ and consequently the root mean square radius of the baryon charge show little dependence on m_χ. On the other hand, the changes in $C(r)$ and the skyrmion mass are recognizable. Inside the skyrmion, especially at the center, $C(r)$ deviates from its vacuum value 1. Note that this change in $C(r)$ is multiplied by f_π^2 in the current algebra term of the Lagrangian. Thus, $C(r) \leq 1$ reduces the *effective* f_π inside the single skyrmion, which implies a partial restoration of the chiral symmetry there. The reduction in the effective pion decay constant is reflected in the single skyrmion mass.

The larger the scalar mass is, the smaller its coupling to the pionic field and the lesser its effect on the single skyrmion. In the limit of $m_\chi \to \infty$, the scalar field

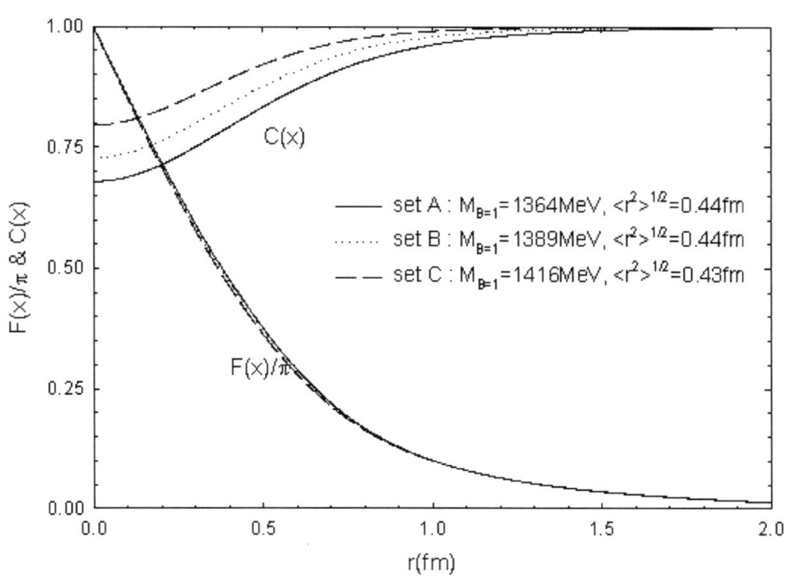

Fig. 5. Profile functions $F(x)$ and $C(x)$ as a function of x.

is completely decoupled from the pions and the model returns back to the original one, where $C(r) = 1$, $M_{sk} = 1479\,\text{MeV}$ and $\langle r^2\rangle^{1/2} = 0.43\,\text{fm}$.

3.4. *Dense skyrmion matter and chiral symmetry restoration*

The second step is to construct a crystal configuration made up of skyrmions with a minimal energy for a given density.

Referring to Refs. 24, 28 and 34 for the full details, we emphasize here the role of the dilaton field in the phase transition scenario for skyrmion matter. Let the dilaton field $\chi(\mathbf{r})$ be a constant throughout the whole space as

$$\chi/f_\chi = X. \tag{8}$$

Then the energy per baryon number of the system for a given density can be calculated and conveniently expressed as[28]

$$E/B(X,L) = X^2(E_2/B) + (E_4/B) + X^3(E_m/B) + (2L^3)\left(X^4\left(\ln X - \frac{1}{4}\right) + \frac{1}{4}\right), \tag{9}$$

where E_2, E_4 and E_m are, respectively, the contributions from the current algebra term, the Skyrme term and the pion mass term of the Lagrangian to the energy of the skyrmion system, described in Sec. 3.1, and $(2L^3)$ is the volume occupied by a single skyrmion.[24]

The quantity $E/B(X,L)$ can be taken as an *in-medium* effective potential for X, modified by the coupling of the scalar to the background matter. Using the parameter values of Ref. 34 for the Skyrme model without the scalar field, the effective potential $E/B(X)$ for few values of L behaves as shown in Fig. 6(a). At low density (large L), the minimum of the effective potential is located close to $X = 1$. As the density increases, the quadratic term in the effective potential $E/B(X)$

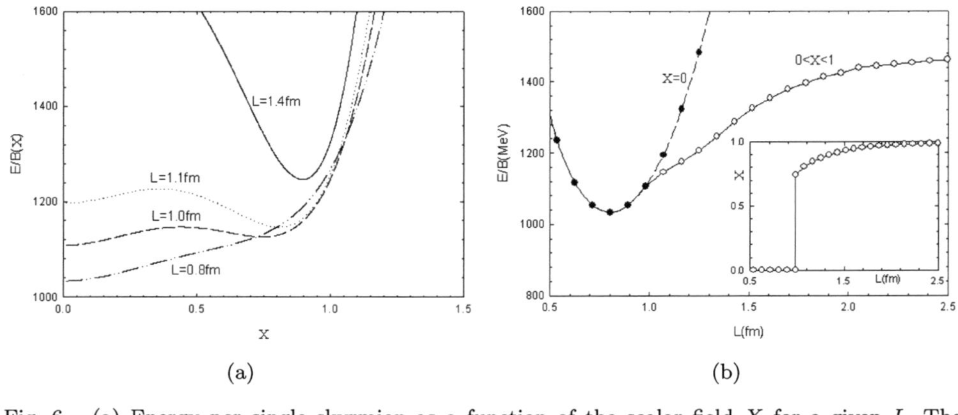

Fig. 6. (a) Energy per single skyrmion as a function of the scalar field X for a given L. The results are obtained with (E_2/B), (E_4/B) and (E_m/B) of Ref. 34 and with the parameter set B; (b) Energy per single skyrmion as a function of L.

develops another minimum at $X = 0$ which is an unstable extremum of the potential $V(X)$ in free space. At $L \sim 1$ fm, the newly developed minimum competes with the one near $X \sim 1$. At higher density, the minimum shifts to $X = 0$ where the system stabilizes.

In Fig. 6(b), we plot $E/B(X_{\min}, L)$ as a function of L, which is obtained by minimizing $E/B(X, L)$ with respect to X for each L. The figure in the small box is the corresponding value of $X_{\min}$ as function of L. There we see the explicit manifestation of a first-order phase transition. Although the present discussion is based on a simplified analysis, it essentially encodes the same physics as in the more rigorous treatment of χ given in Ref. 28.

We show in Fig. 7 the average values $\langle\sigma\rangle$ and $\langle\chi/f_\chi\rangle$ over space for the minimum energy crystal configurations obtained by the complete numerical calculation without any approximation for χ. These data show that a 'structural' phase transition takes place, characterized by $\langle\sigma\rangle = 0$, at lower density then the *genuine* chiral phase transition which occurs when $\langle\chi\rangle = 0$. The value of $\langle\sigma\rangle$ becomes 0 when the structure of the skyrmion crystal undergoes a change from the single skyrmion FCC to the half-skyrmion CC. Thus, the pseudogap phase persists in an intermediate density region, where the $\langle\chi/f_\chi\rangle$ does not vanish while $\langle\sigma\rangle$ does.[35] A similar pseudogap structure has been also proposed in hot QCD.[36]

The two-step phase transition is schematically illustrated in Fig. 7. Let ρ_p and ρ_c be the densities at which $\langle\sigma\rangle$ and $\langle\chi\rangle$ vanish, respectively.

(A) At low density ($\rho < \rho_p$), matter slightly reduces the vacuum value of the dilaton field from that of the baryon-free vacuum. This implies a shrinking of the radius of the chiral circle by the same ratio. Since the skyrmion takes all the values

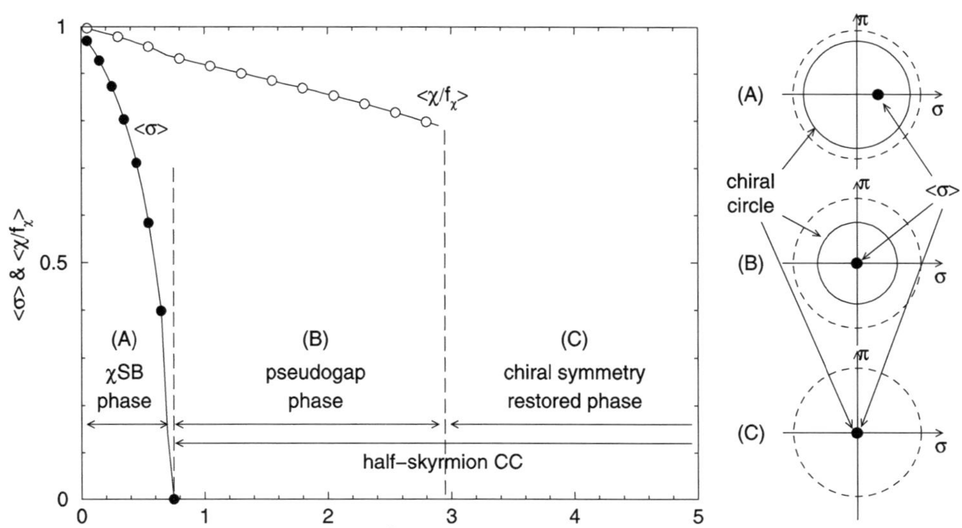

Fig. 7. Average values of $\sigma = \frac{1}{2}\mathrm{Tr}(U)$ and χ/f_χ of the lowest energy crystal configuration at a given baryon number density.

on the chiral circle, the expectation value of σ is not located on the circle but inside the circle. Skyrmion matter at this density is in the chiral symmetry broken phase.

(B) At some intermediate densities ($\rho_p < \rho < \rho_c$), the expectation value of σ vanishes while that of the dilaton field is still nonzero. The skyrmion crystal is in a CC configuration made of half-skyrmions localized at the points where $\sigma = \pm 1$. Since the average value of the dilaton field does not vanish, the radius of the chiral circle is still finite. Here, $\langle\sigma\rangle = 0$ does not mean that chiral symmetry is completely restored. We interpret this as a pseudogap phase.

(C) At higher density ($\rho > \rho_c$), the phase characterized by $\langle\chi/f_\chi\rangle = 0$ becomes energetically favorable. Then, the chiral circle, describing the fluctuating pion dynamics, shrinks to a point.

The density range for the occurrence of a pseudogap phase strongly depends on the parameter choice of m_χ. For small m_χ below 700 MeV, the pseudogap has almost zero size.

In the case of massive pions, the chiral circle is tilted by the explicit (mass) symmetry breaking term. Thus, the exact half-skyrmion CC, which requires a symmetric solution for points with value $\sigma = +1$ and those with $\sigma = -1$, cannot be constructed and consequently the phase characterized by $\langle\sigma\rangle = 0$ does not exist for any density. Thus no pseudogap phase arises. However, $\langle\sigma\rangle$ is always inside the chiral circle and its value drops much faster than that of $\langle\chi/f_\chi\rangle$. Therefore, only if the pion mass is small a pseudogap phase can appear in the model.

4. Conclusions

We used a Lagrangian with quark and meson fields, based on chiral symmetry and scale invariance, to analyze the possible presence of a $B = \frac{1}{2}$ phase at large baryon densities, similarly to what happens in skyrmionic matter.[24] In the Skyrme model, since the baryon number is carried by the skyrmion, the half-skyrmion phase is strictly connected to the topological modification of the pion fields inside the lattice, as the system reaches high densities. Motivated by the so-called quarkyonic phase,[3] we have shown that a similar mechanism occurs in a model like the CDM, which contains explicit fermionic degrees of freedom. In our case the baryon number is carried only by quarks and the $B = \frac{1}{2}$ or half-skyrmion phase is achieved by a delocalization of the baryonic charge, i.e., by a modification of the baryon density profile, as we move to high densities.

The analysis presented here is built upon a series of approximations, but we are certain that the delocalization mechanism will survive more sophisticated analyses because it is physically very intuitive. The main result of our discussion is that inside the quarkyonic phase scenario, within the structural changes that nuclear matter could undergo before deconfinement, the half-skyrmion phase arising from bosonized QCD might take place along with the restoration of QCD symmetries. The fact that we are not considering in our approximations the effect of the long-range tail of the interaction might have an influence on the detailed values of the density at which the phase transition takes place but we do not expect any qualitative change in the mechanism explored.

We have presented an approach to hadronic physics based on Skyrme's philosophy, namely that baryons are solitons of a theory described in terms of meson fields, which can be justified from QCD in the large N_c expansion. We have adopted the basic principles of effective field theory. Given a certain energy domain we describe the dynamics by a Lagrangian defined in terms of the mesonic degrees of freedom active in that domain, we thereafter implement the symmetries of QCD, and describe the baryonic sectors as topological winding number sectors and solve in these sectors the equations derived from the Lagrangian with the appropriate boundary conditions for the sector. Skyrme models also represent a nice tool for understanding low density cold hadronic matter. We have shown that when hadronic matter is compressed Skyrme models provide useful information on the chiral phase transitions. Skyrmion matter is realized as a crystal and we have seen that at low densities it is an FCC crystal made of skyrmions. The phase transition occurs when the FCC crystal transforms into a half-skyrmion CC one. We have discovered the crucial role of the scale dilaton in describing the expected phase transition towards a chiral symmetry restored phase.

The main result of our presentation is the realization that phase transition scenario is not as simple as initially thought but contains many features which make it interesting and phenomenologically appealing. In this line of thought, it is exciting to have unveiled scenarios near the phase transition of unexpected interesting phenomenology in line with proposals of the so-called quarkyonic matter.[3]

Acknowledgments

I would like to thank my collaborators Hee-Jung Lee, Valentina Mantovani-Sarti and Dong-Pil Min whose work has been widely used for this contribution. I have enjoyed a long and intense collaboration with Byung-Yoon Park which has extended beyond the realm of physics. He has been the force behind our dense matter studies. I owe gratitude to Mannque Rho, whom I first met during the days of the "little bag," who inspired much of my research for a long time and motivated these studies during our visits to KIAS and Seoul National University. All of this would not have happened without the encouragement, generosity and scientific guidance of Gerry, G. E. Brown. I am deeply grateful to him for having shown me his approach to physics. Gerry and his work have always been inspiring and motivating.

This work has been supported in part by MINECO (Spain) Grant No. FPA2013-47443-C2-1-P, GVA-PROMETEOII/2014/066 and SEV-2014-0398.

References

1. H. Saito, S. Aoki, K. Kanaya, H. Ohno, S. Ejiri, T. Hatsuda, Y. Maezawa and T. Umeda, The order of the deconfinement phase transition in a heavy quark mass

region, in *Proc. 28th Int. Symp. Lattice Field Theory*, Villasimius, Italy, Paper 212 (PoS, 2010).
2. P. de Forcrand, J. Langelage, O. Philipsen and W. Unger, The phase diagram of strong coupling QCD including gauge corrections, in *Proc. 31st Int. Symp. Lattice Field Theory*, Mainz, Germany, Paper 142 (PoS, 2014).
3. L. McLerran and R. D. Pisarski, *Nucl. Phys. A* **796** (2007) 83.
4. M. G. Alford, K. Rajagopal and F. Wilczek, *Nucl. Phys. B* **537** (1999) 443.
5. Y. Nambu and G. Jona-Lasinio, *Phys. Rev.* **122** (1961) 345.
6. T. H. R. Skyrme, *Proc. R. Soc. Lond. A* **260** (1961) 127.
7. T. Skyrme, *Nucl. Phys.* **31** (1962) 556.
8. I. Zahed and G. E. Brown, *Phys. Rep.* **142** (1986) 1.
9. D. Diakonov, V. Y. Petrov and P. Pobylitsa, *Nucl. Phys. B* **306** (1988) 809.
10. V. M. Sarti and V. Vento, *Phys. Lett. B* **728** (2014) 323.
11. A. Drago and V. M. Sarti, *Phys. Rev. C* **86** 015211.
12. E. K. Heide, S. Rudaz and P. J. Ellis, *Nucl. Phys. A* **571** (1994) 713.
13. G. W. Carter, P. J. Ellis and S. Rudaz, *Nucl. Phys. A* **603** (1996) 367.
14. L. Bonanno and A. Drago, *Phys. Rev. C* **79** (2009) 045801.
15. J. Schechter, *Phys. Rev. D* **D21** (1980) 3393.
16. M. A. Migdal and A. A. Shifman, *Phys. Lett. B* **114** (1982) 445.
17. E. K. Heide, S. Rudaz and P. J. Ellis, *Phys. Lett. B* **293** (1992) 259.
18. G. W. Carter and P. J. Ellis, *Nucl. Phys. A* **628** (1998) 325.
19. M. Birse, J. Rehr and L. Wilets, *Phys. Rev. C* **38** (1988) 359.
20. E. Wigner and F. Seitz, *Phys. Rev.* **43** (1933) 804.
21. N. K. Glendenning and B. Banerjee, Soliton matter as a model of dense nuclear matter, Report No. LBL-20281, Lawrence Berkeley Laboratory, Berkeley, CA, USA (1985).
22. H. Forkel *et al.*, *Nucl. Phys. A* **504** (1989) 818.
23. D. Hahn and N. K. Glendenning, *Phys. Rev. C* **36** (1987) 1181.
24. B.-Y. Park and V. Vento, Skyrmion approach to finite density and temperature, in *The Multifaceted Skyrmion*, eds. G. E. Brown and M. Rho. (World Scientific, Singapore, 2010).
25. A. D. Jackson and J. J. M. Verbaarschot, *Nucl. Phys. A* **484** (1988) 419.
26. H.-J. Lee, B.-Y. Park, D.-P. Min, M. Rho and V. Vento, *Nucl. Phys. A* **723** 427 (2003).
27. Z. Tesanovic, O. Vafek and M. Franz, *Phys. Rev. B* **65** (2002) 180511.
28. H.-J. Lee, B.-Y. Park, M. Rho and V. Vento, *Nucl. Phys. A* **726** (2003) 69.
29. J. R. Ellis and J. Lanik, *Phys. Lett. B* **150** (1985) 289.
30. G. E. Brown and M. Rho, *Phys. Rev. Lett.* **66** (1991) 2720.
31. G. E. Brown, A. D. Jackson, M. Rho and V. Vento, *Phys. Lett. B* **140** (1984) 285.
32. R. J. Furnstahl, H.-B. Tang and B. D. Serot, *Phys. Rev. C* **52** (1995) 1368.
33. C. Song, G. E. Brown, D.-P. Min and M. Rho, *Phys. Rev. C* **56** (1997) 2244.
34. H.-J. Lee, B.-Y. Park, M. Rho and V. Vento, *Nucl. Phys. A* **741** (2004) 161.
35. H. Reinhardt and B. V. Dang, *Phys. Rev. D* **38** (1988) 2881.
36. K. Zarembo, *JETP Lett.* **75** (2002) 59.

Neutral current coherent cross-sections — Implications on detecting SN and earth neutrinos with gaseous spherical TPC's

J. D. Vergados*,‡ and Y. Giomataris†

**ARC Centre of Excellence in Particle Physics at the Terascale and Centre for the Subatomic Structure of Matter (CSSM), University of Adelaide, Adelaide SA 5005, Australia*

†IRFU, CEA, Université Paris-Saclay, F-91191 Gif-Sur-Yvette, France

**vergados@uoi.gr*

†Ioanis.Giomataris@cern.ch

The detection of galactic supernova (SN) neutrinos represents one of the future frontiers of low energy neutrino physics and astrophysics. The neutron coherence of neutral currents (NCs) allows quite large cross-sections in the case of neutron rich targets, which can be exploited in detecting earth and sky neutrinos by measuring nuclear recoils. They are relatively cheap and easy to maintain. These (NC) cross-sections are not dependent on flavor conversions and, thus, their measurement will provide useful information about the neutrino source. In particular, they will yield information about the primary neutrino fluxes and perhaps about the spectrum after flavor conversions in neutrino sphere. They might also provide some clues about the neutrino mass hierarchy. The advantages of large gaseous low threshold and high resolution time projection counters (TPC) detectors are discussed.

Keywords: Neutral currents; supernova neutrinos; Earth neutrinos; Gaseous detectors; spherical TPC; coherent cross section; astrophysics; nuclear recoil; neutrino oscillations; normal hierarchy; inverted hierarchy; Xe, Ar, Ne targets.

1. The Supernova Neutrino Flux

We will assume that the neutrino spectrum can be described as a Fermi Dirac Distribution with a given temperature T and a chemical potential $\mu = aT$. The constants T and a will be treated as free parameters. Thus,

$$f_{\rm sp}(E_\nu, T, a) = \mathcal{N}\frac{1}{1+\exp(E_\nu/T - a)}, \tag{1}$$

where $\mathcal{N}$ is a normalization constant. The temperature T is taken to be 3.5, 5 and 8 MeV for electron neutrinos (ν_e), electron antineutrinos ($\tilde{\nu}_e$) and all other flavors (ν_x), respectively. The parameter a will be taken to be $0 \leq a \leq 5$. The average neutrino energies obtained from this distribution are shown in Table 1.

‡Permanent address: Theoretical Physics, University of Ioannina, Ioannina, Gr 451 10, Greece

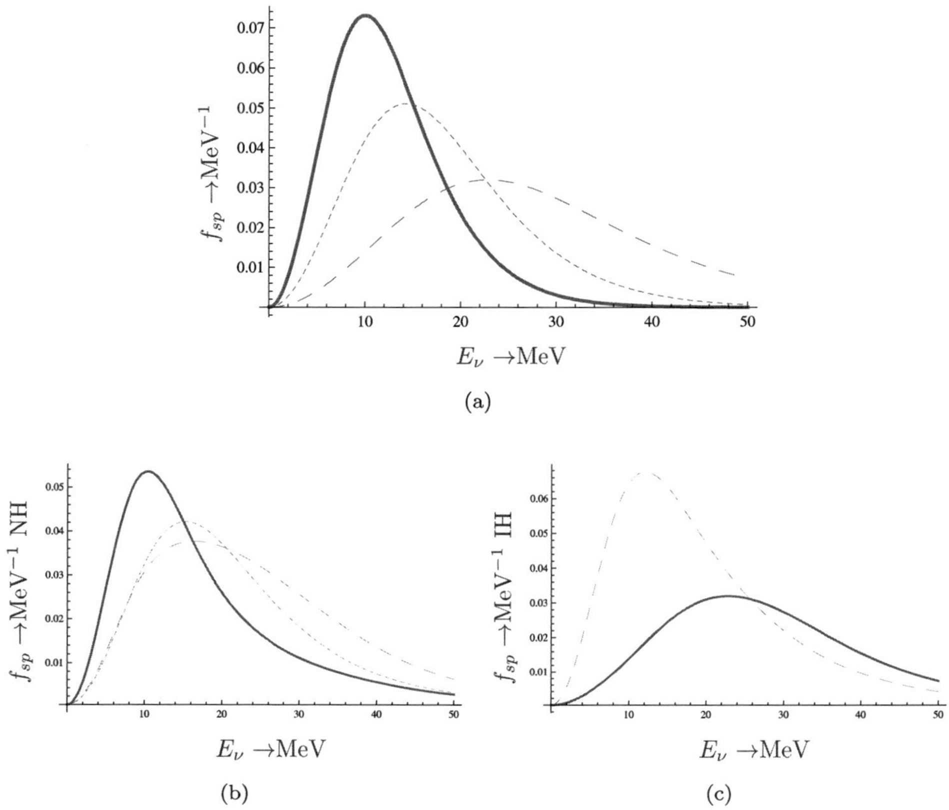

Fig. 1. (a) The normalized to unity SN spectrum for $a = 3$. (b) The modified SN spectra also for $a = 3$ in the normal hierarchy (NH) scenario. (c) The inverted hierarchy (IH) scenario. The continuous, dotted and dashed curves correspond to $T = 3.5$ (ν_e), 5 ($\tilde{\nu}_e$) and 8 (ν_x), respectively. In the case of the inverted hierarchy scenario, the spectra for ν_e and $\tilde{\nu}_e$ coincide.

Table 1. The average SN neutrino energies as a function of the parameters a and T.

	$\prec E_\nu \succ$ (MeV)		
a	ν_e $T = 3.5\,\text{MeV}$	$\tilde{\nu}_e$ $T = 5\,\text{MeV}$	$\sum_x \nu_x$ $T = 8\,\text{MeV}$
0	11.0298	15.7569	25.211
0.75	11.4504	16.3578	26.1724
1.50	12.0787	17.2553	27.6085
2.00	12.6194	18.0277	28.8443
3.00	13.9733	19.9619	31.9391
4.00	15.6313	22.3305	35.7288
5.00	17.5179	25.0255	40.0408

Table 2. The number of primary neutrinos emitted in a typical SN explosion as a function of the parameters a and T in units of 10^{58}.

	$N_\nu/10^{58}$		
a	ν_e $T = 3.5\,\text{MeV}$	$\tilde{\nu}_e$ $T = 5\,\text{MeV}$	$\sum_x \nu_x$ $T = 8\,\text{MeV}$
0	0.282969	0.198079	0.495196
0.75	0.272575	0.190802	0.477006
1.50	0.258397	0.180878	0.452194
2.00	0.247326	0.173128	0.43282
3.00	0.223361	0.156353	0.390882
4.00	0.199669	0.139768	0.349421
5.00	0.178167	0.124717	0.311792

Table 3. The (time integrated) neutrino flux, in units of $10^{12}\,\text{cm}^{-2}$, at a distance 10 kpc from the source.

	$\Phi_\nu/10^{12}\,\text{cm}^{-2}$		
a	ν_e $T = 3.5\,\text{MeV}$	$\tilde{\nu}_e$ $T = 5\,\text{MeV}$	$\sum_x \nu_x$ $T = 8\,\text{Mev}$
0	0.234318	0.164023	0.410057
0.75	0.225711	0.157997	0.394994
1.50	0.213971	0.149779	0.374448
2.00	0.204803	0.143362	0.358405
3.00	0.184958	0.129471	0.323677
4.00	0.16534	0.115738	0.289345
5.00	0.147534	0.103274	0.258185

The number of emitted neutrinos[1] can be obtained from the total emitted energy $U_\nu = 3 \times 10^{53}\,\text{erg}$

$$N_\nu = \frac{U_\nu}{\prec E_\nu \succ}. \tag{2}$$

The obtained results are shown in Table 2. The (time averaged) neutrino flux $\Phi_\nu = N_\nu/(4\pi D^2)$ at a distance $D = 10\,\text{kpc} = 3.1 \times 10^{22}\,\text{cm}$ is given in Table 3.

2. Modification of the SRN Spectra Due to Neutrino Oscillation

Even though the neutral current (NC) detector is neutrino flavor blind, the neutrino oscillations modify the SNS (supernova neutrino spectrum) spectrum. This modification will affect the expected rates since the different flavors have different temperature and chemical potential. This modification has recently been discussed.[2,3] Neutrino oscillations imply:

$$\frac{dN_{\nu_e}}{dE_\nu} = U_{e1}^2 \frac{dN_{\nu_1}}{dE_\nu} + U_{e2}^2 \frac{dN_{\nu_2}}{dE_\nu} + U_{e3}^2 \frac{dN_{\nu_3}}{dE_\nu}, \tag{3}$$

$$\frac{dN_{\nu_x}}{dE_\nu} = U_{x1}^2 \frac{dN_{\nu_1}}{dE_\nu} + U_{x2}^2 \frac{dN_{\nu_2}}{dE_\nu} + U_{x3}^2 \frac{dN_{\nu_3}}{dE_\nu}, \quad x = \mu, \tau \tag{4}$$

for neutrinos and antineutrinos. The superscript zero refers to the primary neutrino spectra. It can be shown that for the NH $m_1 < m_2 < m_3$

$$\frac{dN_{\nu_1}}{dE_\nu} \approx \frac{dN^0_{\nu_e}}{dE_\nu}, \quad \frac{dN_{\nu_2}}{dE_\nu} \approx \frac{dN^0_{\nu_x}}{dE_\nu}, \quad \frac{dN_{\nu_3}}{dE_\nu} \approx \frac{dN^0_{\nu_x}}{dE_\nu} \,\text{NH}, \tag{5}$$

while for the IH $m_3 < m_2 < m_1$, these equations become:

$$\frac{dN_{\nu_1}}{dE_\nu} \approx \frac{dN^0_{\nu_x}}{dE_\nu}, \quad \frac{dN_{\nu_2}}{dE_\nu} \approx \frac{dN^0_{\nu_x}}{dE_\nu}, \quad \frac{dN_{\nu_3}}{dE_\nu} \approx \frac{dN^0_{\nu_e}}{dE_\nu} \,\text{IH}. \tag{6}$$

Combining these equations, we get for NI:

$$\frac{dN_{\nu_e}}{dE_\nu} \approx \frac{2}{3}\frac{dN^0_{\nu_e}}{dE_\nu} + \frac{1}{3}\frac{dN^0_{\nu_x}}{dE_\nu}, \quad \frac{dN_{\nu_x}}{dE_\nu} \approx \frac{1}{6}\frac{dN^0_{\nu_e}}{dE_\nu} + \frac{5}{6}\frac{dN^0_{\nu_x}}{dE_\nu}, \tag{7}$$

while for the IH we find:

$$\frac{dN_{\nu_e}}{dE_\nu} \approx \frac{dN^0_{\nu_x}}{dE_\nu}, \quad \frac{dN_{\nu_x}}{dE_\nu} \approx \frac{1}{2}\left(\frac{dN^0_{\nu_e}}{dE_\nu} + \frac{dN^0_{\nu_x}}{dE_\nu}\right). \tag{8}$$

For completeness, we mention that as the neutrinos continue to propagate outwards, they encounter a further modification by the Mikheyev–Smirnov–Wolfenstein (MSW) effect.[4,5] It has been shown[6] that

$$\begin{aligned} \frac{dN''_{\nu_e}}{dE_\nu} &\approx P_m \frac{dN_{\nu_e}}{dE_\nu} + (1 - P_m)\frac{dN_{\nu_x}}{dE_\nu}, \\ \frac{dN''_{\nu_x}}{dE_\nu} &\approx (1 - P_m)\frac{dN_{\nu_e}}{dE_\nu} + P_m \frac{dN_{\nu_x}}{dE_\nu}. \end{aligned} \tag{9}$$

The parameter P_m depends not only on the mixing angles but also on the crossing probabilities PH (probability high) and PL (probability low) for the neutrino eigenstates at higher and lower resonances. It is also different for neutrinos and antineutrinos, but we will not elaborate further.

3. The TPC Detector

It will be very interesting to see whether one can gain detailed information about the supernova neutrino spectrum (SNS) and perhaps gain information about the neutrino hierarchy from a NC detector with high sensitivity.

To this end we are proposing to use a gaseous spherical time projection counter (TPC) detector dedicated for SN neutrino detection through the neutrino-nucleus coherent process. More specifically to use a gaseous spherical TPC detector, dedicated to SN neutrino detection, exploiting the neutrino-nucleus neutron coherent process. The first idea is to employ a small size spherical TPC detector filled with a high pressure noble gas.[7,8] Today, the spherical detector is used for dark matter search at LSM (0.6 m in diameter 10 bar pressure) underground laboratory and for NEWS experiment a future project at SNOLAB (1.5 m, 10 bar).[9] Dark matter detector is focused in light-WIMP search using gas target of light elements as H, He and Ne are more sensitive in the GeV and sub-GeV range, compared to current experiments using of Xe and Ge. During data taking the threshold was set at 30 eV, a second hint to reach low WIMP mass sensitivity.

For the SN project, we could use a conceptual design of detector based on the existing technology and increase detector diameter and pressure to 50 bar taking into account recent developments. A key issue for such highpressure operation is the use of a sensor ball smaller than the current 6.3 mm in diameter. In recent laboratory investigations, we have successfully used 2 mm ball sensor. This is coming closer to our goal of 1 mm ball, which would be the ideal size for reaching stable operation at 50 bar. Using such small sensors will require lower operation voltage and therefore induce low electric field (E) at large distances:

$$E(r_1) = r_2 \frac{V_0}{r_1^2},$$

which shows that the electric field at the periphery is proportional to the radius of the small ball (r_2) and inversely proportional to the radius square of the external sphere (r_1). Such low field may become a concern for large detector. Clearly, there is a contradiction between large detector and small sensor.

Recently, we have successfully tested, however, a new idea with a multi-ball sensor which is supposed to solve this problem. The multi-ball system is employing many small conductive balls arranged around a larger spherical surface, that could also tune the electric field at large distance and at the same time it could produce a segmentation of the detector. In the multi-ball arrangement, the electric field at large distance is proportional to the distance of the balls from the center of the spherical detector and not to the radius of the balls as it is in the case of existing central single ball detector. Note that such multi-ball system will optimize the electric field for any size of the external sphere and the detector segmentation will help to avoid pile-up events.

An enhancement of the NC component is achieved via the coherent effect of all neutrons in the target. Thus employing, e.g., Xe at 10 Atm, with a feasible threshold energy of about 100 eV in the recoiling nuclei, we show that one may expect, depending on the neutrino hierarchy, between 300 and 500 events for a sphere of radius 3 m. This can go up to 1500 and 2500 events, if the pressure is raised at 50 Atm, something quite feasible today taking into account recent detector progress.

4. The Differential and Total Cross-Section

The differential cross-section for a given neutrino energy E_ν can be cast in the form[10]:

$$\left(\frac{d\sigma}{dT_A}\right)_w (T_A, E_\nu) = \frac{G_F^2 A m_N}{2\pi} (N^2/4) F_{\text{coh}}(T_A, E_\nu) \tag{10}$$

with

$$F_{\text{coh}}(T_A, E_\nu) = F^2(q^2)\left(1 + \left(1 - \frac{T_A}{E_\nu}\right)^2 - \frac{A m_N T_A}{E_\nu^2}\right), \tag{11}$$

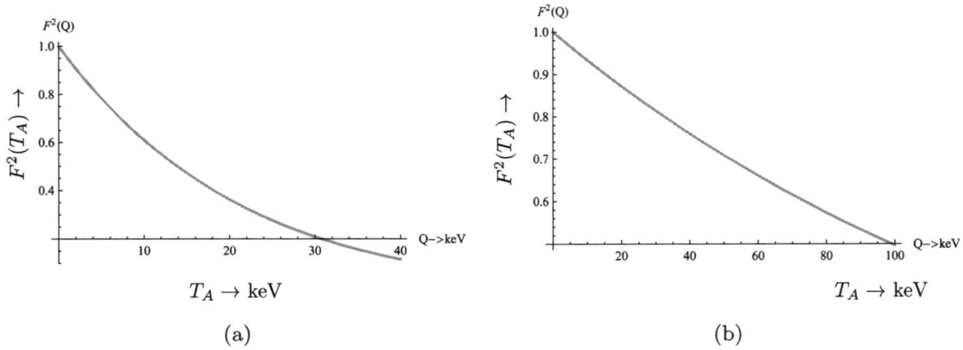

Fig. 2. The square of the nuclear form factor, $F^2(T_A)$, as a function of the recoil energy for (a) $A = 131$ and (b) $A = 40$. Note that the maximum recoil energy is different for each target.

where N is the neutron number and $F(q^2) = F(T_A^2 + 2Am_NT_A)$ is the nuclear form factor. The effect of the nuclear form factor depends on the target (see Fig. 2).

Since the SN source is not "monochromatic" the above equation can be written as

$$\frac{d\sigma}{dT_A} = \int_{E(T_A)}^{(E_\nu)_{\max}} \left(\frac{d\sigma}{dT_A}\right)_w (T_A, E_\nu) f_{\text{sp}}(E_\nu, T, a) dE_\nu, \tag{12}$$

where $(E_\nu)_{\max}$ is the maximum neutrino energy and

$$E(T_A) = \frac{T_A}{2} + \sqrt{\frac{T_A}{2}\left(M_A + \frac{T_A}{2}\right)},$$

where $(E_\nu)_{\max} = \infty$.

Integrating the total cross-section of Fig. 3 from $T_A = E_{\text{th}}$ to infinity, we obtain the total cross-section. The threshold energy E_{th} depends on the detector.

The number of the observed events for each neutrino species is found to be:

$$N_{\text{ev}}(a, T) = \Phi_\nu(a, T)\sigma(a, T, E_{\text{th}})N_N(P, T_0, R), \tag{13}$$

$$N_N(P, T_0, R) = \frac{P}{kT_0}\frac{4}{3}\pi R^3, \tag{14}$$

where N_N is the number of nuclei in the target, which depends on the pressure, (P), the absolute temperature, (T_0), and the radius R of the detector. We find:

$$N_N(P, T_0, R) = 1.04 \times 10^{30} \frac{P}{10\,\text{Atm}} \frac{300^\circ\text{K}}{T_0} \left(\frac{R}{10\,\text{m}}\right)^3. \tag{15}$$

4.1. *Results for the Xe target*

The differential cross-section for neutrino elastic scattering, obtained with the above neutrino spectrum, on the target $^{131}_{54}$Xe is shown in Fig. 3. Integrating the total cross-section of Fig. 3 from $T_A = 0$ to infinity, we obtain the total cross-section

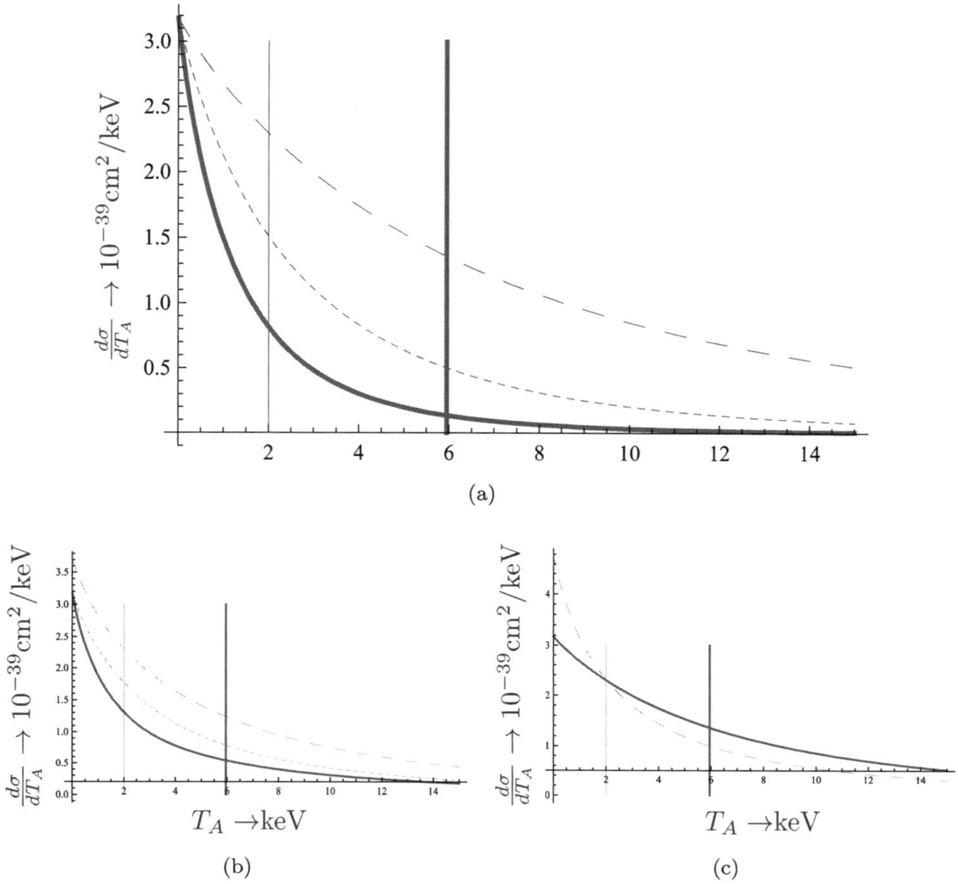

Fig. 3. The differential cross-section for elastic neutrino nucleus scattering in the case of the target $^{131}_{54}$Xe as a function of the recoil energy T_A in keV for (a) the primordial spectrum, (b) the NH and (c) IH scenario. For a threshold energy of 2 keV, only the space on the right of the vertical bar is available. In the presence of quenching, only the space on the right of the thick vertical bar is available. Otherwise, the notation is the same as in Fig. 1.

given in Table 4. The above results refer to an ideal detector operating down to zero energy threshold. In the case of nonzero threshold, the event rate is suppressed as shown in Fig. 4.

Using Eq. (13) and the above total cross-sections, after summing over all neutrino species (i.e., over all T), we obtain the number of events shown in Table 5. It is a surprise for us that the total rate increases somewhat after the modification of the primary neutrino spectra. It is not shown here, but the fraction of ν_x is 77%, 65% and 46% for the S, NH and IH neutrino spectra, respectively, i.e., it decreases as we go from the primary to the modified neutrino spectra. This is not relevant for a NC detector, but it may be an important element for detectors using the charged current process, so long as they are sensitive only to the ν_e and $\bar{\nu}_e$ components.

Table 4. The total neutrino nucleus cross-section in the case of Xe target in units of 10^{-39} cm^2 assuming zero detector threshold.

	$\sigma/10^{-39}$ cm^2			
a	ν_e $T = 3.5$ MeV	$\tilde{\nu}_e$ $T = 5$ MeV	$\sum_x \nu_x$ $T = 8$ MeV	Total
0	4.117	8.312	19.764	32.194
0.75	4.361	8.815	20.921	34.097
1.50	4.749	9.608	22.727	37.083
2.00	5.104	10.330	24.346	39.780
3.00	6.074	12.288	28.621	46.083
4.00	7.408	14.966	34.147	56.521
5.00	9.118	18.364	40.546	68.028

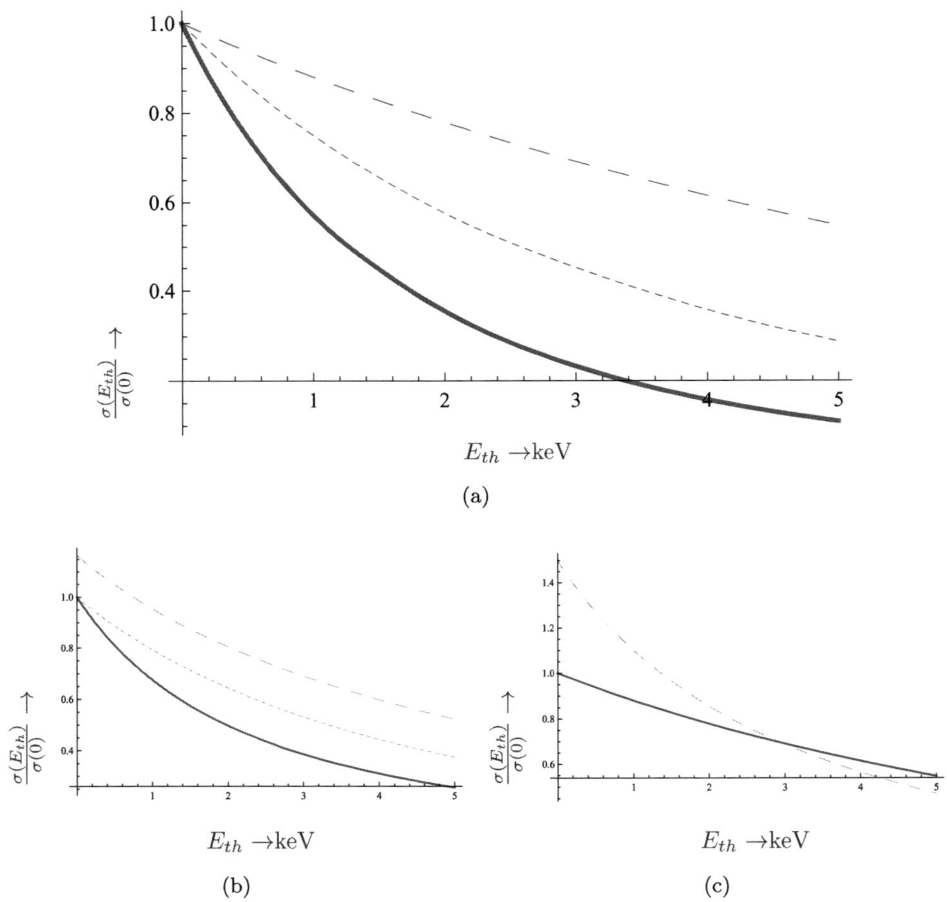

Fig. 4. The ratio of the cross-section at threshold $E_{\rm th}$ divided by that at zero threshold as a function of the threshold energy in keV in the case of a Xe target for (a) the primordial spectrum, (b) the NH and (c) IH scenario. Otherwise the notation is the same as in Fig. 1.

Table 5. The total event rate as a function of a in the case of a gaseous Xe target under a temperature 300°K and various pressures with the indicated spherical detector radii. S, NH and IH stand for neutrino spectra in the standard (primary), modified in the NH and the IH neutrino, respectively. These results were obtained by summing over all neutrino species assuming a zero detector energy threshold. In labeling, the columns 1 $\Leftrightarrow$ ($R = 10\,\text{m}$, $P = 10\,\text{Atm}$), 2 $\Leftrightarrow$ ($R = 3\,\text{m}$, $P = 50\,\text{Atm}$), 3 $\Leftrightarrow$ ($R = 4\,\text{m}$, $P = 10\,\text{Atm}$).

	1 S	1 NH	1 IH	2 S	2 NH	2 IH	3 S	3 NH	3 IH
0	10,872	12,275	15,083	1467	1657	2036	695	785	965
0.8	11,089	12,520	15,383	1497	1690	2076	709	801	984
1.5	11,427	12,901	15,850	1542	1741	2139	731	825	1014
2.0	11,726	13,238	16,262	1583	1787	2195	750	847	1040
3.0	12,482	14,089	17,302	1685	1902	2335	798	901	1107
4.0	13,378	15,093	18,523	1806	2037	2500	856	965	1185
5.0	14,287	16,108	19,749	1928	2174	2666	914	1030	1263

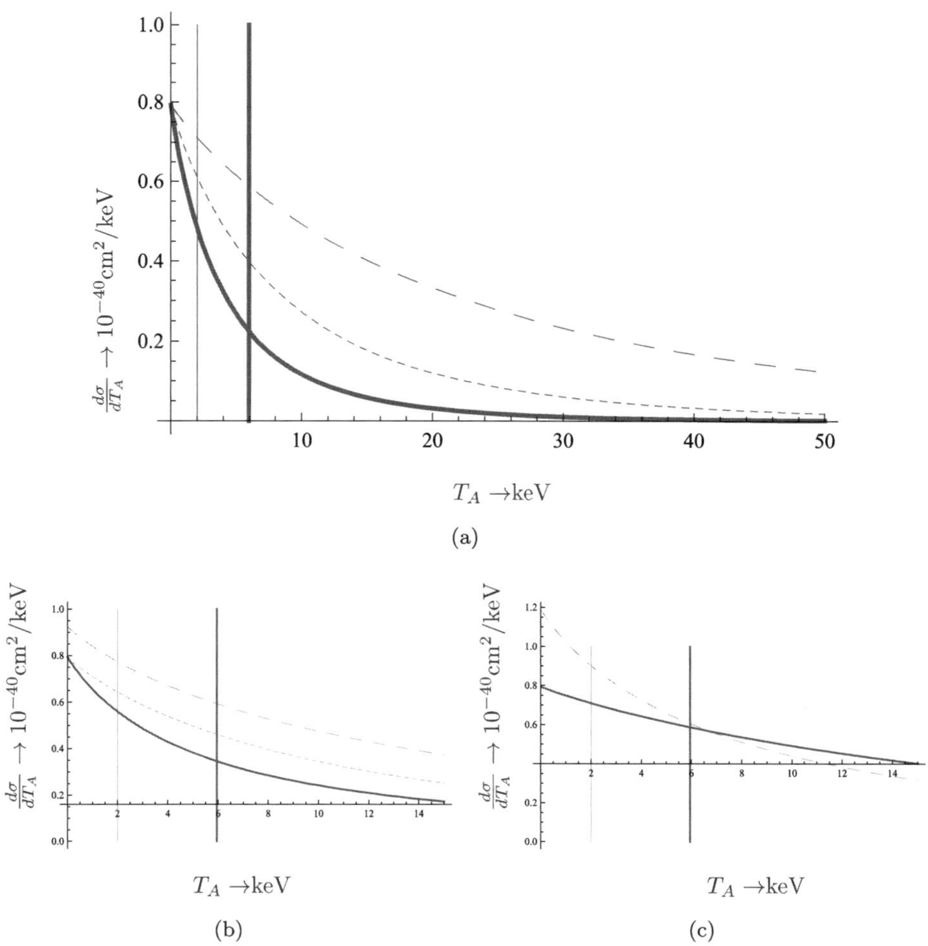

Fig. 5. The same as in 3 in the case of the Ar target.

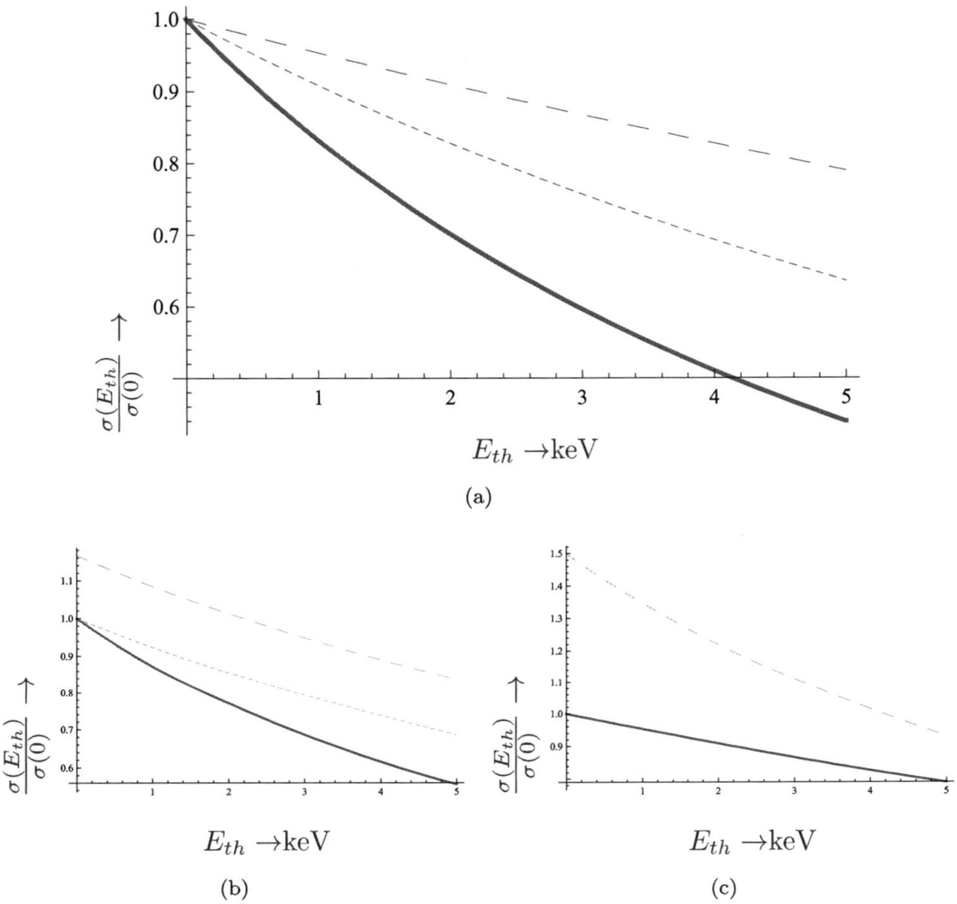

Fig. 6. The same as in Fig. 4 in the case of the Ar target.

Table 6. The total neutrino nucleus cross-section in the case of Ar target in units of 10^{-40} cm^2 assuming zero detector threshold.

	$\sigma/10^{-40}$ cm^2			
a	ν_e $T = 3.5$ MeV	$\tilde{\nu}_e$ $T = 5$ MeV	$\sum_x \nu_x$ $T = 8$ MeV	Total
0	3.324	6.520	13.678	23.521
0.75	3.525	6.908	14.412	24.845
1.50	3.843	7.518	15.528	26.888
2.00	4.133	8.067	16.497	28.693
3.00	4.917	9.537	18.905	33.359
4.00	5.990	11.488	21.690	39.168
5.00	7.353	13.843	24.480	45.676

Table 7. The same as in Table 5 in the case of the target ^{40}Ar.

	1 S	1 NH	1 IH	2 S	2 NH	2 IH	3 S	3 NH	3 IH
0	777	874	1070	104	118	144	49	55	68
0.75	789	889	1087	106	120	146	50	56	69
1.5	808	910	1113	109	122	150	51	58	71
2.0	824	928	1134	111	125	153	52	59	72
3.0	861	968	1182	116	130	159	55	61	75
4.0	895	1005	1225	120	135	165	57	64	78
5.0	920	1031	1254	124	139	169	58	66	80

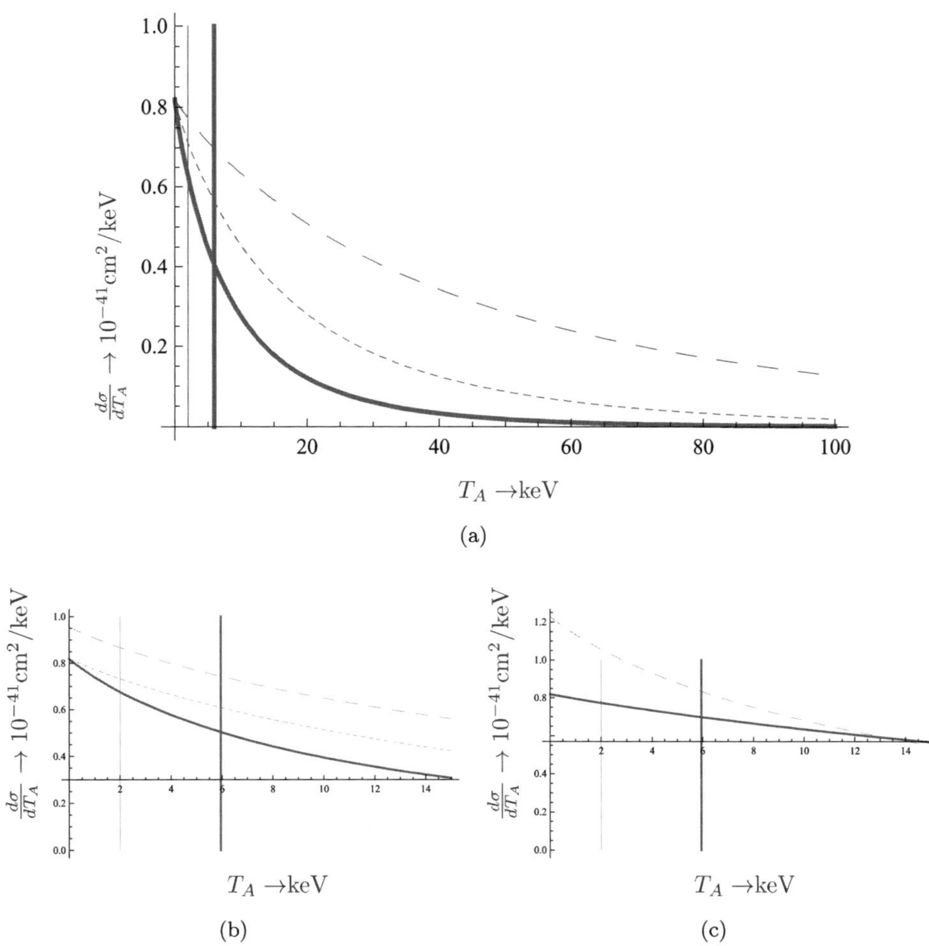

Fig. 7. The same as in 3 in the case of the Ne target.

In the presence of a detector threshold of even 1 keV, the above rates are reduced by about 20% (50%) in the absence (presence) of quenching.

4.2. *The Ar target*

The differential cross-section for neutrino elastic scattering on the target $^{40}_{18}$Ar is shown in Fig. 5. For comparison, we are currently calculating the differential cross-sections to the excited states of ^{40}Ar due to the NC. We are also going to calculate the charged current cross-sections (ν_e, e^-) and $(\tilde{\nu}_e, e^+)$ on ^{40}Ar, which are of interest in the proposal GLACIER, one of the large detectors.[a]

In the presence of a detector threshold of even 1 keV, the above rates are reduced by about 10% (30%) in the absence (presence) of quenching (see Fig. 6).

The total cross section is shown in Table 6 and the total rate in Table 7.

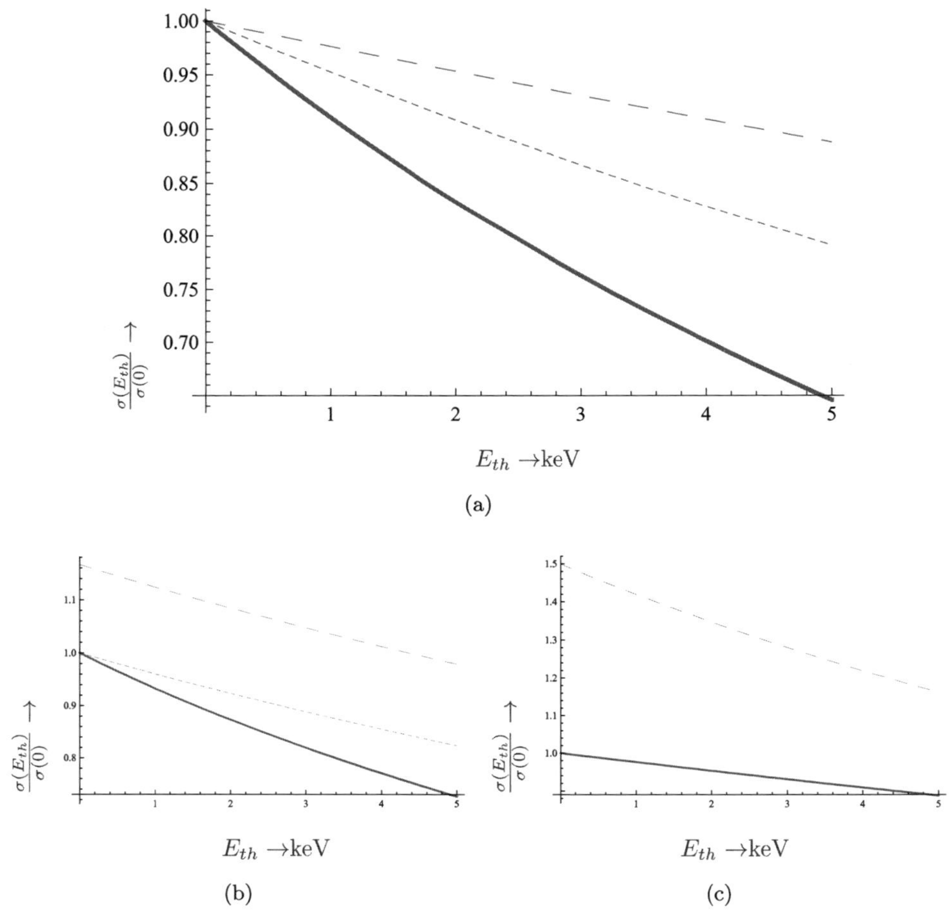

Fig. 8. The same as in Fig. 4 in the case of the Ne target.

[a]V. Tsakstara, private communication.

Table 8. The total neutrino nucleus cross-section in the case of Ne target in units of 10^{-41} cm^2 assuming zero detector threshold.

	$\sigma/10^{-41}$ cm^2			
a	ν_e $T = 3.5$ MeV	$\tilde{\nu}_e$ $T = 5$ MeV	$\sum_x \nu_x$ $T = 8$ MeV	Total
0	6.861	13.456	28.232	48.548
0.75	7.277	14.258	29.747	51.281
1.50	7.934	15.515	32.049	55.497
2.00	8.531	16.649	34.049	59.229
3.00	10.150	19.683	39.021	68.854
4.00	12.364	23.708	44.772	80.844
5.00	15.176	28.568	50.537	94.281

Table 9. The same as in Table 5 in the case of the target ^{20}Ne.

	1 S	1 NH	1 IH	2 S	2 NH	2 IH	3 S	3 NH	3 IH
0	160	180	220	21	24	29	10	11	14
0.75	163	183	224	22	24	30	10	11	14
1.5	166	187	229	22	25	31	10	12	14
2.0	170	191	234	22	25	31	10	12	14
3.0	177	199	243	23	26	32	11	12	15
4.0	184	207	253	24	28	34	11	13	16
5.0	190	213	258	25	28	34	12	13	16

4.3. *The Ne target*

The differential cross-section for neutrino elastic scattering on the target $^{20}_{10}$Ne is shown in Fig. 7 and the total cross-section in Table 8 and the total rate in Table 9. The effect of threshold on the total rate is shown in Fig. 8.

In the presence of a detector threshold of even 1 keV, the above rates are reduced by about 5% (10%) in the absence (presence) of quenching (see Fig. 8).

5. Conclusions

From the above results, one can clearly see the advantages of a gaseous spherical TPC detector dedicated for SN neutrino detection. The first idea is to employ a small size spherical TPC detector filled with a high pressure noble gas. An enhancement of the NC component is achieved via the coherent effect of all neutrons in the target. Thus employing, e.g., Xe at 10 Atm, with a feasible threshold energy of about 100 eV in the recoiling nuclei, one expects, depending on the neutrino hierarchy, between 300 and 500 events for a sphere of radius 3 m. This can go up to 1500 and 2500 events if the pressure is raised at 50 Atm, something quite feasible even today. Employing ^{40}Ar, one expects between 150 and 200 events but it can become larger if the pressure can be increased above 50 Atm, something quite realistic.

The second idea is to build several such low cost and robust detectors and install them in several places over the world, utilizing the small radius spheres filled with Ar under relatively small pressure. The first estimates involving small low pressure spheres show that the required background level is modest and therefore there is no need for a deep underground laboratory. A mere 100 m water equivalent coverage seems to be sufficient to reduce the cosmic muon flux at the required level (in the case of many such detectors in coincidence, a modest shield is sufficient). The maintenance of such systems, quite simple and needed only once every few years, could be easily assured by universities or even by secondary schools, with only specific running programs. Admittedly such a detector scheme, measuring low energy nuclear recoils from neutrino nucleus elastic scattering, will not be able to determine the incident neutrino vector and, therefore, it is not possible to localize the SN this way. This can be achieved by a cluster of such detectors in coincidence by a triangulation technique.

A network of such detectors in coincidence with a sub-keV threshold could also be used to observe unexpected low energy events. This low energy range has never been explored using massive detectors. A challenge of great importance will be the synchronization of such a detector cluster with the astronomical γ-ray burst telescopes to establish whether low energy recoils are emitted in coincidence with the mysterious γ bursts.

In summary, networks of such dedicated gaseous TPC detectors, made out of simple, robust and cheap technology, can be simply managed by an international scientific consortium and operated by students. This network comprises a system, which can be cheaply maintained for several decades (or even centuries). Obviously this is a key point towards preparing to observe few galactic SN explosions.

Thus, with adequate funding and if we are lucky to soon have a SN not much further than 10 kps, we might be able to observe SN neutrinos. This event, in conjunction with organizing an activity to honor Gerry, will be a tribute to G. E. Brown for the memorable work he did on the equation of state of collapsing stars, in collaboration with another giant of physics, H. Bethe. Among other things, they contributed to the understanding of the birth of our SN neutrinos.

Acknowledgments

JDV would like to acknowledge support by ARC Centre of Excellence in Particle Physics at the Terascale, University of Adelaide, and thank Professor A. W. Thomas for his hospitality. He would also like to express his appreciation to Tom Kuo for providing the opportunity to dedicate this work to the unforgettable Gerry.

References

1. R. Tomas, M. Kachellriess, A. G. G. Raffelt, A. Janka and L. Schreck, *J. Cosmol. Astropart. Phys.* **0409** (2004) 015.
2. A. S. Dighe and A. Y. Smirnov, *Phys. Rev. D* **62** (2000) 033007.

3. K. Nakazato, E. Moshida, Y. Nino and H. Suzuki, *Astrophys. J.* **804** (2015) 75, arXiv:1503.01236 (astro-ph.HE).
4. L. Wolfenstein, *Phys. Rev. D* **17** (1978) 2369.
5. S. P. Mikheev and A. Y. Smirnov, *Sov. J. Nucl. Phys.* **42** (1985) 913; *Yad. Fiz.* **42** (1985) 1441.
6. S. H. Chiu, C.-C. Huang and K.-C. Lai, *Prog. Theor. Exp. Phys.* **1** (2015) 063, arXiv:1312.4262 (hep-ph).
7. I. Giomataris *et al.*, *J. Instrum.* **3** (2008) 09007.
8. E. Bourgamont *et al.*, *J. Mod. Phys.* **3** (2012) 57.
9. G. Gerbier *et al.*, arXiv:1401.7902 [astro-ph.IM].
10. Y. Giomataris and J. Vergados, *Phys. Lett. B* **634** (2006) 23.

Permanent electric dipole moments of single-, two- and three-nucleon systems

Andreas Wirzba[*,†,¶], Jan Bsaisou[*] and Andreas Nogga[*,‡,§,‖,**]
[*]*IKP-3 and IAS-4, Forschungszentrum Jülich, D-52425 Jülich, Germany*
[†]*Kavli Institute for Theoretical Physics, University of California, Santa Barbara, CA 93106-4030, USA*
[‡]*Institute of Nuclear and Particle Physics, Ohio University, Athens, OH 45701, USA*
[§]*Department of Physics and Astronomy, Ohio University, Athens, OH 45701, USA*
[¶]*a.wirzba@fz-juelich.de*
[‖]*a.nogga@fz-juelich.de*

A nonzero electric dipole moment (EDM) of the neutron, proton, deuteron or helion, in fact, of any finite system necessarily involves the breaking of a symmetry, either by the presence of external fields (i.e., electric fields leading to the case of *induced* EDMs) or explicitly by the breaking of the discrete parity and time-reflection symmetries in the case of *permanent* EDMs. We discuss two theorems describing these phenomena and report about the cosmological motivation for an existence of CP breaking beyond what is generated by the Kobayashi–Maskawa mechanism in the Standard Model and what this might imply for the permanent EDMs of the nucleon and light nuclei by estimating a window of opportunity for physics beyond what is currently known. Recent — and in the case of the deuteron even unpublished — results for the relevant matrix elements of nuclear EDM operators are presented and the relevance for disentangling underlying New Physics sources is discussed.

Keywords: Electric dipole moments; CP violation; chiral Lagrangians; beyond standard model.

1. The Problem with Permanent Electric Dipole Moments

Gerry Brown was always interested in *magnetic* dipole moments of baryons and nuclei, and especially in confronting the theoretical predictions of quark, chiral bag and Skyrme models with the experimental results. But to our knowledge (compare also with Ref. 1), he never worked on *electric* dipole moments. This choice definitely turned out to be wise in his case, since during his lifetime — and even until the time of writing — all measurements of the electric dipole moment (EDM) of any

[¶]Corresponding author.
[**]On leave of absence from the Forschungszentrum Jülich.

(sub)atomic particle have been compatible with zero — only more and more restrictive upper bounds have been established since the first experiment in the 1950s by Smith, Purcell, and Ramsey[2] for the neutron EDM.

1.1. *The subtle character of EDMs of subatomic particles*

Why are permanent EDMs, which in classical electrodynamics just correspond to (the integrals over) the spatial three-vectors of displaced charges (or in general charge densities), much more subtle in the case of subatomic particles or, generically, in the realm of Quantum Mechanics? The reason is that their existence is tied to the breaking of the discrete symmetries of parity (P) conservation and time-reflection (T) invariance, such that they are intrinsically small.[a] In fact, the order of magnitude of the nucleon EDM ($d_{\rm N}$) can be estimated as follows:[3]

(I) The starting scale is given by the CP and P conserving (magnetic) moment of the nucleon, which is of the order of the nuclear magneton

$$\mu_{\rm N} = \frac{e}{(2m_{\rm p})} \sim 10^{-14}\, e\,\text{cm}, \tag{1}$$

where $e > 0$ is the unit of electric charge and $m_{\rm p}$ the proton mass.

(II) Furthermore, as we will discuss below, a nonzero permanent EDM requires P and CP violation. The cost of P violation can empirically be estimated in terms of Fermi's constant $G_{\rm F} \approx 1.166 \cdot 10^{-5}\,\text{GeV}^{-2}$ times the square of the axial decay constant of the pion, $F_\pi \approx 92.2\,\text{MeV}$, the order parameter of the spontaneous breaking of chiral symmetry of Quantum Chromodynamics (QCD) at low energies.[4] The dimensionless product scales therefore as $G_{\rm F} \cdot F_\pi^2 \sim 10^{-7}$.

(III) The cost related to the *additional* CP violation follows from, e.g., the ratio of the amplitude moduli of $K_{\rm L}^0$ to $K_{\rm S}^0$ decays into two pions[4]:

$$|\eta_{+-}| = \frac{|\mathcal{A}(K_{\rm L}^0 \to \pi^+\pi^-)|}{|\mathcal{A}(K_{\rm S}^0 \to \pi^+\pi^-)|} = (2.232 \pm 0.011) \cdot 10^{-3}. \tag{2}$$

In summary, the modulus of the EDM of the nucleon cannot be larger than

$$|d_{\rm N}| \sim 10^{-3} \times 10^{-7} \times \mu_{\rm N} \sim 10^{-24}\, e\,\text{cm}, \tag{3}$$

which is 10 orders of magnitude smaller than the corresponding magnetic dipole moments, without getting into conflict with known physics — on top of the EDM measurements themselves which are nowadays even more restrictive (see below).

In the standard model (SM) the sole source for CP violation, if the QCD θ term is assumed to be absent, is the Kobayashi–Maskawa[5] (KM) mechanism, which, however, only generates a nonzero CP-violating phase if the Cabbibo–Kobayashi–Maskawa (CKM) quark-mixing matrix involves at least three quark generations.

[a]The breaking of T implies CP violation (in terms of the charge conjugation C symmetry) if CPT is conserved.

This KM generated CP violation is therefore flavor-violating, while the EDMs are, by nature, flavor-diagonal. This means that the SM (without the QCD θ term) is "punished" by the additional cost of a further factor $G_F F_\pi^2 \sim 10^{-7}$ to undo the flavor violation. In summary, the SM prediction for the nucleon EDM, based on the KM mechanism, is therefore much smaller than Eq. (3), namely

$$|d_N^{SM}| \sim 10^{-7} \times 10^{-24}\, e\,\mathrm{cm} \sim 10^{-31}\, e\,\mathrm{cm}. \tag{4}$$

This result agrees in magnitude with the three-loop estimates of Refs. 6 and 7 and also with the two-loop calculations of Refs. 8 and 9 (see also Ref. 10) which include both a strong *penguin* short-range diagram and a long-range pion loop.[b] Even recent *loop-less* calculations[11] involving propagators of charm-flavored sea-quarks give a result of the same order.[c]

From the above estimates one can infer that an EDM of the nucleon measured in the window

$$10^{-24} e\,\mathrm{cm} > |d_N| \gtrsim 10^{-30} e\,\mathrm{cm} \tag{5}$$

could be a clear signal for new physics beyond the KM mechanism of the SM: either *strong CP violation* by a sufficiently large QCD θ term or genuinely new physics, as, e.g., supersymmetric (SUSY) models, multi-Higgs models, left–right symmetric models, etc.

1.2. *Two theorems for the existence of EDMs*

This brings us back to the original question: Why do nonzero EDMs of finite quantum systems necessarily require the breaking of some symmetry? The above statement can be interpreted as a special case of the following theorem which, e.g., is well-known to apply for the case of the chiral symmetry breaking for lattice QCD (see e.g., Refs. 13–15).[d]

Theorem 1. *In* any finite *quantum system in the* absence *of any explicitly broken-symmetry there cannot exist a spontaneously broken ground state.*

The condition of *finiteness* applies to both the spatial extent of the system and to the height of the pertinent quantum "walls". Therefore the tunneling probability from *any* broken-symmetry state to any *alternative* one is non-vanishing. This

[b]Note that one-loop contributions to EDMs resulting from the KM mechanism of the SM have to vanish, since the CP-violating KM matrix element at the first loop vertex is canceled by its Hermitian conjugate at the other side.

[c]The EDM of the electron is even further suppressed by a factor 10^{-7} in the SM, i.e., $|d_e^{SM}| \sim 10^{-38}\, e\,\mathrm{cm}$, which follows from a further weak-interaction insertion and one additional quark/gluon loop.[12] The SM prediction for the muon is slightly larger, namely $|d_\mu^{SM}| \sim 10^{-35}\, e\,\mathrm{cm}$, because of the lepton mass ratio $m_\mu/m_e \sim 200$.

[d]In a finite lattice scenario, even when the lattice size becomes larger and larger, a nonzero value of a quark condensate can only be measured if the mass of the pertinent current quark differs from zero corresponding to an *explicit* breaking of chiral symmetry.

opens up the way for finite systems to form one totally symmetrized state from all the broken-symmetry alternatives which is then the real ground state of the system, while the same tunneling amplitudes induce nonvanishing gaps to suitable antisymmetric combinations of these states which are therefore excited states and nondegenerate to the ground state.

The question might arise why this does not apply to magnetic moments which have known and very well-tabulated[4] nonzero values for the cases of the electron, muon, proton, neutron, other baryons, nuclei, etc. In fact, the theorem applies but the solution is trivial. The nonzero value of the total angular momentum (i.e., the spin in the case of subatomic particles) suffices to induce the (rotational) symmetry breaking since it defines an axis (in the laboratory frame) for the projection of the magnetic moment,[e] which shares the same axial-vector properties as the spin. The appearance of a nonzero magnetic moment for a particle without any spin or angular momentum is forbidden and would indeed come as a surprise.

So what is the difference to the case of an EDM which has the operational definition of the displacement vector of the charges? Why is the existence of a nonzero spin of a particle or finite quantum system not enough to induce the necessary symmetry breaking? The difference is that spins and, in general, angular momenta **J** are *axial* vectors (as the magnetic fields **B**) while the EDM **d** is a *polar* vector:

$$\begin{aligned} \mathbf{J} &\overset{P}{\longleftrightarrow} \ \mathbf{J}, \quad \mathbf{d} \overset{P}{\longleftrightarrow} -\mathbf{d}, \\ \mathbf{J} &\overset{T}{\longleftrightarrow} -\mathbf{J}, \quad \mathbf{d} \overset{T}{\longleftrightarrow} \ \mathbf{d}. \end{aligned} \tag{6}$$

Without the explicit breaking of the discrete T (time-reflection) and P (parity) symmetries, the presence of a nonzero spin or angular momentum would not be enough to define the direction for the projection of the EDM vector, since the sign of this projection would be reversed under the above mentioned discrete symmetries if they are conserved. However, in the rest-frame of any subatomic particle with nonvanishing mass there are simply not any other vectors than the spin and total angular momentum. So it should be clear that at least in these cases there is a need for extra symmetry breaking if these particles are to carry a nonvanishing permanent EDM. The word *permanent* is important here, since the realization of an *induced* EDM is of course possible — without the breaking of T and P — in the presence of a nonvanishing electric field **E** which has the properties of a polar vector as the EDM. Note that, interpreted in this way, our theorem still holds also for the case of *induced* EDMs: the presence of an electric field, which by nature breaks the (rotational) symmetry of the system, is the stated precondition.

But we know that macroscopic and mesoscopic devices (capacitors, batteries, etc.) and even certain molecules (H_2O or NH_3) obviously can have sizable dipole

[e]More precisely, for any nonzero spin and any direction in space it is always possible to find an eigenstate of the spin operator with *nonvanishing* projection (quantum number).

moments which correspond to their spatial extent times the involved charges. Well, most of these systems break a symmetry classically. There, is e.g., a spatial vector pointing from one plate of a capacitor to the other one or from one nucleus in a diatomic polar molecule to the other one (which differs in charge and/or mass). But if these systems are interpreted quantum mechanically, as axially symmetric rods or as (a)symmetric tops, one should keep in mind the difference between body-fixed directions and lab-frame-fixed ones. If a polar symmetry still applies, the projection on a stationary state of fixed angular momentum in the lab-fixed frame suffices to average out the body-fixed (classical or intrinsic) EDM to a vanishing expectation value..[3,16,17] In the (a)symmetric top scenario, the tunneling amplitude from the state pointing in one direction of the lab-frame (spin) axis to the one projected onto the opposite direction would be small, but nonzero,[3] such that even then the theorem applies in principle. A nonzero EDM is measured in practice in the latter cases, since the applied electric fields might be small but nonvanishing, such that the measured EDM has the character of an induced EDM. Alternatively, the pertinent temperature of the system is nonzero or the system, because it might be unstable, it might have finite level widths or the measuring time might be not long enough to resolve the single levels, especially if the tunneling gaps were tiny.[18,19] In this way the resulting *de facto* degeneracy between the parity-even ground state and excited parity-odd states together with the direction defined either by the non vanishing electric field or solely by the total-spin direction would be sufficient to define the orientation for the resulting explicit symmetry breaking. If, however, the system were cooled down to such small temperatures and sufficiently shielded against electric fields and observed for a long enough time that the *de facto* degeneracy would be lifted, a non vanishing EDM could not be measured even in these cases. Only a truly infinite system would escape the consequences of the theorem.

Let us summarize what was stated above by the following theorem which describes the existence of permanent EDMs.[1,18,20]

Theorem 2. *Any nonvanishing coefficient d in the relation of the expectation values $\langle j^P|\mathbf{d}|j^P\rangle = d\langle j^P|\mathbf{J}|j^P\rangle$ of the EDM operator $\mathbf{d} \equiv \int \mathbf{r}\rho(\mathbf{r})d^3r$ (where $\rho(\mathbf{r})$ is the charge density) and the total angular momentum $\vec{J}$ expressed in terms of a stationary state $|j^P\rangle$ of a particle with at least one nonzero generalized "charge", nonzero mass, nonzero total angular momentum j and specified parity P, such that $\langle j^P|\mathbf{J}|j^P\rangle \neq 0$ in general, and no other energy degeneracy than its rotational one is a signal of P and T violation*[f] *and, because of the CPT theorem, of flavor-diagonal CP violation.*

The above particle can be an "*elementary*" particle as a quark, charged lepton, $W^{\pm}$ boson, Dirac neutrino, etc., or a "*composite*" particle as a neutron, proton,

[f]Without the violation of P, $\langle j^P|\mathbf{d}|j^P\rangle$ would just vanish since $\mathcal{P}|j^P\rangle = (-1)^P|j^P\rangle$ and $\mathcal{P}\mathbf{d}\mathcal{P} = -\mathbf{d}$, where $\mathcal{P}$ is the parity operator which has the property $\mathcal{P}^2 = 1$.

nucleus, atom, molecule or even a solid body, as long as it meets the requirements stated in the above theorem. Namely, it is important that

(i) The particle or system should carry a nonvanishing angular momentum to define an axis (excluding therefore scalar and pseudoscalar particles),
(ii) It should not be self-conjugate[g] in order to prevent that the charge-conjugation property of the particle does not even allow a unique orientation in its body-fixed frame,
(iii) It should be in a stationary state (i.e., the observation time and, in case it is a resonance, its lifetime should be so large that the pertinent energy level including its width has not any overlap with the levels of other states of opposite parity),
(iv) There should be no degeneracy (except of the states which only differ in their *magnetic* quantum numbers and which have the same parity of course). Otherwise the ground state of even parity could mix with a state of opposite parity and the directional information coming from the spin would suffice to define the quantization axis for the EDM without the need of explicit P and T breaking.

However, it is well known and especially applies to the case of molecular systems with closely spaced rotational levels or atoms with a sizable octupole moment that the near-degeneracy of the two opposite-parity levels might produce sizable enhancement factors for P- and T- violating quantities, see e.g., Refs. 16 and 17. A similar mechanism is at work in the case of the *induced* EDMs of water or ammonia molecules, see e.g., Ref. 21: In a simplified picture there is a pair of nearly degenerate states of opposite parity $|\pm\rangle$ (where $|+\rangle$ is the ground state) with energy levels which rearrange according to

$$E_{2,1} = \frac{1}{2}(E_- + E_+) \pm \sqrt{\frac{1}{4}(E_- - E_+)^2 + (e\langle \mathbf{r} \rangle \cdot \mathbf{E}\,)^2} \tag{7}$$

when exposed to an electric field $\mathbf{E}$. Here $\langle \mathbf{r} \rangle$ is the *transition* (not a diagonal!) matrix element of the charge displacement vector $\mathbf{r}$ between the states $|+\rangle$ and $|-\rangle$. For a sufficiently large $\mathbf{E}$ field, the second term in the square root dominates and there will be an approximately linear behavior of the levels, $E_{2,1} = \frac{1}{2}(E_- + E_+) \pm |e\langle \mathbf{r} \rangle \cdot \mathbf{E}\,|$, which mimics a linear Stark effect — but note the appearance of an absolute value. For a weak enough $\mathbf{E}$ field and a sufficiently low temperature we would find instead the following behavior (quadratic Stark effect):

$$E_{2,1} = E_{\mp} \pm \frac{(e\langle \mathbf{r} \rangle \cdot \mathbf{E})^2}{E_- - E_+} + \cdots . \tag{8}$$

[g]Examples for *self-conjugate* particles with spin are Majorana neutrinos or the ω, ρ^0 or ϕ^0 mesons, but not their SU(3) partners K^* which carry strangeness quantum numbers.

Thus the molecule has always an *induced* EDM which can be enhanced by the small energy difference between the states of opposite parity.[h]

2. Motivation for EDMs

Why should we be interested in measuring *permanent* EDMs? One reason is of course the window of opportunity which the tiny CP-violating KM mechanism of the SM opens for the search of *New Physics*, see relation (5). The other reason is the CP violation by itself. Independently of how much matter surplus might have originally been created in the Big Bang, after the inflation epoch the primordial baryon–antibaryon (density) asymmetry should have been leveled out to an extremely high precision. However, about 380,000 years later, when the temperature of the Universe had sufficiently decreased such that hydrogen atoms became stable against the radiation pressure and therefore the corresponding photons could not couple any longer to an electron–proton plasma, the ratio of this asymmetry to the photon density n_γ had the following measured value:

$$\left.\frac{n_B - n_{\bar{B}}}{n_\gamma}\right|_{\rm CMB} = (6.05 \pm 0.07) \cdot 10^{-10}. \tag{9}$$

This result was derived from the cosmic microwave background (CMB) measurements by the COBE, WMAP and Planck satellite missions,[22] while the prediction of the SMs of particle physics and cosmology is more than seven orders of magnitude less, see e.g., Ref. 23.

In fact, CP violation is one of the three conditions for the dynamical generation of the baryon–antibaryon asymmetry during the evolution of the universe as formulated by Sakharov in 1967.[24] These conditions can be paraphrased as follows:

(i) There has to exist a mechanism for the generation of baryon charge B in order to depart from the initial value $B = 0$ (after inflation).
(ii) Both C and CP have to be violated such that the production mechanisms and rates of B can be distinguished from the ones of $\bar{B}$ (even then the pertinent helicities are summed).
(iii) Either CPT has to be broken as well[i] or the dynamical generation had to take place during a stage of nonequilibrium (i.e., a sufficiently strong first-order phase transition) to discriminate, in the average, the B production reaction from its back reaction and to escape from the fact that $\langle B \rangle = 0$ holds on the average if CPT symmetry holds, i.e., from the time independence in the equilibrium phase.

[h]The case of a two-level system with a magnetic-moment interaction in the presence of a magnetic field is totally different, since there is always a linear contribution $\langle \pm | \boldsymbol{\mu} | \pm \rangle \cdot \mathbf{B}$, no matter how weak the $\mathbf{B}$ field might be, because these expectation values are diagonal.

[i]This in turn would imply the violation of Lorentz-invariance or locality or hermiticity.

While baryon plus lepton number $(B+L)$ violation can be accommodated by the SM in an early stage of the evolution via sphalerons,[25] the SM cannot satisfy the other two conditions:

(i) The CP breaking by the KM mechanism[5] is far too small; even a $\bar{\theta}$ angle[j] of the order of 10^{-10} which would still be compatible with the empirical bound for the neutron EDM[28] cannot help in generating a sufficient baryon–antibaryon asymmetry: there would be a mismatch of the scales relevant for an electroweak (or even higher) transition on the one hand and the $\sim$ 1–2 GeV regime where QCD becomes sufficiently nonperturbative such that instanton effects are not suppressed any longer on the other hand, see e.g., Ref. 29;

(ii) At vanishing chemical potential the SM, which as a relativistic quantum field theory is of course CPT invariant, shows only a rapid crossover instead of a phase transition of first order.

Therefore, the observed matter–antimatter asymmetry together with the insufficient CP violation of the SM represents one of the few existing indicators that there might be *New Physics* beyond the (BSM) physics which in turn might imply EDM values for subatomic particles that are larger in magnitude than those predicted by the KM mechanism of the SM. Note that this "evidence" for substantial EDM values (especially for hadrons and nuclei) is at best circumstantial and by no means compulsory.

The current status on the experimental bound of the neutron EDM is $|d_n| < 2.9 \cdot 10^{-26}\, e\,\text{cm}$ as measured by the Sussex/RAL/ILL group.[28] It cuts by two orders of magnitude into the new physics window (5), excluding in this way already some simple and minimal variants of the above mentioned New Physics models, especially some variants of (minimal) SUSY. The corresponding bound for the proton, namely $|d_p| < 2.0 \cdot 10^{-25}\, e\,\text{cm}$, is inferred from a theoretical calculation[30] applied to the input from the 2016 EDM bound for the diamagnetic ^{197}Hg atom, $|d_{\text{Hg}}| < 7.4 \cdot 10^{-30}\, e\,\text{cm}$.[31] The same method would predict for the neutron the bound $|d_n| < 1.6 \cdot 10^{-26}\, e\,\text{cm}$ which is even slightly less than the Sussex/RAL/ILL limit[28] but is of course affected by the imponderables in extracting the relevant nuclear matrix elements of the ^{197}Hg nucleus. The EDM bound on the electron is again inferred indirectly, since theoretical calculations[32–34] are needed to deduce it from the corresponding EDM bounds on paramagnetic atoms, e.g.,^{205}Tl with $|d_{\text{TL}}| < 9.4 \cdot 10^{-25}\, e\,\text{cm}$,[35] or on polar molecules, as e.g., YbF[36,37] or the recent ThO measurement by the ACME group.[38] The latter gives the most stringent bound on the electron EDM, $|d_e| < 8.7 \cdot 10^{-29}\, e\,\text{cm}$.

All the measurements mentioned above have in common that they only apply to overall charge-neutral states, since the corresponding particles can be confined

[j]The QCD parameter $\bar{\theta}$ is the sum of the original angle of the QCD θ term and the phase of the determinant of the quark mass matrix: $\theta \to \theta + \arg\det \mathcal{M}_{\text{q}}$. Even if canceled by the Peccei–Quinn mechanism,[26,27] small contributions might reappear generated by BSM physics and by possible "Peccei–Quinn breaking" terms from Planck-scale physics.

at rest in a trap even in the presence of reversible external electric fields (and a weak holding magnetic field) which are needed to extract the EDM signal. In order to trap charged particles (e.g., the proton, deuteron or helion), which would just be accelerated by a constant electric field, storage rings (see e.g., Refs. 39–41) or — as in the case molecular ions — traps with rotating electric fields[42] have to be applied. In fact, as a byproduct of $(g-2)_\mu$ measurements in storage rings, there already exists a very weak bound on the EDM of muon,[43] $|d_\mu| < 1.8 \cdot 10^{-19}\, e\,\text{cm}$, as compared to the SM estimate of $\sim 10^{-35}\, e\,\text{cm}$.

3. EDM Sources Beyond the KM Mechanism

If a nonzero permanent EDM could eventually be inferred from some measurement, we would then know that the source behind the pertinent T violation (or CP violation if CPT holds) would most likely not be the KM mechanism of the SM (recall Eq. (5)), but we would still be unable to pin down the very CP-violating mechanism: it could be genuine New Physics (as SUSY, two-Higgs models, left–right symmetric models) or just the QCD θ term, if the relevant angle were in the window $10^{-10} \gtrsim |\bar\theta| \gtrsim 10^{-14}$ (see below). In the case of genuine New Physics, the scale of the relevant CP-violating operator(s) would have to be larger than the electroweak scale, probably even larger than what is accessible by LHC physics.

However, by matching possible candidate models of CP-violating physics at these high scale(s) to the coefficients of SM operators of dimension-six[k] and higher, the machinery of effective field theories (EFTs) and the renormalization group can be applied.[45] In a repeating chain, the relevant operators, which also mix under this procedure, can perturbatively be run down until a (SM) particle threshold is reached (subsequently the top quark, Higgs boson, $W^\pm$ and Z bosons, and finally the bottom and charm quarks), where the corresponding particle should be integrated out and the coefficients of the operators should be matched to those containing only the remaining active SM degrees of freedom. This cascading perturbative procedure has to stop when the realm of nonperturbative QCD is reached somewhere between 2 GeV and 1 GeV, say. At this chiral scale Λ_χ the pertinent EFT Lagrangian of dimension-six[46] can be written as[l]

$$\begin{aligned}\mathcal{L}^{\not T,\not P} &= -\bar\theta \frac{g_s^2}{64\pi^2}\epsilon^{\mu\nu\rho\sigma} G^a_{\mu\nu} G^a_{\rho\sigma} \\ &\quad -\frac{i}{2}\sum_{q=u,d} d_q \bar q \gamma_5 \sigma_{\mu\nu} F^{\mu\nu} q - \frac{i}{2}\sum_{q=u,d} \tilde d_q \bar q \gamma_5 \frac{1}{2}\lambda^a \sigma^{\mu\nu} G^a_{\mu\nu} q \\ &\quad + \frac{d_W}{6} f_{abc}\epsilon^{\mu\nu\alpha\beta} G^a_{\alpha\beta} G^b_{\mu\rho} G^{c\,\rho}_{\nu} + \sum_{i,j,k,l=u,d} C^{4q}_{ijkl}\bar q_i \Gamma q_j \bar q_k \Gamma' q_l, \end{aligned} \tag{10}$$

[k]The solely existing CP-violating operator of dimension-five is a Majorana mass term which is only relevant for neutrino physics.[44]

[l]In the following, we will concentrate on the EDM contributions to nucleons and light nuclei. Therefore, the terms involving leptons and strange quark contributions are not listed.

where we included, for completeness, the dimension-four QCD θ term as well. The relevant quantities are the quark fields of flavor q, the field strength tensors $F_{\mu\nu}$ and $G^a_{\mu\nu}$ of the photon and the gluon, respectively, the color SU(3) structure constants f_{abc} and Gell-Mann matrices λ^a. The various Lorentz structures of the matrices Γ and Γ' ensure that the dimension-six four-quark operators, which have net zero flavor, violate the P and T symmetries..[46] The coefficients d_q of the quark EDM terms and $\tilde{d}_q$ of the quark chromo-EDM terms scale as $\sim v_{\rm EW}/\Lambda^2_{\not T}$ while the coefficient $d_{\rm W}$ of the gluon chromo-EDM term, the so-called Weinberg term, and the coefficients $C^{4\rm q}_{ijkl}$ of the four-quark EDM terms scale as $\sim 1/\Lambda^2_{\not T}$. $\Lambda_{\not T}$ is the scale of the underlying T (or CP) breaking model and $v_{\rm EW}$ is the electroweak vacuum expectation value which is a relic of the original coupling to the Higgs field which had to be inserted to preserve the SM symmetries.[m] Note that via a chiral $U_A(1)$ transformation the first term on the right-hand side of Eq. (10) can be rotated into the term $-\bar{\theta} m^*_{\rm q} \sum_q \bar{q} i\gamma_5 q$ where $m^*_{\rm q} = m_u m_d/(m_u + m_d)$ is the reduced quark mass. In this way, it is evident that all EDM contributions of the QCD θ term have to vanish in the chiral limit (as it is also the case for the quark and quark chromo-EDM contributions) and that therefore the nucleon EDM induced by the *strong CP* breaking term has to scale as

$$|d^{\bar{\theta}}_{\rm N}| \sim \bar{\theta} \cdot \frac{m^*_{\rm q}}{\Lambda_{\rm QCD}} \cdot \frac{e}{2m_{\rm N}} \sim \bar{\theta} \cdot 10^{-16} e\,{\rm cm}, \qquad (11)$$

such that the window (5) for physics beyond the KM mechanism of the SM together with the current bound on the neutron EDM[28] is compatible with the $10^{-10} \gtrsim |\bar{\theta}| \gtrsim 10^{-14}$ window for searches of strong CP breaking.

The θ term contribution, the Weinberg term and two of three four-quark terms are flavor/isospin symmetric, while the quark and quark chromo-EDMs can be separated into isospin-conserving and isospin-breaking combinations, respectively. The third four-quark operator stems from left–right symmetric physics which breaks isospin and also chiral symmetry at the fundamental level — for more details, see e.g., Ref. 46.

4. EDMs in the Nonperturbative Realm of QCD

At and below the chiral scale Λ_χ, perturbative methods are not applicable any longer in order to continue toward the hadronic, nuclear or even atomic scales relevant for the EDM experiments, since the degrees of freedom change from the quark/gluon ones to hadronic ones. The relevant methods which allow for an estimate of the pertinent uncertainties are lattice QCD and chiral EFT (i.e., chiral perturbation theory and its extension to multi-baryon systems).

[m] Thus the coefficients $d_{u,d}$ and $\tilde{d}_{u,d}$ effectively scale as $\sim m_{u,d}/\Lambda^2_{\not T}$ to ensure that chiral symmetry is preserved in the limit of vanishing current quark masses $m_{u,d}$. The quark and quark chromo-EDM operators are therefore counted as dimension-six operators.

Some progress has been made in lattice QCD calculations of the EDM of the neutron (and in some cases of proton) when it is induced by the θ term, see e.g., Refs. 47 and 48, but the extrapolation to physical pion masses still seems to be problematic for specific lattice methods.[48] First lattice results for the quark EDM scenario relating the nucleon EDM to the tensor charges of the quark flavors are promising[49] but of course not sufficient to constrain realistic models of CP breaking. Recently, there have been attempts to work out the quark flavor contributions to the nucleon EDMs also for the case of quark chromo-EDMs.[50] Lattice estimates of the nucleon EDM resulting from the Weinberg term or even from the four-quark terms are still left for future studies, not to mention lattice computations of the EDMs of (light) nuclei.

What is the situation from the chiral EFT point of view? The first chiral calculations of the nucleon EDM induced by the QCD θ term were already performed in the late seventies of the last century.[51] The results of this calculation and more modern ones[52,53] are that the leading and subleading CP-violating pion-loop contributions to the isovector nucleon EDM could be more and more pinned down while the isoscalar contribution turned out to be more suppressed. In fact, the leading chiral loop diverges, inducing a logarithmic scale dependence and the need for finite counter terms of the same order as the isovector loop contribution to the neutron and proton EDMs. The number of these terms can be constrained to two as shown in Ref. 53 and confirmed by Ref. 54 — even for the three-flavor case. As there do not exist any measurement or theoretical methods (apart from lattice studies for the θ term scenario) to constrain these counter term coefficients other than naive dimensional analysis arguments (which only refer to the magnitude but not to the sign), there is not much predictive power of the chiral EFT approach for the total single-nucleon EDMs. Only when these calculations are coupled with the lattice ones, which have their own problems as mentioned above, there can be predictions for the $\bar{\theta}$-induced nucleon EDMs, see Refs. 47, 54 and 55.

5. EDMs of Light Nuclei

However, the situation is quite different for light nuclei as, e.g., the deuteron or helion. As already observed in the mid-1980s in Ref. 56, the same CP-violating pion exchange that causes the divergence in the loop diagrams appears already at the tree-level order in the two-nucleon contributions to the EDMs of (light) nuclei, such that in this application there is no need for sizable counter terms which are always local contact terms by nature. In fact, the first CP-violating NN-contact terms appear only at next-to-next-to-leading order relative to the contribution of the corresponding pion exchange diagram, see e.g., Ref. 57.

5.1. *The hadronic parameters*

Using chiral symmetry and isospin structure arguments the following chiral EFT Lagrangian for the (leading) P- and T-violating terms including single-nucleon, purely pionic, pion-nucleon and two-nucleon-contact interactions can be postulated:[46,57–62]

$$
\begin{aligned}
\mathcal{L}^{\not T \not P}_{\mathrm{EFT}} = & -d_{\mathrm{n}}\bar{N}(1-\tau_3)S^\mu N v^\nu F_{\mu\nu} - d_{\mathrm{p}}\bar{N}(1+\tau_3)S^\mu N v^\nu F_{\mu\nu} \\
& + (m_{\mathrm{N}}\Delta)\pi_3\pi^2 + g_0\bar{N}\boldsymbol{\tau}\cdot\boldsymbol{\pi}N + g_1\bar{N}\pi_3 N \\
& + C_1\bar{N}N\mathcal{D}_\mu(\bar{N}S^\mu N) + C_2\bar{N}\boldsymbol{\tau}N\cdot\mathcal{D}_\mu(\bar{N}S^\mu\boldsymbol{\tau}N) \\
& + C_3\bar{N}\tau_3 N\mathcal{D}_\mu(\bar{N}S^\mu N) + C_4\bar{N}N\mathcal{D}_\mu(\bar{N}\tau_3 S^\mu N) \\
& + \text{terms of higher order in the chiral expansion,}
\end{aligned}
\tag{12}
$$

where in principle the values of the coefficients of the effective Lagrangian (12) characteristically depend on the coefficients of the Lagrangian (10) (and might eventually be derived by lattice methods). In this way the models for the underlying physics, which again feed with different strength into the coefficients of (10), can in principle be disentangled if sufficiently enough EDM measurements can be matched to sufficiently enough EDM calculations of single- and multi-baryon systems. Currently, this step from the EFT Lagrangian (10) to the chiral EFT Lagrangian (12) exists only in rudimentary form for the θ term case, allowing the determination of the CP-violating three-pion coefficient Δ and, respectively, the isospin-conserving and isospin-violating πNN coefficients g_0 and g_1 as function of $\bar{\theta}$, including uncertainties:[61,62]

$$
\begin{aligned}
\Delta^\theta &= \frac{\epsilon(1-\epsilon^2)}{16F_\pi m_{\mathrm{N}}}\frac{M_\pi^4}{M_K^2-M_\pi^2}\bar{\theta}+\cdots &&= (-0.37\pm 0.09)\cdot 10^{-3}\bar{\theta}, \\
g_1^\theta &= 8c_1 m_{\mathrm{N}}\Delta^\theta + (0.6\pm1.1)\cdot 10^{-3}\bar{\theta} &&= (3.4\pm 1.5)\cdot 10^{-3}\bar{\theta}, \\
g_0^\theta &= \bar{\theta}\frac{\delta m_{\mathrm{np}}^{\mathrm{str}}(1-\epsilon^2)}{4F_\pi\epsilon} &&= (-15.5\pm 1.9)\cdot 10^{-3}\bar{\theta}.
\end{aligned}
\tag{13}
$$

The involved quantities are the pion decay constant, the isospin-averaged masses of the nucleon, pion and kaon,[4] the strong neutron–proton mass splitting $m_{\mathrm{np}}^{\mathrm{str}}$,[63,64] the quark mass ratio m_u/m_d,[65] which feeds into $\epsilon \equiv \frac{m_u-m_d}{m_u+m_d}$, and the ChPT coefficient c_1 (related to the nucleon sigma term).[66] The additional contribution to g_1^θ results from an independent chiral structure[58] and its stated value was estimated in Ref. 57.

However, even in the θ term scenario, the other coefficients, especially the total neutron and proton EDM values d_{n} and d_{p}, but also the isospin-conserving and isospin-breaking NN-contact coefficients $C_{1,2}$ and $C_{3,4}$, respectively, can only be estimated in magnitude but not in sign, either by naive dimensional arguments or by the magnitude of the subleading loop contributions in the case of the C_i coefficients.[61,62] While the coefficients d_{n} and d_{p} can in principle be matched to the corresponding lattice QCD calculations which still — as already mentioned — are problematic, the estimated contributions of the C_i terms have to be treated as systematic uncertainties — even for light nuclei and even in the theoretically most simple θ term scenario.

So far there do not exist similar relations between the other parameters of the Lagrangian (10) and the effective chiral Lagrangian (12) in the case of realistic underlying models. However, for the case of a minimal left–right symmetric model, because of its inherent isospin-breaking nature, a cross relation between the Δ parameter and g_0 and g_1 can be established:[46,61,62]

$$g_1^{\rm LR} = 8c_1 m_{\rm N}\Delta^{\rm LR} = (-7.5 \pm 2.3)\Delta^{\rm LR},$$
$$g_0^{\rm LR} = \frac{\delta m_{\rm np}^{\rm str} m_{\rm N}\Delta^{\rm LR}}{M_\pi^2} = (0.12 \pm 0.02)\Delta^{\rm LR}. \tag{14}$$

5.2. *The EDMs of deuteron, helion and triton*

The EDMs of the deuteron, helion and triton follow from the multiplication of the coefficients of the chiral effective Lagrangian (12) (see (13) and (14) for special cases) and the nuclear matrix elements calculated in Refs. 61 and 62 and listed in Tables 1 and 2, respectively, as

$$\begin{aligned} d_{\rm D} &= d_{\rm p}\cdot\mathcal{F}(d_{\rm p}^{\rm D}) + d_{\rm n}\cdot\mathcal{F}(d_{\rm n}^{\rm D}) + g_1\cdot\mathcal{F}(g_1^{\rm D}) + \Delta\cdot\mathcal{F}(\Delta f_{g_1}^{\rm D}) \\ &\quad + \{C_3\cdot\mathcal{F}(C_3^{\rm D}) + C_4\cdot\mathcal{F}(C_4^{\rm D})\}, \end{aligned} \tag{15}$$

$$\begin{aligned} d_{^3\rm He} &= d_{\rm p}\cdot\mathcal{F}(d_{\rm p}[^3{\rm He}]) + d_{\rm n}\cdot\mathcal{F}(d_{\rm n}[^3{\rm He}]) + \Delta\cdot\mathcal{F}(\Delta[^3{\rm He}]) \\ &\quad + g_0\cdot\mathcal{F}(g_0[^3{\rm He}]) + g_1\cdot\mathcal{F}(g_1[^3{\rm He}]) + \Delta\cdot\mathcal{F}(\Delta f_{g_1}[^3{\rm He}]) \\ &\quad + C_1\cdot\mathcal{F}(C_1[^3{\rm He}]) + C_2\cdot\mathcal{F}(C_2[^3{\rm He}]) \\ &\quad + \{C_3\cdot\mathcal{F}(C_3[^3{\rm He}]) + C_4\cdot\mathcal{F}(C_4[^3{\rm He}])\}, \end{aligned} \tag{16}$$

and the analog of Eq. (16) for the triton case, i.e., $^3{\rm He} \to {}^3{\rm H}$. The terms in curly brackets (the C_3 and C_4 contributions) are of subleading order because of the additional isospin-breaking and can be neglected in all cases, except for an underlying model which is left–right symmetric.

Table 1. Contributions to the deuteron EDM calculated from the N^{2}LO (chiral) χEFT potential,[67,68] the Av_{18} potential[69] and the CD-Bonn potential,[70] respectively, see Refs. 61 and 62 (remember that $e > 0$ here). In addition, unpublished results using the recent χEFT potential[71–73] with terms up to N^{4}LO are presented. The tabulated values still have to be multiplied by the corresponding coefficients from the chiral Lagrangian (12) which are listed as "units". Each C_i has the dimension [fm^3], $d_{\rm n}$ and $d_{\rm p}$ carry the dimension [e fm], while g_1, g_0 and Δ are dimensionless.

Term	N^{2}LO ChPT	N^{4}LO ChPT	Av_{18}	CD-Bonn	Units
$\mathcal{F}(d_{\rm n}^{\rm D})$	0.939 ± 0.009	0.936 ± 0.008	0.914	0.927	$d_{\rm n}$
$\mathcal{F}(d_{\rm p}^{\rm D})$	0.939 ± 0.009	0.936 ± 0.008	0.914	0.927	$d_{\rm p}$
$\mathcal{F}(g_1^{\rm D})$	0.183 ± 0.017	0.182 ± 0.002	0.186	0.186	$g_1\,e\,{\rm fm}$
$\mathcal{F}(\Delta f_{g_1}^{\rm D})$	-0.748 ± 0.138	-0.646 ± 0.023	-0.703	-0.719	$\Delta\,e\,{\rm fm}$
$\mathcal{F}(C_3^{\rm D})$	0.05 ± 0.05	0.033 ± 0.001	—	—	$C_3\,e\,{\rm fm}^{-2}$
$\mathcal{F}(C_4^{\rm D})$	-0.05 ± 0.05	-0.006 ± 0.007	—	—	$C_4\,e\,{\rm fm}^{-2}$

Table 2. Contributions to the helion and trition EDM calculated from the N^{2}LO (chiral) χEFT potential,[67,68] the Av_{18}+UIX potential[69,74] and the CD-Bonn+TM potential,[70,75] see Refs. 61 and 62. Further details as in the caption of Table 1.

Term	Nucleus	N^{2}LO ChPT	Av_{18}+UIX	CD-Bonn+TM	Units
$\mathcal{F}(d_\mathrm{n})$	^{3}He	0.904 ± 0.013	0.875	0.902	d_n
	^{3}H	-0.030 ± 0.007	−0.051	−0.038	d_n
$\mathcal{F}(d_\mathrm{p})$	^{3}He	-0.029 ± 0.006	−0.050	−0.037	d_p
	^{3}H	0.918 ± 0.013	0.902	0.876	d_p
$\mathcal{F}(\Delta)$	^{3}He	-0.017 ± 0.006	−0.015	−0.019	$\Delta\, e\,\mathrm{fm}$
	^{3}H	-0.017 ± 0.006	−0.015	−0.018	$\Delta\, e\,\mathrm{fm}$
$\mathcal{F}(g_0)$	^{3}He	0.111 ± 0.013	0.073	0.087	$g_0\, e\,\mathrm{fm}$
	^{3}H	-0.108 ± 0.013	−0.073	−0.085	$g_0\, e\,\mathrm{fm}$
$\mathcal{F}(g_1)$	^{3}He	0.142 ± 0.019	0.142	0.146	$g_1\, e\,\mathrm{fm}$
	^{3}H	0.139 ± 0.019	0.142	0.144	$g_1\, e\,\mathrm{fm}$
$\mathcal{F}(\Delta f_{g_1})$	^{3}He	-0.608 ± 0.142	−0.556	−0.586	$\Delta\, e\,\mathrm{fm}$
	^{3}H	-0.598 ± 0.141	−0.564	−0.576	$\Delta\, e\,\mathrm{fm}$
$\mathcal{F}(C_1)$	^{3}He	-0.042 ± 0.017	−0.0014	−0.016	$C_1\, e\,\mathrm{fm}^{-2}$
	^{3}H	0.041 ± 0.016	0.0014	0.016	$C_1\, e\,\mathrm{fm}^{-2}$
$\mathcal{F}(C_2)$	^{3}He	0.089 ± 0.022	0.0042	0.033	$C_2\, e\,\mathrm{fm}^{-2}$
	^{3}H	-0.087 ± 0.022	−0.0044	−0.032	$C_2\, e\,\mathrm{fm}^{-2}$
$\mathcal{F}(C_3)$	^{3}He/^{3}H	-0.04 ± 0.03	—	—	$C_3\, e\,\mathrm{fm}^{-2}$
$\mathcal{F}(C_4)$	^{3}He/^{3}H	0.07 ± 0.03	—	—	$C_4\, e\,\mathrm{fm}^{-2}$

The first two terms proportional to d_p and d_n are the single-nucleon contributions to the total EDMs. Since d_n and d_p can independently be determined by separate experiments, these single-nucleon terms can be subtracted from the expressions in (15) and (16) in order to determine the multi-nucleon contributions of the corresponding EDMs — just by using experimental input. The quantities proportional to Δ are either contributing to genuinely irreducible three-body interactions in the helion and triton cases (which numerically, however, turn out to be small) or to finite and momentum-transfer-dependent loop corrections (see Δf_{g_1}) of the isospin-breaking CP-violating pion exchange (proportional to g_1). These corrections are exceptionally large and add up coherently to the g_1 contributions which then factually are governed by *three different* terms with different chiral structures which will be difficult to disentangle without chiral EFT methods.

There appear less terms for the deuteron case since it is only a two-nucleon system (excluding the Δ three-body term) and since the deuteron acts as an isospin filter: the isospin-conserving (CP-violating) g_0 and $C_{1,2}$ terms are excluded since they only induce a transition to the Pauli-allowed $^1P_1$ intermediate states which cannot be undone by the coupling of the photon. The isospin-breaking terms, however, are allowed since the transition to the $^3P_1$ intermediate states can be reversed by the (isovector part of the) photon coupling.

In contrast to the application of phenomenological nuclear potentials,[69,70,74,75] the calculations using chiral potentials[67,68] allow for the specification of uncertainties in addition to central values.[61,62] The latter are mostly compatible with the results of the phenomenological potentials (which agree, where a comparison is possible, with the calculations of other groups[59,76,77]), except for the short-range contact terms. These are very sensitive to the model-dependent specifics of the short-range repulsion of the phenomenological potentials — for more details see Ref. 61. We therefore refrain from showing results of these phenomenological potentials for the isospin-breaking $C_{3,4}$ contributions which are of subleading nature to start with. Finally, in Table 1 also unpublished results for the recent chiral NN potential[71–73] with terms up to order N^{4}LO are reported. The values are compatible with the older N^{2}LO calculations but with reduced uncertainties.

6. Conclusion

Let us conclude by describing a way to identify or exclude the QCD θ term or the left–right symmetric models as the primary candidate for an underlying CP violation beyond the KM mechanism of the SM. This can be achieved solely via measurements of the EDMs of the neutron, proton, deuteron and helion. Note that the exclusive measurements of the single nucleon EDMs will not suffice to achieve this, since any reasonable underlying model will predict d_p and d_n to be approximately of the same magnitude and most probably of opposite sign.

However, if experimental information about $d_{^3\mathrm{He}}$ and d_n can be established, then a fit-value of the $\bar{\theta}$ angle can be extracted from Eq. (16) with input from (13), treating the small contribution of the proton and of the contact terms as systematical uncertainties. With Eq. (15) applied to this result, the nuclear part of the deuteron EDM and, if the proton EDM is measured as well (or calculated by lattice methods), also the total deuteron EDM can be predicted allowing for a test of the θ term scenario (instead or in addition to numerical lattice tests).

Alternatively, a measurement of the neutron, proton and deuteron EDM allows to extract the $\bar{\theta}$ fit-value — again solely from experiment — and for the prediction of the total helion EDM, including uncertainties. This alternative extraction has the advantage that there are not any systematical uncertainties related to the NN-contact interactions, since these are "filtered out" in the deuteron case.

The characteristic signal for the QCD θ term scenario would be

$$d_\mathrm{D} - 0.94(d_\mathrm{p} + d_\mathrm{n}) \approx -(d_{^3\mathrm{He}} - 0.9 d_\mathrm{n}) \approx \frac{1}{2}(d_{^3\mathrm{H}} - 0.9 d_\mathrm{p}). \tag{17}$$

The establishment of the last part of this relation is of course rather unlikely, since a triton EDM measurement would be necessary. At the same time we would have predictions of the coefficients Δ^θ, g_0^θ and g_1^θ (with $g_1^\theta/g_0^\theta \approx -0.2$) which can be used as input for EDM calculations of heavier nuclei.

If the dimension-four QCD θ term can be excluded — this test should always be done as the first one — then the next simplest step is to test the left–right scenario which also has a telling signal. The above described measurements, either the route of $d_{^3\mathrm{He}}$ and d_n or the route of d_D, d_n and d_p, allow to extract the Δ^{LR} parameter and to predict the other alternative. The characteristic signal of the left–right model would be

$$|d_{\mathrm{n}}| \approx |d_{\mathrm{p}}| \ll |d_{\mathrm{D}}| \quad \text{and} \quad d_{\mathrm{D}} \approx d_{^3\mathrm{He}} \approx d_{^3\mathrm{H}}, \tag{18}$$

which is quite distinct from the θ term scenario. Furthermore, the ratio $-g_1^{\mathrm{LR}}/g_0^{\mathrm{LR}} \gg 1$ is very different from its θ term counterpart.

If both models can be excluded, then the measured values of d_{n}, d_{p} and d_{D} still allow to extract an effective coefficient g_1 which includes the Δf_{g_1} modification. Using this as an input, a further measurement of $d_{^3\mathrm{He}}$ would then allow to isolate the value of the coefficient g_0. The ratio g_1/g_0 of these values should be rather different from those predicted in the θ term and in the left–right symmetric scenarios, as otherwise one of these cases could not be excluded any longer. The extracted g_1 and g_0 values can be used to predict EDMs for other nuclei, namely light nuclei as the triton or heavier ones as measured in the case of diamagnetic atoms if the calculation of the nuclear matrix elements of these heavy nuclei can eventually be done with less than 50% uncertainty, say. More details can be found in Refs. 46, 61 and 62.

Acknowledgments

We would like to thank Tom Kuo and Ismail Zahed for the invitation to write this paper and express our gratitude to our collaborators and colleagues Christoph Hanhart, Ulf-G. Meißner, Jordy de Vries, Evgeny Epelbaum, Kolya Nikolaev, Werner Bernreuther and Wouter Dekens for sharing their insights into the topics presented here. A. Wirzba would like to thank the Kavli Institute for Theoretical Physics for hospitality and Barry Holstein and Jörg Pretz for comments on an earlier version of the manuscript. This research was supported by the DFG (Grant No. DFG/TR-110), the NSFC (Grant No. 11261130311) through funds provided to the Sino-German CRC 110 Symmetries and the Emergence of Structure in QCD, and in part by the National Science Foundation under Grant No. NSF PHY-1125915. The resources of the Jülich Supercomputing Center at the Forschungszentrum Jülich in Germany, namely the supercomputers JUROPA, JURECA and JUQUEEN, have been instrumental in the computations reported here.

References

1. A. Wirzba, *Nucl. Phys A.* **928** (2014) 116, arXiv:1404.6131.
2. J. H. Smith, E. M. Purcell and N. F. Ramsey, *Phys. Rev.* **108** (1957) 120.
3. I. B. Khriplovich and S. K. Lamoreaux, *CP Violation without Strangeness* (Springer, 1997).
4. Particle Data Group (K. A. Olive *et al.*), *Chin. Phys. C* **38** (2014) 090001.
5. M. Kobayashi and T. Maskawa, *Prog. Theor. Phys.* **49** (1973) 652.
6. I. B. Khriplovich, *Phys. Lett. B* **173** (1986) 193.
7. A. Czarnecki and B. Krause, *Phys. Rev. Lett.* **78** (1997) 4339.
8. M. B. Gavela *et al.*, *Phys. Lett. B* **109** (1982) 215.

9. I. Khriplovich and A. Zhitnitsky, *Phys. Lett. B* **109** (1982) 490.
10. J. O. Eeg and I. Picek, *Nucl. Phys. B* **244** (1984) 77.
11. T. Mannel and N. Uraltsev, *J. High Eenergy Phys.* **1303** (2013) 064, arXiv:1205.0233.
12. M. E. Pospelov and I. B. Khriplovich, *Sov. J. Nucl. Phys.* **53** (1991) 638.
13. H. Leutwyler, *Phys. Lett. B* **189** (1987) 197.
14. J. Gasser and H. Leutwyler, *Nucl. Phys. B* **307** (1988) 763.
15. D. J. Gross, *Proc. Natl. Acad. Sci. USA* **93** (1996) 1456.
16. N. Auerbach, V. Flambaum and V. Spevak, *Phys. Rev. Lett.* **76** (1996) 4316.
17. V. Spevak, N. Auerbach and V. V. Flambaum, *Phys. Rev. C* **56** (1997) 1357.
18. C. H. Greene, A. S. Dickinson and H. R. Sadeghpour, *Phys. Rev. Lett.* **85** (2000) 2458.
19. W. Li *et al.*, *Science* **334** (2011) 1110.
20. W. Bernreuther, *Talk at the EDM Seminar*, Aachen/Jülich, April 14 2012.
21. W. Bernreuther and M. Suzuki, *Rev. Mod. Phys.* **63** (1991) 313.
22. Planck (P. A. R. Ade *et al.*), *Astron. Astrophys.* **594** (2016) A13.
23. W. Bernreuther, *Lect. Notes Phys.* **591** (2002) 237, arXiv:hep-ph/0205279.
24. A. D. Sakharov, *Pisma Zh. Eksp. Teor. Fiz.* **5** (1967) 32.
25. V. Kuzmin, V. Rubakov and M. Shaposhnikov, *Phys. Lett. B* **155** (1985) 36.
26. R. D. Peccei and H. R. Quinn, *Phys. Rev. Lett.* **38** (1977) 1440.
27. R. D. Peccei and H. R. Quinn, *Phys. Rev. D* **16** (1977) 1791.
28. C. A. Baker *et al.*, *Phys. Rev. Lett.* **97** (2006) 131801, arXiv:hep-ex/0602020.
29. V. Kuzmin, M. Shaposhnikov and I. Tkachev, *Phys. Rev. D* **45** (1992) 466.
30. V. F. Dmitriev and R. A. Sen'kov, *Phys. Atom. Nucl.* **66** (2003) 1940.
31. B. Graner, Y. Chen, E. G. Lindahl and B. R. Heckel, *Phys. Rev. Lett.* **116** (2016) 161601, arXiv:1601.04339.
32. E. R. Meyer and J. L. Bohn, *Phys. Rev. A* **78** (2008) 010502.
33. V. Dzuba, V. Flambaum and C. Harabati, *Phys. Rev. A* **84** (2011) 052108.
34. L. V. Skripnikov, A. N. Petrov and A. V. Titov, arXiv:1308.0414v2.
35. B. C. Regan, E. D. Commins, C. J. Schmidt and D. DeMille, *Phys. Rev. Lett.* **88** (2002) 071805.
36. J. J. Hudson *et al.*, *Nature* **473** (2011) 493.
37. D. M. Kara *et al.*, *New J. Phys.* **14** (2012) 103051, arXiv:1208.4507.
38. ACME (J. Baron *et al.*), *Science* **343** (2014) 269, arXiv:1310.7534.
39. F. J. M. Farley *et al.*, *Phys. Rev. Lett.* **93** (2004) 052001.
40. EDM Collab. (Y. K. Semertzidis *et al.*), *AIP Conf. Proc.* **698** (2004) 200.
41. J. Pretz, *Hyperfine Interact.* **214** (2013) 111, arXiv:1301.2937.
42. A. E. Leanhardt *et al.*, *J. Mol. Spectrosc.* **270** (2011) 1.
43. Muon (g-2) Collab. (G. Bennett *et al.*), *Phys. Rev. D* **80** (2009) 052008.
44. S. Weinberg, *Phys. Rev. Lett.* **43** (1979) 1566.
45. W. Dekens and J. de Vries, *J. High Energy Phys.* **1305** (2013) 149, arXiv:1303.3156.
46. W. Dekens *et al.*, *J. High Energy Phys.* **1407** (2014) 069, arXiv:1404.6082.
47. F. K. Guo *et al.*, *Phys. Rev. Lett.* **115** (2015) 062001, arXiv:1502.02295.
48. E. Shintani, T. Blum, T. Izubuchi and A. Soni, *Phys. Rev. D* **93** (2016) 094503.
49. T. Bhattacharya, V. Cirigliano, R. Gupta, H.-W. Lin and B. Yoon, *Phys. Rev. Lett.* **115** (2015) 212002, arXiv:1506.04196.
50. T. Bhattacharya, V. Cirigliano, R. Gupta, E. Mereghetti and B. Yoon, *PoS* **LATTICE2015** (2016) 238, arXiv:1601.02264.
51. R. J. Crewther, P. Di Vecchia, G. Veneziano and E. Witten, *Phys. Lett. B* **88** (1979) 123.
52. A. Pich and E. de Rafael, *Nucl. Phys. B* **367** (1991) 313.
53. K. Ottnad, B. Kubis, U.-G. Meißner and F.-K. Guo, *Phys. Lett. B* **687** (2010) 42.

54. F.-K. Guo and U.-G. Meißner, *J. High Energy Phys.* **1212** (2012) 097, arXiv:1210.5887.
55. T. Akan, F.-K. Guo and U.-G. Meißner, *Phys. Lett. B* **736** (2014) 163.
56. V. V. Flambaum, I. B. Khriplovich and O. P. Sushkov, *Sov. Phys. JETP* **60** (1984) 873.
57. J. Bsaisou *et al.*, *Eur. Phys. J. A* **49** (2013) 31, arXiv:1209.6306.
58. E. Mereghetti, W. Hockings and U. van Kolck, *Ann. Phys.* **325** (2010) 2363.
59. J. de Vries *et al.*, *Phys. Rev. C* **84** (2011) 065501, arXiv:1109.3604.
60. J. de Vries, E. Mereghetti, R. G. E. Timmermans and U. van Kolck, *Ann. Phys.* **338** (2013) 50, arXiv:1212.0990.
61. J. Bsaisou *et al.*, *J. High Energy Phys.* **1503** (2015) 104, arXiv:1411.5804, Erratum **1505** (2015) 083.
62. J. Bsaisou, U.-G. Meißner, A. Nogga and A. Wirzba, *Ann. Phys.* **359** (2015) 317.
63. A. Walker-Loud, *PoS* **LATTICE2013** (2014) 013, arXiv:1401.8259.
64. S. Borsanyi *et al.*, *Science* **347** (2015) 1452, arXiv:1406.4088.
65. S. Aoki *et al.*, *Eur. Phys. J. C* **74** (2014) 2890, arXiv:1310.8555.
66. V. Bernard, N. Kaiser and U.-G. Meißner, *Int. J. Mod. Phys. E* **4** (1995) 193.
67. E. Epelbaum, W. Glöckle and U.-G. Meißner, *Nucl. Phys. A* **747** (2005) 362.
68. E. Epelbaum, H.-W. Hammer and U.-G. Meißner, *Rev. Mod. Phys.* **81** (2009) 1773.
69. R. B. Wiringa, V. G. J. Stoks and R. Schiavilla, *Phys. Rev. C* **51** (1995) 38.
70. R. Machleidt, *Phys. Rev. C* **63** (2001) 024001, arXiv:nucl-th/0006014.
71. E. Epelbaum, H. Krebs and U.-G. Meißner, *Eur. Phys. J. A* **51** (2015) 53.
72. E. Epelbaum, H. Krebs and U.-G. Meißner, *Phys. Rev. Lett.* **115** (2015) 122301.
73. LENPIC (S. Binder *et al.*), *Phys. Rev. C* **93** (2016) 044002, arXiv:1505.07218.
74. B. S. Pudliner, V. R. Pandharipande, J. Carlson, S. C. Pieper and R. B. Wiringa, *Phys. Rev. C* **56** (1997) 1720, arXiv:nucl-th/9705009.
75. S. A. Coon and H. K. Han, *Few Body Syst.* **30** (2001) 131.
76. Y.-H. Song, R. Lazauskas and V. Gudkov, *Phys. Rev. C* **87** (2013) 015501.
77. N. Yamanaka and E. Hiyama, *Phys. Rev. C* **91** (2015) 054005.

Hidden local symmetry and beyond

Koichi Yamawaki
Kobayashi-Maskawa Institute for the Origin of Particles and the Universe (KMI), Nagoya University, Chikusa-ku, Nagoya 464-8602, Japan
yamawaki@kmi.nagoya-u.ac.jp

Gerry Brown was a godfather of our hidden local symmetry (HLS) for the vector meson from the birth of the theory throughout his life. The HLS is originated from very nature of the nonlinear realization of the symmetry G based on the manifold G/H, and thus is universal to any physics based on the nonlinear realization. Here, I focus on the Higgs Lagrangian of the Standard Model (SM), which is shown to be equivalent to the nonlinear sigma model based on $G/H = \mathrm{SU}(2)_L \times \mathrm{SU}(2)_R/\mathrm{SU}(2)_V$ with additional symmetry, the nonlinearly-realized scale symmetry. Then, the SM does have a dynamical gauge boson of the $\mathrm{SU}(2)_V$ HLS, "SM ρ meson", in addition to the Higgs as a pseudo-dilaton as well as the NG bosons to be absorbed in to the W and Z. Based on the recent work done with Matsuzaki and Ohki, I discuss a novel possibility that the SM ρ meson acquires kinetic term by the SM dynamics itself, which then stabilizes the skyrmion dormant in the SM as a viable candidate for the dark matter, what we call "dark SM skyrmion (DSMS)".

Keywords: Hidden local symmetry; standard model; dark matter; nonlinear realization; non-pertubative dynamics; skyrmion.

1. Introduction

HLS song to Gerry:
Hearty thanks to you Gerry picking me the blueberry
Long time passing a summer at garden of Brown Gerry
Sweet and sour memory everlasting with a smile of Grand Gerry

Gerry Brown was enthusiastic about the hidden local symmetry (HLS)[1,2] and the associated vector manifestation (VM) of chiral symmetry,[3,4] probably even more than myself and my HLS collaborators.[5] He as an editor of the Physics Reports encouraged (sometimes urged) us to publish the HLS review articles in Physics Reports, the result being one at tree-level[2] and another at loop-level.[4,a] In contrast to Gerry who was working on the hadron physics as well as nuclear physics, my main interest was not the hadron physics itself but the physics hidden behind the Standard Model (SM) Higgs, although I mostly worked on the concrete realization of the HLS in the hadron physics. HLS in fact proved very successful in describing

[a]The relevant one-loop results were obtained in Refs. 6, 7, for the case without and with the quadratic divergence, respectively.

the experimental facts in hadron physics, which then could be an excellent monitor of any theory giving rise to the nonlinear realization, or the spontaneous breakdown of the symmetry.

The HLS as it stands is very universal, not just in hadron physics and particle physics, but in any system described by the nonlinear realization of the symmetry G spontaneously broken down to $H(\in G)$. The nonlinear sigma model based on G/H is always gauge equivalent to the HLS model having a larger symmetry $G_{\rm global} \times H_{\rm local}$, the $H_{\rm local}$ being the HLS which is a spontaneously broken gauge symmetry. At classical level, the gauge boson of the HLS is merely an auxiliary field as a static composite of the Nambu–Goldstone (NG) bosons living on G/H, and thus can be solved away so that the theory is reduced back to the original model without HLS. However, the HLS gauge boson at quantum level develops the kinetic term due to the own dynamics of the nonlinear sigma model, or those of the underlying theory if any (see Refs. 2, 4–7 for discussions and concrete calculations).

This is analogous to the dynamical generation of the composite scalar meson in the Nambu–Jona-Lasinio (NJL) model, where the composite scalar is introduced as the auxiliary field at classical level and then the quantum theory develops the kinetic term as well as the quartic self coupling (corresponding to the kinetic term of the Yang–Mills field for HLS), the result being equivalent to the linear sigma model with Yukawa coupling (corresponding to the gauge coupling to matter in the HLS theory).[8] This method was fully developed[9] in the top quark condensate model,[10,b] based on the explicit (gauged) NJL model.

Apart from the dynamical generation of the HLS gauge boson, the concept of the HLS as the effective theory of the underlying QCD-like theories has been developed: a manifestation of the Seiberg duality as the magnetic gauge symmetry,[4,6,7,13,14] an infinite tower of HLS ("Moose") of the deconstructed/latticized gauge theory in the extra dimensions,[15] further a basis of the holographic QCD,[16,17] and so on.

Here, I focus on novel aspects of the SM Higgs Lagrangian with Higgs mass 125 GeV, which to our surprise has far richer physics than ever recognized, without recourse to the UV completion[18]: We have recently found[19] (see also Ref. 12) that the SM Higgs Lagrangian is straightforwardly rewritten in the form of the scale-invariant nonlinear sigma model similar to the effective theory[20] of the walking technicolor,[21] i.e., the nonlinear sigma model for both the chiral $SU(2)_L \times SU(2)_R$ symmetry and the scale symmetry realized nonlinearly, near the "conformal limit" (similar to the BPS limt[22]), with the quartic coupling $\lambda \to 0$ with the Higgs VEV $v =$ fixed, where the SM Higgs boson ϕ is nothing but a pseudo-dilaton, pseudo-NG boson of the spontaneously broken scale symmetry. Accordingly, it has an $SU(2)_V$ HLS and its dynamical gauge boson, "SM rho meson", as an analogue of the QCD rho meson. The resultant theory takes the same form as the scale-invariant HLS model[23] considered for the waking technicolor up to the explicit scale-symmetry breaking potential.

[b]Precise equivalence of Ref. 9 to this formulation is explicitly shown in Ref. 11.

We then have found[18] that the SM rho meson can be dynamically-generated by the SM dynamics itself without recourse to the UV completion beyond the SM, based on the one-loop calculation[4,6,7] of the the kinetic term of the HLS gauge boson in the general nonlinear sigma model. This led us to an amazing fact[18] that the dark matter candidate does already exist *inside but not outside* the SM ("dark side" of the SM), namely the dynamically-generated HLS gauge boson, the SM rho meson, can stabilize the *skyrmion*, "dark SM skyrmion (DSMS)" denoted as X_s.

This is in an analogous way to the well-known mechanisms[24–26] that the QCD skyrmion (say, nucleon with $I = J = 1/2$) is stabilized by the HLS rho meson in QCD, up to a notable difference that the kinetic term of the QCD rho meson is already generated by the underlying QCD not by the nonlinear sigma model own dynamics.[27] Here, we consider a scalar DSMS with $I = J = 0$ having a topological charge $Q_{X_s} = 1$ of $U(1)_{X_s}$. The idea to identify the skyrmion as a dark matter in a certain generalization of the SM Higgs sector (some new physics beyond the SM) is not new (see e.g., Ref. 28 and references cited therein.) In contrast, it is really novel to find the dark matter candidate inside the SM as it is.

A salient feature of the DSMS is the coupling to the Higgs as a pseudo-dilaton, which is unambiguously determined in the low energy limit due to the low energy theorem of the scale symmetry as is well known for the pseudo-dilaton in a different context.[29] Accordingly, together with the nature of the soliton extended object of strong coupling system, the DSMS yields a novel dark matter phenomenology.[18] Thus the dormant new physics inside the SM awaken!

This is in sharp contrast to the current view that the dark matter is definitely originated from the physics "beyond the SM". In fact the SM Higgs Lagrangian written in the form of the linear sigma model is usually understood as something different from the nonlinear sigma model regarded as the strong coupling (heavy Higgs mass) limit with the Higgs decoupled, which is in obvious disagreement with the reality of 125 GeV Higgs, and hence regarded irrelevant to the HLS as well. However, even for the light Higgs it has obviously the symmetry $G = \mathrm{SU}(2)_L \times \mathrm{SU}(2)_R$ spontaneously broken down to $H = \mathrm{SU}(2)_{L+R} = \mathrm{SU}(2)_V$ and thus G can be realized nonlinearly by the NG bosons living on the manifold G/H, and hence does have an $\mathrm{SU}(2)_V$ HLS. As mentioned above, such a light SM Higgs near the conformal/BPS limit simply becomes a pseudo-dilaton instead of being decoupled, giving rise to the same nonlinear realization of G/H plus the nonlinearly realized scale symmetry, the result being the scale-invariant HLS Lagrangian similarly to that[23] considered in the walking technicolor as mentioned above.

It is further well known in the hadron physics[24] that the gauge boson of the HLS can stabilize the skyrmion, with the kinetic term becoming precisely equal to the Skyrme term in the heavy mass limit of the rho meson where the rho meson field behaving like the auxiliary field composed of the nonlinear pions. It has also been shown[25,26,28] that including the scalar meson (corresponding to the SM Higgs as a pseudo-dilaton in our case) does not invalidate the skyrmion, but rather makes the skyrmion mass lighter, as was particularly discussed[25] in essentially the same form as our scale-invariant HLS model of the SM Higgs Lagrangian (up to the potential term).

We found[18] that such a lighter mass shift of the skyrmion due to the pseudo-dilaton for DSMS is in accord with the current direct detection experimental limit LUX 2006,[30] which yields unambiguously, through the characteristic low energy theorem, the upper bound (instead of lower bound) of the mass of the DSMS to be very light $M_{X_s} \lesssim 38\,\text{GeV}$. As such a light particle, it is further constrained by the most stringent LHS data on the Higgs invisible decay[31] as $M_{X_s} \lesssim 18\,\text{GeV}$, which is contrasted with most of the WIMP dark matter candidates having mass of order of 100 GeVs. Furthermore, it is crucial that the annihilation cross-section can be roughly evaluated by the extended size of the soliton DSMS, which we estimated in the limit of heavy SM rho mass limit (Skyrme term limit) in rough consistency with the presently observed relic abundance.[32]

In the next two sections, I will recapitulate Ref. 19 (see also Ref. 12) showing that the SM Higgs Lagrangian usually written in the form of *linear* sigma model actually can be straightforwardly rewritten in to the *nonlinear* sigma model based on G/H with an additional symmetry, the scale symmetry, which is also spontaneously broken and realized nonlinearly by another (pseudo-) NG boson, the (pseudo-) dilaton, which is nothing but the SM Higgs with mass of 125 GeV. Then, it will be further shown to be gauge equivalent to the scale-invariant version of the HLS Lagrangian. In Sec. 4, before discussing the dynamical generation of the HLS gauge boson à la Refs. 4–7. In Sec. 5, I will discuss a well-known good example of the dynamical generation of the kinetic term of the auxiliary field, namely the composite Higgs in the NJL model, where the auxiliary Higgs field in fact becomes dynamical at the quantum level and the system becomes equivalent to the Higgs–Yukawa model (linear sigma model)[8] (as to the conformal/BPS limit, see Ref. 12). In Sec. 6, I will discuss the recent result[18] that the skyrmion in the SM does exist, stabilized by the SM rho meson, and it is a viable candidate for the dark matter, DSMS, totally within the SM without physics beyond the SM. Section 7 is devoted to summary and discussions, where some possible UV completion of the SM such as the walking technicolor will also be addressed in the context of the DSMS.

2. SM Higgs as a Scale-Invariant Nonlinear Sigma Model

The SM Lagrangian takes the form[19]:

$$\mathcal{L}_{\text{Higgs}} = |\partial_\mu h|^2 - \mu_0^2 |h|^2 - \lambda |h|^4 \tag{1}$$

$$= \frac{1}{2}[(\partial_\mu \hat{\sigma})^2 + (\partial_\mu \hat{\pi}_a)^2] - \frac{1}{2}\mu_0^2[\hat{\sigma}^2 + \hat{\pi}_a^2] - \frac{\lambda}{4}[\hat{\sigma}^2 + \hat{\pi}_a^2]^2, \tag{2}$$

where we have rewritten the conventional form in Eq. (1) in to the $\text{SU}(2)_L \times \text{SU}(2)_R$ linear sigma model in Eq. (2) through

$$h = \begin{pmatrix} \phi^+ \\ \phi^0 \end{pmatrix} = \frac{1}{\sqrt{2}} \begin{pmatrix} i\hat{\pi}_1 + \hat{\pi}_2 \\ \hat{\sigma} - i\hat{\pi}_3 \end{pmatrix} \tag{3}$$

with the potential in the form:

$$V(\hat{\sigma},\hat{\pi}) = \frac{1}{2}\mu_0^2[\hat{\sigma}^2+\hat{\pi}_a^2] + \frac{\lambda}{4}[\hat{\sigma}^2+\hat{\pi}_a^2]^2 = \frac{1}{2}\mu_0^2\sigma^2 + \frac{\lambda}{4}\sigma^4, \tag{4}$$

$$\sigma^2(x) \equiv \hat{\sigma}^2(x) + \hat{\pi}_a^2(x), \tag{5}$$

which has a minimum at the chiral-invariant circle

$$\langle\sigma^2(x)\rangle = \frac{-\mu_0^2}{\lambda} \equiv v^2 = (246\,\mathrm{GeV})^2. \tag{6}$$

The SM Higgs Lagrangian can be further rewritten in to

$$\mathcal{L}_{\mathrm{Higgs}} = \frac{1}{2}\mathrm{tr}(\partial_\mu M\partial^\mu M^\dagger) - \left[\frac{\mu_0^2}{2}\mathrm{tr}(MM^\dagger) + \frac{\lambda}{4}(\mathrm{tr}(MM^\dagger))^2\right] \tag{7}$$

with the 2×2 matrix M

$$M = (i\tau_2 h^*, h) = \frac{1}{\sqrt{2}}(\hat{\sigma}\cdot 1_{2\times 2} + 2i\hat{\pi}), \quad \left(\hat{\pi}\equiv\hat{\pi}_a\frac{\tau_a}{2}\right), \tag{8}$$

which transforms under $G = \mathrm{SU}(2)_L\times\mathrm{SU}(2)_R$ as:

$$M \to g_L\, M\, g_R^\dagger, \quad (g_{R,L}\in\mathrm{SU}(2)_{R,L}). \tag{9}$$

Now any complex matrix M can be decomposed in to the Hermitian (always diagnonalizable) matrix H and unitary matrix U as $M = HU$ ("polar decomposition")[2]:

$$M(x) = H(x)\cdot U(x), \quad H(x) = \frac{1}{\sqrt{2}}\begin{pmatrix}\sigma(x) & 0\\ 0 & \sigma(x)\end{pmatrix}, \quad U(x) = \exp\left(\frac{2i\pi(x)}{F_\pi}\right) \tag{10}$$

with $\pi(x) = \pi^a(x)\frac{\tau^a}{2}$ $(a = 1,2,3)$ and $F_\pi = v = \langle\sigma(x)\rangle$. The chiral transformation of M is carried by U, while H is a chiral-singlet such that:

$$U \to g_L\, U\, g_R^\dagger, \quad H\to H, \tag{11}$$

where $g_{L/R}\in\mathrm{SU}(2)_{L/R}$. Note that the radial mode σ is a chiral-singlet in contrast to $\hat{\sigma}$ which is a chiral nonsinglet transformed in to the chiral partner $\hat{\pi}_a$ by the chiral rotation.

We can parametrize $\sigma(x)$ as the nonlinear base of the scale transformation:

$$\sigma(x) = v\cdot\chi(x), \quad \chi(x) = \exp\left(\frac{\phi(x)}{F_\phi}\right), \tag{12}$$

where $F_\phi = v$ is the decay constant of the dilaton ϕ as the SM Higgs.[c]

[c] The scale (dilatation) transformations for these fields are

$$\delta_D\sigma = (1+x^\mu\partial_\mu)\sigma, \quad \delta_D\chi = (1+x^\mu\partial_\mu)\chi, \quad \delta_D\phi = F_\phi + x^\mu\partial_\mu\phi\,. \tag{13}$$

Note that $\langle\sigma(x)\rangle = v\langle\chi(x)\rangle = v\neq 0$ breaks spontaneously the scale symmetry, but not the chiral symmetry, since $\sigma(x)$ ($\chi(x)$ and ϕ as well) is a chiral singlet. This is a nonlinear realization of the scale symmetry: the $\phi(x)$ is a dilaton, NG boson of the spontaneously broken scale symmetry.

With Eqs. (10) and (12), we can straightforwardly rewrite the SM Higgs Lagrangian Eq. (7) in to a form of the nonlinear sigma model[19]:

$$\mathcal{L}_{\rm Higgs} = \left[\frac{F_\phi^2}{2}(\partial_\mu\chi)^2 + \frac{F_\pi^2}{4}\chi^2\cdot {\rm tr}(\partial_\mu U\partial^\mu U^\dagger)\right] - V(\phi)$$

$$= \chi^2(x)\cdot\left[\frac{1}{2}(\partial_\mu\phi)^2 + \frac{F_\pi^2}{4}{\rm tr}(\partial_\mu U\partial^\mu U^\dagger)\right] - V(\phi),$$

$$V(\phi) = \frac{\lambda}{4}v^4[(\chi^2(x)-1)^2-1] = \frac{M_\phi^2F_\phi^2}{8}[(\chi^2(x)-1)^2-1],$$

$$F_\phi = F_\pi = v,\quad M_\phi^2 = 2\lambda v^2. \tag{14}$$

This is nothing but the scale-invariant nonlinear sigma model, similar to the effective theory of the walking technicolor,[20] apart from the form of the explicit scale-symmetry breaking potential: $V(\phi) = \frac{M_\phi^2F_\phi^2}{4}\chi^4(\ln\chi - \frac{1}{4})$ with $F_\phi \neq F_\pi \neq v$ in general, instead of that in Eq. (14).

The scale symmetry is explicitly broken only by the potential $V(\phi)$ such that $\delta_D V(\phi) = \lambda v^4\chi^2 = -\theta_\mu^\mu$ whose scale dimension $d_\theta = 2$ (originally the tachyon mass term), namely, the scale symmetry is broken only by the dimension 2 operator instead of dimension 4 in the walking technicolor arising from the trace anomaly of quantum mechanical origin: This yields the mass of the (pseudo-)dilaton as the Higgs $M_\phi^2 = 2\lambda v^2$, which is in accord with the partially conserved dilatation current (PCDC) for $\partial^\mu D_\mu = \theta_\mu^\mu$:

$$M_\phi^2F_\phi^2 = -\langle 0|\partial^\mu D_\mu|\phi\rangle F_\phi = -d_\theta\langle\theta_\mu^\mu\rangle = 2\lambda v^4\langle\chi^2(x)\rangle = 2\lambda v^4 \tag{15}$$

with $F_\phi = v$, where D_μ is the dilatation current: $\langle 0|D_\mu(x)|\phi\rangle = -iq_\mu F_\phi e^{-iqx}$, or equivalently $\langle 0|\theta_{\mu\nu}|\phi(q)\rangle = F_\phi(q_\mu q_\nu - q^2 g_{\mu\nu}/3)$.

Hence the SM Higgs as it stands is a (pseudo-) dilaton, with the mass arising from the dimension 2 operator in the potential, which vanishes for $\lambda \to 0$:

$$M_\phi^2 = 2\lambda v^2 \to 0 \quad \left(\lambda\to 0, v = \sqrt{\frac{-\mu_0^2}{\lambda}} = \text{fixed} \neq 0\right) \tag{16}$$

("conformal limit",[19] which corresponds to the so-called "Bogomol'nyi–Prasad–Sommerfield (BPS) limit" of 't Hooft–Polyakov monopole in the Georgi–Glashow model, similar to the SUSY flat direction[22]).[d] In fact, the Higgs mass 125 GeV

Although χ is a dimensionless field, it transforms as that of dimension 1, while ϕ having dimension 1 transforms as the dimension 0, instead. The physical particles are ϕ and π which are defined by the nonlinear realization, in contrast to the tachyons $\hat\sigma$ and $\hat\pi_a$.

[d]With vanishing potential, $V(\phi)\to 0$, this limit still gives an interacting theory where the physical particles π and ϕ have derivative coupling in the same sense as in the nonlinear chiral Lagrangian Eq. (20). It should be contrasted to the triviality limit, $\lambda \to 0$ without fixing $v = \sqrt{\frac{-\mu_0^2}{\lambda}} \neq 0$, which yields only a free theory of tachyons $\hat\pi$ and $\hat\sigma$. The interaction of course generates the trace anomaly of dimension 4 just like that of the walking technicolor, even in the conformal/BPS limit

implies that the SM Higgs is in near conformal/BPS limit $\lambda \to 0$ with $v =$ fixed:

$$\lambda = \frac{1}{2}\left(\frac{M_\phi}{v}\right)^2 \simeq \frac{1}{2}\left(\frac{125\,\text{GeV}}{246\,\text{GeV}}\right)^2 \simeq \frac{1}{8} \ll 1. \tag{17}$$

Note that mass term of all the SM particles except the Higgs is scale-invariant.

By the electroweak gauging as usual; $\partial_\mu U \Rightarrow \mathcal{D}_\mu U = \partial_\mu U - ig_2 W_\mu U + ig_1 U B_\mu$ in Eq. (14), we see that the mass term of W/Z is scale-invariant thanks to the dilaton factor χ, and so is the mass term of the SM fermions f: $g_Y \bar{f} h f = (g_Y v/\sqrt{2})(\chi \bar{f} f)$, all with the scale dimension 4. This implies that the couplings of the SM Higgs as a pseudo-dilaton to all the SM particles are written in the scale-invariant form and thus obey the low energy theorem of the scale symmetry in perfect agreement with the experiments:

The low energy theorem for the pseudo-dilaton $\phi(q_\mu)$ coupling to the canonical matter filed X at $q_\mu \to 0$ reads

$$g_{\phi X^\dagger X} = \frac{2M_X^2}{F_\phi}, \quad g_{\phi \bar{X} X} = \frac{M_X}{F_\phi}, \quad (F_\phi = v) \tag{18}$$

for complex scalar and spin 1/2 fermion, respectively,[29] which can also be read from the scale invariance of the mass term;

$$\begin{aligned} M_X^2 \cdot \chi^2 X^\dagger X &= M_X^2 X^\dagger X + \frac{2M_X^2}{v}\phi X^\dagger X + \cdots, \\ M_X \cdot \chi \bar{X} X &= M_X \bar{X} X + \frac{M_X}{v}\phi \bar{X} X + \cdots \end{aligned} \tag{19}$$

for the respective canonical field with the canonical dimension (For the neutral scalar, we have $g_{\phi XX} = M_X^2/F_\phi$). See Refs. 33 and 34 for the general form of the low energy theorem of the scale symmetry including the anomalous dimension.

On the other hand, if we take the limit $\lambda \to \infty$, then the SM Higgs Lagrangian goes over to the usual nonlinear sigma model *without scale symmetry*:

$$\mathcal{L}_{\text{NL}\sigma} = \frac{F_\pi^2}{4}\text{tr}(\partial^\mu U \partial_\mu U^\dagger), \quad \left(\lambda \to \infty,\, F_\pi = v = \sqrt{\frac{-\mu_0^2}{\lambda}} = \text{fixed} \neq 0\right), \tag{20}$$

where the potential is decoupled with $\chi(x)$ frozen to the minimal point $\chi(x) \equiv 1$ ($\phi(x) \equiv \langle\phi(x)\rangle = v \neq 0$), so that the scale symmetry breaking is transferred from the potential to the kinetic term, which is no longer transformed as the dimension 4 operator.

The $\lambda \to \infty$ limit is known to be a good effective theory (chiral perturbation theory) of the ordinary QCD which in fact lacks the scale symmetry at all, perfectly consistent with the nonlinear sigma model, Eq. (20).

where the tree-level potential vanishes. This limit should also be distinguished from the popular limit $\mu_0^2 \to 0$ with $\lambda =$ fixed $\neq 0$, where the Coleman–Weinberg potential as the explicit scale symmetry breaking is generated by the trace anomaly (dimension 4 operator) due to the quantum loop in somewhat different context.

Absence of the scale symmetry in QCD corresponding to $\lambda \to \infty$ is also consistent with the well-known failure of the old idea to regard the lightest scalar $f_0(500)$ ("σ") as the pseudo-dilaton[29]: if it were the pseudo-dilaton, the low energy theorem of the scale symmetry in Eq. (18) would uniquely determine the low energy limit dilaton couplings to the matter (including massive pion) in units of the dilaton decay constant $f_\sigma (\geq f_\pi = 93\,\mathrm{MeV})$. We may take the coupling ratio which is free from the unknown parameter f_σ, and see a typical case of couplings to π and the nucleon[e]: $g_{\sigma\pi\pi}/g_{\sigma NN} = m_\pi/m_N$ for $\mathcal{L} = m_\pi g_{\sigma\pi\pi} \cdot \sigma\pi^a\pi_a,\ g_{\sigma NN} \cdot \bar{N}\sigma N$, which predicts $g^2_{\sigma\pi\pi} \simeq (m_\pi/m_N)^2 g^2_{\sigma NN} \simeq 2$ for the observed value $g_{\sigma NN} \simeq 10$ (the value consistent with the low energy theorem $g_{\sigma NN} = m_N/f_\sigma$, if $f_\sigma = f_\pi$). Then, the pseudo-dilaton width would be

$$\Gamma_\sigma \simeq 3 \times \frac{m_\pi^2 g^2_{\sigma\pi\pi}}{8\pi m_\sigma}\left(1 - \frac{4m_\pi^2}{m_\sigma^2}\right)^{1/2} \sim 7 - 8\,\mathrm{MeV}(m_\sigma = 500 - 600\,\mathrm{MeV}), \quad (21)$$

which is compared with the experiment $\Gamma_{f_0} = 400 - 700\,\mathrm{MeV}$, roughly two orders magnitude smaller (unless the on-shell couplings are drastically distorted from the low energy limit values, though it is another symptom of the absence of the scale symmetry anyway)

Thus, there is no remnants of scale symmetry in the QCD in the free space (There could be some emergent scale symmetry for the hot and/or dense QCD, however[35]). This is in sharp contrast to the SM Higgs whose couplings to the SM particles (quark/lepton and $W/Z/\gamma$) all respect the low energy theorem of the scale symmetry even on the SM Higgs on-shell away from the low energy limit, in perfect agreement with the experiments as mentioned before.

3. HLS in the SM Higgs Lagrangian, the "SM Rho Meson"

The SM Higgs Lagrangian was further shown[19] to be *gauge equivalent* to the *scale-invariant* version[23] of the HLS Lagrangian,[1,2,4,6,7] which contains possible new vector boson, "SM rho", hidden behind the SM Higgs Lagrangian, as an analogue of the QCD ρ meson.[f]

As usual, the HLS can be made explicit by dividing $U(x)$ in to two parts[1,2,4,6,7]:

$$U(x) = \xi_L^\dagger(x) \cdot \xi_R(x), \quad (22)$$

where $\xi_{R,L}(x)$ transform under $G_{\mathrm{global}} \times H_{\mathrm{local}}$ as

$$\xi_{R,L}(x) \to h(x) \cdot \xi_{R,L}(x) \cdot {g'}^\dagger_{R,L}, \quad U(x) \to \hat{g}_L U(x) {g'}^\dagger_R,$$
$$(h(x) \in H_{\mathrm{local}}, {g'}_{R,L} \in G_{\mathrm{global}}). \quad (23)$$

The H_{local} is a gauge symmetry of group H arising from the redundancy (gauge symmetry) how to divide U in to two parts.

[e]The low energy theorem coupling $g_{\phi X^\dagger X} = 2M_X^2/F_\phi$ in Eq. (18) corresponds to $2m_\pi g_{\sigma\pi\pi}$ with the conventional dimensionless coupling $g_{\sigma\pi\pi} = m_\pi/f_\sigma$ used here.

[f]The s-HLS model was also discussed in a different context, ordinary QCD in medium.[35]

Then, we can introduce the HLS gauge boson, "SM rho" meson, $\rho_\mu(x)$ by covariant derivative as

$$D_\mu \xi_{R,L}(x) = \partial_\mu \xi_{R,L}(x) - i\rho_\mu(x)\xi_{R,L}(x), \tag{24}$$

which transform in the same way as $\xi_{R,L}$. Then, we have two covariant objects transforming homogeneously under H_{local}:

$$\begin{aligned}
&\{\hat{\alpha}_{\mu,R,L}, \hat{\alpha}_{\mu,||,\perp}\} \to h(x) \cdot \{\hat{\alpha}_{\mu,R,L}, \hat{\alpha}_{\mu,||,\perp}\} \cdot h^\dagger(x), \\
&\quad \hat{\alpha}_{\mu,R,L} \equiv \frac{1}{i} D_\mu \xi_{R,L} \cdot \xi^\dagger_{R,L} = \frac{1}{i}\partial_\mu \xi_{R,L} \cdot \xi^\dagger_{R,L} - \rho_\mu, \\
&\quad \hat{\alpha}_{\mu,||,\perp} \equiv \frac{1}{2}(\hat{\alpha}_{\mu,R} \pm \hat{\alpha}_{\mu,L}) = \begin{cases} \alpha_{\mu||} - \rho_\mu, \\ \alpha_{\mu\perp}, \end{cases}
\end{aligned} \tag{25}$$

where

$$\begin{aligned}
\alpha_{\mu||} &= \frac{1}{2i}\left(\partial_\mu \xi_R \cdot \xi^\dagger_R + \partial_\mu \xi_L \cdot \xi^\dagger_L\right) = \frac{1}{F_\rho}\partial_\mu \check{\rho} - \frac{i}{2v^2}[\partial_\mu \pi, \pi] + \cdots, \\
\alpha_{\mu\perp} &= \frac{1}{2i}\left(\partial_\mu \xi_R \cdot \xi^\dagger_R - \partial_\mu \xi_L \cdot \xi^\dagger_L\right) \\
&= \frac{1}{2i}\xi_L \cdot \partial_\mu U \cdot \xi^\dagger_R = \frac{1}{2i}\xi_R \partial_\mu U^\dagger \cdot \xi^\dagger_L
\end{aligned} \tag{26}$$

with $\check{\rho}$ and F_ρ being the NG boson to be absorbed into the longitudinal ρ_μ and its decay constant, respectively, as introduced by $\xi_{R,L} = e^{i\check{\rho}/F_\rho} \cdot e^{\pm i\pi/v}$ (see the discussions below).[g]

We thus have two independent invariants under the larger symmetry $G_{\text{global}} \times H_{\text{local}}$:

$$\mathcal{L}_A = v^2 \cdot \text{tr}\hat{\alpha}^2_\perp(x) = v^2 \cdot \text{tr}\alpha^2_{\mu\perp}(x) = \frac{v^2}{4} \cdot \text{tr}(\partial_\mu U \partial^\mu U^\dagger), \tag{27}$$

$$\begin{aligned}
\mathcal{L}_V &= v^2 \cdot \text{tr}\hat{\alpha}^2_{\mu||}(x) = v^2 \cdot \text{tr}\,(\rho_\mu(x) - \alpha_{\mu||}(x))^2, \\
&= v^2 \cdot \text{tr}\left(\left(\rho_\mu(x) - \frac{1}{F_\rho}\partial_\mu \check{\rho}\right) - \frac{i}{2v^2}[\partial_\mu \pi, \pi] + \cdots\right)^2,
\end{aligned} \tag{28}$$

where $\mathcal{L}_A$ is the original nonlinear sigma model on *G/H*, a part of the SM Higgs Lagrangian in the form of Eq. (14). Then, the scale-invariant $G_{\text{global}} \times H_{\text{local}}$ model takes the form:

$$\mathcal{L}_s - \text{HLS} = \chi^2(x) \cdot \left(\frac{1}{2}(\partial_\mu \phi)^2 + \mathcal{L}_A + a\mathcal{L}_V\right) \tag{29}$$

with a being an arbitrary parameter. The kinetic term of $\check{\rho}$ is normalized as $F^2_\rho = aF^2_\pi = av^2$.

[g]In the HLS papers,[1,2,4,6,7] $\check{\rho}$ was denoted by σ. In order to avoid confusion, we will use $\check{\rho}$ in this paper.

We now fix the gauge of HLS as $\xi_L^\dagger = \xi_R = \xi = e^{i\pi/v}$ such that $U = \xi^2$ (unitary gauge $\check{\rho} = 0$). Then, $H_{\rm local}$ and $H_{\rm global}(\subset G_{\rm global})$ get simultaneously broken spontaneously (Higgs mechanism), leaving the diagonal subgroup $H = H_{\rm local} + H_{\rm global}$, which is nothing but the subgroup H of the original G of G/H: $H \subset G$. Then, the extended symmetry $G_{\rm global} \times H_{\rm local}$ is simply reduced back to the original nonlinear realization of G on the manifold G/H, both are gauge equivalent to each other.

Thus, the SM Lagrangian in the form of Eq. (14) is gauge equivalent to

$$\mathcal{L}_{\rm Higgs-HLS} = \chi^2(x) \cdot \left(\frac{1}{2} (\partial_\mu \phi)^2 + \frac{v^2}{4} \cdot {\rm tr}(\partial_\mu U \partial^\mu U^\dagger) \right) - V(\phi)$$
$$+ \chi^2(x) \cdot F_\rho^2 \cdot {\rm tr}(\rho_\mu(x) - \alpha_{\mu||}(x))^2, \quad (F_\rho^2 = a v^2), \tag{30}$$

where the second line $\chi^2 a \mathcal{L}_V$ is the extra term which has the rho field ρ_μ as the auxiliary field. In the path integral language, it is a Gaussian integral which would not change the physics. In fact, at classical level in the absence of the kinetic term for the HLS gauge boson ρ_μ, it is simply solved away to yield

$$\chi^2(x) \cdot F_\rho^2 \cdot {\rm tr}(\rho_\mu(x) - \alpha_{\mu||}(x))^2 = 0, \tag{31}$$

after using the equation of motion $\rho_\mu = \alpha_{\mu||}$. Then $\mathcal{L}_{\rm Higgs-HLS}$ in Eq. (30) is simply reduced back to the original SM Higgs Lagrangian $\mathcal{L}_{\rm Higgs}$ in nonlinear realization, Eq. (14).

As we shall discuss later, however, the auxiliary field ρ_μ actually acquires the kinetic term,

$$\mathcal{L}^{(\rho)}_{\rm kinetic} = -\frac{1}{2g^2} {\rm tr}\, \rho_{\mu\nu}^2, \tag{32}$$

by the quantum corrections with g being the gauge coupling of HLS. Then, the quantum theory for the SM Higgs takes the form:

$$\mathcal{L}_{\rm Higgs-HLS} = \chi^2 \cdot \left(\frac{1}{2} (\partial_\mu \phi)^2 + \frac{v^2}{4} \cdot {\rm tr}(\partial_\mu U \partial^\mu U^\dagger) \right) - V(\phi)$$
$$+ \chi^2 \cdot F_\rho^2 \cdot {\rm tr}(\rho_\mu - \alpha_{\mu||})^2 - \frac{1}{2g^2} {\rm tr}\, \rho_{\mu\nu}^2 + \cdots, \tag{33}$$

where "$\cdots$" stands for other induced terms at quantum level.

When it happens, after rescaling the kinetic term of ρ_μ, $\rho_\mu(x) \to g\, \rho_\mu(x)$ to the canonical one $-\frac{1}{2} {\rm tr}\, \rho_{\mu\nu}^2$, the $\chi^2\, a\mathcal{L}_V$ term yields the scale-invariant mass term of ρ_μ,

$$\chi^2\, a\mathcal{L}_V = \chi^2 F_\rho^2 \cdot {\rm tr}(g\rho_\mu - \alpha_{\mu,||})^2 = M_\rho^2\, {\rm tr}(\rho_\mu)^2 + g_{\rho\pi\pi} \cdot 2i{\rm tr}(\rho^\mu [\partial_\mu \pi, \pi]) + \cdots,$$
$$M_\rho^2 = g^2 F_\rho^2 = a (gv)^2, \quad g_{\rho\pi\pi} = \frac{F_\rho^2}{2v^2}\, g = \frac{a}{2} g \tag{34}$$

with the mass acquired by the Higgs mechanism mentioned above, which provides the standard KSRF I relation for $a = F_\rho^2/F_\pi^2 = F_\rho^2/v^2 = 2$[1,2]:

$$M_\rho^2 = \left(\frac{4F_\pi^2}{F_\rho^2}\right) g_{\rho\pi\pi}^2 F_\pi^2 = \frac{4}{a} g_{\rho\pi\pi}^2 v^2. \tag{35}$$

Note that the HLS gauge boson acquires the scale-invariant mass term thanks to the dilaton factor χ^2, the nonlinear realization of the scale symmetry, in sharp contrast to the Higgs (dilaton) which acquires mass only from the explicit breaking of the scale symmetry.

The electroweak gauge bosons ($\in \mathcal{R}_\mu(\mathcal{L}_\mu)$) are introduced by extending the covariant derivative of Eq. (24) this time by gauging $G_{\rm global}$, which is *independent of* $H_{\rm local}$ in the HLS extension:

$$D_\mu \xi_{R,L}(x) \Rightarrow \hat{D}_\mu \xi_{R,L}(x) \equiv \partial_\mu \xi_{R,L}(x) - i\rho_\mu(x)\, \xi_{R,L}(x) + i\xi_{R,L}(x)\, \mathcal{R}_\mu(\mathcal{L}_\mu). \tag{36}$$

We then, finally, have a gauged s-HLS version of the Higgs Lagrangian (gauged-s-HLS)[h]:

$$\mathcal{L}_{\rm Higgs-HLS}^{\rm gauged} = \chi^2(x) \cdot \left[\frac{1}{2}(\partial_\mu \phi)^2 + \hat{\mathcal{L}}_A + a\hat{\mathcal{L}}_V\right] - V(\phi) + \mathcal{L}_{\rm kinetic}^{(\rho,\mathcal{L},\mathcal{R})} + \cdots \tag{37}$$

with

$$\hat{\mathcal{L}}_{A,V} = \mathcal{L}_{A,V}(D_\mu \xi_{R,L}(x) \Rightarrow \hat{D}_\mu \xi_{R,L}(x)). \tag{38}$$

This yields, besides Eq. (34), a notable a-independent relation (KSRF II) between the $\rho - \gamma$ mixing strength g_ρ and $g_{\rho\pi\pi}$ from the mass term $\chi^2 a\mathcal{L}_V$[1,2]:

$$g_\rho = gF_\rho = 2F_\pi^2 g_{\rho\pi\pi} = 2v^2 g_{\rho\pi\pi}, \tag{39}$$

and its extension to $W/Z - \rho$ mixing strength (low energy theorem of HLS: Proof in Ref. 36) should be intact even when the mass term becomes dimension 4 with the extra factor χ^2 which introduces additional symmetry, the scale symmetry). As usual in the Higgs mechanism, the gauge bosons of gauged$-H_{\rm global}(\subset$ gauged$-G_{\rm global})$ get mixed with the gauge bosons of HLS, leaving only the gauge bosons of the unbroken diagonal subgroup (gauged$-H$) $= H_{\rm local}+$(gauged$-H_{\rm global}$) be massless after mass diagonalization.

Again note that the mass terms including the couplings of all the SM particles except for the Higgs mass term $V(\phi)$ are dimension 4 operators and thus are scale-invariant.

[h] This form of the Lagrangian is the same as that of the effective theory of the one-family ($N_F = 8$) walking technicolor,[23] except for the shape of the scale-violating potential $V(\phi)$ which has a scale dimension 4 (trace anomaly) in the case of the walking technicolor instead of two of the SM Higgs case (Lagrangian mass term).

4. Dynamical Generation of HLS Gauge Boson I: Lesson from the Dynamical Higgs in the NJL Model

Before discussing the dynamical generation of the HLS gauge boson, kinetic term and Yang–Mills self-couplings as well, here, we recapitulate the well-known formulation[8,9] to show the dynamical generation of the composite Higgs model, kinetic term and quartic self-coupling, based on the strong coupling phase $G > G_{\rm cr} \neq 0$ in the NJL model. The NJL Lagrangian for the N_C-component two-flavored fermion ψ takes the form[12]:

$$\mathcal{L}_{\rm NJL} = \bar\psi i\gamma^\mu\partial_\mu\psi + \frac{G}{2}\left[(\bar\psi\psi)^2 + (\bar\psi i\gamma_5\tau^a\psi)^2\right]. \tag{40}$$

We can add the auxiliary field term which does not change the physics (Gaussian term trivially integrated out in terms of the path integral):

$$\mathcal{L}_{\rm aux} = -\frac{1}{2G}(\hat\sigma - G\bar\psi\psi)^2 - \frac{1}{2G}(\hat\pi^a - G\bar\psi i\gamma_5\tau^a\psi)^2. \tag{41}$$

This in fact yields zero $\mathcal{L}_{\rm aux} = 0$ when equation of motion $\hat\sigma = G\bar\psi\psi$ and $\hat\pi^a = G\bar\psi i\gamma_5\tau^a\psi$ are used. These auxiliary field terms just correspond to the HLS term in Eq. (30) which also yields zero when the equation of motion for ρ_μ is used. Then, the resultant Lagrangian read

$$\mathcal{L}_{\rm NJL} + \mathcal{L}_{\rm aux} = \bar\psi(i\gamma^\mu\partial_\mu + \hat\sigma + i\gamma_5\tau^a\hat\pi_a)\psi - \frac{1}{2}\frac{1}{G}(\hat\sigma^2 + \hat\pi_a^2), \tag{42}$$

where we can again check that the equations of motion of the auxiliary fields $\hat\sigma \sim G\bar\psi\psi$ and $\hat\pi^a \sim G\bar\psi i\gamma_5\tau^a\psi$ are plugged back in the Lagrangian resulting in the original Lagrangian.

However, at quantum level, a "miracle" takes place[i]: in the large N_C limit ($N_C \to \infty$ with $N_C G \neq 0$ fixed), we may integrate the loop contribution from the cutoff scale Λ down to some infrared scale μ to have dynamical generated kinetic term and quartic coupling of $(\hat\sigma, \hat\pi^a)$ in the sense of the Wilsonian renormalization-group as

$$\mathcal{L}_{\rm induced} = \frac{1}{2}Z_\phi[(\partial_\mu\hat\sigma)^2 + (\partial_\mu\hat\pi)^2] + \frac{1}{4}Z_\phi[(\hat\sigma)^2 + (\hat\pi^2)]^2, \tag{43}$$

$$Z_\phi = \frac{N_C}{8\pi^2}\ln\frac{\Lambda^2}{\mu^2}, \tag{44}$$

together with the mass shift

$$\mathcal{L}_{\rm mass} = \frac{1}{2}\left(\frac{1}{G} - \frac{N_C}{4\pi^2}(\Lambda^2 - \mu^2)\right)(\hat\sigma^2 + \hat\pi_a^2). \tag{45}$$

When we take $\mu \to \Lambda$, all the induced terms disappear and we get back to the original bare Lagrangian Eq. (42). The condition $Z_\phi(\mu) \to 0$ for $\mu \to \Lambda$ is

[i]Actually, it is not an miracle nor magic, since the following formulation yields precisely the same result as the traditional direct large N_C computation with gap equation and Bethe–Salpeter equation, giving in fact the bound state Higgs dynamically-generated.

the so-called compositeness condition.[9] After rescaling the induced kinetic term to the canonical one, $Z_\phi^{1/2}(\hat{\sigma}, \hat{\pi}^a) \to (\hat{\sigma}, \hat{\pi}^a)$, the quantum theory for $\hat{\sigma}$ and $\hat{\pi}$ sector yields precisely the same form as the SM Higgs Eq. (2),[8] with

$$\mu_0^2 = \left(\frac{1}{G} - \frac{N_C}{4\pi^2}(\Lambda^2 - \mu^2)\right) Z_\phi^{-1} = -\lambda v^2 < 0,$$

$$\lambda = Z_\phi Z_\phi^{-2} = Z_\phi^{-1} = \left[\frac{N_C}{8\pi^2} \ln \frac{\Lambda^2}{\mu^2}\right]^{-1}, \tag{46}$$

where we have used the relation $F_\pi^2 = v^2 = -\mu_0^2/\lambda$ in Eq. (6). The quadratic running in mass is understood to be renormalized in to the bare mass term $1/G$ so as to keep $\mu_0^2 < 0$ (spontaneous breaking of the chiral symmetry) as usually done by the gap equation. On the other hand, there is no bare term of the quartic coupling (unless we introduce eight-fermion operator), the induced quartic coupling has no renormalization such that $\lambda(\mu)$ has a Landau pole at $\mu = \Lambda$ in the SM language.[j]

5. Dynamical Generation of HLS Gauge Boson II: SM Rho Meson

Now, we discuss[18] that the kinetic term of ρ_μ is generated dynamically by the quantum loop,[2,4,6,7] in the same sense as that of the dynamical generation of the kinetic term (and the quartic coupling as well) of the composite Higgs in the NJL model, which is an auxiliary field at the tree-level or at composite scale.[8]

In order to discuss the off-shell ρ (in space-like momentum) relevant to the skyrmion stabilization to be discussed in Sec. 6, we adopt the background field gauge as in Refs. 4–7 (where Feynman diagrams are explicitly given). The relevant diagrams for the two-point function of the SM rho at one-loop are given in Fig. 10 of Refs. 4–7. By integrating out the high frequency modes from the cutoff scale Λ (composite scale) to the scale μ in the Wilsonian sense, we have the dynamically-generated kinetic term as given in Eq. (4.194) of Refs. 4–7 ($N_f = 2$, in the case at hand)[k]:

$$\mathcal{L}_{\text{kinetic}}^{(\rho)} = -\frac{1}{2g(\mu)^2} \text{tr}\rho_{\mu\nu}^2, \tag{47}$$

$$\frac{1}{g(\mu)^2} = \frac{1}{(4\pi)^2} \frac{a^2}{12} \ln \frac{\Lambda^2}{\mu^2}, \quad (\mu^2 \ll M_\rho^2),$$

$$= \frac{1}{(4\pi)^2} \left(\frac{a^2}{12} + \left(-\frac{22}{3} + \frac{1}{12}\right)\right) \ln \frac{\Lambda^2}{\mu^2}, \quad (\mu^2 \gg M_\rho^2). \tag{48}$$

[j] There of course exists a conformal/BPS limit of the NJL model written in the form of the SM Higgs Lagrangian. See Ref. 12.

[k] There is a discontinuity between the results for $\mu \gg M_\rho$ and $\mu \ll M_\rho$ near $\mu \sim M_\rho$, which is an artifact of disregarding the finite parts of the loop integral,[4,6,7] and irrelevant to the discussions here.

For $\mu^2 \ll M_\rho^2$[l], we have only the loop contribution of π (longitudinal W/Z when the electroweak gauging switched on) with $g_{\rho\pi\pi} = a/2$, which is characterized by $a^2/12$ in front of the log in Eq. (48). The resultant kinetic term includes the self-couplings of the SM rho as it should (in a way consistent with the gauge invariance), similar to the composite Higgs self-coupling as well as the kinetic term in the NJL model. The kinetic term vanishes $\frac{1}{g^2(\mu)} \to 0$ for $\mu \to \Lambda$, in a way similar to composite Higgs in the NJL model.[8] This is the simplest case when $M_\rho = \mathcal{O}(\Lambda)$.

On the other hand, for $\mu^2 \gg M_\rho^2$, there arise additional contributions: the gauge loop from the dynamically-generated SM rho self-coupling, with the usual factor, $-22/3$, together with $+1/12$ from the loop of the would-be NG boson $\check{\rho}$ (the longitudinal SM rho) having the ρ coupling $1/2$, adds up to the characteristic factor $(a^2 - 87)/12$ instead of $a^2/12$. For the gauge coupling $g(\mu)^2$ to have a Landau pole at $\mu = \Lambda$ in conformity of the compositeness condition, we would need

$$a > \sqrt{87} \simeq 9.3 \quad (M_\rho \ll \Lambda). \tag{49}$$

Although Eq. (48) is similar to that of the QCD rho meson, an outstanding difference is that the QCD rho kinetic term is already generated by the underlying QCD and hence the standard QCD value,[1,2] $a = 2$, has no problem about the HLS framework of the QCD rho meson, while in the SM rho case for $M_\rho^2 \ll \Lambda^2$, we generally need a large value of a to get the kinetic term to be dynamically-generated by the SM dynamics alone without recourse to the UV completion.

Of course there is a possibility that even in the SM rho meson the bare kinetic term may be provided by some underlying theory beyond the SM, such as the walking technicolor[21] whose low energy effective theory is the scale-invariant HLS model[23] similar to the SM Higgs Lagrangian as discussed before. In such a case, a large a would not be needed as in the QCD rho meson.

In the next section, we shall only discuss the simplest scenario $M_\rho = \mathcal{O}(\Lambda)$ without such a UV completion as a benchmark for the SM skyrmion for the dark matter.

By the loop effects, the M_ρ^2 would develop (potentially large) imaginary parts in the time-like region for decaying to the $\pi\pi$ ($W_L W_L, W_L Z_L$ when the electroweak couplings switched on) if $M_\rho > 2M_{W/Z}$. However, this would not affect the skyrmion physics which is relevant to the space-like ρ. Thus, we will confine ourself to the kinetic term hereafter, although it would give rise to interesting collider physics if $M_\rho = \mathcal{O}$ (TeV).

6. SM Skyrmion as a Dark Matter, "Dark SM Skyrmion"

Now that we have discussed the dynamical generation of the kinetic term of the SM rho meson, we can discuss the skyrmion living inside the SM Higgs Lagrangian in

[l]Rescaling the kinetic term to the canonical one by $\rho_\mu \to g\rho_\mu$, we have the off-shell mass $M_\rho^2(\mu) = g^2(\mu)F_\rho^2 = ag^2(\mu)v^2$ which behaves as $M_\rho(\mu) \to \infty(0)$ as $\mu \to \Lambda(0)$. We have defined M_ρ as $M_\rho^2 \equiv M_\rho^2(\mu = M_\rho) = g^2(\mu = M_\rho)F_\rho^2$.

the form of Eq. (33), which does have a skyrmion stabilized by the SM rho meson as a viable candidate for the dark matter, "dark SM skyrmion (DSMS)" denoted as X_s..[18]

It is in fact well-known in the hadron physics[24] that the QCD rho meson stabilizes the skyrmion (nucleon) with $I = J = 1/2, 3/2, \ldots$, with a numerical result giving rise to the skyrmion mass somewhat smaller than that in the original Skyrme model. It was further shown[24] that the Skyrme term may be regarded as the heavy mass limit of the HLS rho meson, $M_\rho^2 = ag^2v^2 \gg v^2$, such that $a \to \infty,\ g = \text{const.}$, where the rho field configuration in the term $\chi^2 a\mathcal{L} = a\chi^2(\rho_\mu - \alpha_{\mu,||})^2$ in Eq. (33) is restricted to the auxiliary field configuration as a composite of the π in the nonlinear base:

$$\rho_\mu \to \alpha_{\mu||} = \frac{1}{2i}(\partial_\mu \xi_R(x) \cdot \xi_R^\dagger(x) + \partial_\mu \xi_L(x) \cdot \xi_L^\dagger(x)), \quad (a \to \infty,\ g = \text{const.}), \tag{50}$$

and hence

$$\rho_{\mu\nu} \to \rho_{\mu\nu}|_{\rho_\mu = \alpha_{\mu||}} = i[\hat{\alpha}_{\mu\perp}, \hat{\alpha}_{\nu\perp}], \quad \hat{\alpha}_{\mu\perp} = \frac{\xi_L(\partial_\mu U)\xi_R^\dagger}{2i}, \tag{51}$$

$$\mathcal{L}_{\text{kinetic}}^{(\rho)}(\rho_\mu) \to -\frac{1}{2g^2}\text{tr}(i[\hat{\alpha}_{\mu\perp}, \hat{\alpha}_{\nu\perp}])^2 = \frac{1}{32g^2}\text{tr}[[\partial_\mu U U^\dagger, \partial_\nu U U^\dagger]^2], \tag{52}$$

namely, the kinetic term becomes precisely equal to the Skyrme term with $e = g$ in this limit.

It has also been shown[25,26,28] that including the scalar meson (corresponding to the SM Higgs as a dilaton in our case) does not invalidate the skyrmion, but rather makes the skyrmion mass lighter. Particularly, the skyrmion in QCD was discussed[25] in essentially the same form as our scale-invariant HLS form of the SM Higgs Lagrangian Eq. (33) except for the potential term $V(\phi)$ (dimension 4 instead of our case with dimension 2) and adopted values of the parameters F_π, F_ϕ, a, g, etc.

Thus the DSMS X_s emerges as a soliton solution from the Lagrangian Eq. (33) to be a topological bosonic matter carrying the topological number, which we call $U(1)_{X_s}$.[18] The $U(1)_{X_s}$ symmetry protects the decay of the DSMS (X_s) completely, so the X_s can be a dark matter candidate. Here, we consider a complex scalar DSMS with $I = J = 0$.

The DSMS is essentially generated by the scale-invariant part of Eq. (33) and hence its coupling is dictated by the nonlinear realization of the scale symmetry. In the low energy limit, $q^2 \ll v^2 \simeq (246\,\text{GeV})^2$, the DSMS ($X_s$) coupling to the SM Higgs ϕ as a pseudo-dilaton is unambiguously determined by the low-energy theorem of the scale symmetry,[29] as described in Eq. (18)[18]

$$g_{\phi X_s X_s} = \frac{2M_{X_s}^2}{v} = 2v\lambda_{X_s}, \quad \lambda_{X_s} = \frac{M_{X_s}^2}{v^2}, \tag{53}$$

which is relevant to the dark matter detection experiments for weakly-interacting massive-particle (WIMP) such as the LUX experiments.[30]

Through the Higgs (ϕ) exchange at zero-momentum transfer as given in Eq. (53), the relevant spin-independent (SI) elastic scattering cross-section of the DSMS X_s

with the target nucleus, (Xe : $Z = 54, A = 131.293, u = 0.931\,\text{GeV}$) per nucleon ($N = p, n$), can be calculated as

$$\sigma_{\text{SI}}^{\text{elastic/nucleon}}(X_s N \to X_s N) = \frac{\lambda_{X_s}^2}{\pi M_\phi^4}\left[\frac{Z \cdot m_*(p, X_s) g_{\phi pp} + (A - Z) \cdot m_*(n, X_s) g_{\phi nn}}{A}\right]^2, \tag{54}$$

where $m_*(N, X_s) = \frac{M_{X_s} m_N}{M_{X_s} + m_N}$ and $g_{\phi pp(nn)} = \sum_q \sigma_q^{(p(n))}/v \simeq 0.248(0.254)$ GeV$/v$.[37,38]

Using $m_{p(n)} \simeq 938(940)\,\text{MeV}$ together with the electroweak scale $v \simeq 246\,\text{GeV}$ and the Higgs mass $M_\phi \simeq 125\,\text{GeV}$, we can numerically evaluate the cross-section as a function of the DSMS mass M_{X_s}. We find[18] (see Fig. 1) that the currently strongest exclusion limit from the latest LUX2016 experiment[30] implies

$$M_{X_s} \lesssim 38\,\text{GeV}. \tag{55}$$

Note that we have the upper bound instead of the lower bound in contrast to conventional WIMP models due to the characteristic dilatonic coupling proportional to $M_{X_s}^2$ as in Eq. (53).

Since the LUX2016 limit in Eq. (55) implies $M_{X_s} < M_\phi/2 \simeq 63\,\text{GeV}$, further constraint on the mass of the DSMS will be placed through the Higgs invisible decay limit at collider experiments. The on-shell coupling of the SM Higgs as a pseudo-dilaton to the $X_s \bar{X}_s$, relevant to the invisible decay process, should be the same as that determined by the low-energy theorem, i.e., $q \sim M_\phi \ll v$ in Eq. (53), as it is true for the couplings of the SM Higgs as a pseudo-dilaton to all the SM particles as mentioned before. The partial decay width of the Higgs ϕ to the $X_s \bar{X}_s$ is thus

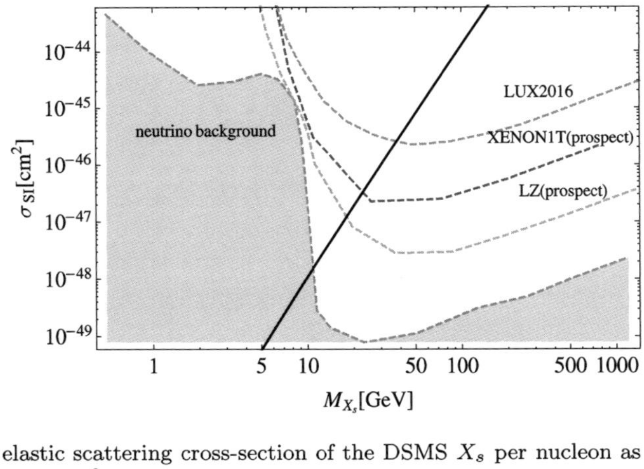

Fig. 1. The SI elastic scattering cross-section of the DSMS X_s per nucleon as a function of the mass M_{X_s} in units of cm^2 (solid curve). The most stringent constraint at present from the latest LUX2016 experiment[30] are shown and also shown are the projected experiments with the xenon target by the end of this decade.[39] The gray domain, surrounded by the dashed curve on the bottom, stands for the atmospheric and astrophysical neutrino background.[40]

unambiguously computed as

$$\Gamma(\phi \to X_s \bar{X}_s) = \frac{\lambda_{X_s}^2 v^2}{4\pi M_\phi} \sqrt{1 - \frac{4M_{X_s}^2}{M_\phi^2}}. \tag{56}$$

The branching ratio is then constructed as $\mathrm{Br}[\phi \to X_s \bar{X}_s] = \Gamma(\phi \to X_s \bar{X}_s)/\Gamma_\phi^{\rm tot} = \Gamma(\phi \to X_s \bar{X}_s)/[\Gamma_\phi^{\rm SM} + \Gamma(\phi \to X_s \bar{X}_s)]$ with the total SM Higgs width (without the $X_s \bar{X}_s$ decay mode) $\Gamma_\phi^{\rm SM} \simeq 4.1\,\mathrm{MeV}$ at the mass of 125 GeV.[41] Currently, the most stringent upper limit on the Higgs invisible decay has been set by the CMS Collaboration combined with the run II data set with the luminosity of $2.3\,\mathrm{fb}^{-1}$.[31] Figure 2 shows the exclusion limit on the X_s mass at 95% C.L., $\mathrm{Br}_{\rm invisible} \lesssim 0.2$.[31] From the figure, we find[18]

$$M_{X_s} \lesssim 18\,\mathrm{GeV}. \tag{57}$$

In Fig. 2, the future prospected 95% C.L. limits in the LHC and ILC experiments[42] are also shown.

We should emphasize that such a characteristic mass range of the dark matter candidate is a salient feature of the scale symmetry of the DSMS coupled to the SM Higgs as a pseudo-dilaton as the low energy theorem of the scale symmetry.

Now, we come to the relic abundance of the DSMS in the thermal history of the Universe.[18] In the thermal history of the Universe, the DSMS emerges after the electroweak phase transition at the temperature $T = \mathcal{O}(v)$. Below the freeze-out temperature $T < T_f$ ($x = M_{X_s}/T > x_f = M_{X_s}/T_f \simeq 20$), the DSMS number density evolves merely according to the adiabatic expansion of the universe, and

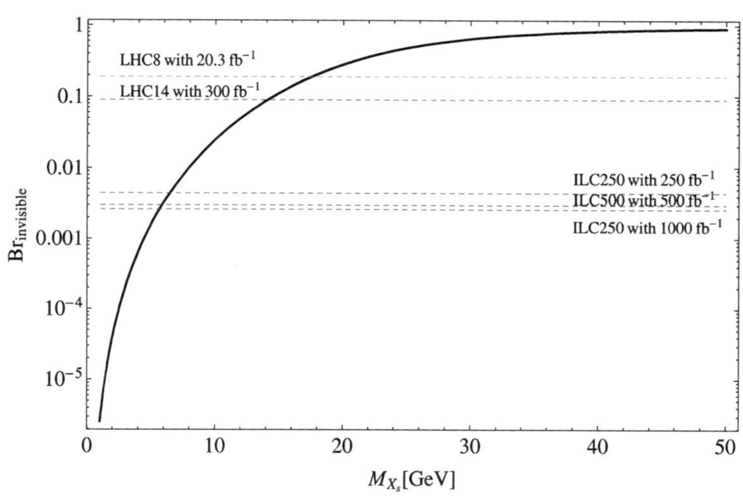

Fig. 2. The branching fraction of the Higgs decaying to the DSMS pair as a function of the mass M_{X_s} (solid curve) compared with the most stringent (95% C.L.) constraint at present from the LHC run I combined with the early stage of the 13 TeV run reported by the CMS group.[31] Also, plotted are the expected 95% C.L. limits in the projected collider experiments including the 14 TeV LHC and ILC.[42]

the DSMS is cooled down to become a cold dark matter just like WIMPs, with the relic abundance observed in the Universe today. Such a relic abundance can be estimated by the standard procedure, so-called the freeze-out thermal relic[43]:

$$\Omega_{X_s} h^2 = \frac{2 \times (1.07 \times 10^9) x_f}{g_*(T_f)^{1/2} M_{\rm Pl}\,{\rm GeV} \langle \sigma_{\rm ann} v_{\rm rel} \rangle}, \tag{58}$$

where $M_{\rm Pl}$ stands for the Planck mass scale $\simeq 10^{19}\,{\rm GeV}$, $\langle \sigma_{\rm ann} v_{\rm rel}\rangle$ is the thermal average of the annihilation cross-section times the relative velocity of $X_s \bar{X}_s$, $v_{\rm rel}$, and $g_*(T_f)$ denotes the effective degrees of freedom for relativistic particles at $T = T_f$. The prefactor 2 comes from counting both X_s-particle and $\bar{X}_s$-anti-particle present today. The freeze-out temperature T_f can be determined by $x_f = \ln[2 \times 0.038 \times [g_*(T_f) x_f]^{-1/2} M_{\rm Pl} \cdot M_{X_s} \cdot \langle \sigma_{\rm ann} v_{\rm rel}\rangle]$.

In evaluating the cross-section, we expect that, until the freeze-out time, the $X_s \bar{X}_s$ annihilation in to the $U(1)_{X_s}$ current will predominantly take place. Here, we note that the DSMS is a soliton, an extended particle with a finite radius. Such an annihilation process can be viewed as the classical $X_s \bar{X}_s$ collision with the $U(1)_{X_s}$ charge radius $R_{X_s} = (\langle r^2_{X_s} \rangle_{X_s})^{1/2}$, so

$$\langle \sigma_{\rm ann} v_{\rm rel} \rangle = \mathcal{O}(\pi R^2_{X_s}) = \mathcal{O}(\pi \cdot \langle r^2_{X_s} \rangle_{X_s}). \tag{59}$$

(Similar observation was made in Ref. 28). Thus, we need to evaluate the size of $\langle r^2_{X_s} \rangle_{X_s}$, in the standard skyrmion calculation with a scalar meson in the literature.[25,26,28]

To get a rough idea of the DSMS as a dark matter, we here discuss[18] the simplest case, the heavy rho (rho decoupled) limit mentioned above, where the rho kinetic term may be replaced by the Skyrme term Eq. (52). The Lagrangian Eq. (33) reads:

$$\begin{aligned} \mathcal{L}_{\rm Higgs-Skyrme} &= \chi^2(x) \cdot \left(\frac{1}{2} (\partial_\mu \phi)^2 + \frac{v^2}{4} \cdot {\rm tr}(\partial_\mu U \partial^\mu U^\dagger) \right) - V(\phi) \\ &\quad + \frac{1}{32 g^2} {\rm tr}[[\partial_\mu U U^\dagger, \partial_\nu U U^\dagger]^2], \end{aligned} \tag{60}$$

which keeps the scale-invariance of Eq. (33), except for the potential $V(\phi)$.

Then, the skyrmion system is essentially the same as the one analyzed in Ref. 28 as to the soliton solution and numerical results in spite of the lack of scale symmetry in their Lagrangian and the associated scale-noninvariant form of the Higgs profile. Using the standard spherically symmetric hedgehog ansatz, $U(\mathbf{x}) = \exp(i \boldsymbol{\tau} \hat{\mathbf{r}}\, F(\mathbf{r}))$, we have[18]

$$\begin{aligned} M_{X_s} &= \frac{2\pi v}{g} \int_0^\infty dr \Bigg[\chi^2(r)(r^2 F'(r)^2 + 2 \sin^2(F(r))) \\ &\quad + \sin^2(F(r)) \left(\frac{\sin^2(F(r))}{r^2} + 2F'(r)^2 \right) \\ &\quad + \chi^2(r) r^2 \phi'(r)^2 + \frac{M_\phi^2}{4 g^2 v^2} r^2 (\chi(r)^4 - 2\chi(r)^2 + 1) \Bigg] \simeq 35\, \frac{v}{g}, \end{aligned}$$

$$\langle r^2_{X_s}\rangle_{X_s} = \int_0^\infty dr\, r^2(4\pi r^2)J^0_{X_s} = \frac{2}{\pi}\int_0^\infty dr\, r^2 \sin^2(F(r))F'(r)$$
$$\simeq (2.2)^2\frac{1}{g^2v^2}, \tag{61}$$

for $g \gtrsim 478(\gg 1)$ with $M_{X_s} \lesssim 18\,\text{GeV}$ in Eq. (57),[m] where we have defined a dimensionless parameter $\tilde{r} = gvr$ and $F(r) = \tilde{F}(\tilde{r})$, and renamed $\tilde{r} \to r$ and $\tilde{F} \to F$ in the final expression for notational convenience. $J^0_{X_s}$ is the topological $U(1)_{X_s}$ current is defined by

$$J^\mu_{X_s} = \frac{1}{24\pi^2}\epsilon^{\mu\nu\rho\sigma}\text{tr}\left(U^\dagger\partial_\nu U\cdot U^\dagger\partial_\rho U\cdot U^\dagger\partial_\sigma U\right) \tag{62}$$

and we have taken the topological charge $Q_{X_s} = 1$.

Combining all of them together in to the formulas given above, we find

$$\Omega_{X_s}h^2 = \mathcal{O}(0.1) \tag{63}$$

at $M_{X_s} = 18\,\text{GeV}$ (and $T_f \simeq 1\,\text{GeV}$), which is roughly consistent with the presently observed dark matter relic $\simeq 0.12$.[32]

7. Summary and Discussions

We have discussed a nobel role of the HLS in the SM Higgs Lagrangian. The SM Higgs Lagrangian was shown to be cast in to precisely the scale-invariant nonlinear sigma model, with the SM Higgs being the pseudo-dilaton, which was further shown to be gauge equivalent to the scale-invariant version of the HLS Lagrangian. Then the dynamical gauge boson "SM rho meson" of the HLS stabilizes the skyrmion, "dark Standard Model skyrmion (DSMS)" X_s, a novel candidate for the dark matter without recourse to beyond the SM.

A salient feature of DSMS is the nonlinearly realized scale invariance of the whole dynamics, which unambiguously determines the couplings of the DSMS to the SM Higgs as a pseudo-dilaton in terms of the low energy theorem of the spontaneously broken scale invariance. This imposes a definite constraint on the mass $M_{X_s} \lesssim 38\,\text{GeV}$ from the direct detection experiments LUX2016. With such a mass smaller than half of the SM Higgs mass, we have further constraint $M_{X_s} \lesssim 18\,\text{GeV}$ from the SM Higgs invisible decay data also definitely constrained by the low-energy theorem. Based on this salient constraint, we have discussed that the DSMS in a benchmark case of the heavy SM rho mass limit (scale-invariant Skyrme model limit) is roughly

[m] The equation of motion of $F(r)$ and $\phi(r)$ are given by $-[\chi^2r^2 + 2\sin^2 F]\cdot F'' = \chi^2(-\sin(2F) + 2rF') + 2\chi\chi' r^2F' - \sin^2 F\sin(2F)/r^2 + \sin(2F)(F')^2$ and $\phi'' = (F')^2 + 2\sin^2(F)/r^2 - (\phi')^2 - 2\phi'/r + [M_\phi^2/(2g^2v^2)](\chi^2 - 1)$, with the boundary conditions $F(0) = \pi\,, F(\infty) = 0$ and $\phi'(0) = 0$, $\phi(\infty) = 0$. The solution implies that $\chi(0) \to 0\,(\phi(0) \to -\infty)$, i.e., symmetry restoration at the origin, in the conformal/BPS limit Eq. (16), similarly to the BPS limit of the 't Hooft–Polyakov monopole in the Georgi–Glashow model.[22]

consistent with the relic abundance $\Omega_{X_s} h^2 \simeq 0.12$ in view of the extended size of the soliton.

If this scenario within the SM is established, then we should look for the beyond-SM physics somewhere other than the dark matter which is currently regarded as the biggest motivation for going beyond the SM.

The discussion here is straightforwardly applied to some UV completion, in which case the bare kinetic term of the HLS gauge boson is already generated by the underlying theory, the situation similar to the QCD rho meson and the QCD skyrmion stabilized by the QCD rho. A typical example for such a UV completion is the walking technicolor having the approximate scale symmetry and thus the technidilaton,[21] which is described by essentially the same type of the scale-invariant nonlinear chiral Lagrangian as the present theory (see e.g., Ref. 19), with technidilaton identified as 125 GeV Higgs, though having somewhat larger decay constant $F_\phi > v$. With the HLS gauge bosons being the technirho and the skyrmion the technibaryon and largeness of $F_\phi > v$, the mass upper bound of the dark matter technibaryon from the LUX constraint (see Fig. 1) shifts to the larger mass range, say $F_\phi > v$ for a typical walking technicolor case $F_\phi \simeq 5v$,[20] so that the Higgs invisible decay constraint becomes irrelevant. Also, the technibaryon usually with mass on order of TeV can be light when including the effects of the scalar meson (pseudo-dilaton in our case) on the skyrmion solution as already observed in the literature.[25,26,28]

We also comment[18] on the possibility for the DSMS as an asymmetric dark matter, which generates the current relic abundance of the DSMS through the electroweak sphaleron process collaborating with both the DSMS and the SM Higgs nonlinear dynamics of HLS with a "HLS sphaleron".

In summary, we have discussed a novel possibility that the HLS together with skyrmion, another pet idea of Gerry,[27] will give a great impact in the sense somewhat other than that Gerry was anticipating but I hope he would certainly enjoy it with a nice smile.

Acknowledgments

I would like to thank Shinya Matsuzaki and Hiroshi Ohki for collaborations and helpful discussions on the DSMS. I also thank Mannque Rho for stimulating correspondence on the scale symmetry and pseudo-dilaton in the gauge theory and the effective theory.

References

1. M. Bando, T. Kugo, S. Uehara, K. Yamawaki and T. Yanagida, *Phys. Rev. Lett.* **54** (1985) 1215; M. Bando, T. Kugo and K. Yamawaki, *Nucl. Phys. B* **259** (1985) 493; T. Fujiwara, T. Kugo, H. Terao, S. Uehara and K. Yamawaki, *Prog. Theor. Phys.* **73** (1985) 926; M. Bando, T. Fujiwara and K. Yamawaki, *Prog. Theor. Phys.* **79** (1988) 1140.

2. M. Bando, T. Kugo and K. Yamawaki, *Phys. Rept.* **164** (1988) 217.
3. M. Harada and K. Yamawaki, *Phys. Rev. Lett.* **86** (2001) 757.
4. M. Harada and K. Yamawaki, *Phys. Rept.* **381** (2003) 1.
5. G. E. Brown and M. Rho, *Phys. Rept.* **363** (2002) 85.
6. M. Harada and K. Yamawaki, *Phys. Lett. B* **297** (1992) 151.
7. M. Harada and K. Yamawaki, *Phys. Rev. Lett.* **83** (1999) 3374.
8. T. Eguchi, *Phys. Rev. D* **14** (1976) 2755; T. Kugo, *Prog. Theor. Phys.* **55** (1976) 2032.
9. W. A. Bardeen, C. T. Hill and M. Lindner, *Phys. Rev. D* **41** (1990) 1647.
10. V. A. Miransky, M. Tanabashi and K. Yamawaki, *Phys. Lett. B* **221** (1989) 177; *Mod. Phys. Lett. A* **4** (1989) 1043.
11. K. Yamawaki, *Prog. Theor. Phys. Suppl.* **123** (1996) 19.
12. K. Yamawaki, arXiv:1511.06883 [hep-ph]; arXiv:1605.01951 [hep-ph].
13. Z. Komargodski, *J. High Energy Phys.* **1102** (2011) 019.
14. R. Kitano, *J. High Energy Phys.* **1111** (2011) 124.
15. N. Arkani-Hamed, A. G. Cohen and H. Georgi, *Phys. Rev. Lett.* **86** (2001) 4757; C. T. Hill, S. Pokorski and J. Wang, *Phys. Rev. D* **64** (2001) 105005.
16. D. T. Son and M. A. Stephanov, *Phys. Rev. D* **69** (2004) 065020; J. Erlich, E. Katz, D. T. Son and M. A. Stephanov, *Phys. Rev. Lett.* **95** (2005) 261602.
17. T. Sakai and S. Sugimoto, *Prog. Theor. Phys.* **113** (2005) 843.
18. S. Matsuzaki, H. Ohki and K. Yamawaki, arXiv:1608.03691 [hep-ph].
19. H. S. Fukano, S. Matsuzaki, K. Terashi and K. Yamawaki, *Nucl. Phys. B* **904** (2016) 400.
20. S. Matsuzaki and K. Yamawaki, *J. High Energy Phys.* **1512** (2015) 053.
21. K. Yamawaki, M. Bando and K. I. Matumoto, *Phys. Rev. Lett.* **56** (1986) 1335.
22. J. A. Harvey, arXiv:hep-th/9603086.
23. M. Kurachi, S. Matsuzaki and K. Yamawaki, *Phys. Rev. D* **90** (2014) 055028.
24. Y. Igarashi, M. Johmura, A. Kobayashi, H. Otsu, T. Sato and S. Sawada, *Nucl. Phys. B* **259** (1985) 721; U. G. Meissner and I. Zahed, *Phys. Rev. Lett.* **56** (1986) 1035; U. G. Meissner, N. Kaiser, A. Wirzba and W. Weise, *Phys. Rev. Lett.* **57** (1986) 1676; F. R. Klinkhamer, *Z. Phys. C* **31** (1986) 623.
25. B. Y. Park, M. Rho and V. Vento, *Nucl. Phys. A* **736** (2004) 129; *Nucl. Phys. A* **807** (2008) 28.
26. Y. L. Ma, M. Harada, H. K. Lee, Y. Oh, B. Y. Park and M. Rho, *Phys. Rev. D* **90** (2014) 034015; B. R. He, Y. L. Ma and M. Harada, *Phys. Rev. D* **92** (2015) 076007.
27. I. Zahed and G. E. Brown, *Phys. Rept.* **142** (1986) 1.
28. R. Kitano and M. Kurachi, *J. High Energy Phys.* **1607** (2016) 037.
29. P. Carruthers, *Phys. Rept.* **1** (1971) 1.
30. D. S. Akerib *et al.*, arXiv:1608.07648 [astro-ph.CO]; A. Tan *et al.* [Pandax-II Collaboration], *Phys. Rev. Lett.* **117**(12), 121303 (2016).
31. CMS Collab., CMS-PAS-HIG-16-016.
32. Planck Collab. (P. A. R. Ade *et al.*), *Astron. Astrophys.* **571** (2014) A16.
33. M. Bando, K. I. Matumoto and K. Yamawaki, *Phys. Lett. B* **178** (1986) 308.
34. S. Matsuzaki and K. Yamawaki, *Phys. Rev. D* **86** (2012) 035025.
35. H. K. Lee, W. G. Paeng and M. Rho, *Phys. Rev. D* **92** (2015) 125033; W. G. Paeng, T. T. S. Kuo, H. K. Lee and M. Rho, *Phys. Rev. C* **93** (2016) 055203; W. G. Paeng and M. Rho, arXiv:1604.06889 [nucl-th].
36. M. Harada, T. Kugo and K. Yamawaki, *Phys. Rev. Lett.* **71** (1993) 1299; *Prog. Theor. Phys.* **91** (1994) 801.
37. JLQCD Collab. (H. Ohki *et al.*), *Phys. Rev. D* **78** (2008) 054502; *Phys. Rev. D* **87** (2013) 034509.

38. J. Hisano, K. Ishiwata and N. Nagata, *J. High Energy Phys.* **1506** (2015) 097.
39. J. L. Feng *et al.*, arXiv:1401.6085 [hep-ex].
40. J. Billard, L. Strigari and E. Figueroa-Feliciano, *Phys. Rev. D* **89** (2014) 023524.
41. https://twiki.cern.ch/twiki/bin/view/; LHCPhysics/CERNYellowReportPageBR# TotalWidthAnchor.
42. H. Baer *et al.*, arXiv:1306.6352 [hep-ph].
43. E. W. Kolb and M. S. Turner, *The Early Universe* (Addison-Wesley, Redwood City, CA, 1990).